Grasses of the Intermountain Region

Grasses of the Intermountain Region

Edited by   Laurel K. Anderton and Mary E. Barkworth

Illustrated by   Cindy Talbot Roché, Linda Ann Vorobik, Sandy Long, Annaliese Miller, Bee F. Gunn, and Christine Roberts

Intermountain Herbarium
Utah State University
Logan, Utah 84322

Intermountain Herbarium
Logan, Utah 84322-5305

The account of *Danthonia* is reproduced by permission of Stephen J. Darbyshire
for the Department of Agriculture and Agri-Food, Government of Canada,
©Minister of Public Works and Government Services, Canada, 2002.
The account of *Avena* is reproduced by permission of Bernard R. Baum for the
Department of Agriculture and Agri-Food, Government of Canada,
©Minister of Public Works and Government Services, Canada, 2007.
The accounts of *Schizachne* and *Vahlodea* are reproduced by permission
of Jacques Cayouette and Stephen J. Darbyshire for the Department of
Agriculture and Agri-Food, Government of Canada, ©Minister of Public
Works and Government Services, Canada, 2007.
The accounts of *Leucopoa* and *Schedonorus* are reproduced by permission of Stephen
J. Darbyshire for the Department of Agriculture and Agri-Food, Government of
Canada, ©Minister of Public Works and Government Services, Canada, 2007.

Manufactured in the United States of America

Cover design by Barbara Yale-Read
Cover photo by Mark Elzey

Library of Congress Cataloging-in-Publication Data
Grasses of the Intermountain Region
edited by Laurel K. Anderton and Mary E. Barkworth;
illustrated by Cindy Talbot Roché ... [et al.].
p. cm.
"Grasses of the Intermountain Region is a modification of the two grass
volumes of the Flora of North America"–Pref.
Includes index.

ISBN 978-0-87421-765-0(pbk.)
1. Grasses–Great Basin–Identification.
I. Anderton, Laurel K. II. Barkworth, Mary E. III. Roché, Cindy Talbot
QK495.G74G744 2010
584'90979–dc22

Contents

Preface

Grasses of the Intermountain Region is a modification of the two grass volumes of the *Flora of North America* (FNA). It is designed for identifying members of the *Poaceae* in the region between the Sierra Nevada and Rocky Mountains. The reduction in the number of taxa treated has reduced the length of the keys and made it possible to include, in a single volume, descriptions and illustrations for all taxa treated as well as provide distribution maps for species that are established in the area.

Region: In this volume, the Intermountain Region is interpreted somewhat more broadly than in the *Intermountain Flora* (see Cronquist et al. 1972), differing primarily by 1) including, in addition to the regions covered by that flora, the southern tip of Nevada; 2) using the crest of the Sierra Nevada, rather than the lower slopes, as the western boundary; 3) using the Colorado River as the southern boundary in Arizona; and 4) using county boundaries (rather than geographic boundaries) to define its limits in Oregon, Idaho, and Wyoming. This last explains the rather jagged boundaries of the distribution maps.

The reason for using county boundaries was pragmatic. Information on which species grow in the region comes from plants preserved as herbarium specimens, the labels of which provide information on when, where, and by whom they were collected. This information almost always includes the name of the county in which the specimen was collected but, because most were collected before development of inexpensive GPS units, obtaining more precise information is a major task. This situation is, however, changing rapidly thanks to the major push by herbaria to georeference their collections.

Maps: The distribution maps are based on information from herbarium specimens and published sources. Two levels of data are represented. The gray areas represent counties in which the species has been collected; the dots reflect the actual collection location. Many of these dots were obtained by retrospectively georeferencing information on specimen labels. As such, their accuracy varies but they provide better information than relying entirely on county records. It should also be borne in mind that many of these specimens are old; some of the localities shown have since become part of a town, golf course, highway, field, or other constructed landscape.

For determining whether a species from Arizona, California, or Oregon was collected within the region, we consulted http://ucjeps.berkeley.edu/consortium/, http://swbiodiversity.org/seinet/index.php, and http://www.oregonflora.org/atlas.php. If we were still unable to determine whether a species was inside or outside the region, we included it.

The maps in this volume incorporate some updates from the information in the FNA volumes and the *Manual of Grasses for North America*, but we have not reviewed them all. The maps are also available online at http://herbarium.usu.edu/webmanual. This online resource provides additional information about the source of individual records and is updated at irregular intervals. For information about a particular record or to provide additional records, contact mary@biology.usu.edu.

Organization: To conserve space, the descriptions, illustrations, and distribution maps are in three separate sections and abbreviations have been used in the descriptions. The order in which the taxa are treated reflects their phylogenetic relationships, as these are now understood. Grouping the treatments in this manner makes it easier to locate and compare the descriptions and illustrations of similar genera and species than does an alphabetical arrangement.

A bipartite number is used to indicate the location of each genus, the part before the dot indicating the tribe and the part after the dot indicating the genus. These numbers also form part of the page headers. In the generic treatments, the names of the species are followed by two page numbers; the italicized number refers to the illustration page, the underlined number to the map page. Tripartite numbers are associated with each illustration and map. The first two parts of these numbers correspond to the tribal and generic numbers used as page headers on the treatment pages; the third part of each number indicates the position of the species within the text material for its genus.

The literature cited section includes all the works cited in this volume. For a more extensive listing, readers should consult the *Flora* volumes or go online at http://utc.usu.edu/grassbib.htm.

Numbers: There are 482 species described in this volume. They represent five subfamilies, 15 tribes, and 134 genera. These numbers include six hybrid genera, mostly in the *Triticeae*, and six hybrid species. Of the 482 species, 286 are native and known to be in the region, 101 are introduced species that are established in the region, and 44 are cultivated. Most of the cultivated species are grown as ornamentals but some, such as wheat (*Triticum aestivum*) and corn (*Zea mays*), are grown for food or forage. Of the established species some, such as crested wheatgrass (*Agropyron cristatum*), were deliberately introduced; the introduction of others, such as medusahead (*Taeniatherum caput-medusae*), was accidental. The status of the other 51 species in the region (including the hybrid species) is in doubt for one reason or another. Some of them are species known to grow in adjacent areas, but whether they actually extend into the region is not known. For other species, it is doubtful whether they are native or introduced to the region from other parts of North America.

A few species that are not thought to be established have been included. Some of these, such as *Hygroryza aristata*, seem likely to become aggressive weeds if they escape. A few species, such as *Glyceria declinata* and *Ehrharta calycina*, are known only from old collections and are probably no longer present in the region. *Piptochaetium lasianthum* is included because it was, tentatively, reported from the region by Welsh et al. (2003, 2007). The specimen proved to be an *Achnatherum hymenoides* hybrid.

Taxonomic changes: There are some differences between the taxonomic treatments in this volume and its two predecessors, the grass volumes in the *Flora of North America* series and the *Manual of Grasses for North America*. The changes include recognition of both *Podagrostis humilis* and *P. thurberiana*, treatment of *Buchloë dactyloides* as a species of *Bouteloua*, and recognition of *Distichlis spicata* subsp. *spicata* based on evidence from a 2009 abstract. The additional taxa have been illustrated by Dr. C.T. Roché. She has also prepared new illustrations for some of the original plates in which the habit was illustrated by a taxon from outside the region.

The name *Thinopyrum ponticum*, which is based on *Triticum ponticum* Podp., needs to be changed. Tsvelev (1993) made it a synonym of *Elytrigia obtusiflora*, based on *Triticum obtusiflorum* DC. While checking type specimens, it became evident that there may be earlier epithets available. In order to avoid adding to the confusion by combining "obtusiflorum" with *Thinopyrum* and then finding an earlier epithet, we have continued to use *Thinopyrum ponticum* in this volume.

Synonyms: Many of the taxa treated in this volume are treated under other names in older floras. The index includes all such names, showing the name used in this volume and the page on which its description is located.

Authorship: Most of the these treatments are drawn nearly directly from those in the *Flora of North America* volumes, and their authorship is usually the same as in those volumes. The FNA authors were sent copies of the modified versions for approval. Some suggested ways in which the modified versions could be improved, for which we thank them. In a few cases, there has been a change of authorship either because the treatment has been changed or the author no longer wishes to be involved. The citation for the volume is:

Anderton, L.K. and M.E. Barkworth (Eds). 2009. *Grasses of the Intermountain Region*. Intermountain Herbarium. 559 pp.

Acknowledgments:. In addition to the the authors of the treatments, we thank Dr. Kanchi Gandhi, Harvard University, for his rapid and careful responses to all our nomenclatural questions, Dr. Cindy Roché for the additional illustrations, Chris Garrard for modifying our mapping program to create the maps used in this volume, and John Lowry for developing the maps for the front end pages. We also thank SEINet, the California Consortia of Herbaria, and the Flora of Oregon project for making maps and detailed specimen information freely available over the web.

Authors

We thank the following treatment authors for their contribution to this volume.

Charles M. Allen
Center for Environmental
    Management of Military Lands
Fort Polk, Louisiana

Kelly W. Allred
New Mexico State University
Las Cruces, New Mexico

Sharon J. Anderson
U.S. Department of Agriculture-
    Agricultural Research Service
Fort Detrick, MD

Laurel K. Anderton
Utah State University
Logan, Utah

Carol R. Annable
Carpinteria, California

Claus Baden†
The Royal Veterinary and
    Agricultural University
Frederiksberg, Denmark

Nigel P. Barker
Rhodes University
Grahamstown, South Africa

Mary E. Barkworth
Utah State University
Logan, Utah

Bernard R. Baum
Agriculture and Agri-Food Canada
Ottawa, Ontario

Christine M. Bern
Colorado State University
Fort Collins, Colorado

Roland von Bothmer
Swedish University of Agricultural
    Sciences
Alnarp, Sweden

David M. Brandenburg
The Dawes Arboretum
Newark, Ohio

Paul B.H. But
Chinese University of Hong Kong
Hong Kong, China

Christopher S. Campbell
University of Maine
Orono, Maine

Julian J.N. Campbell
The Nature Conservancy
Lexington, Kentucky

Jack R. Carlson
Natural Resources Conservation
    Service
Washington, D.C.

Jacques Cayouette
Agriculture and Agri-Food Canada
Ottawa, Ontario

Lynn G. Clark
Iowa State University
Ames, Iowa

Laurie L. Consaul
Canadian Museum of Nature
Ottawa, Ontario

William J. Crins
Ontario Ministry of Natural
    Resources
Peterborough, Ontario

Thomas F. Daniel
California Academy of Sciences
San Francisco, California

Stephen J. Darbyshire
Agriculture and Agri-Food Canada
Ottawa, Ontario

Patricia Dávila Aranda
Escuela Nacional de Estudios
    Profesionales, Iztacala
Tlalnepantla, México

Jerrold I. Davis
Cornell University
Ithaca, New York

Melvin Duvall
Northern Illinois University
DeKalb, Illinois

Robert W. Freckmann
University of Wisconsin-Stevens
    Point
Stevens Point, Wisconsin

Signe Frederiksen
Københavns Universitet
Copenhagen, Denmark

Mark L. Gabel
Black Hills State University
Spearfish, South Dakota

Grass Phylogeny Working Group
Nigel P. Barker
Lynn G. Clark
Jerrold I. Davis
Melvin Duvall
Gerald F. Guala
Catherine Hsiao
Elizabeth A. Kellogg
H. Peter Linder
Roberta J. Mason-Gamer
Sarah Y. Mathews
Mark P. Simmons
Robert J. Soreng
Russell E. Spangler

Craig W. Greene†
College of the Atlantic
Bar Harbor, Maine

Gerald F. Guala
Fairchild Tropical Garden
Miami, Florida

David W. Hall
David W. Hall Consultant, Inc.
Gainesville, Florida

Barry E. Hammel
Missouri Botanical Garden
St. Louis, Missouri and
Instituto Nacional de Biodiversidad
Santo Domingo de Heredia, Costa
    Rica

M.J. Harvey
Victoria, British Columbia

Stephan J. Hatch
Texas A&M University
College Station, Texas

Richard J. Hebda
Royal British Columbia Museum
Victoria, British Columbia

Khidir W. Hilu
Virginia Polytechnic Institute and
    State University
Blacksburg, Virginia

Catherine Hsiao
U.S. Department of Agriculture-
    Agricultural Research Service
Logan, Utah

Hugh H. Iltis
University of Wisconsin-Madison
Madison, Wisconsin

Niels H. Jacobsen
The Royal Veterinary and
    Agricultural University
Frederiksberg, Denmark

Elizabeth A. Kellogg
University of Missouri-St. Louis
St. Louis, Missouri

Michel G. Lelong
University of South Alabama
Mobile, Alabama

H. Peter Linder
University of Cape Town
Rondebosch, South Africa

Robert I. Lonard
University of Texas-Pan American
Edinburg, Texas

Kendrick L. Marr
Royal British Columbia Museum
Victoria, British Columbia

Roberta J. Mason-Gamer
University of Illinois
Chicago, Illinois

Sarah Y. Mathews
University of Missouri
Columbia, Missouri

Peter W. Michael
University of Sydney and National
    Herbarium of New South Wales
Sydney, Australia

Laura A. Morrison
Oregon State University
Corvallis, Oregon

Leon E. Pavlick†
Royal British Columbia Museum
Victoria, British Columbia

Paul M. Peterson
Smithsonian Institution
Washington, D.C.

Grant L. Pyrah
Missouri State University
Springfield, Missouri

Charlotte G. Reeder
University of Arizona
Tucson, Arizona

John R. Reeder†
University of Arizona
Tucson, Arizona

James M. Rominger
Northern Arizona University
Flagstaff, Arizona

Björn Salomon
Swedish University of Agricultural
    Sciences
Alnarp, Sweden

Robert B. Shaw
Texas A&M University
College Station, Texas

Mark P. Simmons
Ohio State University
Columbus, Ohio

James P. Smith, Jr.
Humboldt State University
Arcata, California

Neil W. Snow
Bishop Museum
Honolulu, Hawaii

Robert J. Soreng
Smithsonian Institution
Washington, D.C.

Russell E. Spangler
University of Minnesota
St. Paul, Minnesota

Lisa A. Standley
Vanasse Hangen Brustlin, Inc.
Watertown, Massachusetts

Michael T. Stieber
Morton Arboretum
Lisle, Illinois

Edward E. Terrell
Smithsonian Institution
Washington, D.C.

John W. Thieret†
Northern Kentucky University
Highland Heights, Kentucky

Rahmona A. Thompson
East Central University
Ada, Oklahoma

Gordon C. Tucker
Eastern Illinois University
Charleston, Illinois

Jesús Valdés-Reyna
Universidad Autonoma Agraria
    Antonio Narro
Saltillo, México

Alan S. Weakley
University of North Carolina-Chapel
    Hill
Chapel Hill, North Carolina

Robert D. Webster
U.S. Department of Agriculture-
    Agricultural Research Service
Beltsville, Maryland

J.K. Wipff
Barenbrug USA
Albany, Oregon

Abbreviations

abx . . . . abaxial
adx . . . . . adaxial, adaxially
ann . . . . annual
anth . . . . anthers
apc . . . . apices
asex . . . . asexual
aur . . . . auricles
bas . . . . basal, basally
bisex . . . . bisexual
bld . . . . blades
br . . . . branches
brchd . . . . branched
brchg . . . . branching
cal. . . . . calluses
car . . . . caryopses
cent . . . . central
ces . . . . cespitose
clm . . . . culms
clstgn . . . . cleistogenes
col . . . . collars
dis . . . . disarticulation,
        disarticulating
emb . . embryos, embryonic
emgt . . . . emarginate
epdm . . . . epidermes
exvag . . . . extravaginal
flt . . . . .florets
fnctl . . . . functional
ftl . . . . fertile
glab . . . . glabrous
glm . . . . glumes
infl . . . . inflorescences

infvag . . . . infravaginal
intnd . . . . internodes
invag . . . . intravaginal
jnct . . . . junction
lat . . . . lateral, laterally
lf . . . . leaf
lfy . . . . leafy
lig . . . . ligules
lm . . . . lemmas
lo . . . . lower
lod . . . . lodicules
lvs . . . . leaves
memb . . . . . . membranous
mid . . . . middle
mrg . . . . margins
mrgl . . . . marginal
nd . . . . nodes
occ . . . . occasionally
ov . . . . ovaries
pal . . . . paleas
pan . . . panicles, paniculate
ped . . . . pedicels
pedlt . . . . pedicellate
per . . . . perennial
pist . . . . pistillate
pl . . . . plants
pluricsp . . . . pluricespitose
pri . . . . primary
psdlig . . . . . .pseudoligules
psdpet . . . . pseudopetioles
        pseudopetiolate
psdspklt . pseudospikelets

psdinvag . . . .
. . . . . . pseudointravaginal
rchl . . . . rachillas
rchs . . . . rachises
rcm . . . . racemes, racemose
rcmly . . . . racemosely
rdcd . . . . reduced
rdmt . . . . . rudiments,
        rudimentary
rebrchg . . . . . . rebranching
rhz . . . . rhizomes,
        rhizomatous
sec . . . . secondary
shth . . . . sheaths
smt . . . . sometimes
spklt . . . . spikelets
sta . . . . stamens
stln . . stolons, stoloniferous
stmt . . . . staminate
strl . . . . sterile
sty . . . . styles
sex . . . . sexual, sexually
subglab . . . . . . subglabrous
subtm . . . . subterminal
tml . . . terminal, terminally
unisex . . . . unisexual
up . . . . upper
usu . . . . usually

Taxonomic Treatments

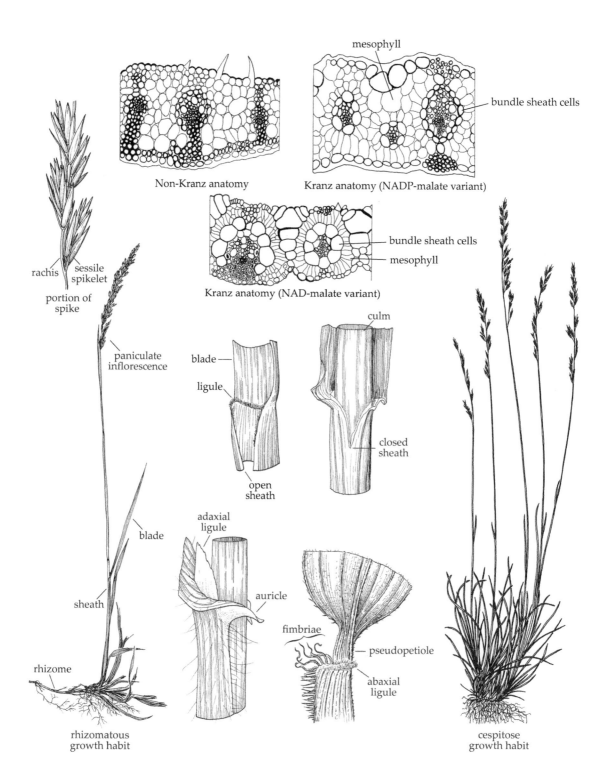

mesophyll

bundle sheath cells

Non-Kranz anatomy

Kranz anatomy (NADP-malate variant)

bundle sheath cells

mesophyll

Kranz anatomy (NAD-malate variant)

rachis

sessile
spikelet

portion of
spike

paniculate
inflorescence

culm

blade

ligule

open
sheath

closed
sheath

blade

sheath

rhizome

rhizomatous
growth habit

adaxial
ligule

auricle

fimbriae

pseudopetiole

abaxial
ligule

cespitose
growth habit

POACEAE Barnhart
GRAMINEAE Adans., alternate name

● Grass Family   Lynn G. Clark and Elizabeth A. Kellogg

Pl ann or per; usu terrestrial, smt aquatic; tufted, mat-forming, csp, pluricsp, or with solitary *culms* (flower stems), rhz and stln often well developed. **Clm** ann or per, herbaceous or woody, usu erect or ascending, smt prostrate or decumbent for much of their length, occ climbing, rarely floating; **nd** prominent, smt concealed by lf shth; **intnd** hollow or solid, bases meristematic; **brchg** from the bas nd only or from bas, mid, and up nd; **bas brchg** exvag or invag; **up brchg** invag, exvag, or infvag. **Lvs** alternate, 2-ranked, each composed of a shth and bld encircling the clm or br; **shth** usu open, smt closed with mrg fused for all or part of their length; **aur** (lobes of tissue extending beyond the margins of the sheaths on either side) smt present; **lig** usu present at the shth-bld jnct, particularly on the adx surface, abx lig common in the *Bambusoideae*, memb, smt ciliate, adx lig usu present, of memb to hyaline tissue, a line of hairs, or a ciliate membrane; **bld** usu linear to lanceolate, occ ovate to triangular, bases smt *pseudopetiolate* (having a petiole-like constriction), venation usu parallel, smt with evident cross veins, occ divergent. **Infl** (*synflorescences*) usu compound, composed of simple or complex aggregations of pri infl, aggregations pan, spicate, or rcm or of spikelike br, often with an evident *rachis* (central axis), pri infl *spikelet*, *pseudospikelet*, or spklt equivalents; **infl br** usu without obvious bracts. **Spklt** with (0–1)2(3–6) *glumes* (empty bracts) subtending 1–60 flt, glm and flt distichously attached to the *rachilla* (central axis); **psdspklt** with bud-subtending bracts below the glm. **Glm** usu with an odd number of veins, smt awned. **Flt** bisex, stmt, or pist, often lat or dorsally compressed, smt round in cross section, usu composed of a *lemma* (lower bract) and *palea* (upper bract), lod, and reproductive organs; **lm** usu with an odd number of veins, often awned, bases frequently thick and hard, forming a cal, backs rounded or keeled over the midvein, awns usu 1(–3), arising bas to tml; **pal** usu with 2 major veins, with 0 to many additional veins between the major veins, smt also in the mrg, often keeled over the major veins; **lod** (0)2–3, inconspicuous, usu without veins, bases swelling at anthesis; **sta** usu 3, smt 1(2) or 6+, filaments capillary, anth versatile, usu all alike within a flt, smt 1 or 2 evidently longer than the others; **ov** 1-loculed, with (1)2–3(4) sty or sty br, stigmatic region usu plumose. **Fruits** car, pericarp usu dry and adhering to the seeds, smt fleshy or dry and separating from the seeds at maturity or when moistened; **emb** ⅓ as long as to almost equaling the car, highly differentiated with a *scutellum* (absorptive organ), a shoot with lf primordium covered by the *coleoptile* (shoot sheath), and a root covered by the *coleorhiza* (root sheath); **hila** punctate to linear. $x = 5, 6, 7, 9, 10, 11, 12$. The formulaic name, *Poaceae*, is based on *Poa*, the largest grass genus; the alternate name, *Gramineae*, comes from the Latin, '*gramen*' grass.

 The *Poaceae* or grass family includes approximately 700 genera and 11,000 species. This volume treats 8 subfamilies, 15 tribes, 142 genera, and 483 species. Of these, all the subfamilies and tribes, 88 genera, and 330 species are native to the Intermountain Region. The remaining taxa include introductions, both established and not established, and species cultivated as ornamentals, forage, or food.

 Grasses constitute the fourth largest plant family in terms of number of species. Nevertheless, the family is clearly more significant than any other plant family in terms of geographic, ecological, and economic importance. Grasses grow in almost all terrestrial environments, including dense forests, open deserts, and freshwater streams and lakes. There are no truly marine grasses, but some species grow within reach of the highest tides. In addition to being widely distributed, grasses are often dominant or co-dominant over large areas. This is reflected in the many words that exist for grasslands, words such as meadow, palouse, pampas, prairie, savanna(h), steppe, and veldt. Not surprisingly, grasses are of great ecological importance as soil stabilizers and as providers of shelter and food for many different animals.

 The economic importance of grasses to humans is almost impossible to overestimate. The wealth of individuals and countries is dependent on the availability of such sources of grain as *Triticum* (wheat), *Oryza* (rice), *Zea* (corn or maize), *Hordeum* (barley), *Avena* (oats), *Secale* (rye), *Eragrostis* (tef), *Zizania* (wild rice), and *Sorghum*. Most countries invest heavily in research programs designed to develop better strains of these grasses and the many other grasses that are used for livestock, soil stabilization, and revegetation. Developing improved grasses for recreation areas, such as playing fields, golf courses, and parks, is also a major industry. Increasing recognition of the aesthetic value of grasses is reflected in their prominence in horticultural catalogs.

 There are, of course, grasses that are considered undesirable, but even the most obnoxious grasses may be well regarded over a portion of their range. For instance, *Bromus tectorum* (cheatgrass) is regarded as a noxious, fire-prone invader of

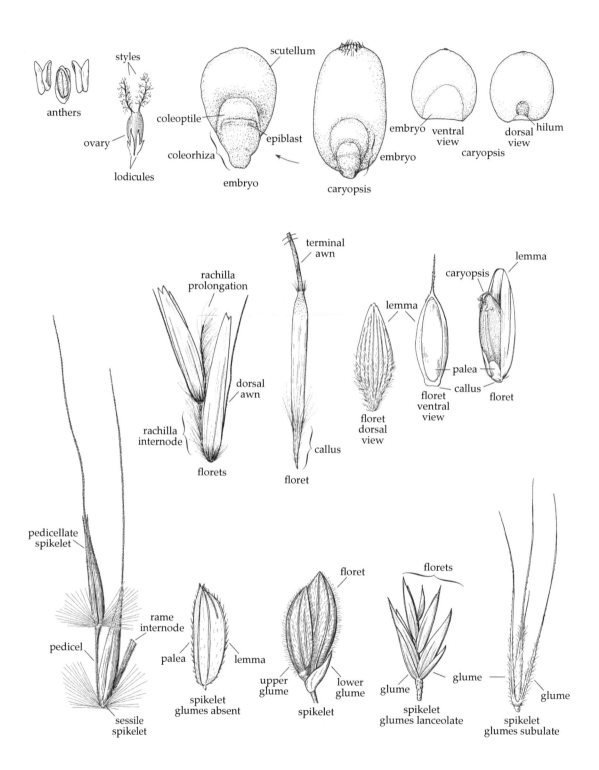

anthers

styles

ovary

lodicules

scutellum

coleoptile

coleorhiza

epiblast

embryo

embryo

embryo

ventral
view

dorsal
view

hilum

caryopsis

caryopsis

rachilla
prolongation

dorsal
awn

rachilla
internode

florets

terminal
awn

callus

floret

lemma

floret
dorsal
view

lemma

caryopsis

palea

callus

floret
ventral
view

floret

pedicellate
spikelet

pedicel

sessile
spikelet

rame
internode

palea

lemma

spikelet
glumes absent

upper
glume

floret

lower
glume

spikelet

florets

glume

glume

spikelet
glumes lanceolate

florets

glume

glume

spikelet
glumes subulate

Intermountain Region ecosystems, but it is also welcomed as a source of early spring feed in some parts of the United States. *Cynodon dactylon* (bermudagrass) is listed as a noxious weed in some jurisdictions; in others it is valued as a lawn grass.

Although grasses are widespread and often dominant in open areas, all evidence points to their having originated in forests, probably in the Southern Hemisphere, by the late Cretaceous to early Paleocene (70–55 mya). Recent evidence from phytoliths (isolated silica bodies commonly produced inside the epidermal cells of grasses and some other plants) embedded in fossil coprolites strongly suggests that grasses evolved earlier in the Cretaceous than previously thought. Living representatives of the three earliest lineages of the grass family, together comprising about 30 species, are perennial, broad-leaved plants of relatively small stature, native to tropical or subtropical forests in South America, Africa, southeast Asia, some Pacific Islands, and northern Australia. The major diversification of the family probably occurred in the Oligocene (34–24 mya), and was associated with climatic changes that produced more open habitats. All major lineages of the grass family were present by the middle of the Miocene (24–5.5 mya); C_4 photosynthesis in grasses had also evolved by then.

Grasses exhibit three variants of C_4 photosynthesis. They are associated with Kranz leaf anatomy, in which the vascular bundles are surrounded by a cylinder of cells (bundle sheath cells) that, because they contain starch, stain dark with iodine. When seen in cross-section, the cylinders form rings or wreaths (Kranz in German) around the vascular bundles. C_4 grasses also have three or fewer mesophyll cells between adjacent vascular bundles. C_3 grasses do not have starch-containing bundle sheath cells and have more than three mesophyll cells between the vascular bundles. Both these features can be seen in hand-made cross-sections (Hattersley and Watson 1976); they are, admittedly, more useful in the herbarium than in the field.

Key to Tribes
Mary E. Barkworth

1. Culms perennial, woody (except in greenhouse plants) often developing complex branching systems from the upper nodes; leaves on the upper portion of the culm or distal on the branches usually pseudopetiolate (see p. 7) . . . . *Bambusoideae*
1. Culms usually annual, sometimes overwintering, rarely woody, sometimes branching from the upper culm nodes but branching not complex; leaves usually not pseudopetiolate.
  2. Spikelets almost always with two dissimilar florets, the lower floret sterile or staminate, frequently reduced to a lemma, sometimes missing, the upper floret usually bisexual, sometimes unisexual or sterile; upper floret well developed, usually with a leathery to hard, usually unawned, lemma or reduced, with a hyaline, often awned, lemma, awn attached at or near the top of the lemma; rachilla not prolonged beyond the base of the upper floret (*Panicoideae*).
    3. Glumes flexible, membranous, the lower glume usually shorter than the upper glume, the upper glume subequal to or exceeded by the upper floret; lower lemma similar in texture to the glumes; upper lemma usually leathery to hard; spikelets usually single or in pairs; disarticulation below the glumes . . . . 14. *Paniceae*
    3. Glumes usually stiff, leathery to indurate, often subequal, at least 1 and usually both exceeding the upper floret (excluding the awn); both lemmas hyaline; most spikelets in pairs or triplets, at least one spikelet in the group usually sessile; disarticulation often in the inflorescence branches, below the sessile spikelet, sometimes below the glumes . . . . 15. *Andropogoneae*
  2. Spikelets with other than 2 florets or, if with 2, the lower floret bisexual or the upper floret awned from the back or base of the lemma or the spikelets bulbiferous; glumes usually membranous; rachilla often prolonged beyond the base of the distal floret.
    4. All or most spikelets bulbiferous . . . . 8. *Poeae* (in part)
    4. All or most spikelets sexual.
      5. Spikelets with 1 floret; lemma terminating in a 3-branched awn, the lateral branches sometimes greatly reduced; callus well developed; ligules of hairs or a ciliate membrane, the cilia longer than the membranous base. . . . . 13. *Aristideae*
      5. Spikelets with more than 1 sexual floret or, if with only 1, the lemma not terminating in a 3-branched awn; callus sometimes well developed; ligules membranous, of hairs, or a ciliate membrane.
        6. Spikelets with 1 sexual floret; glumes absent or less than ¼ the length of the adjacent floret; lower glume, if present, without veins; upper glume, if present, without veins or 1-veined.
          7. Inflorescence 1-sided, spikelike; spikelets triangular in cross section . . . . 3. *Nardeae*
          7. Inflorescence paniculate; spikelets laterally compressed or terete.
            8. Culms spongy; plants growing in wet places, often emergent; lemmas of the bisexual or pistillate florets with 3–14 veins . . . . 2. *Oryzeae*
            8. Culms not spongy; plants of wet or dry habitats but not emergent; lemmas of bisexual or pistillate florets with 1–3 veins. . . . . 10. *Cynodonteae* (in part)
        6. Spikelets usually with more than 1 sexual floret, usually with 2 glumes, 1 or both glumes more than ¼ the length of the adjacent floret and/or with more than 1 vein (glumes always longer than ¼ the length of the adjacent floret in taxa with only 1 sexual floret).

9. Lemmas with 9 veins extending into equal, plumose awns . . . . 11. *Pappophoreae*
9. Lemmas awned or unawned, apices entire, mucronate, bilobed, or bifid, occasionally 4-lobed or 4–5-toothed, sometimes erose.
  10. Cauline leaf sheaths closed for at least ½ their length; glumes usually exceeded by the distal floret, sometimes greatly so.
    11. Spikelets 5–80 mm long, not bulbiferous; lemmas usually awned, often bilobed or bifid, veins convergent distally; ovary apices hairy . . . . 6. *Bromeae*
    11. Spikelets 0.7–60 mm long, sometimes bulbiferous, lemmas often unawned, not both bilobed/bifid and with convergent veins; ovary apices usually glabrous.
      12. Lemma veins (4)5–15, usually prominent, parallel distally; spikelets 2.5–60 mm long, not bulbiferous . . . . 4. *Meliceae*
      12. Lemma veins 1–9, often inconspicuous, usually convergent distally; spikelets 0.7–18(20) mm long, sometimes bulbiferous . . . . 8. *Poeae* (in part)
  10. Cauline leaf sheaths open for at least ½ their length; glumes sometimes exceeded by the distal floret, sometimes exceeding it.
    13. Spikelets with 1 floret; lemmas terminally awned, the junction of the lemma and awn conspicuous; rachilla not prolonged beyond the base of the floret . . . . 5. *Stipeae*
    13. Spikelets with 1–60 florets; lemmas unawned or awned, awns basal, subterminal, or terminal, if terminal, the junction with the lemma not conspicuous; rachilla often prolonged beyond the base of the distal floret.
      14. Ligules, at least of the basal leaves, composed of a line of hairs or a ciliate membrane or ridge with cilia longer than the basal membrane or ridge; leaves usually hairy on either side of the ligule; auricles absent.
        15. Lemmas of the fertile florets hairy all over, with 3–11 inconspicuous veins, if lemma margins hairy on the lower portion, hairs without papillose bases; lemma apices usually bilobed or bifid and awned or mucronate from the sinus; awns twisted, at least on the lower portion . . . . . 12. *Danthonieae*
        15. Lemmas of the fertile florets glabrous all over or hairy over the veins, usually with 1–3 conspicuous veins, sometimes with 3 inconspicuous veins or 5–11 veins, marginal hairs often with papillose bases; lemma apices entire, bilobed, bifid, or 4-lobed; awns usually not twisted.
          16. Lemmas 1–11-veined, veins glabrous or hairy, marginal hairs not papillose-based; neither rachillas nor calluses pilose; basal internodes of the culms not persistent . . . . 10. *Cynodonteae* (in part)
          16. Lemmas 3(5)-veined, veins glabrous, marginal hairs, if present, papillose-based; rachillas or calluses pilose or the basal internodes of the culms persistent . . . . 9. *Arundineae*
      14. Ligules membranous or absent, if ciliate, the cilia shorter than the membranous base; leaves usually glabrous on either side of the ligule; auricles present or absent.
        17. Inflorescences panicles or 1-sided racemes, not spikelike, without spikelike branches; spikelets solitary, with only 1 sexual floret, this distal to any sterile or staminate florets.
          18. Lemmas of the lower florets membranous, sometimes reduced to a small hairy strip much shorter than the sexual floret . . . . 8. *Poeae* (in part)
          18. Lemmas of the lower florets hard, at least the upper sterile floret exceeding the sexual floret . . . . 1. *Ehrharteae*
        17. Inflorescences panicles, racemes, or spikes; spikelets solitary, paired, or in triplets, often with more than 1 sexual floret; sterile florets, if present, distal to the sexual floret.

19. Lemmas with 1–3 or 9–11 conspicuous veins; sheaths open; blade cross-sections exhibiting Kranz leaf anatomy . . . . 10. *Cynodonteae* (in part)
19. Lemmas with (1)3–15 veins, these often inconspicuous; sheaths open or closed; blades not exhibiting Kranz leaf anatomy.
   20. Inflorescences panicles . . . . 8. *Poeae* (in part)
   20. Inflorescences spikes or spikelike; spikelets 1–5+ per node, often with 1 sessile spikelet.
      21. Inflorescences spikes or spikelike, with 1–5 spikelets per node, if 3, usually 1 sessile and the other 2 shortly pedicellate, if with only 1 spikelet per node, the spikelets embedded in the rachis or tangential to it, in either case two glumes present on all spikelets . . . . 7. *Triticeae*
      21. Inflorescences spikelike panicles with highly reduced branches, these often fused to the rachis, or spikes with only 1 spikelet per node, the spikelets radial to the rachis, but the terminal spikelet having only 1 glume . . 8. *Poeae* (in part)

1. **BAMBUSOIDEAE** Luerss.

Grass Phylogeny Working Group

Pl usu per, rarely ann, rhz. **Clm** woody or herbaceous, hollow or solid; often developing complex vegetative brchg; **lvs** distichous, if complex vegetative brchg present, lvs of the clm (*culm leaves*) differing from those of the vegetative br (*foliage leaves*); **aur** often present; **abx lig** rarely present on the clm lvs, usu present on the foliage lvs; **adx lig** memb or chartaceous, ciliate or not; **psdpet** smt present on the clm lvs, usu present on the foliage lvs; **bld** usu relatively broad, venation parallel, often with evident cross venation; **mesophyll** nonradiate; **adx palisade layer** usu absent; **fusoid cells** usu well developed, large; **arm cells** usu well developed and highly invaginated; **Kranz anatomy** not developed; **midribs** complex or simple; **stomates** with dome-shaped, triangular, or more rarely parallel-sided subsidiary cells; **adx bulliform cells** present; **bicellular microhairs** present, tml cells tapered; **papillae** common and abundant. **Infl** spicate, rcm, or pan, comprising spklt or psdspklt, the spklt lacking subtending bracts and prophylls, completing their development during 1 period of growth, the psdspklt having subtending bracts, prophylls, and bas bud-bearing bracts developing 2 or more orders of true spklt with different phases of maturity. **Spklt** bisex or unisex, with 1 to many flt. **Glm** absent or 1–2+; **lm** without uncinate hairs, smt awned, awns single; **pal** well developed; **lod** (0)3(6+), memb, vascularized, often ciliate; **anth** usu 2, 3, or 6, rarely 10–120; **ov** glab or hairy, smt with an apical appendage; **haustorial synergids** absent; **sty** or **sty br** 1–4. **Car: hila** linear, usu as long as the car; **endosperm** hard, without lipid, containing compound starch grains; **emb** small relative to the car; **epiblasts** present; **scutellar cleft** present; **mesocotyl intnd** absent; **emb lf mrg** overlapping. $x = 7, 9, 10, 11, 12$.

Members of the *Bambusoideae* are native to temperate and tropical forests, high montane grasslands, river banks, and savannahs of Asia, North America, and South America. They have traditionally been placed in two tribes, *Bambuseae* and *Olyreae*, woody species being placed in the *Bambuseae* and herbaceous species in the *Olyreae*. Recent work (Sarawood et al. 2009) indicates that the woody species should be divided into two tribes, *Arundinarieae*, most of whose species grow in temperate habitats, and the *Bambuseae*, most of whose species grow in tropical habitats.

No members of the *Bambusoideae* are native in the Intermountain Region, but they are increasingly popular as ornamentals. Identification of the introduced species is problematic because the taxonomy of the wild species from which they are derived is poorly known. Another factor is that there is no record of which species are being grown in the region. Descriptions and illustrations of three species, representing three genera, are presented in order to provide a taste of the characteristics used to distinguish bamboos.

1. **Bambusa multiplex** (Lour.) Raeusch. *ex* Schult. & Schult. f. Hedge Bamboo [p. 277]
Pl densely clumping, without thorny br. **Clm** 0.5–7 m tall, 1–2.5 cm thick, emerging at an angle, broadly arching above, usu thin-walled and hollow, solid in some cultivars; **nd** not swollen; **intnd** all similar, 3–60 cm. **Br** to 20 per nd, erect to spreading, the cent br slightly

dominant, often becoming densely congested and forming tangled clusters of rhz, aborted shoots, and stunted roots, branchlets of the lo br not thornlike. **Clm lvs** 12–15 cm, narrowly triangular, tardily deciduous, initially light green, becoming reddish brown to stramineous, glab; **aur** and **fimbriae** developed; **bld** 1–2 cm, initially appressed to the clm, initially antrorsely hispid on both surfaces, becoming glab. **Foliage lvs: shth** glab; **lig** to 0.5 mm; **aur** absent; **fimbriae** smt present; **bld** 7–15 cm long, 1–2 cm wide, abx surfaces glaucous and slightly pubescent, adx surfaces dark green and glab. **Psdspklt** 30–40 mm, with up to 10 flt. $2n = 72$.

Bambusa multiplex is native to southeast Asia. It is now widely planted around the world. The dense foliage with many leaves on each branchlet makes it well suited to hedging. A large number of cultivars are available, some with striped culms and leaves, others with greatly reduced stature and leaf size suitable for bonsai culture or hedging. The tangled branch clusters allow natural dispersal and easy propagation in hot, humid climates. Plants listed as *B. glaucescens* (Willd.) Sieb. *ex* Munro in North America probably belong to *B. multiplex*.

2. Phyllostachys bambusoides Siebold & Zucc. Giant Timber Bamboo, Madake [p. 278]

Clm to 22 m tall and 15 cm thick, erect or leaning towards the light, base sinuous in some cultivars; **intnd** glab, usu green, in cultivars golden yellow, or with yellow and green stripes, lustrous; **nodal ridges** usu prominent (scarcely discernible in 'Crookstem' forms); **shth scars** thin, not strongly flared, glab. **Clm lvs: shth** glab or pubescent, greenish to ruddy-buff, more or less densely dark-brown-spotted; **aur** absent from the bas shth, narrow to broadly ovate or falcate on the up shth; **fimbriae** greenish, crinkled; **lig** rounded and ciliolate to truncate and ciliate with coarse hairs; **bld** short, lanceolate, reflexed and crinkled on the lo lvs, those above longer and recurved, green or variously striped. **Foliage lvs: aur** and **fimbriae** usu well developed; **lig** well developed; **bld** to 20 cm long and 3.2 cm wide, usu puberulent to subglab. $2n = 48$.

Phyllostachys bambusoides, a widely cultivated species, is hardy to -17°C. Several cultivars are available, differing in the color of their culms and leaves.

3. Pseudosasa japonica (Siebold & Zucc. *ex* Steud.) Makino *ex* Nakai Japanese Arrow Bamboo, Metake, Yadake [p. 279]

Clm 1–3(5) m tall, to 1.5 cm thick, erect or nodding, finely ridged; **nd** slightly raised; **shth scar** large; **intnd** long, finely mottled, with a light ring of wax below the nd. **Clm shth** to 25 cm, bas glab, distally appressed-hispid, persistent; **aur** and **fimbriae** absent; **bld** 2–5 cm, erect, abx surfaces glab. **Br** usu 1 per nd, with no bas buds or br, smt rebrchg from more distal nd. **Foliage lvs: shth** glab, edges memb; **aur** absent or small and erect; **fimbriae** absent or scarce, erect; **lig** long, oblique, erose, slightly pubescent, abx lig glab to finely ciliate; **bld** 15–35 cm long, 1.5–5 cm wide, glab or abx surfaces sporadically shortly red-brown tomentose, light green to glaucous, adx surfaces dark green, glossy, glab. **Spklt** 3.5–10 cm, narrowly cylindrical, curved, with 5–20(25) flt. **Lm** 1.2–1.5 cm, glab, often mucronate, mucros about 2 mm; **pal** nearly equaling the lm, glab, keels finely cilate.

Pseudosasa japonica is a widely cultivated ornamental species that used to be grown for arrows in Japan. There are no known wild populations. It forms a tough and effective screen, and has become naturalized in British Columbia and the eastern United States. A shorter cultivar with partially ventricose culms, 'Tsutsumiana', and cultivars with variegated leaves are also available.

2. EHRHARTOIDEAE Link   Grass Phylogeny Working Group

Pl ann or per. **Clm** ann, smt woody, hollow or solid. **Lvs** distichous; **shth** open; **aur** smt present; **abx lig** absent; **adx lig** memb, scarious, or of hairs; **psdpet** smt present; **bld** rarely cordate or sagittate at the base, venation parallel; **mesophyll** not radiate; **adx palisade layer** usu absent; **fusoid cells** smt present; **arm cells** absent or present; **Kranz anatomy** not developed; **midribs** simple or complex; **adx bulliform cells** present; **stomates** with dome-shaped or triangular subsidiary cells; **bicellular microhairs** present, tml cells tapered; **papillae** smt present. **Infl** pan, rcm, or spikes, rarely with bracts other than those of the spklt; **dis** usu above the glm, smt beneath the spklt or at the base of the pri br. **Spklt** bisex or unisex, with 1 pist or bisex flt, smt with 1–2 strl flt below the fnctl flt. **Glm** absent or 2; **lm** without uncinate hairs, smt tml awned, awns single; **pal** well developed, lacking in strl flt; **lod** 2, usu memb, rarely fleshy, heavily vascularized; **anth** (1)3–6(16); **ov** glab, without an apical appendage; **sty** 2, free to the base to fused throughout, 2-brchd. **Fruits** car or achenes; **hila** long-linear; **endosperm** without lipid, usu containing compound starch grains, rarely with simple starch grains; **emb** to ⅓ the length of the car; **epiblasts** usu present; **scutellar cleft** usu present; **mesocotyl intnd** absent or very short; **emb lf mrg** usu overlapping. *x* = 12 (10, 15, 17).

The *Ehrhartoideae* encompasses 120 species and three tribes. Its members grow in forests, open hillsides, and aquatic habitats. One tribe, the *Oryzeae*, is native to North America; the *Ehrharteae* is introduced. Morphologically, the *Oryzeae* and the *Ehrharteae* are characterized by spikelets that have a distal unisexual or bisexual floret with up to two proximal sterile florets, and, frequently, six stamens in the staminate or bisexual florets.

1. Spikelets with 2 sterile florets below the functional floret, both well developed, at least the upper sterile floret as long as or longer than the functional floret; glumes from ½ as long as the spikelets to exceeding the florets; culms not aerenchymatous; plants of dry to damp habitats . . . . 1. *Ehrharteae*
1. Spikelets with 0–2 sterile florets below the functional floret, when present, sterile florets ⅛–⁹⁄₁₀ as long as the functional floret; glumes absent or highly reduced; culms aerenchymatous; plants of wet habitats . . . . 2. *Oryzeae*

1. EHRHARTEAE Nevski   Mary E. Barkworth

Pl ann or per. **Clm** ann, (1)6–200 cm, smt woody, not aerenchymatous, smt brchd above the base. **Shth** open, usu rounded on the back, glab or not, smt scabrous; **col** frequently with tuberculate hairs; **aur** usu present, often ciliate; **lig** usu memb, smt a memb rim or of hairs; **psdpet** not present; **bld** linear, venation parallel, cross venation not evident, abx surfaces with microhairs and variously shaped silica bodies, cross sections non-Kranz; **1st seedling lvs** with well-developed, erect bld. **Infl** tml, pan or unilateral rcm; **dis** above the glm, flt falling as a cluster. **Spklt** solitary, terete or lat compressed, with 3 flt, lo 2 flt strl, tml flt bisex, at least the up strl flt as long as or longer than the bisex flt; **rchl** smt shortly prolonged beyond the base of the bisex flt. **Glm** 2, from ½ as long as to exceeding the flt, (3)5–7-veined. **Stl flt: lm** coriaceous, 5–7-veined, awned or unawned; **pal** lacking. **Bisex flt: lm** lanceolate or rectangular, firmly cartilaginous to coriaceous, 5–7-veined, veins inconspicuous, apc entire, unawned; **pal** 0–2(5)-veined; **lod** 2, free; **anth** 1–6; **sty** 2, fused or free to the base, stigmas linear, plumose. **Car** ellipsoid; **hila** linear, at least ½ as long as the car; **emb** up to ⅓ the length of the car, waisted, without an epiblast, with a scutellar tail and a minute mesocotyl intnd. *x* = 12.

The number of genera recognized in the *Ehrharteae* varies from one to four (Willemse 1982; Edgar and Connor 2000; Wheeler et al. 2002). The largest genus, *Ehrharta*, is native to Africa, the other three being Australasian. Only one genus, *Ehrharta*, has been introduced to North America.

1.01 EHRHARTA Thunb.   Mary E. Barkworth

Pl ann or per; synoecious. **Clm** 6–200 cm, smt woody, erect to decumbent, smt brchd above the base, usu pubescent; **bas brchg** invag. **Lvs** bas or bas and cauline; **shth** terete, open; **aur** present, often ciliate; **lig** 0.5–3 mm, truncate, memb or of hairs; **bld** linear to lanceolate, smt dis from the shth. **Infl** rcm or pan; **pri br** spreading to ascending; **dis** above the glm, not between the flt. **Spklt** 2–17 mm, solitary, pedlt, terete or lat compressed, with 3 flt, lo 2 flt strl, at least the up equaling or exceeding the distal flt, distal flt bisex. **Glm** from about ½ as long as to exceeding the flt, (3)5–7-veined. **Stl flt** consisting only of lm; **stl lm** firmer than the glm, glab or pubescent, stipitate or non-stipitate, smooth to rugose, unawned or awned, lowest lm often with lat earlike appendages at the base, up lm subequal to or longer than the distal flt. **Bisex flt: lm** often indurate at maturity, glab, 5–7-veined, keeled, unawned, smt mucronate; **pal** thinner than the lm, 1–2(5)-veined; **anth** (1–5)6, yellow; **sty** 2, fused or free to the base, white or brown. **Car** lat compressed. *x* = 12. Named for Jakob Friedrich Ehrhart (1742–1795), a German botanist of Swiss origin who studied under Linnaeus.

Ehrharta is a genus of approximately 25 species, most of which are native to southern Africa. Only one species has been found in the Intermountain Region.

1. **Ehrharta calycina** Sm. PERENNIAL
   VELDTGRASS [p. *280*]
Pl per; ces, often rhz. **Clm** 30–75(180) cm, erect, glab. **Shth** finely striate, smooth, smt densely pubescent, with short hairs between the veins, usu purplish; **aur** ciliate; **lig** about 1 mm, lacerate, glab; **bld** 2–9 cm long, 2–7 mm wide, flat or involute, surfaces glab, smt scabridulous, mrg hairy, wavy. **Pan** 7–22 cm, smt partially enclosed in the up lf shth, smt nodding; **ped** curved or bent,

smt straight. **Spklt** 4–9 mm, U-shaped, purplish. **Glm** subequal, 3–8 mm long, ¾–⁹⁄₁₀ the length of the spklt, 7-veined; **strl lm** hairy, smooth, lo strl lm from ⅔ the length of to equaling the up strl lm, bases with earlike appendages, apc of both lm mucronate or shortly awned; **bisex lm** slightly shorter than the up strl lm, 5–7-veined, glab or sparsely pubescent; **pal** shorter than the lm, 2-veined; **anth** 6, 2.5–3.5 mm. **Car** about 3 mm. $2n = 24$–28, 30.

Ehrharta calycina is native to southern Africa. It was introduced to Davis, California, as a drought-resistant grass for rangelands, but it is unable to withstand heavy grazing. It was collected at a nursery in Humboldt County, Nevada, but it has not persisted in the Intermountain Region. Four varieties have been described; they are not treated here.

2. **ORYZEAE** Dumort.

Edward E. Terrell

Pl ann or per; synoecious or monoecious. **Clm** ann, 20–500 cm tall, aerenchymatous, smt floating. **Lvs** aerenchymatous; **aur** present or absent; **lig** memb or scarious, smt absent; **psdpet** smt present; **bld** with parallel veins, cross venation not evident; **abx bld epdm** with microhairs and transversely dumbbell-shaped silica bodies; **1st seedling lf** without a bld. **Infl** usu pan, smt rcm or spikes; **dis** below the spklt, not occurring in cultivated taxa. **Spklt** lat compressed or terete, with 1 bisex or unisex flt, if unisex, pist and stmt spklt in the same or different pan, smt with 2 strl flt below the sex flt, these no more than ½(⁹⁄₁₀) the length of the ftl flt; **rchl** not prolonged. **Glm** absent or highly rdcd, forming an annular ring or lobes at the ped apc; **stl flt** ⅛–½(⁹⁄₁₀) as long as the spklt; **ftl lm** 3–14-veined, memb or coriaceous, apc entire, unawned or with a tml awn; **pal** similar to the lm, 3–10-veined, 1-keeled; **lod** 2; **anth** usu 6(1–16); **sty** 2, bases fused or free, stigmas linear, plumose. **Fruits** usu car, smt achenes, ovoid, oblong, or cylindrical; **emb** of the F+FP or F+PP type, small or elongate, with or without a scutellar tail; **hila** usu linear. $x = 12, 15, 17$.

The *Oryzeae* include about 10–12 genera and 70–100 species. Its members are native to temperate, subtropical, and tropical regions. *Oryza sativa*, rice, is one of the world's most important crop species. One genus, *Leersia*, is native to the Intermountain Region.

1. Plants rooted in soil, native in the Intermountain Region. . . . . 2.01 *Leersia*
1. Plants floating, growing in aquaria in the Intermountain Region . . . . 2.02 *Hygroryza*

2.01 **LEERSIA** Sw.

Grant L. Pyrah

Pl usu per, rarely ann; terrestrial or aquatic; rhz or csp; synoecious. **Clm** 20–150 cm (occ longer in floating mats), erect or decumbent, often rooting at the nd, brchd or unbrchd. **Lvs** equitably distributed along the clm; **shth** open; **aur** absent; **lig** memb; **psdpet** absent; **bld** aerial, linear to broadly lanceolate, flat or folded, smt involute when dry. **Infl** tml pan, usu exserted, axillary pan smt present; **dis** beneath the spklt. **Spklt** bisex, with 1 flt; **flt** lat compressed, linear to suborbicular in sideview. **Glm** absent; **cal** not stipelike, glab; **lm** and **pal** subequal, chartaceous to coriaceous, ciliate-hispid or glab, tightly clasping along the mrg; **lm** 5-veined, obtuse or acute to acuminate, smt mucronate, usu unawned; **pal** 3-veined, unawned; **lod** 2; **anth** 1, 2, 3, or 6; **sty** 2, bases fused, stigmas lat exserted, plumose. **Car** lat compressed; **emb** about ⅓ as long as the car; **hila** linear. $x = 12$. Named for Johann Daniel Leers (1727–1774), a German botanist and pharmacist.

Leersia is a genus of about 17 aquatic to mesophytic species that grow primarily in tropical and warm-temperate regions. One species is native to the Intermountain Region. *Leersia* is closely allied to *Oryza*.

1. **Leersia oryzoides** (L.) Sw. Rᴵᴄᴇ Cᴜᴛɢʀᴀss [p. 280, 493]

Pl per; rhz, rhz elongate, scaly, scales not imbricate. **Clm** 35–150 cm tall, 1–3 mm thick, brchg, decumbent, sprawling, rooting at the nd, tml portions erect. **Shth** scabrous; **lig** 0.5–1 mm; **bld** 7–30 cm long, 5–15 mm wide, spreading to slightly ascending, both surfaces usu scabrous. **Pan** 10–30 cm, tml, also axillary, exserted or enclosed at maturity, spreading on exserted pan, usu 2 or more br at the lo nd, 1 at the up nd; **br** 4–10 cm, the lo ⅓ naked, spklt imbricate. **Spklt** (4)4.2–6.5 mm long, 1.3–1.7 mm wide, elliptic. **Lm** and **pal** usu ciliate on the keels and mrg, glab or puberulent elsewhere; **anth** 3, 1.5–2(3) mm in chasmogamous spklt, 0.4–0.7 mm in cleistogamous spklt. **Car** 2–3.5 mm, asymmetrically pyriform to obovoid, whitish to dark brown. $2n = 48$.

Leersia oryzoides grows in wet, heavy, clay or sandy soils, and is often aquatic. It is found across most of southern Canada, extending south throughout the contiguous United States into northern Mexico, and flowers from July to October. It has also become established in Europe and Asia.

2.02 **HYGRORYZA** Nees

J.K. Wipff

Pl per; aquatic, producing long, floating clm; synoecious. **Clm** 50–150 cm, spongy, developing adventitious roots at the nd, brchd; **br** erect, lfy. **Lvs** cauline, glab, veins tessellate; **shth** open, inflated, serving as floats; **lig** absent or hyaline; **psdpet** present; **bld** elliptic, ovate, ovate-lanceolate, or oblong. **Infl** tml pan, aerial, lowermost br whorled; **dis** beneath the spklt cal. **Spklt** bisex, lat compressed, with 1 flt. **Glm** absent or an annular rim; **cal** (1)2–10 mm, stipelike, glab, jnct with the ped marked by a tan constriction; **lm** 5-veined, mrg clasping the pal, apc acuminate, awned, awns tml, antrorsely scabridulous; **pal** similar to the lm, 3-veined, 1-keeled, acute-acuminate, unawned; **lod** 2, glab; **anth** 6; **sty** 2, bases not fused, stigmas lat exserted, plumose. **Car** terete, fusiform; **emb** small; **hila** linear, almost as long as the emb. $x = 12$. Name from the Greek *hygros*, 'wet' or 'moist', and *oryza*, 'rice', referring to its aquatic habit and similarity to rice.

Hygroryza is a monospecific Asian genus that grows in India, Sri Lanka, and throughout southeast Asia. It forms floating masses, often of considerable extent, in lakes and slow-moving streams, and is sometimes a weed in rice.

1. **Hygroryza aristata** (Retz.) Nees ASIAN WATERGRASS, WATER STARGRASS [p. *281*]

Clm 50–150 cm, floating, brchd, flexuous; **nd** rooting, roots feathery, whorled. **Shth** glab, open, inflated; **lig** absent or 0.5–0.8 mm, hyaline, truncate; **psdpet** shorter than 1 mm; **bld** 2–8 cm long, 5–20 mm wide, flat, bases rounded to cordate, apc blunt to rounded. **Pan** 3–8 cm, pyramidal, with 4–9 br; **lo br** 2–4 cm, whorled, spreading or deflexed, glab; **ped** 0.2–2 mm, smt absent. **Flt** (6)7–18 mm (including the stipelike cal), narrowly lanceolate; **cal** (1)2–10 mm, stipelike; **lm** 5–8 mm, chartaceous, keeled, veins and mrg with hairs, glab or pubescent between the veins, awns 5–14 mm; **pal** as long as the lm, chartaceous, glab, keels ciliate or scabrous; **anth** 6, about 3.5 mm. **Car** about 3.5 mm. $2n = 24$.

Hygroryza aristata is sold for ponds and aquaria, where its long, feathery, adventitious roots have a decorative effect. It is included in this treatment because it has the potential to become a significant weed problem.

3. **POOIDEAE** Benth.

Grass Phylogeny Working Group

Pl ann or per; smt matlike, smt ces, smt stln, smt rhz. **Clm** usu hollow, smt solid. **Lvs** distichous; **shth** usu open to the base, varying to closed for nearly their full length; **aur** present or absent; **abx lig** absent; **adx lig** scarious or memb, smt puberulent or scabridulous, usu not ciliate, cilia smt shorter than the base; **psdpet** rarely present; **bld** usu linear, smt broadly so, venation parallel; **cross sections** non-Kranz, mesophyll nonradiate, adx palisade layer absent, fusoid and arm cells usu absent; **midribs** usu simple; **adx bulliform cells** present; **stomates** with parallel-sided subsidiary cells; **epdm** usu lacking bicellular microhairs, smt with unicellular microhairs, papillae usu absent, when present, rarely more than 1 per cell. **Infl** usu tml, pan, spikes, or rcm, usu ebracteate; **dis** usu below the flt, smt below the glm, at the rchs nd, or at the infl bases. **Spklt** usu bisex, infrequently unisex or mixed, usu lat compressed or not compressed, occ dorsally compressed, with 1–30 sex flt, distal flt(s) often rdcd, infrequently spklt with 1–2 rdcd or stmt bas flt and a single tml sex flt. **Glm** usu 2, up or lo glm smt absent, rarely both glm absent; **lm** without uncinate hairs, awned or not, awns single, bas to apical; **pal** usu well developed, smt rdcd or absent; **lod** 2(3), usu lanceolate and broadly memb distally, rarely truncate and fleshy, usu not veined or obscurely veined, smt distinctly veined, smt ciliate; **anth** (1, 2)3; **ov** glab or smt hairy distally, smt with an apical appendage; **haustorial synergids** absent; **sty** (1)2(–4), bases close together, smt fused. **Car**: **hila** linear, elliptic, ovate, or punctate; **endosperm** usu hard, smt soft or liquid, with or without lipids, starch grains compound or simple; **emb** less than ½ the length of the car; **epiblasts** usu present; **scutellar cleft** usu absent; **mesocotyl intnd** usu absent; **emb lf mrg** overlapping. $x = 7, 10$.

The subfamily *Pooideae* includes approximately 3300 species, making it the largest subfamily in the *Poaceae*. It reaches its greatest diversity in cool temperate and boreal regions, extending across the tropics only in high mountains.

The circumscription and relationships of tribes within the *Pooideae* are unsettled (see, for example, Catalán 2004, Quintanar et al. 2007, Doring et al. 2007, Bouchenak-Khelladi et al. 2008, Schneider et al. in press). In this flora, some previously recognized tribes have been combined within the *Poeae*. Further work will undoubtedly support the breakup of this expanded *Poeae* into additional tribes, but there is as yet no clear indication as to what the boundaries of such tribes should be.

1. Inflorescences 1-sided spikes, the spikelets radial to and partially embedded in the rachises; spikelets with 1 floret each . . . . 3. *Nardeae*
1. Inflorescences panicles, racemes, or 2-sided spikes, the spikelets radial or tangential to and sometimes embedded in the rachises but never both radial and embedded; spikelets with 1–30 florets.
  2. Cauline leaf sheaths closed for at least ¾ their length; lemmas longer than 4.5 mm and/ or awned and/or with prominent, parallel veins.
    3. Lemmas with parallel veins, often also with a purplish transverse band in the distal half, usually unawned; plants perennial; ovary apices glabrous; bases of style branches strongly recurved . . . . 4. *Meliceae*
    3. Lemmas with convergent veins, without a purplish transverse band in the distal half, usually awned; plants annual or perennial; ovary apices hairy; bases of style branches not recurved . . . . 6. *Bromeae*
  2. Cauline leaf sheaths open for all or most of their length, if closed the lemmas shorter than 7 mm, unawned, and with inconspicuous, convergent veins.
    4. Inflorescences spikes or spikelike.
      5. Inflorescences without branches, with 1–5 spikelets per node, if 1, the spikelets tangential to the rachis and all spikelets with 2 glumes . . . . 7. *Triticeae*
      5. Inflorescences with or without branches, if branched, the branches shorter than 3 mm, often fused to the rachis, if 1, the spikelets radial to the rachis and all but the terminal spikelet with only 1 glume. . . . . 8. *Poeae* (in part)
    4. Inflorescences panicles or racemes.
      6. Spikelets with 1 floret; lemmas with 1 terminal awn, the junction of the lemma and the awn abrupt, evident; glumes subequal to or longer than the floret (excluding the awn). . . . . 5. *Stipeae*
      6. Spikelets with 1-22 florets; lemmas unawned or awned, the awn basal, dorsal or terminal, if terminal the transition from the midvein to the awn gradual, not evident; glumes from shorter to longer than the adjacent florets or absent . . . 8. *Poeae* (in part)

3. **NARDEAE** W.D.J. Koch
Mary E. Barkworth

Pl per; ces. **Clm** ann, to 60 cm; **intnd** hollow. **Shth** open, mrg not fused; **col** glab, without tufts of hair at the sides; **aur** absent; **lig** scarious, not ciliate, those of the up and lo cauline lvs usu similar; **psdpet** not present; **bld** filiform, venation parallel, cross venation not evident, sec veins parallel to the midvein; **cross sections** non-Kranz, without arm or fusoid cells, **adx epdm** with bicellular microhairs, not papillate. **Infl** tml spikes, 1-sided, spklt solitary, radial to the rchs; **rchs** with the spklt partially embedded. **Spklt** not compressed, triangular in cross section, with 1 flt, flt bisex; **rchl** not prolonged beyond the flt base; **dis** above the glm, beneath the flt. **Glm** absent or vestigial; **lo glm** a cupular rim; **up glm** absent or vestigial; **flt** 5–10 mm, not compressed; **cal** poorly developed, glab; **lm** chartaceous, 3-veined, angled over the veins, most strongly so over the lat veins, apc entire, awned, awns tml, not brchd, lm-awn transition gradual, not evident; **pal** subequal to the lm, hyaline, 2-keeled; **lod** absent; **anth** 3; **ov** glab; **sty** 1. **Car** fusiform, sty bases not persistent; **hila** linear, more than ½ as long as the car; **emb** about ⅙ the length of the car. *x* = 13.

The *Nardeae* include two genera, *Nardus* and *Lygeum*. Embryo characters, ligule texture, and DNA sequence data place them in the *Pooideae*. Only *Nardus* is found in the Intermountain Region. Its inclusion here in the *Pooideae* reflects the findings of the Grass Phylogeny Working Group (2001) and Bouchenak-Khelladi et al. (2008).

3.01 **NARDUS** L.
Mary E. Barkworth

Pl per; ces. **Clm** 3–60 cm, erect; **bas brchg** invag. **Lvs** mostly bas; **shth** open; **aur** absent; **lig** memb, entire, rounded; **bld** filiform, tightly convolute, epdm with bicellular microhairs. **Infl** tml spikes, 1-sided, spklt in 2 rows, loosely to closely imbricate; **rchs** terminating in a bristle; **dis** below the flt. **Spklt** triangular in cross section, with 1 flt, flt bisex. **Lo glm** a highly rdcd, cupular rim; **up glm** absent or vestigial; **flt** 5–10 mm; **lm** linear-lanceolate to lanceolate-oblong, chartaceous, enveloping the pal, 3-veined, awned; **pal** hyaline, 2-veined, 2-keeled; **lod** absent; **anth** 3; **sty** 1. $x = 13$. Name from the Greek *nardos*, referring to spikenard, an aromatic herb. It is not clear why the name was applied to this genus; its only species is not scented.

Nardus is a monospecific European genus.

1. **Nardus stricta** L. MATGRASS [p. *281*, <u>493</u>]

Clm (3)10–40(60) cm, stiff, wiry, frequently gray-green; **nd** 1(2) per clm, restricted to the lo portion of the clm, pubescent; **intnd** glab. **Shth** smooth, whitish, tough; **lig** 0.5–1(2) mm, blunt; **bld** 4–30 cm long, 0.5–1 mm wide, stiff, tightly convolute, abx surfaces hispid, hairs about 0.3 mm, adx surfaces scabridulous, ribbed over the veins, apc sharply acute. **Spikes** (1)3–8 cm, terminating in a bristle, bristle to 1 cm. **Spklt** 5–10 mm, narrowly linear, triangular in cross section, bluish or purplish; **lm** 5–10 mm, 2–3-keeled, awned, awns 1–4.5 mm; **pal** slightly shorter than the lm; **anth** 1–4 mm. **Car** 3–4.5 mm, tightly enclosed by the lm and pal. $2n = 26$.

Nardus stricta is a widespread xerophytic and glycophytic species in Europe, usually growing in open areas on sandy or peaty soils. In North America, it is found in scattered locations from upper Michigan to Newfoundland and Greenland, and in Oregon and Idaho, where it is listed as a state noxious weed. The stiff, sharp leaves make it unpalatable; hence it tends to survive in areas of heavy grazing. This, combined with its broad ecological range, makes its potential for spreading in western rangelands a matter of concern.

4. **MELICEAE** Endl.

Mary E. Barkworth

Pl usu per, smt ann; ces, smt rhz. **Clm** ann, not woody, not brchg above the base; **intnd** hollow. **Shth** closed for their whole length or almost so; **col** without tufts of hair on the sides; **aur** smt present; **lig** hyaline, glab, often lacerate, occ ciliate, those of the lo and up cauline lvs usu similar; **psdpet** absent; **bld** linear to narrowly lanceolate, venation parallel, cross venation smt evident; **cross sections** non-Kranz, without arm or fusoid cells; **epdm** without microhairs, smt papillate. **Infl** tml pan or rcm; **dis** above the glm and beneath the flt or below the glm. **Spklt** 2.5–60 mm, not viviparous, slightly to strongly lat compressed, with 1–30 flt, proximal flt bisex, distal 1–3 flt usu strl, smt pist, smt rdcd and amalgamated into a knob- or club-shaped *rudiment*; **rchl** prolonged beyond the base of the distal flt. **Glm** exceeded by the distal flt, shorter than to longer than the adjacent lm, mostly memb, scarious distally, 1–11-veined, apc usu rounded to acute; **flt** lat or dorsally compressed; **cal** blunt, glab or with hairs; **lm** of sex flt rectangular or ovate, mostly memb, scarious distally, often with a purplish band adjacent to the scarious apc, (4)5–15-veined, veins not converging distally, often prominent, unawned or awned, awns not brchd, apc entire to bilobed or bifid, awns straight, subtml or from the sinuses; **pal** from shorter than to longer than the lm, similar in texture, 2-veined, veins keeled, smt winged; **lod** 2, fleshy, usu connate into a single structure, without a memb wing, truncate, not ciliate, not or scarcely veined; **anth** 1, 2, or 3; **ov** glab; **sty** 2-brchd, bases persistent, br plumose distally. **Car** ovoid to ellipsoid, longitudinally grooved or not; **hila** usu linear; **emb** less than ⅓ as long as the car. $x = (8)9, 10$.

There are approximately 130 species and 8 or 9 genera in the *Meliceae*. Four of the genera are monotypic. *Melica* and *Glyceria*, the two largest genera, are well represented in North America. *Pleuropogon* and *Schizachne* are primarily North American, but extend into eastern Asia. Four genera are native to the Intermountain Region.

Molecular studies (e.g., Grass Phylogeny Working Group 2001; Bouchenak-Khelladi et al. 2008) show the tribe to be monophyletic and somewhat basal within the *Pooideae*. Members of the tribe are most easily recognized by the combination of closed leaf sheaths, scarious lemma apices, and non-converging lemma veins. The tribe also differs from other tribes in the *Pooideae* in having 2 unwinged lodicules that are usually connate into a single structure, and a base chromosome number of 9 or 10. *Catabrosa* and *Briza*, whose inclusion in the tribe was suggested by the preliminary results of Mejia-Saulés and Bisby (2000), have more membranous lemma margins and free, winged lodicules. *Briza* also has open leaf sheaths and more convergent lemma veins. Their inclusion is not supported by the molecular data.

1. Calluses hairy; lemmas awned, awns 8–15 mm long, twisted on the lower portion, divergent to slightly geniculate . . . . 4.03 *Schizachne*
1. Calluses glabrous; lemmas unawned or awned, awns to 12 mm long, straight
   2. Inflorescences racemes; palea keels winged, the wings notched and awned . . . . 4.04 *Pleuropogon*
   2. Inflorescences usually panicles, racemes in depauperate specimens; palea keels not winged or the wings entire and unawned.

4.01 GLYCERIA R. Br.   Mary E. Barkworth and Laurel K. Anderton

Pl usu per, rarely ann; rhz. **Clm** (10)20–250 cm, erect or decumbent, freely rooting at the lo nd, not cormous based. **Shth** closed for at least ¾ their length, often almost entirely closed; **lig** scarious, erose to lacerate; **bld** flat or folded. **Infl** tml, usu pan, smt rcm in depauperate specimens, br appressed to divergent or reflexed. **Spklt** cylindrical and terete or oval and lat compressed, with 2–16 flt, tml flt in each spklt strl, rdcd; **dis** above the glm, below the flt. **Glm** much smaller than to equaling the adjacent lm, 1-veined, obtuse or acute, often erose; **lo glm** 0.3–4.5 mm; **up glm** 0.6–7 mm; **cal** glab; **lm** memb to thinly coriaceous, rounded over the back, smooth or scabrous, glab or hairy, hairs to about 0.1 mm, 5–11-veined, veins usu evident, often prominent and ridged, not or scarcely converging distally, apical mrg hyaline, smt with a purplish band below the hyaline portion, apc acute to rounded or truncate, entire, erose, or irregularly lobed, unawned; **pal** shorter than to longer than the lm, keeled, keels smt winged; **lod** thick, smt connate, not winged; **anth** (1)2–3; **ov** glab; **sty** 2-brchd, br divergent to recurved, plumose distally. $x = 10$. Name from the Greek *glukeros*, 'sweet', the caryopses of the type species being sweet.

Glyceria includes approximately 35 species, all of which grow in wet areas. All but five species are native to the Northern Hemisphere. The genus is represented in the Intermountain Region by four native and two introduced species.

All native species of *Glyceria* are palatable to livestock. They are rarely sufficiently abundant to be important forage species. Some grow in areas that are soon degraded by grazing. *Glyceria maxima* can cause cyanide poisoning in cattle.

Glyceria resembles *Puccinellia* in the structure of its spikelets and its preference for wet habitats; it differs in its inability to tolerate highly alkaline soils, and its usually more flexuous panicle branches, closed leaf sheaths, and single-veined upper glumes. Some species are apt to be confused with *Torreyochloa pallida*, another species associated with wet habitats but one that, like *Puccinellia*, has open leaf sheaths.

Glyceria includes several species that appear to intergrade. In some cases, the distinctions between such taxa are more evident in the field, particularly when they are sympatric. Recognition of such taxa at the specific level is merited unless it can be shown that all the distinctions between them are inherited as a group.

Culm thickness is measured near midlength of the basal internode; it does not include leaf sheaths. Unless otherwise stated, ligule measurements reflect both the basal and upper leaves. Ligules of the basal leaves are usually shorter than, but similar in shape and texture to, those of the upper leaves. The number of spikelets on a branch is counted on the longest primary branches, and includes all the spikelets on the secondary (and higher order) branches of the primary branch. Pedicel lengths are measured for lateral spikelets on a branch, not the terminal spikelet. Lemma characteristics are based on the lowest lemmas of most spikelets in a panicle. There is often, unfortunately, considerable variation within a panicle.

1. Spikelets cylindrical, terete except at anthesis, their length more than 5 times their width.
    2. Lemmas 2.7–5.4 mm long, apices usually acute, sometimes obtuse, entire or almost so; blades of mid-cauline leaves densely papillose on the adaxial (upper) surface . . . . 5. *G. borealis*
    2. Lemmas (3.5)4–6 mm long, apices acute, with a well-developed lobe on one or both sides; adaxial surface of mid-cauline leaves not papillose. . . . . 6. *G. declinata*
1. Spikelets laterally compressed, their length 1–4 times their width.
    3. Upper glumes 2.5–5 mm long, longer than wide . . . . 2. *G. maxima*
    3. Upper glumes 0.6–3.7 mm long, if longer than 3 mm, shorter than wide.
        4. Lemma apices almost flat; anthers 3; vein of one or both glumes usually extending to the apices . . . . 1. *G. grandis*
        4. Lemma apices prow-shaped; anthers 2; veins of both glumes terminating before the apices.
            5. Blades 2–6 mm wide; anthers 0.2–0.6 mm long; culms 1.5–3.5 mm thick. . . . . 3. *G. striata*
            5. Blades 6–15 mm wide; anthers 0.5–0.8 mm long; culms 2.5–8 mm thick . . . . 4. *G. elata*

1. **Glyceria grandis** S. Watson AMERICAN GLYCERIA, AMERICAN MANNAGRASS [p. *282*, 493]

Pl per. **Clm** 50–150 (200) cm tall, 8–12 mm thick, erect or decumbent and rooting at the base. **Shth** smooth or scabridulous, keeled; **lig** 1–5 (7) mm, truncate to rounded, lig of the lo lvs stiff at the base, lig of the up lvs flexible throughout; **bld** 25–43 cm long, 4.5–15 mm wide. **Pan** 16–42 cm long, 12–20 cm wide, open; **br** (7)10–18 cm, lax, widely divergent to drooping, with 35–80+ spklt; **ped** 1–15 mm. **Spklt** 3.2–10 mm long, 2–3 mm wide, somewhat lat compressed, oval to elliptic in side view, with 4–10 flt. **Glm** mostly hyaline, usu the midvein of 1 or both glm extending to the apc,

apc acute; **lo glm** 1–2.3 mm; **up glm** 1.5–2.7 mm; **rchl intnd** 0.5–0.8 mm; **lm** 1.8–3 mm, prominently (5)7-veined, veins often scabridulous, intercostal regions smooth, apc rounded to truncate, smt erose, almost flat at maturity; **pal** from shorter than to slightly longer than the lm, lengths more than 3 times widths, keels not winged, ciliolate, tips not strongly incurved, truncate to notched between the keels; **anth** 3, 0.5–1.2 mm. **Car** 1–1.5 mm. $2n = 20$.

Glyceria grandis grows on banks and in the water of streams, ditches, ponds, and wet meadows, from Alaska to Newfoundland and south in the mountains to California, Arizona, and New Mexico in the western United States, and to Virginia and Tennessee in the eastern United States. It is sometimes confused with *G. elata* and *Torreyochloa pallida*. It differs from the former in having acute glumes with long veins, more evenly dark florets, flatter lemma apices, and paleal keel tips that do not point towards each other. It differs from *Torreyochloa pallida* in its closed leaf sheaths and 1-veined glumes.

Glyceria grandis S. Watson var. **grandis** GIANT GLYCERIA, GIANT MANNAGRASS [p. 282]
Spklt 3.2–6.4 mm, with 4–8 flt.

Glyceria grandis var. *grandis* is the more widespread of the two varieties, growing throughout the range of the species. It is the only variety to grow in the Intermountain Region.

2. **Glyceria maxima** (Hartm.) Holmb. TALL GLYCERIA, ENGLISH WATERGRASS [p. 282]
Pl per. **Clm** 60–250 cm tall, 6–12 mm thick, erect. **Shth** scabridulous, keeled; **lig** 1.2–6 mm, rounded or with a cent point, lig of the lo lvs thick, stiff, and opaque, lig of the up lvs thinner and translucent; **bld** 30–60 cm long, 6–20 mm wide, both surfaces smooth or adx surfaces scabridulous. **Pan** 15–45 cm long, to 30 cm wide, open; **br** 8–20 cm, lax, strongly divergent or drooping at maturity, scabridulous, pri br with 50+ spklt; **ped** 0.8–10 mm. **Spklt** 5–12 mm long, 2–3.5 mm wide, somewhat lat compressed, oval in side view, with 4–10 flt. **Glm** unequal, usu the midvein of 1 or both reaching to the apc; **lo glm** 2–3 mm; **up glm** 3–4 mm, longer than wide; **rchl intnd** 0.5–1 mm; **lm** 3–4 mm, 7-veined, veins scabridulous, apc broadly acute to rounded, slightly prow-shaped; **pal** subequal to the lm, lengths more than 3 times widths, keels not winged, ciliate, tips not strongly incurved, curved to broadly notched between the keels; **anth** 3, (1)1.2–2 mm. **Car** 1.5–2 mm. $2n = 60$.

Glyceria maxima is native to Eurasia. It grows in wet areas, including shallow water, at scattered locations in the Intermountain Region. It is also sometimes grown as an ornamental in the region, but this should be discouraged because it might become invasive. At some sites, the species appears to be spreading, largely vegetatively. It is easily confused with large specimens of *G. grandis*, but differs in its firmer, more prow-tipped lemmas as well as its larger lemmas and, usually, larger anthers.

3. **Glyceria striata** (Lam.) Hitchc. RIDGED GLYCERIA [p. 283, 493]
Pl per. **Clm** 20–80 (100) cm tall, (1.5)2–3.5 mm thick, not or only slightly spongy, smt rooting at the lo nd. **Shth** smooth to scabridulous, keeled, smt weakly so; **lig** 1–4 mm, usu rounded, smt acute to mucronate, erose-lacerate; **bld** 12–30 cm long, 2–6 mm wide, abx surfaces smooth or scabridulous, adx surfaces scabridulous to scabrous. **Pan** 6–25 cm long, 2.5–21 cm wide, pyramidal, open, nodding; **br** 5–13 cm, straight to lax, lo br usu strongly divergent to drooping at maturity, smt ascending, with 15–50 spklt, these often confined to the distal ⅔; **ped** 0.5–7 mm. **Spklt** 1.8–4 mm long, 1.2–2.9 mm wide, lat compressed, oval in side view, with 3–7 flt. **Glm** ovate, 1–1.5 times longer than wide, narrowing from midlength or above, veins terminating below the apical mrg, apc often splitting with age; **lo glm** 0.5–1.2 mm, rounded to obtuse; **up glm** 0.6–1.2 mm, acute or rounded; **rchl intnd** 0.1–0.6 mm; **lm** 1.2–2 mm, ovate in dorsal view, veins raised, scabridulous over and between the veins, apc acute, prow-shaped; **pal** slightly shorter than to equaling the lm, lengths 1.5–3 times widths, keeled, keels not winged, tips pointing towards each other, apc narrowly notched between the keels; **anth** 2, (0.2)0.4–0.6 mm, purple or yellow. **Car** 0.5–2 mm. $2n = 20$ [reports of $2n = 28$ are questionable].

Glyceria striata grows in bogs, along lakes and streams, and in other wet places. Its range extends from Alaska to Newfoundland and south into Mexico. Plants from the eastern portion of the range have sometimes been treated as *G. striata* var. *striata*, and those from the west as *G. striata* var. *stricta* (Scribn.) Fernald. Eastern plants tend to have somewhat narrower leaves and thinner culms than western plants, but the variation appears continuous. In the west, larger specimens are easy to confuse with *G. elata*. The two species are sometimes found growing together without hybridizing; this and molecular data (Whipple et al. 2007) support their recognition as separate species. The differences between the two in growth habit and stature are evident in the field; they are not always evident on herbarium specimens.

4. **Glyceria elata** (Nash) M.E. Jones TALL MANNAGRASS [p. 283, 493]
Pl per. **Clm** 75–150 cm tall, 2.5–8 mm thick, spongy, decumbent and rooting at the lo nd. **Shth** scabridulous or hirtellous, not or weakly keeled; **lig** 2.5–4(6) mm, truncate to acute, erose, puberulent; **bld** 19–40+ mm long, 6–12(15) mm

wide, abx surfaces smooth or scabridulous, adx surfaces usu scabrous, smt scabridulous. **Pan** 15–30 cm long, 12–30 cm wide, pyramidal, open; **br** 12–17 cm, divergent to drooping, lax, with 30–50+ spklt; **ped** 0.3–5 mm. **Spklt** 3–6 mm long, 1.5–2.8 mm wide, lat compressed, oval in side view, with 3–4(6) flt. **Glm** 1–1.5(2) times longer than wide, narrowing beyond midlength, veins terminating below the apical mrg, apc obtuse to rounded; **lo glm** 0.7–1.5 mm; **up glm** 1–1.5 mm; **rchl intnd** 0.5–0.6 mm; **lm** 1.7–2.2 mm, oval in dorsal view, 5–7-veined, veins raised throughout, scabridulous, apc rounded, prow-shaped; **pal** subequal to or often slightly longer than the lm, lengths 2.4–3 times widths, oval in dorsal view, keels not winged, tips pointing towards each other, apc narrowly notched between the keels; **anth** 2, 0.5–0.8 mm. **Car** 0.8–1.5 mm long, 0.5–0.7 mm wide; **hila** as long as the car. $2n = 20$.

Glyceria elata grows in wet meadows and shady moist woods, from British Columbia east to Alberta and south to California and New Mexico. It is very similar to, and sometimes confused with, *G. striata*, but the two sometimes grow together and show no evidence of hybridization. Their differences in growth habit and stature are evident in the field. Molecular data (Whipple et al. 2007) confirm that *G. elata* and *G. striata* are distinct, closely related entities.

Glyceria elata is also sometimes confused with *G. grandis*. It differs in having rounded glumes with veins that terminate below the apices, more readily disarticulating florets, and greener lemmas with more prow-shaped apices, as well as in having paleal keel tips that point towards each other.

5. **Glyceria borealis** (Nash) Batch. Boreal Glyceria, Boreal Mannagrass [p. *284*, 493]
Pl per. **Clm** 60–100 cm tall, 1.5–5 mm thick, often decumbent and rooting at the lo nd. **Shth** glab, keeled; **lig** 4–12 mm; **bld** 9–25 cm long, 2–7 mm wide, often floating, abx surfaces smooth, adx surfaces of the midcauline lvs densely papillose, glab. **Pan** 18–40(50) cm long, 0.5–2(5) cm wide, arching, usu narrow, open at anthesis, bases often enclosed in the up lf shth at maturity; **br** 5–10(15) cm, usu 1–3(5) per nd, usu appressed to strongly ascending, occ spreading, longer br with 3–6 spklt; **ped** 1.2–5 mm. **Spklt** 9–22 mm long, 0.8–2.5 mm wide, cylindrical and terete, except at anthesis when slightly lat compressed, rectangular in side view, with 8–12 flt. **Glm** elliptic, apc rounded to obtuse, smt erose; **lo glm** 1.2–2.2 mm; **up glm** 2–3.8 mm, rounded; **rchl intnd** 0.6–3.5 mm; **lm** 2.7–5.4 mm, veins raised, scabridulous or smooth, intercostal

regions usu smooth, smt scabridulous, midvein terminating about (0.1)0.2 mm short of the apical mrg, apc usu acute, smt obtuse, entire or almost so; **pal** usu shorter than to equaling the lm, smt exceeding them by up to 0.5 mm, keels narrowly winged, apc bifid, teeth to 0.2 mm, parallel to weakly incurved; **anth** 3, 0.4–1.5 mm. **Car** 1.2–2 mm. $2n = 20$.

Glyceria borealis is a widespread native species that grows in the northern portion of North America, extending southward through the western mountains into northern Mexico. It grows along the edges and muddy shores of freshwater streams, lakes, and ponds. In the southern portion of its range, *G. borealis* is restricted to subalpine and alpine areas. The midcauline leaves of *G. borealis* almost always have densely papillose upper leaf surfaces. Voss (1972) stated that such surfaces are non-wettable and develop on the floating leaves.

6. **Glyceria declinata** Bréb. Low Glyceria [p. *284*, 493]
Pl usu per, rarely ann. **Clm** (10)20–92 cm tall, 1.5–2.5 mm thick, ascending to erect from a decumbent, brchg base. **Shth** glab, keeled; **lig** 4–9 mm; **bld** (2)3–12 cm long, 4–8 mm wide, adx surfaces not papillose, apc abruptly acute. **Pan** 6–30 cm long, 1–2.5 cm wide; **br** 1.5–9.5 cm, ascending, with 1–5 spklt; **ped** 1–2.5 mm. **Spklt** 11–24 mm long, 1.3–3 mm wide, cylindrical and terete, except slightly lat compressed at anthesis, rectangular in side view, with 8–15 flt. **Glm** oval; **lo glm** 1.4–3.5 mm; **up glm** 2.5–4.9 mm; **rchl intnd** 1.2–1.8 mm; **lm** (3.5)4–6 mm, 7-veined, veins and intercostal regions scabridulous, prickles about 0.05 mm, midveins extending to within 0.1 mm of the apical mrg, apc acute, with a well-developed lobe on one or both sides opposite the lat veins, entire to crenulate between the lat lobes; **pal** exceeding the lm by 0.2–1(1.5) mm, keels winged, apc bifid, teeth 0.3–0.5 mm; **anth** 0.5–1.4 mm, usu purple. **Car** 1.8–2.5 mm. $2n = 20$.

Glyceria declinata is a European species that is now established in North America. It has been found once in northeastern Nevada, but it is not known if the population has persisted. In Europe, it grows in low-calcium, acidic soils and tolerates drier conditions than other European species of *Glyceria* (Conert 1992). In Denmark, it tends to grow in areas that are highly trampled (Niels Jacobsen and Signe Frederiksen, pers. comm.).

In western North America, *G. declinata* has been confused with *G. ×occidentalis*. The most reliable distinguishing characteristics are the lateral lemma lobes of *G. declinata* and its rather short, straight panicle branches. The two species also differ in their ploidy level, *G. declinata* being diploid and *G. ×occidentalis* tetraploid (Church 1949).

4.02 **MELICA** L.   Mary E. Barkworth

Pl per; csp or soboliferous, not or only shortly rhz. **Clm** (4)9–250 cm, smt forming a bas corm; **nd** and intnd usu glab. **Shth** closed almost to the top; **aur** smt present; **lig** thinly memb, erose to lacerate, usu glab, those of the lo lvs shorter than those of the up lvs; **bld** flat or folded, glab or hairy, particularly on the adx surfaces, smt scabrous. **Infl** tml pan; **pri br** often appressed; **sec br** appressed or divergent; **ped** either more or less straight or sharply bent below the spklt, scabrous to strigose distally; **dis** below the glm in species with sharply bent ped, above the glm in other species. **Spklt** with 1–7 bisex flt, terminating in a strl structure, the rdmt, composed of 1–4 strl flt; **rdmt** smt morphologically distinct from the bisex flt, smt similar but smaller. **Glm** memb or chartaceous, distal mrg wide, translucent; **lo glm** 1–9-veined; **up glm** 1–11-veined; **cal** glab; **lm** memb bas, smt becoming coriaceous at maturity, glab or with hairs, (4)5–15-veined, usu unawned, smt awned, awns to 12 mm, straight; **pal** from ½ as long as to almost equaling the lm, keels usu ciliate; **lod** fused into a single, collarlike structure extending ½–⅔ around the base of the ov; **anth** (2)3. **Car** usu 2–3 mm, smooth, glab, longitudinally furrowed, falling from the flt when mature. $x = 9$. From the Latin *mel*, 'honey', a classical name for an unknown, but presumably sweet, plant.

   Melica includes approximately 80 species, which grow in all temperate regions of the world except Australia, usually in shady woodlands on dry stony slopes (Mejia-Saulés and Bisby 2003). The species are relatively nutritious, but are rarely sufficiently abundant to be important as forage. Nine species are native to the Intermountain Region. Two European species are grown as ornamentals in the region; some of the native species merit such use.

   In the following key and descriptions, unless otherwise stated, comments on the panicle branches apply to the longest branches within the panicle; glume widths are measured from side to side, at the widest portion; lemma descriptions are for the lowest floret in the spikelets; and rachilla internode comments apply to the lowest internode in the spikelets.

1. Spikelets disarticulating below the glumes; pedicels sharply bent just below the spikelets.
   2. Lemmas conspicuously hairy, the hairs 3.5–5 mm long . . . . 11. *M. ciliata*
   2. Lemmas not hairy, sometimes scabridulous or scabrous.
      3. Rudiments clublike, not resembling the lower florets; culms 60–250 cm tall; plants cultivated . . . . 12. *M. altissima*
      3. Rudiments acute to acuminate, similar to but smaller than the lower florets; culms 9–100 cm tall; plants native.
         4. Spikelets broadly V-shaped when mature, 5–13 mm wide; upper glumes 6–18 mm long . . . . 9. *M. stricta*
         4. Spikelets parallel-sided when mature, 1.5–5 mm wide; upper glumes 5–8 mm long . . . . 10. *M. porteri*
1. Spikelets disarticulating above the glumes; pedicels not sharply bent below the pedicels.
   5. Rudiments truncate to acute, not resembling the lowest florets.
      6. Bisexual florets 1(2); paleas almost as long as the lemmas . . . . 1. *M. imperfecta*
      6. Bisexual florets 2–5; paleas about ¾ the length of the lemmas . . . . 5. *M. californica*
   5. Rudiments tapering, smaller than but otherwise similar to the lowest florets in shape.
      7. Lemmas awned . . . . 6. *M. harfordii* (in part)
      7. Lemmas unawned.
         8. Lemmas strongly tapering and acuminate, the veins usually hairy . . . . 7. *M. subulata*
         8. Lemmas acute to obtuse, the veins hairy or not.
            9. Lemmas pubescent, the hairs on the marginal veins clearly longer than the hairs elsewhere . . . . 6. *M. harfordii* (in part)
            9. Lemmas glabrous, scabrous, or pubescent, never with clearly longer hairs on the marginal veins.
               10. Rachilla internodes swollen when fresh, wrinkled when dry . . . . 8. *M. fugax*
               10. Rachilla internodes not swollen when fresh, not wrinkled when dry.
                  11. Paleas about ½ the length of the lemmas; lower panicle branches bearing 5–15 spikelets; culms not forming corms . . . . 4. *M. frutescens*
                  11. Paleas ⅔–¾ the length of the lemmas; lower panicle branches bearing 1–5 spikelets; culms sometimes forming corms.
                     12. Glumes about (½)⅔ the length of the spikelets; ligules 2–6 mm long; corms almost sessile on the rhizomes . . . . 3. *M. bulbosa*
                     12. Glumes usually less than ½ as long as the spikelets; ligules 0.1–2mm long; corms connected to the rhizomes by a root-like structure . . . . 2. *M. spectabilis*

1. **Melica imperfecta** Trin. LITTLE CALIFORNIA
   MELIC [p. *285*, 493]

Pl densely ces, not rhz. **Clm** 35–120 cm, not
forming corms; **intnd** scabridulous immediately
above the nd. **Shth** glab or pilose; **lig** 0.8–6.5
mm; **bld** 1–6 mm wide, abx surfaces glab or
puberulent, adx surfaces with hairs. **Pan** 5–36
cm; **br** 2.5–9 cm, appressed to reflexed, straight or
flexuous, with 5–30 spklt; **ped** not sharply bent;
dis above the glm. **Spklt** 3–7 mm, with 1(2) bisex
flt; **rchl intnd** 0.3–0.6 mm. **Lo glm** 2–5 mm long,
1–2 mm wide, 1-veined; **up glm** 2.5–6 mm long,
1.5–2.5 mm wide, 1-veined; **lm** 3–7 mm, glab,
smt scabrous, with 7+ veins, veins prominent,
apc rounded to acute, unawned; **pal** almost as
long as the lm; **anth** 1.5–2.5 mm; **rdmt** 1–4 mm,
not resembling the lo flt, longer and thicker than
the tml rchl intnd, truncate to obtuse. $2n = 18$.

Melica imperfecta grows from sea level to 1500 m, on stable
coastal dunes, dry, rocky slopes, and in open woods, from
California and southern Nevada south to Baja California,
Mexico. Plants vary with respect to size, panicle shape, and
pubescence, but no infraspecific taxa merit recognition.

2. **Melica spectabilis** Scribn. PURPLE
   ONIONGRASS [p. *285*, 493]

Pl loosely ces, rhz. **Clm** 45–100 cm, forming
corms, corms connected to the rhz by a rootlike,
10–30 mm structure, which usu remains attached
to the corm; **intnd** smooth. **Shth** usu glab, often
pilose at the throat and col; **lig** 0.1–2 mm; **bld**
2–5 mm wide, abx surfaces scabridulous over the
veins, adx surfaces usu glab. **Pan** 5–26 cm; **br** 2–5
cm, usu appressed, smt divergent and flexuous,
with 2–3 spklt; **ped** not sharply bent; **dis** above
the glm. **Spklt** 7–19 mm, with 3–7 bisex flt, base
of the distal flt concealed at anthesis; **rchl intnd**
1–2 mm, not swollen when fresh, not wrinkled
when dry. **Glm** usu less than ½ the length of the
spklt; **lo glm** 3.5–6.4 mm long, 1.5–3 mm wide,
1–3-veined; **up glm** 5–7 mm long, 2.3–3.5 mm
wide, 5–7-veined; **lm** 6–9 mm, glab, scabridulous,
5–11-veined, veins inconspicuous, apc rounded to
acute, unawned; **pal** about ⅔ the length of the lm;
anth 1.5–3 mm; **rdmt** 1.5–3.5 mm, acute, distinct
from the bisex flt, smt surrounded by a small strl
flt similar in shape to the bisex flt. $2n = 18$.

Melica spectabilis grows in moist meadows, flats, and open
woods, from 1200–2600 m, primarily in the Pacific Northwest
and the Rocky Mountains. It is often confused with *M. bulbosa*,
differing in its shorter glumes, "tailed" corm, and the more
marked and evenly spaced purplish bands of its spikelets.

3. **Melica bulbosa** Geyer *ex* Porter & J.M. Coult.
   ONIONGRASS [p. *285*, 493]

Pl loosely ces, rhz. **Clm** 29–100 cm, forming
corms, corms almost sessile on the connecting

rhz; **intnd** scabridulous above the nd. **Shth** usu
scabridulous, smt sparsely pilose; **lig** 2–6 mm;
bld 1.5–5 mm wide, abx surfaces scabridulous,
adx surfaces with hairs. **Pan** 7–30 cm; **br** 2–6.5
cm, appressed, usu straight, with 1–5 spklt; **ped**
straight; **dis** above the glm. **Spklt** 6–24 mm, with
4–7 bisex flt, base of the distal flt concealed at
anthesis; **rchl intnd** 1–2 mm, not swollen when
fresh, not wrinkled when dry. **Glm** from (½)⅔
as long as to equaling the spklt; **lo glm** 5.5–10.5
mm long, 2–3 mm wide, 3–5-veined; **up glm**
6–14 mm long, 2.3–3.5 mm wide, 5–7-veined; **lm**
6–12 mm, glab, smooth or scabrous, 7–11-veined,
veins prominent, apc emgt to acute, unawned;
pal about ¾ the length of the lm; **anth** 3, 1.5–4
mm; **rdmt** 1.5–5 mm, truncate to tapering, smt
resembling the bisex flt in shape. $2n = 18$.

Melica bulbosa grows from 1370–3400 m, mostly in open
woods on dry, well-drained slopes and along streams. It is
the most widespread species of *Melica* in the Intermountin
region.

Melica bulbosa differs from *M. spectabilis* in its sessile corms
and longer glumes. In addition, in *M. bulbosa* the spikelets
have purplish bands which appear to be concentrated
towards the apices; in *M. spectabilis* the bands appear more
regularly spaced. *Melica bulbosa* differs from *M. fugax* in not
having swollen rachilla internodes.

4. **Melica frutescens** Scribn. WOODY MELIC
   [p. *286*, 493]

Pl densely ces, not rhz. **Clm** 60–200 cm, not
forming corms, often brchd from the lo nd;
intnd smooth. **Shth** glab, smt scabridulous, smt
purplish; **lig** 2.5–9 mm; **bld** 2–5 mm wide, abx
sufaces scabridulous, adx surfaces puberulent.
Pan 12–40 cm; **br** 3.5–9 cm, appressed, with 5–15
spklt; **ped** straight; **dis** above the glm. **Spklt** 9–18
mm, with 3–5 bisex flt; **rchl intnd** 1–1.3 mm, not
swollen when fresh, not wrinkled when dry. **Lo**
glm 7–12 mm long, 2–3 mm wide, 5–7-veined; **up**
glm 8–15 mm long, 2.5–3.5 mm wide, 5–7-veined;
lm 8–11 mm, glab, chartaceous for the distal ⅓
or more, 7–9-veined, smt purplish bas, veins
inconspicuous, apc rounded to acute, unawned;
pal about ½ the length of the lm; **anth** 3, 1–2
mm; **rdmt** 2–6 mm, blunt, enclosed in empty lm
resembling those of the bisex flt. $2n = 18$.

Melica frutescens grows from 300–1500 m in the dry hills
and canyons of southern California, Arizona, and adjacent
Mexico.

5. **Melica californica** Scribn. CALIFORNIA
   MELIC [p. *286*, 493]

Pl densely ces, not rhz. **Clm** 50–130 cm, not
forming corms; **lo nd** strigose; **intnd** usu smooth,
smt puberulent below the nd, lo 2–3 intnd usu
swollen. **Shth** glab or pilose; **lig** 1.5–4 mm; **bld**

1.5–5 mm wide, strigose on both surfaces. **Pan** 4–30 cm; **br** 3–6 cm, appressed, straight, with 4–15 spklt; **ped** straight; **dis** above the glm. **Spklt** 5–15 mm, with 2–5 bisex flt; **rchl intnd** 1.1–1.6 mm. **Lo glm** 3.5–12 mm long, 2.5–3 mm wide, 3–5-veined; **up glm** 5–13 mm long, 2–2.5 mm wide, 5–7-veined; **lm** 5–9 mm, glab, smooth to scabrous, 7–9-veined, veins inconspicuous, apc rounded to broadly acute, unawned; **pal** about ¾ the length of the lm; **anth** 3, 1.8–3 mm; **rdmt** 1.4–3 mm, clublike, not resembling the bisex flt, truncate to acute. $2n = 18$.

Melica californica grows from sea level to 2100 m, in a wide range of habitats, from dry, rocky, exposed hillsides to moist woods. Its range extends from Oregon to California. It differs from *M. bulbosa* in its more obtuse spikelets and less strongly colored lemmas, as well as in not having corms.

Melica californica var. *nevadensis* Boyle supposedly differs from var. *californica* in having shorter spikelets (averaging 8, rather than 10, mm), more acute glumes and lemmas, blunter rudiments, and in being restricted to the lower Sierra Nevada; the two varieties intergrade, both morphologically and geographically.

Boyle (1945) obtained vigorous sterile hybrids from crosses between *M. californica* and *M. imperfecta*, but found no natural hybrids.

6. Melica harfordii Bol. HARFORD MELIC [p. *286*, 493]

Pl ces, not rhz. **Clm** 35–120 cm, not forming corms; **intnd** smooth. **Shth** glab or pilose, often most pilose at the throat and col; **lig** 0.5–1.5 mm; **bld** 1.5–4.5 mm wide, abx surfaces smooth, adx surfaces scabridulous, glab or puberulent. **Pan** 6–25 cm; **br** 3–8 cm, appressed, with 2–6 spklt; **ped** straight; **dis** above the glm. **Spklt** 7–20 mm, with 2–6 bisex flt; **rchl intnd** 2–2.4 mm. **Glm** obtuse to subacute; **lo glm** 4–10 mm long, 1.5–2.5 mm wide, 3–5-veined; **up glm** 5–11 mm long, 1.8–2.5 mm wide, 5–7-veined; **lm** 6–16 mm, hairy, hairs to 0.75 mm on the back, 0.7–1.3 mm on the mrg, 9–11-veined, veins inconspicuous, apc mucronate to rounded, usu awned, awns 0.5–3 mm, fragile; **pal** about ¾ as long as to nearly equaling the length of the lm; **anth** 3, 2.2–4 mm; **rdmt** 2.5–6 mm, tapering, resembling the bisex flt. **Car** about 5 mm. $2n = 18$.

Melica harfordii grows primarily in the Pacific coast ranges from Washington to California, as well as in the Sierra Nevada and a few other inland locations, usually on dry slopes or in dry, open woods. The awns in *M. harfordii* often escape attention because they do not always extend beyond the lemma.

7. Melica subulata (Griseb.) Scribn. ALASKAN ONIONGRASS, TAPERED ONIONGRASS [p. *286*, 493]

Pl ces, rhz. **Clm** 55–125 cm, forming corms, corms attached to the rhz; **intnd** scabridulous bas. **Shth** usu scabridulous, smt glab or pilose; **lig** 0.4–5 mm, to 1.5 mm on the lo lvs, to 5 mm on the up lvs; **bld** 2–10 mm wide, abx surfaces smooth or scabridulous, adx surfaces scabridulous, glab or with hairs. **Pan** 8–25 cm, lax; **br** 1.7–9 cm, usu appressed to ascending, occ divergent, with 1–5 spklt; **ped** not sharply bent; **dis** above the glm. **Spklt** 10–28 mm, with 2–5 bisex flt; **rchl intnd** 1.8–2 mm. **Lo glm** 4–8 mm long, 1.3–2.2 mm wide, 1–3-veined; **up glm** 5.5–11.5 mm long, 2–3 mm wide, 3–5-veined; **lm** 5.5–18 mm, usu strigose over the veins, hairs longest towards the base, 7–9-veined, veins prominent, apc strongly tapering and acuminate, unawned; **pal** ½–¾ the length of the lm; **anth** 1.5–2.5 mm; **rdmt** 4–9 mm, tapering, resembling the bisex flt. **Car** 4–5 mm. $2n = 18$.

Melica subulata grows from sea level to 2300 m in mesic, shady woods. Its range extends from the Aleutian Islands of Alaska through British Columbia to California, east to Lawrence County, South Dakota, and into Colorado.

8. Melica fugax Bol. LITTLE MELIC [p. *287*, 493]

Pl ces, not rhz. **Clm** 10–60 cm, forming corms; **intnd** smooth or scabridulous. **Shth** scabridulous to scabrous; **lig** 0.5–2.6 mm; **bld** 1.2–5 mm wide, smt pilose on both surfaces. **Pan** 4.5–18 cm; **br** 0.8–4 cm, appressed to ascending, with 1–5 spklt; **ped** straight. **Spklt** 4–17 mm, with 2–5 bisex flt; **rchl intnd** 2.1–2.3 mm, swollen when fresh, wrinkled when dry; **dis** above the glm. **Lo glm** 3–5 mm long, 1.5–2.5 mm wide, 1–3-veined; **up glm** 3.5–7 mm long, 2.5–3.5 mm wide, 5-veined; **lm** 4–7 mm, glab or scabrous, 4–11-veined, veins inconspicuous, apc rounded to acute, unawned; **pal** almost as long as the lm; **anth** 3, 1–2 mm; **rdmt** 1.5–3.5 mm, tapering, resembling the bisex flt. $2n = 18$.

Melica fugax grows at elevations to 2200 m on dry, open flats, hillsides, and woods, from British Columbia to California and east to Idaho and Nevada. It is usually found on soils of volcanic origin, and rarely below 1300 m.

Melica fugax is often confused with *M. bulbosa*, but its rachilla internodes are unmistakable and unique among the species in the Intermountain Region, being swollen when fresh and wrinkled when dry. One specimen, C.L. Hitchcock 15521 [WTU 114265] from Elmore County, Idaho, appears to be a hybrid. It has shrunken caryopses and combines the rachilla of *M. fugax* with the lemma pubescence, size, and overall appearance of *M. subulata*, but lacks corms.

9. Melica stricta Bol. ROCK MELIC [p. *287*, 493]

Pl densely ces, not rhz. **Clm** 9–85 cm, not forming corms; bas intnd often thickened; **intnd** smooth. **Shth** scabridulous; **lig** 2.5–5 mm; **bld** 1.5–5 mm wide, abx surfaces glab, scabridulous, adx surfaces smt strigose, smt glab or scabridulous. **Pan** 3–30 cm; **br** 0.5–10 cm, appressed, with

1–5 spklt; **ped** sharply bent below the spklt; **dis** below the glm. **Spklt** 6–23 mm long, 5–13 mm wide, broadly V-shaped when mature, with 2–4 bisex flt; **rchl intnd** 1.8–2.1 mm. **Lo glm** 6–16 mm long, 3.5–5 mm wide, 4–7-veined; **up glm** 6–18 mm long, 3–5 mm wide, 5–9-veined; **lm** 6–16 mm, glab, scabridulous, 5–9-veined, veins inconspicuous, apc acute, unawned; **pal** ½–¾ the length of the lm; **anth** 1–3 mm; **rdmt** 2–7 mm, resembling the lo flt, acute to acuminate. **Car** 4–5 mm. 2*n* = 18.

Melica stricta grows from 1200–3350 m on rocky, often dry slopes, sometimes in alpine habitats. Its range extends from Oregon and California to Utah. Boyle (1945) recognized two varieties, more on their marked geographical separation than on their morphological divergence. Only one variety grows in the Intermountain Region.

Melica stricta Bol. var. **stricta** [p. 287]

Shth usu purplish, becoming dark brown. **Glm** 3–4 mm wide; **pal** about ½ the length of the lm; **anth** 1–2 mm.

Melica stricta var. *stricta* is the more widespread of the two varieties, growing throughout the range of the species except in the mountains of southern California. It is the only variety found in the Intermountain Region.

10. Melica porteri Scribn. PORTER'S MELIC [p. 287, 493]

Pl not or loosely ces, shortly rhz. **Clm** 55–100 cm, not forming corms; **intnd** smooth, bas intnd not thickened. **Shth** often scabrous on the keels, otherwise smooth; **lig** 1–7 mm; **bld** 2–5 mm wide, both surfaces glab, scabridulous. **Pan** 13–25 cm; **br** 1–9 cm, straight and appressed or flexible and ascending to strongly divergent, with 1–12 spklt; **ped** sharply bent below the spklt; **dis** below the glm. **Spklt** 8–16 mm long, 1.5–5 mm wide, parallel-sided when mature, with 2–5 bisex flt; **rchl intnd** 1.9–2.1 mm. **Glm** green, pale, or purplish tinged; **lo glm** 3.5–6 mm long, 2–3 mm wide, 3–5-veined; **up glm** 5–8 mm long, 2–3 mm wide, 5-veined; **lm** 6–10 mm, glab, chartaceous on the distal ⅓, 5–11-veined, veins conspicuous, apc rounded to acute, unawned; **pal** about ⅔ the length of the lm; **anth** 1–2.5 mm; **rdmt** 1.8–5 mm, acute to acuminate, resembling the bisex flt. 2*n* = 18.

Melica porteri grows on rocky slopes and in open woods, often near streams. It grows from Colorado and Arizona to central Texas and northern Mexico. One variety has been found in the Intermountain Region. Living plants are sometimes confused with *Bouteloua curtipendula*; the similarity is superficial.

Melica porteri Scribn. var. **porteri** [p. 287]

Pan narrow; **br** straight, appressed. **Glm** green or pale.

Melica porteri var. *porteri* grows from northern Colorado to Arizona and central Texas, and south to the Sierra Madre Occidental, Mexico.

11. Melica ciliata L. CILIATE MELIC, SILKY-SPIKE MELIC, HAIRY MELIC [p. 288]

Pl ces, smt shortly rhz. **Clm** 20–60(100) cm, not forming corms. **Shth** glab or shortly and sparsely pubescent; **lig** 1–4 mm; **bld** 7–15 cm long, 1–4 mm wide, usu involute. **Pan** 4–8(25) cm, narrowly cylindrical, lax, pale; **br** 1.5–4 cm, appressed to ascending, with 3–12(15) spklt; **ped** sharply bent below the spklt; **dis** below the glm. **Spklt** 6–8 mm, with 1 bisex flt, smt purple-tinged. **Lo glm** 4–6 mm long, 1.5–2.5 mm wide, ovate, 1–5-veined, acute; **up glm** 6–8 mm long, about 1.5 mm wide, lanceolate, acute to acuminate; **lm** 4–6.5 mm, lanceolate, 7–9-veined, papillose, mrg and mrgl veins pubescent, hairs 3.5–5 mm, not twisted; **rdmt** 1–1.7 mm, ovoid, not resembling the bisex flt. 2*n* = 18, 36.

Melica ciliata is grown as an ornamental in North America and is not known to have escaped. It is native to Europe, northern Africa, and southwestern Asia, where it grows on damp to somewhat dry soils.

12. Melica altissima L. TALL MELIC, SIBERIAN MELIC [p. 288]

Pl loosely ces. **Clm** 60–250 cm, not forming corms, scabrous below the pan. **Shth** retrorsely scabridulous; **lig** 3–5 mm; **bld** to 20 cm long, 5–15 mm wide, flat, lax. **Pan** 10–20 cm long, 1–2(5) cm wide, cylindrical, pale or purplish; **br** about 3 cm, strongly ascending to appressed, often with 15+ spklt; **ped** sharply bent below the spklt; **dis** below the glm. **Spklt** 7–11 mm, with 1–2(3) bisex flt. **Glm** subequal in length and similar in shape, 7–10.5 mm long, 3–4 mm wide, glab, ovate-elliptic, obtuse to acute, ivory or purple, 7-veined; **lm** 7–11 mm, glab, scabridulous, 9–13-veined, scarious, apc acute; **pal** about ⅔ the length of the lm; **rdmt** 2.5–3 mm, pyriform. **Car** about 3 mm. 2*n* = 18.

Melica altissima is native to Eurasia. It is grown as an ornamental in North America. In its native region, it grows on the moist soils of shrubby thickets and forest edges, and on rocky slopes. Plants with dark purple glumes and lemmas can be called *M. altissima* var. *atropurpurea* Host.

4.03 **SCHIZACHNE** Hack.

Jacques Cayouette and Stephen J. Darbyshire

Pl per; loosely csp. **Clm** 30–110 cm, glab, often decumbent at the base; **nd** glab, becoming dark. **Shth** closed almost to the top; **lig** memb, mrg often united; **bld** folded or loosely involute, glab or pilose. **Infl** pan or rcm, with 4–20 spklt; **br** straight and appressed to lax and drooping. **Spklt** slightly lat compressed, with 3–6 flt; **dis** above the glm and beneath the flt. **Glm** exceeded by the lowest lm in each spklt, chartaceous, often anthocyanic below, the up ⅓ hyaline; **cal** rounded, with hairs; **lm** chartaceous, slightly scabrous, 7–9-veined, veins parallel, conspicuous, apc scarious, bifid, awned from below the teeth, awns 8–15 mm, divergent or slightly geniculate; **pal** shorter than the lm, 2-veined, veins ciliate, keeled; **lod** truncate; **anth** 3; **ov** glab. **Car** 3.2–3.8 mm, smooth, shiny, falling free of the lm and pal. $x = 10$. Name from the Greek *schizo*, 'split', and *achne*, 'chaff', alluding to the split (bifid) lemmas.

Schizachne is a monospecific genus that extends across North America as well as from the Ural Mountains of Russia to Kamchatka and Japan.

1. **Schizachne purpurascens** (Torr.) Swallen FALSE MELIC [p. *289*, <u>493</u>]

Clm (30)50–80(110) cm, smt slightly decumbent at the base, otherwise erect. **Lig** 0.5–1.5 mm; **bld** 2–4(5) mm wide, glab or adx surfaces pilose. **Pan** 7–13(17) cm, open or closed, often rcm in depauperate pl. **Spklt** 11.5–17 mm. **Glm** glab, acute; **lo glm** 4.2–6.2 mm, faintly (1)3–5-veined; **up glm** 6–9 mm, faintly (3)5-veined; **lm** 8–10.5(12) mm; **awns** 8–15 mm, as long as or longer than the lm bodies, somewhat twisted and divergent or slightly geniculate; **anth** 1.4–2 mm. $2n = 20$.

In North America, *Schizachne purpurascens* grows in moist to mesic woods, from south of the tree line in Alaska and northern Canada through the Rocky Mountains to New Mexico in the west, and to Kentucky and Maryland in the east.

4.04 **PLEUROPOGON** R. Br.

Paul P.H. But

Pl ann or per; ces or rhz. **Clm** 5–160 cm, erect or geniculate at the base, glab; **bas brchg** exvag. **Shth** closed almost to the top; **lig** memb; **bld** flat to folded, adx surfaces with prominent midribs. **Infl** tml, rcm, rarely pan. **Spklt** lat compressed, with 5–20(30) flt, up flt rdcd; **dis** above the glm and beneath the flt. **Glm** unequal to subequal, shorter than the adjacent lm, memb to subhyaline, mrg scarious; **lo glm** 1-veined; **up glm** 1–3-veined; **rchl intnd** in some species swollen and glandular bas, the glandular portion turning whitish when dry; **cal** rounded, glab; **lm** thick, herbaceous to memb, 7(9)-veined, veins parallel, mrg scarious, apc scarious, entire or emgt, midvein smt extended into an awn, awns straight; **pal** subequal to the lm, 2-veined, keeled over each vein, keels winged, with 1 or 2 awns or a flat triangular appendage; **lod** 2, completely fused; **anth** 3, opening by pores; **ov** glab. $x = 8, 9, 10$. Name from the Greek, *pleura*, 'side', and *pogon*, 'beard', a reference to the awns on the sides of the palea in some species.

Pleuropogon is a genus of five hydrophilous species, one circumboreal in the arctic, the other four restricted to the Pacific coast of North America, extending from southern British Columbia to central California. One species extends into the Intermountain Region.

The flat, triangular paleal appendages differ from bristly or flattened awns in being wider at the base, and smooth rather than scabrous.

1. **Pleuropogon oregonus** Chase OREGON SEMAPHOREGRASS [p. *289*, <u>493</u>]

Pl per; not ces, rhz. **Clm** 40–95 cm tall, 2–3.5 mm thick, erect. **Shth** glab, smooth or scabridulous; **lig** 5–10 mm, rounded or acute, often erose; **bld** 5–17 cm long, 4–9 mm wide, smooth or scabridulous over the veins, apc spinose. **Rcm** 13–20 cm, with 6–7 spklt; **lo intnd** 3.5–7.2 cm; **up intnd** shorter; **ped** 2–5(12) mm. **Spklt** 20–40(50) mm, with 7–14 flt, lo flt bisex, up flt pist, tml flt usu strl. **Glm** lanceolate to ovate, acute, erose; **lo glm** 2–3 mm; **up glm** 2.5–4.5 mm; **rchl intnd** 2–3 mm long, 0.1–0.2 mm thick, without a glandular swelling at the base; **lm** 5.5–7 mm, scabridulous, 7-veined, veins prominent, apc truncate, smt erose, awned, awns 5–12 mm; **paleal keels** awned, awns 3–9 mm, inserted ⅓–½ of the way from the base; **anth** about 4 mm. **Car** 2.5–3 mm. $2n = $ unknown.

Pleuropogon oregonus grows in swampy ground, wet meadows, and stream banks. It is known, even historically, from only a few locations in Union and Lake counties, Oregon. In 1975 it was thought to be extinct, but a population has since been discovered at one location in Lake County. The species is listed as threatened by the state of Oregon.

5. **STIPEAE** Dumort.

Mary E. Barkworth

Pl usu per; usu tightly to loosely csp, smt rhz. **Clm** ann or per, not woody, smt brchg at up nd. **Lvs** bas concentrated to evenly distributed; **shth** open, mrg not fused, smt ciliate distally, bas shth smt concealing axillary pan (*cleistogenes*), smt wider than the bld; **col** smt with tufts of hair at the sides; **aur** absent; **lig** scarious, often ciliate, cilia usu shorter than the membranous base, lig of the lo and up cauline lvs smt differing in size and vestiture; **psdpet** absent; **bld** linear to narrowly lanceolate, venation parallel, cross venation not evident, cross sections non-Kranz, without arm or fusoid cells; **epdm** of adx surfaces smt with unicellular microhairs, cells not papillate. **Infl** usu tml pan, occ rdcd to rcm in depauperate pl, smt 2–3 pan developing from the highest cauline nd. **Spklt** usu with 1 flt, smt with 2–6 flt, lat compressed to terete; **rchl** not prolonged beyond the base of the flt in spklt with 1 flt, prolonged beyond the base of the distal flt in spklt with 2–6 flt, prolongation hairy, hairs 2–3 mm; **dis** above the glm and beneath the flt. **Glm** usu exceeding the flt(s), always longer than ¼ the length of the adjacent flt, 1–10-veined, narrowly lanceolate to ovate, hyaline or memb, flexible; **flt** usu terete, smt lat or dorsally compressed; **cal** usu well developed, rounded or blunt to sharply pointed, often antrorsely strigose; **lm** lanceolate, rectangular, or ovate, memb to coriaceous or indurate, 3–5-veined, veins inconspicuous, apc entire, bilobed, or bifid, awned, lm-awn jnct usu conspicuous, awns 0.3–30 cm, not brchd, usu tml and centric or eccentric, smt subtml, caducous to persistent, not or once- to twice-geniculate, if geniculate, proximal segment(s) twisted, distal segment straight, flexuous, or curled, not or scarcely twisted; **lod** 2 or 3; **anth** 1 or 3, smt differing in length within a flt; **ov** glab throughout or pubescent distally; **sty** 2(3–4)-brchd. **Car** ovoid to fusiform, not beaked, pericarp thin; **hila** linear; **emb** less than ⅓ the length of the car. $x = 7, 8, 10, 11, 12$.

The tribe *Stipeae* includes 15–25 genera and 550–610 species. It grows in Africa, Australia, South and North America, and Eurasia. In Australia, South America, and Asia, it is often the dominant grass tribe over substantial areas. Most species grow in arid or seasonally arid, temperate regions.

In a presentation to the Fourth International Conference on The Comparative Biology of the Monocotyledons, Romaschenko et al. (2008) argued that all the North American *Stipeae* are more closely related to the South American *Stipeae* than to the Asian *Stipeae*, but they have not as yet published the names that reflect their findings. Of the genera native to the Intermountain Region, the two that will be affected are *Achnatherum* and *Piptatherum*.

1. Paleas sulcate, longer than the lemmas; lemma margins involute, fitting into the paleal groove; lemma apices not lobed . . . . 5.04 *Piptochaetium*
1. Paleas flat, from shorter than to longer than the lemmas; lemma margins convolute or not overlapping; lemma apices often lobed or bifid.
    2. Lemma margins strongly overlapping their whole length at maturity, lemma bodies usually rough throughout, lemma apices not lobed.
        3. Awns persistent; upper leaf blades longer than 2 cm . . . . 5.06 *Nassella*
        3. Awns deciduous; upper leaf blades shorter than 1.5 cm . . . . 5.05 *Oryzopsis*
    2. Lemma margins usually not overlapping their whole length or only slightly so, strongly overlapping in some species with smooth, glabrous lemma bodies; lemma bodies usually smooth at least on the lower portion, apices 1–2-lobed; paleas from ⅓ as long as to equaling or slightly exceeding the lemmas, 2-veined, at least on the lower portion, usually hairy or both lemma and palea glabrous.
        4. Awns once-geniculate, the basal section hairy with hairs 3–8 mm long, the distal section straight, scabrous or scabridulous; ligules of basal leaves densely hairy, those of upper leaves glabrous or almost so . . . . 5.07 *Pappostipa*
        4. Awns not, once-, or twice-geniculate, the basal portion scabridulous, scabrous, or with hairs shorter than 2 mm, the distal portion straight or wavy, scabridulous to scabrous, sometimes with hairs up to 3 mm long; ligules of basal and upper leaves glabrous.
            5. Calluses 1.5–6 mm long, sharply pointed; awns 65–225 mm long; paleas about as long as the lemmas, with a "pinched" top . . . . 5.03 *Hesperostipa*
            5. Calluses 0.1–2 mm long, blunt to sharply pointed; awns 1–70 mm long; paleas usually shorter than the lemmas, if as long as and with a "pinched" top, the awns shorter than 65 mm.
                6. Florets usually dorsally compressed at maturity, sometimes terete; lemma margins separate their whole length at maturity; paleas as long as or longer than the lemmas . . . . 5.02 *Piptatherum*

6. Florets terete or laterally compressed at maturity; lemma margins separate for part or all their length at maturity; paleas often shorter than the lemmas . . . . 5.01 *Achnatherum*

5.01 **ACHNATHERUM** P. Beauv.   Mary E. Barkworth

Pl per; tightly to loosely csp. **Clm** 10–250 cm, erect, not brchg at the up nd; **bas brchg** exvag or invag. **Shth** open, mrg often ciliate distally; clstgn not present in bas lf shth; **col** smt with hairs on sides; **aur** absent; **lig** hyaline to memb, glab or pubescent, smt ciliate; **bld** flat, convolute, or involute, apc acute, flexible, bas bld not overwintering, flag lf bld more than 10 mm long. **Infl** pan, usu contracted, smt 2+ at tml nd; **br** usu straight, smt flexuous. **Spklt** usu appressed, with 1 flt; **rchl** not prolonged; **dis** beneath flt. **Glm** exceeding flt, usu lanceolate, 1–7-veined, acute, acuminate, or obtuse; **flt** usu terete, fusiform or globose, smt somewhat lat compressed; **cal** 0.1–4 mm, blunt to sharp, usu strigose; **lm** stiffly memb to coriaceous, smooth, usu hairy, smt glab, hairs on the lm body to 6 mm, evenly distributed, hairs on the up ¼ smt somewhat longer than those below, not both markedly longer and more divergent, apical hairs to 7 mm, lm mrg usu not or only weakly overlapping, firmly overlapping in some species with glab lm, usu with 0.05–3 mm lobes, smt unlobed, lobes usu memb and flexible, smt thick, apc awned, lm-awn jnct evident; awn 3–80 mm, centric, readily deciduous to persistent, usu scabrous to scabridulous, smt hairy in whole or in part, if shorter than 12 mm, usu deciduous, not or once-geniculate and scarcely twisted, if longer than 12 mm, usu persistent, once- or twice-geniculate and twisted below, tml segment usu straight, smt flexuous; **pal** from ⅓ as long as to slightly longer than the lm, usu pubescent, 2-veined, not keeled over veins, flat between veins, veins usu terminating below apc, smt prolonged 1–3 mm, apc usu rounded; **lod** 2 or 3, memb, not lobed; **anth** 3, 1.5–6 mm, smt penicillate; **ov** with 2 sty br, br fused at base. **Car** fusiform, not ribbed, sty bases persistent; **hila** linear, almost as long as car; **emb** ⅕–⅓ the length of the car. x = 10 or 11. Name from the Greek *achne*, 'scale', and *ather*, 'awn', a reference to the awned lemmas.

As interpreted here, *Achnatherum* is one of the larger and more widely distributed genera in the *Stipeae*. Fourteen species are native to the Intermountain Region. Romaschenko et al. (2008) argue that the North American species are only distantly related to *Achnatherum* and other Asian genera. The new generic name for most (not all) North American species will probably be *Eriocoma*. They are retained in *Achnatherum* pending publication of the necessary combinations.

In the key, glume widths are the distance between the midvein and the margin. Floret lengths include the callus, but not the apical lobes. Floret thickness refers to the thickest part of the floret.

1. Awns persistent, basal segment pilose, at least some hairs longer than 0.5 mm.
    2. Flag leaves with ligules 3–8 mm long; lemmas with 1 thick, inflexible apical lobe up to 0.1 mm long. . . . . 6. *A. thurberianum*
    2. Flag leaves with ligules 0.3–3 mm long; lemmas usually with 2 thin, flexible apical lobes up to 1 mm long, sometimes without an apical lobe.
        3. Calluses 0.5–0.7 mm long; paleas ½–¾ as long as the lemmas; palea apices with hairs about 1 mm long. . . . . 2. *A. nevadense*
        3. Calluses 0.8–1.2 mm long; paleas ⅖–⅗ as long as the lemmas; palea apices usually with hairs shorter than 1 mm . . . . 3. *A. occidentale*
1. Awns persistent or deciduous, basal segment scabridulous, scabrous, or with hairs shorter than 0.5 mm.
    4. Lemmas with evenly distributed hairs, those at midlength 1.2–6 mm long, usually not conspicuously longer than those at the apices.
        5. Awns persistent.
            6. Plants sterile; anthers indehiscent, the pollen grains with collapsed walls . . . . 16. Hybrids involving *A. hymenoides* (in part)
            6. Plants fertile; anthers dehiscent, the pollen grains mostly spherical, not collapsed.
                7. Awns once-geniculate; sheaths becoming flat and ribbonlike with age . . . . 7. *A. parishii* (in part)
                7. Awns twice-geniculate; sheaths not becoming flat and ribbonlike with age . . . . 11. *A. pinetorum*
        5. Awns rapidly deciduous.
            8. Florets at least 4.5 mm long, fusiform; anthers dehiscent or indehiscent.
                9. Anthers dehiscent, the pollen grains spherical . . . . 12. *A. webberi*

9. Anthers indehiscent, the pollen grains collapsed . . . .
. . . . .16. Hybrids involving *A. hymenoides* (in part)
8. Florets 2.5–4.5 mm long, usually ovoid to obovoid, sometimes fusiform; anthers dehiscent.
   10. Panicle branches terminating in a pair of spikelets on conspicuously divaricate pedicels, the shorter pedicel usually more than ½ as long as the longer pedicel, sometimes subequal to it . . . . 14. *A. hymenoides*
   10. Panicle branches terminating in a pair of spikelets on loosely appressed pedicels, the shorter pedicel usually less than ½ as long as the longer pedicel . . 15. *A. arnowiae*
4. Lemmas glabrous or with hairs 0.2–1.5(2) mm long, lemma apices glabrous or with hairs evidently longer than those at midlength.
   11. Lemma apices with hairs 2–7 mm long, usually more than 1 mm longer than those at midlength.
      12. Calluses sharp; paleas ⅓–½ as long as the lemmas. . . . . 9. *A. scribneri*
      12. Calluses blunt to acute, not sharp; paleas ½–⁹⁄₁₀ as long as the lemmas . . . . 7. *A. parishii* (in part)
   11. Lemma apices glabrous or with hairs up to 2.2 mm long, usually less than 1 mm longer than those at midlength.
      13. Awns 5–6 mm long, readily deciduous. . . . . 13. *A. swallenii*
      13. Awns 10–80 mm long, persistent.
         14. Terminal awn segment flexuous . . . . 8. *A. aridum*
         14. Terminal awn segment straight or slightly arcuate.
            15. Lemma apices with a single, thick, inflexible lobe about 0.1 mm long . . . 5. *A. lemmonii*
            15. Lemma apices unlobed or with 1–2 thin, flexible lobes up to 1.2 mm long.
               16. Glumes unequal, the lower glume exceeding the upper glume by 1–4 mm . . . . 10. *A. perplexum*
               16. Glumes subequal, the lower glume exceeding the upper glume by less than 1 mm.
                  17. Paleas ⅗–⁹⁄₁₀ as long as the lemmas, the apical hairs clearly exceeding the palea apex; blades 0.5–2 mm wide; awns 12–25 mm long . . . . 1. *A. lettermanii*
                  17. Paleas ⅓–⅔ as long as the lemmas, the apical hairs usually not exceeding the palea apex; blades (0.5)1.2–5 mm wide; awns 19–45 mm long . . . . 4. *A. nelsonii*

1. **Achnatherum lettermanii** (Vasey) Barkworth
LETTERMAN'S NEEDLEGRASS [p. *290*, <u>493</u>]
Pl tightly ces, not rhz. **Clm** 15–90 cm tall, 0.5–0.8 mm thick, usu glab, smt puberulent to 5 mm below the lo nd; **nd** 2–3. **Bas shth** smooth, glab, mrg not ciliate; **col**, including the sides, glab or sparsely pubescent, col of the flag lvs glab; **lig** 0.2–1.5(2) mm, without tufts of hair on the sides, truncate to rounded; **bld** 0.5–2 mm wide, abx surfaces smooth to scabridulous, adx surfaces scabrous or puberulent. **Pan** 7–19 cm long, 0.5–1 cm wide; **br** straight, appressed to strongly ascending, longest br 1.2–2.5 cm. **Spklt** appressed to the br. **Glm** 6.5–9 mm, subequal; **lo glm** 1(3)-veined; **up glm** to 0.5 mm shorter than the lo glm, 0.6–1 mm wide, 1-veined; **flt** 4.5–6 mm long, 0.8–1 mm thick, fusiform, terete, widest below midlength; **cal** 0.4–1 mm, blunt; **lm** evenly hairy, hairs at midlength about 0.5 mm, apical hairs 0.7–1.5(2) mm, apical lobes 0.3–0.8 mm, memb, flexible; **awns** 12–25 mm, persistent, twice-geniculate, scabrous, tml segment straight;

pal 3–4 mm, ¾–⅘(⁹⁄₁₀) as long as the lm, veins terminating at or before the apc, apc round, flat, apical hairs 0.5–1 mm, extending beyond the pal body; **anth** 1.5–2 mm, dehiscent, not penicillate. **Car** about 4 mm, fusiform. 2*n* = 32.

Achnatherum lettermanii grows in meadows and on dry slopes, from sagebrush to subalpine habitats, at 1700–3400 m. Its range extends from Oregon and Montana to southern California, Arizona, and New Mexico. When sympatric with *A. nelsonii*, *A. lettermanii* tends to grow in shallower or more disturbed soils. It can be distinguished from that species by its generally finer leaves and more tightly cespitose growth habit, as well as its blunter calluses and longer paleas. Its relatively long paleas also distinguish *A. lettermanii* from *A. perplexum*.

2. **Achnatherum nevadense** (B.L. Johnson) Barkworth NEVADA NEEDLEGRASS [p. *290*, <u>494</u>]
Pl ces, not rhz. **Clm** 20–85 cm tall, 0.8–1.2 mm thick, usu retrorsely pubescent below the lo nd, smt glab, smt pubescent over the whole of the intnd; **nd** 3–4. **Bas shth** glab or pubescent, smt scabridulous, usu glab at the throat, becoming

brown to gray-brown; **col**, including the sides, glab; **bas lig** 0.2–0.7 mm, truncate; **up lig** 0.3–1 mm, often wider than the bld; **bld** usu 10–25 cm long, 1–3 mm wide, usu involute, abx surfaces glab, adx surfaces more or less puberulent. **Pan** 6–25 cm long, 0.4–1.5 cm wide; **br** appressed, lo br 2–7.5 cm. **Spklt** appressed to the br. **Glm** subequal, 7–14 mm; **lo glm** 0.6–0.9 mm wide; **flt** 5–6.5 mm long, 0.6–1 mm thick, fusiform, terete; **cal** 0.5–0.7 mm, sharp, dorsal boundary of the glab tip with the cal hairs rounded to acute; **lm** evenly hairy, hairs 0.5–1 mm at midlength, apical hairs 0.8–2 mm, longer than those at midlength and those at the base of the awn, apical lobes memb, about 0.2 mm; **awns** 20–35 mm, persistent, twice-geniculate, first 2 segments pilose, with hairs 0.5–1.5 mm and of mixed lengths, tml segment scabridulous to smooth; **pal** 2.8–4.2 mm, ½–¾ as long as the lm, pubescent, distal hairs usu about 1 mm, extending well beyond the apc, apc rounded; **anth** about 2.5 mm, dehiscent, not penicillate. **Car** 3–5.5 mm, fusiform. $2n = 68$.

Achnatherum nevadense grows in sagebrush and open woodlands, from Washington to south-central Wyoming and south to California and Utah. Johnson (1962) argued that it is an alloploid derivative of *A. occidentale* and *A. lettermanii*.

The apical lemma hairs of *Achnatherum nevadense* appear longer than the lowermost awn hairs. This difference is not reflected in the lengths shown because a few of the basal awn hairs may be as long as those at the top of the lemma, but the majority are shorter. This is the best character for distinguishing *A. nevadense* and *A. occidentale* subsp. *californicum* from *A. occidentale* subsp. *pubescens*. In addition, in *A. nevadense* and *A. occidentale* subsp. *californicum*, the hairs on the first awn segments tend to look untidy because of their varied lengths and the different angles they make with the awn; those of *A. occidentale* subsp. *occidentale* and subsp. *pubescens* have basal awn segments with tidier looking hairs.

Differentiating between *Achnatherum nevadense* and *A. occidentale* subsp. *californicum* can be difficult, but they differ in the shape of the boundary between the glabrous and strigose portions of the callus. In addition, *A. nevadense* is usually pubescent below the lower cauline nodes, and has paleas that are longer in relation to the lemmas.

3. Achnatherum occidentale (Thurb.) Barkworth [p. 291, 494]

Pl tightly ces, not rhz. **Clm** 14–120(180) cm tall, 0.3–2 mm thick, intnd glab or puberulent to densely pubescent; **nd** 2–4, glab or pubescent. **Bas shth** glab or puberulent to densely pubescent, often ciliate at the throat; **col** often with tufts of hair at the sides; **lig** 0.2–1.5 mm, often ciliate; **bld** 0.5–3 mm wide and flat, or convolute and 0.1–0.8 mm in diameter, lax to straight. **Pan** 5–30 cm long, 0.5–1.5 cm wide; **br** appressed, straight, longest br 1–7 cm. **Spklt** appressed to the br. **Glm** subequal, 9–15 mm long, 0.6–0.9 mm wide;

flt 5.5–7.5 mm long, 0.5–0.9 mm thick, fusiform, terete; **cal** 0.8–1.2 mm, sharp, dorsal boundary of the glab tip with the cal hairs narrowly acute; **lm** evenly hairy, hairs 0.5–1.5 mm at midlength, apical hairs somewhat longer than those below, smt similar in length to those at the base of the awns, smt longer, apical lobes 0.3–0.5 mm, memb; **awns** 15–55 mm, twice-geniculate, usu the first 2 segments evidently hairy and the tml segment scabrous, glab, or partly to wholly pilose, smt all three segments scabrous; **pal** 2.2–3.5 mm, ⅖–⅗ as long as the lm, hairs at the tip usu shorter than 1 mm, frequently extending beyond the apc, apc rounded; **anth** 2.5–3.5 mm, dehiscent, not penicillate. **Car** 4–6 mm, fusiform. $2n = 36$.

Achnatherum occidentale, which extends from British Columbia to California, Utah, and Colorado, varies considerably in pubescence and size. All three subspecies grow in the Intermountain Region.

1. Terminal awn segment usually pilose; culms 0.3–1 mm thick, glabrous even on the basal internodes; glumes often purplish . . . . subsp. *occidentale*
1. Terminal awn segment usually scabrous or glabrous, occasionally pilose at the base; culms 0.5–2 mm thick; glumes usually green.
   2. First awn segment pilose, the hairs of different lengths intermixed, or scabrous; apical lemma hairs longer than the basal awn hairs . . . . subsp. *californicum*
   2. First awn segment pilose, the hairs gradually and evenly becoming shorter towards the first geniculation; apical lemma hairs similar in length to the basal awn hairs . . . . subsp. *pubescens*

Achnatherum occidentale subsp. californicum (Merr. & Burtt Davy) Barkworth CALIFORNIA NEEDLEGRASS [p. 291]

Clm 30–100(180) cm tall, 0.5–2 mm thick, intnd glab or pubescent, smt densely pubescent; **nd** 2–4, glab or pubescent. **Shth** glab, pubescent, or pilose; **col** usu with tufts of hair at the sides; **bld** 0.8–2 mm wide, usu erect or ascending, adx surfaces pilose. **Pan** 8–30 cm; **br** appressed; **longest br** to 7 cm. **Glm** usu green; **lo glm** 0.6–0.8 mm wide; **lm** with hairs to 0.8 mm at midlength, apical hairs to 1.8 mm, longer than the bas awn hairs; **awns** 18–55 mm, scabrous or pilose on the first 2 segments, with hairs to 0.5 mm and of mixed lengths, tml segment usu scabrous, occ pilose at the base. $2n = 36$.

Achnatherum occidentale subsp. *californicum* grows from Washington through Idaho to southwestern Montana and south to California and Nevada, with disjunct records from south-central Wyoming and southwestern Utah. Its elevation range is 2000–4000 m.

Johnson (1962) postulated that *Achnatherum occidentale* subsp. *californicum* is a hybrid derivative of *A. nelsonii* and *A. occidentale*; it intergrades with both. The scattering of longer hairs among shorter hairs on the basal awn segments, combined with the long apical lemma hairs, give florets of subsp. *californicum* a more untidy appearance than those of the other two subspecies. It resembles *A. nevadense* in this respect, but differs from that species in the shape of the boundary between the glabrous and strigose portions of the callus, in usually being glabrous below the lower cauline nodes, and in having paleas that are shorter in relation to the lemmas. Plants with scabrous awns are often confused with *A. nelsonii* subsp. *nelsonii*; they differ in having sharper calluses, a more elongated extension of the glabrous callus area into the strigose portion of the callus, and, usually, longer awns.

Achnatherum occidentale (Thurb.) Barkworth subsp. occidentale WESTERN NEEDLEGRASS [p. 291]

Clm 14–50 cm tall, 0.3–1 mm thick, intnd glab. **Bas shth** mostly glab, often ciliate or sparsely tomentose at the throat; **col** usu without tufts of hair at the sides; **bld** 0.3–2 mm wide when flat, usu convolute and 0.1–0.2 mm in diameter. **Pan** 5–20 cm; **longest br** 1.5–4 cm. **Glm** often purplish; **lm** with apical pubescence similar in length to the bas awn pubescence; **awns** 15–42 mm, first 2 segments always pilose, tml segment usu pilose, hairs becoming shorter distally, occ scabridulous or smooth.

Achnatherum occidentale subsp. *occidentale* grows above 2400 m, primarily in California. It differs from *A. occidentale* subsp. *pubescens* in having culms that are glabrous throughout and awns with terminal segments that are usually pilose. One specimen from Idaho has been seen; it has the palea/lemma ratio of subsp. *californicum* but the pilose terminal awn segments, slender habit, and purplish coloration of subsp. *occidentale*.

Achnatherum occidentale subsp. pubescens (Vasey) Barkworth COMMON WESTERN NEEDLEGRASS [p. 291]

Clm 32–120 cm tall, 0.8–1.3(2) mm thick, bas intnd puberulent to pubescent. **Bas shth** usu pubescent, smt densely pubescent, occ glab; **col** often with tufts of hair at the sides; **bld** 1–3 mm wide, adx surfaces pubescent to pilose. **Pan** 10–30 cm; **longest br** 2–7 cm. **Glm** usu green; **lm** with apical pubescence similar in length to the bas awn pubescence; **awns** 24–50 mm, pilose on the first 2 segments, with hairs gradually becoming shorter distally, tml segment scabrous or glab. $2n = 36$.

Achnatherum occidentale subsp. *pubescens* grows from Washington to California and eastward to Wyoming, at 1300–1600 m. It is the most widespread and variable subspecies of *A. occidentale*, intergrading with subsp. *californicum*, *A. nelsonii*, and *A. lettermanii*. It differs from the latter two in its shorter paleas and its pilose awns.

4. Achnatherum nelsonii (Scribn.) Barkworth [p. 292, <u>494</u>]

Pl ces, not rhz. **Clm** 40–175 cm tall, 0.7–2.4 mm thick, lo cauline intnd usu glab, smt slightly pubescent below the lo nd; **nd** 2–5. **Bas shth** glab or sparsely to densely pubescent, mrg smt ciliate; **col** glab or somewhat pubescent, without tufts of hair on the sides, col of the flag lvs glab or sparsely pubescent; **bas lig** 0.2–0.7 mm, memb, truncate to rounded, usu not ciliate; **up lig** 1–1.5 mm, acute; **bld** (0.5)1.2–5 mm wide. **Pan** 9–36 cm long, 0.8–2 cm wide; **br** ascending to appressed, straight. **Spklt** appressed to the br. **Glm** 6–12.5 mm long, 0.7–1.1 mm wide; **lo glm** exceeding the up glm by 0.2–0.8 mm; **flt** 4.5–7 mm long, 0.6–0.9 mm thick, fusiform; **cal** 0.2–1 mm, blunt to sharp, dorsal boundary of the glab tip with the cal hairs almost rounded to acute; **lm** evenly hairy, hairs at midlength 0.5–1 mm, hairs at the apc to 2 mm, erect to ascending, apical lobes 0.1–0.4 mm, memb, flexible; **awns** 19–45 mm, persistent, twice-geniculate, first 2 segments scabrous or with hairs shorter than 0.5 mm, tml segment straight; **pal** 2–4 mm, ⅓–⅔ as long as the lm, pubescent, hairs usu not exceeding the apc, veins terminating before the apc, apc rounded; **anth** 2–3.5 mm, dehiscent, not penicillate. **Car** 3–4 mm, fusiform. $2n = 36, 44$.

Achnatherum nelsonii grows in meadows and openings, from sagebrush steppe and pinyon-juniper woodlands to subalpine forests, at 500–3500 m. It differs from *A. perplexum* in having more dense inflorescences, straighter, more equal glumes, and in flowering in late spring to early summer rather than early fall. It is sometimes sympatric with *A. lettermanii*, from which it differs in its shorter paleas and wider leaves, and its tendency to grow in deeper or less disturbed soils. It differs from *A. lemmonii* in having wider leaf blades, shorter paleas, and membranous lemma lobes, and from *A. nevadense* and *A. occidentale* in its scabrous awns and the truncate to acute boundary of the glabrous tip of the callus with the callus hairs.

The two subspecies intergrade to some extent. There is also intergradation with *Achnatherum occidentale*, possibly as a result of hybridization and introgression.

1. Calluses blunt, dorsal boundary of the glabrous tip and the callus hairs almost straight to rounded; awns 19–31 mm long . . . . subsp. *dorei*
1. Calluses sharp, dorsal boundary of the glabrous tip and the callus hairs acute; awns 19–45 mm long . . . . subsp. *nelsonii*

Achnatherum nelsonii subsp. dorei (Barkworth & J.R. Maze) Barkworth DORE'S NEEDLEGRASS [p. 292]

Cal blunt, glab tips 0.02–0.06 mm, dorsal boundary of the glab tip and the cal hairs almost straight to rounded; **awns** 19–31 mm.

Achnatherum nelsonii subsp. *dorei* grows from the southern Yukon Territory to southern Nevada and Utah, but it is most common in the north. In regions where both subspecies grow, subsp. *dorei* is at higher elevations than subsp. *nelsonii*.

Some of the reports of *Achnatherum nelsonii* subsp. *dorei* (identified as *Stipa columbiana* Macoun by many authors) from New Mexico and Arizona are probably based on *A. perplexum*, which differs in having sparse, narrow inflorescences and slightly recurved glumes. The two also differ in flowering time, *A. nelsonii* subsp. *dorei* flowering in late spring to early summer and *A. perplexum* in the fall. *Achnatherum nelsonii* subsp. *nelsonii* is also present in New Mexico.

Achnatherum nelsonii (Scribn.) Barkworth
subsp. **nelsonii** NELSON'S NEEDLEGRASS [p. 292]

Cal sharp, acute to acuminate, glab tips 0.05–0.15 mm, dorsal boundary of the glab tip and the cal hairs acute; **awns** 19–45 mm.

Achnatherum nelsonii subsp. *nelsonii* is the more common of the two subspecies in the Intermountain Region. Within the region, it tends to grow at lower elevations than subsp. *dorei*. It intergrades with subsp. *dorei* in Montana and Wyoming, and with *A. occidentale* subsp. *pubescens* in California.

5. Achnatherum lemmonii (Vasey) Barkworth
LEMMON'S NEEDLEGRASS [p. 292, 494]

Pl tightly ces, not rhz. **Clm** 15–90 cm tall, 0.7–1 mm thick, glab, pubescent, or tomentose; **nd** 3–4. **Bas shth** glab, pubescent, or tomentose; **col**, including the sides, glab or sparsely pubescent, hairs shorter than 0.5 mm; **bas lig** 0.5–1.2 mm, hyaline, glab, truncate to acute; **up lig** to 2.5 mm; **bas bld** 0.5–1.5 mm wide, folded to convolute, abx surfaces smooth, glab, adx surfaces prominently ribbed, often with 0.3–0.5 mm hairs, smt glab; **up bld** to 2.5 mm wide, otherwise similar to the bas bld. **Pan** 7–21 cm long, about 1 cm wide; **br** straight, strongly ascending to appressed, longest br 4–5 cm. **Spklt** appressed to the br. **Glm** subequal, 7–11.5 mm; **lo glm** 0.9–1.1 mm wide, 4–5-veined; **up glm** 3-veined; **flt** 5.5–7 mm long, 0.8–1.3 mm thick, fusiform, somewhat lat compressed; **cal** 0.4–1.2 mm, blunt; **lm** coriaceous, evenly pubescent, hairs 0.4–1 mm, apc 1-lobed, lobe about 0.1 mm long, thick, stiff, apical lm hairs 0.4–0.8 mm; **awns** 16–30 mm, persistent, (once)twice-geniculate, all segments scabrous, tml segment straight; **pal** 4.5–6.5 mm, from ¾ as long as to equaling the lm, sparsely to moderately pubescent, hairs not exceeding the apc, veins terminating below the apc, apc flat or pinched; **anth** 2.3–3.5 mm, dehiscent, not penicillate. **Car** 4–5 mm, fusiform. 2n = 34.

Achnatherum lemmonii grows in sagebrush and yellow pine associations, from southern British Columbia to California and east to Utah. It differs from *A. nelsonii* in having narrower leaves, laterally compressed florets with a thick apical lobe, and longer paleas. There are two subspecies, but only subsp. *lemmonii* grows in the Intermountain Region.

Achnatherum lemmonii (Vasey) Barkworth
subsp. **lemmonii** [p. 292]

Shth and clm glab or pubescent, hairs to 0.2 mm, smt varying within a population.

Achnatherum lemmonii subsp. *lemmonii* grows throughout the range shown on the map, on both serpentine and non-serpentine soils.

6. Achnatherum thurberianum (Piper)
Barkworth THURBER'S NEEDLEGRASS [p. 293, 494]

Pl ces, not rhz. **Clm** 30–75 cm tall, 0.5–1.7 mm thick, intnd pubescent or glab, pubescence more common on the lo intnd, particularly just below the nd; **nd** 2–3, lo nd retrorsely pubescent, up nd glab or pubescent. **Bas shth** glab, usu smooth, brown or gray-brown; **col** glab, without tufts of hair at the sides; **bas lig** 1.5–6 mm, hyaline, rounded to acute, lacerate; **up lig** to 8 mm, hyaline, acute, glab; **bld** 0.5–2 mm wide, convolute, abx surfaces scabrous, adx surfaces scabrous or hairy, hairs about 0.3 mm. **Pan** 7–15 cm long, 0.5–2.5 cm wide, often included in the up lf shth at the start of anthesis; **br** 1.5–6 cm, appressed to strongly ascending, with 1–6 spklt. **Glm** often purplish; **lo glm** 10–15 mm long, 1.2–2 mm wide; **up glm** to 2 mm shorter; **flt** 6–9 mm long, 0.7–1.2 mm thick, fusiform, terete; **cal** 0.9–1.5 mm, sharp; **lm** coriaceous, evenly pubescent or the back glabrate distally, hairs 0.5–0.8 mm, apc lobed on 1 mrg, lobe about 0.1 mm long, thick, apical lm hairs 0.5–0.8 mm; **awns** 32–56 mm, twice-geniculate, first 2 segments pilose, hairs 0.8–2 mm, tml segment glab, often scabrous; **pal** 4.6–6.1 mm, ¾–⁹⁄₁₀ as long as the lm, sparsely pubescent towards the base; **anth** 2.5–3.5 mm, dehiscent, not penicillate. **Car** 5–7 mm, fusiform. 2n = 34.

Achnatherum thurberianum grows in canyons and foothills, primarily in sagebrush desert and juniper woodland associations, from Washington to southern Idaho and southwestern Montana and from California to Utah, at 900–3000 m. Its long ligules and pilose awns make it one of the easier North American species of *Achnatherum* to identify.

7. Achnatherum parishii (Vasey) Barkworth
[p. 293, 494]

Pl tightly ces, not rhz. **Clm** 14–80 cm tall, 0.8–2 mm thick, intnd glab or pubescent below the nd; **nd** 3–5, glab. **Bas shth** mostly glab, smt pubescent at the base, flat and ribbonlike with age, mrg smt hairy distally, hairs adjacent to the lig 0.5–3 mm; **col** glab; **lig** truncate, abx surfaces pubescent, ciliate, cilia as long as or longer than the bas membrane, lig of bas lvs 0.3–0.8 mm, of up lvs 0.5–1.5 mm, asymmetric; **bld** 4–30+ cm long, 1–4.2 mm wide, usu flat and more or less straight, smt tightly convolute and arcuate. **Pan** 7–15 cm long, 1.5–4 cm wide; **br** strongly ascending at maturity, longest br 1.5–4 cm. **Glm** unequal to subequal, narrowly lanceolate, 3–5-veined; **lo glm** 9–15 mm long, 0.9–1.2 mm wide; **up glm** 8–15 mm; **flt** 4.8–6.5 mm long, 0.8–1 mm thick, fusiform, terete; **cal** 0.2–0.8 mm, acute; **lm** evenly and densely hairy, hairs 1.5–3.5 mm at midlength, apical hairs 2.5–5 mm; **awns** 10–35 mm, persistent, once-geniculate, first segment scabrous or strigose, hairs to 0.3 mm, tml segment straight; **pal** 2.5–4.5 mm, ½–⅘ times the length of the lm, hairy between the veins, hairs often as long as those on the lm but not as dense, apc usu rounded, occ somewhat pinched; **anth** 2.3–4.5 mm, dehiscent, not penicillate. **Car** 3–6 mm, fusiform. 2*n* = unknown.

Achnatherum parishii grows from the coastal ranges of California to Nevada and Utah, south to Baja California, Mexico, and to the Grand Canyon in Arizona. It differs from *A. scribneri* in its shorter, blunter calluses and more abundant lemma hairs; and from *A. perplexum* in having longer hairs on its lemmas.

1. Basal sheath margins glabrous or hairy distally, hairs to 0.5 mm long; culms 14–35 cm tall . . . . subsp. *depauperatum*
1. Basal sheath margins hairy distally, hairs 1–3.2 mm long; culms 20–80 cm tall . . . . subsp. *parishii*

Achnatherum parishii subsp. depauperatum
(M.E. Jones) Barkworth Low Needlegrass
[p. 293]

Clm 14–35 cm tall, 0.8–1.7 mm thick, glab. **Bas shth mrg** glab or hairy distally, hairs to 0.5 mm; **bld** 4–15 cm long, 1–2.5 mm wide, usu 0.5–1.5 mm in diameter and tightly convolute, conspicuously arcuate distally, abx surfaces smooth, glab, adx surfaces scabrous. **Pan** 7–12 cm long, 1.5–2.5 cm wide. **Flt** 4.8–6 mm; **pal** densely hairy between the veins, some hairs as long as those on the lm; **awns** 10–17 mm; **anth** 2.3–2.8 mm. **Car** 3–4 mm, fusiform. 2*n* = unknown.

Achnatherum parishii subsp. *depauperatum* grows in gravel and on rocky slopes, in juniper and mixed desert shrub associations, from central Nevada to western Utah. It differs from *A. webberi* in its persistent awns and thicker leaves that tend to curl when dry, and from *A. parishii* subsp. *parishii* in its smaller stature, glabrous or shortly hairy sheath margins, and densely hairy paleas.

Achnatherum parishii (Vasey) Barkworth
subsp. **parishii** Parish's Needlegrass
[p. 293]

Clm 20–80 cm tall, 1.5–2 mm thick, mostly glab, pubescent below the nd. **Bas shth mrg** hairy distally, hairs 1–3.2 mm; **bld** 11–30+ cm long, 2.5–4.2 mm wide, usu flat or only partly closed,

smt completely convolute, straight to somewhat arcuate distally. **Pan** 11–15 cm long, 2–4 cm wide. **Flt** 5.5–6.5 mm; **awns** 15–35 mm; **pal** sparsely hairy between the veins, hairs about ½ as long as the lm hairs, apc usu rounded, occ somewhat pinched; **anth** 3.5–4.5 mm, glab. **Car** 5–6 mm, fusiform. 2*n* = unknown.

Achnatherum parishii subsp. *parishii* grows on dry, rocky slopes, in desert shrub and pinyon-juniper associations, from the coastal ranges of California to northeastern Nevada, eastern Utah, and the Grand Canyon in Arizona. It differs from subsp. *depauperatum* in its longer culms, hairy sheath margins, and sparsely hairy paleas.

8. Achnatherum aridum (M.E. Jones)
Barkworth Mormon Needlegrass
[p. 294, 494]

Pl ces, not rhz. **Clm** 35–85 cm tall, 0.9–2.5 mm thick, usu glab and smooth, smt scabridulous or puberulent; **nd** 2–3. **Bas shth** glab, up shth mrg hyaline distally; **col** of the bas shth occ with a small tuft of 0.8 mm hair on the sides, col of the up lvs glab, scabridulous, or sparsely puberulent; **lig** 0.2–1.5 mm, truncate to rounded, erose, smt ciliate, cilia about 0.05 mm; **bld** 0.9–3 mm wide, abx surfaces smooth or scabridulous, glab, adx surfaces hirtellous, hairs to 0.5 mm. **Pan** 5–17 cm long, 1–1.5 cm wide, contracted, bases often enclosed at anthesis; **br** appressed or strongly ascending, straight, lo br 1.5–4 cm. **Lo glm** 8–15 mm long, 0.6–0.8 mm wide; **up glm** 1–5 mm shorter; **flt** 4–6.5 mm long, 0.6–1.1 mm thick, fusiform, terete; **cal** 0.2–1 mm, sharp; **lm** evenly hairy on the lo portion, hairs 0.2–0.5 mm, the distal ⅕–¼ often glab, apical hairs absent or fewer than 5, to 1.5 mm; **awns** 40–80 mm, persistent, obscurely once-geniculate, scabridulous, tml segment flexuous; **pal** 2–3.2 mm, ½–¾ as long as the lm, pubescent, hairs exceeding the apc, apc rounded, flat; **anth** 2–3.5 mm, dehiscent, not penicillate. 2*n* = unknown.

Achnatherum aridum grows on rocky outcrops, in shrub-steppe and pinyon-juniper associations, from southeastern California to Colorado and New Mexico, at 1200–2000 m.

9. Achnatherum scribneri (Vasey) Barkworth
Scribner's Needlegrass [p. 294, 494]

Pl ces, not rhz. **Clm** 25–90 cm tall, 0.5–1.6 mm thick, glab; **nd** 3. **Bas shth** becoming flat and papery, mrg ciliate distally; **col** glab, with tufts of hair on the sides, hairs on the bas lvs to 1.5 mm, hairs on the flag lvs 1–2.5 mm; **bas lig** 0.3–0.8 mm, truncate, erose, ciliate, cilia 0.2–0.4 mm; **up lig** to 1.5 mm, asymmetric, obliquely truncate for most of their width, abruptly longer on 1 side; **bld** to 30 cm long, 2–5 mm wide, flat or involute, long-tapering. **Pan** 7–21 cm long, 0.5–1 cm wide; **br** appressed to ascending, straight. **Lo glm** 10–17 mm long, 0.7–1.2 mm wide, exceeding the up glm by 2.5–4.5 mm, apc tapering, often slightly recurved; **flt** 6–9.5 mm long, 0.6–1.1 mm thick, fusiform, terete, widest at or below midlength; **cal** 0.5–1.5 mm, sharp; **lm** evenly hairy, hairs at midlength to 1 mm, apical hairs 2–3 mm, ascending, apical lobes 0.3–0.5 mm; **awns** 13–25 mm, persistent, usu once-geniculate, first segment scabrous, tml segment straight; **pal** 2.5–3.5 mm, $\frac{1}{3}$–$\frac{1}{2}$ as long as the lm, pubescent, hairs not exceeding the apc, apc rounded; **anth** 3–5 mm, dehiscent, not penicillate. **Car** 5–6 mm, fusiform. $2n = 40$.

Achnatherum scribneri grows on rocky slopes, in pinyon-juniper and ponderosa pine associations at 1500–2700 m, from southeastern Wyoming through Colorado to Arizona, New Mexico, western Oklahoma, and Texas, and in southern Utah. The Utah populations appear to be disjunct from the species' primary range; this may reflect a lack of collecting. *Achnatherum scribneri* is similar to *A. parishii* and *A. perplexum*, differing from them in its sharp calluses.

10. **Achnatherum perplexum** Hoge & Barkworth PERPLEXING NEEDLEGRASS [p. 294, 494]

Pl ces, not rhz. **Clm** 35–90 cm tall, 0.7–2.2 mm thick, lo intnd glab, puberulent to 5 mm below the nd; **nd** 2–3. **Bas shth** mostly glab, mrg ciliolate distally; **col** glab, including the sides; **bas lig** 0.2–0.5 mm, truncate, ciliolate, cilia to 0.1 mm; **up lig** 0.2–3.5 mm, rounded to acute; **bld** to 30 cm long, 1–3 mm wide. **Pan** 10–25 cm long, 0.5–1.5 cm wide; **br** ascending to appressed, straight. **Spklt** appressed to the br. **Glm** unequal; **lo glm** 10–15 mm long, 0.5–1.1 mm wide, exceeding the up glm by 1–3(4) mm; **flt** 5.5–11 mm long, 0.7–1 mm thick, fusiform, terete, widest at or below midlength; **cal** 0.4–0.6 mm, blunt; **lm** evenly hairy, hairs at midlength about 1 mm, apical hairs 1–2 mm, ascending to divergent, apical lobes 0.2–0.5 mm, memb, flexible; **awns** 10–19 mm, persistent, once(twice)-geniculate, bas segments scabrous, tml segments straight; **pal** 2.8–5.6 mm, $\frac{1}{2}$–$\frac{2}{3}$ as long as the lm, hairy, hairs not or scarcely exceeding the apc, veins terminating at or before the apc, apc acute to rounded; **anth** 2.5–4 mm, dehiscent, not penicillate. **Car** 3–6 mm, fusiform. $2n =$ unknown.

Achnatherum perplexum grows on slopes in pinyon-pine associations of the southwestern United States and adjacent Mexico, at 1500–1700 m. It flowers in late summer to early fall. It has generally been confused with *A. scribneri* and *A. nelsonii*. It differs from *A. scribneri* in the glabrous collar

margins of its basal leaves and its blunt calluses; from *A. nelsonii* and *A. lettermanii* in its unequal glumes; and from *A. lettermanii* in its relatively short paleas.

11. **Achnatherum pinetorum** (M.E. Jones) Barkworth PINEWOODS NEEDLEGRASS [p. 295, 494]

Pl tightly ces, not rhz. **Clm** 14–50(80) cm tall, 0.4–0.9 mm thick, mostly glab, lo intnd often puberulent or pubescent, particularly below the nd; **nd** 2–3. **Bas shth** not becoming flat and ribbonlike with age, usu glab, throat smt with a few hairs, hairs about 0.2 mm; **col** glab, including the sides; **bas lig** 0.2–0.8 mm, truncate to rounded, memb, glab; **up lig** to 2 mm, rounded; **bld** usu involute and 0.2–0.4 mm in diameter, 0.5–1 mm wide if flat, often arcuate distally, abx surfaces scabridulous, adx surfaces hairy, hairs about 0.1 mm. **Pan** 4.5–20 cm long, 0.5–1 cm wide, contracted; **br** appressed, lo br 1–5 cm, with 2–7 spklt. **Glm** subequal, 7–11 mm long, 0.6–0.9 mm wide, lanceolate, not saccate; **flt** 3.5–5.5 mm long, 0.6–0.8 mm thick, fusiform, terete; **cal** 0.4–0.6 mm, sharp; **lm** densely and evenly pilose, hairs at midlength 1.5–3.5 mm, apical hairs to 5 mm, apical lobes 0.3–2 mm, thin; **awns** 13–25 mm, persistent, twice-geniculate, first 2 segments scabrous; **pal** 2.5–4 mm, from $\frac{2}{3}$ as long as to equaling the lm, hairy; **anth** 1.8–2.6 mm, dehiscent, not penicillate. **Car** 2.5–4 mm, fusiform. $2n = 32$.

Achnatherum pinetorum usually grows on rocky soil, in pinyon-juniper to subalpine associations, at 2100–3300 m. Its range extends from Oregon, Idaho, and Montana south to California, Nevada, and Colorado. It differs from *A. webberi* in its longer, persistent awns, and from *A. lettermanii* in its sharp calluses and longer lemma hairs.

12. **Achnatherum webberi** (Thurb.) Barkworth WEBBER'S NEEDLEGRASS [p. 295, 494]

Pl tightly ces, not rhz. **Clm** 12–35 cm tall, 0.4–0.7 mm thick, smooth or antrorsely scabridulous; **nd** 2–3. **Bas shth** glab, smooth or scabridulous; **col** glab, without tufts of hair on the sides; **bas lig** 0.1–1 mm, truncate to rounded; **up lig** 1–2 mm, acute; **bld** 0.5–1.5 mm wide when flat, usu folded to involute and about 0.5 mm in diameter, stiff, abx surfaces smooth or scabrous, adx surfaces scabrous. **Pan** 2.5–7 cm long, 0.5–2 cm wide, contracted; **br** appressed, longest br 1–2 cm. **Glm** subequal, 6–10 mm long, 0.6–0.9 mm wide, lanceolate, not saccate; **flt** 4.5–6 mm long, 0.7–1 mm thick, fusiform, terete; **cal** 0.3–0.8 mm, blunt; **lm** evenly and densely pilose, hairs 2.5–3.5 mm, apical lobes 0.6–1.9 mm, memb; **awns** 4–11 mm, readily deciduous, straight to once-geniculate, scabrous; **pal** 4–5.6 mm, from as long as to slightly longer than the lm; **anth** 1.6–2 mm, dehiscent, not penicillate. **Car** 3.5–4.5 mm, fusiform. $2n = 32$.

Achnatherum webberi grows in dry, open flats and on rocky slopes, often with sagebrush, at 1500–2500 m. It grows at scattered locations from Oregon and Idaho to California and Nevada. It differs from *A. hymenoides* in its cylindrical floret and non-saccate glumes, and from *A. pinetorum* and *A. parishii* subsp. *parishii* in its shorter, deciduous awns. It also has narrower blades than *A. parishii* subsp. *depauperatum*.

13. Achnatherum swallenii (C.L. Hitchc. & Spellenb.) Barkworth SWALLEN'S NEEDLEGRASS [p. 295, 494]

Pl tightly ces, not rhz. **Clm** 15–25 cm tall, 0.5–1 mm thick, glab; **nd** 2–3. **Bas shth** mostly glab, pubescent at the base, throats glab; **col**, including the sides, glab; **bas lig** 0.2–0.3 mm, obtuse to rounded, glab; **up lig** to 0.5 mm, rounded to broadly acute; **bld** 0.4–0.7 mm wide, arcuate, abx surfaces smooth or scabridulous, adx surfaces with hairs shorter than 0.5 mm. **Pan** 3–6.5 cm long, 0.3–0.7 cm wide; **br** appressed, lo br 1–5 cm, with 1–5 spklt. **Glm** subequal, 4–5.5 mm, not saccate, apc narrowly acute to acuminate, midveins often prolonged into an awnlike tip; **lo glm** 0.6–1 mm wide, apc narrowly acute; **flt** 2.5–3.5 mm long, 0.6–0.8 mm thick, fusiform, terete; **cal** 0.1–0.2 mm, blunt; **lm** evenly hairy, hairs 0.3–0.5 mm, all similar in length, apical lobes 0.3–0.5 mm, thickly memb; **awns** 5–6 mm, once-geniculate, readily deciduous, bas segment scabridulous; **pal** 2.2–2.5 mm, slightly shorter than the lm; **anth** 1.5–2 mm, dehiscent, not penicillate. **Car** 2–3 mm, ovoid. $2n = 34$.

Achnatherum swallenii grows on open, rocky sites, frequently with low sagebrush, in Idaho and western Wyoming, at 1500–2200 m. It is a dominant species in parts of eastern Idaho, although it is poorly represented in collections.

14. Achnatherum hymenoides (Roem. & Schult.) Barkworth INDIAN RICEGRASS [p. 296, 494]

Pl tightly ces, not rhz. **Clm** 25–70 cm tall, 0.7–1.3 mm thick, glab or partly scabridulous; **nd** 3–4. **Shth** glab or scabridulous, smt puberulent on the distal mrg, hairs to 0.8 mm; **col** glab, smt with tufts of hair on the sides, hairs to 1 mm; **bas lig** 1.5–4 mm, hyaline, glab, acute; **up lig** to 2 mm; **bld** usu convolute, 0.1–1 mm in diameter, abx surfaces smooth or scabridulous, adx surfaces pubescent. **Pan** 9–20 cm long, 8–14 cm wide; **br** ascending to strongly divergent, longest br 3–15 cm; **ped** paired, conspicuously divaricate,

shorter ped in each pair usu at least ½ as long as the longer ped. **Glm** subequal, 5–9 mm long, 0.8–2 mm wide, saccate below, puberulent, hairs about 0.1 mm, tapering above midlength, apc acuminate; **lo glm** 5-veined at the base, 3-veined at midlength; **up glm** 5–7-veined at the base; **flt** 3–4.5 mm long, 1–2 mm thick, obovoid; **cal** 0.4–1 mm, sharp; **lm** indurate, densely and evenly pilose, hairs 2.5–6 mm, easily rubbed off, apc not lobed; **awns** 3–6 mm, rapidly deciduous, not geniculate, scabrous; **pal** subequal to the lm in length and texture, glab, apc pinched; **anth** 1.5–2 mm, penicillate, dehiscent, well filled. **Car** 2–3 mm. $2n = 46, 48$.

Achnatherum hymenoides grows in dry, well-drained soils, primarily in the western United States and Canada, and in northern Mexico. The roots of *A. hymenoides* are often surrounded by a rhizosheath formed by mucilaginous secretions to which soil particles attach. This rhizosheath harbors nitrogen-fixing organisms that probably contribute to the success of the species as a colonizer.

Native Americans used the seeds of *Achnatherum hymenoides* for food. It is also one of the most palatable native grasses for livestock. Several cultivars have been developed for use in restoration work, and it is becoming increasingly available for use as an ornamental.

Achnatherum hymenoides forms natural hybrids with other members of the *Stipeae*. See discussion on p. 31.

15. Achnatherum arnowiae (S.L. Welsh & N.D. Atwood) Barkworth ARNOW'S RICEGRASS [p. 296, 494]

Pl tightly ces, not rhz. **Clm** 15–75 cm tall. **Shth** smooth, mostly glab, mrg ciliate; **col** glab, with or without tufts of hair at the sides, hairs 0.7–2 mm; **lig** 1–4 mm, glab or sparsely hairy, acute, smt ciliate; **bld** 1–2 mm wide, involute, 0.5–1 mm in diameter, abx surfaces scabridulous or smooth, adx surfaces densely hairy, hairs about 0.2 mm. **Pan** 5–20 cm long, 0.5–2.8 cm wide, loosely contracted; **br** strongly ascending, longest br 0.5–2.5(5) cm. **Spklt** evenly distributed along the br; **ped** loosely appressed to the br, paired, unequal, shorter ped in each pair usu less than ½ as long as the longer ped. **Glm** slightly unequal, saccate below, tapering from about midlength, veins and smt also the intercostal regions puberulent, hairs to 0.1 mm, apc acute to acuminate; **lo glm** 5.1–6.1 mm; **up glm** 4.3–5.2 mm; **flt** 2.8–4.2 mm, ovoid; **cal** 0.2–0.4 mm, acute; **lm** indurate, dark gray-brown, smooth, densely pilose, hairs at midlength and at the apc similar, 2–3 mm, easily rubbed off, apc not lobed; **awns** 3–4.4 mm, rapidly deciduous, not geniculate, scabrous; **pal** similar to the lm in length and texture, glab, apc pinched; **anth** about

1 mm, dehiscent, penicillate. **Car** 1–1.7 mm long, 0.8–1 mm in diameter, globose to obovoid. 2*n* = unknown.

Achnatherum arnowiae grows in pinyon-juniper, sagebrush, and mixed desert shrub communities in Utah, at 1400–2000 m. Welsh et al. (2008) state that specimens belonging to *A. arnowiae* are often filed as *A.* ×*bloomeri*, and they suggest that this species may also be hybrid in origin, possibly with *Hesperostipa comata* as one of its parents. Another possibility is that it is a derivative of *A. hymenoides* that is adapted to particular soil types.

16. Hybrids involving *A. hymenoides* [p. 296, 494]

Numerous natural hybrids exist between *Achnatherum hymenoides* and other members of the *Stipeae*. Johnson (1945, 1960, 1962, 1963, 1972) described several of these; all are sterile. Using the treatment adopted here, Johnson's hybrids have as the second parent *A. occidentale* (all subspecies), *A. thurberianum*, *A. scribneri*, *Pappostipa speciosa*, and *Nassella viridula*. Evidence from herbarium specimens suggests that *A. hymenoides* also forms sterile hybrids with other species of *Achnatherum*. The name *Achnatherum* ×*bloomeri* applies only to hybrids between *A. hymenoides* and *A. occidentale*, but plants keying here may include any of the interspecific hybrids. They all differ from *A. hymenoides* in having more elongated florets and awns 10–20 mm long, and from their other parent, in most instances, in having longer lemma hairs and more saccate glumes. Identification of the second parent is best made in the field by noting which other species of *Stipeae* are present, bearing in mind that species that are not in anthesis at the same time in one year might have sufficient overlap for hybridization in other years.

No binomial has been proposed for the hybrid with *Pappostipa speciosa*.

Sterile hybrids have anthers that do not dehisce, and contain few, poorly formed pollen grains. They also fail to form good caryopses, but this is also true of some non-hybrid plants. In the case of non-hybrid plants, failure to form good caryopses can result from failure to capture pollen or from incompatibility between the pollen grain and the pistillate plant.

5.02 **PIPTATHERUM** P. Beauv.   Mary E. Barkworth

Pl per; csp or soboliferous, smt rhz. **Clm** 10–140(150) cm, erect, usu glab, usu smooth; **nd** 1–6; **brchg** invag or exvag at the base, not brchg above the base; **prophylls** concealed by the lf shth. **Lvs** smt bas concentrated; **clstgn** not present; **shth** open, glab, smooth to scabrous; **aur** absent; **lig** 0.2–15 mm, memb to hyaline; **bld** 0.5–16 mm wide, flat, involute, valvate, or folded, often tapering in the distal ⅓, apc acute to acuminate, not stiff, bas bld not overwintering, smt not developed, flag lf bld well developed, longer than 1 cm. **Infl** 3–40 cm, tml pan, open or contracted; **br** straight or flexuous, usu scabrous, rarely smooth; **ped** often appressed to the br. **Spklt** 1.5–7.5 mm, with 1 flt; **rchl** not prolonged beyond the flt; **dis** above the glm, beneath the flt. **Glm** from 1 mm shorter than to exceeding the flt, subequal or the lo glm longer than the up glm, memb, 1–9-veined, veins evident, apc obtuse to acute or acuminate; **flt** 1.5–10 mm, usu dorsally compressed, smt terete; **cal** 0.1–0.6 mm, glab or with hairs, blunt; **lm** 1.2–9 mm, smooth, coriaceous or stiffly memb, tawny or light brown to black at maturity, 3–7-veined, mrg flat, separated and parallel for their whole length at maturity, apc not lobed or lobed, glab or hairy, hairs about 0.5 mm, not spreading, awned, lm-awn jnct evident; **awns** 1–18(20) mm, centric, often caducous, almost straight to once- or twice-geniculate, scabrous; **pal** as long as or slightly longer than the lm, similar in texture and pubescence, 2(3)-veined, not keeled over the veins, flat between the veins, veins terminating near the apc, apc often pinched; **anth** 3, 0.6–5 mm, smt penicillate; **sty** 2 and free to their bases, or 1 with 2–3 br. **Car** glab, ovoid to obovoid; **hila** ½ as long as to equaling the car. *x* = 11, 12. Name from the Greek *pipto*, 'fall', and *ather*, 'awn'.

As treated here, *Piptatherum* has approximately 30 species, most of which are Eurasian. They extend from lowland to alpine regions, and grow in habitats ranging from mesic forests to semideserts. Romaschenko et al. (2008) suggested that no North American species should be included in the genus. No changes have been made, pending publication of the necessary name changes.

The pistils in *Piptatherum* exhibit variability in the development of the styles, a feature that can be seen only in florets shortly before or at anthesis. This variability is reported in the descriptions, but the number of specimens examined per species is low, sometimes only two.

1. Panicle branches appressed; florets 3–6 mm long; lemmas 2-lobed, lobes thick; awn persistent, once geniculate . . . . 1. *P. exiguum*
1. Panicle branches strongly divergent; florets 1.5–4.1 mm long; lemmas not lobed; awns caducous, not geniculate.
   2. Lemmas and calluses usually glabrous, sometimes sparsely hairy; awns 4–8 mm long; ligules 0.4–2.5 mm long, truncate. . . . . 2. *P. micranthum*
   2. Lemmas and calluses hairy, hairs to 0.5 mm; awns 1–2.5 mm long; ligules 1.8–5.5 mm long, acute . . . . 3. *P. shoshoneanum*

1. **Piptatherum exiguum** (Thurb.) Dorn LITTLE
   PIPTATHERUM [p. 297, 494]

Pl tightly ces, not rhz. **Clm** 12–40 cm, smt not
exceeding the bas lvs, scabridulous; **bas brchg**
mostly invag. **Lvs** bas concentrated; **shth** mostly
smooth, smt scabridulous distally; **lig** 1.2–3.5
mm, acute; **bas bld** 9–30 cm long, 0.6–1.4 mm
wide when open, usu valvate and 0.4–0.8
mm in diameter, both surfaces scabrous. **Pan**
3.5–9 cm, lo nd with 1–2 br; **br** straight, tightly
appressed to the rchs, lo br 1–2 cm, with 1–2
spklt. **Glm** subequal, 3.5–6 mm long, 1.8–2.2
mm wide, ovate, apc acute; **flt** 3–6 mm, terete;
cal 0.2–0.5 mm, hirsute, dis scars circular; **lm**
evenly pubescent, tan or gray-brown at maturity,
mrg not overlapping at maturity, apical lobes 2,
0.3–0.4 mm, thick; **awns** 3.9–7 mm, persistent,
strongly once-geniculate, bas segment twisted;
pal not exceeding the lm lobes, similar in texture
and pubescence; **anth** 1.5–3 mm; **ov** with a
conelike extension bearing a 3-brchd sty. **Car**
about 2.5 mm long, 1 mm thick; **hila** linear, ⁹⁄₁₀ as
long as to equaling the car. $2n = 22$.

Piptatherum exiguum grows on rocky slopes and outcrops
in upper montane habitats, from central British Columbia
to southwestern Alberta and south to northern California,
Nevada, Utah, and northern Colorado.

2. **Piptatherum micranthum** (Trin. & Rupr.)
   Barkworth SMALL-FLOWERED PIPTATHERUM
   [p. 297, 494]

Pl loosely ces, not rhz. **Clm** 20–85 cm, glab; **bas
brchg** exvag. **Lvs** bas concentrated; **shth** glab;
lig 0.4–1.5(2.5) mm, truncate; **bld** 5–16 cm long,
0.5–2.5 mm wide, usu involute. **Pan** 5–20 cm, lo
nd with 1–3 br; **br** 2–6 cm, divergent to reflexed
at maturity, with 3–10(15) spklt, sec br appressed
to the pri br. **Glm** 2.5–3.5 mm, acute; **lo glm**
1(3)-veined; **up glm** 3-veined; **flt** 1.5–2.5 mm,
dorsally compressed; **cal** 0.1–0.2 mm, glab or
sparsely hairy, dis scars circular; **lm** usu glab, smt
sparsely pubescent, brownish, shiny, 5-veined,
mrg not overlapping at maturity; **awns** 4–8 mm,
straight or almost so, caducous; **anth** 0.6–1.2 mm,

not penicillate; **ov** truncate to rounded, bearing 2
separate sty. **Car** about 1.2 mm long, about 0.8
mm wide; **hila** linear, ¾–⁹⁄₁₀ as long as the car.
$2n = 22$.

Piptatherum micranthum grows on gravel benches, rocky
slopes, and creek banks, from British Columbia to Manitoba
and south to Arizona, New Mexico, and western Texas.
The combination of small, dorsally compressed florets and
appressed pedicels distinguishes this species from all other
native North American *Stipeae*.

3. **Piptatherum shoshoneanum** (Curto &
   Douglass M. Hend.) P.M. Peterson & Soreng
   SHOSHONE PIPTATHERUM [p. 298, 494]

Pl tightly ces, not rhz. **Clm** 20–50 cm, intnd
smooth to scabridulous; **nd** glabrate; **bas brchg**
invag. **Lvs** bas concentrated; **lig** 1.8–5.5 mm,
hyaline, acute, often lacerate; **bld** 4–16 cm
long, 1–2.5 mm wide when flat, 0.6–1 mm in
diameter when involute, abx surfaces smooth
or scabridulous, adx surfaces scabridulous.
Pan 3.3–22 cm, lo nd with 1–2(4) br; **br** 1.8–11.6
cm, flexuous, initially appressed, becoming
strongly divergent to reflexed, sec and tertiary
br appressed to the pri br. **Glm** subequal, 3.2–
5.3 mm, exceeding the flt by 0.2–1.5 mm, ovate
to broadly lanceolate, 1–9-veined, apc acute to
acuminate; **flt** 2.4–4.1 mm, dorsally compressed;
cal about 0.3 mm, hairy, dis scars round; **lm**
coriaceous, evenly pubescent, hairs to 0.5 mm,
becoming tawny with age, mrg not overlapping
at maturity; **awns** 1–2.5 mm, straight or slightly
arcuate, caducous; **pal** 2.1–3.6 mm, similar to the
lm in texture and pubescence; **anth** 1.7–2.2 mm,
penicillate; **ov** truncate, bearing 2 sty. **Car** 1.8–2
mm long, about 0.8 mm thick; **hila** linear, about
⁹⁄₁₀ the length of the car. $2n = 20$.

Piptatherum shoshoneanum is known best from eastern
Idaho, where it grows in the canyons of the Middle Fork of
the Salmon River and its tributaries. It has also been found
750 km to the southwest in the Belted Range of southwestern
Nevada. So far, field work in the intervening mountain ranges
has not revealed additional populations. It usually grows
in moist crevices of igneous, metamorphic, or sedimentary
cliffs and rock walls.

5.03 HESPEROSTIPA (M.K. Elias) Barkworth   Mary E. Barkworth

Pl per; csp, not rhz. **Clm** 12–110 cm, erect, not brchg at the up nd; prophylls shorter than the shth. **Lvs**
not overwintering, not bas concentrated; **clstgn** not developed; **shth** smooth; **aur** absent; **lig** memb,
frequently ciliate; **bld** 4–40 cm long, 0.5–4.5 mm wide, usu tightly involute, adx surfaces conspicuously
ridged, apc narrowly acute, not sharp. **Infl** tml pan, contracted or open. **Spklt** 15–60 mm, with 1 flt;
rchl not prolonged beyond the base of the flt; **dis** above the glm and beneath the flt. **Glm** 15–60 mm
long, 2–4 mm wide, tapering from near the base to a hairlike tip; **flt** 7–25 mm, narrowly cylindrical;
cal 2–6 mm, sharp, densely strigose distally; **lm** indurate, smooth, mrg flat, slightly overlapping
at maturity, the up portion fused into a papillose, ciliate crown, awned, lm-awn jnct distinct; **awn**
50–225 mm, persistent, twice-geniculate, often weakly so, lo segments twisted and scabrous to pilose,

tml segment not twisted, usu scabridulous or pilose; **pal** equal to the lm, flat, pubescent, coriaceous, 2-veined, veins terminating at the apc, apc indurate, prow-tipped; **anth** 3, 1.2–9 mm. **Car** fusiform, not ribbed. $x = 11$. Name from the Greek *hesperos*, 'west', and the generic name *Stipa*.

Hesperostipa is a North American endemic that resembles the Eurasian *Stipa sensu stricto* in overall morphology, but is more closely related to the primarily South American genera *Piptochaetium* and *Nassella*. There are five species in the genus, three of which have been found in the Intermountain Region.

1. Lemmas with lines of hairs extending from the base, glabrous between the lines; terminal segment of awn straight . . . . 1. *H. spartea*
1. Lemma hairy all over; terminal segment of the awn usually flexuous, sometimes straight.
    2. Basal segment of the awn scabrous to strigose, the hairs shorter than 0.5 mm; ligules of lower leaves 1–6.5 mm long . . . . 2. *H. comata*
    2. Basal segment of the awn pilose, the hairs 1–3 mm long; ligules of lower leaves 0.5–1 mm long. . . . . 3. *H. neomexicana*

1. **Hesperostipa spartea** (Trin.) Barkworth
PORCUPINEGRASS [p. 298, 494]

Clm 45–90 (145) cm; **lo nd** usu crossed by lines of pubescence, occ glab. **Lo shth** usu with ciliate mrg; **lig of lo lvs** 0.3–3 mm, stiff, truncate to rounded, usu entire; **lig of up lvs** 3–7.5 mm, thin, acute, often lacerate; **bld** 1.5–4.5 mm wide. **Pan** 10–25 cm. **Glm** 22–45 mm, subequal; **flt** 15–25 mm; **cal** 3.5–6 mm; **lm** unevenly pubescent, densely pubescent on the mrg and in lines on the lo portion of the lm, glab distally, hairs brown at maturity; **awns** 90–190 mm, tml segment straight, scabridulous. $2n = 44, 46$.

Hesperostipa spartea grows at elevations of 200–2600 m, primarily in the grasslands of the central plains and southern prairies of the United States and Canada. It was once a common species, but its habitat is now intensively cultivated. Native Americans used bundles of the florets for combs. Its presence in the Intermountain Region probably reflects introductions. It is not known to have persisted in the region.

2. **Hesperostipa comata** (Trin. & Rupr.)
Barkworth NEEDLE-AND-THREAD [p. 299, 494]

Clm 12–110 cm; **lo nd** glab or pubescent. **Lo shth** glab or pubescent, not ciliate; **lig of lo lvs** 1–6.5 mm, scarious, usu acute, smt truncate, often lacerate; **lig of up lvs** to 7 mm; **bld** 0.5–4 mm wide, usu involute. **Pan** 10–32 cm, contracted. **Glm** 16–35 mm, 3–5-veined; **lo glm** 18–35 mm; **up glm** 1–3 mm shorter; **flt** 7–13 mm; **cal** 2–4 mm; **lm** evenly pubescent, hairs about 1 mm, white, smt glab immediately above the cal; **awns** 65–225 mm, first 2 segments scabrous to strigose, hairs shorter than 1 mm, tml segment scabridulous.

Hesperostipa comata is found primarily in the cool deserts, grasslands, and pinyon-juniper forests of western North America. Both subspecies are primarily cleistogamous.

1. Terminal awn segment 40–120 mm long, sinuous to curled at maturity; lower cauline nodes usually concealed by the sheaths; panicles often partially enclosed in the uppermost sheath at maturity . . . . subsp. *comata*

1. Terminal awn segment 30–80 mm long, straight; lower cauline nodes usually exposed; panicles usually completely exserted at maturity . . . . subsp. *intermedia*

Hesperostipa comata (Trin. & Rupr.) Barkworth subsp. **comata** [p. 299]

Lo cauline nd usu concealed by the shth. **Pan** often partially included in the uppermost shth at maturity. **Awns** 75–225 mm, tml segment 40–120 mm, sinuous to curled. $2n = 38, 44, 46$.

Hesperostipa comata subsp. *comata* grows on well-drained soils of cool deserts, grasslands, and sagebrush associations, at elevations of 200–2500 m. It is widespread and often abundant in western and central North America, particularly in disturbed areas. Intermediates to *H. neomexicana* exist but are not common.

Hesperostipa comata subsp. **intermedia** (Scribn. & Tweedy) Barkworth [p. 299]

Lo cauline nd usu exposed. **Pan** usu fully exserted at maturity. **Awns** 65–130 mm, tml segment 30–80 mm, straight. $2n = 44–46$.

Hesperostipa comata subsp. *intermedia* is found in pinyon-juniper woodlands, at elevations of 2175–3075 m, in the Sierra Nevada and Rocky Mountains, from southern Canada to New Mexico.

3. **Hesperostipa neomexicana** (Thurb.)
Barkworth NEW MEXICAN NEEDLEGRASS [p. 299, 495]

Clm 40–100 cm; **lo nd** glab. **Lo shth** glab or puberulent, not ciliate; **lig of lo lvs** 0.5–1 mm, thickly memb, rounded; **lig** of up lvs to 3 mm, scarious, acute; **bld** 0.5–1 mm wide. **Pan** 10–30 cm. **Glm** subequal, 30–60 mm; **flt** 15–18 mm; **cal** 4–5 mm; **lm** evenly pubescent, hairs shorter than 1 mm; **awns** 120–220 mm, first 2 segments hairy, hairs mostly 0.2–1 mm, tml segment flexible, pilose, hairs 1–3 mm. $2n = 44$.

Hesperostipa neomexicana grows in grassland, oak, and pinyon pine associations, from 800–2400 m, usually in well-drained, rocky areas in the southwestern United States.

5.04 **PIPTOCHAETIUM** J. Presl   Mary E. Barkworth

Pl per; ces, not rhz. **Clm** 4–150 cm, usu erect, smt decumbent, glab, not brchd above the base; **bas brchg** invag; **prophylls** shorter than the shth, mostly glab, keels usu with hairs, apc bifid, teeth 1–3 mm; **clstgn** not developed. **Shth** open to the base, mrg glab; **lig** memb, decurrent, truncate to acute, smt highest at the sides, smt ciliate; **bld** convolute to flat, translucent between the veins, often sinuous distally. **Infl** tml pan, open or contracted, spklt usu confined to the distal ½ of each br. **Spklt** 4–22 mm, with 1 flt; **rchl** not prolonged beyond the base of the flt; **dis** above the gl, beneath the flt. **Glm** subequal, longer than the flt, lanceolate, 3–7(8)-veined; **flt** globose to fusiform, terete to lat compressed; **cal** well developed, sharp or blunt, glab or antrorsely strigose, hairs yellow to golden brown; **lm** coriaceous to indurate, glab or pubescent, striate, particularly near the base, smooth, papillose, or tuberculate, often smooth on the lo portion and papillate to tuberculate distally, mrg involute, fitting into the grooved pal, apc fused into a crown, awned, lm often narrowed below the crown, crowns usu ciliate; **awns** caducous to persistent, usu twice-geniculate, first 2 segments usu twisted and hispid, tml segment straight and scabridulous; **pal** longer than the lm, similar in texture, glab, sulcate between the veins, apc prow-tipped; **lod** 2 or 3, memb, glab, blunt or acute; **anth** 3; **ov** glab; **sty** 2. **Car** terete to globose or lens-shaped. $x = 11$. Name from the Greek *pipto*, 'fall', and *chaite*, 'long hair'.

Piptochaetium is primarily South American, being particularly abundant in Argentina. It has 27 species. One species, *P. pringlei*, is native in the Intermountain Region; *P. lasianthum*, an Argentinean species, was reported to be present by Welsh et al. (2003). Its description is based on that in Cialdella and Arriaga (1998).

1. Florets 6.5–10 mm long, 1.5–2.5 mm wide; panicles with 10–25 spikelets . . . . 1. *P. pringlei*
1. Florets 2.3–3 mm long, about 1 mm wide; panicles with 30–40 spikelets . . . . 2. *P. lasianthum*

1. **Piptochaetium pringlei** (Beal) Parodi
PRINGLE'S SPEARGRASS [p. *300*, <u>495</u>]

Clm 50–125 cm, mostly glab, pubescent below the nd; **nd** 2–3, dark, glab or slightly pubescent. **Shth** smooth to scabridulous; **lig** of bas lvs 0.5–2.8 mm, truncate to rounded, of up lvs 1–3.5 mm, rounded to acute; **bld** 10–30 cm long, 1–3.5 mm wide, 3–5-veined, abx surfaces glab, smooth, adx surfaces smooth or scabrous over the veins, mrg smooth or scabrous. **Pan** 6–20 cm, open, with 10–25 spklt; **br** ascending, flexuous; **ped** to 1 mm, flattened, hispid. **Glm** subequal, 9–12 mm long, 2.5–3.5 mm wide; **lo glm** 5–7-veined; **up glm** 7-veined; **flt** 6.5–10 mm long, 1.5–2.1 mm thick, terete to somewhat lat compressed; **cal** 0.6–1.9 mm, blunt to acute, strigose; **lm** golden brown to dark brown at maturity, shiny or not, smooth to spiny-tuberculate distally or for almost their entire length, pubescent, hairs tawny to golden brown, evenly distributed or somewhat more abundant on the bas ½, apc tapering to the crown; **crowns** 0.5–0.6 mm, inconspicuous, straight, hairy, hairs 0.5–1 mm; **awns** 19–27(35) mm, persistent, twice-geniculate, smt inconspicuously so; **pal** 6.3–9.5 mm; **lod** 2, 1–1.5 mm, acute; **anth** 3.5–5.5 mm, smt penicillate. **Car** about 7 mm, fusiform. $2n = 42$.

Piptochaetium pringlei grows in oak woodlands, often on rocky soils, in the southwestern United States and northwestern Mexico.

2. **Piptochaetium lasianthum** Griseb. [p. *300*]
Clm (40)80–100 cm tall, usu glab; **nd** 3–4, glab or lo nd slightly hairy. **Shth** glab, smt scabrous; **lig** of bas lvs 1–2 mm, obtuse; **bld** (4.5)15–25 cm long, about 1 mm wide, 3(5)-veined, abx surfaces glab or slightly hairy, adx surfaces glab, smt scabridulous near the base. **Pan** 12–25 cm long, somewhat open, with 30–40 spklt; **br** ascending, flexuous, scabrous, ped 2–7 mm, terete, hispidulous. **Glm** subequal, 5.5–7.5(10) mm, acuminate; **lo glm** 5(7)-veined; **up glm** 7-veined; **flt** 2.3–3 mm long, about 1 mm wide, slightly lat compressed; **cal** 0.5–0.8 mm, blunt, strigose; **lm** golden brown at maturity, dull, densely antrorsely strigose, the hairs to 4 mm long, evenly distributed, persistent, golden brown, exceeding the crown by 2–4 mm; **crowns** inconspicuous, slightly constricted at the base, hairy; **awns** 15–30 mm long, persistent, twice-geniculate; **pal** 2–2.5 mm; **lod** 2, about 0.5 mm.

Piptochaetium lasianthum is native to Argentina, Brazil, and Uruguay. Welsh et al. (2003) reported finding it in Zion National Park, but admitted that the identification was tentative. Examination of the voucher specimen showed it to be a hybrid having *Achnatherum hymenoides* as one parent (see p. 30). The description and illustration of *P. lasianthum* is included here to help prevent further confusion. Its most evident distinction from the hybrids involving *A. hymenoides* is the golden-brown color of its hairs and its glabrous, grooved, palea.

5.05 **ORYZOPSIS** Michx. Mary E. Barkworth

Pl per; csp, not rhz. **Clm** 25–65 cm, erect or spreading, **Bas brchg** exvag; prophylls not visible; **nd** glab. **Lvs** mostly bas; **clstgn** not developed; **shth** open, glab; **aur** absent; **lig** memb, longest at the sides or rounded, ciliate; **bld** of bas lvs 30–90 cm, remaining green over winter, erect when young, recumbent in the fall, bases twisted, placing the abx surfaces uppermost, cauline lf bld rdcd, flag lf bld 2–12 mm, conspicuously narrower than the top of the shth. **Infl** pan, contracted. **Spklt** 5–7.5 mm, with 1 flt; **rchl** not prolonged beyond the base of the flt; **dis** above the glm, beneath the flt. **Glm** subequal, 6–10-veined, apc mucronate; **flt** terete to lat compressed; **cal** usu shorter than ⅕ the length of the flt, blunt, distal portions pilose; **lm** coriaceous, pubescent at least bas, 3–5(9)-veined, mrg strongly overlapping at maturity, awned, lm-awn jnct conspicuous, lobed, lobes 0.1–0.2 mm; awn more or less straight, deciduous; **pal** similar to the lm in length, texture, and pubescence, concealed by the lm, 2-veined, flat between the veins; **lod** 2, free, memb, 2-veined; **anth** 3; **sty** 1, with 2 br; **ov** glab. **Car** falling with the lm and pal. *x* = 11, 12. Name from the Greek *oryza*, 'rice', and *opsis*, 'appearance', in reference to a supposed resemblance to rice.

Oryzopsis is treated here as a monospecific genus that is restricted to North America. The North American species previously included in *Oryzopsis* are here included in *Achnatherum* and *Piptatherum*.

1. **Oryzopsis asperifolia** Michx. Roughleaf
   Ricegrass, Winter Grass, [p. *301*, <u>495</u>]
Clm 25–65 cm. **Bas lig** 0.2–0.7 mm, rounded, smt longest at the sides; **bld** of bas lvs 30–90 cm long, 4–9 mm wide, flag lf bld 8–12 mm. **Pan** 3.5–13 cm, contracted. **Glm** subequal, 5–7.5 mm long, 2.5–4 mm wide, 6–10-veined, apc mucronate; **flt** 5–7 mm; **cal** 0.8–2 mm, blunt, distal portions with a dense ruff of soft hairs; **lm** coriaceous or indurate at maturity, pale green, white, or yellowish, smt purple-tinged, glossy or dull, pubescent at least bas, mrg overlapping, concealing the pal; **awns** 7–15 mm; **pal** similar to the lm; **anth** 2–4 mm, usu penicillate. **Car** 4–6.5 mm. 2*n* = 46.

Oryzopsis asperifolia grows in both deciduous and coniferous woods, usually on open, rocky ground in areas with well-developed duff. It is found from the Yukon and Northwest Territories south to New Mexico along the Rocky Mountains, and from British Columbia east to Newfoundland and Maryland.

Leaf development in *Oryzopsis asperifolia* is unusual in that the leaves start to develop in midsummer, the blades growing upright. As the year progresses, the blades bend over, but stay alive and green through winter and the following spring. The part of the sheaths that remains below the level of the duff is usually bright purple (Dore and McNeill 1980).

5.06 **NASSELLA** (Trin.) E. Desv. Mary E. Barkworth

Pl usu per, rarely ann; usu csp, occ rhz. **Clm** 10–175(210) cm, smt brchd at the up nd, br flexible; **prophylls** not evident, shorter than the shth. **Lvs** mostly bas, not overwintering; **shth** open; **clstgn** smt present; **aur** absent; **lig** memb, smt pubescent or ciliate; **bld** of bas lvs 3–60 cm long, 0.2–8 mm wide, apc narrowly acute to acute, not sharp, flag lf bld 1–80 mm, bases about as wide as the top of the shth. **Infl** tml pan, smt partially included at maturity. **Spklt** 3–22 mm, with 1 flt; **rchl** not prolonged beyond the base of the flt; **dis** above the glm, beneath the flt. **Glm** longer than the flt, narrowly lanceolate or ovate, bas portion usu purplish at anthesis, color fading with age, (1)3–5-veined, smt awned; **flt** usu terete, smt slightly lat compressed; **cal** blunt or sharp, glab or antrorsely strigose; **lm** usu papillose or tuberculate, at least distally, smt smooth throughout, glab or variously hairy, strongly convolute, wrapping 1.2–1.5 times around the car, apc not lobed, fused distally into crowns, these often evident by their pale color and constricted bases; **crowns** mostly glab, rims often bearing hairs with bulbous bases; **awns** tml, centric or eccentric, deciduous or persistent, usu twice-geniculate, second geniculation often obscure; **pal** up to ½ as long as the lm, glab, without veins, flat; **lod** 2 or 3, if 3, the third somewhat shorter than the other 2; **anth** 1 or 3, if 3, often of 2 lengths, penicillate; **ov** glab; **sty** 2, bases free. **Car** glab, not ribbed; **hila** elongate; **emb** to ⅖ as long as the car. *x* = 7, 8. Name not explained by Desvaux (1854), but possibly from the Latin *nassa*, a narrow-necked basket for catching fish.

Nassella includes at least 116 species (Barkworth and Torres 2001), the majority of which are South American. Only two species grow in the Intermountain Region. The strongly convolute lemmas distinguish *Nassella* from all other genera of *Stipeae* in the Americas and, in combination with the reduced, ecostate, glabrous paleas, from all other genera in the tribe worldwide. Molecular data (Jacobs et al. 2006; Romaschenko et al. 2008; Barkworth et al. 2008) support the expanded interpretation of *Nassella* to include *Amelichloa*.

Many species of *Nassella* develop both cleistogamous and chasmogamous florets in the terminal panicle. The cleistogamous florets have 1–3 anthers that are less than 1 mm long; the chasmogamous florets have 3 anthers that are significantly longer. In addition, some species develop cleistogenes, i.e., panicles in the axils of their basal sheaths. Spikelets of cleistogenes have reduced or no glumes, and florets with no or very short awns.

1. Florets 7.5–11.5 mm long; lemmas constricted below the crown; awns 38–100 mm long . . . . 1. *N. pulchra*
1. Florets 3.4–5.5 mm long; lemmas not constricted below the crown; awns 19–32 mm long . . . . . . 2. *N. viridula*

1. **Nassella pulchra** (Hitchc.) Barkworth Purple Nassella, Purple Needlegrass [p. *301*, 495]

Pl per; ces, not rhz. **Clm** 35–100 cm tall, 1.8–3.1 mm thick, erect or geniculate at the lowest nd, smt scabrous below the pan, intnd mostly glab, lo intnd smt pubescent below the nd; **nd** 2–3, pubescent. **Shth** glab or hairy, smt mostly glab, smt the distal mrg ciliate, varying within a pl; **col** with tufts of hair at the sides, hairs 0.5–0.8 mm; **lig** 0.3–1.2 mm, glab, truncate to rounded; **bld** 10–20 cm long, 0.8–3.5 mm wide, flat to convolute, abx surfaces glab or sparsely pilose. **Pan** 18–60 cm, open; **br** 3–9 cm, spreading, flexuous, often pilose at the axils, with 2–6 spklt; **ped** 3–10 mm. **Glm** subequal, 12–20 mm long, 1.1–2.2 mm wide, narrowly lanceolate, glab; **flt** 7.5–11.5 mm long, about 1.2 mm thick, terete; **cal** 1.8–3.5 mm, sharp, strigose; **lm** papillose, evenly pubescent at maturity, constricted below the crown; **crowns** 0.6–1.1 mm long, 0.5–0.7 mm wide, straight-sided to slightly flared, rims with 0.8–0.9 mm hairs; **awns** 38–100 mm long, 0.3–0.45 mm thick at the base, strongly twice-geniculate, tml segment straight; **anth** 3.5–5.5 mm, penicillate. **Car** 4.5–6 mm. 2*n* = 64.

Nassella pulchra grows in oak chaparral and grassland communities of the coast ranges and Sierra foothills of California, extending south into Mexico. It probably never formed extensive grasslands (Hamilton 1997), flourishing primarily in moderately disturbed areas.

2. **Nassella viridula** (Trin.) Barkworth Green Nassella, Green Needlegrass [p. *301*, 495]

Pl per; ces, not rhz. **Clm** 35–120 cm, erect or geniculate bas, intnd mostly glab, pubescent below the lo nd; **nd** 2–3, glab. **Shth** mostly glab, mrg usu ciliate; **col** of bas lvs hispidulous, with tufts of hair at the side s, hairs 0.5–1.8 mm, col of flag lvs glab or sparsely pubescent; **lig** 0.2–1.2 mm, glab, truncate to rounded; **bld** 10–30 cm long; 1.5–3 mm wide, flat to convolute, abx surfaces scabridulous, glab, adx surfaces glab. **Pan** 2.9–7.2 cm, loosely contracted; **br** 1–4 cm, appressed or ascending, with 3–7 spklt; **ped** 1–9 mm. **Glm** subequal, 6.8–13 mm long, 1–2.1 mm wide, narrowly lanceolate, glab, apiculate; **flt** 3.4–5.5 mm long, 1–1.2 mm wide, terete; **cal** 0.7–1.4 mm, moderately sharp, strigose; **lm** papillose, evenly pubescent, not constricted below the crown; **crowns** 0.4–0.5 mm long, 0.3–0.5 mm wide, not conspicuous, straight-sided, rims with 0.5–0.75 mm hairs; **awns** 19–32 mm, evidently twice-geniculate, tml segment straight; **anth** (0.8)2–3 mm, smt penicillate. **Car** about 3.5 mm. 2*n* = 82, 88.

Nassella viridula grows in grasslands and open woods, frequently on sandy soils. It is the most widespread species of *Nassella* in North America. Its morphology, distribution, and high chromosome number suggest that it may be an alloploid between *Nassella* and *Achnatherum*. It is included in *Nassella* because it resembles *Nassella* more than *Achnatherum* in the characters distinguishing the two genera.

5.07 **PAPPOSTIPA** Speg.

Mary E. Barkworth

Pl per; tightly ces, smt stln. **Clm** 10–70 cm tall; **bas brchg** invag; **nd** 2-3. **Shth** whitish to reddish brown or purplish; **lig** 0.5–3 mm long, lo lig often densely hairy, smt merely ciliate, smt glab, up lig glab or sparsely hairy; **bld** 0.5–2.5 mm wide, convolute, apc pointed, abx surfaces smt scabrous, usu glab, smt with hairs, adx surfaces hairy. **Pan** 3–24 cm long, often partially included in the up lf shth at maturity. **Glm** subequal, 15–45 mm long, longer than the flt, hyaline, with 1–3(5) veins, acuminate. **Flt** 4–17 mm, fusiform; **cal** 1–4 mm; **lm** evenly hairy with hairs 0.5–1 mm long or glab on the distal portion of the back, apc not fused into a crown, lobed, lobes to 1 mm; **awns** 20–70(195) cm long, persistent, strongly once-geniculate, first segment 15–33 mm, densly hairy, hairs 4–9 mm long, tml segment 12–45(170) mm long; **pal** usu subequal to the lm, smt less than ½ as long, usu hairy, smt glab.

Pappostipa is native to South America and most abundant in Argentina. Romaschenko et al. (2008) recognize 14 species in the genus of which one, *P. speciosa*, grows in both South and North America. The genus is readily distinguished by its single florets and sharply once-geniculate awns with their plumose column. The generic description is based on that provided by Romaschenko et al. (2008)

1. **Pappostipa speciosa** (Trin. & Rupr.)

Romasch. DESERT NEEDLEGRASS [p. *302*, *495*]
Pl tightly ces, not rhz. **Clm** 30–60 cm, bases
orange-brown; **nd** 3–6; **bas brchg** invag. **Shth**
mostly glab, throats densely ciliate, bas shth
reddish brown, flat and ribbonlike with age;
lig varying within a pl, lig of lo lvs 0.3–1 mm,
densely hairy and ciliate, hairs 0.2–1 mm, often
longer than the bas membrane, lig of up lvs to 2.5
mm, hyaline to scarious, glab or hairy, usu less
hairy than the lo lig, smt ciliate; **bld** 10–30 cm
long, 0.5–2 mm wide when flat, usu rolled, to 1
mm in diameter, abx surfaces glab, smooth, adx
surfaces pilose. **Pan** 10–15 cm, dense, frequently
partially included in the up lf shth at maturity;
br ascending. **Spklt** 16–24 mm. **Glm** linear-
lanceolate, glab, tapering from below midlength
to the narrowly acute apc; **lo glm** 16–24 mm,
1-veined; **up glm** 13–19 mm, 3–5-veined; **flt** (6)8–
10 mm; **cal** 0.8–1.6(3) mm, sharp; **lm** densely
and evenly hairy, hairs about 0.5 mm, without
a pappus; **awns** 35–45(80) mm, once-geniculate,
first segment pilose, hairs 3–8 mm, tml segment
glab, smooth; **pal** 3.2–5.1 mm, ⅖–⅔(⅘) the length
of the lm, usu hairy, hairs about 0.5 mm. $2n = 66$,
68, about 74.

Pappostipa speciosa grows on rocky slopes in canyons of
arid and semiarid regions of the southwestern United States
and northern Mexico, and in Chile and northern to central
Argentina. Several varieties are recognized in South America.
It is not clear to which of these varieties, if any, the North
American plants belong.

The reddish brown leaf bases, differing lower and upper
ligules, and the pilose, once-geniculate awns make *Pappostipa
speciosa* an easy species to recognize in North America. It is
also an attractive species, well worth cultivating. It prefers
open areas with well-drained soils. The growth of young
shoots and flowering is stimulated by fire.

6. **BROMEAE** Dumort.

Mary E. Barkworth

Pl ann or per; usu csp, smt rhz. **Clm** ann, not woody, not brchg above the base; **intnd** usu hollow,
rarely solid. **Shth** closed, mrg united for most of their length; **col** without tufts of hair on the sides;
aur smt present; **lig** memb, smt shortly ciliate, those of the up and lo cauline lvs usu similar; **psdpet**
absent; **bld** linear to narrowly lanceolate, venation parallel, cross venation not evident, without arm
or fusoid cells, cross sections non-Kranz, epdm without microhairs, not papillate. **Infl** usu tml pan,
smt rdcd to rcm in depauperate pl; **dis** above the glm and beneath each flt. **Spklt** 5–80 mm, not
viviparous, terete to lat compressed, with 3–30 bisex flt, distal flt smt rdcd; **rchl** prolonged beyond
the bases of the distal flt. **Glm** usu unequal, rarely more or less equal, exceeded by the distal flt, usu
longer than ¼ the length of the adjacent flt, lanceolate, 1–9(11)-veined; **flt** terete to lat compressed;
cal glab, not well developed; **lm** lanceolate to ovate, rounded or keeled over the midvein, herbaceous
to coriaceous, 5–13-veined, veins converging somewhat distally, apc usu minutely bilobed to bifid,
rarely entire, usu awned, smt unawned, awns unbrchd, tml or subtml, usu straight, smt geniculate;
pal usu shorter than the lm; **lod** 2, glab, not veined; **anth** 3; **ov** with hairy apc; **sty** 2, bases free. **Car**
narrowly ellipsoid to linear, longitudinally grooved; **hila** linear; **emb** about ⅙ the length of the car.
$x = 7$.

There are three genera in the *Bromeae*. One genus, *Bromus*, grows in the Intermountain Region. It is most closely related to
the *Triticeae*, which it resembles in the pubescent apices of its ovaries, simple endosperm starch grains, nucleic acid sequences,
and seedling development. These data do not support a close relationship between the *Bromeae* and *Brachypodium*, a genus that
has sometimes been included in the tribe.

6.01 **BROMUS** L.

Leon E. Pavlick† and Laurel K. Anderton

Pl per, ann, or bien; usu csp, smt rhz. **Clm** 5–190 cm. **Shth** closed to near the top, usu pubescent;
aur smt present; **lig** memb, to 6 mm, usu erose or lacerate; **bld** usu flat, rarely involute. **Infl** pan,
smt rcm in depauperate specimens, erect to nodding, open to dense, occ 1-sided; **br** usu ascending
to spreading, smt reflexed or drooping. **Spklt** 5–70 mm, terete to lat compressed, with 3–30 flt; **dis**
above the glm, beneath the flt. **Glm** unequal, usu shorter than the adjacent lm, always shorter than the
spklt, glab or pubescent, usu acute, rarely mucronate; **lo glm** 1–7(9)-veined; **up glm** 3–9(11)-veined;
lm 5–13-veined, rounded to keeled, glab or pubescent, apc entire, emgt, or toothed, usu tml or subtml
awned, smt with 3 distinct awns or unawned; **pal** usu shorter than the lm, ciliate on the keels, adnate
to the car; **anth** (2)3. $x = 7$. Name from the Greek *bromos*, an ancient name for 'oats', which was based
on *broma*, 'food'.

Bromus grows in temperate and cool regions. It is estimated to include 100–400 species, the number depending on how
the species are circumscribed. Of the 29 species in the Intermountain Region, 14 are native and 15 are introduced. The native

perennial species provide considerable forage for grazing animals, with some species being cultivated for this purpose. The introduced species, most of which are annuals, range from sporadic introductions to well-established members of the region's flora. Many are weedy and occupy disturbed sites. Some are used for hay; others have sharp, pointed florets and long, rough awns that can injure grazing animals.

This treatment is based on one submitted by Pavlick, who died before it could be reviewed and edited. It has been substantially revised by Anderton. The majority of Pavlick's taxonomic concepts are retained, despite the necessity for overlap in many key leads.

We thank Hildemar Scholz of the Botanic Garden and Botanical Museum Berlin-Dahlem, Free University Berlin, for providing accurately identified specimens of the weedy European species for use in preparing the illustrations, and for his helpful suggestions for the keys and descriptions. In the keys and descriptions, the distances from the bases of the subterminal lemma awns to the lemma apices are measured on the most distal florets in a spikelet.

1. Lemmas strongly keeled, at least distally; spikelets strongly laterally compressed; lower glumes 3–7(9)-veined . . . . sect. *Ceratochloa*
1. Lemmas rounded over the midvein; spikelets terete to moderately laterally compressed; lower glumes 1–5-veined.
   2. Awns, if present, arising less than 1.5 mm below the lemma apices; lemma apices entire, emarginate, or with teeth less than 1 mm long.
      3. Lower glumes 1–3-veined; upper glumes 3–5-veined; plants perennial or annual, if annual, the lower glumes 1-veined and the upper glumes 3-veined . . . . sect. *Bromopsis*
      3. Lower glumes 3–5-veined; upper glumes 5–9-veined; plants annual or biennial, if biennial, the upper glumes 7-veined and/or the lateral veins of the lemmas prominently ribbed. . . . . sect. *Bromus* (in part)
   2. Awns arising 1.5 mm or more below the lemma apices, lemma apices entire, emarginate, or with teeth to 5 mm long.
      4. Awns usually geniculate, sometimes only divaricate, lemma teeth 2–3 mm long, usually aristate, sometimes only acuminate . . . . sect. *Neobromus*
      4. Awns straight, arcuate, or divaricate, not geniculate, sometimes absent; lemma teeth absent or to 5 mm long, acuminate.
         5. Lower glumes 1–3-veined; upper glumes 3–5-veined; spikelets with parallel or diverging sides in outline, often widening distally; lemma apices bifid, teeth (0.8)1–5 mm long . . . . sect. *Genea*
         5. Lower glumes 3–5-veined; upper glumes 5–9-veined; spikelets with parallel or converging sides in outline; lemma apices entire to bifid, teeth less than 1 mm long, apices sometimes split and teeth appearing longer. . . . . sect. *Bromus* (in part)

Bromus sect. **Ceratochloa**

1. Lemmas unawned or with awns to 3.5 mm long; lemmas usually glabrous, sometimes pubescent distally, veins prominent for most of their length . . . . 1. *B. catharticus* (in part)
1. Lemmas awned, awns (2)4–17 mm long; lemmas pubescent or glabrous, veins obscure or prominent.
   2. Lower panicle branches to 20 cm long, with 1–3 spikelets on the distal ½, sometimes confined to the tips; culms 3–5 mm thick . . . . 2. *B. sitchensis*
   2. Lower panicle branches to 10 cm long, with 1–5 spikelets variously distributed; culms less than 4 mm thick.
      3. Upper glume about as long as the lowest lemma in each spikelet; lemmas glabrous or pubescent distally or throughout, the marginal hairs, if present, longer than those elsewhere . . . . 3. *B. arizonicus*
      3. Upper glume shorter than the lowest lemma in each spikelet; lemmas glabrous or pubescent only on the margins or throughout, if throughout, the marginal hairs similar in length to those elsewhere.
         4. Lemmas and sheath throats glabrous. . . . . 5. *B. polyanthus*
         4. Lemmas and/or sheath throats with hairs.
            5. Lemmas 9–13-veined, veins often raised and riblike distally or throughout . . . . 1. *B. catharticus* (in part)
            5. Lemmas 7–9-veined, veins usually not raised or riblike . . . . 4. *B. carinatus*

Bromus sect. **Bromopsis**

1. Plants rhizomatous.
    2. Lemma backs sparsely to densely hairy throughout, or on the lower portion and margins, or along the marginal veins and keel; cauline nodes and leaf blades pubescent or glabrous; awns usually present, to 7.5 mm long, sometimes absent . . . . 7. *B. pumpellianus*
    2. Lemma backs usually glabrous, occasionally sparsely puberulent at the base and sometimes on the margins; cauline nodes and leaf blades usually glabrous, rarely hairy; awns absent or to 3 mm long . . . . 6. *B. inermis*
1. Plants not rhizomatous.
    3. Most lower glumes within a panicle 3-veined, sometimes some 1-veined.
        4. Most upper glumes within a panicle 5-veined, sometimes some 3-veined . . . . 8. *B. laevipes*
        4. Most upper glumes within a panicle 3-veined, sometimes some 5-veined.
            5. Leaf blades often glaucous; glumes usually glabrous, rarely slightly pubescent
             . . . . 9. *B. frondosus*
            5. Leaf blades not glaucous; glumes usually pubescent, rarely glabrous . . . . 10. *B. porteri* (in part)
    3. Most lower glumes within a panicle 1-veined, sometimes some 3-veined.
        6. Awns (4)6–12 mm long; ligules 2–6 mm long . . . . 11. *B. vulgaris*
        6. Awns to 5 mm long; ligules to 2.5 mm long.
            7. Panicle branches appressed to slightly spreading; culm nodes 1–4 . . . . 12. *B. suksdorfii*
            7. Panicle branches ascending to drooping; culm nodes (1)2–8.
                8. Glumes usually pubescent, rarely glabrous
                    9. Upper glume mucronate . . . . 13. *B. mucroglumis*
                    9. Upper glume not mucronate . . . . 10. *B. porteri* (in part)
                8. Glumes usually glabrous, sometimes pubescent.
                    10. Lemma margins and backs usually pubescent, sometimes nearly glabrous; awns 2–4 mm long; anthers 1.8–4 mm long . . . . 14. *B. lanatipes*
                    10. Lemma margins conspicuously hirsute or densely pilose, at least along the lower ½, the backs glabrous at least on the lower lemmas in a spikelet; awns 3–5 mm long; anthers 1–2.7 mm long.
                        11. Backs of all lemmas glabrous; anthers 1–1.4 mm long; upper glumes 7.1–8.5 mm long . . . . 15. *B. ciliatus*
                        11. Backs of the upper lemmas in a spikelet hairy; anthers 1.6–2.7 mm long; upper glumes 8.9–11.3 mm long . . . . 16. *B. richardsonii*

Bromus sect. **Neobromus**

This section includes one species, 17. *Bromus berteroanus*.

Bromus sect. **Genea**

1. Lemmas 20–35 mm long . . . . 18. *B. diandrus*
1. Lemmas 9–20 mm long.
    2. Spikelets longer than the panicle branches; panicle branches ascending to spreading, never drooping . . . . 21. *B. rubens*
    2. Spikelets usually shorter than the panicle branches; panicle branches ascending to spreading or drooping.
        3. Lemmas 14–20 mm long; panicles with spreading, ascending, or drooping branches, rarely any branches with more than 3 spikelets . . . . 19. *B. sterilis*
        3. Lemmas 9–12 mm long; panicles with drooping branches, often 1 or more branches with 4–8 spikelets . . . . 20. *B. tectorum*

Bromus sect. **Bromus**

1. Lemmas inflated, 6–8 mm wide; unawned or with awns up to 1 mm long; spikelets ovate . . . . 22. *B. briziformis*
1. Lemmas not inflated, 1.2–5 mm wide; awns 3–13 mm long, rarely absent; spikelet shape various.
    2. Lemma margins inrolled at maturity; floret bases visible at maturity; rachilla internodes visible at maturity; caryopses sometimes thick, strongly inrolled.

3. Lower leaf sheaths glabrous or loosely pubescent and glabrate; lemmas 6.5–8.5(10) mm long, margins evenly rounded; awns straight or flexuous . . . . 23. *B. secalinus*
3. Lower leaf sheaths evenly covered with stiff hairs; lemmas 8–11.5 mm long, margins bluntly angled; awns straight . . . . 24. *B. commutatus* (in part)

2. Lemma margins not inrolled at maturity; floret bases concealed at maturity; rachilla internodes concealed at maturity; caryopses thin, weakly inrolled or flat.

4. Awns arising less than 1.5 mm below the lemma apices, erect or weakly divaricate, not twisted at the base.

5. Panicle branches shorter than the spikelets; lemmas chartaceous, with prominent ribs over the veins, often concave between the veins; anthers 0.6–1.5 mm long . . . . 25. *B. hordeaceus* (in part)

5. At least some panicle branches longer than the spikelets; lemmas coriaceous, veins obscure or distinct, not ribbed; anthers 0.7–3 mm long.

6. Anthers 0.7–1.7 mm long; rachilla internodes 1.5–2 mm long; lemmas 8–11.5 mm long, margins bluntly angled . . . . 24. *B. commutatus* (in part)

6. Anthers 1.5–3 mm long; rachilla internodes 1–1.5 mm long; lemmas 6.5–8 mm long, margins rounded . . . . 26. *B. racemosus*

4. Awns arising 1.5 mm or more below the lemma apices, erect to strongly divaricate, often twisted at the base.

7. Panicle branches shorter than the spikelets, slightly curved or straight, panicles erect . . . . 25. *B. hordeaceus* (in part)

7. At least some panicle branches as long as or longer than the spikelets, sometimes sinuous; panicles nodding.

8. Lower glumes 7–10 mm long; upper glumes 8–12 mm long; panicle branches conspicuously sinuous; awns erect to weakly spreading; lemma margins rounded . . . . 27. *B. arenarius*

8. Lower glumes 4.5–7 mm long; upper glumes 5–8 mm long; panicle branches sometimes sinuous; awns erect to strongly divergent; lemma margins slightly to strongly angled above the middle.

9. Lemmas with hyaline margins 0.3–0.6 mm wide, slightly angled above the middle; branches somewhat drooping, sometimes sinuous, often with more than 1 spikelet . . . . 28. *B. japonicus*

9. Lemmas with hyaline margins 0.6–0.9 mm wide, strongly angled above the middle; branches not drooping or sinuous, usually with 1 spikelet . . . . 29. *B. squarrosus*

Bromus sect. **Ceratochloa** (P. Beauv.) Griseb.

Pl ann, biennial, or per. **Spklt** elliptic to lanceolate, strongly lat compressed, with 3–12 flt. **Lo glm** 3–7(9)-veined; **up glm** 5–9(11)-veined; **lm** lanceolate, lat compressed, strongly keeled, at least distally, apc entire or with acute teeth, teeth shorter than 1 mm; **awns** straight, erect to slightly divaricate.

Bromus sect. *Ceratochloa* is native to North and South America, and contains about 25 species. It is marked by polyploid complexes; the major one in North America is the *Bromus carinatus* complex. This treatment recognizes four species in the complex: *B. arizonicus*, *B. carinatus*, *B. polyanthus*, and *B. sitchensis*. The lowest chromosome number known for members of this complex is 2n = 28, found in *B. carinatus*; the highest is 2n = 84, found in *B. arizonicus*. *Bromus polyanthus* is octoploid with 2n = 56. One other species in the section, *B. catharticus*, has been introduced from South America and is also part of a polyploid complex.

There is morphological intergradation among the species recognized here, and some evidence that these intermediates are sometimes partially fertile (Harlan 1945a, 1945b; Stebbins and Tobgy 1944; Stebbins 1947). Stebbins and Tobgy (1944) commented that partial hybrid sterility between plants placed in different species on the basis of their morphology "supports the recognition of more than one species among the octoploid members of the complex," but Stebbins (1981) later stated that "...all the North American octoploids ... should be united into a single species, in spite of the barriers of hybrid sterility that separate them."

1. **Bromus catharticus** Vahl Rescue Grass [p. *302*, <u>*495*</u>]

Pl ann, biennial, or per; loosely ces or tufted. **Clm** 30–120 cm tall, 2–4 mm thick, erect or decumbent. **Shth** usu densely, often retrorsely, hairy, hairs smt confined to the throat; **aur** absent; **lig** 1–4 mm, glab or pilose, obtuse, lacerate to erose; **bld** 4–30 cm long, 3–10 mm wide, flat, glab or hairy on both surfaces. **Pan** 9–28 cm, usu open, erect or nodding; **lo br** shorter than 10 cm, 1–4 per nd, spreading or ascending, with up to 5 spklt variously distributed. **Spklt** (17)20–40 mm, shorter than at least some ped and br, elliptic to lanceolate, strongly lat compressed, not crowded or overlapping, with 4–12 flt. **Glm** smooth or scabrous, glab or pubescent; **lo glm** 7–12 mm, 5–7(9)-veined; **up glm** 9–17 mm, 7–9(11)-veined, shorter than the lowest lm; **lm** 11–20 mm, lanceolate, lat compressed, strongly keeled, usu glab, smt pubescent distally, smooth or scabrous, 9–13-veined, veins often raised and riblike, mrg smt conspicuous, hyaline, whitish or partly purplish, apc entire or toothed, teeth acute, shorter than 1 mm; **awns** absent or to 10 mm; **anth** 0.5–1 mm in cleistogamous flt, 2–5 mm in chasmogamous flt. 2*n* = 42.

One variety has been found in the Intermountain Region.

Bromus catharticus Vahl var. **catharticus**
[p. 302]

Pl ann or biennial; tufted. **Clm** 30–120 cm, erect or decumbent. **Shth** usu densely, often retrorsely, hairy, hairs smt confined to the throat; **lig** 1–4 mm, glab or pilose, erose; **bld** 4–26 cm long, 3–10 mm wide, glab or hairy on both surfaces. **Pan** 9–28 cm, open, erect or nodding; **lo br** 1–4 per nd, spreading or ascending, with 1–5 spklt. **Spklt** 20–30 mm, with 6–12 flt. **Lo glm** 5–7-veined; **up glm** 9–13 mm, (7)9(11)-veined; **lm** 11–20 mm, glab or scabrous, smt pubescent distally, (9)11–13-veined; **awns** absent or to 3.5 mm; **anth** about 0.5 mm in cleistogamous flt, 2–4 mm in chasmogamous flt. 2*n* = 42.

Bromus catharticus var. *catharticus* is native to South America. It has been widely introduced in North America as a forage crop and is now established, particularly in the southern half of the United States. It usually grows on disturbed soils.

2. **Bromus sitchensis** Trin. SITKA BROME, ALASKA BROME [p. 303, 495]
Pl per; loosely ces. **Clm** 120–180 cm tall, 3–5 mm thick, erect. **Shth** glab or sparsely pilose; **aur** absent; **lig** 3–4 mm, glab or hairy, obtuse, lacerate; **bld** 20–40 cm long, 2–9 mm wide, flat, sparsely pilose adx or on both surfaces. **Pan** 25–35 cm, open; **lo br** to 20 cm, 2–4(6) per nd, spreading, often drooping, with 1–3 spklt on the distal ½, smt confined to the tips. **Spklt** 18–38 mm, elliptic to lanceolate, strongly lat compressed, with (5)6–9 flt. **Glm** glab, smt scabrous; **lo glm** 6–10 mm, 3–5-veined; **up glm** 8–11 mm, 5–7-veined; **lm** 12–14(15) mm, lanceolate, lat compressed, 7–11-veined, strongly keeled at least distally, usu glab, smt hirtellous, mrg smt sparsely pilose, apc entire or with acute teeth, teeth shorter than 1 mm; **awns** 5–10 mm; **anth** to 6 mm. 2*n* = 42, 56.

Bromus sitchensis grows on exposed rock bluffs and cliffs, and in meadows, often in the partial shade of forests along the ocean edge, and on road verges and other disturbed sites. Its range extends from the Aleutian Islands and Alaska panhandle through British Columbia to southern California.

3. **Bromus arizonicus** (Shear) Stebbins
ARIZONA BROME [p. 303, 495]
Pl ann; tufted. **Clm** 30–90 cm tall, to 3 mm thick, erect. **Shth** retrorsely pilose, smt mostly glab, throats smt with hairs; **aur** absent; **lig** 1–4 mm, usu glab, obtuse, erose; **bld** 8–18 cm long, 3–9 mm wide, flat, sparsely pilose on both surfaces or the abx surfaces glab. **Pan** 12–25 cm, somewhat contracted or open; **lo br** shorter than 10 cm, 2–3(5) per nd, initially erect to ascending, spreading at maturity, with 1–2 spklt variously distributed. **Spklt** 18–25 mm, elliptic to lanceolate, strongly lat compressed, with 4–8 flt. **Glm** subequal, smooth or scabrous; **lo glm** 8–12.5 mm, 3-veined; **up glm** 9.5–14 mm, 7-veined, about as long as the lowest lm; **lm** 9.5–14 mm, lanceolate, lat compressed, prominently 7-veined, strongly keeled at least distally, glab or pubescent distally or throughout, mrgl hairs, if present, longer than those elsewhere, apc entire or with acute teeth shorter than 1 mm; **awns** 6–13 mm, smt slightly geniculate; **anth** 0.4–0.5 mm. 2*n* = 84.

Bromus arizonicus grows in dry, open areas and disturbed ground of the southwest, usually below 2000 m. Its range extends from California and southern Nevada into Arizona, New Mexico, and northern Mexico.

Stebbins et al. (1944) demonstrated that, like *Bromus carinatus* var. *carinatus*, *B. arizonicus* obtained three of its genomes from *B. catharticus* or a close relative, but the remaining three genomes are not homologous with those in *B. carinatus*, probably being derived from a species in a section other than *Ceratochloa*. The small anthers of *B. arizonicus* strongly suggest that most seed is produced by selfing.

4. **Bromus carinatus** Hook. & Arn. [p. 304, 495]
Pl ann, biennial, or per; loosely ces. **Clm** 45–120(180) cm tall, usu less than 3 mm thick, erect. **Shth** mostly glab or retrorsely soft pilose, throats usu hairy; **aur** smt present on the lo lvs; **lig** 1–3.5(4) mm, glab or sparsely hairy, acute to obtuse, lacerate or erose; **bld** 8–30 cm long, 1–12 mm wide, flat or becoming involute, glab or sparsely pilose to pubescent on 1 or both surfaces. **Pan** 5–40 cm, lax, open or erect; **lo br** usu shorter than 10 cm, 1–4 per nd, ascending to strongly divergent or reflexed, with 1–4 spklt variously distributed. **Spklt** 20–40 mm, shorter than at least some ped and br, elliptic to lanceolate, strongly lat compressed, not crowded or overlapping, smt purplish, with 4–11 flt. **Glm** glab or pubescent; **lo glm** 7–11

mm, 3–7(9)-veined; **up glm** 9–13 mm, shorter than the lowest lm, 5–9(11)-veined; **lm** 10–16(17) mm, lanceolate, lat compressed, strongly keeled distally, usu more or less uniformly pubescent or pubescent on the mrg only, smt glab or scabrous, 7–9-veined, veins usu not raised or riblike, apc entire or with acute teeth shorter than 1 mm; **awns** 4–17 mm, smt slightly geniculate; **anth** 1–6 mm. 2*n* = 28, 42, 56.

Bromus carinatus is native from British Columbia to Saskatchewan and south to Mexico. One variety, var. *marginatus*, grows in the Intermountain Region. It differs from var. *carinatus*, which is primarily coastal, in having awns that are mostly 4–7 mm long rather than 8–17 mm long. It is often treated as a species, *B. marginatus* Nees *ex* Steud.

Bromus carinatus var. marginatus (Nees) Barkworth & Anderton Mountain Brome [p. 304]

Pl per. **Clm** 45–120(180) cm tall. **Shth** usu sparsely retrorsely pilose throughout, ranging from densely pilose to glab, except at the throat, throats always pilose; **aur** smt present on the lo lvs; **lig** 1–3.5 mm, sparsely hairy, acute to obtuse, erose or lacerate; **bld** 8–25 cm long, 1–12 mm wide, flat or involute, glab or sparsely pilose to pubescent on 1 or both surfaces. **Pan** 5–20(30) cm, erect; **lo br** 1–4 per nd, erect or ascending. **Spklt** with 4–9 flt. **Lo glm** 7–11 mm, 3–7(9)-veined; **up glm** 9–13 mm, 5–9(11)-veined; **lm** 10–14(17) mm, pubescent on the backs and mrg, on the mrg only, or glab, 7–9-veined; **awns** 4–7 mm; **anth** 1–6 mm. 2*n* = 42.

Bromus carinatus var. *marginatus* is primarily an inland species and grows on open slopes, grass balds, shrublands, meadows, and open forests, in montane and subalpine zones. It grows from British Columbia to Saskatchewan, south through the western United States and into northern Mexico. Its elevational range is 350–2200 m in the northern part of its distribution, and 1500–3300 m in the south.

Bromus carinatus var. *marginatus* is variable and intergrades with *B. carinatus* var. *carinatus* to the west and *B. polyanthus* to the southeast. As treated here, *B. carinatus* var. *marginatus* includes *B. luzonensis* J. Presl, which has been recognized mainly on the basis of its canescent sheaths and blades; this trait is highly variable and may be environmentally determined.

5. **Bromus polyanthus** Scribn. Colorado Brome, Great-Basin Brome [p. 304, 495]

Pl per; loosely ces. **Clm** 60–120 cm tall, to 3 mm thick, erect, glab or puberulent. **Shth** usu smooth or scabrous, smt hairy except at the throat; **aur** absent; **lig** (1)2–2.5 mm, glab, obtuse, erose; **bld** 10–31 cm long, 2–9 mm wide, flat, smt scabrous, usu glab, rarely puberulent to pubescent near the col. **Pan** 10–20 cm, open to somewhat contracted; **lo br** shorter than 10 cm, (1)2–3 per nd, erect, ascending or spreading, with 1–2 spklt variously distributed. **Spklt** 20–35 mm, shorter than at least some ped and br, elliptic to lanceolate, strongly lat compressed, not crowded or overlapping, with 6–11 flt. **Glm** smooth or scabrous; **lo glm** (5.5)7–10(11.5) mm, 3-veined; **up glm** (7.5) 9–11(12.5) mm, 5–7-veined, shorter than the lowest lm; **lm** 12–15 mm, lanceolate, lat compressed, strongly keeled at least distally, glab, smt scabrous, 7–9-veined, veins usu not raised or riblike, apc entire or with acute teeth, teeth shorter than 1 mm; **awns** 4–7 mm; **anth** 1–5 mm. 2*n* = 56.

Bromus polyanthus grows on open slopes and in meadows. It is found primarily in the central Rocky Mountains, but the limits of its range include British Columbia in the north, California in the west, and Arizona, New Mexico, and western Texas in the south. It intergrades with *B. carinatus* var. *marginatus*. Plants with an erect, contracted panicle and awns 4–6 mm long can be called *B. polyanthus* Scribn. var. *polyanthus*; those with with an open, nodding panicle and awns up to 8 mm long can be called *B. polyanthus* var. *paniculatus* Shear. Because the variation in both characters is continuous, the varieties are not recognized here.

Bromus sect. Bromopsis Dumort.

Pl usu per, smt ann. **Spklt** elliptic to lanceolate, more or less terete initially, smt becoming lat compressed at anthesis, with (3)4–14(16) flt. **Lo glm** 1–3-veined; **up glm** 3–5-veined; **lm** elliptic to lanceolate, rounded over the midvein, apc subulate, acute, obtuse or rounded, entire or slightly emgt; **awns** straight, arising less than 1.5 mm below the lm apc.

Bromus sect. *Bromopsis* is native to Eurasia as well as to North and South America, and has about 90 species.

6. **Bromus inermis** Leyss. Smooth Brome, Hungarian Brome [p. 305, 495]

Pl per; rhz, rhz short to long-creeping. **Clm** 50–130 cm, erect, single or a few together; **nd** (2)3–5(6), usu glab, rarely pubescent; **intnd** usu glab, rarely pubescent. **Shth** usu glab, rarely pubescent or pilose; **aur** smt present; **lig** to 3 mm, glab, truncate, erose; **bld** 11–35(42) cm long, 5–15 mm wide, flat, usu glab, rarely pubescent or pilose. **Pan** 10–20 cm, open, erect; **br** ascending or spreading. **Spklt** 20–40 mm, elliptic to lanceolate, terete to moderately lat compressed, smt purplish, with (5)8–10 flt. **Glm** glab; **lo glm** (4)6–8(9) mm, 1(3)-veined; **up glm** (5)7–10 mm, 3-veined; **lm** 9–13 mm, elliptic to lanceolate, rounded over the midvein, usu glab and smooth, smt scabrous, mrg smt sparsely puberulent, the bas part of the backs less frequently so, apc acute to obtuse, entire; **awns** absent or to 3 mm, straight, arising less than 1.5 mm below the lm apc; **anth** 3.5–6 mm. $2n = 28, 56$.

Bromus inermis is native to Eurasia, and is now found in disturbed sites in Alaska, Greenland, and most of Canada as well as south throughout most of the contiguous United States except the southeast. It has also been used for rehabilitation, and is planted extensively for forage in pastures and rangelands from Alaska and the Yukon Territory to Texas.

Bromus inermis is similar to *B. pumpellianus*, differing mainly in having glabrous lemmas, nodes, and leaf blades, but lack of pubescence is not a consistently reliable distinguishing character.

7. Bromus pumpellianus Scribn. ARCTIC BROME [p. *305*, 495]

Pl per; usu rhz, smt ces, rhz short to long-creeping. **Clm** 50–135 cm, erect or ascending, smt geniculate, usu single or few together, smt clumped; **nd** 2–7, pubescent or glab; **intnd** glab or pubescent. **Shth** pilose, villous, or glab; **aur** smt present on the lo lvs; **lig** to 4 mm, glab, truncate or obtuse, erose; **bld** 7–30 cm long, 2.5–8.5(9) mm wide, flat, pubescent or glab on both surfaces, smt only the adx surface pubescent. **Pan** 10–24 cm, open or contracted, erect or nodding; **br** erect to spreading. **Spklt** 16–32(45) mm, elliptic to lanceolate, terete to moderately lat compressed, smt purplish, with 4–14 flt. **Glm** glab or hairy; **lo glm** (4)5–10 mm, 1(3)-veined; **up glm** (5)7.5–13 mm, 3-veined; **lm** 9–16 mm, lanceolate, rounded over the midvein, sparsely to densely hairy throughout, or on the mrg and lo portion of the back, or along the mrgl veins and keel, apc subulate to acute, entire or slightly emgt, lobes shorter than 1 mm; **awns** usu present, smt absent, to 7.5 mm, straight, arising less than 1.5 mm below the lm apc; **anth** 3.5–7 mm. $2n = 28, 56$.

The range of *Bromus pumpellianus* extends from Asia to North America, where it includes Alaska, the western half of Canada, the western United States as far south as New Mexico, and a few other locations eastward. It grows on sandy and gravelly stream banks and lake shores, sand dunes, meadows, dry grassy slopes, and road verges. Pavlick recognized two subspecies, only one of which, subsp. *pumpellianus*, is present in the Intermountain Region. *Bromus pumpellianus* is sometimes treated as a subspecies of *B. inermis*. It differs from that species primarily in its tendency to have pubescent lemmas, nodes, and leaf blades.

8. Bromus laevipes Shear CHINOOK BROME [p. *306*, 495]

Pl per; not rhz. **Clm** 50–150 cm, erect or bas decumbent, often rooting from the lo nd; **nd** 3–5(6), pubescent; **intnd** usu glab, often

puberulent-pubescent just below the nd, rarely puberulent throughout. **Shth** glab, smt slightly pubescent near the throat, smt with hairs in the aur position; **aur** absent or vestigial on the bas lvs; **lig** 2–4.2 mm, glab, obtuse, lacerate; **bld** 13–26 cm long, 4–10 mm wide, light green or glaucous, flat, glab, smt scabrous on both surfaces. **Pan** 10–20 cm, open, nodding; **br** ascending to spreading, often drooping. **Spklt** 23–35 mm, elliptic to lanceolate, terete to moderately lat compressed, with 5–11 flt. **Glm** glab, smt scabrous, mrg often bronze-tinged; **lo glm** 6–9 mm, 3-veined; **up glm** 8–12 mm, 5-veined; **lm** 12–16 mm, elliptic to lanceolate, rounded over the midvein, backs sparsely pilose, pubescent, or scabrous, mrg densely pilose, at least on the lo ½, often bronze-tinged, apc acute to obtuse, entire, rarely slightly emgt, lobes shorter than 1 mm; **awns** 4–6 mm, straight, arising less than 1.5 mm below the lm apc; **anth** 3.5–5 mm. $2n = 14$.

Bromus laevipes grows from northern Oregon to southern California. It grows in shaded woodlands and on exposed brushy slopes, at 300–1500 m.

9. Bromus frondosus (Shear) Wooton & Standl. WEEPING BROME [p. *306*, 495]

Pl per; not rhz. **Clm** 50–100 cm, erect to spreading; **nd** 3–5, usu glab, rarely pubescent; **intnd** glab. **Shth** usu glab, smt pubescent or pilose, especially the lo shth, midrib of the culm lvs usu narrowed just below the col; **aur** absent; **lig** 1–3 mm, glab, truncate to obtuse, laciniate; **bld** 10–20 cm long, 3–6 mm wide, flat, often glaucous, usu glab, smt scabrous, bas bl often pubescent. **Pan** 10–20 cm, open; **br** ascending and spreading or declining and drooping. **Spklt** 15–30 mm, elliptic to lanceolate, terete to moderately lat compressed, with (4)5–10 flt. **Glm** usu glab, rarely slightly pubescent, 3-veined; **lo glm** 5.5–8 mm; **up glm** 6.5–9 mm, often mucronate; **lm** 8–12 mm, elliptic to lanceolate, rounded over the midvein, backs pubescent or glab, mrg usu with longer hairs, apc subulate to obtuse, entire; **awns** 1.5–4 mm, straight, arising less than 1.5 mm below the lm apc; **anth** 1.5–3.5 mm. $2n = 14$.

Bromus frondosus grows in open woods and on rocky slopes, at 1500–2500 m. Its range extends from Colorado, Arizona, and New Mexico into Mexico.

10. Bromus porteri (J.M. Coult.) Nash NODDING BROME [p. *307*, 495]

Pl per; not rhz. **Clm** 30–100 cm, erect; **nd** (2)3–4(5), glab or pubescent; **intnd** mostly glab, puberulent near the nd. **Shth** glab or pilose, midrib of the clm lvs not abruptly narrowed just below the col; **aur**

absent; **lig** to 2.5 mm, glab, truncate or obtuse, erose or lacerate; **bld** (3)10–25(35) cm long, 2–5(6) mm wide, flat, not glaucous, both surfaces usu glab, smt the adx surface pilose. **Pan** 7–20 cm, open, nodding, often 1-sided; **br** slender, ascending to spreading, often recurved and flexuous. **Spklt** 12–38 mm, elliptic to lanceolate, terete to moderately lat compressed, with (3)5–11(13) flt. **Glm** usu pubescent, rarely glab; **lo glm** 5–7(9) mm, usu 3-veined, smt 1-veined; **up glm** 6–10 mm, 3-veined, not mucronate; **lm** 8–14 mm, elliptic, rounded over the midvein, usu pubescent or pilose, mrg often with longer hairs, backs and mrg rarely glab, apc acute or obtuse to truncate, entire; **awns** (1)2–3(3.5) mm, straight, arising less than 1.5 mm below the lm apc; **anth** (1)2–3 mm. $2n = 14$.

Bromus porteri grows in montane meadows, grassy slopes, mesic steppes, forest edges, and open forest habitats, at 500–3500 m. It is found from British Columbia to Manitoba, and south to California, western Texas, and Mexico. It is closely related to *B. anomalus*, and has often been included in that species. It differs chiefly in its lack of auricles, and in having culm leaves with midribs that are not narrowed just below the collar.

11. **Bromus vulgaris** (Hook.) Shear COMMON
    BROME [p. *307*, <u>495</u>]

Pl per; not rhz. **Clm** 60–120 cm, erect or spreading; **nd** (3)4–6(7), usu pilose; **intnd** glab. **Shth** pilose or glab; **aur** absent; **lig** 2–6 mm, glab, obtuse or truncate, erose or lacerate; **bld** 13–25(33) cm long, to 14 mm wide, flat, abx surfaces usu glab, smt pilose, adx surfaces usu pilose, smt glab. **Pan** 10–15 cm, open; **br** ascending to drooping. **Spklt** 15–30 mm, elliptic to lanceolate, terete to moderately lat compressed, with (3)4–9 flt. **Glm** glab or pilose; **lo glm** 5–8 mm, 1(3)-veined; **up glm** 8–12 mm, 3-veined; **lm** 8–15 mm, lanceolate, rounded over the midvein, backs sparsely hairy or glab, mrg usu coarsely pubescent, smt glab, apc subulate to acute, entire; **awns** (4)6–12 mm, straight, arising less than 1.5 mm below the lm apc; **anth** 2–4 mm. $2n = 14$.

Bromus vulgaris grows in shaded or partially shaded, often damp, coniferous forests along the coast, and inland in montane pine, spruce, fir, and aspen forests, from sea level to about 2000 m. Its range extends from coastal British Columbia eastward to southwestern Alberta and southward to central California, northern Utah, and western Wyoming.

Varieties have been described within *Bromus vulgaris*; because their variation is overlapping, none are recognized here.

12. **Bromus suksdorfii** Vasey SUKSDORF'S
    BROME [p. *307*, <u>495</u>]

Pl per; not rhz. **Clm** 50–100 cm, erect; **nd** 2–3(4), glab, intnd glab or puberulent just below the nd.

Shth glab; **aur** absent; **lig** to 1 mm, glab, truncate; **bld** (8)12–19(24) cm long, 4–8(14) mm wide, flat, glab, mrg scabrous. **Pan** 6–14 cm, erect, contracted; **br** erect or ascending. **Spklt** 15–30 mm, elliptic to lanceolate, terete to moderately lat compressed, with (3)5–7 flt. **Glm** glab or sparsely pubescent; **lo glm** 7–11 mm, 1(3)-veined; **up glm** 9–12 mm, 3-veined; **lm** 12–15 mm, elliptic, rounded over the midvein, backs and mrg pubescent or nearly glab, apc obtuse, entire; **awns** 2–5 mm, straight, arising less than 1.5 mm below the lm apc; **anth** 2–3.5 mm. $2n = 14$.

Bromus suksdorfii grows on open slopes and in open subalpine forests, at about 1300–3300 m, from southern Washington to southern California.

13. **Bromus mucroglumis** Wagnon
    SHARPGLUME BROME [p. *308*, <u>495</u>]

Pl per; not rhz. **Clm** 50–100 cm, erect or spreading; **nd** 5–7, pilose or pubescent; **intnd** glab. **Bas shth** pubescent or pilose, throats pilose; **up shth** pubescent or glab, midrib of the clm lvs not abruptly narrowed just below the col; **aur** absent; **lig** 1–2 mm, glab, truncate or obtuse; **bld** 20–30 cm long, (4)7–11 mm wide, flat, both surfaces pilose or the abx surface glab. **Pan** 10–20 cm, open, nodding; **br** ascending or spreading. **Spklt** 20–30 mm, elliptic to lanceolate, terete to moderately lat compressed, with 5–10 flt. **Glm** usu pilose or pubescent, rarely glab; **lo glm** 6–8 mm, 1-veined; **up glm** 8–8.5 mm, 3-veined, mucronate; **lm** 10–12 mm, elliptic to lanceolate, rounded over the midvein, backs and mrg pilose or pubescent, apc acute to obtuse, entire; **awns** 3–5 mm, straight, arising less than 1.5 mm below the lm apc; **anth** 1.5–3 mm. $2n = 28$.

Bromus mucroglumis grows at 1500–3000 m in the southwestern United States and northern Mexico.

14. **Bromus lanatipes** (Shear) Rydb. WOOLY
    BROME [p. *308*, <u>495</u>]

Pl per; not rhz. **Clm** 40–90 cm, erect; **nd** 3–5(7), mainly pubescent; **intnd** mostly glab, puberulent near the nd. **Bas shth** densely pilose or glab; **up shth** glab or almost so, midrib of the clm lvs not abruptly narrowed just below the col; **aur** absent; **lig** 1–2 mm, glab, truncate or obtuse, smt lacerate; **bld** 5–20 cm long, to 7 mm wide, flat, both surfaces glab, smt scabrous. **Pan** 10–25 cm, open, nodding; **br** ascending to spreading. **Spklt** 10–30 mm, elliptic to lanceolate, terete to moderately lat compressed, with 7–12(16) flt. **Glm** usu glab, smt pubescent; **lo glm** 5–6.5(7) mm, 1(3)-veined; **up glm** (6)7–9 mm, 3-veined, not mucronate; **lm** 8–11 mm, elliptic, rounded over the midvein, backs and mrg pubescent, smt nearly glab, apc truncate or obtuse, entire, rarely emgt, lobes shorter than 1 mm; **awns** 2–4 mm, straight, arising less than 1.5 mm below the lm apc; **anth** 1.8–4 mm. 2*n* = 28.

Bromus lanatipes grows in a wide range of habitats at 800–2500 m, from Wyoming through the southwestern United States to northern Mexico.

15. **Bromus ciliatus** L. FRINGED BROME [p. *308*, 496]

Pl per; not rhz. **Clm** 45–120(150) cm, erect; **nd** (3)4–7(8), all pubescent or the lo nd smt glab; **intnd** glab. **Bas shth** usu retrorsely pilose, smt glab; **up shth** glab, throats glab or pilose, midrib of the clm lvs not abruptly narrowed just below the col; **aur** smt present; **lig** 0.4–1.4 mm, usu glab, rarely pilose, truncate, erose; **bld** 13–25 cm long, 4–10 mm wide, flat, abx surfaces usu glab, smt pilose, adx surfaces usu pilose, smt glab. **Pan** 10–20 cm, open, nodding; **br** ascending, spreading, or drooping. **Spklt** 15–25 mm, elliptic to lanceolate, terete to moderately lat compressed, with 4–9 flt. **Glm** glab; **lo glm** 5.5–7.5 mm, 1(3)-veined; **up glm** 7.1–8.5 mm, 3-veined, not mucronate; **lm** 9.5–14 mm, elliptic to lanceolate, rounded over the midvein, backs glab, smt scabrous, mrg conspicuously hirsute on the lo ½–⅔, apc obtuse to acute, entire; **awns** 3–5 mm, straight, arising less than 1.5 mm below the lm apc; **anth** 1–1.4 mm. 2*n* = 14.

Bromus ciliatus grows in damp meadows, thickets, woods, and stream banks across almost all of northern North America except the high arctic, extending further south mainly through the western United States to Mexico. The variation in pubescence of leaf sheaths is continuous and does not merit taxonomic recognition.

16. **Bromus richardsonii** Link RICHARDSON'S BROME [p. *309*, 496]

Pl per; not rhz. **Clm** 50–110(145) cm, erect to spreading; **nd** (3)4–5(6), usu glab, smt pubescent; **intnd** usu glab. **Bas shth** often retrorsely pilose; **clm shth** glab, often tufted-pilose near the aur position, midrib of the clm lvs not abruptly narrowed just below the col; **aur** absent; **lig** 0.4–2 mm, glab, rounded, erose, ciliolate; **bld** 10–35 cm long, 3–12 mm wide, flat, glab. **Pan** 10–20(25) cm, open, nodding; **br** ascending to spreading or drooping, filiform. **Spklt** 15–25(40) mm, elliptic to lanceolate, terete to moderately lat compressed, with (4)6–10(15) flt. **Glm** usu glab, smt pubescent; **lo glm** 7.5–12.5 mm, 1(3)-veined; **up glm** 8.9–11.3 mm, 3-veined, often mucronate; **lm** 9–14(16) mm, elliptic, rounded over the midvein, mrg more or less densely pilose on the lo ½ or ¾, lo lm in a spklt glab across the backs, uppermost lm with appressed hairs on the backs, apc obtuse, entire; **awns** (2)3–5 mm, straight, arising less than 1.5 mm below the lm apc; **anth** 1.6–2.7 mm. 2*n* = 28.

Bromus richardsonii grows in meadows and open woods in the upper montane and subalpine zones, at 2000–4000 m in the southern Rocky Mountains, and at lower elevations northwards. Its range extends from southern Alaska to southern California and northern Baja California, Mexico; it is found as far east as Saskatchewan, South Dakota, and western Texas. Specimens with pubescent nodes and glumes are apparently confined to the southwestern United States.

Bromus sect. **Neobromus** (Shear) Hitchc.

Pl ann. **Spklt** elliptic to lanceolate, more or less terete, with 3–9 flt. **Lo glm** 1(3)-veined; **up glm** 3(5)-veined; **lm** lanceolate to linear-lanceolate, rounded over the midvein, apc acuminate, bifid, teeth aristate or acuminate; **awns** geniculate, divaricate, arising 1.5 mm or more below the lm apc.

Bromus sect. *Neobromus* has two species, both of which are native to South America. *Bromus berteroanus* has become established in the Intermountain Region.

17. **Bromus berteroanus** Colla CHILEAN CHESS [p. *309*, 496]

Pl ann; often tufted. **Clm** 30–60 cm, slender. **Shth** pilose-pubescent to nearly glab; **bld** 7–28 cm long, 2–9 mm wide, pilose or glab. **Pan** 10–20 cm long, 3–9 cm wide, erect, dense; **br** appressed to spreading, smt flexuous. **Spklt** 15–20 mm, elliptic to lanceolate, more or less terete, with 3–9 flt. **Glm** glab, acuminate; **lo glm** 8–10 mm, 1-veined; **up glm** 12–16 mm, 3(5)-veined; **lm** 11–14 mm, lanceolate to linear-lanceolate, sparsely pubescent, 5-veined, rounded over the midvein, apc acuminate, bifid, teeth 2–3 mm, usu aristate, smt acuminate; **awns** 13–20 mm, geniculate, strongly to moderately twisted in the bas portion, arising 1.5 mm or more below the lm apc; **anth** 2–2.5 mm. 2*n* = unknown.

Bromus berteroanus is from Chile, and can now be found in dry areas in western North America, from British Columbia to Baja California, Mexico.

Bromus sect. **Genea** Dumort.

Pl ann. **Spklt** with parallel or diverging sides in outline, terete to moderately lat compressed, with 4–11 flt. **Lo glm** 1–3-veined; **up glm** 3–5-veined; **lm** lanceolate to linear-lanceolate, rounded over the midvein, apc acuminate, teeth 0.8–5 mm; **awns** straight or arcuate, arising 1.5 mm or more below the lm apc.

Bromus sect. *Genea* is native to Europe and northern Africa; four of its six species are established in the Intermountain region.

18. **Bromus diandrus** Roth GREAT BROME, RIPGUT GRASS [p. *310*, 496]

Pl ann. **Clm** 20–90 cm, erect or decumbent, puberulent below the pan. **Shth** softly pilose, hairs often retrorse or spreading; **aur** absent; **lig** 2–3 mm, glab, obtuse, lacerate or erose; **bld** 3.5–27 cm long, 1–9 mm wide, both surfaces pilose. **Pan** 13–25 cm long, 2–12 cm wide, erect to spreading; **br** 1–7 cm, stiffly erect to ascending or spreading, with 1 or 2 spklt. **Spklt** 25–70 mm, sides parallel or diverging distally, moderately lat compressed, with 4–11 flt. **Glm** smooth or scabrous, mrg hyaline; **lo glm** 15–25 mm, 1–3-veined; **up glm** 20–35 mm, 3–5-veined; **lm** 20–35 mm, linear-lanceolate, scabrous, 7-veined, rounded over the midvein, mrg hyaline, apc bifid, acuminate, teeth 3–5 mm; **awns** 30–65 mm, straight, arising 1.5 mm or more below the lm apc; **anth** 0.5–1 mm. $2n = 42, 56$.

Bromus diandrus is native to southern and western Europe. It is now established in North America, where it grows in disturbed ground, waste places, fields, sand dunes, and limestone areas. It occurs from southwestern British Columbia to Baja California, Mexico, and eastward to Montana, Colorado, and Texas. The common name 'ripgut grass' indicates the effect it has on animals if they consume the sharp, long-awned florets of this species.

Bromus diandrus, as treated here, includes *B. rigidus* Roth. Sales (1993) reduced these two taxa to varietal rank, pointing out that the differences between them in panicle morphology and callus and scar shape are subtle enough that identification of many specimens beyond *B. diandrus sensu lato* is often impossible.

19. **Bromus sterilis** L. BARREN BROME [p. *310*, 496]

Pl ann. **Clm** 35–100 cm, erect or geniculate near the base, glab. **Shth** densely pubescent; **aur** absent; **lig** 2–2.5 mm, glab, acute, lacerate; **bld** 4–20 cm long, 1–6 mm wide, pubescent on both surfaces. **Pan** 10–20 cm long, 5–12 cm wide, open; **br** 2–10 cm, spreading, ascending or drooping, rarely with more than 3 spklt. **Spklt** 20–35 mm, usu shorter than the pan br, sides parallel or diverging distally, moderately lat compressed, with 5–9 flt. **Glm** smooth or scabrous, mrg hyaline; **lo glm** 6–14 mm, 1(3)-veined; **up glm** 10–20 mm, 3(5)-veined; **lm** 14–20 mm, narrowly lanceolate, pubescent or puberulent, 7(9)-veined, rounded over the midvein, mrg hyaline, apc acuminate, bifid, teeth 1–3 mm; **awns** 15–30 mm, straight, arising 1.5 mm or more below the lm apc; **anth** 1–1.4 mm. $2n = 14, 28$.

Bromus sterilis is native to Europe, growing from Sweden southward. In the Intermountain Region, it grows in road verges, waste places, fields, and overgrazed rangeland. It is widespread in western and eastern North America, but is mostly absent from the Great Plains and the southeastern states.

20. **Bromus tectorum** L. CHEATGRASS, DOWNY CHESS [p. *310*, 496]

Pl ann. **Clm** 5–90 cm, erect, slender, puberulent below the pan. **Shth** usu densely and softly retrorsely pubescent to pilose, up shth smt glab; **aur** absent; **lig** 2–3 mm, glab, obtuse, lacerate; **bld** to 16 cm long, 1–6 mm wide, both surfaces softly hairy. **Pan** 5–20 cm long, 3–8 cm wide, open, lax, drooping distally, usu 1-sided; **br** 1–4 cm, drooping, usu 1-sided and longer than the spklt, usu at least 1 br with 4–8 spklt. **Spklt** 10–20 mm, usu shorter than the pan br, sides parallel or diverging distally, moderately lat compressed, often purplish tinged, not densely crowded, with 4–8 flt. **Glm** villous, pubescent, or glab, mrg hyaline; **lo glm** 4–9 mm, 1-veined; **up glm** 7–13 mm, 3–5-veined; **lm** 9–12 mm, lanceolate, glab or pubescent to pilose, 5–7-veined, rounded over the midvein, mrg hyaline, often with some hairs longer than those on the backs, apc acuminate, hyaline, bifid, teeth 0.8–2(3) mm; **awns** 10–18 mm, straight, arising 1.5 mm or more below the lm apc; **anth** 0.5–1 mm. $2n = 14$.

Bromus tectorum is a European species that is well established in the Intermountain Region and other parts of the world. It grows in disturbed sites, such as overgrazed rangelands, fields, sand dunes, road verges, and waste places. In the southwestern United States, *Bromus tectorum* is considered a good source of spring feed for cattle, at least until the awns mature. It is highly competitive and dominates rapidly after fire, especially in sagebrush areas. The resulting dense, fine fuels permanently shorten the fire-return interval, further hindering reestablishment of native species. It now dominates large areas of the sagebrush ecosystem of the Intermountain Region.

Specimens with glabrous spikelets have been called *Bromus tectorum* f. *nudus* (Klett & Richt.) H. St. John. They occur throughout the range of the species, and are not known to have any other distinguishing characteristics. For this reason, they are not given formal recognition in this treatment.

21. Bromus rubens L. FOXTAIL CHESS, RED BROME [p. 311, 496]

Pl ann. Clm 10–40 cm, erect or ascending, often puberulent below the pan. Shth softly pubescent to pilose; aur absent; lig 1–3(4) mm, pubescent, obtuse, lacerate; bld to 15 cm long, 1–5 mm wide, flat, pubescent on both surfaces. Pan 2–10 cm long, 2–5 cm wide, erect, dense, often reddish brown; br 0.1–1 cm, ascending, never drooping, not readily visible, with 1 or 2 spklt. Spklt 18–25 mm, much longer than the pan br, densely crowded, subsessile, with parallel sides or widening distally, moderately lat compressed, with 4–8 flt. Glm pilose, mrg hyaline; lo glm 5–8 mm, 1(3)-veined; up glm 8–12 mm, 3–5-veined; lm 10–15 mm, linear-lanceolate, pubescent to pilose, 7-veined, rounded over the midvein, mrg hyaline, apc acuminate, teeth 1–3 mm; awns 8–20 mm, straight, reddish, arising 1.5 mm or more below the lm apc; anth 0.5–1 mm. 2n =14, 28.

Bromus rubens is native to southern and southwestern Europe. It now grows in North America in disturbed ground, waste places, fields, and rocky slopes, from southern Washington to southern California, eastward to Idaho, New Mexico, and western Texas.

Bromus L. sect. Bromus

Pl usu ann, smt biennial. Spklt with parallel or converging sides in outline, terete to moderately lat compressed, with 4–30 flt. Lo glm 3–5-veined; up glm 5–9-veined; lm elliptic, lanceolate, obovate, or rhombic, rounded over the midvein, apc subulate, acute, acuminate, or obtuse, notched, minutely bifid, or toothed, teeth shorter than 1 mm, apc smt split and the teeth appearing longer; awns (0)1(3), straight or flexuous, recurved or divaricate.

Bromus sect. *Bromus* has about 40 species that are native to Eurasia, northern Africa, and Australia; 8 species have been introduced to the Intermountain Region.

22. Bromus briziformis Fisch. & C.A. Mey. RATTLESNAKE BROME [p. 311, 496]

Pl ann. Clm 20–62 cm, erect or ascending. Shth densely pilose; lig 0.5–2 mm, hairy, obtuse, erose; bld 3–13 cm long, 2–4 mm wide, pilose to pubescent on both surfaces. Pan 5–15 cm long, 3–7 cm wide, open, secund, nodding; br smt longer than the spklt, curved to reflexed. Spklt 15–27 mm long, 8–12 mm wide, ovate, lat compressed; flt 7–15, bases concealed at maturity; rchl intnd concealed at maturity. Glm smooth or scabridulous; lo glm 5–6 mm, 3–5-veined; up glm 6–8 mm, 7–9-veined; lm 9–10 mm long, 6–8 mm wide, inflated, obovate or rhombic, coriaceous, smooth or scabridulous, obscurely 9-veined, rounded over the midvein, mrg hyaline, 1–1.3 mm wide, abruptly angled, not inrolled at maturity, apc acute to obtuse, bifid, teeth shorter than 1 mm; awns usu absent, smt to 1 mm, arising less than 1.5 mm below the lm apc; anth 0.7–1 mm. Car equaling or shorter than the pal, thin, weakly inrolled or flat. 2n = 14.

Bromus briziformis grows in waste places, road verges, and overgrazed areas. It is native to southwest Asia and Europe, and is adventive in North America, occurring from southern British Columbia to as far south as New Mexico, and in scattered locations eastward. The unique shape of its spikelets has led to its use in dried flower arrangements and as a garden ornamental. The common name may refer to the similarity of the spikelets to a rattlesnake's tail.

23. Bromus secalinus L. RYEBROME [p. 311, 496]

Pl ann. Clm 20–80 (120) cm, erect. Lo shth glab or loosely pubescent and glabrate; lig 2–3 mm, glab, obtuse; bld 15–30 cm long, 2–4 mm wide, abx surfaces pilose or glab, adx surfaces pilose. Pan 5–23 cm long, 2.5–12 cm wide, open, nodding; br spreading to ascending; lo br slightly drooping, often secund after anthesis, not sinuous. Spklt 10–20 mm, shorter than at least some pan br, ovoid-lanceolate or ovate, lat compressed, not purple-tinged; flt 4–9(10), ascending-spreading after flowering, bases visible at maturity; rchl intnd visible at maturity. Glm scabrous or glab; lo glm 4–6 mm, 3–5-veined; up glm 6–7 mm, 7-veined; lm 6.5–8.5(10) mm long, 1.7–2.5 mm wide, elliptic, coriaceous, obscurely 7-veined, rounded over the midvein, backs usu glab, smt pubescent, scabrous to puberulent on the mrg and near the apc, mrg evenly rounded, inrolled at maturity, apc acute to obtuse, bifid, teeth shorter than 1 mm; awns (0)3–6(9.5) mm, straight or flexuous, arising less than 1.5 mm below the lm apc; anth 1–2 mm. Car equaling the pal, thick, strongly inrolled at maturity. 2n = 28.

Bromus secalinus is native to Europe. It is widespread in North America, where it grows in fields, on waste ground, and along roadsides. Specimens with pubescent spikelets may be called *B. secalinus* var. *velutinus* (Schrad.) W.D.J. Koch.

24. **Bromus commutatus** Schrad. Meadow
    Brome, Hairy Chess [p. *312*, 496]

Pl ann. **Clm** 40–120 cm, erect or ascending. **Lo
shth** densely hairy, hairs stiff, often retrorse; **up
shth** pubescent or glab; **lig** 1–2.5 mm, glab or
pilose, obtuse, ciliolate; **bld** 9–18 cm long, 2–4
mm wide, pilose on both surfaces. **Pan** 7–16
cm long, 3–6 cm wide, open, erect to ascending;
br smt longer than the spklt, slender, ascending
to spreading. **Spklt** 14–18(30) mm, oblong-
lanceolate, terete to moderately lat compressed,
not purple-tinged; **flt** 4–9(11), bases concealed
or visible at maturity; **rchl intnd** 1.5–2 mm,
concealed or visible at maturity. **Glm** usu glab,
smt scabrous or pubescent; **lo glm** 5–7 mm,
5-veined; **up glm** 6–9 mm, 7(9)-veined; **lm** 8–11.5
mm long, 1.7–2.6 mm wide, elliptic to lanceolate,
coriaceous, backs usu glab, distinctly 7(9)-veined,
not ribbed, rounded over the midvein, mrg
scabrous or pubescent, bluntly angled, inrolled
or not at maturity, apc acute to obtuse, bifid,
teeth shorter than 1 mm; **awns** 3–10 mm, straight,
arising less than 1.5 mm below the lm apc, awn
of the lowest lm shorter than the others; **anth**
0.7–1.7 mm. **Car** equaling or shorter than the pal,
weakly to strongly inrolled. 2*n* = 14, 28, 56.

Bromus commutatus grows in fields, waste places, and
road verges. It is native to Europe and the Baltic region.
Hildemar Scholz (pers. comm.) recognizes three subspecies
of *B. commutatus* in Europe; no attempt has been made to
determine which subspecies are present in the Intermountain
Region.

25. **Bromus hordeaceus** L. Lopgrass [p. *312*,
    496]

Pl ann or biennial. **Clm** 2–70 cm, erect or
ascending. **Lo shth** densely, often retrorsely,
pilose; **up shth** pubescent or glab; **lig** 1–1.5 mm,
hairy, obtuse, erose; **bld** 2–19 cm long, 1–4 mm
wide, abx surfaces glab or hairy, adx surfaces
hairy. **Pan** 1–13 cm long, 1–4 cm wide, erect,
usu ovoid, open, becoming dense, occ rdcd to 1
or 2 spklt; **br** shorter than the spklt, ascending
to erect, straight or almost so. **Spklt** (11)14–
20(23) mm, lanceolate, terete to moderately lat
compressed; **flt** 5–10, bases concealed at maturity;
rchl intnd concealed at maturity. **Glm** pilose or
glab; **lo glm** 5–7 mm, 3–5-veined; **up glm** 6.5–8
mm, 5–7-veined; **lm** 6.5–11 mm long, 3–5 mm
wide, lanceolate, chartaceous, antrorsely pilose
to pubescent, or glab proximally or throughout,
7–9-veined, lat veins prominently ribbed,
rounded over the midvein, hyaline mrg abruptly
or bluntly angled, not inrolled at maturity, apc
acute, bifid, teeth shorter than 1 mm; **awns** 6–8

mm, usu arising less than 1.5 mm below the lm
apc, straight to recurved at maturity; **anth** 0.6–1.5
mm. **Car** equaling or shorter than the pal, thin,
weakly inrolled to flat. 2*n* = 28.

Bromus hordeaceus is native to southern Europe and
northern Africa. It is weedy, growing in disturbed areas
such as roadsides, fields, sandy beaches, and waste places,
and can be found in many locations in North America. Its
origin is obscure. Ainouche et al. (1999) reviewed various
suggestions, and concluded that at least one of its diploid
ancestors may have been an extinct or undiscovered species
related to *B. caroli-henrici* Greuter, a diploid species.

The three subspecies below are usually morphologically
distinct. Ainouche et al. (1999), however, found no evidence
of genetic differentiation among them.

1. Lemmas 6.5–8(9) mm long, glabrous or pubescent.
   .... subsp. *pseudothominei*
1. Lemmas (7)8–11 mm long, usually pubescent or
   pilose.
   2. Awns more than 0.1 mm wide at the base,
      straight, erect; culms (3)10–70 cm long.......
      .... subsp. *hordeaceus*
   2. Awns less than 0.1 mm wide at the base, often
      divaricate or recurved at maturity; culms 15–
      25(60) cm long .... subsp. *molliformis*

Bromus hordeaceus L. subsp. **hordeaceus**
[p. *312*]

Clm (3)10–70 cm. **Pan** (3)5–10 cm, usu with
more than 1 spklt. **Lm** (7)8–11 mm, usu pilose
or pubescent, mrg bluntly angled; **awns** more
than 0.1 mm wide at the base, straight, erect. **Car**
shorter than the pal.

Bromus hordeaceus subsp. *hordeaceus* grows throughout the
range of the species, being most prevalent in southwestern
British Columbia, the western United States, and the
northeastern coast.

Bromus hordeaceus subsp. **molliformis** (J.
Lloyd *ex* Billot) Maire & Weiller [p. *312*]

Clm 15–25(60) cm. **Pan** to 10 cm, usu with more
than 1 spklt. **Lm** (7)8–11 mm, pubescent, mrg
rounded; **awns** less than 0.1 mm wide at the
base, often divaricate or recurved at maturity.
Car shorter than the pal.

Bromus hordeaceus subsp. *molliformis* grows in California
and other scattered locations, including Idaho, New Mexico,
and southern Michigan.

Bromus hordeaceus subsp. **pseudothominei** (P.
M. Sm.) H. Scholz [p. *312*]

Clm (3)10–70 cm. **Pan** to 10 cm, usu with more
than 1 spklt. **Lm** 6.5–8(9) mm, usu glab, mrg
often abruptly angled; **awns** straight, erect. **Car**
usu as long as the pal.

Bromus hordeaceus subsp. *pseudothominei* grows
sporadically throughout the range of the species in North
America. Hitchcock (1951) included it in *B. racemosus*.

26. Bromus racemosus L. Smooth Brome
[p. *313*, <u>496</u>]

Pl ann. **Clm** 20–110 cm, erect or ascending. **Lo shth** densely hairy, hairs stiff, often retrorse; **up shth** glab or pubescent; **lig** 1–2 mm, glab or hairy, erose; **bld** 7–18 cm long, 1–4 mm wide, pilose on both surfaces. **Pan** 4–16 cm long, 2–3 cm wide, erect, open; **br** smt longer than the spklt, slender, usu ascending, slightly curved or straight. **Spklt** 12–20 mm, lanceolate, terete to moderately lat compressed; **flt** 5–6, bases concealed at maturity; **rchl intnd** 1–1.5 mm, concealed at maturity. **Glm** smooth to scabrous; **lo glm** 4–6 mm, (3)5-veined; **up glm** 4–7 mm, 7-veined; **lm** 6.5–8 mm long, 3–4.5 mm wide, elliptic to lanceolate, coriaceous, backs smooth, distinctly 7(9)-veined, not ribbed, rounded over the midvein, mrg scabrous, rounded, not inrolled at maturity, apc acute to obtuse, bifid, teeth shorter than 1 mm; **awns** 5–9 mm, all more or less equal in length, straight, arising less than 1.5 mm below the lm apc; **anth** 1.5–3 mm. **Car** shorter than the pal, thin, weakly inrolled or flat. $2n = 28$.

Bromus racemosus grows in fields, waste places, and road verges. It is native to western Europe and the Baltic region, and occurs throughout much of southern Canada and the United States. Hitchcock (1951) included *B. hordeaceus* subsp. *pseudothominei* in *B. racemosus*.

27. Bromus arenarius Labill. Australian
Brome [p. *313*, <u>496</u>]

Pl ann. **Clm** 20–40 cm, erect to ascending. **Shth** densely retrorsely pilose; **lig** 1.5–2.5 mm, glab or pilose, obtuse, lacerate; **bld** 7–8 cm long, 3–6 mm wide, pilose on both surfaces. **Pan** (4)10–15 cm long, 4–7 cm wide, open, nodding; **br** smt longer than the spklt, spreading or ascending, sinuous. **Spklt** 10–20 mm, lanceolate, terete to moderately lat compressed; **flt** 5–9(11), bases concealed at maturity; **rchl intnd** concealed at maturity. **Glm** densely pilose; **lo glm** 7–10 mm, 3-veined; **up glm** 8–12 mm, (5)7-veined; **lm** 9–11(13) mm long, 1–1.8 mm wide, lanceolate, densely pilose, distinctly 7-veined, rounded over the midvein, mrg rounded, not inrolled at maturity, apc acute, bifid, teeth shorter than 1 mm; **awns** 10–16 mm, straight to weakly spreading, arising 1.5 mm or more below the lm apc; **anth** 0.7–1 mm. **Car** equaling or shorter than the pal, thin, weakly inrolled. $2n =$ unknown.

Bromus arenarius grows in dry, often sandy slopes, fields, and waste places. Native to Australia, it is now widely scattered throughout California, and is also recorded from Oregon, eastern Nevada, Arizona, New Mexico, Texas, and Pennsylvania.

28. Bromus japonicus Thunb. Japanese Brome
[p. *313*, <u>496</u>]

Pl ann. **Clm** (22)30–70 cm, erect or ascending. **Shth** usu densely pilose; **up shth** smt pubescent or glab; **lig** 1–2.2 mm, pilose, obtuse, lacerate; **bld** 10–20 cm long, 2–4 mm wide, usu pilose on both surfaces. **Pan** 10–22 cm long, 4–13 cm wide, open, nodding; **br** usu longer than the spklt, ascending to spreading or somewhat drooping, slender, flexuous, smt sinuous, often with more than 1 spklt. **Spklt** 20–40 mm, lanceolate, terete to moderately lat compressed; **flt** 6–12, bases concealed at maturity; **rchl intnd** concealed at maturity. **Glm** smooth or scabrous; **lo glm** 4.5–7 mm, (3)5-veined; **up glm** 5–8 mm, 7-veined; **lm** 7–9 mm long, 1.2–2.2 mm wide, lanceolate, coriaceous, smooth proximally, scabrous on the distal ½, obscurely (7)9-veined, rounded over the midvein, mrg hyaline, 0.3–0.6 mm wide, obtusely angled above the mid, not inrolled at maturity, apc acute, bifid, teeth shorter than 1 mm; **awns** 8–13 mm, strongly divergent at maturity, smt erect, twisted, flattened at the base, arising 1.5 mm or more below the lm apc; **anth** 1–1.5 mm. **Car** equaling or shorter than the pal, thin, weakly inrolled or flat. $2n = 14$.

Bromus japonicus grows in fields, waste places, and road verges. It is native to central and southeastern Europe and Asia, and is distributed throughout much of the United States and southern Canada.

29. Bromus squarrosus L. Squarrose Brome
[p. *314*, <u>496</u>]

Pl ann. **Clm** 20–60 cm, erect or geniculately ascending. **Lo shth** densely pilose; **lig** 1–1.5 mm, hairy, obtuse, erose, ciliolate; **bld** 5–15 cm long, 4–6 mm wide, densely pilose on both surfaces. **Pan** 7–20 cm long, 4–8 cm wide, rcm, open, nodding, often with few spklt, usu secund; **br** smt longer than the spklt, ascending-spreading, flexuous, slightly curved, usu with 1 spklt. **Spklt** 15–70 mm, broadly oblong or ovate-lanceolate, terete to moderately lat compressed; **flt** 8–30, bases concealed at maturity; **rchl intnd** concealed at maturity. **Glm** smooth or scabrous; **lo glm** 4.5–7 mm, 3–5(7)-veined; **up glm** 6–8 mm, 7-veined; **lm** 8–11 mm long, 2–2.4 mm wide, lanceolate, chartaceous, smooth or scabridulous, 7–9-veined, rounded over the midvein, mrg hyaline, 0.6–0.9 mm wide, strongly angled above the mid, not inrolled at maturity, apc acute, bifid, teeth shorter than 1 mm; **awns** 8–10 mm, flattened and smt twisted at the base, divaricate at maturity, arising 1.5 mm or more below the lm apc; **anth** 1–1.3 mm. **Car** equaling the pal, thin, weakly inrolled or flat. $2n = 14$.

Bromus squarrosus grows in overgrazed pastures, fields, waste places, and road verges. Native to central Russia and southern Europe, it can be found mainly in southern Canada and the northern half of the United States.

7. TRITICEAE Dumort.   Mary E. Barkworth

Pl ann or per; smt csp, smt rhz. **Clm** ann, not woody, usu erect, not brchg above the base; **intnd** hollow or solid. **Shth** usu open, those of the bas lvs smt closed; **col** without tufts of hair on the sides; **aur** usu present; **lig** memb or scarious, smt ciliolate, those of the up and lo cauline lvs usu similar; **psdpet** absent; **bld** linear to narrowly lanceolate, venation parallel, cross venation not evident, without arm or fusoid cells, surfaces without microhairs, not papillate, cross sections non-Kranz. **Infl** usu spikes or spikelike rcm, with 1–5 sessile or subsessile spklt per nd, occ pan, smt with morphologically distinct strl and bisex spklt within an infl; **ped** absent or to 4 mm; **dis** usu above the glm and beneath the flt, smt in the rchs, smt at the infl bases. **Spklt** usu lat compressed, smt terete, with 1–16 bisex flt, the distal (or only) flt smt strl; **rchl** smt prolonged beyond the base of the distal flt. **Glm** unequal to equal, shorter than to longer than the adjacent flt, subulate, lanceolate, rectangular, ovate, or obovate, 1–5-veined, absent or vestigial in some species; **flt** lat compressed to terete; **cal** glab or hairy; **lm** lanceolate to rectangular, stiffly memb to coriaceous, smt keeled, 5(7)-veined, veins not converging distally, inconspicuous, apc entire, lobed, or toothed, unawned or awned, awns tml, unbrchd, lm-awn jnct not evident; **pal** usu subequal to the lm, smt considerably shorter or slightly longer than the lm; **lod** 2, without venation, usu ciliate; **anth** 3; **ov** with hairy apc; **sty** 2, bases free. **Car** ovoid to fusiform, longitudinally grooved, not beaked, pericarp thin; **hila** linear; **emb** about ⅓ as long as the car. $x = 7$.

The *Triticeae* are primarily north-temperate in distribution. The tribe includes 400–500 species, among which are several important cereal, forage, and range species. Its generic treatment is contentious. This reflects the prevalence of natural hybridization, introgression, polyploidy, and reticulate relationships among its species. These factors preclude the circumscription of monophyletic groups, and mitigate against the delineation of morphologically coherent groups. Tsvelev (1975) argued that these same factors contribute to the tribe's success by maintaining a "generalist" genome.

The following key does not include intergeneric hybrids; they are treated in the text on the following pages: ×*Triticosecale* (p. 56), ×*Pseudelymus* (p. 60), ×*Elyhordeum* (p. 61), ×*Elyleymus* (p. 73), and ×*Leydeum* (p. 78). In the field, they can usually be detected by their intermediate morphology and sterility. In sterile plants, the anthers are indehiscent, somewhat pointed, and tend to remain on the plants. Measurements of rachis internodes and spikelets should be made at midspike.

1. Spikelets 2–7 at all or most nodes.
    2. Spikelets 3 at each node, the central spikelets sessile, the lateral spikelets usually pedicellate, sometimes all 3 spikelets sessile in cultivated plants; spikelets with 1 floret, usually only the central spikelet with a functional floret, the florets of the lateral spikelets usually sterile and reduced, in cultivated plants all florets functional or those of the lateral spikelets functional and those of the central spikelet reduced . . . . 7.01 *Hordeum*
    2. Spikelets usually other than 3 at each node, if 3, all 3 sessile; spikelets with 1–11 florets, if 1 floret, additional reduced or sterile florets present distal to the functional floret in at least 1 spikelet per node.
        3. Plants annual, weedy; spikelets with only 1 seed-forming floret . . . . 7.03 *Taeniatherum*
        3. Plants perennial, usually not weedy; spikelets usually with more than 1 seed-forming floret.
            4. Lemma awns (0)1–120 mm long; anthers 0.9–6 mm long; blades with well-spaced, unequally prominent veins on the adaxial surfaces. . . . . 7.12 *Elymus* (in part)
            4. Lemmas usually unawned or with awns up to 7 mm long, if awns 16–35 mm long, anthers 6–8 mm; blades usually with closely spaced, equally prominent veins on the adaxial surfaces.
                5. Disarticulation in the spikelets, beneath the florets; plants sometimes cespitose, often rhizomatous . . . . 7.15 *Leymus* (in part)
                5. Disarticulation tardy, in the rachises; plants cespitose, not rhizomatous . . . . 7.17 *Psathyrostachys*
1. Spikelets 1 at all or most nodes.
    6. Spikelets usually more than 3 times the length of the middle rachis internodes, usually divergent, sometimes ascending; rachis internodes 0.2–5.5 mm long.

7. Lemmas strongly keeled, keels conspicuously scabrous distally, scabrules 0.5–0.8 mm long . . . . 7.04 *Secale*
7. Lemmas rounded proximally, sometimes keeled distally, keels not or inconspicuously scabrous distally.
   8. Plants annual; anthers 0.4–1.4 mm long; spikes 0.8–4.5 cm long . . . . 7.02 *Eremopyrum*
   8. Plants perennial; anthers 3–5 mm long; spikes 1.3–15 cm long . . . . 7.08 *Agropyron*
6. Spikelets ½–3 times the length of the middle rachis internodes, appressed or ascending; rachis internodes 3–28 mm long.
   9. Glumes subulate to lanceolate, if lanceolate tapering from near or below midlength, 1–3(4)-veined at midlength.
      10. Glume midveins curving sideways distally; glumes lanceolate, tapering to acuminate apices from near midlength or below; plants always rhizomatous. . . . 7.14 *Pascopyrum*
      10. Glume midveins straight; glumes subulate to lanceolate, tapering from below midlength, keels straight or almost so; plants often rhizomatous. . . . . 7.15 *Leymus* (in part)
   9. Glumes lanceolate, rectangular, ovate, or obovate, narrowing beyond midlength, usually in the distal ¼, (1)3–5(7)-veined at midlength.
      11. Plants annual; glumes often with lateral teeth or awns, midveins smooth throughout.
         12. Glumes rounded over the midveins; plants weedy. . . . . 7.06 *Aegilops*
         12. Glumes keeled over the midveins; plants cultivated, sometimes escaping . . . . . . 7.07 *Triticum*
      11. Plants perennial; glumes without lateral teeth or awns, midveins sometimes scabrous.
         13. Glumes stiff, truncate, obtuse, or acute, unawned; glume keels smooth proximally, usually scabrous distally. . . . . 7.18 *Thinopyrum*
         13. Glumes flexible, acute to acuminate, sometimes awn-tipped; glume keels usually uniformly smooth or scabrous their whole length, sometimes smooth proximally and scabrous distally.
            14. Spikelets distant, not or scarcely reaching the base of the spikelet above on the same side of the rachis; anthers 4–8 mm long. . . . . 7.09 *Pseudoroegneria*
            14. Spikelets usually more closely spaced, reaching midlength of the spikelet above on the same side of the rachis; anthers 0.7–7 mm long . . . . 7.12 *Elymus* (in part)

7.01 HORDEUM L.   Roland von Bothmer, Claus Baden†, and Niels H. Jacobsen

Pl summer or winter ann or per; csp, smt shortly rhz. **Clm** to 135(150) cm, erect, geniculate, or decumbent; **nd** glab or pubescent. **Shth** open, pubescent or glab; **aur** present or absent; **lig** hyaline, truncate, erose; **bld** flat to more or less involute, more or less pubescent on both sides. **Infl** usu spikelike rcm, smt spikes, all customarily called spikes, with 3 spklt at each nd, cent spklt usu sessile, smt pedlt, ped to 2 mm, lat spklt usu pedlt, ped curved or straight, smt all 3 spklt sessile in cultivated pl; **dis** usu in the rchs, the spklt falling in triplets, cultivated forms generally not dis. **Spklt** with 1 flt. **Glm** awnlike, usu exceeding the flt. **Lat spklt** usu strl or stmt, often bisex in cultivated forms; **flt** pedlt, usu rdcd; **lm** awned or unawned. **Cent spklt** bisex; **flt** sessile; **rchl** prolonged beyond the flt; **lm** ovate, glab to pubescent, 5-veined, usu awned, rarely unawned; **pal** almost equal to the lm, narrowly ovate, keeled; **lod** 2, broadly lanceolate, mrg ciliate; **anth** 3, usu yellowish. **Car** usu tightly enclosed in the lm and pal at maturity. 2n = 14, 28, 42. Name from the old Latin name for barley.

Hordeum is a genus of 32 species that grow in temperate and adjacent subtropical areas, at elevations from 0–4500 m. The genus is native to Eurasia, the Americas, and Africa, and has been introduced to Australasia. The species are confined to rather moist habitats, even on saline soils. The annual species occupy seasonally moist habitats that cannot sustain a continuous grass cover.

Some species of *Hordeum*, such as *H. murinum*, are cosmopolitan weeds. *Hordeum vulgare* is widely cultivated for feed, malt, and flour. Archeological records suggest that *Hordeum* and *Triticum* were two of the earliest domesticated crops.

Eight species of *Hordeum* grow in the Intermountain Region: five are native, two are established weeds, and one is cultivated and occasionally persists as a weed. This treatment differs from Bothmer et al. (2007) in specifically recognizing *H. geniculatum* as distinct from *H. marinum* (Jakob et al. 2007).

Spike measurements and lemma lengths, unless stated otherwise, do not include the awns.

1. Plants perennial.
    2. Glumes of the central spikelet flattened near the base . . . . 5. *H. arizonicum* (in part)
    2. Glumes of the central spikelet setaceous, rarely slightly flattened near the base.
        3. Glumes 15–18 mm long, divergent to strongly divergent at maturity . . . . 4. *H. jubatum* (in part)
        3. Glumes 7–19 mm long, divergent or not at maturity . . . . 3. *H. brachyantherum*
1. Plants annual.
    4. Auricles to 8 mm long, well developed even on the upper leaves; lemmas of the lateral florets 6–15 mm long.
        5. Rachises usually not disarticulating at maturity; glumes of the central spikelets pubescent; lemmas of the central florets at least 3 mm wide, unawned or with awns 30–180 mm long; usually 1 or both lateral spikelets forming seeds . . . . 8. *H. vulgare*
        5. Rachises disarticulating at maturity; glumes of the central spikelet ciliate; lemmas of the central florets to 2 mm wide, with awns 20–40 mm long; lateral spikelets staminate . . . . 7. *H. murinum*
    4. Auricles absent or to 0.3 mm long; lemmas of the lateral florets 1.7–8.5 mm long.
        6. Glumes bent, strongly divergent at maturity.
            7. Glumes of the central spikelets slightly flattened near the base, 11–28 mm long
                 . . . . 5. *H. arizonicum* (in part)
            7. Glumes of the central spikelets setaceous, not flattened near the base, (15)35–85 mm long . . . . 4. *H. jubatum* (in part)
        6. Glumes straight, ascending to slightly divergent at maturity.
            8. Lemmas of the lateral spikelets with awns 3–8 mm long . . . . 6. *H. geniculatum*
            8. Lemmas of the lateral spikelets unawned or with awns shorter than 3 mm.
                9. Glumes of the central spikelets distinctly flattened near the base. . . . . 1. *H. pusillum*
                9. Glumes of the central spikelets setaceous to slightly flattened near the base.
                    10. Spikes 4–8 mm wide; lemmas of the central spikelets with awns 3–12 mm long; ligules 0.3–0.8 mm long . . . . 2. *H. depressum*
                    10. Spikes 6–20 mm wide; lemmas of the central spikelets with awns 10–22 mm long; ligules 0.6–1.8 mm long . . . . 5. *H. arizonicum* (in part)

1. Hordeum pusillum Nutt. LITTLE BARLEY [p. 314, 496]

Pl ann; loosely tufted. **Clm** 10–60 cm, erect, geniculate, or ascending; **nd** glab. **Shth** glab or slightly pubescent; **lig** 0.2–0.8 mm; **aur** absent; **bld** to 10.5 cm long, to 4.5 mm wide, sparsely to densely pubescent on both sides. **Spikes** 2–9 cm long, 3–7 mm wide, erect, often partially enclosed at maturity, pale green. **Glm** straight, not divergent at maturity. **Cent spklt: glm** 8–17 mm long, 0.5–1.5 mm wide, distinctly flattened near the base; **lm** 5–8.5 mm, usu glab, smt sparsely to densely pubescent, awned, awns 3.5–9.5 mm; **anth** 0.7–1.8 mm. **Lat spklt** usu strl; **glm** to 18 mm; **lo glm** distinctly flattened, more or less winged bas; **lm** 2.5–5.7 mm, usu awned, awns to 1.8 mm, rarely unawned; **anth** 0.6–1.2 mm. 2*n* = 14.

Hordeum pusillum grows in open grasslands, pastures, and the borders of marshes, and in disturbed places such as roadsides and waste places, often in alkaline soil. It is more common east and south of the Rocky Mountains than in the Intermountain Region.

2. Hordeum depressum (Scribn. & J.G. Sm.) Rydb. LOW BARLEY [p. 314, 496]

Pl ann; loosely tufted. **Clm** 10–55 cm, erect; **nd** glab. **Bas shth** pubescent; **lig** 0.3–0.8 mm;

aur absent; **bld** to 7.5(13.5) cm long, to 4.5 mm wide, sparsely to densely pubescent on both sides. **Spikes** 2.2–7 cm long, 4–8 mm wide, often partially enclosed at maturity, pale green or with a reddish tinge to the glm and awns. **Glm** straight, ascending to slightly divergent at maturity. **Cent spklt: glm** 5.5–20.5 mm long, to 0.5 mm wide, setaceous to slightly flattened near the base; **lm** 5–9 mm, glab, awned, awns 3–12 mm; **anth** 0.5–1.5 mm. **Lat spklt** strl or stmt, occ bisex; **glm** 5–20 mm; **lo glm** slightly flattened near the base; **up glm** setaceous throughout; **lm** 1.8–8.5 mm, unawned or awned, awns to 1 mm. 2*n* = 28.

Hordeum depressum grows in vernal pools and ephemeral habitats, often in alkaline soil. It is restricted to the western United States.

3. Hordeum brachyantherum Nevski [p. 315, 496]

Pl per; loosely to densely ces. **Clm** to 90 cm, erect to geniculate, not bulbous; **nd** glab. **Shth** glab or densely pubescent; **aur** absent; **bld** to 19 cm long, to 8 mm wide, glab or with hairs on both surfaces, hairs smt of mixed lengths. **Spikes** 3–8.5 cm, green to somewhat purple. **Glm** 7–19 mm, ascending to slightly divergent at maturity.

Cent spklt: glm 9–19 mm long, about 0.2 mm wide, setaceous throughout, rarely flattened near the base; **lm** 5.5–10 mm, usu glab, rarely pubescent, awned, awns 3.5–14 mm; **anth** 0.8–4 mm. **Lat spklt** stmt; **glm** 7–19 mm, setaceous; **lo glm** smt flattened near the base; **lm** rdmt to well developed, awns to 7.5 mm, rarely absent; **anth** 0.8–4 mm. $2n = 14, 28, 42$.

Hordeum brachyantherum is native to the Kamchatka Peninsula and western North America, and has been introduced to a few locations in the eastern United States. It grows in salt marshes, pastures, woodlands, subarctic woodland meadows, and subalpine meadows. One subspecies is present in the Intermountain Region.

Hordeum brachyantherum Nevski subsp.
brachyantherum MEADOW BARLEY,
NORTHERN BARLEY [p. 315]

Pl densely ces. **Clm** 30–95 cm, often robust, smt slender. **Bas shth** usu glab, smt sparsely pubescent; **bld** to 19 cm long, to 8 mm wide, both sides usu glab, smt with hairs to 0.5 mm on both surfaces. **Glm** 7–17 mm, usu straight at maturity; **lm** usu awned, awns to 6.5 mm, usu straight at maturity; **anth** 0.8–3.5 mm. $2n = 28, 42$.

Hordeum brachyantherum subsp. *brachyantherum* grows in pastures and along streams and lake shores, from sea level to 4000 m. Its range extends from Kamchatka through western North America to Baja California, Mexico. One population from California is known to be hexaploid.

4. **Hordeum jubatum** L. [p. 316, 496]

Pl per, smt appearing ann; ces. **Clm** 20–80 cm, geniculate to straight, not bulbous-based; **nd** glab. **Shth** glab or pubescent; **lig** to 0.8 mm; **aur** absent; **bld** to 15 cm long, to 5 mm wide, scabrous, smt hairy. **Spikes** 3–15 cm, usu nodding, whitish green to light purplish. **Glm** 15–85 mm long, conspicuous, bent, divergent to strongly divergent at maturity. **Cent spklt: glm** (15)35–85 mm, setaceous throughout, strongly spreading at maturity; **lm** 4–8.5 mm, glab, awned, awns 11–90 mm, straight to ascending; **pal** 5.5–8 mm; **anth** 0.6–1.2 mm. **Lat spklt** stmt or strl; **glm** 17–83 mm, setaceous; **lm** 4–6.5 mm, awned; **awns** 2–15 mm, divergent; **anth** 1–1.5 mm. $2n = 28$.

Hordeum jubatum grows in meadows and prairies around riverbeds and seasonal lakes, often in saline habitats, and along roadsides and in other disturbed sites. It is native from eastern Siberia through most of North America to Mexico, growing at elevations of 0–3000 m. It is sometimes grown as an ornamental.

Hordeum jubatum shows a wide range of variation in almost all characters; most such variation is not taxonomically significant. *Hordeum jubatum* subsp. *intermedium* is considered to be a subspecies of *H. jubatum* because no clear-cut discontinuities exist in the characters used to distinguish it from *H. jubatum* subsp. *jubatum*.

1. Glumes of the central spikelet 15–35 mm long; lemma awns of the central spikelets 11–35 mm long . . . . subsp. *intermedium*
1. Glumes of the central spikelet 35–85 mm long; lemma awns of the central spikelets 35–90 mm long . . . . subsp. *jubatum*

Hordeum jubatum subsp. **intermedium**
Bowden INTERMEDIATE BARLEY [p. 316]

Cent spklt: glm 15–35 mm, spreading at maturity; **lm** awned, awns 11–35 mm.

Hordeum jubatum subsp. *intermedium* is most abundant in the dry prairies of the northern Rocky Mountains and northern plains, growing at 0–3000 m. It is sometimes treated as a species, either as *H. intermedium* Hausskn. or *H. ×intermedium* Hausskn., the latter reflecting a suspected hybrid origin involving *H. jubatum* and *H. brachyantherum*.

Hordeum jubatum L. subsp. **jubatum** FOXTAIL
BARLEY, SQUIRRELTAIL BARLEY, SQUIRRELTAIL
GRASS [p. 316]

Cent spklt: glm 35–85 mm, strongly spreading at maturity; **lm** awned, awns 35–90 mm.

Hordeum jubatum subsp. *jubatum* is the more widespread of the two subspecies, extending from eastern Siberia through most of North America to northern Mexico. It grows in moist soil along roadsides and other disturbed areas, as well as in meadows, the edges of sloughs and salt marshes, and on grassy slopes.

5. **Hordeum arizonicum** Covas ARIZONA
BARLEY [p. 317, 496]

Pl ann to biennial, per under favorable conditions; forming small tufts. **Clm** 21–75 cm, rarely geniculate, not bulbous; **nd** glab. **Lo shth** pubescent; **up shth** glab; **lig** 0.6–1.8 mm; **aur** absent; **bld** to 13 cm long, to 4 mm wide, flat, glaucous, both surfaces scabrous, hairy, smt only sparsely hairy. **Spikes** 5–12 cm long, 6–10 mm wide, erect, often partially enclosed at maturity, pale green. **Glm** bent, strongly divergent at maturity. **Cent spklt: glm** 11–28 mm long, 0.2–0.4 mm wide, flattened near the base, setaceous above the mid; **lm** 5–9 mm, glab, awned, awns 10–22 mm; **anth** 1–2.2 mm. **Lat spklt** strl; **glm** to 27 mm long, 0.2–0.4 mm wide, flattened near the base; **lm** 2.5–5 mm, unawned or awned, awns to 3 mm, lm and awn together to 7 mm, the transition from the lm to the awn gradual. $2n = 42$.

Hordeum arizonicum grows in saline habitats, along irrigation ditches, canals, and ponds in the southwestern United States and northern Mexico. It is a segmental allopolyploid between *H. jubatum* and either *H. pusillum* or *H. intercedens*.

6. **Hordeum geniculatum** All. [p. 317, 497]

Pl summer ann, loosely tufted. **Clm** to 50 cm, strongly geniculate; **nd** glab. **Bas shth** hairy; **lig** 0.2–0.5 mm; **aur** absent or to 0.3 mm; **bld** to 8 cm long, 1.5–4 mm wide, flat or involute, sparsely to densely hairy on both sides. **Spikes** 1.5–7 cm long, 5–10(20) mm wide, dense, green or the awns and glm somewhat purplish; **rchs** dis at maturity, triplets breaking up after rchs dis. **Glm** straight, ascending to slightly divergent at maturity. **Cent spklt: glm** 14–26 mm long, 0.2–1.1 mm wide at the base, setaceous distally; **lm** 5–8 usu smooth or scabrous, smt pubescent, awned, awns 6–18 mm; **anth** 0.8–1.3 mm, yellowish. **Lat spklt** strl; **lo glm** usu setaceous, occ slightly widened near the base, not winged. 2*n* = 14, 28.

Hordeum geniculatum is a Mediterranean species that is now established in North America, including the Intermountain Region. It grows in grassy fields, waste places, and open ground, often in saline soils. Bothmer et al. (2007) treated this species as *H. marinum* subsp. *gussoneanum* (Parl.) Thell.

7. **Hordeum murinum** L. [p. *318*, <u>497</u>]

Pl ann; loosely tufted. **Clm** to 110 cm, usu erect, smt almost prostrate; **nd** glab. **Lo shth** often completely surrounding the clm, glab or somewhat pilose; **lig** 1–4 mm; **aur** to 8 mm, well developed even on the up lvs; **bld** to 28 cm, usu flat, occ with involute mrg, glab or sparsely pilose, smt scabrous. **Spikes** 3–8 cm long, 7–16 mm wide, pale green to distinctly reddish, especially the awns; **rchs** dis at maturity. **Cent spklt** sessile, flt sessile or pedlt, ped to 2 mm; **glm** 11–25 mm long, 0.8–1.8 mm wide, flattened, mrg usu distinctly ciliate; **lm** 8–14 mm long, to 2 mm wide, more or less smooth, awned, awns 20–40 mm; **lod** glab or with 1+ cilia; **anth** 0.2–3.2 mm, gray to yellow, smt with purple spots. **Lat spklt** stmt, flt sessile; **glm** flattened, mrg ciliate; **lm** 8–15 mm, awned, awns 20–50 mm; **pal** 8–15 mm; **rchl** 2.5–6.5 mm, slender or gibbous, yellow. 2*n* = 14, 28, 42.

Hordeum murinum is native to Eurasia, where it is a common weed in areas of human disturbance. It is thought to have originated around seasides, sandy riverbanks, and animal watering holes. It is now an established weed in the southwestern United States and other scattered locations. Prostrate plants are associated with grazing. Three subspecies are recognized.

1. Central spikelets sessile to subsessile; lemmas of the central florets subequal to those of the lateral florets, the awns longer than those of the lateral florets; paleas of the lateral florets almost glabrous
.... subsp. *murinum*
1. Central spikelets pedicellate; lemmas of the central florets from subequal to shorter than those of the lateral florets, the awns from shorter to longer than those of the lateral florets; paleas of the lateral florets scabrous to hairy.
2. Lemmas of the central florets much shorter than those of the lateral florets; paleas of the

lateral florets scabrous on the lower ½; anthers of the central and lateral florets similar in size
.... subsp. *leporinum*
2. Lemmas of the central florets about equal to those of the lateral florets; paleas of the lateral florets distinctly pilose on the lower ½; anthers of the central florets 0.2–0.6 mm long, those of the lateral florets 1.2–1.8 mm long ... subsp. *glaucum*

Hordeum murinum subsp. **glaucum** (Steudel) Tzvelev Smooth Barley [p. *318*]

Pl winter ann; loosely tufted. **Clm** 15–40 cm, usu erect, smt almost prostrate; **nd** glab. **Lo shth** often completely surrounding the clm, glaucous, glab or sparsely pilose; **lig** 1–4 mm; **aur** to 8 mm, well developed, even on the up lvs; **bld** to 28 cm, usu flat, occ with involute mrg, glab or sparsely pilose, smt scabrous. **Spikes** 3–8 cm long, 7–16 m wide, smt glaucous, often brownish when mature; **rchs** dis at maturity. **Cent spklt** sessile; **flt** ped; **lm** 8–14 mm long, to 2 mm wide, subequal to those of the lat flt, more or less smooth, awned, awns 20–40 mm, as long as or longer than those of the lat flt; **anth** 0.2–0.6 mm, shorter than those of the lat flt, more or less purple spotted. **Lat spklt** ped, stmt; **flt** sessile; **glm** flattened, mrg ciliate; **pal** 8–5 mm, pilose; **anth** 1.2–1.8 mm. 2*n* = 14.

Hordeum murinum subsp. *glaucum* grows in grasslands, fields, and waste places in arid areas. It is native to the eastern Mediterranean.

Hordeum murinum subsp. **leporinum** (Link) Arcang. Mouse Barley [p. *318*]

Pl winter ann; loosely tufted. **Clm** 30–110 cm, usu erect, smt almost prostrate; **nd** glab. **Lo shth** often completely surrounding the clm, not glaucous, glab or sparsely pilose; **lig** 1–4 mm; **aur** to 8 mm, well developed, even on the up lvs; **bld** to 28 cm, usu flat, occ with involute mrg, glab or sparsely pilose, smt scabrous. **Spikes** 3–8 cm long, 7–16 mm wide, green initially, often becoming purplish at maturity; **rchs** dis at maturity. **Cent spklt** sessile; **flt** ped; **lm** 8–14 mm long, to 2 mm wide, shorter than those of the lat flt, more or less smooth, awned, awns 20–40 mm, shorter than those of the lat flt; **anth** 0.9–3 mm. **Lat spklt** ped, stmt; **flt** sessile; **glm** flattened, mrg ciliate; **pal** 8–5 mm, scabrous on the lo portion; **anth** 1.2–3.2 mm. 2*n* = 14, 28, 42.

Hordeum murinum subsp. *leporinum* grows in waste places, roadsides, and disturbed areas in arid regions. It is native to the Mediterranean region.

Hordeum murinum L. subsp. **murinum** Wall Barley, Farmer's Foxtail, Way Barley [p. *318*]

Pl winter ann. **Clm** 30–60 cm. **Lvs** green. **Spikes** green. **Cent flt** sessile to subsessile; **lm** subequal to those of the lat flt, awns longer than those of the lat flt; **anth** 0.8–1.4 mm. **Lat spklt: pal** almost glab; **anth** 0.8–1.4 mm; **rchl** about 0.15 mm, pale. 2*n* = 28.

Hordeum murinum subsp. *murinum* grows in waste places that are somewhat moist. It is native to Europe. In North America, it has the most restricted distribution of the three subspecies, being found from Washington to Arizona, and in scattered locations from Maine to Virginia.

8. **Hordeum vulgare** L. BARLEY [p. *319*, <u>497</u>]
Pl summer or winter ann; loosely tufted. **Clm** to 100(150) cm, usu erect; **nd** glab. **Lo shth** pilose; **up shth** glab; **aur** to 6 mm, well developed even on the up lvs; **bld** to 30 cm long, 5–15 mm wide, flat, scabrous or glab. **Spikes** 5–10 cm long, 0.8–2 cm wide, green to purplish or blackish; **nd** 10–30, with 3 spklt per nd, 0–2 lat spklt, in addition to the cent spklt, forming seed at maturity (resulting in 2-, 4-, and 6-rowed barley); **rchs** usu not dis at maturity. **Cent spklt** sessile; **glm** 10–30 mm,

pubescent, flattened near the base; **lm** 6–12 mm long, 3+ mm wide, glab, smt scabrous, particularly distally, unawned or awned, awns 30–180 mm, usu scabrous; **anth** 6–10 mm, yellowish. **Lat spklt** usu sessile if seed-forming, pedlt if strl; **ped** to 3 mm; **lm** usu 6–15 mm, awned when ftl, obtuse to acute when strl. 2*n* = 14 (28).

Hordeum vulgare is native to Eurasia. Plants in the Intermountain Region belong to the cultivated subspecies, *H. vulgare* L. subsp. *vulgare*. The progenitor of cultivated barley, *H. vulgare* subsp. *spontaneum* (K. Koch) Thell., has a brittle rachis, tough awn, and, often, shrunken seeds. It does not grow in the Intermountain Region.

Hordeum vulgare L. subsp. *vulgare* was first domesticated in western Asia. It is now grown in most temperate parts of the world. In the Intermountain Region, it occurs as a cultivated species that is often found as an adventive in fields, roadsides, and waste places throughout the region, not just at the locations shown on the map. There are many distinctive, but interfertile, forms. Bothmer et al. (1995) presented an artificial classification of such forms.

7.02 EREMOPYRUM (Ledeb.) Jaub. & Spach

Signe Frederiksen

Pl ann. **Clm** 3–40 cm, geniculate. **Shth** open for most of their length; **aur** present, often inconspicuous; **lig** 0.4–2 mm, memb, truncate; **bld** 1–6 mm wide, flat, linear. **Infl** distichous spikes, 0.8–4.5 cm, with 1 spklt per nd, usu erect when mature; **rchs** intnd flat, mrg glab or with hairs, hairs white; **mid intnd** 0.5–3 mm; **dis** in the rchs, at the nd beneath each spklt, or at the base of each flt. **Spklt** 6–25 mm, including the awns, more than 3 times the length of the intnd, divergent, lat compressed, with 2–5 bisex flt, strl flt distal or absent. **Glm** equal, 4–19 mm, including the awns, coriaceous, becoming indurate, 1-keeled initially, smt 2-keeled at maturity, keels glab or hairy, never with tufts of hair, bases slightly connate, apc tapering to a sharp point or straight awn; **lm** 5–24 mm, coriaceous, rounded bas, keeled distally, 5-veined, unawned or shortly awned; **pal** usu shorter and thinner than the lm, 2-keeled, ciliate or scabrous distally, keels smt prolonged into 2 toothlike appendages; **anth** 3, 0.4–1.3 mm, yellow; **ov** pubescent; **sty** 2, free to the base. *x* = 7. **Haplomes** F, Xe. Name from the Greek *eremia*, 'desert', and *pyros*, 'wheat'.

Eremopyrum includes 5–10 species that grow in steppes and semidesert regions from Turkey to central Asia and Pakistan. Three species have been found in North America; only *E. triticeum* grows in the Intermountain Region. In their native ranges, species of *Eremopyrum* are valuable fodder on ephemeral spring pastures.

1. **Eremopyrum triticeum** (Gaertn.) Nevski
ANNUAL WHEATGRASS [p. *319*, <u>497</u>]
Clm to 30 cm, mostly glab, puberulent below the spikes. **Shth** of up lvs inflated; **bld** 1–3(6) mm wide, scabrous or shortly pilose distally. **Spikes** 1.3–2.4 cm long, 0.8–2 cm wide, elliptic, ovate, or nearly circular in outline; **dis** beneath each flt, smt at the base of the spikes, not in the rchs. **Spklt** 6–12 mm, with 2–3 flt. **Glm** 4–7.5 mm, glab, 1-veined and -keeled, becoming 2-keeled by the development of a ridge adjacent to the vein,

bases prominently inflated and curved; **lm** 5–7.5 mm, prominently keeled towards the subulate apc, lowest lm in each spklt pubescent on the proximal ½, hairs 0.1–0.15 mm, glab distally, the other lm glab; **pal keels** not prolonged. 2*n* = 14.

Eremopyrum triticeum is known primarily from scattered disturbed sites in western North America. Like most weeds, it is probably more widely distributed than herbarium records indicate. It is tolerant of alkaline soils and is summer-dormant.

7.03 TAENIATHERUM Nevski

J.K. Wipff

Pl ann. **Clm** (5)10–55(70) cm, erect, glab; **nd** 3–6. **Lvs** evenly distributed; **shth** open, usu glab; **aur** 0.1–0.5 mm, rarely absent; **lig** memb, truncate; **bld** flat to involute. **Infl** spikes, erect; **nd** 4–24(28), each with 2(3, 4) spklt; **intnd** 0.5–3.5 mm. **Spklt** with 2(3) flt, the lowest flt in each spklt bisex, the distal flt(s) highly rdcd, strl; **dis** above the glm. **Glm** 5–80 mm, equal, awnlike, erect to spreading or reflexed, bases connate. **Bisex flt: lm** 5-veined, glab or scabrous, mrg flat, scabrous, apc tml awned, awns 20–110 mm, longer than the lm, divergent, often cernuous; **pal** as long as the lm, keels antrorsely ciliate, apc truncate; **lod** 2, lobed, ciliate. **Rdcd flt: lm** 3-veined, awned; **pal** absent; **anth** 3, yellow to purple. **Car** narrowly elliptic, with an adx groove, apc pubescent. $x = 7$. Haplome **Ta**. Name from the Greek *taenia*, 'ribbon', and *ather*, 'awn'.

  Taeniatherum includes only one species. It is native to Eurasia.

1. **Taeniatherum caput-medusae** (L.) Nevski
  Medusahead [p. *320*, <u>497</u>]

Clm (5)10–55(70) cm. **Aur** 0.1–0.5 mm, rarely absent; **lig** 0.2–0.6 mm; **bld** (0.2) 0.7–2.5 mm wide, flat to involute. **Spikes** 1.2–6 cm. **Spklt** 6–45 mm; **glm** (5)7–80 mm, awnlike, erect to reflexed. **Bisex flt: lm** 5.5–8 mm, awns (20) 30–110 mm, divergent; **anth** 0.8–1 mm. **Car** 4–5.2 mm. $2n = 14$.

  Taeniatherum caput-medusae is native from Portugal and Morocco east to Kyrgyzstan. It usually grows on stony soils,

and flowers from May–June (July). It is an aggressive invader of disturbed sites in the western United States, where it has become a serious problem on rangelands. It is listed as a noxious weed by the U.S. Department of Agriculture.

  Frederiksen (1986) recognized three subspecies within *Taeniatherum caput-medusae*, distinguishing among them on the basis of morphology and geography. All North American plants belong to *Taeniatherum caput-medusae* (L.) Nevski subsp. *caput-medusae*, which differs from the other two subspecies in its longer glumes and shorter lemmas.

7.04 **SECALE** L.   Mary E. Barkworth

Pl ann, bien, or short-lived per; csp when per. **Clm** 25–120(300) cm. **Shth** open; **aur** usu present, 0.5–1 mm; **lig** memb, truncate, often lacerate; **bld** flat or involute. **Infl** lat compressed, distichous spikes; **mid intnd** 2–4 mm, with 1 spklt per nd, spklt strongly ascending; **dis** in the rchs, below the spklt, rchs not or tardily dis in cultivated strains. **Spklt** 10–18 mm, with 2(3) flt; **flt** bisex. **Glm** 8–20 mm, shorter than the adjacent lm, linear to subulate, scabrous, mrg hyaline, 1-veined, keeled, keels terminating in an awn, awns to 35 mm; **lm** 8–19 mm, strongly lat compressed, strongly keeled, keels conspicuously scabrous distally, scabrules 0.6–1.3 mm, lm apc tapering to a scabrous awn, awns 2–50 mm; **anth** 3, 2.3–12 mm, yellow. $x = 7$. Haplome **R**. *Secale* is the classical Latin name for rye.

  Secale is a genus of three species. All are native to the Mediterranean region and western Asia. One species is established in the Intermountain Region. All three species are diploids.

  ×Triticosecale is an artificially derived hybrid between *Triticum* and *Secale* that is now widely cultivated (see next).

1. **Secale cereale** L. Rye [p. *321*, <u>497</u>]

Pl ann or biennial. **Clm** (35)50–120(300) cm. **Bld** (3)4–12 mm wide, usu glab. **Spikes** (2) 4.5–12(19) cm, often nodding when mature; **dis** tardy, in the rchs, at the nd, or not occurring. **Glm** 8–20 mm, keels scabrous, terminating in awns, awns 1–3 mm; **lm** 14–18 mm, awns 7–50 mm; **anth** about 7 mm. $2n = 14, 21, 28$.

  Secale cereale is cultivated as a cereal species and, particularly in Canada, for whisky. It is also frequently used for soil stabilization along roadsides where it frequently becomes established. When dry, the spike is often nodding. Frederiksen and Petersen (1998) placed cultivated plants with a non-disarticulating rachis into *Secale cereale* L. subsp. *cereale*, and wild or weedy plants with more fragile rachises into *S. cereale* subsp. *ancestrale* Zhuk. Remains of cultivated rye dating to 6000 B.C. have been found in Turkey.

7.05 **×TRITICOSECALE** Wittm. *ex* A. Camus Triticale [p. *321*]   Mary E. Barkworth

Pl ann. **Clm** to 130 cm, erect, straight or geniculate at the lowest nd. **Lvs** mainly cauline; **lig** 2–4 mm, memb, truncate to rounded. **Infl** tml, distichous spikes, with solitary spklt; **intnd** 3–5 mm, densely pilose, at least on the edges. **Spklt** 10–17 mm, with 2–4 flt, distal flt usu rdcd. **Glm** 9–12 mm, asymmetrically keeled, keels stronger and smt conspicuously ciliate distally, apc retuse to acute, awned, awns 3–4 mm; **lm** 10–15 mm, lat compressed, keeled, keels smt ciliate distally, tml awned, awns 3–50 mm; **anth** 3, yellow. $x = 7$.

  ×Triticosecale comprises hybrids between *Secale* and *Triticum*. Natural hybrids between the two genera are rare, but Triticale, which consists of cultivars derived from artificial hybrids between *S. cereale* and *T. aestivum*, is becoming an increasingly important cereal crop. The existing binomials in *×Triticosecale* do not apply to Triticale because they involve different species of *Triticum*.

  Triticale cultivars often have a complex ancestry, involving multiple hybridizations, backcrossings, and artificially induced chromosome doubling. Their genetic material varies from being derived almost entirely from *Triticum* to being derived almost entirely from *Secale*. For this reason, they are best identified as such, e.g., *×Triticosecale* 'Newton' or *×Triticosecale* 'Bokolo'. A hybrid formula would misrepresent their genetic and morphological diversity.

7.06 AEGILOPS L.

Mary E. Barkworth

Pl ann. **Clm** 14–80 cm, usu glab, erect or geniculate at the base, with (1)2–4(5) nd. **Shth** open; **aur** ciliate; **lig** 0.2–0.8 mm, memb, truncate; **bld** 1.5–10 mm wide, linear to linear-lanceolate, flat, spreading. **Infl** tml spikes, with 2–13 spklt, usu with 1–3 additional rdmt spklt at the base; **intnd** 6–12 mm; **dis** either at the base of the spikes or in the rchs, the spklt falling attached to the intnd above or below. **Spklt** solitary at each nd, ½–2(3) times the length of the intnd, tangential to the rchs, appressed or ascending, the up spklt(s) smt strl; **ftl spklt** 5–15 mm, with 2–7 flt, the distal flt often strl. **Glm** ovate to rectangular, rounded on the back, scabrous or pubescent, with several prominent veins, midveins smooth throughout, apc truncate, toothed, or awned, smt indurate at maturity; **lm** rounded on the back, apc toothed, frequently awned; **pal** chartaceous, 2-keeled, keels ciliate; **anth** 3, 1.5–4 mm, not penicillate; **ov** with pubescent apc. **Car** lanceolate to lanceolate-ovate. $x = 7$. **Haplomes B, C, D, S, T, U, M, N**. The name is derived from from the Greek *aigilops*, a word which has multiple etymological interpretations (Slageren 1994, pp. 19–20, 118–119), including 'wild oats' and several variants related to the Greek *aigos*, 'goat'.

Aegilops has about 23 species and is native to the Canary Islands, and from the Mediterranean region to central Asia. It is sometimes included in *Triticum* because the two form natural hybrids and both are involved in the evolution of the cultivated wheats, including *T. aestivum*. They are treated as distinct genera here, in keeping both with past practice and with their differing ecological attributes, *Aegilops* being a weedy genus. Two species have been found in the Intermountain Region. This treatment is derived from one submitted by Sandra M. Saufferer for the *Flora of North America*, vol. 24.

1. Lemmas of lower fertile spikelets with apices mucronate or awned, awns 0.1–0.5 cm . . . . 1. *A. cylindrica*
1. Lemmas of lower fertile spikelets with 2–3 teeth, if 3-toothed, the central tooth the longest, sometimes extending into a 10 mm awn . . . . 2. *A. triuncialis*

1. **Aegilops cylindrica** Host Jointed Goatgrass [p. 322, 497]

Clm 14–50 cm, erect to decumbent at the base, usu with many tillers. **Shth** with hyaline mrg, smt ciliate; **bld** 3–15 cm long, 2–5 mm wide. **Spikes** 2.2–12 cm long, about 0.3 cm wide, narrowly cylindrical, 10–45 times longer than wide, with (2)3–8(12) ftl spklt; **rdmt spklt** absent or 1–2; **dis** initially at the base of the spikes and secondarily in the rchs, the spklt remaining attached to the intnd above. **Ftl spklt** 9–12 mm, narrowly cylindrical, scabrous or pubescent, with 3–5 flt, the lo (1)2–3 flt ftl. **Glm of ftl spklt** awned, awns of the lo spklt 2–5 mm; **glm** of apical spklt 7–9 mm, scabrid, awned, awns 3–6 cm, usu flanked by 2 lat teeth; **lm of lo ftl spklt** 9–10 mm, adx surfaces velutinous distally, apc mucronate or awned, awns 0.1–0.5 cm; **lm of apical spklt** 1-awned, awns (2)4–8 cm, flanked by 2 teeth. **Car** 6–7 mm, adhering to the lm and pal. **Haplomes DC.** $2n = 28$.

Aegilops cylindrica is a widespread weed in North America, being particularly troublesome in winter wheat. It usually grows in disturbed sites, such as roadsides, fields, and along railroad tracks. It is native to the Mediterranean region and central Asia, and is adventive in other temperate countries.

Hybrids between *Aegilops cylindrica* and *Triticum aestivum*, called ×*Aegilotriticum sancti-andreae* (Degen) Soó [p. 322], have been found in various parts of North America. They often have a few functional seeds which, on maturing into reproductive plants, can backcross to either parent. For this reason, *A. cylindrica* is considered a serious weed in many wheat-growing areas.

2. **Aegilops triuncialis** L. Barbed Goatgrass [p. 322, 497]

Clm 17–60 cm, geniculate to semiprostrate at the base, usu with several tillers. **Shth** with hyaline mrg, lo cauline shth mrg usu ciliate; **bld** 1.5–7 cm long, 2–3 mm wide. **Spikes** 2.2–6 cm long, bases 0.4–0.5 cm wide, narrowly ellipsoid, becoming subcylindrical distally, with 2–7 ftl spklt; **rdmt spklt** (2)3; **dis** at the base of the spikes. **Lo ftl spklt** 7–13 mm, lanceolate-ovate, with 3–5 flt, the first 1–2 flt ftl; **up spklt** 7–9 mm, rdcd. **Glm of lo ftl spklt** 6–10 mm, 2–3-awned, awns 1.5–6 cm, glab, scabrous, or velutinous, if 3-awned, the cent awn often shorter than the lat awns, smt rdcd to a tooth; **glm of apical spklt** 6–8 mm, 3-awned or with 1 awn and 2 lat teeth, awns 2.5–8 cm, if 3-awned, the cent awn the longest; **lm of lo ftl spklt** 7–11 mm, with 2–3 teeth, if 3-toothed, the cent tooth the longest, smt extending into a 10 mm awn. **Car** 5–8 mm, falling free from the lm and pal. **Haplomes UC.** $2n = 28$.

North American collections of *Aegilops triuncialis* are from disturbed sites, mostly roadsides and railroads. The native range of the species extends from the Mediterranean area east to central Asia and south to Saudi Arabia. Specimens from North America belong to *Aegilops triuncialis* L. var. *triuncialis*, in which the glumes of the apical spikelets have a 5–8 cm central awn flanked by shorter (1–3 cm) lateral awns, and the glumes of the lower fertile spikelets have 2–3 awns of 1.5–6 cm.

7.07 **TRITICUM** L.   Laura A. Morrison

Pl ann. **Clm** 14–180 cm, solitary or brchd at the base; **intnd** usu hollow throughout in hexaploids, usu solid for about 1 cm below the spike in diploids and tetraploids, even if hollow below. **Shth** open; **aur** present, often deciduous at maturity; **lig** memb; **bld** flat, glab or pubescent. **Infl** usu tml spikes, distichous, with 1 spklt per nd, occ brchd; **intnd** (0.5)1.4–8 mm; **dis** in the rchs, the spklt usu falling with the intnd below to form a wedge-shaped diaspore, smt falling with the adjacent intnd to form a barrel-shaped diaspore, domesticated taxa usu non-dis, or dis only under pressure. **Spklt** 10–25(40) mm, usu 1–3 times the length of the intnd, appressed to ascending, with 2–9 flt, the distal flt often strl. **Glm** subequal, ovate, rectangular, or lanceolate, chartaceous to coriaceous, usu stiff, tightly to loosely appressed to the lo flt, with 1 prominent keel, at least distally, keels often winged and ending in a tooth or awn, a second keel or prominent lat vein present in some taxa; **lm** keeled, chartaceous to coriaceous, 2 lowest lm usu awned, awns 3–23 cm, scabrous, distal lm unawned or awned, awns to 2 cm; **pal** hyaline-memb, splitting at maturity in diploid taxa; **anth** 3. **Car** tightly (hulled wheats) or loosely (naked wheats) enclosed by the glm and lm, lm and pal not adherent; **Endosperm** flinty or mealy. $x = 7$. **Haplomes A, B, D**, and **G**. *Triticum* is the classical Latin name for wheat.

Triticum is a genus of approximately 25 wild and domesticated species. It was first cultivated in western Asia at least 9000 years ago and is now the world's most important crop, being planted more widely than any other genus. Three species are cultivated in North America, of which *T. aestivum* is by far the most important.

1. Culms partially or complete solid for 1 cm below the spikes; glumes with 1 fully developed keel . . . . .1. *T. durum*
1. Culms usually hollow to the base of the spikes; glumes 1-keeled but this often developed only in the upper portion of the glumes.
   2. Glumes loosely appressed to the lower florets; rachises not disarticulating, even under pressure . . . . .2. *T. aestivum*
   2. Glumes tightly appressed to the lower florets; rachises disarticulating under pressure . . . . .3. *T. spelta*

1. **Triticum durum** Desf. Durum Wheat, Macaroni Wheat, Hard Wheat [p. 323]

Clm 60–160 cm; **nd** glab; **intnd** mostly hollow, solid for 1 cm below the spikes. **Bld** 7–16 mm wide, usu glab. **Spikes** 4–11 cm, about as wide as thick, never brchd; **rchs** ciliate to partially ciliate at the nd and mrg, not dis; **intnd** 3–6 mm. **Spklt** 10–15 mm, with 5–7 flt, 2–4 seed-forming. **Glm** 8–12 mm, coriaceous, loosely appressed to the lo flt, with 1 prominent keel, terminating in a tooth, tooth to 0.3 mm; **lm** 10–12 mm, lo 2 lm awned, awns to 23 cm; **pal** not splitting at maturity. **Endosperm** usu flinty, smt mealy. **Haplomes AuB**. $2n = 28$.

Triticum durum is a domesticated spring wheat that is grown in temperate climates throughout the world. It is typically used for macaroni-type pastas, semolina, bulghur, flat breads, and pita. It imparts a yellowish color to bread.

2. **Triticum aestivum** L. Wheat, Bread Wheat, Common Wheat, Soft Wheat [p. 323]

Clm 14–150 cm; **nd** glab or pubescent; **intnd** usu hollow, even immediately below the spikes. **Bld** 6–15(20) mm wide, glab or pubescent. **Spikes** (3.5)6–18 cm, usu thicker than wide to about as thick as wide, wider than thick in compact forms; **rchs** shortly ciliate at the nd and mrg, not dis. **Spklt** 10–15 mm, appressed or ascending, with 3–9 flt, 2–5 seed-forming. **Glm** 6–12 mm, coriaceous, loosely appressed to the lo flt, usu keeled in the distal ½, smt prominently keeled to the base, terminating in a tooth or awn, awns to 4 cm; **lm** 10–15 mm, toothed or awned, awns to 12 cm; **pal** not splitting at maturity. **Endosperm** mealy to flinty. **Haplomes AuBD**. $2n = 42$.

Triticum aestivum is the most widely cultivated wheat. In addition to being grown for bread flour, *T. aestivum* cultivars are used for pastry-grade flour, Oriental-style soft noodles, and cereals.

3. **Triticum spelta** L. Spelt, Dinkel [p. 323]

Clm 80–120 cm; **nd** glab or pubescent; **intnd** hollow, even immediately below the spikes. **Bld** 12–20 mm wide, sparsely pubescent. **Spikes** 6–20 cm, about as wide as thick, slender, almost cylindrical, narrowing distally; **rchs** glab or sparsely hairy at the nd and mrg, dis with pressure, dis units barrel-shaped or wedge-shaped. **Spklt** 12–16 mm, with 3–5 flt, 1–3 seed-forming. **Glm** 5–10 mm, coriaceous, tightly appressed to the lo flt, truncate, with 1 prominent keel, keel winged to the base, terminating in a tooth; **lm** 8–12 mm, toothed or awned, awns on the lo 2 lm to 10 cm, the third lm smt awned, awns to 2 cm; **pal** not splitting at maturity. **Endosperm** usu flinty. **Haplomes AuBD**. $2n = 42$.

Triticum spelta is grown in North America for the specialty food and feed grain markets. It is known for yielding a pastry-grade flour not suitable for bread making unless mixed with *T. aestivum*, the bread-quality flour. Modern plant breeding programs are improving its gluten profile to upgrade its bread-making quality. Consequently, claims that *T. spelta* is a safe option for consumers with gluten intolerance should be treated with caution.

7.08 **AGROPYRON** Gaertn.

Mary E. Barkworth

Pl per; densely to loosely csp, smt rhz. **Clm** 25–110 cm, geniculate or erect. **Shth** open; **aur** usu present; **lig** memb, often erose. **Infl** spikes, usu pectinate; **mid intnd** 0.2–3(5.5) mm, bas intnd often somewhat longer. **Spklt** solitary, usu more than 3 times as long as the intnd, usu divergent or spreading from the rchs, with 3–16 flt; **dis** above the glm and beneath the flt. **Glm** shorter than the adjacent lm, lance-ovate to lanceolate, 1–5-veined, asymmetrically keeled, a sec keel smt present on the wider side, keels glab or with hairs, hairs not tufted, apc acute and entire, smt awned, awns to 6 mm; **lm** 5–7-veined, asymmetrically keeled, acute to awned, awns to 4.5 mm; **pal** from slightly shorter than to exceeding the lm, bifid; **anth** 3, 3–5 mm, yellow. **Car** usu falling with the lm and pal attached. $x = 7$. Haplome **P**. Name from the Greek *agrios*, 'wild', and *pyros*, 'wheat'.

Agropyron, it is now agreed, should be restricted to perennial species of *Triticeae* with keeled glumes, i.e., *A. cristatum* and its allies, or the "crested wheatgrasses". The excluded species are distributed among *Pseudoroegneria*, *Thinopyrum*, *Elymus*, *Eremopyrum*, and *Pascopyrum*. This leaves *Agropyron* as a Eurasian genus that includes diploid, tetraploid, and hexaploid plants, all of which contain a single genome, designated the **P** genome by the International Triticeae Consortium. The genus is now widespread in western North America, frequently being used for soil stabilization on degraded rangeland and abandoned cropland, because it is highly tolerant of grazing and provides good spring forage.

This treatment recognizes two species within the Intermountain Region, a very broadly interpreted *Agropyron cristatum*, which includes Dewey's *A. cristatum* and *A. desertorum*, and a traditionally interpreted *A. fragile*. *Agropyron cristatum* in North America reflects a process that might be called de-speciation.

1. Lemmas usually awned, awns 1–6 mm long; spikelets diverging from the rachises at angles of 30–95°; spikes narrowly to broadly lanceolate, rectangular, or ovate in outline . . . . 1. *A. cristatum*
1. Lemmas unawned, sometimes mucronate; spikelets diverging from the rachises at an angle of less than 30(35)°; spikes linear to narrowly lanceolate in outline . . . . 2. *A. fragile*

1. **Agropyron cristatum** (L.) Gaertn. Crested Wheatgrass [p. 324, <u>497</u>]

Pl occ rhz. **Clm** 25–110 cm, smt geniculate. **Lig** to 1.5 mm; **bld** 1.5–6 mm wide, glab or pubescent. **Spikes** 1.3–10.5(15) cm long, 5–25 mm wide, narrowly to broadly lanceolate, rectangular, or ovate, smt tapering distally; **intnd** (0.2)0.7–5(8) mm, glab or pilose, smt all more or less equal, smt short and long intnd alternating within a spike, bas intnd often longer than those at midlength. **Spklt** 7–16 mm, diverging at angles of 30–95° at maturity, with 3–6(8) flt. **Glm** 3–6 mm, glab or with coarse hairs on the keels, acute, usu awned, awns 1.5–3 mm; **lm** 5–9 mm, glab or with hairs, keeled, keels smt scabrous distally, apc acute, usu awned, awns 1–6 mm; **anth** 3–5 mm. $2n = 14, 28, 42$.

Agropyron cristatum is native from central Europe and the eastern Mediterranean to Mongolia and China. According to Tsvelev (1976), the most widely distributed taxon outside the Soviet Union is *A. cristatum* subsp. *pectinatum*. In North America, the reticulate genetic history of crested wheatgrass and the absence of any native populations argue against attempting recognition of subspecies.

Among the more commonly encountered variants of *Agropyron cristatum* in the Intermountain Region are the cultivar 'Fairway', which was considered by Dillman (1946) and Dewey (1986) to belong to *A. cristatum* rather than *A.*

desertorum, and its derivatives 'Parkway' and 'Ruff'. The name "Fairway" is also widely used in agricultural circles to refer to any crested wheatgrass that looks like the cultivar 'Fairway'. 'Standard' crested wheatgrass, which Dewey (1986) and others placed in *A. desertorum*, originally referred to a particular seed lot (S.P.I. 19537) that the Montana Wheatgrowers' Association decided to use as a standard against which to compare the performance of other crested wheatgrass strains. The term is now applied by agronomists to all crested wheatgrasses that are less leafy and have more lanceolate spikes than 'Fairway' crested wheatgrasses. There are numerous cultivars of crested wheatgrass available.

Because it is easy to establish, *Agropyron cristatum* has often been used to restore productivity to areas that have been overgrazed, burned, or otherwise disturbed. This ability, combined with its high seed production, tends to prevent establishment of most other species, both native and introduced.

2. **Agropyron fragile** (Roth) P. Candargy Siberian Wheatgrass [p. 324]

Pl not rhz. **Clm** 30–100 cm, rarely geniculate. **Lig** to 1 mm; **bld** 1.5–6 mm wide. **Spikes** (5)8–15 cm long, 5–13 mm wide, linear to narrowly lanceolate; **intnd** 1.5–5 mm. **Spklt** 7–16 mm, appressed or diverging up to 30(35)° from the rchs. **Glm** 3–5 mm, glab or hairy, often awned, awns 1–3 mm; **lm** 5–9 mm, keels scabrous distally, apc unawned, smt mucronate, mucros up to 0.5 mm; **anth** 4–5 mm. $2n = (14), 28, (42)$.

Agropyron fragile is native from the southern Volga basin through the Caucasus to Turkmenistan and Mongolia. It is more drought-tolerant than *A. cristatum*. It is rarely found outside of experimental plantings. This may change as more cultivars become available. Hybrids with *A. cristatum* are fertile.

7.09 PSEUDOROEGNERIA (Nevski) Á. Löve   Jack R. Carlson

Pl per; usu csp, smt rhz. **Clm** 30–100 cm, usu erect, smt decumbent or geniculate. **Lvs** evenly distributed; **shth** open; **aur** well developed; **lig** memb; **bld** flat to loosely involute. **Infl** tml spikes, erect, with 1 spklt per nd; **intnd** (7)10–20(28) mm at midlength, lo intnd often longer than those at midlength. **Spklt** (8)12–25 mm, 1.1–1.5(2) times the length of the intnd, usu appressed, smt slightly divergent, with 4–9 flt; **dis** above the glm and beneath the flt. **Glm** unequal, from shorter than to slightly longer than the lowest lm in the spklt, lanceolate to oblanceolate, (3)4–5(7)-veined, usu acute to obtuse, occ truncate, narrowing beyond midlength, veins prominent; **lm** inconspicuously 5-veined, unawned or tml awned, awns straight to strongly bent and divergent; **anth** 4–8 mm. $x = 7$. Haplome **St**. Name from the Greek *pseudo*, 'false', and the genus *Roegneria*, an Asian taxon often included in *Elymus*.

Pseudoroegneria includes 15–20 species, one of which is North American and the remainder either Eurasian or Asian. All species currently included in the genus are obligate outcrossers, and almost all are diploids or autotetraploids (Jensen et al. 1992) containing the **St** haplome (designation by the International Triticeae Consortium). This genome is the most widely distributed in the *Triticeae*, being found in all species of *Elymus sensu lato* as well as some species of *Thinopyrum*.

1. Pseudoroegneria spicata (Pursh) Á. Löve
BLUEBUNCH WHEATGRASS [p. 324, <u>497</u>]

Pl loosely ces, smt rhz. **Clm** 30–100 cm tall, 0.5–2 mm thick, smt glaucous. **Lig** truncate, 0.1–0.4 mm on the lo lvs, 0.2–0.4 mm on the up lvs; **bld** 2–6 mm wide, involute when dry, flag lf bld strongly divergent when dry, abx surfaces smooth, glab, adx surfaces scabrous or hirsute. **Spikes** 8–15 cm long, 3–8(10) mm wide excluding the awns; **mid intnd** 7–20(25) mm, glab, scabrous on the angles. **Spklt** 8–22(25) mm, with 4–9 flt. **Glm** 6–13 mm long, 0.9–2.2 mm wide, about ½ the length of the spklt, glab, smt scabrous over the veins, acute; **lm** 9–14 mm, unawned or with a tml, strongly divergent awn, awns to 25 mm. $2n = 14, 28$.

Pseudoroegneria spicata is primarily a western North American species, extending from the east side of the coastal mountains to the western edge of the Great Plains, and from the Yukon Territory to northern Mexico. It grows on medium-textured soils in arid and semiarid steppe, shrub-steppe, and open woodland communities. It is an important forage plant in the northern portion of the Intermountain Region. Several cultivars have been developed.

Rhizomatous plants are favored in relatively moist habitats, and cespitose plants in dry habitats (Daubenmire 1960). Daubenmire noted that rhizomatous plants produce few inflorescences and, possibly for this reason, are collected less frequently than cespitose plants. He also found that awn length varies continuously within plants grown from seed. He concluded that the ability to produce rhizomes and unawned plants is heritable, that the two characters are not linked, and that the form which becomes dominant at a local site is determined by environmental conditions.

Based on informal observations, plant breeders working with *Pseudoroegneria spicata* consider that awn presence is determined by a single major gene, and modified by some minor genes. The unawned condition is apparently dominant, as seed from crosses of heterozygotic, diploid, unawned parents gives rise to around 50% awned offspring.

The above observations make it clear that the awned and unawned phases of *Pseudoroegneria spicata* are of little taxonomic significance, despite their evident morphological difference. If it is considered necessary to distinguish between them, the awned phase can be called *Pseudoroegneria spicata* (Pursh) Á. Löve f. *spicata* and the unawned phase *P. spicata* f. *inermis* (Scribn. & J.G. Sm.) Barkworth.

Pseudoroegneria spicata has been suggested as one of the parents in numerous natural hybrids with species of *Elymus*. These hybrids are usually mostly sterile, but development of even a few viable seeds permits introgression to occur, as well as the formation of distinctive populations. It is often difficult to detect such hybrids, particularly if they involve the unawned form of *Pseudoroegneria*. The named hybrids are treated under ×*Pseudelymus* (see below). Others are discussed under the *Elymus* parent.

7.10 ×PSEUDELYMUS Barkworth & D.R. Dewey   Mary E. Barkworth

Pl per; smt rhz. **Clm** 50–80 cm, erect, glab. **Lvs** not bas concentrated; **shth** glab or puberulent; **aur** present; **lig** truncate. **Infl** distichous spikes, with 1(2) spklt(s) per nd. **Spklt** appressed, with 3–5 flt. **Glm** unequal, linear-lanceolate to lanceolate; **lm** awned or unawned; **pal** slightly shorter than to slightly longer than the lm; **anth** indehiscent.

×*Pseudelymus* comprises hybrids between *Pseudoroegneria* and *Elymus*. Only one species is treated here. Another species, *E. albicans*, is thought to be a similar hybrid, but it is treated as a species because it is frequently highly fertile.

1. **×Pseudelymus saxicola** (Scribn. & J.G. Sm.)
   Barkworth & D.R. Dewey [p. 325]
Pl not rhz. **Clm** 0.8–1 mm thick; **lig** 0.2–0.4 mm;
bld 1.5–2 mm wide. **Spikes** 7–14 cm, with 1 spklt
per nd; **intnd** 7–9 mm. **Spklt** 12–15 mm excluding
the awns; **dis** beneath the flt, smt also in the rchs.
Glm 15–21 mm including the awns, glm bodies
6–8 mm long, 0.7–1 mm wide, (1)3–4-veined; **lm**
10–11 mm excluding the awns, (14)18–37 mm
including the awns, apc often bifid; **anth** about
2.5 mm.

×Pseudelymus saxicola consists of hybrids between
Pseudoroegneria spicata and *Elymus elymoides*. It is a rather
common hybrid in western North America. It differs from *E.
albicans*, which is thought to be derived from hybrids between
P. spicata and *E. lanceolatus*, in lacking rhizomes, having longer
awns on its glumes and lemmas, and having disarticulating
rachises. It is more likely to be confused with *E. ×saundersii*,
but differs in its longer glume and lemma awns.

7.11 ×ELYHORDEUM Mansf. *ex* Tsitsin & K.A. Petrova   Mary E. Barkworth

Pl per; usu csp, occ shortly rhz. **Infl** tml, spikes or spikelike, with 1–3(7) spklt per nd, lat spklt usu
shortly pedlt, cent spklt sessile or nearly so; **dis** tardy, at the rchs nd and beneath the flt. **Spklt**
with 1–4 flt. **Glm** subulate to narrowly lanceolate, usu awned; **lm** usu awned; **anth** strl. **Car** rarely
formed.

×Elyhordeum is the name given to hybrids between *Elymus* and *Hordeum*. These hybrids are fairly common. All appear to be
sterile, i.e., they do not produce good pollen or set seed. The descriptions should be treated with reservation because, in some
instances, only the type specimen has been examined. For that reason, no key is provided. Only named hybrids are described
and illustrated.

Interspecific hybrids between *Elymus elymoides* or *E. multisetus* and other species of *Elymus* resemble the *×Elyhordeum*
hybrids in having tardily disarticulating, spikelike inflorescences and awned glumes and lemmas, but are more likely to have
solitary spikelets, even at the lowest node. Distinguishing between them and *×Elyhordeum* hybrids, without knowledge of
other species of *Triticeae* at a site, is challenging.

Inflorescence measurements, unless stated otherwise, do not include the awns.

1. **×Elyhordeum macounii** (Vasey) Barkworth
   & D.R. Dewey [p. 325]
Clm 50–100 cm. **Shth** usu glab; **lig** truncate; **bld**
9–16 cm long, 2–5 mm wide, stiff, ascending,
scabrous. **Spikes** 4–13 cm long, about 5 mm wide,
erect, lo nd with 1–2 spklt, up nd with 1 spklt, the
spklt imbricate. **Spklt** with 1–3 flt, those at the
lo nd frequently with 3 glm. **Glm** 6–9 mm, not
indurate at the base, awned, awns as long as or
longer than the glm bodies; **lm** 6–11 mm, oblong-
lanceolate, glab or smt scabrous distally, awned,
awns 10–20 mm. 2*n* = 28.

×Elyhordeum macounii consists of hybrids between *Elymus
trachycaulus* and *Hordeum jubatum*. It is quite common in
the Intermountain Region. Backcrosses to *E. trachycaulus*
may have non-disarticulating rachises; they are likely to
be identified as *E. trachycaulus*, falling between subsp.
trachycaulus and subsp. *subsecundus*.

2. **×Elyhordeum stebbinsianum** (Bowden)
   [p. 325]
Clm 60–80 cm. **bld** 3–5 mm wide, abx surfaces
smooth, adx surfaces scabridulous. **Infl** spikelike,
10–16 cm, with 3 spklt per nd; **intnd** averaging
3.4 mm. **Spklt** with 1–3 flt, the lat spklt sessile
or pedlt, ped to 3 mm. **Glm** 0.5–1 mm wide,
terminating in a 4–7 mm awn; **lm** glab to scabrous,
awned, awns 5–11 mm; **anth** indehiscent. 2*n* =
unknown.

×Elyhordeum stebbinsianum consists of hybrids between
Elymus glaucus and *Hordeum brachyantherum*. Bowden (1958)

reported that they appear to be completely sterile. They have
been found at scattered locations in western North America;
although not reported from the Intermountain Region, they
can be expected because both parents are present.

3. **×Elyhordeum californicum** (Bowden)
   Barkworth [p. 325]
Clm 30–40 cm, glab throughout or hairy on the
lo portion. **Shth** glab; **aur** to 0.5 mm; **lig** 0.4–1.5
mm, rounded; **bld** to about 15 cm long, 1–2(3)
mm wide, abx surfaces glab or sparsely hairy,
adx surfaces scabridulous, all veins equally
prominent. **Infl** spikelike, 2.5–5 cm long, 7–10
mm wide excluding the awns, 20–30 mm wide
including the awns, straight, nd with 3 spklt;
intnd 2.2–4 mm, bases about ½ as wide as the
apc. **Spklt** 8–10 mm excluding the awns, 25–28
mm including the awns, with 1–2 flt, the lat spklt
shortly pedlt, the cent spklt sessile. **Glm** 15–35
mm including the awns, setaceous, straight; **lm**
7–8 mm, smooth throughout or scabrous distally,
awned, awns 10–20 mm, straight or slightly
divergent at maturity; **anth** about 1.5 mm.

×Elyhordeum californicum consists of hybrids between
Elymus elymoides or *E. multisetus* and *Hordeum brachyantherum*
subsp. *brachyantherum*. It was described by Bowden (1958) on
the basis of specimens collected in California. It is included
because it seems probable that it will be found at many other
locations where the two parents grow together.

7.12 **ELYMUS** L.   Mary E. Barkworth, Julian J.N. Campbell, and Björn Salomon

Pl per; smt csp, smt rhz, smt stln. **Clm** 8–180(220) cm, usu erect to ascending, smt strongly decumbent to prostrate, usu glab. **Lvs** usu evenly distributed, smt somewhat bas concentrated; **shth** open for most of their length; **aur** often present; **lig** memb, usu truncate or rounded, smt acute, entire or erose, often ciliolate; **bld** 1–24(25) mm wide, abx surfaces usu smooth or scabrous, smt with hairs, adx surfaces scabrous or with hairs, particularly over the veins, usu with unequal, not strongly ribbed, widely spaced veins, smt with equal, strongly ribbed, closely spaced veins. **Infl** spikes, usu exserted, with 1–3(5) spklt per nd, intnd (1.5)2–26 mm; **rchs** with scabridulous, scabrous, or ciliate edges. **Spklt** usu appressed to ascending, smt strongly divergent or patent, with 1–11 flt, the lowest flt usu fnctl, strl and glmlike in some species, the distal flt often rdcd; **dis** usu above the glm and beneath each flt, smt also below the glm or in the rchs. **Glm** usu 2, absent or highly rdcd in some species, usu equal to subequal, smt unequal, usu linear-lanceolate to linear, setaceous, or subulate, smt oblanceolate to obovate, (0)1–7-veined, smt keeled over 1 vein, not necessarily the cent vein, keel vein smt extending into an awn; **lm** linear-lanceolate, obscurely 5(7)-veined, apc acute, often awned, smt bidentate, teeth to 0.2 mm, smt with bristles, bristles to 10 mm, awns tml or from the sinus, straight or arcuately divergent, not geniculate; **pal** from shorter than to slightly longer than the lm, keels scabrous or ciliate, at least in part; **anth** 3, 0.7–7 mm. **Car** with hairy apc. $x = 7$. **Haplomes St, H, Y.** Name from the Greek *elyo*, 'rolled up', the caryopses being tightly embraced by the lemmas and paleas.

As interpreted here, *Elymus* is a widespread, north-temperate genus of about 150 species. It includes *Sitanion* Raf., but moves some taxa that others include in *Elymus* to *Leymus, Pascopyrum, Pseudoroegneria,* and *Thinopyrum*. Twelve species of *Elymus* are native to the Intermountain Region. Three named, naturally occurring, interspecific hybrids are described at the end of the treatment. They are not included in the key. Other interspecific hybrids undoubtedly exist. Because many of the hybrids are partially fertile, backcrossing and introgression occurs. Intergeneric hybrids are treated under ×*Elyhordeum* (see previous), ×*Elyleymus* (p. 73), and ×*Pseudelymus* (p. 60); most are sterile.

The complex patterns of morphological diversity within *Elymus* in North America probably reflect a combination of multiple origins involving different progenitors, introgression, hybridization both within the genus and with other members of the tribe, and morphological plasticity. Little is known concerning the relative importance of these factors. Two infraspecific ranks have been used to aid in circumscribing the known variation. In general, infraspecies taxa that show great morphological and ecological distinction are treated as subspecies; others, as varieties.

All species of *Elymus* are alloploids that combine one copy of the **St** haplome present in *Pseudoroegneria* with at least one other haplome. So far as is known, all species that are native to North America, as well as many species native to northern Eurasia, are tetraploids with one additional haplome, the **H** genome from *Hordeum* sect. *Critesion*. *Elymus repens* and *E. hoffmannii*, the other two hexaploid species in this treatment, basically combine two copies of the **St** haplome with one of the **H** haplome, but the molecular data for *E. repens* point to a more complex situation (Mason-Gamer 2001). For further discussion of generic delimitation in the *Triticeae*, see Barkworth (2000), Yen et al. (2005), and Barkworth and von Bothmer (2005).

In the key and descriptions, unless otherwise stated, the following conventions are observed: the number of culm nodes refers to the number of nodes above the base; measurements of spikes include the awns, while measurements of spikelets, glumes, and lemmas do not; rachis internodes are measured in the middle of the spike; glume widths of lanceolate to linear glumes are measured at the widest point, and those of linear to setaceous glumes about 5 mm above the base of the glumes; the number of florets in a spikelet includes the distal reduced, sterile florets; dates of anthesis, when provided, are for the central range of each species.

The curvature of the lemma awns is often important in identifying individual species. The curvature increases with maturity, and may vary within a spike. If a plant appears to have at least some strongly curved lemma awns, it should be taken through the "strongly curved" side of the key.

1. Spikelets 2–3(3) at all or most nodes; glume bodies linear-lanceolate to setaceous or subulate, margins usually firm, sometimes hyaline or scarious; lemmas usually awned, awns to 120 mm long.
    2. Rachises disarticulating at maturity; glumes 10–135 mm long (including the awns), sometimes longitudinally split, flexuous or outcurving from the base; lowest floret in each spikelet often sterile, glumelike; blades 1–6 mm wide.
        3. Glumes entire or split into 2–3 divisions; lemma awns about 0.4 mm wide at the base; rachis internodes 3–10(15) mm long. . . . . 6. *E. elymoides*
        3. Glumes split into 3–9 divisions; lemma awns about 0.2 mm wide at the base; rachis internodes 3–5 mm long . . . . 5. *E. multisetus*
    2. Rachises not disarticulating at maturity; glumes 7–40 mm long (including the awns), entire, straight; lowest floret in each spikelet functional; blades 2–18 wide.
        4. Glume bases strongly bowed out, without evident veins and more or less terete for at least 1 mm; lemma awns shorter than 5 mm . . . . 1. *E. curvatus*

    4. Glume bases more or less straight, usually with evident veins and more or less flat to
       within 1 mm of the base; lemma awns usually longer than 5(3)6–40(50) mm.
       5. Spikes strongly drooping; glume margins abrupt, neither hyaline nor scarious. . . . . 2. *E. canadensis*
       5. Spikes erect or nodding; glume margins hyaline or scarious.
          6. Glume bodies (6)9–14(19) mm long; lemmas 8–16 mm long, awns usually
             straight to flexuous; auricles usually present, to 2.5 mm long . . . . 3. *E. glaucus* (in part)
          6. Glume bodies 6–9 mm long; lemmas (5)7–11 mm long, awns outcurving
             from near the base; auricles minute or absent . . . . 4. *E. dahuricus*
1. Spikelets 1 at all or most nodes; glume bodies linear-lanceolate to ovate or obovate, margins
  hyaline, scarious, or chartaceous; lemmas awned or unawned.
    7. Anthers 3–7 mm long; plants often strongly rhizomatous, sometimes not or only weakly
      so.
      8. At least some lemmas with strongly divergent, outcurving, or reflexed awns.
         9. Culms prostrate to decumbent and geniculate, 20–50 cm tall; plants not
            rhizomatous, growing in subalpine and alpine habitats . . . . 12. *E. sierrae* (in part)
         9. Culms erect or decumbent only at the base, (15)40–130 cm tall; plants often
            rhizomatous, growing in valley and midmontane habitats.
           10. Plants strongly rhizomatous; lemma awns 4–12 mm long; blades 1–3 mm
              wide . . . . 14. *E. albicans*
           10. Plants not or weakly rhizomatous; lemma awns 9–28 mm long; blades
              1.7–5 mm wide . . . . .13. *E. wawawaiensis*
      8. Lemmas unawned or with straight to flexuous awns.
        11. Glumes keeled distally, keels smooth and inconspicuous on the lower portion,
          conspicuous and scabrous on the distal portion; lemmas glabrous.
          12. Adaxial surfaces of the blades usually sparsely pilose, sometimes glabrous,
             veins smooth, the primary veins separated by secondary veins; plants
             strongly rhizomatous . . . . .15. *E. repens*
          12. Adaxial surfaces of the blades glabrous, veins smooth or scabrous, all
             veins more or less equally prominent; plants slightly to moderately
             rhizomatous . . . . 16. *E. hoffmannii*
        11. Glumes not keeled or keeled throughout their length, keels smooth or scabrous
          their whole length, sometimes hairy, conspicuous or not; lemmas glabrous or
          hairy.
          13. Plants strongly rhizomatous; lemmas acute to awn-tipped, the awns up to
             2 mm long, glumes 5–14 mm long, ½—¾ the length of the adjacent lemmas
             . . . . 9. *E. lanceolatus* (in part)
          13. Plants not or weakly rhizomatous; lemmas usually awned, awns 1–35 mm
             long; glumes 6–19 mm long, from ¾ as long as to equaling the length of the
             adjacent lemmas.
            14. Spikelets usually twice as long as the internodes; midspike internodes
              4–12 mm long; glumes often awned, sometimes unawned; blades
              usually lax . . . . 3. *E. glaucus* (in part)
            14. Spikelets from shorter than to almost twice as long as the internodes;
              midspike internodes 9–27 mm long; glumes unawned; blades usually
              straight . . . . 10. *E. stebbinsii*
    7. Anthers 0.7–3 mm long; plants usually not or weakly rhizomatous, sometimes strongly
      rhizomatous.
      15. Culms prostrate to decumbent and geniculate, 20–50 cm tall; lemma awns 15–30
        mm long; plants of subalpine and alpine habitats.
        16. Anthers 1–1.6 mm long; internodes 2.5–5(7) mm long; disarticulation initially
          in the rachises; spikelets appressed to ascending . . . . .11. *E. scribneri*
        16. Anthers 2–3.5 mm long; internodes 5–15 mm long; rachises not disarticulating;
          spikelets ascending to divergent . . . . 12. *E. sierrae* (in part)
      15. Culms usually ascending to erect, sometimes geniculate or weakly decumbent at
        the base; lemmas unawned or with awns to 40 mm long; plants of valley to alpine
        habitats.
        17. Plants strongly rhizomatous; anthers (2.5)3–6 mm long . . . . 9. *E. lanceolatus* (in part)
        17. Plants not or weakly rhizomatous; anthers 0.7–3.5 mm long.

18. Glumes 3(5)-veined; glume margins unequal, the wider margins 0.3–1 mm wide, usually widest in the distal third; lemma awns 0.5–3 mm long.... 8. *E. violaceus*
18. Glumes 3–7-veined; glume margins about equal, 0.1–0.5 mm wide, widest at or slightly beyond midlength; lemmas unawned or with awns to 40 mm long.
    19. Glumes 1.8–2.3 mm wide, margins 0.2–0.3 mm wide............... 7. *E. trachycaulus*
    19. Glumes 0.4–1.5(2) mm wide, margins 0.1–0.2 mm wide .... 3. *E. glaucus* (in part)

1. Elymus curvatus Piper AWNLESS WILDRYE [p. 326, 497]

Pl ces, not rhz, often glaucous. **Clm** 60–110 cm, stiffly erect, or the base smt geniculate; **nd** 6–9, concealed or exposed, glab. **Lvs** evenly distributed; **shth** glab, often reddish brown; **aur** to 1 mm, smt absent; **lig** shorter than 1 mm, ciliolate; **bld** 5–15 mm wide, lo bld usu lax, shorter, narrower, and senescing earlier than up bld, up bld usu ascending and somewhat involute, adx surfaces smooth or scabridulous, occ scabrous. **Spikes** 9–15 cm long, (0.5)0.7–1.3 cm wide, erect, exserted or the bases slightly sheathed, with 2 spklt per nd; **intnd** 2.5–4.5 mm long, about 0.25–5 mm thick at the thinnest sections, smooth or scabrous beneath the spklt. **Spklt** 10–15 mm, appressed, often reddish brown at maturity, with (2)3–4(5) flt, lowest flt fnctl; **dis** below the glm and beneath the flt, or the lowest flt falling with the glm. **Glm** equal or subequal, the bas 2–3 mm terete, indurate, strongly bowed out, without evident venation, glm bodies 7–15 mm long, 1.2–2.1 mm wide, linear-lanceolate, widening above the base, 3–5-veined, usu glab or scabrous, occ hispidulous, rarely hirsute on the veins, mrg firm, awns 0–3(5) mm; **lm** 6–10 mm, glab or scabrous, rarely hirsute, awns (0.5)1–3(4) mm, rarely 5–10 mm on the lm of the distal spklt, straight; **pal** 6–10 mm, obtuse, often emgt; **anth** 1.5–3 mm. **Anthesis** late June to mid-August. 2*n* = 28, 42.

Elymus curvatus grows in moist or damp soils of open forests, thickets, grasslands, ditches, and disturbed ground, especially on bottomland. It is widespread from British Columbia and Washington, through the Intermountain Region and northern Rockies, to the northern Great Plains. It is infrequent or rare in the midwest, the Great Lakes region, and the northeast, and is virtually unknown in the southeast.

2. Elymus canadensis L. GREAT PLAINS WILDRYE [p. 326, 497]

Pl loosely ces, rarely with rhz to 4 cm long and 1–2 mm thick, often glaucous. **Clm** (40)60–150(180) cm, erect or decumbent; **nd** 4–10, mostly concealed by the lf shth, glab. **Lvs** evenly distributed; **shth** smooth or scabridulous, glab or hirsute, often reddish brown; **aur** 1.5–4 mm, brown or purplish black; **lig** to 1(2) mm, truncate, ciliolate; **bld** (3)4–15(20) mm wide, usu firm, often ascending and somewhat involute, usu dull green, drying to grayish, adx surfaces usu smooth or scabridulous and glab, rarely sparsely hispid to villous. **Spikes** 6–30 cm long, 3–7 cm wide, usu nodding, smt pendent or almost erect, usu with 2(3) spklt per nd, occ to 5 at some nd, rarely with 1 at some nd but never throughout; **intnd** (2)3–5(7) mm long, or 5–10 mm long towards the base, 0.2–0.35 mm thick at the thinnest sections, glab or with a few hairs below the spklt. **Spklt** 12–20 mm excluding the awns, more or less divergent, with (2)3–5(7) flt, lowest flt fnctl; **dis** usu above the glm and beneath each flt, rarely also below the glm. **Glm** usu equal, occ subequal, 11–40 mm including the awns, the bas 0–1 mm subterete and slightly indurate, glm bodies 6–13 mm long, 0.5–1.6 mm wide, linear-lanceolate to subsetaceous, entire, widening or parallel-sided above the base, 3–5-veined, glab to scabrous-ciliate, rarely villous on the veins, mrg firm, awns (5)10–25(27) mm, straight to outcurving; **lm** 8–15 mm, glab, scabrous, hispid, or uniformly villous with the hairs generally appressed, awns (10)15–40(50) mm, moderately to strongly outcurving, often contorted at the spike bases; **pal** 7–13 mm, acute, usu bidentate; **anth** 2–3.5 mm. **Anthesis** May to July. 2*n* = 28, rarely 42.

Elymus canadensis grows on dry to moist or damp, often sandy or gravelly soil on grasslands, dunes, stream banks, ditches, roadsides, and disturbed ground. Two of its varieties grow in the Intermountain Region.

1. Lemmas usually smooth or scabridulous, occasionally hirsute; spikes usually nodding, occasionally almost erect; internodes 3–4 mm long, not strongly glaucous.............. var. *brachystachys*
1. Lemmas usually villous or hispid; spikes nodding to almost pendent; internodes 4–7 mm long, often strongly glaucous .... var. *canadensis*

Elymus canadensis var. brachystachys (Scribn. & C.R. Ball) Farw. [p. 326]

Spikes 6–20 cm, nodding, not strongly glaucous, often becoming yellowish or pale reddish brown, rarely with 3 spklt per nd; **intnd** mostly 3–4 mm. **Glm** not clearly indurate or bowed out at the base, awns 10–20 mm; **lm** smooth or scabridulous, awns usu 20–30 mm, moderately outcurving.

Elymus canadensis var. *brachystachys* is widespread in the southern Great Plains from Nebraska to Mexico, where anthesis is from March to early June. It also occurs sporadically as far north as southern Canada, from British Columbia to Quebec.

Elymus canadensis L. var. **canadensis** [p. 326]
Spikes (6)10–25(30) cm, nodding to almost pendent, often strongly glaucous, often with 3 spklt per nd; **intnd** 4–7 mm. **Glm** not clearly indurate or bowed out at the base, awns 10–25 mm; **lm** villous or hispid, awns 15–40 mm, moderately to strongly outcurving.

Elymus canadensis var. *canadensis* is most abundant in the eastern portion of the Intermountain Region, but even there it is not common.

3. **Elymus glaucus** Buckley COMMON WESTERN WILDRYE, BLUE WILDRYE [p. 327, 497]
Pl densely to loosely ces, smt weakly rhz, often glaucous. **Clm** 30–140 cm, erect or slightly decumbent; **nd** 4–7, mostly exposed, usu glab, smt puberulent. **Lvs** evenly distributed; **shth** scabrous or smooth, glab or, particularly those of the lo lvs, retrorsely puberulent to hirsute, often purplish; **aur** usu present, to 2.5 mm, often purplish; **lig** to 1 mm, truncate, erose-ciliolate or entire; **bld** 2–13(17) mm wide, usu lax, smt slightly involute, adx surfaces glab, scabrous, or strigose on the veins, smt pilose to villous. **Spikes** 5–21 cm long, (0.2)0.5–2 cm wide, erect to slightly nodding, rarely somewhat pendent, usu with 2 spklt per nd, smt with 1 at all or most nd, rarely with 3 at some nd; **intnd** 4–8(12) mm long, 0.15–0.5 mm thick at the thinnest sections, angles scabrous, glab below the spklt. **Spklt** 8–25 mm, smt purplish at higher latitudes and elevations, appressed to slightly divergent, with (1)2–4(6) flt, lowest flt fnctl; **dis** above the glm and beneath each flt. **Glm** subequal, ¾ as long as or equaling the adjacent lm, bases often overlapping, usu flat and thin with evident venation, glm bodies (6)9–14(19) mm long, 0.6–1.5(2) mm wide, linear-lanceolate, entire, widening above the base, (1)3–5(7)-veined, 2–3 veins extending to the apc, glab, veins smooth or evenly scabrous, mrg 0.1–0.2 mm wide, whitish hyaline, tapering towards the apc, unawned or awned, awns to 5(9) mm, straight; **lm** (8)9–14(16) mm, glab, scabrous, or short-hirsute, awns (0)1–30(35) mm, usu straight to flexuous, smt slightly curving; **pal** 7–13 mm, keels straight or slightly concave, usu scabrous

to ciliate, apc often bidentate; **anth** 1.5–3.5 mm. **Anthesis** from May to July. $2n = 28$.

Elymus glaucus grows in moist to dry soil in meadows, thickets, and open woods. It is widespread in western North America, including the Intermountain Region. Populations can differ greatly in morphology, especially in rhizome development, leaf width, pubescence, and the prevalence of solitary spikelets. Their crossing relationships are partly correlated with such variation (Snyder 1950, 1951; Stebbins 1957, Wilson et al. 2001). Rhizome development and the production of solitary spikelets may also be environmental responses. Rhizomatous plants are more common on unstable slopes or sandy soils. Plants with solitary spikelets are more common on poor soil or in shade. In the herbarium, specimens with one spikelet per node are sometimes difficult to distinguish from *E. trachycaulus*, but there is usually little problem in the field, *E. glaucus* having more evenly leafy culms, laxer and wider blades, more tapered glumes that are always awn-tipped or awned, and shorter anthers. Only one subspecies of *E. glaucus* grows in the Intermountain Region.

Elymus glaucus hybridizes with several other species of *Elymus*, including *E. elymoides* (p. 66), *E. multisetus* (p. 66) (see *E.* ×*hansenii*, p. 73), *E. trachycaulus* (p. 67), and *E. stebbinsii* (p. 70). These hybrids often appear at least partially fertile. *Elymus glaucus* can also form intergeneric hybrids with *Leymus* and *Hordeum* (see ×*Elyleymus*, p. 73, and ×*Leydeum*, p. 78).

Elymus glaucus Buckley subsp. **glaucus** [p. 327]
Shth glab, scabrous or pubescent; **bld** 4–17 mm wide, adx surfaces glab or strigose, occ pilose to hirsute with hairs of fairly uniform length. **Glm** awns (0.5)1–5(9) mm; **lm** awns (5)10–25(35) mm.

Elymus glaucus subsp. *glaucus* grows throughout the range of the species, from sea level to 2500 m.

4. **Elymus dahuricus** Turcz. *ex* Griseb. [p. 327]
Pl ces, not rhz, often glaucous. **Clm** 30–130 cm, erect; **nd** 4–7, mostly exposed, usu glab, occ short-hairy. **Lvs** evenly distributed; **shth** glab; **aur** minute or absent; **lig** 0.5–1 mm; **bld** 3–18 mm wide, lax, usu pale green, smt glaucous, adx surfaces usu smooth or scabrous on the veins, smt sparsely pilose. **Spikes** 7–23 cm long, 1–2.5 cm wide, usu slightly nodding, smt erect, usu with 2 spklt per nd, occ with 1 spklt at some nd; **intnd** 3–6 mm long, 0.2–0.8 mm thick at the thinnest sections, angles usu with scattered hairs. **Spklt** 10–15 mm, appressed to divergent, often purplish, with (2)3–4(5) flt, lowest flt fnctl; **dis** above the glm, beneath each flt. **Glm** equal, the bases flat, not indurate, veins evident, glm bodies 6–9 mm long, 1–1.5 mm wide, linear-lanceolate, entire, widening or parallel-sided above the base, (1)3–5(7)-veined, veins scabrous, mrg hyaline or scarious, awns (0)1–5 mm, straight or outcurving; **lm** (5)7–11 mm, usu glab

and smooth throughout, smt scabrous to hispid distally and on the mrg, mrgl hairs not markedly longer than those elsewhere, awns (3)6–17(20) mm, usu somewhat outcurving from near the base; **pal** 7–11 mm, keels spinose-ciliate, apc obtuse or truncate; **anth** 1.5–3.5 mm. **Anthesis** from May to July. $2n = 42$.

Elymus dahuricus is widespread in temperate central and eastern Asia. It is a hexaploid with an **StYH** genome constitution. It has been introduced for reclamation in some parts of western North America, possibly including the Intermountain Region. It is most likely to be confused with *E. glaucus* (see previous), from which it differs in its palea shape. Several varieties have been described in Asia; only *Elymus dahuricus* Turcz. *ex* Griseb. var. *dahuricus* has been introduced to North America.

5. Elymus multisetus (J.G. Sm.) Burtt Davy
BIG SQUIRRELTAIL [p. *328*, 497]

Pl ces, not rhz. **Clm** 15–65 cm, erect to ascending, usu puberulent; **nd** 4–6, mostly concealed, glab. **Lvs** evenly distributed; **shth** glab or white-villous; **aur** usu present, 0.5–1.5 mm; **lig** to 1 mm, truncate, entire or lacerate; **bld** 1.5–4(5) mm wide, often ascending and involute, adx surfaces scabrous, pilose, or villous. **Spikes** 5–20 cm long, 5–15 cm wide, erect, smt partially enclosed at the base, with 2 spklt per nd, rarely with 3–4 at some nd; **intnd** 3–5(8) mm long, 0.1–0.3 mm thick at the thinnest sections, glab beneath the spklt. **Spklt** 10–15 mm, divergent, with 2–4 flt, lowest flt strl and glmlike in 1 or both spklt at each nd; **dis** initially at the rchs nd, subsequently beneath each flt. **Glm** subequal, (10)30–100 mm including the awns, the bases indurate and glab, glm bodies (2)5–10 mm long, 1–2 mm wide, setaceous, 2–3-veined, mrg firm, awns (8)25–90 mm, each split into 3–9 unequal divisions, scabrous, flexuous to outcurving from near the glm bases at maturity; **ftl lm** 8–10 mm, smooth or scabrous near the apc, 2 lat veins extending into bristles to 10 mm, awns (10)20–110 mm long, about 0.2 mm wide at the base, divergent to arcuate; **pal** 7–9 mm, veins usu extending into about 1 mm bristles, apc acute to truncate; **anth** 1–2 mm. **Anthesis** from late May to June. $2n = 28$.

Elymus multisetus grows in dry, often rocky, open woods and thickets on slopes and plains in western North America at elevations up to 2000 m. It usually grows in less arid habitats than *E. elymoides* subsp. *elymoides* (p. 67), but the two taxa are sometimes sympatric.

Wilson (1963) reported a wide belt of introgression between *Elymus multisetus* and *E. elymoides* subsp. *elymoides* from southeastern California to southern Nevada, but not in other areas where they are sympatric. There are also probable hybrids with *E. glaucus* (p. 65) and *Pseudoroegneria spicata* (p. 60).

6. Elymus elymoides (Raf.) Swezey [p. *328*, *329*, 497]

Pl ces, often glaucous, not rhz. **Clm** 8–65 (77) cm, erect or geniculate to slightly decumbent, smt puberulent; **nd** 4–6, mostly concealed, usu glab, smt pubescent. **Lvs** evenly distributed; **shth** glab, scabrous, puberulent, or densely white-villous; **aur** usu present, to about 1 mm, often purplish; **lig** shorter than 1 mm, truncate, entire or lacerate; **bld** (1)2–4(6) mm wide, spreading or ascending, often involute, smt folded, abx surfaces glab to puberulent, adx surfaces scabrous, puberulent, hirsute, or white-villous. **Spikes** 3–20 cm long, 5–15 cm wide, erect to sub-flexuous, with 2–3 spklt per nd, rarely with 1 at some nd; **intnd** 3–10(15) mm long, 0.1–0.4 mm thick at the thinnest sections, usu glab, smt puberulent beneath the spklt. **Spklt** 10–20 mm, divergent, smt glaucous, at least 1 spklt at a nd with 2–4(5) flt, 1–4(5) flt ftl, smt all flt strl in the lat spklt; **dis** initially at the rchs nd, subsequently beneath each flt. **Glm** subequal, 20–135 mm including the often undifferentiated awns, the bases indurate and glab, glm bodies 5–10 mm long, 1–3 mm wide, linear to setaceous, 1–3-veined, mrg firm, awns 15–125 mm, scabrous, smt split into 2–3 unequal divisions, flexuous to outcurving from near the base at maturity; **ftl lm** 6–12 mm, glab, scabrous, or appressed-pubescent, 2 lat veins extending into bristles to 10 mm, awns 15–120 mm long, about 0.4 mm wide at the base, often reddish or purplish, scabrous, flexuous to curved near the base; **pal** 6–11 mm, veins often extending into bristles to 2(5) mm, apc acute to truncate; **anth** 0.9–2.2 mm. **Anthesis** from late May to July. $2n = 28$.

Elymus elymoides grows in dry, often rocky, open woods, thickets, grasslands, and disturbed areas, from sagebrush deserts to alpine tundra. It is widespread in western North America, from British Columbia to northern Mexico and the western Great Plains. It is often dominant in overgrazed pinyon-juniper woodlands. Although palatable early in the season, the disarticulating, long-awned spikes irritate grazing animals later in the year. All four subspecies grow in the Intermountain Region.

Elymus elymoides intergrades with *E. multisetus* (see previous) in parts of its southern range (Wilson 1963). It is sometimes confused with *E. scribneri* (p. 70), but differs in having more than one spikelet per node, narrower glumes, and less tardily disarticulating rachises. Hybrids with several other species in the *Triticeae* are known; they can often be recognized by their tardily disarticulating rachises. Named interspecific hybrids (pp. 72–73) (and the other parent) are *E. ×saundersii* (*E. trachycaulus*) and possibly *E. ×hansenii* (*E. elymoides* or *E. multisetus* × *E. glaucus*). Hybrids with *E. sierrae* have not been named; they are common where the two species are sympatric. They have broader glume bases, shorter glume awns, and longer anthers than *E. elymoides*.

1. Rachis nodes with 3 spikelets, the central spikelet usually with 2 fertile florets, the florets of the lateral spikelets rudimentary to awnlike; lemma awns 15–30 mm long. subsp. *hordeoides*
1. Rachis nodes usually with 2 spikelets, each spikelet usually with (1)2–4(5) fertile florets; lemma awns 15–120 mm long.
   2. No spikelets appearing to have 3 glumes, the lowermost floret in each spikelet well developed; paleas rarely with the veins extended as bristles. . . . . subsp. *brevifolius*
   2. One or more of the spikelets at most nodes appearing to have 3 glumes, the lowest 1–2 florets sterile and glumelike; paleas usually with the veins extended as bristles.
      3. Glumes with awns 15–70 mm long, all glumes entire . . . . subsp. *californicus*
      3. Glumes with awns 35–85 mm long, one of the glumes at most nodes with the awn split into 2 or 3 divisions. . . . . subsp. *elymoides*

Elymus elymoides subsp. brevifolius (J.G. Sm.) Barkworth LONGLEAF SQUIRRELTAIL [p. 328]

Clm 25–65(77) cm, erect. **Bld** usu puberulent abx, smt glab. **Spikes** 7–20 cm, usu exserted, usu with 2 spklt per nd. **Spklt** with (1)2–4(5) flt, lowermost flt fnctl. **Glm** awns 50–125 mm, entire; **lm** awns 50–120 mm; **pal** rarely with the veins extended as bristles.

Elymus elymoides subsp. *brevifolius* has a wide ecological and elevation range, extending from the arid Sonoran Desert to subalpine habitats, from 600–3500 m.

Elymus elymoides subsp. californicus (J.G. Sm.) Barkworth CALIFORNIA SQUIRRELTAIL [p. 328]

Clm 8–40 cm, erect or decumbent. **Bld** usu glab abx, smt puberulent. **Spikes** 3–10 cm, often partly included, usu with 2 spklt per nd. **Spklt** with (1)2–3 ftl flt, lowest 1–2 flt strl and glmlike. **Glm** awns 15–40(70) mm, entire; **lm** awns 25–70 mm, usu exceeding those of the glm; **pal** often with the veins extended as 1–2 mm bristles.

Elymus elymoides subsp. *californicus* grows in midmontane to arctic-alpine habitats, at elevations of 1500–4200 m. Plants transitional to subsp. *elymoides* occur where the two are sympatric.

Elymus elymoides (Raf.) Swezey subsp. elymoides COMMON SQUIRRELTAIL [p. 329]

Clm 15–45 cm, erect to decumbent. **Bld** usu puberulent abx, smt glab. **Spikes** 4–15 cm, exserted or partly included, usu with 2 spklt per nd. **Spklt** with (1)2–3(4) ftl flt, lowest 1–2 flt strl and glmlike. **Glm** awns 35–85 mm, often split into 2, smt 3, unequal divisions; **lm** awns 25–75 mm, usu exceeded by those of the glm; **pal** with the veins extended as bristles.

Elymus elymoides subsp. *elymoides* grows in desert and shrub-steppe areas of western North America. It is frequently associated with disturbed sites and is the most commonly encountered subspecies.

Elymus elymoides subsp. hordeoides (Suksd.) Barkworth [p. 329]

Clm 10–20 cm, erect. **Bld** glab or puberulent abx. **Spikes** 3–6 cm, exserted or partly included, with 3 spklt per nd. **Spklt** in the cent position usu with 2 ftl flt, the lat spklt usu with rdmt to awnlike flt. **Glm** awns 15–50 mm, usu entire; **lm** awns of ftl flt 15–30 mm; **pal** with or without distinct bristles.

Elymus elymoides subsp. *hordeoides* grows in dry, rocky, often shallow soils, particularly in *Artemisia rigida–Poa secunda* communities, from eastern Washington and Idaho to northern California and Nevada. It resembles some *Elymus–Hordeum* hybrids. It has been found in two locations in the Intermountain Region, one in Elmore County, Idaho, and one in Pershing County, Nevada.

7. Elymus trachycaulus (Link) Gould [p. 330, 497]

Pl usu ces, smt weakly rhz. **Clm** 30–150 cm, ascending to erect; **nd** usu glab. **Lvs** somewhat bas concentrated; **shth** usu glab, smt markedly retrorsely hirsute or villous; **aur** absent or to 1 mm; **lig** 0.2–0.8 mm, truncate; **bld** 2–5(8) mm wide, flat to involute, usu straight and ascending, abx surfaces usu smooth and glab, smt hairy, adx surfaces usu glab, smt conspicuously hairy. **Spikes** 4–25 cm long, 0.4–1 cm wide, erect, with 1 spklt at all or most nd; **intnd** (4)7–9(12) mm, edges scabrous, both surfaces smooth and glab. **Spklt** 9–17(20) mm long, usu at least twice as long as the intnd, 3–6 mm wide, appressed, with 3–9 flt, lowest flt fnctl; **rchl** glab or hairy, hairs to 0.3 mm; **dis** above the glm, beneath each flt. **Glm** subequal, 5–17 mm long, from ¾ as long as to longer than the adjacent lm, 1.8–2.3 mm wide, lanceolate to narrowly ovate, widest about midlength, usu green, purple at higher latitudes and elevations, flat or asymmetrically keeled for their full length, 3–7-veined, the keel vein usu scabrous, the others smooth or scabrous, only 1 vein extending to the apex, adx surfaces glab, mrg hyaline or scarious, usu more or less equal, 0.2–0.5 mm wide, widest at or slightly beyond midlength, apc acute to awned, awns to 11 mm; **lm** 6–13 mm, glab, usu smooth proximally, often scabridulous distally over the veins, apc acute, usu awned, awns to 40 mm, usu straight, smt weakly curved if shorter than 10 mm; **pal** subequal to the lm, keels straight or slightly outwardly curved below the apc, tapering to the apc, apc truncate, 0.15–0.3 mm wide, keel veins often extending beyond the intercostal region, smt forming teeth; **anth** (0.8)1.2–2.5 (3) mm. 2n = 28.

Elymus trachycaulus grows from sea level to 3300 m, usually in open or moderately open areas, but sometimes in forests. Its range extends from the boreal forests of North America east through Canada to Greenland and south into Mexico. It exhibits considerable variability in the presence or absence of rhizomes, the length and density of the spike, awn development on the glumes and lemmas, and glume venation. The variability in these features has often been used to circumscribe infraspecific taxa, but most such taxa, even though locally distinctive, appear to intergrade. Some of the features appear to be strongly influenced by environmental factors. For instance, plants growing in forested areas of northwestern North America tend to be slightly rhizomatous, more gracile, and later-flowering that those in adjacent, more exposed areas; whether they constitute a distinct taxon or merely a forest ecotype is not clear. Plants growing at higher elevations tend to have glumes with more widely spaced veins and broader, often unequal margins, resembling *E. violaceus* in these respects. Whether this reflects ecotypic differentiation, hybridization with *E. violaceus* (see next), or greater genetic continuity than is suggested by their placement in different species is not clear.

Jozwik (1966) recognized four groups within *Elymus trachycaulus*. Group I comprised unawned or shortly awned specimens; group II a polymorphic assemblage of awned specimens; group III a rather homomorphic group of specimens with secund spikes and relatively long awns; and group IV a relatively homomorphic group of unawned, high-elevation specimens. He concluded that group II consists of hybrids and backcrosses between *E. trachycaulus* and other species of *Triticeae*. He based this conclusion on consideration of field observations, artificial hybrids, the polymorphism of the specimens, and the geographic distribution of the group. This last was similar to that of unawned specimens of *E. trachycaulus*, but the populations were highly scattered within the area concerned. Jozwik's group III is treated here as *E. trachycaulus* subsp. *subsecundus*. His group IV is treated here as *E. violaceus*.

Elymus trachycaulus has been implicated in several interspecific and intergeneric hybrids. Named interspecific hybrids (pp. 72–73) (and the other parent) are *E. ×pseudorepens* (*E. lanceolatus*), and *E. ×saundersii* (*E. elymoides*). There is one named intergeneric hybrid in the Intermountain Region, *×Elyhordeum macounii* (p. 61) (*Hordeum jubatum*). Hybrids with *Elymus elymoides*, *E. multisetus*, and *Hordeum jubatum* have brittle rachises and tend to be awned. Others are harder to recognize.

1. Lemma awns 17–40 mm long, longer than the lemma body, straight; spikes somewhat 1-sided
. . . . subsp. *subsecundus*
1. Lemmas unawned or with awns to 24 mm long, shorter or longer than the lemma body, straight or curved; spikes 2-sided.
    2. Lemma awns 9–24 mm long . . *E. trachycaulus* hybrids
    2. Lemmas unawned or with awns to 9 mm long, the awns sometimes curved . . . . subsp. *trachycaulus*

Elymus trachycaulus subsp. **subsecundus** (Link) Á. Löve & D. Löve ONE-SIDED WHEATGRASS [p. *330*]
Clm 40–110 cm. **Spikes** 7–25 cm, somewhat 1-sided. **Spklt** with 3–7 flt, the bases usu visible.

Glm 11–17 mm, long-acuminate or awned, awns to 11 mm; **lm** awned, awns 17–40 mm, longer than the lm body, straight.

Elymus trachycaulus subsp. *subsecundus* grows primarily in the Great Plains. It differs from plants of *E. glaucus* (p. 65) with solitary spikelets, in its 1-sided spike and stiffer, more basally concentrated leaves. It may comprise derivatives of *E. trachycaulus* subsp. *trachycaulus* × *Hordeum jubatum* (p. 53) hybrids that are adapted to moist prairies. The unilateral spike is particularly characteristic of artificial hybrids between the two species, and is uncommon in other hybrids (Jozwik 1966).

Elymus trachycaulus (Link) Gould subsp. **trachycaulus** SLENDER WHEATGRASS [p. *330*]
Clm 30–150 cm. **Spikes** (4)8–30 cm long, 0.5–0.8 cm wide, 2-sided; **intnd** 8–15 mm. **Spklt** with 3–9 flt, the bases usu visible. **Glm** 5–17 mm, at least 1 vein scabrous to near the base, smt all veins scabrous, unawned or with straight awns shorter than 2 mm; **lm** unawned or awned, awns to 5 mm, straight.

Elymus trachycaulus subsp. *trachycaulus* grows throughout the habitat and range of the species, and exhibits considerably more variation than subsp. *subsecundus*. Two aspects of the variation that seem particularly worthy of further study are the glume venation and the spacing of spikelets in the spikes. Plants with glumes having 5–7 well-developed, narrowly spaced veins are restricted to lower elevations and the southern portion of the subspecies range; northern plants and plants at higher elevations generally have 3–5 weakly developed and widely spaced veins. Glumes of the former resemble those of *E. glaucus*; the glumes of the latter, those of *E. violaceus* (see next). This may reflect introgression with *E. glaucus* and *E. violaceus*, both of which are often sympatric with *E. trachycaulus* subsp. *trachycaulus*. Spikelet spacing also varies considerably. In at least some instances, plants with widely spaced spikelets appear to be associated with more shady habitats.

8. **Elymus violaceus** (Hornem.) Feilberg ARCTIC WHEATGRASS [p. *330*, <u>497</u>]
Pl ces, not rhz. **Clm** 18–75 cm, often decumbent or geniculate; **nd** usu glab. **Shth** glab; **aur** about 0.5 mm; **lig** 0.5–1 mm, truncate; **bld** 3–4 mm wide, flat, glab or hairy, abx surfaces less densely hairy and with shorter hairs than the adx surfaces, apc acute. **Spikes** 5–12 cm long, 0.4–0.7 cm wide excluding the awns, erect, with 1 spklt per nd; **intnd** 4–5.5 mm, edges ciliate. **Spklt** 11–19 mm, appressed, with (3)4–5 flt; **rchl** hairy, hairs about 0.4 mm; **dis** above the glm, beneath each flt. **Glm** 8–12 mm long, 1.2–2 mm wide, about ¾ as long as to equaling the adjacent lm, narrowly ovate to obovate, often purplish, glab, smt scabrous, flat or equally keeled the full length, keels and other veins usu smooth, smt scabrous, 3(5)-veined, adx surfaces glab, mrg usu unequal, the wider mrg 0.3–1 mm wide, usu widest in the distal ⅓, apc

acute to rounded, often awned, awns to 2 mm; **lm** glab or pubescent, hairs flexible, all similar, apc usu awned, awns 0.5–3 mm, straight; **pal** subequal to the lm, tapering to the apc, apc about 0.4 mm wide; **anth** 0.7–1.3 mm. $2n = 28$.

Elymus violaceus grows in arctic, subalpine, and alpine habitats, on calcareous or dolomitic rocks, from Alaska through arctic Canada to Greenland, and south in the Rocky Mountains to southern New Mexico. In western North America, it forms intermediates with *E. scribneri* (p. 70), *E. trachycaulus* (see previous), and *E. alaskanus* (Scribn. & Merr.) Á. Löve. It is treated here as including *E. alaskanus* subsp. *latiglumis* [≡ *Agropyron latiglume*], *E. alaskanus* being restricted to plants with relatively short glumes that are often found in valleys and at lower elevations than *E. violaceus*. Western plants of *E. violaceus* tend to be more glaucous, have shorter spikes and spikelets, and more obovate glumes than plants from Greenland but, until more is known about the extent and genetic basis of the variation in and among *E. violaceus*, *E. alaskanus*, and *E. trachycaulus*, formal taxonomic recognition seems inappropriate.

9. Elymus lanceolatus (Scribn. & J.G. Sm.) Gould [p. 331, 497]

Pl strongly rhz, smt glaucous. **Clm** 22–130 cm, erect; **nd** glab. **Lvs** often mostly bas, smt more evenly distributed; **shth** glab or pubescent; **aur** usu present on the lo lvs, 0.5–1.5 mm; **lig** 0.1–0.5 mm, erose, smt ciliolate; **bld** 1.5–6 mm wide, generally involute, abx surfaces usu glab, adx surfaces strigose, ribs subequal in size and spacing. **Spikes** 3.5–26 cm long, 0.5–1 cm wide, erect to slightly nodding, usu with 1 spklt per nd, smt with 2 at a few nd; **intnd** 3.5–15 mm long, 0.1–0.8 mm wide, glab or hairy. **Spklt** 8–31 mm, 1.5–3 times longer than the intnd, appressed, with 3–11 flt; **rchl** glab or hairy, hairs to 1 mm; **dis** above the glm, beneath each flt. **Glm** subequal, 5–14 mm long, ½–¾ the length of the adjacent lm, 0.7–1.3 mm wide, lanceolate, glab or hairy, smooth or scabrous, 3–5-veined, flat or weakly, often asymmetrically keeled, keels straight, mrg narrow, tapering from the base or from beyond midlength, apc acute to acuminate, smt mucronate or shortly awned; **lm** 7–12 mm, glab or hairy, hairs all alike, smt scabrous, acute to awn-tipped, awns to 2 mm, straight; **pal** about equal to the lm, keels straight below the apc, smooth or scabrous proximally, smt hairy, scabrous distally, intercostal region glab or with hairs, apc 0.2–0.3 mm wide; **anth** (2.5)3–6 mm. $2n = 28$.

Elymus lanceolatus grows in sand and clay soils and dry to mesic habitats. It is found primarily in western North America between the coastal mountains and 95° W longitude, with the exception of *E. lanceolatus* subsp. *psammophilus*, which extends to the Great Lakes. All three subspecies grow in the Intermountain Region.

Elymus lanceolatus is primarily outcrossing, and hybridizes with several species of Triticeae. *Elymus albicans* (p. 71) is thought to be derived from hybridization with the awned phase of *Pseudoroegneria spicata* (p. 60). Judging from specimens of controlled hybrids, hybridization with *E. trachycaulus* (p. 67) and unawned plants of *P. spicata* probably occur, but would be almost impossible to detect without careful observation in the field. Experimental hybrids are partially fertile, and capable of backcrossing to either parent (Dewey 1965, 1967, 1968, 1975, 1976).

1. Lemmas with hairs, not scabrous . . . . . subsp. *lanceolatus*
1. Lemmas smooth, sometimes scabrous distally, mostly glabrous, sometimes the lemma margins hairy proximally . . . . subsp. *riparius*

Elymus lanceolatus (Scribn. & J.G. Sm.) Gould subsp. lanceolatus THICKSPIKE WHEATGRASS [p. 331]

Clm 60–130 cm. **Spikes** 10–22 cm; **intnd** 7–15 mm, smooth, scabrous, or hairy distally. **Spklt** 10–28 mm. **Lm** not scabrous, moderately hairy, hairs stiff, shorter than 1 mm.

Elymus lanceolatus subsp. *lanceolatus* grows in clay, sand, loam, and rocky soils, and is widely distributed in western North America. It is most likely to be confused with the octoploid *Pascopyrum smithii* (p. 74); it differs morphologically from that species in having more evenly distributed leaves and acute glumes that tend to taper from midlength or higher, rather than acuminate glumes that tend to taper from below midlength. In addition, the midvein of the glumes of *E. lanceolatus* is straight, whereas that of *P. smithii* "leans" to the side distally.

Elymus lanceolatus subsp. psammophilus (J.M. Gillett & H. Senn) Á. Löve SAND-DUNE WHEATGRASS [p. 331]

Clm 20–95 cm. **Spikes** 4–26 cm; **intnd** 3.5–13 mm, hairy at least distally. **Spklt** 9–31 mm. **Lm** densely hairy, hairs flexible, usu many longer than 1 mm; **pal** hairy between the keels, keels hairy proximally.

Elymus lanceolatus subsp. *psammophilus* tends to grow in sandy soils. It was originally described from around the Great Lakes, but plants with similar vestiture have been found scattered throughout the western range of the species, almost always in association with sandy soils.

Elymus lanceolatus subsp. riparius (Scribn. & J.G. Sm.) Barkworth STREAMBANK WHEATGRASS [p. 331]

Clm usu 22–60 cm. **Spikes** 6–10 cm; **intnd** 3.5–10 mm, glab, smt scabrous. **Spklt** 10–17 mm. **Lm** smooth, smt scabrous distally, mostly glab, mrg smt hairy proximally.

Elymus lanceolatus subsp. *riparius* grows throughout most of the western part of the range of *E. lanceolatus*, being more common in mesic habitats and clay soils than the other two subspecies.

10. Elymus stebbinsii Gould [p. 332, <u>497</u>]

Pl ces or shortly rhz. **Clm** 60–140 cm; **nd** glab or retrorsely pubescent. **Lvs** evenly distributed; **shth** glab or pubescent; **aur** usu present, 0.5–2 mm; **lig** 0.3–3.5 mm, truncate to acute, smt long-ciliate; **bld** 4–6.5 mm wide, flat or the mrg involute, straight. **Spikes** 15–31 cm long, 0.4–1.5 cm wide including the awns, 0.4–0.8 cm wide excluding the awns, erect, with 1 spklt per nd; **intnd** 9–27 mm long, 1–1.3 mm wide, glab, smooth. **Spklt** 13–29 mm long, from shorter than to almost twice as long as the intnd, 2.5–5 mm wide, appressed, with 5–7 flt; **rchl** glab; **dis** above the glm and beneath each flt. **Glm** subequal, 7.5–12 mm long, 1.2–1.5 mm wide, lanceolate, widest at about midlength, flat or rounded on the back, 5-veined, veins smooth, scabrous or just the midvein scabridulous, mrg widest at about midlength, apc acute, unawned; **lm** 9–12 mm, glab, smt scabrous, acute, unawned or awned, awns to 28 mm, straight; **pal** subequal to the lm, tapering, apc 0.2–0.3 mm wide; **anth** (3.5)4–7 mm. $2n = 28$.

Elymus stebbinsii is restricted to California, where it grows on dry slopes, chaparral, and wooded areas, at elevations below 1600 m. It differs from other *Elymus* species primarily in its combination of long anthers and solitary spikelets. It is often confused with *E. glaucus* (p. 65) and *E. trachycaulus* (p. 67) with solitary spikelets. It differs from both in its longer anthers, and from most representatives of *E. glaucus* in its acute, but unawned, glumes.

1. Lemmas awned, awns 8–28 mm long; lower leaf sheaths rarely pubescent; spikelets 13–22 mm long
 . . . . subsp. *septentrionalis*
1. Lemmas unawned or with awns to 8(12) mm long; lower leaf sheaths pubescent or glabrous; spikelets 17–29 mm long . . . . subsp. *stebbinsii*

Elymus stebbinsii subsp. **septentrionalis**
 Barkworth NORTHERN STEBBINS'
 WHEATGRASS [p. 332]

Lowest visible cauline nd usu glab, rarely pubescent. **Lo lf shth** usu glab, rarely pubescent. **Spike intnd** 9–21 mm. **Spklt** 13–22 mm. **Lm** awned, awns 8–28 mm.

Elymus stebbinsii subsp. *septentrionalis* grows primarily in the Sierra Nevada. Its range extends from near the Oregon border to Tulare County, and includes the coastal mountains north of San Francisco Bay.

Elymus stebbinsii Gould subsp. **stebbinsii**
 STEBBINS' WHEATGRASS [p. 332]

Lowest visible cauline nd often pubescent. **Lo lf shth** pubescent or glab. **Spike intnd** 16.3–27 mm. **Spklt** 17–29 mm. **Lm** unawned or awned, awns to 8(12) mm.

Elymus stebbinsii subsp. *stebbinsii* is best known from the coastal mountains south of San José. It also grows at scattered locations from the central Sierra Nevada south to the Transverse Mountains.

11. Elymus scribneri (Vasey) M.E. Jones
 SCRIBNER'S WHEATGRASS [p. 332, <u>497</u>]

Pl ces, not rhz. **Clm** 15–35(55) cm, prostrate to strongly decumbent, at least at the base; **nd** glab. **Shth** glab or shortly pilose; **aur** usu present, 0.5–1 mm; **lig** 0.2–0.4(0.7) mm, usu truncate, occ acute, entire to erose; **bld** 1.5–4 mm wide, usu involute, adx surfaces prominently ribbed. **Spikes** 3.5–10 cm long, 0.8–1.2 cm wide excluding the awns, 3–6 cm wide including the awns, usu with 1 spklt per nd, occ with 2 spklt at the lo nd; **intnd** 2.5–5(7) mm long, 0.5–1 mm wide, glab, mostly smooth, edges scabrous. **Spklt** 9–15 mm long, 6–12 mm wide, appressed to ascending, with 3–6 flt; **rchl intnd** 0.8–1.3 mm, scabridulous; **dis** initially at the rchs nd, subsequently beneath each flt. **Glm** 4–9 mm long, 0.5–1 mm wide, mostly glab, midveins scabrous, 3–5-veined, entire, tapering into a divergent, 12–30 mm awn; **lm** 7–10 mm, usu glab, occ scabridulous, awned, awns 15–30 mm, divergent, scabridulous; **pal** usu longer than the lm, apc ciliate, truncate or the veins extending into teeth, teeth about 0.5 mm; **anth** 1–1.6 mm. $2n = 28$.

Elymus scribneri grows in rocky areas in open subalpine and alpine regions, at 2500–3200 m, often in windswept locations. It is often confused with *E. elymoides* (p. 66), but differs from that species in having only one spikelet per node, wider glumes, and more tardily disarticulating rachises. It also resembles *E. sierrae* (see next), from which it differs in its disarticulating rachises, denser spikes, and shorter anthers.

Several taxonomists have suggested that *E. scribneri* consists of fertile hybrids between *E. violaceus* (p. 68) and *E. elymoides*. This suggestion is supported by the frequency with which the three taxa are sympatric, the morphological variation exhibited by *E. scribneri*, and cytogenetic data (Dewey 1967).

12. Elymus sierrae Gould SIERRA WHEATGRASS
 [p. 333, <u>498</u>]

Pl ces, not rhz. **Clm** 20–50 cm, prostrate or decumbent and geniculate; **nd** 1–2, exposed, glab. **Lvs** bas concentrated; **shth** glab; **aur** usu present, to 1 mm on the lo lvs; **lig** 0.2–0.5 mm, erose; **bld** 1–5 mm wide, flat, abx surfaces smooth, glab, adx surfaces prominently ridged over the veins, with scattered hairs, hairs to 0.2 mm, veins closely spaced. **Spikes** 5–15 cm long, 1.5–2.5 cm wide including the awns, 0.7–1.2 cm wide excluding the awns, flexuous, erect to nodding distally, with 1 spklt at most nd, occ some of the lo nd with 2 spklt; **intnd** 5–15 mm long, 0.2–0.5 mm wide, both surfaces glab, edges ciliate, not scabrous. **Spklt** 15–20 mm, ascending to divergent, with 3–7 flt; **rchl** glab; **dis** above the glm, beneath each flt. **Glm** subequal, 6–9 mm long, 0.7–1 mm wide, lanceolate, glab, the bases evidently veined, apc entire, tapering into a 3–10 mm awn; **lm** 12–16 mm, glab, smt scabridulous, apc bidentate, awned, awns 15–30 mm, arcuately diverging to strongly recurved; **pal** subequal to the lm, apc about 0.4 mm wide; **anth** 2–3.5 mm. $2n = 28$.

Elymus sierrae is best known from rocky slopes and ridgetops in the Sierra Nevada, at 2100–3400 m, and is also found in Washington and Oregon. It resembles *E. scribneri* (see previous), differing in its non-disarticulating rachises, longer rachis internodes, and longer anthers. Hybrids with *E. elymoides* (p. 66) have glumes with awns 15+ mm long, and some spikelets with narrower glume bases and shorter anthers. Specimens with wide-margined glumes suggest hybridization with *E. violaceus* (p. 68).

13. Elymus wawawaiensis J.R. Carlson & Barkworth SNAKERIVER WHEATGRASS [p. 333, 498]

Pl ces, smt weakly rhz. **Clm** (15)50–130 cm, erect, mostly glab; **nd** usu glab, smt slightly pubescent. **Lvs** more or less evenly distributed; **bas shth** glabrate, mrg not evidently ciliate; **aur** absent or to 1.2 mm; **lig** 0.1–1.1 mm; **bld** to 28 cm long, 1.7–5 mm wide, involute when dry, adx surfaces usu densely pubescent, rarely sparsely pubescent. **Spikes** 5–20 cm long, 2.5–3 cm wide including the awns, erect to slightly nodding, with 1 spklt per nd; **intnd** 5–12 mm long, about 0.2 mm thick, about 0.3 mm wide, glab beneath the spklt. **Spklt** 10–22 mm long, about twice as long as the intnd, 2–8.5 mm wide, appressed, with 4–10 flt; **rchl** glab; **dis** above the glm, beneath each flt. **Glm** 4–10 mm long, 0.5–1.3 mm wide, narrowly lanceolate, widest at or below midlength, glab, often glaucous, 1–3-veined, flat or weakly keeled, mrg 0.1–0.2 mm wide, widest near midlength, apc usu acuminate, awned or unawned, awns to 6 mm; **lm** 6–12 mm, smooth or slightly scabrous, mrg often sparsely pubescent proximally, apc awned, longest awns in the spklt 9–28 mm, strongly divergent; **pal** 7.2–10.5 mm, keels scabrous distally, tapering to the 0.2–0.3 mm wide apc; **anth** 3.5–6 mm. $2n = 28$.

Elymus wawawaiensis grows primarily in shallow, rocky soils of slopes in coulees and reaches of the Salmon, Snake, and Yakima rivers of Washington, northern Oregon, and Idaho. There are also a few records from localities at some distance from the Snake River and its tributaries. These probably reflect deliberate introductions. C.V. Piper, who worked for the U.S. Department of Agriculture in southeastern Washington from 1892–1902, frequently distributed seed to farmers in the region from populations that he considered superior; he considered *E. wawawaiensis* to be a superior form of what is here called *Pseudoroegneria spicata* (p. 60). Another source of introduced populations is 'Secar', a cultivar of *E. wawawaiensis* that is recommended as a forage grass for arid areas of the northwestern United States.

Elymus wawawaiensis resembles a vigorous version of *Pseudoroegneria spicata*, and was long confused with that species. It differs in its more imbricate spikelets and narrower, stiff glumes. In its primary range, *E. wawawaiensis* is often sympatric with *P. spicata*, but the two tend to grow in different habitats, *E. wawawaiensis* growing in shallow, rocky soils and *P. spicata* in medium- to fine-textured loess soil. The two species also differ cytologically, *E. wawawaiensis* being an allotetraploid, and *P. spicata* consisting of diploids and autotetraploids.

14. Elymus albicans (Scribn. & J.G. Sm.) Á. Löve MONTANA WHEATGRASS [p. 333, 498]

Pl strongly rhz. **Clm** 40–100 cm, erect or decumbent only at the base, glab. **Lvs** somewhat bas concentrated; **shth** glab; **aur** usu present, to 0.8 mm; **lig** 0.2–0.5 mm, ciliolate; **bld** 1–3 mm wide, usu involute, adx surfaces scabrous to strigose. **Spikes** 4–14 cm long, 1.5–2.5 cm wide including the awns, 0.3–0.8 cm wide excluding the awns, erect, with 1 spklt per nd; **intnd** 6–14 mm long, 0.2–0.4 mm wide, glab or pubescent beneath the spklt. **Spklt** 10–18 mm, 1.5–2 times longer than the intnd, appressed to ascending, with 3–7 flt; **rchl** strigillose; **dis** above the glm, beneath each flt. **Glm** subequal, ½ as long as to almost equaling the adjacent lm, glab or hairy, weakly keeled, keels and adjacent veins smooth to evenly and strongly scabrous from the base to the apc, mrg 0.2–0.3 mm wide, apc acute, acuminate, or shortly awned; **lo glm** 4–8 mm; **up glm** 4.5–8 mm; **lm** 7.5–9.5 mm, glab or densely hairy, awns 4–12 mm, at least some strongly divergent; **pal** subequal to the lm, tapering to the 0.1–0.3 mm wide apc; **anth** 3–5 mm. $2n = 28$.

Elymus albicans grows primarily in the central Rocky Mountains and the western portion of the Great Plains. It tends to grow in shallow, rocky soils on wooded or sagebrush-covered slopes, rather than in deep loams. It is derived from hybrids between *Pseudoroegneria spicata* (p. 60) and *E. lanceolatus* (p. 69). In practice, it is probably restricted to hybrids involving the awned variant of *Pseudoroegneria spicata*, because the hybrid origin of plants involving the unawned variant would probably not be recognized.

Populations of *Elymus albicans* differ in their reproductive abilities (Dewey 1970). In some, most plants yield good seed; in others, most plants are sterile. Some fertile populations appear to be self-perpetuating; others appear to consist of recent hybrids and some backcrosses. Although treated here as a species, *E. albicans* could equally well be treated as a hybrid in ×*Pseudelymus* (p. 60), but the combination has not been published. Plants with glabrous lemmas, presumed to be derived from crosses with glabrous individuals of *E. lanceolatus*, have sometimes been treated as a distinct taxon; they are not formally recognized here.

15. **Elymus repens** (L.) Gould QUACKGRASS, COUCH-GRASS [p. *334*, 498]

Pl strongly rhz, smt glaucous. **Clm** 50–100 cm. **Lvs** smt somewhat bas concentrated; **shth** pilose or glab proximally; **aur** 0.3–1 mm; **lig** 0.25–1.5 mm; **bld** 6–10 mm wide, usu flat, abx surfaces glab or sparsely pilose, adx surfaces usu sparsely pilose over the veins, smt glab, veins smooth, widely spaced, pri veins prominent, separated by the sec veins. **Spikes** 5–15 cm long, 0.5–1.5 cm wide, erect, usu with 1 spklt per nd, occ with 2 at a few nd; **intnd** 4–6(9.5) mm long, 0.5–1.2 mm wide, smooth or scabrous, glab, evenly puberulent, or sparsely pilose, hairs to 0.3 mm. **Spklt** 10–27 mm, appressed to ascending, with 4–7 flt; **dis** above the glm, beneath each flt. **Glm** oblong, glab, keeled distally, keels inconspicuous and smooth proximally, scabrous and conspicuous distally, lat veins inconspicuous, hyaline mrg present in the distal ½, apc acute, unawned or awned, awns to 3 mm; **lo glm** 8.8–11.4 mm, 3–6-veined; **up glm** 7–12 mm, 5–7-veined; **lm** 8–12 mm, glab, mostly smooth, smt scabridulous distally, unawned or with a 0.2–4(10) mm awn, awns straight; **pal** 7–9.5 mm, keels ciliate from ½ to almost the entire length, apc emgt, truncate, or rounded; **anth** 4–7 mm. 2*n* = 22, 42.

Elymus repens is native to Eurasia; it is now established as a weed throughout much of the world, including the Intermountain Region. It grows well in disturbed sites, spreading rapidly via its long rhizomes, as well as by seed. It is also drought tolerant. Although listed as a noxious weed in several states, it provides good forage. It differs from *E. hoffmannii* (see next) in having widely spaced, unequally prominent leaf veins and, usually, shorter awns.

16. **Elymus hoffmannii** K.B. Jensen & Asay HOFFMANN'S WHEATGRASS [p. *334*]

Pl slightly to moderately rhz. **Clm** 54–135 cm, glab. **Lvs** evenly distributed; **shth** glab; **aur** absent or to 1 mm; **lig** 0.6–1 mm, truncate, erose; **bld** 5–13 mm wide, flat to involute, abx surfaces smooth, glab, adx surfaces glab, veins closely spaced, all more or less equally prominent, smooth or scabrous. **Spikes** 10–50 cm long, 0.8–1.8 cm wide, with 1 spklt per nd, glab below the spklt; **intnd** 5–8 mm long, about 0.2 mm thick, about 0.3 mm wide, both surfaces hairy, hairs 0.2–0.4 mm. **Spklt** 15–27 mm, appressed to ascending, with 5–7 flt; **rchl** scabridulous; **dis** above the glm, beneath the flt. **Glm** equal, 5–11 mm long, 1.3–1.8 mm wide, stiff, lanceolate to linear-lanceolate, strongly rounded to keeled distally, keels inconspicuous and smooth on the proximal ⅓–½, conspicuous and with a few

teeth distally, lat veins inconspicuous, hyaline mrg 0.1–0.2 mm wide, apc acuminate to awned, awns to 8 mm; **lm** 7–12 mm, glab, smooth, apc unawned or awned, awns to 12 mm, straight; **pal** ciliate on the keels, apc about 0.6 mm wide; **anth** 4–7 mm. 2*n* = 42.

Elymus hoffmannii was described from a breeding line of plants developed from seeds collected in Erzurum Province, Turkey by J.A. Hoffmann and R.J. Metzger (Jensen & Asay 1996). There is no information available about its native distribution. As indicated in the key, *E. hoffmannii* differs from *E. repens* (see previous) primarily in its evenly prominent, closely spaced leaf veins and, usually, in having longer awns.

The description of *Elymus hoffmannii* was explicitly written to encompass the cultivar 'NewHy' that is derived from an artificial cross between *E. repens* and *Pseudoroegneria spicata* (p. 60). Because of its morphological similarity to plants obtained from the Turkish seed, Jensen and Asay suggested that *E. hoffmannii* had a similar parentage. 'NewHy' was released as a cultivar in the 1980s. Its distribution within the Intermountain Region is not known.

Named hybrids

Elymus is notorious for its ability to hybridize. Most of its interspecific hybrids are partially fertile, permitting introgression between the parents. The descriptions provided below are restricted to the named interspecific hybrids. They should be treated with caution and some skepticism; some are based solely on the type specimen, because little other reliably identified material was available. Moreover, as the descriptions of the non-hybrid species indicate, many other interspecific hybrids exist.

The parentage of all hybrids is best determined in the field. Perennial hybrids, such as those in *Elymus*, can persist in an area after one or both parents have died out, but the simplest assumption is that both are present. Interspecific hybrids of *Elymus* that have disarticulating rachises presumably have *E. elymoides* or *E. multisetus* as one of their parents.

17. **Elymus ×pseudorepens** (Scribn. & J.G. Sm.) Barkworth & D.R. Dewey FALSE QUACKGRASS [p. *334*]

Pl rhz. **Clm** 30–100 cm, ascending or erect, glab, mostly smooth, smt scabrous below the nd. **Lvs** evenly distributed; **shth** glab; **lig** to 0.5 mm; **bld** 10–25 cm long, 2–7 mm wide, involute when dry, both surfaces scabrous, adx surfaces sparsely pilose, hairs 0.7–1 mm. **Spikes** 5.5–13 cm long, 0.4–0.6 cm wide, with 1 spklt per nd; **intnd** 3.5–5 mm. **Glm** 8–18 mm, equaling or exceeded by the adjacent lm, more or less flat, 5–9-veined, mrg unequal, the wider mrg to about 0.3 mm, narrowly acute, smt awned, awns to 1 mm; **lm** 7.5–15 mm, smooth and glab proximally, scabrous distally, mucronate or awned, awns to 3 mm; **anth** 1.5–2 mm.

Elymus ×pseudorepens consists of hybrids between *E. lanceolatus* (p. 69) and *E. trachycaulus* (p. 67). It does not appear to be common in the Intermountain Region.

18. Elymus ×saundersii Vasey [p. *335*]

Pl ces, not rhz. **Clm** 50–80 cm, erect, glab or pilose. **Lvs** somewhat bas concentrated; **shth** retrorsely hairy; **aur** to 1 mm; **lig** 0.5–1 mm; **bld** 10–15 cm long, 4–5 mm wide, flat, becoming involute when dry, tapering to the apc. **Spikes** 10–25 cm long, 1–2.5 cm wide, with 1 spklt per nd; **intnd** 4–5 mm; **dis** at the rchs nd. **Spklt** 10–25 mm excluding the awns, with 3–6 flt. **Glm** 6–8 mm, linear-lanceolate to lanceolate, 3–5-veined, veins scabrous, apc smt toothed, awned, awns 8–13 mm; **lm** 10–13 mm, glab, smooth proximally, scabrous distally, awned, awns 20–45 mm, outcurving; **anth** 1.5–2 mm, usu indehiscent.

Elymus ×*saundersii* comprises hybrids between *E. trachycaulus* (p. 67) and *E. elymoides* (p. 66). Such hybrids are found throughout much of the western portion of the contiguous United States, mostly in disturbed areas. The hybrids are generally sterile and, as in all hybrids involving *E. elymoides* or *E. multisetus* (p. 66), the rachises disarticulate at maturity.

19. Elymus ×hansenii Scribn. [p. *335*]

Pl ces, not rhz. **Clm** 60–120 cm. **Lvs** evenly distributed; **shth** smooth; **lig** to 1 mm; **bld** 10–30 cm long, 2–8 mm wide, flat or the mrg involute. **Spikes** 5–20 cm, straight or nodding, with 2+ spklt per nd; **intnd** about 10 mm; **dis** in the rchs. **Spklt** about 15 mm, with 3–5 flt. **Glm** narrowly lanceolate, 2–3-veined, awned, awns 25–35 mm; **lm** 10–12 mm, awned, awns 40–50 mm, outcurving; **pal** subequal to the lm, truncate or bidentate.

Elymus ×*hansenii* refers to hybrids between *E. glaucus* (p. 65) and either *E. elymoides* or *E. multisetus* (p. 66). It is not clear which of the latter two species is involved. It is a fairly common hybrid in those parts of western North America where both parents grow. The glumes of the type specimen are as wide as those in *E. glaucus*, and some are divided longitudinally, as in *E. elymoides* and *E. multisetus*. As in other hybrids involving *E. elymoides* and *E. multisetus*, the rachis of *E.* ×*hansenii* disarticulates at maturity.

7.13 ×ELYLEYMUS B.R. Baum   Mary E. Barkworth

Pl per; smt rhz. **Clm** 40–235 cm, erect. **Infl** usu spikes, smt spikelike rcm, 5–35 cm, erect, with 1–3 spklt per nd, ped, when present, to 3 mm. **Spklt** with 2–8 flt; **dis** usu above the glm and beneath the flt, smt below the glm, smt in the rchs, usu tardy. **Glm** linear to lanceolate, often awn-tipped; **lm** 6–25 mm, glab or hairy, usu awned, awns to 15 mm; **anth** 1.5–5 mm.

    ×*Elyleymus* consists of hybrids between *Elymus* and *Leymus*. So far as is known, they are completely sterile, having thin anthers (usually less than 0.5 mm thick) and failing to develop mature caryopses. Only the named hybrids are accounted for here. Each of the entities appears to be distinct, but identification of the parents is, in some instances, tentative. The descriptions are offered with considerable reservation. The illustrations are based on type specimens. Unless stated otherwise, measurements of the spikes include the awns; measurements of the spikelets, glumes, and lemmas do not.

1. ×Elyleymus aristatus (Merr.) Barkworth & D.R. Dewey [p. *335*]

Pl not or shortly rhz. **Clm** 60–130 cm, glab. **Lvs** evenly distributed on the clm; **shth** smooth, glab; **aur** poorly developed, to 0.5 mm; **lig** 1–2.5 mm, scarious, rounded; **bld** about 5.5 mm wide, abx surfaces glab, mostly smooth, scabrous near the mrg, adx surfaces scabridulous, pri veins separated by about 3 sec veins. **Infl** spikes, 6–15 cm long, 10–15 mm wide including the awns, 7–10 mm wide excluding the awns, erect, with 2–3 sessile or subsessile spklt per nd; **intnd** 4–7 mm, concealed by the spklt; **dis** tardy, in the rchs and beneath the flt. **Spklt** 10–20 mm excluding the awns, to 18 mm including the awns, with 3–4 flt. **Glm** 8–15 mm long, 0.3–0.5 mm wide, subequal to unequal, scabrous; **lm** 7.5–12 mm, glab, smooth or scabrous, smt only scabrous distally, midveins prominent and scabrous distally, awns 4–5 mm; **anth** 2.2–2.4 mm.

    Dewey and Holmgren (1962) argued that ×*Elyleymus aristatus* comprises hybrids between *Elymus elymoides* and *Leymus cinereus* or *L. triticoides*.

2. ×Elyleymus hirtiflorus (Hitchc.) Barkworth [p. *336*]

Pl rhz. **Clm** 40–90 cm tall, 2–2.5 mm thick, glab. **Lvs** somewhat bas concentrated; **shth** smooth, glab; **aur** 0.3–0.5 mm; **lig** 0.2–0.5 mm, truncate; **bld** 5–20 mm long, 2–4 mm wide, usu involute, smt flat, abx surfaces smooth, glab or sparsely hairy, particularly towards the base, veins not prominent, adx surfaces scabridulous or scabrous, varying within a pl, all veins equally prominent, apc narrowly acute. **Infl** spikes, 5–18 cm long, 8–10 mm wide, with 1–2 sessile or subsessile spklt per nd; **intnd** 4–5 mm, partially exposed on the sides, hairy. **Spklt** 11–14 mm, to 23 mm including the awns, with 3–6 flt; **dis** above the glm, beneath the flt. **Glm** 12–17 mm long including the awnlike apc, 1–3 mm wide, widest at about ¼ length, 1(3)-veined, keeled, sparsely to densely hairy, hairs 0.3–0.5 mm; **lm** 8.5–10 mm, sparsely to densely hairy, hairs 0.3–0.5 mm, apc awned, awns 5–10 mm, straight; **anth** 1.8–2 mm long.

Bowden (1967) suggested that ×*Elyleymus hirtiflorus* consisted of hybrids between *Elymus trachycaulus* and *Leymus innovatus*, and included in it plants from British Columbia. The name, however, is based on collections from the banks of the Green River, Wyoming, where neither putative parent grows. The more likely parents are *E. lanceolatus* and *L. simplex*. Admittedly, the short anthers argue for *E. trachycaulus* rather than *E. lanceolatus* as the *Elymus* parent.

7.14 PASCOPYRUM Á. Löve   Mary E. Barkworth

Pl per; rhz. **Clm** 20–100 cm. **Lvs** bas concentrated; **shth** striate when dry, smooth, usu glab, rarely pilose; **aur** present; **lig** memb. **Infl** tml, distichous spikes, spklt usu 1 per nd, occ in pairs at the lo nd, spklt at the lo 4–6 nd often strl; **lowest intnd** to 26 mm, 2 times as long as the mid intnd. **Spklt** 12–26(30) mm, 1–3 times the length of the intnd, straight, usu ascending, not appressed, with 2–12 flt; **dis** above the glm, beneath the flt. **Glm** 5–15 mm, ½–⅔ the length of the spklt, usu narrowly lanceolate, stiff, tapering from midlength or below, slightly curving to the side distally, not keeled, 3–5-veined bas, 1-veined distally, apc acuminate; **lm** lanceolate, rounded on the back, acute, mucronate to awned, awns to 5 mm, straight; **pal** slightly shorter than the lm; **anth** 3, 2.5–6 mm. **Car** 4–5 mm, falling with the lm and pal. $2n = 56$. Haplomes **St**, **H**, **Ns**, and **Xm**. Name from the Latin *pascuum*, 'pasture', and the Greek *pyros*, 'wheat'.

Pascopyrum is a North American allooctoploid genus with one species. It combines the genomes of *Leymus* with those of *Elymus*. There are no other species that combine these two tetraploid genomes, although there are many species of both *Elymus* and *Leymus* in Eurasia and North America.

1. **Pascopyrum smithii** (Rydb.) Barkworth & D.R. Dewey WESTERN WHEATGRASS [p. 337, 498]

Clm 20–100 cm, glab. **Aur** 0.2–1 mm, often purple; **lig** about 0.1 mm; **bld** 2–26 cm long, 1–4.5 mm wide, decreasing in length upwards, spreading, rigid, adx surfaces with prominent veins. **Spikes** 5–17 cm; **mid intnd** 4.5–11 mm. **Spklt** 12–26(30) mm, with 2–12 flt; **lowest rchl intnd** in each spklt 0.8–2 mm long, 0.5–0.9 mm wide at the top. **Glm** 5–15 mm, lo glm usu exceeded by the up glm; **lo glm** 0.15–0.8 mm wide at ¾ length; **lm** 6–14 mm, unawned or awned, awns 0.5–5 mm. $2n = 56$.

Pascopyrum smithii is native to sagebrush deserts and mesic alkaline meadows, growing in both clay and sandy soils. It is probably derived from a *Leymus triticoides–Elymus lanceolatus* cross (Dewey 1975) and is frequently confused with both. *Leymus triticoides* differs from *P. smithii* in usually having 2 spikelets per node and glumes that are narrower at the base. In *E. lanceolatus*, the leaves tend to be more evenly distributed and the glumes have straight midveins, become narrow beyond midlength, and tend to be wider at ¾ length (0.35–1.6 mm). In addition, the first rachilla internodes of *E. lanceolatus* are often longer and narrower (the length/width ratio averaging 2.6, versus 1.8 in *P. smithii*).

7.15 LEYMUS Hochst.   Mary E. Barkworth

Pl per; smt csp, often rhz. **Clm** 10–350 cm, erect, with exvag brchg. **Lvs** bas or evenly distributed; **shth** open; **aur** usu present; **lig** memb, truncate to rounded; **bld** often stiff, adx surfaces usu with subequal, closely spaced, prominently ribbed veins, smt with unequal, widely spaced, not prominently ribbed veins. **Infl** usu distichous spikes with 1–8 spklt per nd, smt pan with (2)3–35 spklt associated with each rchs nd; **rchs** with scabrous or ciliate edges; **intnd** 3.5–12(15) mm. **Spklt** ½–3 ¾ times the length of the rchs intnd, usu sessile, smt pedlt, ped to 5 mm, appressed to ascending, with 2–12 flt, the tml flt usu rdcd; **dis** above the glm, beneath the flt. **Glm** usu 2, usu equal to subequal, the lo or both glm smt rdcd or absent, lanceolate and narrowing in the distal ¼, or lanceolate to subulate and tapering from below midlength, pilose or glab, smt scabrous, 0–3(7)-veined, veins evident at least at midlength, smt keeled, keels straight or almost so, apc acute, acuminate, or tapering to an awnlike tip, if distinctly awned, awns to 4 mm; **lm** glab or with hairs, smt scabrous distally, inconspicuously 5–7-veined, rounded over the back proximally, smt keeled distally, keels not conspicuously scabrous distally, apc acute, unawned or awned, awns usu to 7 mm, smt 16–33 mm, straight; **pal** slightly shorter than to slightly longer than the lm, keels usu scabrous or ciliate on the distal portion, smt throughout; **lod** 2, shortly hairy, lobed; **anth** 3, 2.5–10 mm. **Car** with hairy apc. $x = 7$. **Haplomes NsNs or NsXm** (see below). Name an anagram of *Elymus*.

Leymus is a genus of approximately 50 species; most are native to temperate regions in the Northern Hemisphere, but there are 2–3 species in South America. It is most abundant in eastern Asia, with North America being a secondary center.

Most species of *Leymus*, including most North American species, grow well in alkaline soils. They are used for soil stabilization and forage. All the species are self-incompatible, outcrossing polyploids. One of the haplomes present is the **Ns** genome; this genome is also found in *Psathyrostachys*, most species of which are diploids. There is disagreement concerning the second haplome. Wang and Jensen (1994) argued that there are two different haplomes present, the origin of the second one being unknown and designated **Xm**. Bödvarsdóttir and Anamthawat-Jónsson (2003) found no molecular probes that would distinguish between the two genera, from which they argued that *Leymus* is a segmental allopolyploid with only one basic haplome, **Ns**. Morphologically, *Psathyrostachys* and *Leymus* are very similar, the major differences being that *Psathyrostachys* is never rhizomatous, has disarticulating rachises, and, usually, distinctly awned lemmas.

In most species of *Leymus*, at least some of the spikelets are on pedicels up to 2 mm long. Despite this, it is customary to identify the inflorescence of such species as a spike rather than a raceme, as is done in this treatment. Culm thicknesses are measured on the lower internodes. Descriptions of rachis nodes, unless stated otherwise, apply to the internodes at midspike.

1. Lemmas densely hairy, hairs 2–3 mm long . . . . 7. *L. flavescens*
1. Lemmas usually wholly or partially glabrous, if hairy, the hairs shorter than 0.8 mm.
   2. Plants cespitose, not or weakly rhizomatous; culms several to many together.
      3. Spikes with 2–7 spikelets per node; blades 3–12 mm wide; culms (7)100–270 cm tall . . . . . 5. *L. cinereus*
      3. Spikes with 1 spikelet, at least at the distal nodes, sometimes at all nodes or with
         2(3) spikelets at the lower nodes; blades 1–6 mm wide; culms 35–140 cm tall . . . . 6. *L. salina*
   2. Plants strongly rhizomatous; culms solitary or few together.
      4. Culms 2.5–5 mm thick; glumes 10–30 mm long . . . . 1. *L. racemosus*
      4. Culms 1–3 mm thick; glumes 4–16 mm long.
         5. Spikes with 1 spikelet at all or most nodes; lemma awns 2.3–6.5 mm long;
         culms 35–55 cm tall . . . . 2. *L. simplex*
         5. Spikes with 2 or more spikelets at most nodes; lemma awns to 3 mm long; culms
         45–125 cm tall.
            6. Adaxial surfaces of the blades usually with closely spaced, prominently
            ribbed, subequal veins; calluses usually glabrous, occasionally with a few hairs
            about 0.1 mm long . . . . 3. *L. triticoides*
            6. Adaxial surfaces of the blades usually with widely spaced, not prominently
            ribbed veins, the primary veins evidently larger than the intervening
            secondary veins; calluses with hairs about 0.2 mm long . . . . 4. *L. multicaulis*

1. **Leymus racemosus** (Lam.) Tzvelev
MAMMOTH WILDRYE [p. 338, <u>498</u>]

Pl not or only weakly ces, strongly rhz, often glaucous. **Clm** 50–100 cm tall, 8–12 mm thick, solitary or few together, mostly smooth and glab, scabridulous or pubescent below the spikes, hairs to 0.5 mm. **Lvs** exceeded by the spikes; **lig** 1.5–2.5 mm; **bld** 20–40 cm long, 8–20 mm wide. **Spikes** 15–35 cm long, 10–20 mm wide, dense, with 3–8 spklt per nd; **intnd** 8–11 mm, surfaces hairy, hairs to 1 mm, on the edges to 1.5 mm. **Spklt** 12–25 mm, sessile, with 4–6 flt. **Glm** 12–25 mm long, to 2 mm wide, usu exceeding the lm, linear-lanceolate at the base, tapering from below midlength, stiff, glab at least at the base, the cent portion thicker than the mrg, keeled and subulate distally, 1-veined, veins inconspicuous at midlength; **lm** 15–20 mm, pubescent proximally, glab distally, tapering to an awn, awns 1.5–2.5 mm; **pal** keels usu glab, smt ciliate distally; **anth** about 5 mm, dehiscent. $2n = 28$.

Leymus racemosus is native to Europe and central Asia, where it grows on dry, sandy soils. It has been introduced into the Intermountain Region, and collected at various locations, particularly in the northwestern contiguous United States; it is not clear how many of the populations represented by these specimens are still extant. Tsvelev (1976) recognized 4 subspecies. Because there are few North American specimens, and these are incomplete, no attempt has been made to determine to which subspecies the North American plants belong.

2. **Leymus simplex** (Scribn. & T.A. Williams)
D.R. Dewey ALKALI WILDRYE [p. 338, <u>498</u>]

Pl not ces, strongly rhz, often glaucous. **Clm** 35–75 cm tall, 1–2.5 mm thick, solitary or few together, glab or sparsely pubescent near the nd. **Lvs** exceeded by the spikes; **shth** glab, smooth; **aur** infrequently present, to 0.8 mm, the auricular location often with hairs to 2 mm; **lig** 0.3–0.5 mm, truncate, erose; **bld** 4–29 cm long, 1–2(5) mm wide, flat, becoming involute when dry, stiff, adx surfaces scabrous, with scattered hairs to 2 mm, veins 7–11, subequal, prominently ribbed. **Spikes** 1.5–27 cm long, 4–15 mm wide, with 1 spklt per nd at midspike, smt with 2 at the lo nd; **intnd** 7–20 mm, surfaces glab or strigillose, edges ciliate, cilia to 1 mm. **Spklt** 16–25 mm, pedlt, ped 1–2(5) mm, with 3–12 flt. **Glm** subequal, 8–12 mm long, 0.5–1.5 mm wide, subulate, tapering from about ¼ of their length, stiff, glab at least at the base, the cent portion thicker than the mrg, keeled, 0–1(3)-veined, veins inconspicuous at midlength; **lm** 7–12 mm, glab, awned, awns 2.3–6.5(12) mm; **anth** 3.7–4.5 mm, dehiscent. $2n = 28$.

Leymus simplex is found in meadows and drifting sand in southern Wyoming, and along the Green River in northeastern Utah.

1. Culms 55–75 cm tall; spikes 10–27 cm long; internodes 10–20 mm . . . . var. *luxurians*
1. Culms 35–55 cm tall; spikes 1.5–13 cm long; internodes 7–9 mm . . . . var. *simplex*

Leymus simplex var. **luxurians** (Scribn. & T.A. Williams) Beetle [p. *338*]

Clm 55–75 cm tall, 2–2.5 mm thick. **Spikes** 10–27 cm long; **intnd** 10–20 mm. **Spklt** 20–25 mm, with 6–12 flt.

Leymus simplex var. *luxurians* grows at a few locations in Wyoming. It sometimes grows close to var. *simplex*. It may represent clones that have access to more water and/or more nutrients, but the absence of intermediate plants suggests a genetic distinction.

Leymus simplex (Scribn. & T.A. Williams) D.R. Dewey var. **simplex** [p. *338*]

Clm 35–55 cm tall, 1–2 mm thick. **Spikes** 1.5–13 cm long; **intnd** 7–9 mm. **Spklt** 16–22 mm, with 3–6 flt.

Leymus simplex var. *simplex* is found throughout the range of the species, sometimes in close proximity to var. *luxurians*. The two may be environmentally induced variants, but the lack of intermediates suggests a genetic distinction.

3. **Leymus triticoides** (Buckley) Pilg. Beardless Wildrye [p. *339*, 498]

Pl not ces, strongly rhz. **Clm** 45–125 cm tall, 1.8–3 mm thick, solitary or few together. **Lvs** exceeded by the spikes, often bas concentrated; **shth** glab or hairy, hairs 0.5–1 mm; **aur** to 1 mm; **lig** 0.2–1.3 mm, truncate, erose; **bld** 10–35 cm long, 3.5–10 mm wide, flat to involute, usu stiffly ascending, adx surfaces usu scabrous, often also sparsely hairy, hairs to 0.8 mm, most abundant proximally, veins 11–27, closely spaced, subequal, prominently ribbed. **Spikes** 5–20 cm long, 5–15 mm wide, with 2 spklt at midspike, smt 1 or 3 at other nd; **intnd** 5–11.5 mm, usu mostly smooth and glab, smt strigillose distally, edges ciliate, cilia to 0.4 mm. **Spklt** 10–22 mm, with 3–7 flt. **Glm** 5–16 mm long, 0.5–1.2 mm wide, bases not overlapping, glab and smooth proximally, scabrous distally, tapering from below midlength to the subulate apc, stiff, keeled, the cent portion thicker than the mrg, 1(3)-veined, veins inconspicuous at midlength; **cal** usu glab, occ with a few hairs, hairs about 0.1 mm; **lm** 5–12 mm, usu glab, occ sparsely hairy, hairs to 0.3 mm, apc acute, usu awned, awns to 3 mm; **anth** 3–6 mm, dehiscent. $2n = 28$.

Leymus triticoides grows in dry to moist, often saline meadows. Its range extends from southern British Columbia to Montana, south to California, Arizona, and New Mexico, but its populations are widely scattered. There is considerable variation within the species, but no pattern of variation suggesting the existence of infraspecific taxa is known. It is very similar to *L. multicaulis*, strains of which were initially released as *L. triticoides* by the U.S. Department of Agriculture. The most consistent differences between them appear to be in the venation of the leaf blades and the vestiture of the calluses. *Leymus triticoides* is also very similar to *L. simplex*, differing from it in the number of spikelets at the midspike nodes.

Leymus triticoides hybridizes with *L. cinereus*, but these hybrids have not been formally named. Plants identified as *Elymus arenicolus* Scribn. & J.G. Sm. are here included in *L. flavescens*, but may represent hybrids between *L. triticoides* and *L. flavescens*.

4. **Leymus multicaulis** (Kar. & Kir.) Tzvelev Many-stem Wildrye [p. *339*]

Pl somewhat ces, rhz. **Clm** 50–80 cm tall, 1.5–3 mm thick, usu few together, glab, mostly smooth, scabrous beneath the spikes. **Lvs** exceeded by the spikes; **shth** glab, smooth; **aur** to 1 mm; **lig** 1–2 mm; **bld** 3–8 mm wide, flat or the mrg slightly involute, grayish green, smt glaucous, abx surfaces smooth, adx surfaces glab, with both pri and sec veins, pri veins 5–7, not prominently ribbed. **Spikes** 5–14 cm long, 6–13 mm wide, with 2–4(6) spklt per nd; **intnd** 4–6 mm, glab or strigillose, hairs about 0.1 mm, edges ciliate, cilia to 0.4 mm. **Spklt** 8–15 mm, with 2–6 flt. **Glm** 4–10 mm long, to 1 mm wide, stiff, keeled, glab, scabrous, the cent portion thicker than the mrg, bases not overlapping, tapering from below midlength to the subulate apc, inconspicuously 1-veined at midlength; **cal** usu with at least some hairs, hairs about 0.2 mm; **lm** 5–9 mm, mostly glab and smooth, scabrous distally, apc tapering to an awn, awns 2–3 mm, scabrous; **anth** 3–4 mm, dehiscent. $2n = 42$.

Leymus multicaulis is native to Eurasia, extending from the Volga River delta in Russia to Xinjiang, China. In its native range, it grows in alkaline meadows and saline soils, and as a weed in fields, near roads, and around human habitations. It is very similar to *L. triticoides*, and hybrids with that species are highly fertile. A cultivar of *L. multicaulis*, 'Shoshone', that was originally thought to be a productive strain of *L. triticoides*, has been widely distributed for forage. *Leymus multicaulis* differs from *L. triticoides* primarily in having both primary and secondary veins in its blades, and small hairs on its calluses. Because it has only recently been realized that *L. multicaulis* has been introduced, its distribution in the Intermountain Region is not known.

5. **Leymus cinereus** (Scribn. & Merr.) Á. Löve Great Basin Wildrye [p. *339*, 498]

Pl strongly ces, weakly rhz, usu bright green, not glaucous. **Clm** 70–270 cm tall, 2–5 mm thick, many together, lowest nd often pubescent, smt pubescent up to 1.5 cm below the infl. **Lvs**

exceeded by the spikes; **shth** glab or hairy; **aur** to 1.5 mm; **lig** 1.5–8 mm; **bld** 15–45 cm long, 3–12 mm wide, strongly involute to flat, abx surfaces glab, adx surfaces scabrous, 11–25-veined, veins subequal, prominently ribbed. **Spikes** 10–29 cm long, 8–17 mm wide, with 14–28 nd and 2–7 spklt per nd; **intnd** 4–9 mm. **Spklt** 9–25 mm, with 3–7 flt. **Glm** 8–18 mm long, 0.5–2.5 mm wide, subulate distally, stiff, keeled, the cent portion thicker than the mrg, tapering from below midlength, smooth or scabrous, 0–1(3)-veined, veins inconspicuous at midlength; **lm** 6.5–12 mm, glab or hairy, hairs 0.1–0.3 mm, apc acute or awned, awns to 3 mm; **anth** 4–7 mm, dehiscent. 2*n* = 28, 56.

Leymus cinereus grows along streams, gullies, and roadsides, and in gravelly to sandy areas in sagebrush and open woodlands. It is widespread and common in western North America. It resembles *Psathyrostachys juncea*, differing in its larger size, non-disarticulating rachises, larger spikelets with more florets, and longer ligules. Spontaneous hybridization between *L. cinereus* and *L. triticoides* is known; the hybrids do not have a scientific name. The rhizomes found in some specimens may reflect introgression from *L. triticoides* through such hybrids.

6. **Leymus salina** (M.E. Jones) Á. Löve [p. 340, 498]

Pl ces, smt weakly rhz. **Clm** 35–140 cm tall, 1.5–3 mm thick, several together. **Lvs** exceeded by the spikes; **aur** to 1 mm; **lig** 0.1–1 mm, truncate; **bld** 1–5 mm wide, flat to strongly involute, adx surfaces glab or sparsely to densely hirsute, with 5–9 prominently ribbed, subequal veins. **Spikes** 4–14 cm long, 4–11 mm wide, nd below midspike with 1–2(3) spklt, distal nd with 1 spklt; **intnd** 3.5–9 mm, surfaces glab, edges scabrous or strigillose. **Spklt** 9–21 mm, pedlt, ped to 1 mm, with 3–6 flt. **Glm** unequal to subequal, to 12.5 mm long, 0.5–3.2 mm wide, subulate, stiff, keeled, the cent portion thicker than the mrg, tapering from below midlength, 0–1(3)-veined, veins inconspicuous at midlength; **lo glm** 0–12 mm; **up glm** 3.5–12.5 mm; **lm** 7–12.5 mm, usu glab, smt sparsely strigillose, unawned or awned, awns to 2.5 mm; **anth** 2.5–7.5 mm, dehiscent.

The three subspecies of *Leymus salina* differ in their pubescence and geographic distribution, with subsp. *salina* being the most common of the three. Two subspecies grow in the Intermountain Region. The specific epithet comes from the locality of the type collection: Salina Pass, Utah.

1. Basal sheaths glabrous; blades usually glabrous on the abaxial surfaces. . . . . subsp. *salina*
1. Basal sheaths and blades conspicuously hairy on the abaxial surfaces. . . . . subsp. *salmonis*

Leymus salina (M.E. Jones) Barkworth subsp. **salina** Salina Wildrye [p. 340]

Clm 39–102 cm. **Bas shth** glab; **bld** strongly involute, abx surfaces glab, adx surfaces pubescent, usu densely hairy above the lig. **Spikes** with 1 spklt at most nd, including those at midspike. 2*n* = 28.

Leymus salina subsp. *salina* grows on rocky hillsides, primarily in eastern Utah and western Colorado, extending into southern Wyoming and northern Arizona and New Mexico.

Leymus salina subsp. **salmonis** (C.L. Hitchc.) R.J. Atkins Salmon Wildrye [p. 340]

Clm 60–140 cm. **Bas shth** conspicuously hairy; **bld** open to involute, abx surfaces conspicuously hairy, adx surfaces evenly strigillose to strigose. **Spikes** with 1–2 spklt at the cent nd, usu 1 at the distal nd. 2*n* = 28.

Leymus salina subsp. *salmonis* grows at scattered locations on rocky hillsides in the mountains of southern Idaho, Nevada, and western Utah.

7. **Leymus flavescens** (Scribn. & J.G. Sm.) Pilg. Yellow Wildrye [p. 340, 498]

Pl smt ces, strongly rhz. **Clm** 40–120 cm tall, 2–4 mm thick, pubescent beneath the nd. **Lvs** exceeded by the spikes; **shth** glab; **aur** absent, smt with a few hairs in the auricular position; **lig** 0.3–1.5 mm; **bld** 3–4 mm wide, usu involute, adx surfaces scabrous, smt with scattered hairs, hairs to 1 mm, with about 15 closely spaced, subequal, mostly prominently ribbed veins. **Spikes** 10–20 cm long, 12–20 mm thick, with 12–20 nd and 2 spklt per nd; **intnd** 7–10 mm, densely hairy. **Spklt** 13.5–25 mm, with 4–9 flt. **Glm** 8.5–16 mm long, 0.5–2.5 mm wide, stiff, keeled distally, the cent portion thicker than the mrg, tapering from below midlength to the subulate apc, hairy, 0–1(3)-veined, veins inconspicuous at midlength; **lo glm** 8.5–13.5 mm; **up glm** 10–16 mm; **cal** poorly developed; **lm** 10.5–15 mm, densely villous, hairs 2–3 mm, apc unawned or awned, awns to 2 mm; **anth** 4.5–7 mm, dehiscent. 2*n* = 28.

Leymus flavescens grows on sand dunes and open sandy flats, and ditch- and roadbanks, of the Snake and Columbia river valleys.

Plants identified as *Elymus arenicolus* Scribn. & J.G. Sm. are included here, but they may represent hybrids between *Leymus flavescens* and *L. triticoides*. Leckenby, the collector of the type specimen, noted that they grew on sand or sand drifts along the Columbia River, but could not withstand flooding. He could find no seed.

7.16 ×LEYDEUM Barkworth

Mary E. Barkworth

Pl per; rhz, smt shortly so. **Clm** to 140 cm tall, 1–3 mm thick. **Spikes** 10–15 cm long, 5–12 mm wide excluding the awns, erect, smt lax, nd with 2–3 spklt; **intnd** 3–5 mm; **dis** in the rchs, smt delayed. **Spklt** appressed, with 1–3 flt. **Glm** equal or unequal, 10–25 mm long, 0.2–1.5 mm wide, tapering from below midlength or subulate from the base; **lm** glab or hairy, awned, awns 1–10 mm; **anth** 1.8–3 mm long, 0.1–0.3 mm thick. **Car** not developed.

×*Leydeum* consists of hybrids between *Hordeum* and *Leymus*. The number of named ×*Leydeum* hybrids is substantially lower than that for hybrids between *Hordeum* and *Elymus*. This probably reflects the lower likelihood of such hybrids being formed, because *Leymus* does not incorporate the **H** genome of *Hordeum*, whereas *Elymus* does. It is also possible that such hybrids are less likely to be recognized, because inland species of *Leymus*, like *Hordeum*, have narrow glumes. The only species likely to be found in the Intermountain Region, ×*Leydeum piperi*, differs from *Leymus* in its disarticulating rachises, and from *Hordeum* in having 2 spikelets per node and 2–3 florets in the larger spikelets.

1. ×**Leydeum piperi** (Bowden) Barkworth [p. 340]

Pl shortly rhz. **Clm** to 80 cm tall, about 1.5 mm thick. **Lvs** evenly distributed; **shth** smooth; **aur** to 1.5 mm; **lig** 0.5–0.8 mm, truncate to rounded; **bld** 10–20 cm long, 1.5–2.5 mm wide, tapering from near the base, abx surfaces scabridulous, adx surfaces scabrous, with about 12 usu more or less equally prominent veins, apc narrowly acute. **Spikes** 10–15 cm long, 10–15 mm wide including the awns, 5–7 mm wide excluding the awns, lax, nd with 2(3) spklt, 1 (cent if 3) spklt sessile, 1(2) spklt(s) pedlt, ped about 0.2 mm; **intnd** 3–5 mm, completely or mostly concealed by the spklt; **dis** tardy, in the rchs. **Spklt** about 15 mm including

the awns, about 10 mm excluding the awns, with 1–3 flt, the larger or cent spklt with 2–3 flt, the lat or smaller spklt with 1(2) flt(s). **Glm** 10–15 mm long, about 0.2 mm wide, subulate, scabrous; **lm** glab, mostly smooth, scabrous distally, the largest lm of the larger spklt 6– 7 mm, with awns about 9 mm, the largest lm of the smaller or lat spklt about 5 mm, with awns about 5.5 mm; **anth** 1.8–2.2 mm long, about 0.3 mm thick.

Covas (1949) and Bowden (1967) agreed that that the parents of ×*Leydeum piperi* are *Hordeum jubatum* and *Leymus triticoides*. It is not known how common or widespread it is. It differs from ×*Elyhordeum macounii* in its subulate, rather than narrowly linear, glumes.

7.17 PSATHYROSTACHYS Nevski

Claus Baden†

Pl per; csp, forming dense to loose clumps, smt stln, smt rhz. **Clm** 15–120 cm, erect or decumbent. **Shth** of the bas lvs closed, becoming fibrillose, of the up cauline lvs open; **aur** smt present; **lig** 0.2–0.3 mm, memb; **bld** with prominently ribbed veins on the adx surfaces. **Infl** spikes, with 2–3 spklt per nd; **dis** in the rchs. **Spklt** appressed to ascending, with 1–2(3) flt, often with additional rdcd flt distally. **Glm** equal to unequal, (3.5)4.2–48.5(65) mm including the awns, subulate, stiff, scabrous to pubescent, obscurely 1-veined, not united at the base; **lm** 5.5–14.3 mm, narrowly elliptic, rounded, glab or pubescent, 5–7-veined, veins often prominent distally, apc sharply acute to awned, smt with a minute tooth on either side of the awn base, awns 0.8–34 mm, straight, ascending to slightly divergent, smt violet-tinged; **pal** equaling or slightly longer than the lm, memb, scabrous or pilose on and smt also between the keels, bifid; **anth** 3, 2.5–6.8(7) mm, yellow or violet; **lod** 2, acute, entire, ciliate. **Car** pubescent distally, tightly enclosed by the lm and pal at maturity. $x = 7$. Name from the Greek *psathyros*, 'fragile', and *stachys*, 'spike'.

Psathyrostachys has eight species, all of which are native to arid regions of central Asia, from eastern Turkey to eastern Siberia, Russia, and Xinjiang Province, China. One species, *P. juncea*, was introduced to North America as a potential forage species, and is now established in the Intermountain Region.

Psathyrostachys is very similar to *Leymus*, particularly the cespitose species of *Leymus*. The major differences are that *Psathyrostachys* has disarticulating rachises and, usually, distinctly awned lemmas.

1. **Psathyrostachys juncea** (Fisch.) Nevski
RUSSIAN WILDRYE [p. 341, 498]

Pl densely ces. **Clm** (20)30–80(120) cm, erect or decumbent at the base, mostly glab, pubescent below the spikes. **Bas shth** glab, grayish brown, old shth more or less persistent; **aur** 0.2–1.5 mm; **bld** (1)2.5–18 (30) cm long, (1)5–20 mm wide, flat

or involute, abx surfaces smooth or scabridulous, often glaucous. **Spikes** (3)6–11(16) cm long, 5–17 mm wide, erect, with (2)3 spklt per nd; **rchs** hirsute on the mrg, puberulent elsewhere; **intnd** 3.5–6 mm. **Spklt** 7–10(12) mm excluding the awns, strongly overlapping, lat spklt slightly larger than the cent spklt. **Glm** (3.5)4.2–9.4 mm, subulate, scabrous or with 0.3–0.8 mm hairs; **lm** 5.5–7.5 mm, lanceolate, glab or with 0.3–0.8 mm hairs, sharply acute or awned, awns 0.8–3.5 mm; **pal** 5.8–7.6 mm, scabrous, acute; **anth** 2.5–5.1 mm; **lod** 1.3–1.5 mm. **Car** 4.3–5 mm. $2n = 14$, rarely 28.

Psathyrostachys juncea is native to central Asia, primarily to the Russian and Mongolian steppes. It has become established at various locations, from Alaska to Arizona and New Mexico. It is drought-resistant and tolerant of saline soils. In its native range, it grows on stony slopes and roadsides, at elevations to 5500 m.

Psathyrostachys juncea closely resembles *Leymus cinereus*, differing primarily in being shorter and in having shorter ligules and a rachis that breaks up at maturity. Immature plants can be identified by the more uniform appearance of the spikelets. *Psathyrostachys juncea* also tends to have smaller spikelets with fewer florets than *L. cinereus*. Plants with pilose florets have been treated as a distinct taxon; such recognition is not merited.

7.18 THINOPYRUM Á. Löve

Mary E. Barkworth

Pl per; csp or not, smt rhz. **Clm** 10–250 cm, usu erect. **Shth** open, glab or ciliate; **aur** 0.2–1.8 mm or absent; **lig** memb; **bld** convolute or flat. **Infl** tml, distichous spikes, usu not dis at maturity, with 1 spklt at all or most nd; **intnd** 5–30 mm. **Spklt** 1–3 times the length of the mid intnd, solitary, appressed to ascending, often diamond-shaped in outline and arching outwards at maturity; **dis** tardy, usu beneath the flt, smt in the rchs. **Glm** rectangular to lanceolate, narrowing beyond midlength, stiff, indurate to coriaceous, glab or with hairs, keeled or rounded at the base, usu more strongly keeled distally than proximally, 4–9-veined, midveins usu scabrous distally, mrg often hyaline, apc truncate to acute, smt mucronate, unawned, without lat teeth; **lm** 5-veined, coriaceous, glab or with hairs, truncate, obtuse, or acute, smt mucronate or awned, awns to 3 cm; **anth** 3, 2.5–12 mm. $x = 7$. Name from the Greek *thino*, a shore weed, and *pyros*, 'wheat'.

Thinopyrum includes approximately ten species, most of which are alkaline tolerant. It is native from the Mediterranean region to western Asia. Two species are established in the Intermountain Region. The genus is sometimes included in *Elytrigia* Desv. or *Elymus*.

Thinopyrum differs from the other *Triticeae* in its thick, stiff glumes and lemmas. These are several cells thick, even between the veins (Jarvie and Barkworth 1992b).

In the descriptions, measurements and comments about rachis internodes refer to the internodes at midspike. The lowest internodes of a spike are usually 2–4 times as long as those at midspike.

1. Plants rhizomatous; glumes obliquely truncate or obtuse to acute, midveins usually slightly longer and more prominent than the lateral veins. . . . . 1. *T. intermedium*
1. Plants not rhizomatous; glumes truncate, midveins about equal in length and prominence to the lateral veins . . . . 2. *T. ponticum*

1. **Thinopyrum intermedium** (Host) Barkworth & D.R. Dewey [p. *341*, <u>498</u>]

Pl rhz, often glaucous. **Clm** 50–115 cm, glab or hairy, smt hairy only on the nd; **lowest intnd plus shth** 3–5 mm thick. **Shth** mostly glab, often ciliate on the mrg; **aur** 0.5–1.8 mm; **lig** 0.1–0.8 mm; **bld** 2–8 mm wide, flat, abx surfaces glab, adx surfaces usu sparsely strigose, smt with hairs of mixed lengths, with 7–30 ribs, ribs not prominent, mrg whitish, thicker than the veins. **Spikes** 8–21 cm, erect or lax; **intnd** 7–12 mm; **rchs** glab or with hairs, scabrous on the edges, particularly distally, not dis at maturity. **Spklt** 11–18 mm, with 3–10 flt; **dis** beneath the flt. **Glm** oblong, glab and mostly smooth, or strigose with 1–1.5 mm hairs, hairs usu evenly distributed, weakly keeled distally, keels scabrous, at least distally, midvein usu more prominent and longer than the lat veins, mrg not hyaline or hyaline near the apc, apc obliquely truncate or obtuse to acute, smt mucronate; **lo glm** 4.5–7.5 mm long, 1.5–2.5 mm wide, 5–6-veined; **up glm** 5.5–8.5 mm long, 2–3 mm wide, 5–7-veined; **lm** 7.5–10 mm, glab or with 1–1.5 mm hairs, hairs usu evenly distributed, smt only on the outer portion of the lm, apc occ awned, awns to 5 mm; **pal** 7–9.5 mm, keels usu scabrous for ½ their length; **anth** 5–7 mm. $2n = 42, 43$.

Thinopyrum intermedium is native to Europe and western Asia. It is widely established in western North America, having been introduced for erosion control, revegetation, forage, and hay. One of its advantages for erosion control and revegetation is that it establishes rapidly in many different habitats. In Europe, it forms sterile hybrids with *Elymus repens*; no such hybrids are known from North America.

Several subspecies have been recognized within *Thinopyrum intermedium*, usually based on differences in the vestiture of the glumes and lemmas, the presence or absence of lemma awns, and the color of the plants. Assadi (1994) commented that there was little correlation between the different character states. He grew seeds from several wild plants and, even when most of the offspring resembled the parent plant, there was often segregation of some variants. Crossing experiments showed that hybrids between the morphological variants were fertile, and usually had regular meiosis. He noted, however, that the plants with glabrous spikelets tended to grow in mesophytic habitats, those with hairy glumes and lemmas on dry slopes, and those with ciliate glumes and lemmas at the edges of fields and in wet places. This difference in habitat preference was reiterated by Ogle (2001). Because of this ecological distinction, they are formally recognized here as subspecies. Plants with hairs only near the lemma margins are included under *T. intemedium* subsp. *intermedium*. They may be derived from crosses between the hairy and glabrous plants, a possibility that has not been experimentally evaluated. There seems to be little correlation between spikelet vestiture and that of the leaves and stems.

1. Lemmas and glumes glabrous . . . . . subsp. *intermedium*
1. Lemmas with hairs, sometimes only on the margins, hairs 1–1.5 mm long; glumes usually hairy throughout, sometimes glabrous but scabrous over the veins. . . . . subsp. *barbulatum*

Thinopyrum intermedium subsp. **barbulatum** (Schur) Barkworth & D.R. Dewey HAIRY WHEATGRASS [p. 341]

Glm usu hairy throughout, smt merely scabrous on the veins; lm hirsute throughout, smt only on the mrg, hairs 1–1.5 mm.

There is no known difference in geographic distribution between subsp. *barbulatum* and subsp. *intermedium* in the Intermountain Region. Ogle (2001) states that *T. intermedium* subsp. *barbulatum* is adapted to areas with 11–12 inches of rainfall per year.

Thinopyrum intermedium (Host) Barkworth & D.R. Dewey subsp. **intermedium** INTERMEDIATE WHEATGRASS [p. 341]

Glm usu glab, smooth or scabrous over 1 or more veins; lm usu glab, smt with hairs on the mrg.

There is no known difference in geographic distribution between subsp. *intermedium* and subsp. *barbulatum* in the Intermountain Region. Ogle (2001) states that *T. intermedium* subsp. *intermedium* is adapted to areas with 12–13 inches of rainfall per year.

2. Thinopyrum ponticum Barkworth & D.R. Dewey TALL WHEATGRASS, RUSH WHEATGRASS [p. 342, 498]

Pl ces, not rhz. Clm 50–200 cm, glab; **lowest intnd plus shth** about 3.5 mm thick. Shth ciliate on the lo mrg; **aur** 0.2–1.5 mm; **lig** 0.3–1.5 mm; **bld** 2–6.5 mm wide, generally convolute, adx surfaces with 1–8 ribs, ribs rounded, prominent, spinulose, mrg usu thinner than the ribs. Spikes 10–42 cm, erect; **intnd** 9–19 mm; **rchs** glab, not **dis** at maturity. Spklt 13–30 mm, with 6–12 flt; **dis** beneath the flt. Glm oblong, glab, 5–9-veined, midveins about equal in length and prominence to the lat veins, mrg about 0.5 mm wide, hyaline, apc truncate; **lo glm** 6.5–10 mm, midveins occ scabrous distally; **up glm** 7–10 mm; **lm** 9–12 mm, glab; **pal** 7.5–11 mm, keeled, keels ciliate; **anth** 4–6 mm. $2n = 69, 70$.

Thinopyrum ponticum is native to southern Europe and western Asia. In its native range, it grows in dry and/or saline soils. In the Intermountain Region, *T. ponticum* is planted along roadsides for soil stabilization, and is spreading naturally because of its tolerance of the saline conditions caused by salting roads in winter.

8. POEAE R. Br.

Mary E. Barkworth

Pl ann or per; csp, rhz, or stln. Clm ann, not woody, not brchg above the base; **intnd** usu hollow. Shth usu open for most of their length, smt closed; **col** without tufts of hair on the sides; **aur** usu absent; **lig** memb to hyaline, smt ciliate, those of the up and lo cauline lvs usu similar; **psdpet** not developed; **bld** linear to narrowly lanceolate, venation parallel, cross venation not evident, without arm or fusoid cells, epdm without microhairs, not papillate, cross sections non-Kranz. Infl tml, usu pan, smt spikes, pan smt spikelike or rdcd to rcm in depauperate specimens; **dis** usu above the glm and beneath the flt, smt below the glm. Spklt 0.7–50 mm, lat compressed, smt weakly so, smt viviparous, usu with 2–22 flt, smt with 1, strl flt usu distal to the reproductively fnctl flt, smt with 1 or 2 stmt or strl flt below a bisex flt, strl flt often rdcd in size; **rchl** smt prolonged beyond the base of the distal flt. Glm (0,1)2, equal or unequal, shorter or longer than the adjacent flt, smt exceeding the distal flt; **flt** lat compressed; **cal** glab or hairy, not well developed; **lm** lanceolate to ovate, 1–7(9)-veined, unawned or awned, veins usu converging distally, smt parallel, awns from bas to tml on the lm, straight or bent; **pal** 2-keeled, from shorter than to longer than the lm, smt absent or minute; **lod** 2, memb, not or weakly veined; **anth** 3; **ov** usu glab, smt hairy distally; **sty** 2, bases free. Car longitudinally grooved or not, not beaked, pericarp thin; **hila** punctate to linear; **emb** from ¼–⅓ as long as the car. $x = 7$.

The *Poeae* constitute the largest tribe of grasses, encompassing around 115 genera and 2500 species. The species are primarily cool-temperate to arctic in their distribution. In the Intermountain Region, there are 36 non-hybrid genera with 146 species, and 1 hybrid genus with one species. Many of the tribe's species are well known as lawn and pasture grasses, for example, *Poa pratensis* (Kentucky bluegrass), *Dactylis glomerata* (orchard grass), and *Phleum pratense* (timothy).

The tribe's circumscription and its infratribal taxonomy are unclear. It is interpreted here as including generic groups that are, or have been, treated in other works as tribes (e.g., *Agrostideae* Dumort., *Aveneae* Dumort., *Hainardeae* Greut., and

Phalarideae Dumort.). Some of these are sometimes recognized as subtribes, often with modified circumscriptions. Recent studies (e.g., Catalán et al. 2004, Bouchenak-Khelladi et al. 2008) indicate that there are some infratribal groupings that, based on chloroplast DNA data, appear stable; other groupings do not. In addition, there is little support for the monophyly of some genera, notably *Festuca* and its allies.

The following key does not include the hybrid genus ×*Agropogon* (*Agrostis* × *Polypogon*). It is described on page 146. In the key that follows, branch measurements include spikelets, but not awns.

1. All or most spikelets asexual, producing plantlets with a bulblike base, rather than florets with a lemma and palea . . . . 8.10 *Poa* (in part)
1. All or most spikelets developing sexual florets, each floret having a lemma and palea.
  2. Inflorescences spikelike, with 1–2(4) spikelets per node.
    3. Spikelets with 1 functional floret . . . . 8.22 *Scribneria*
    3. Spikelets with 2–25 functional florets.
      4. Lower lemmas in each spikelet awned from about midlength, awns 10–26 mm long, twisted on the lowest section; upper lemmas more shortly awned . . . . 8.25 *Helictotrichon* (in part)
      4. Lemmas unawned or terminally awned, awns straight.
        5. Spikelets sessile, most spikelets with 1 glume, only the terminal spikelet with 2 glumes . . . . 8.05 *Lolium*
        5. Spikelets subsessile to pedicellate, all spikelets with 2 glumes.
          6. Plants perennial . . . . 8.01 *Festuca* (in part)
          6. Plants annual.
            7. Inflorescences usually exceeded by the leaves; lemmas 3.4–7 mm long, apices round to emarginate, not bifid; culms usually prostrate or procumbent . . . . 8.07 *Sclerochloa* (in part)
            7. Inflorescences usually exceeding the leaves; lemmas 2–3 mm long, apices acute to obtuse, sometimes bifid; culms procumbent to erect . . . . 8.20 *Desmazeria* (in part)
  2. Inflorescences panicles or racemes with more than 1 spikelet at the lower nodes.
    8. Inflorescences racemes or spikelike panicles, all inflorescence branches shorter than 1 cm.
      9. Leaves usually exceeding the inflorescences; culms usually prostrate to procumbent; lemmas indurate at maturity . . . . 8.07 *Sclerochloa* (in part)
      9. Leaves usually exceeded by the inflorescences; culms usually erect or decumbent at the base; lemmas usually membranous or papery, sometimes coriaceous, not indurate.
        10. Spikelets disarticulating below the glumes or, if the spikelets are attached to stipes (which resemble calluses), at the base of the stipe; glume bases sometimes fused.
          11. Spikelets weakly laterally compressed, borne on stipes that fall with the spikelet (and resemble an elongated callus); glumes usually awned, glume bases not fused . . . . 8.16 *Polypogon* (in part)
          11. Spikelets strongly laterally compressed, without stipes; glumes unawned or awned, glume bases sometimes fused.
            12. Lemmas dorsally awned; spikelets oval in outline; glumes often fused at the base, often winged distally, keels sometimes ciliate, apices never abruptly truncate . . . . 8.36 *Alopecurus* (in part)
            12. Lemmas usually unawned, occasionally subterminally awned; spikelets U-shaped in outline; glumes not fused at the base, strongly ciliate on the keels, abruptly truncate to an awnlike apex . . . . . . 8.18 *Phleum* (in part)
        10. Spikelets disarticulating above the glumes; glume bases not fused.
          13. Spikelets with 2–25 bisexual florets, the sterile or staminate florets, if present, distal to the bisexual florets.
            14. Sheaths closed for at least ½ their length; lemmas with 3–5 excurrent veins . . . . 8.19 *Sesleria*
            14. Sheaths open for all or most of their length.
              15. Lemmas coriaceous at maturity, unawned . . . . 8.20 *Desmazeria* (in part)
              15. Lemmas membranous at maturity, apically awned, awns 0.3–2.2 mm long . . . . 8.04 *Vulpia* (in part)

13. Spikelets with 1 bisexual floret, sometimes with 1–2 sterile florets below the bisexual floret, these from larger than to much smaller than the bisexual floret, sometimes reduced to pubescent lemmas and easily mistaken for tufts of callus hairs.

    16. Spikelets with 1–2 sterile or staminate florets below the bisexual florets; these from larger than to much smaller than glumes sometimes winged distally. . . . . 8.34 *Phalaris* (in part)

    16. Spikelets without sterile or staminate florets below the bisexual floret; glumes not winged distally.

        17. Spikelet bases U-shaped; glumes equal, strongly keeled, keels strongly ciliate . . . . 8.18 *Phleum* (in part)

        17. Spikelet bases cuneate; glumes unequal, not strongly keeled, midveins not ciliate.

            18. Lemmas unawned or with awns up to 4 mm long; plants perennial or annual . . . . 8.15 *Agrostis* (in part)

            18. Lemma awns 4–16 mm long; plants annual . . . . 8.37 *Apera* (in part)

8. Inflorescences panicles, dense to open, sometimes compact, usually at least some branches longer than 1 cm.

    19. Panicle branches appearing 1-sided; spikelets strongly imbricate, subsessile.

        20. Culms usually prostrate or procumbent; glumes obtuse to emarginate . . . .
. . . . 8.07 *Sclerochloa* (in part)

        20. Culms erect or ascending; glumes apiculate to awn-tipped.

            21. Lemmas awned, awns of the lowest lemmas in the spikelet 0.3–22 mm long . . . . 8.04 *Vulpia* (in part)

            21. Lemmas unawned, sometimes awn-tipped.

                22. Spikelets circular to ovate or obovate in outline, with 1–2 florets; glumes almost entirely concealing the sides of the florets; disarticulation below the glumes . . . . 8.09 *Beckmannia*

                22. Spikelets oval in outline, longer than wide, with 2–6 florets; glumes partially exposing the sides of the florets; disarticulation above the glumes. . . . . 8.08 *Dactylis*

    19. Panicle branches not appearing 1-sided; spikelets usually widely spaced to somewhat imbricate, usually clearly pedicellate, sometimes subsessile, sometimes on stipes.

        23. All or most spikelets in an inflorescence with 1 bisexual floret, sometimes 1–2 sterile or staminate florets below the bisexual floret, the sterile florets sometimes resembling tufts of hair . . . . *Poeae* Subkey I

        23. All or most florets in an inflorescence with 2-25 bisexual florets, usually all florets bisexual or the distal florets sterile or unisexual, sometimes all florets unisexual, sometimes the plants unisexual. . . . . *Poeae* Subkey II

Poeae Subkey I

Synoecious or monoecious grasses with panicles having at least some branches longer than 1 cm and spikelets with 1 bisexual floret, sometimes with 1–2 sterile or staminate florets below the bisexual floret.

1. Spikelets with 1–2 staminate or sterile florets below the bisexual floret, sterile florets sometimes knoblike or resembling tufts of hair.

    2. Spikelets with 2 florets of similar size, the lower floret staminate; lower lemmas awned, upper lemmas unawned or awned . . . . 8.30 *Arrhenatherum*

    2. Spikelets with 2–3(4) florets, the lower 1–2 florets staminate or sterile, sometimes knoblike or resembling tufts of hair; lemmas of the lower florets awned or unawned, lemma of the terminal floret unawned.

        3. Spikelets brown at maturity; lower sterile florets 2, larger than the bisexual floret; fresh leaves sweet smelling when crushed . . . . 8.33 *Anthoxanthum*

        3. Spikelets green to purplish at maturity; lower sterile florets 1–2, varying from knoblike projections to on the callus of the bisexual floret to linear or lanceolate lemmas up to ¾ as long as the bisexual floret . . . . 8.34 *Phalaris* (in part)

1. Spikelets without sterile or staminate florets below the bisexual floret.

4. Spikelets 15–50 mm long; lemmas usually dorsally awned, awns 20–90 mm long,
   sometimes unawned . . . . 8.28 *Avena* (in part)
4. Spikelets 1–15 mm long; lemmas unawned or awned, awns to 18 mm long, basal, dorsal,
   subterminal, or terminal.
   5. Lemmas awned, awns longer than 2 mm.
      6. Disarticulation below the glumes.
         7. Spikelets borne on calluslike stipes; disarticulation at the base of the stipes;
            lemmas 0.5–2 mm long; glumes usually awned . . . . 8.16 *Polypogon* (in part)
         7. Spikelets borne on pedicels; disarticulation immediately below the glumes;
            lemmas 1.5–7.5 mm long; glumes usually unawned.
            8. Paleas absent or greatly reduced; lemma awns attached at midlength
               or below; glume bases often fused; rachillas not prolonged beyond the
               floret base . . . . 8.36 *Alopecurus* (in part)
            8. Paleas at least ¾ as long as the lemmas; lemma awns subterminal;
               glume bases not fused; rachillas usually prolonged beyond the base of
               the distal floret as a stub or slender bristle . . . . 8.35 *Cinna* (in part)
      6. Disarticulation above the glumes.
         9. Rachillas not prolonged beyond the base of the distal floret; paleas absent,
            minute or subequal to the lemmas; lemmas 0.5–4 mm long . . . . 8.15 *Agrostis* (in part)
         9. Rachillas prolonged beyond the base of the distal floret; paleas at least ½ as
            long as the lemmas; lemmas 1–8 mm long.
            10. Plants annual; calluses glabrous or sparsely hairy; lemma apices entire;
               marginal veins not excurrent; lemma awns subterminal . . . . 8.37 *Apera* (in part)
            10. Plants perennial; calluses usually abundantly, sometimes sparsely hairy,
               hairs 0.2–6.5 mm long; lemma apices denticulate; awn attachment from
               nearly basal to subterminal. . . . . 8.26 *Calamagrostis* (in part)
   5. Lemmas unawned or, if awned, awns shorter than 2 mm.
      11. Disarticulation below the glumes, the spikelets falling intact.
         12. Disarticulation of the spikelets 0.1–0.7 mm below the glumes, at the base of
            the stipe; glumes usually awned, awns flexuous. . . . . 8.16 *Polypogon* (in part)
         12. Disarticulation of the spikelets immediately below the glumes; glumes
            unawned or with stiff awns.
            13. Lemma awns subterminal; glume bases not fused; paleas at least ¾ the
               length of the lemmas; rachillas prolonged beyond the distal floret for
               0.1–0.3 mm . . . . 8.35 *Cinna* (in part)
            13. Lemma awns attached at midlength or below; glume bases often fused;
               paleas absent or less that ¾ the length of the lemmas; rachillas not
               prolonged beyond the distal floret. . . . . 8.36 *Alopecurus* (in part)
      11. Disarticulation above the glumes, below the florets.
         14. Paleas often absent or highly reduced, sometimes subequal to the lemmas,
            always lacking veins; rachilla not prolonged beyond the distal floret; lemmas
            usually unawned, if awned, awn attachment from basal to terminal. . . 8.15 *Agrostis* (in part)
         14. Paleas at least ½ as long as the lemmas, 2-veined; rachillas prolonged
            beyond the distal floret for at least 0.1 mm; lemmas often awned, awns
            usually attached to the lower ½ of the lemmas.
            15. Calluses hairy, hairs 0.5–4.5 mm long; lemmas usually awned, awns
               attached to the lower ½ of the lemmas, if higher, the awn hairs longer
               than 2 mm. . . . . 8.26 *Calamagrostis* (in part)
            15. Calluses glabrous or with sparse hairs up to 0.5 mm long; lemmas
               usually unawned, sometimes subapically awned, awns to 1.3 mm
               long . . . . 8.24 *Podagrostis*

Poeae Subkey II

Synoecious, monoecious, or dioecious grasses with spikelets having 2–22 florets, the lower or all
florets sexual, the distal florets sometimes sterile.

1. One or both glumes exceeding the adjacent lemmas, sometimes exceeding the distal
   floret.
   2. All lemmas within a spikelet unawned or with awns shorter than 2 mm.

    3. Spikelets usually with 2 florets; lemmas of the lower florets unawned, lemmas of the upper florets awned, the awns strongly curved or hooked . . . . 8.29 *Holcus* (in part)
    3. Spikelets with 2–33 florets; all lemmas unawned or, if awned, the awns straight.
        4. Glumes 15–50 mm long; plants annual . . . . 8.28 *Avena* (in part)
        4. Glumes 0.4–9 mm long; plants annual or perennial.
            5. Rachilla internodes hairy, hairs at least 1 mm long on the distal portion.
                6. Lemma apices truncate, erose or 2–4-toothed . . . . 8.14 *Deschampsia* (in part)
                6. Lemma apices acute, bifid . . . . 8.31 *Trisetum* (in part)
            5. Rachilla internodes glabrous or with hairs shorter than 1 mm on the distal portion.
                7. Plants strongly rhizomatous; glumes 5–9 mm long. . . . . 8.27 *Scolochloa* (in part)
                7. Plants not or weakly rhizomatous; glumes 0.4–9 mm long.
                    8. Panicle branches densely hairy, hairs 0.1–0.2 mm long; lemma apices entire, sometimes mucronate; lemma veins converging distally . . . . . 8.32 *Koeleria* (in part)
                    8. Panicle branches glabrous, sometimes scabrous; lemma apices entire or serrate to erose, not mucronate; lemma veins more or less parallel distally. . . . . 8.06 *Puccinellia* (in part)
  2. One or all lemmas within a spikelet awned, the awns at least 2 mm long.
    9. Lemmas 14–40 mm long; glumes 7–11-veined. . . . . 8.28 *Avena* (in part)
    9. Lemmas 1.3–16 mm long; glumes 1–9-veined.
      10. Lemmas 7–10 mm long . . . . 8.25 *Helictotrichon* (in part)
      10. Lemmas 1.3–7 mm long.
        11. Disarticulation below the glumes
            12. Spikelets usually with 2 florets, the lower floret bisexual and with an unawned lemma, the upper floret staminate or sterile and with an awned lemma . . . . 8.29 *Holcus* (in part)
            12. Spikelets with 2–5 florets, all florets bisexual or the distal floret(s) sterile; all lemmas awned. . . . . 8.31 *Trisetum* (in part)
        11. Disarticulation above the glumes.
            13. Lowest lemma within a spikelet unawned or with a straight awn up to 9 mm long, the distal lemmas within a spikelet always awned, awns 10-16 mm long, geniculate. . . . . 8.21 *Ventenata* (in part)
            13. All lemmas within a spikelet similarly awned or the awns of the lower lemmas longer than those of the upper lemmas, or the upper lemmas with awns shorter than 10 mm.
                14. Callus hairs about ½ the length of the lemmas; rachillas not prolonged or prolonged for no more than 0.5 mm beyond the base of the distal floret; plants loosely cespitose . . . . 8.23 *Vahlodea*
                14. Calluses usually glabrous or the hairs much shorter than ½ as long as the lemmas, if about ½ as long, the rachillas prolonged more than 0.5 mm beyond the base of the distal floret and the plants usually densely cespitose.
                    15. Panicle branches densely hairy, not scabrous; rachilla internodes glabrous or with hairs shorter than 1 mm on the distal portion . . . . 8.32 *Koeleria* (in part)
                    15. Panicle branches usually glabrous, sometimes scabrous; rachilla internodes hairy, hairs at least 1 mm long on the distal portion.
                      16. Lemma apices truncate, erose or 2–4-toothed . . . 8.14 *Deschampsia* (in part)
                      16. Lemma apices acute, bifid . . . . 8.31 *Trisetum* (in part)
1. Both glumes shorter than or subequal to the adjacent lemmas.
  17. Upper lemma(s) in a spikelet with hooked or geniculate awns, awns 2–16 mm long; lowest lemma in each spikelet unawned or terminally awned, awns to 4 mm long, straight.
    18. Spikelets 9–15 mm long, with 2–20 florets; awns of the distal florets 10–16 mm long, geniculate . . . . 8.21 *Ventenata* (in part)
    18. Spikelets 3–7 mm long, with 2 florets; awns of the distal florets 2–5 mm long, hooked. . . . . 8.29 *Holcus* (in part)
  17. Lemmas all similarly awned or unawned.

19. Lower lemmas with awns longer than 2 mm.
    20. Calluses hairy; rachillas prolonged beyond the base of the distal floret; glumes subequal to the adjacent lemmas. . . . . 8.31 *Trisetum* (in part)
    20. Calluses glabrous or sparsely hairy; rachillas sometimes prolonged beyond the base of the distal floret; one or both glumes often shorter than the adjacent lemmas.
        21. Plants annual; anthers 1 . . . . 8.04 *Vulpia* (in part)
        21. Plants perennial; anthers 3.
            22. Leaves without auricles; blades flat, conduplicate, involute or convolute. . . . . 8.01 *Festuca* (in part)
            22. Leaves with auricles; blades flat . . . . 8.03 *Schedonorus* (in part)
19. Lower lemmas unawned, mucronate, or with awns up to 2 mm long.
    23. Lemma apices rounded, truncate, obtuse, or emarginate.
        24. Lemmas conspicuously 3-veined; lower glumes 0–3-veined . . . . 8.12 *Catabrosa*
        24. Lemmas (3)5–9-veined, the veins often inconspicuous; lower glumes 1–5-veined.
            25. Inflorescences usually exceeded by the leaves; lemmas indurate at maturity; pedicels 0.5–0.8 mm thick; culms usually prostrate to procumbent, sometimes ascending; upper glumes 2.6–6.2 mm long . . . . 8.07 *Sclerochloa* (in part)
            25. Inflorescences exceeding the leaves at maturity; lemmas usually membranous at maturity, sometimes coriaceous; pedicels less than 0.5 mm thick; culms usually erect to ascending; upper glumes 0.7–4.5(9) mm long.
                26. Lower glumes about as long as the upper glumes but no more than ½ as wide; disarticulation below the glumes . . . . 8.13 *Sphenopholis* (in part)
                26. Lower glumes shorter than the upper glumes or subequal and more than ½ as wide; disarticulation above the glumes.
                    27. Panicle branches stiff; lemmas coriaceous at maturity; plants annual; culms to 60 cm tall . . . . 8.20 *Desmazeria* (in part)
                    27. Panicle branches flexible; lemmas usually membranous at maturity, sometimes coriaceous; plants perennial or annual; culms 2–145 cm tall.
                        28. Lemma veins excurrent, making the lemma apices indistinctly 3-lobed or toothed; plants strongly rhizomatous; rhizomes succulent . . . . 8.27 *Scolochloa* (in part)
                        28. Lemma veins not excurrent; lemma apices entire, serrate, or erose; plants sometimes rhizomatous, rhizomes not succulent.
                            29. Lemmas with (3)5(7) veins; plants of saline and alkaline habitats; panicle branches usually straight. . . . . 8.06 *Puccinellia* (in part)
                            29. Lemmas with (5)7–9 veins; plants of non-saline and non-alkaline habitat; panicle branches usually flexuous . . . . 8.11 *Torreyochloa* (in part)
    23. Lemma apices acute to acuminate, sometimes mucronate or shortly awn-tipped.
        30. Lemmas (3)5–9-veined, the veins more or less parallel distally, conspicuous.
            31. Lemmas with (3)5(7) veins; plants not truly rhizomatous, sometimes the culms rooting at buried lower nodes, growing in saline and alkaline habitats; panicle branches usually straight . . . . 8.06 *Puccinellia* (in part)
            31. Lemmas with (5)7–9 veins; plants rhizomatous, growing in non-saline and non-alkaline habitats; panicle branches usually flexuous . . . 8.11 *Torreyochloa* (in part)
        30. Lemmas 3–9-veined, veins converging distally, usually inconspicuous, sometimes conspicuous.
            32. Disarticulation below the glumes; lower glumes about equaling the upper glumes in length but only ½ as wide . . . . 8.13 *Sphenopholis* (in part)
            32. Disarticulation above the glumes, sometimes above the basal floret; lower glumes shorter than the upper glumes or, if subequal, more than ½ as wide.

33. Panicle branches hairy, hairs soft. . . . . 8.32 *Koeleria* (in part)
33. Panicle branches smooth, scabrous, or with stiff hairs.
    34. Basal leaves with auricles. . . . . 8.03 *Schedonorus* (in part)
    34. No leaves with auricles.
        35. Lemma veins parallel distally; plants of saline and alkaline
            habitats . . . . 8.06 *Puccinellia* (in part)
        35. Lemma veins converging distally; plants of many habitats,
            including saline and alkaline habitats.
            36. Leaf blades with a translucent line on either side
                of the midvein, blade apices often prow-tipped;
                lemmas often with a tuft of hairs at the base of the
                midvein; hila round to oval . . . . 8.10 *Poa* (in part)
            36. Leaf blades without translucent lines on either side
                of the midvein, blade apices not prow-tipped, often
                flat; lemmas without a tuft of hairs at the base of the
                midvein; hila usually linear.
                37. Plants perennial.
                    38. Plants and florets bisexual; glumes not
                        translucent; caryopses obovoid to oblong . . .
                        . . . . 8.01 *Festuca* (in part)
                    38. Plants and florets unisexual; glumes trans-
                        lucent; caryopses fusiform . . . . 8.02 *Leucopoa*
                37. Plants annual.
                    39. Ligules up to 1 mm long; lemma apices
                        mucronate or awned . . . . 8.04 *Vulpia* (in part)
                    39. Ligules 1–4 mm long; lemma apices never
                        awned, sometimes mucronate . . . . 8.20 *Desmazeria* (in part)

8.01 FESTUCA L.   Stephen J. Darbyshire and Leon E. Pavlick†

Pl per; bisex; usu densely to loosely csp, with or without rhz, occ stln. **Clm** 5–150(275) cm, usu glab and smooth throughout, smt scabrous or densely pubescent below the infl. **Shth** from open to the base to closed almost to the top, in some species shth of previous years persisting and the bld usu deciduous, in other species the senescent shth rapidly shredding into fibers and decaying between the veins and the bld not deciduous; **col** inconspicuous, usu glab; **aur** absent; **lig** 0.1–2(8) mm, memb, smt longest at the mrg, usu truncate, smt acute, usu ciliate, smt erose; **bld** flat, conduplicate, involute, or convolute, smt glaucous or pruinose, abx surfaces usu glab or scabrous, smt puberulent or pubescent, rarely pilose, adx surfaces usu scabrous, smt hirsute or puberulent, with or without ribs over the major veins; **abx sclerenchyma tissue** varying from longitudinal strands at the mrg and opposite the midvein to adjacent to some or all of the lat veins, longitudinal strands smt lat confluent with other strands into an interrupted or continuous band, smt reaching to the veins and forming pillars; **adx sclerenchyma tissue** smt present in strands opposite the veins at the epdm, the strands smt extending to the veins and, in combination with the abx sclerenchyma, forming girders of sclerenchyma tissue extending from one epdm to the other at some or all of the veins. **Infl** usu open or contracted pan, smt rdcd to rcm, usu with 1–2(3) br at the lo nd; **br** usu erect, spreading to widely spreading at anthesis, smt the lo br reflexed; **Spklt** with (1)2–10 mostly bisex flt, distal flt rdcd or abortive; **rchl** usu scabrous or pubescent, smt smooth and glab; **dis** above the glm, beneath the flt. **Glm** subequal or unequal, usu exceeded by the flt, ovate to lanceolate, acute to acuminate; **lo glm** from shorter than to about equal to the adjacent lm, 1(3)-veined; **up glm** 3(5)-veined; **cal** usu wider than long, usu glab and smooth, smt scabrous, occ pubescent; **lm** usu chartaceous, smt coriaceous, bases more or less rounded dorsally, slightly or distinctly keeled distally, veins 5(7), prominent or obscure, apc acute to attenuate, smt minutely bidentate, usu tml or subtml awned or mucronate; **pal** from shorter than to slightly longer than the lm, veins sparsely to densely scabrous-ciliate, intercostal region usu smooth and glab at the base, usu scabrous and/or puberulent distally, bidentate; **anth** 3; **ov** glab or with hispidulous apc, hairs persisting on the mature car. **Car** obovoid-oblong, adx grooved, usu free of the lm and pal, smt adhering along the groove, smt adhering more broadly; **hila** linear, from ½ as long as to almost as long as the car. $x = 7$. Name from the Latin *festuca*, 'stalk', 'stem', or 'straw' — a name used by Pliny for a weed.

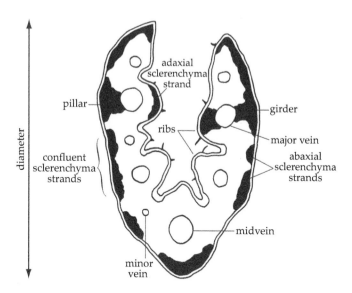

Figure 1. *Festuca* blade cross section

Festuca is a widespread genus, probably having more than 500 species. The species grow in alpine, temperate, and polar regions of all continents except Antarctica. Of the 21 species treated below, 16 are native to the Intermountain Region. One species, *F. rubra*, is represented by both native and introduced subspecies. Many native species provide good forage in western North American grasslands and montane forests. Important cultivated species include *Festuca rubra*, grown for forage and as a turf grass, and *F. trachyphylla*, used as a turf grass and for erosion control. A number of species are cultivated as ornamentals. Of these, only *F. amethystina* and *F. glauca* are included in this account.

The taxonomy of the genus is problematic and contentious, and this treatment is far from definitive. Keying the species ultimately relies on characters that are sometimes difficult to detect on herbarium specimens, such as ovary pubescence and leaf blade sclerenchyma patterns. The key provided here does not rely exclusively upon such characters, but due to intraspecific variability, combinations of overlapping characters must be employed for identification. Any resulting ambiguity in the key may need to be resolved by consulting the species descriptions and illustrations, which retain the information on the more difficult characters.

The distribution of some taxa that are grown for turf, revegetation, and, to a lesser extent, horticulture—such as *Festuca rubra* subsp. *rubra*, *F. trachyphylla*, and *F. valesiaca*—is continually expanding because of their wide commercial availability. The occurrence of these in the Intermountain Region is no doubt more extensive than current herbarium collections indicate.

The distribution of sclerenchyma tissue within the vegetative shoot leaves is often an important diagnostic character. It should be observed in cross sections made from mature, but not senescent, leaves of vegetative shoots, ¼ to halfway up the blades; the sections can be made freehand, with a single-edged razor blade. Sections are best viewed at 40×, and with transmitted light (polarized if possible). Important features seen in the blade cross section are identified in Fig. 1, above.

The main sclerenchyma distribution patterns in *Festuca* are shown in Fig. 2, p. 88 (not all of these patterns are found in the species of the Intermountain Region). Almost all species have a strand of sclerenchyma tissue along the margins and opposite the midvein against the abaxial epidermis (Fig. 2C). Strands may be narrow (about as wide as the adjacent veins or narrower, Fig. 2D) to broad (wider than the adjacent veins, Fig. 2E). Additional strands are often present at the abaxial surface opposite the veins; these strands may be confluent (Figs. 1, 2E), sometimes combining to form a cylinder around the leaf and appearing as a continuous ring or band in the cross sections (Figs. 2F, I). Some species have additional strands on the adaxial surface opposite some or all of the veins (Figs. 2G, H). Another variant is for the abaxial sclerenchyma strands to extend inwards to some or all of the vascular bundles (veins), forming pillars in the cross sections (Figs. 2H, I). If both the abaxial and adaxial strands extend inward to the vascular bundles, they are said to form girders (Figs. 2A, B, I).

Some of the patterns described may co-occur within a leaf. For instance, some veins may be associated with pillars, others with girders; some sclerenchyma strands within a leaf may be confluent, whereas others are not. Although there may be considerable variation in the extent of sclerenchyma development, the general pattern within a species is usually constant. It is this that makes such patterns useful diagnostic characters, particularly for those needing to identify plants in vegetative condition.

Descriptions of leaf blades are based on the leaves of the basal vegetative shoots, where present. For those without basal tufts of vegetative shoots, the cauline leaves are described. Width measurements are provided for leaves that are usually flat, or almost so, when encountered in the field or herbarium. "Diameter" is given for leaves that are usually folded or conduplicate when encountered; for leaves that are oval in cross section when folded, it is the largest diameter (or widest dimension).

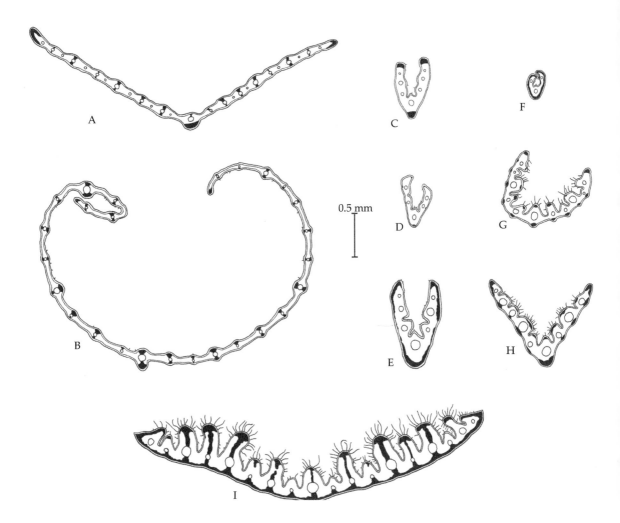

Figure 2. *Festuca* sclerenchyma distribution patterns. A–Leaf blade flat, ribs indistinct, sclerenchyma girders at most veins (*F. subverticillata**, similar to *F. subulata* in the Intermountain Region); **B**–Leaf blade loosely convolute, ribs indistinct, sclerenchyma girders at most veins (*F. subverticillata**, similar to *F. subulata* in the Intermountain Region); **C**–Leaf blade conduplicate, ribs indistinct, sclerenchyma in broad strands at margins and midvein (*F. lenensis**); **D**–Leaf blade conduplicate, ribs indistinct to distinct, sclerenchyma in narrow abaxial strands opposite veins (*F. brachyphylla*); **E**–Leaf blade conduplicate, ribs distinct, sclerenchyma in broad, mostly confluent strands (*F. trachyphylla*); **F**–Leaf blade conduplicate, ribs indistinct, sclerenchyma in a continuous abaxial cylinder or band (*F. filiformis**); **G**–Leaf blade loosely rolled, ribs distinct, sclerenchyma in narrow abaxial and adaxial strands (*F. dasyclada*); **H**–Leaf blade loosely folded; ribs distinct, sclerenchyma in broad abaxial and adaxial strands forming pillars (*F. viridula*); **I**–Leaf blade flat, ribs distinct, sclerenchyma in a continuous abaxial cylinder or band, forming girders at most veins (*F. californica**).

* Species from outside the Intermountain Region

Closure of the leaf sheaths should be checked on young leaves, because the sheaths often split with age, leading to underestimations of the extent of their closure. The fraction of the leaf sheath that is closed varies within and between species of *Festuca*, but the species can be divided into three categories in this regard: those such as *F. rubra*, in which the leaves are closed for at least ¾ their length; those such as *F. saximontana*, in which they are closed from ⅓ to slightly more than ½ their length; and those such as *F. trachyphylla*, in which they are not closed or closed for less than ¼ their length. Lemma awns tend to be longer, and should be measured, on the distal florets within a spikelet.

A multiaccess, interactive key is available at http://utc.usu.edu/keys/IMRFestuca.

1. Primary and secondary panicle branches stiffly and strongly divaricate at maturity, their angles densely scabrous or ciliate; culms densely scabrous or pubescent below the inflorescence; spikelets 5.5–8 mm . . . . 21. *F. dasyclada*
1. Primary and secondary panicle branches lax or stiff, erect to ascending or spreading at maturity, at least the secondary branches not stiffly divaricate, branch angles smooth

or scabrous; culms smooth or scabrous, glabrous or pubescent below the inflorescence; spikelets (2.5)3–15(19) mm.

2. Lemma awns (2.5)5–15(20) mm; inflorescences open, 10–40 cm long, branches lax, usually spreading, sometimes reflexed; blades usually flat or loosely convolute, 3–10 mm wide, veins 13–29. . . . . 1. *F. subulata*

2. Lemmas unawned or with awns to 12 mm, if awns longer than 5 mm, then blades conduplicate and veins (3)5(7); inflorescences open or contracted, up to 20(40) cm long, branch habit varied, from lax to stiff, erect to spreading or reflexed; blades conduplicate or flat, veins 3-25.

  3. Ligules 2–5(9) mm long; lemmas unawned or with mucros to 0.2 mm . . . . 4. *F. thurberi*

  3. Ligules to 1.5(2) mm long; lemmas unawned, mucronate, or with awns up to 6(7) mm long.

    4. Anthers (3.3)4.5–6 mm long; lower glumes 4.5–7.5(8.5) mm long; upper glumes about as long as lower lemmas; leaf blades densely scabrous abaxially. . . . . 3. *F. campestris*

    4. Anthers 0.3–4.5(5) mm long; lower glumes 1–5.5(7) mm long; upper glumes shorter than lower lemmas; leaf blades smooth to slightly scabrous abaxially.

      5. Blades flat, 3–6(10) mm wide; veins 13–25; sclerenchyma girders often present . . . . . 2. *F. sororia*

      5. Blades conduplicate, convolute or loosely folded and (0.2)0.3–1.3 mm in diameter, or flat and up to 7 mm wide; veins 3–9(13); sclerenchyma girders absent, abaxial pillars sometimes present.

        6. Plants rhizomatous; if rhizomes lacking then sheaths closed more than ½ their length; densely or loosely caespitose; sheaths closed ½ their length or more; sheaths of basal leaves shredding into fibers.

          7. Inflorescences 7–15 cm long; spikelets 7–14.5(17) mm; anthers 2.4–4.5 mm; ovary apices glabrous; adaxial sclerenchyma sometimes present . 5. *F. rubra* (in part)

          7. Inflorescences 3-5(8) cm long; spikelets (4.5)5–6.5(7) mm; anthers 0.6–0.9 (1.4) mm; ovary apices pubescent; adaxial sclerenchyma absent. . . . . 6. *F. earlei* (in part)

      6. Plants not rhizomatous; mostly densely caespitose; sheaths usually closed less than ½ their length; sheaths of basal leaves persistent or shredding into fibers.

        8. Culms densely scabrous to pubescent below the inflorescences.

          9. Lemmas 5.5–9 mm; ligules 0.5–1.5(2) mm long; anthers (2)3–4(4.2) mm long; plants not of alpine habitats . . . . 18. *F. arizonica*

          9. Lemmas 3.5–6 mm long; ligules 0.1–0.3 mm; anthers 0.3–0.7(1.1) mm long; plants of alpine habitats . . . . 14. *F. baffinensis*

        8. Culms usually smooth and glabrous below the inflorescences, sometimes sparsely or minutely scabrous.

          10. Lemma awns 3–12 mm, more than half as long as lemma body; inflorescences open, branches lax, widely spreading to reflexed; anthers (1)1.5–2(3) mm; ovary apices densely pubescent; plants of montane habitats . . . . 16. *F. occidentalis*

          10. Lemmas unawned or with awns 0.2–6(7) mm, usually less than half as long as lemma body or if longer then ovary apices glabrous; inflorescences open or contracted, branches lax or stiff, spreading to erect; anthers 0.3–4.5(5); ovary apices glabrous or if pubescent then plants of alpine and subalpine habitats; plants of various habitats.

            11. Anthers (0.3)0.6–1.7(2) mm; plants of alpine and subalpine habitats.

              12. Lemmas (2)2.2–3.5(4) mm; spikelets (2.5)3–5 mm; anthers (0.4)0.6–1.2 mm . . . . 15. *F. minutiflora*

              12. Lemmas 3–4.5(5.6) mm; spikelets (3)4.4–8.8(10) mm; anthers (0.5)0.6–1.7(2) mm.

                13. Plants usually with short rhizomes; ovary apices pubescent . . . . 6. *F. earlei* (in part)

                13. Plants without rhizomes; ovary apices glabrous.

                  14. Abaxial sclerenchyma in broad bands (wider than adjacent veins) or a continuous ring; anthers (0.8)1.2–1.8(2) mm . . . . 13. *F. saximontana* (in part)

14. Abaxial sclerenchyma in narrow bands (as wide as adjacent veins or less); anthers (0.5)0.6–1.1(1.4) mm . . . . 12. *F. brachyphylla* (in part)

11. Anthers (1.4)2–4.5(5) mm, if 2 mm or less then plants not of alpine or subalpine habitats.

    15. Plant loosely or not at all caespitose; sheaths closed more than ½ their length, basal sheaths shredding into fibers . . 5. *F. rubra* (in part)

    15. Plants densely caespitose; sheaths closed less than ½ their length, basal sheaths persistent and not shredding.

        16. Lemma awns (1.5)2–6(7) mm; native plants common in the Intermountain Region and often dominant in their community . . . . 19. *F. idahoensis*

        16. Lemmas unawned or with awns 0.2–2.5(3) mm; introduced or uncommon native plants.

            17. Ovary apices pubescent, sometimes sparsely so; native plants not common in the Intermountain Region and/or not likely to be dominant in their community.

                18. Abaxial sclerenchyma absent; blades of lower cauline leaves not reduced to stiff horny points; lemmas (3.8)4–6 mm; spikelets (6)7–8(11) mm; ovary apices sparsely pubescent; plants of grasslands and open montane forests. . . . . 17. *F. calligera*

                18. Abaxial sclerenchyma present (Fig. 2H); blades of lower cauline leaves usually reduced to stiff horny points; lemmas (4.8)6–8.5 mm; spikelets 9–15 mm; ovary apices densely pubescent; plants of alpine and subalpine habitats . . . . 20. *F. viridula*

            17. Ovary apices glabrous; introduced plants grown as ornamentals turf or soil stabilization or sometimes escaping.

                19. Lemmas not awned or mucronate; inflorescence (3)8–18(25) cm; anthers (2)3–4 mm; introduced plants grown as ornamentals . . . . 8. *F. amethystina*

                19. Lemma awns 0.5–2(3) mm; inflorescence (2)2.5–13(16) cm; anthers (1.4)2–3.4 mm; introduced plants grown for various purposes.

                    20. Anthers (1.8)2.3–3.4 mm; abaxial sclerenchyma in broad confluent bands (Fig. 2E), rarely a continuous ring; commonly planted for soil stabilization, turf and pasture . . . . 11 *F. trachyphylla*

                    20. Anthers (1.4)2–3 mm; abaxial sclerenchyma in 3 distinct bands or a continuous ring, rarely otherwise; rarely planted.

                        21. Abaxial sclerenchyma usually in 3 distinct bands at mid-rib and margins (Fig. 2C), rarely small bands adjacent to other veins; adaxial ribs distinct. . . . . 7. *F. valesiaca*

                        21. Abaxial sclerenchyma in a continuous or almost continuous ring (Fig. 2F); adaxial ribs indistinct.

22. Plants not glaucous or pruinose; lemmas (2.6)3–4(5) mm; plants persisting in old lawns . . . . 9. *F. ovina*
22. Plants usually glaucous or pruinose; lemmas (3.5)4–6(6.2) mm; plants grown as ornamentals, rarely escaping . . . . 10. *F. glauca*

1. Festuca subulata Trin. Bearded Fescue [p. 343, 498]

Pl loosely ces, without rhz, with short exvag tillers. **Clm** (35)50–100(120) cm, erect or decumbent at the base, scabrous. **Shth** closed for less than ⅓ their length, glab or sparsely pubescent, shredding into fibers; **col** glab; **lig** 0.2–0.6(1) mm; **bld** 3–10 mm wide, flat or loosely convolute, abx and adx surfaces scabrous or puberulent, veins 13–29, ribs obscure; **sclerenchyma** in narrow abx and adx strands; **girders** or **pillars** formed at the major veins. **Infl** 10–40 cm, open, with 1–2(5) br per nd; **br** lax, usu spreading, smt reflexed. **Spklt** 6–12 mm, with (2)3–5(6) flt. **Glm** sparsely scabrous towards the apc, acuminate to subulate; **lo glm** (1.8)2–3(4) mm; **up glm** (2)3–6 mm; **cal** wider than long, glab, smooth or slightly scabrous; **lm** 5–9 mm, glab, smt sparsely scabrous, lanceolate, apc entire, acute to acuminate, awned, awns (2.5) 5–15(20) mm, tml, straight, smt curved or kinked; **pal** about as long as or slightly longer than the lm, intercostal region puberulent distally; **anth** 1.5–2.5(3) mm; **ov apc** pubescent. 2*n* = 28.

Festuca subulata grows on stream banks and in open woods, meadows, shady forests, and thickets, to about 2800 m. Its range includes the northern portion of the Intermountain Region. It is an attractive species.

2. Festuca sororia Piper Ravine Fescue [p. 343, 498]

Pl loosely ces, without rhz, with short exvag tillers. **Clm** 60–100 (150) cm, erect, glab; **nd** usu exposed. **Shth** closed for less than ⅓ their length, glab or scabrous, shredding into fibers; **lig** 0.3–1.5 mm; **bld** 3–6(10) mm wide, flat, lax, mrg scabrous, abx and adx surfaces glab, adx surfaces smt scabrous, veins 13–25, ribs obscure; **abx sclerenchyma** in narrow strands; **adx sclerenchyma** developed; **girders** or **pillars** formed at the major veins. **Infl** 10–20(40) cm, open or somewhat contracted, with 1–2(3) br per nd; **br** lax, more or less spreading, spklt borne towards the ends of the br. **Spklt** 7–12 mm, with (2)3–5 flt. **Glm** lanceolate, scabrous at least on the midvein, acute to acuminate; **lo glm** (1.5) 2.5–4(4.5) mm; **up glm** (3)4–6.5 mm; **cal** wider than long, glab, smooth or slightly scabrous; **lm** (5)6–8(9) mm, lanceolate, scabrous or puberulent, acuminate, unawned or awned, awns to 2 mm; **pal** about as long as or shorter than the lm, intercostal region puberulent distally; **anth** 1.6–2.5 mm; **ov apc** pubescent. 2*n* = unknown.

Festuca sororia grows in open woods and on shaded slopes and stream banks, at 2000–3000 m. Its range extends from central Utah and Colorado to Arizona and New Mexico.

3. Festuca campestris Rydb. Mountain Rough Fescue [p. 344, 498]

Pl densely ces, usu without rhz, occ with short rhz. **Clm** (30)40–90(140) cm, scabrous near the infl; **nd** usu not exposed. **Shth** closed for less than ⅓ their length, glab or scabrous, persistent; **col** glab; **lig** 0.1–0.5 mm; **bld** 0.8–2 mm in diameter, usu conduplicate, rarely convolute, gray-green, deciduous, abx surfaces scabrous, adx surfaces scabrous or puberulent, veins (8)11–15(17), ribs (6)7–11; **abx sclerenchyma** usu forming a more or less continuous band; **adx sclerenchyma** developed; **girders** at the 5–7 major veins; **pillars** at some of the other veins. **Infl** (5)9–18(25) cm, open or loosely contracted, with(1)2(3) br per nd; **br** erect to stiffly spreading. **Spklt** 8–13(16) mm, with (3)4–5(7) flt. **Glm** exceeded by the distal flt; **lo glm** 4.5–7.5(8.5) mm, shorter than or about equaling the adjacent lm; **up glm** 5.3–8.2(9) mm; **lm** (6.2)7–8.5(10) mm, chartaceous to somewhat coriaceous, scabrous, backs rounded below the mid, veins more or less obscure, apc mucronate or shortly awned, awns to 1.5 mm; **pal** somewhat shorter than the lm, intercostal region puberulent distally; **anth** (3.3) 4.5–6 mm; **ov apc** pubescent. 2*n* = 56.

Festuca campestris is a common species in prairies and montane and subalpine grasslands, at elevations to about 2000 m. Within the Intermountain Region, it is known only from southeastern Oregon. It is highly palatable and provides nutritious forage.

4. Festuca thurberi Vasey Thurber's Fescue [p. 344, 498]

Pl densely ces, without rhz. **Clm** (45) 60–100(120) cm, glab, smooth or scabrous below the infl. **Shth** closed for less than ⅓ their length, smooth or scabrous, persistent; **col** glab; **lig** 2–5(9) mm, entire or lacerate, not ciliate; **bld** 1.5–3 mm wide, 0.8–1.8 mm in diameter when conduplicate, deciduous, abx surfaces scabrous, adx surfaces scabrous or pubescent, veins 9–15, ribs 7–13; **abx sclerenchyma** a more or less continuous band; **adx sclerenchyma** present; **girders** usu formed at the major veins, smt only pillars present. **Infl** (7)10–15(17) cm, open, with 1–2(3) br per nd; **br** 4.5–9 cm, lax, erect or spreading, spklt borne towards the ends of the br. **Spklt** (8)10–14 mm, with (3)4–5(6) flt. **Glm** unequal to subequal, ovate-lanceolate, scabrous or smooth, acute; **lo glm** (2)3.5–5.5 mm; **up glm** (2.5)4.5–6.5(7) mm; **lm** 6–10 mm, lanceolate to ovate-lanceolate, scabrous or smooth, unawned, smt mucronate, mucros to 0.2 mm; **pal** shorter than to as long as the lm, intercostal region puberulent distally; **anth** 3–4.5 mm; **ov apc** densely pubescent. 2*n* = 28, 42.

Festuca thurberi is a large bunchgrass of dry, rocky slopes and hills, open forests, and meadows in montane and subalpine regions, at (1000)2000–3500 m. Its range extends from eastern Utah to southern Wyoming, through Colorado to northern New Mexico.

5. Festuca rubra L. Red Fescue [p. 344, 345, 498]

Pl usu rhz, usu loosely to densely ces, clm smt single and widely spaced, smt stln. **Clm** (8)10–120 (130) cm, erect or decumbent, glab and smooth. **Shth** closed for about ¾ their length when young, readily splitting with age, usu pubescent, at least distally, hairs retrorse or antrorse, smt glab, not persistent, shth of older vegetative shoots shredding into fibers; **col** glab; **lig** 0.1–0.5 mm; **bld** usu conduplicate or convolute and 0.3–2.5 mm in diameter, smt flat and 1.5–7 mm wide, abx surfaces glab, smooth or scabrous, adx surfaces scabrous or pubescent, veins 5–9(13), ribs (3)5–7(9), usu conspicuous; **abx sclerenchyma** in 5–9(13) discrete or partly confluent strands, rarely forming a complete band; **adx sclerenchyma** smt present in fascicles opposite the veins; **girders** and **pillars** not developed. **Infl** (2)3.5–25(30) cm, usu open or loosely contracted pan, occ rcm, with 1–3 br per nd, lo br with 2+ spklt; **br** erect or spreading, stiff or lax, glab, scabrous, or pubescent. **Spklt** (6)7–17 mm, with 3–10 flt. **Glm** ovate-lanceolate to lanceolate, exceeded by the distal flt; **lo glm** (1.5)2–6(7) mm; **up glm** (3)3.5–8.5 mm; **lm** 4–9.5 mm, usu glab and smooth, smt scabrous towards the apc, smt densely pubescent throughout, attenuate or acuminate in side view, awned,

awns (0.1)0.4–4.5 mm; **pal** slightly shorter than to about equaling the lm, intercostal region puberulent distally; **anth** 1.8–4.5 mm; **ov apc** glab. 2*n* = 28, 42, 56, 70.

Festuca rubra is interpreted here as a morphologically diverse polyploid complex that is widely distributed in the arctic and temperate zones of Europe, Asia, and North America. Its treatment is complicated because Eurasian material has been introduced in other parts of the world. In addition, hundreds of forage and turf cultivars have been developed, many of which have also been widely distributed.

Within the complex, morphologically, ecologically, geographically, and/or cytogenetically distinct taxa have been described, named, and given various taxonomic ranks. In some cases these taxa represent extremes, and in other cases they are morphologically intermediate between other taxa. Moreover, hybridization and/or introgression between native taxa, and between native and non-native taxa, may be occurring.

Overlap in morphological characters between most taxa in the complex has led some taxonomists to ignore the variation within the complex, calling all its members *Festuca rubra* without qualification. This obscures what is known about the complex, and presents an extremely heterogeneous assemblage of plants as a single "species" — or a mega-species. The following account attempts to reflect the genetic diversity of the *F. rubra* complex in the Intermountain Region. All the taxa are recognized as subspecies, but they are not necessarily equivalent in terms of their distinction and genetic isolation. Much more work on the taxonomy of the *F. rubra* complex is needed before the boundaries of individual taxa can be firmly established.

A strongly rhizomatous alpine and subalpine form, *F. rubra* subsp. *vallicola* (Rydb.) Pavlick, is distinctive in its non-cespitose habit with widely spaced culms. It is known to the north and northeast of the Intermountain Region and may be expected there.

Festuca earlei (see next) is sometimes confused with *F. rubra*. It differs in having pubescent ovary apices and shorter anthers.

1. Plants not rhizomatous, densely cespitose . . . .
. . . . subsp. *commutata*
1. Plants rhizomatous, usually loosely to densely cespitose, sometimes with solitary culms.
    2. Vegetative shoot blades usually flat or loosely conduplicate; plants strongly rhizomatous; adaxial sclerenchyma strands always present
. . . . subsp. *fallax*
    2. Vegetative shoot blades usually conduplicate, sometimes flat; plants strongly or weakly rhizomatous; adaxial sclerenchyma strands sometimes present . . . . subsp. *rubra*

Festuca rubra subsp. commutata Gaudin
Chewing's Fescue [p. 344]

Pl without rhz, usu densely ces. **Clm** 25–90 cm. **Shth** red-brown, scarious near the base, puberulent or pubescent, slowly shredding into fibers; **bld** 0.3–0.7(1) in diameter, conduplicate, smt glaucous, abx surfaces scabrous or smooth; **abx sclerenchyma** in narrow to broad strands;

adx sclerenchyma rarely present. **Infl** 4–13(30) cm, more or less contracted, often secund; **br** spreading at anthesis, scabrous on the angles. **Spklt** 7–11 mm, with 3–9 flt. **Glm** ovate-lanceolate, acute; **lo glm** 2.5–4 mm; **up glm** 3.5–5 mm; **lm** 4.5–6 mm, green or reddish violet distally, glab, smooth, awned, awns 1–3.3 mm; **anth** 1.8–2.2(3) mm. 2*n* = 28, 42.

Festuca rubra subsp. *commutata* is extensively used for lawns and road verges. It is native to Europe, growing from southern Sweden southward, but is widely introduced elsewhere in the world. It is common in North America south of Alaska, Yukon Territory, and the Northwest Territories.

Festuca rubra subsp. **fallax** (Thuill.) Nyman
FLATLEAF RED FESCUE [p. 344]
Pl ces, strongly rhz, stln, smt with solitary clm. **Clm** 50–90(130) cm. **Shth** of innovations red-brown at the base, scarious, pubescent on the up parts, shredding into fibers; **bld** usu flat and 2–7 mm wide, or loosely conduplicate and 0.8–2.5 mm in diameter; **abx sclerenchyma** in 5–9 broad strands; **adx sclerenchyma** always present. **Infl** 9–15 cm, open, lax or erect. **Spklt** 8–14(17) mm, with (4)6–10 flt. **Glm** ovate-lanceolate to lanceolate, acute to acuminate; **lo glm** 3.5–4.5(7) mm; **up glm** 5–6.5(8.5) mm; **lm** 5–7.5(9.5) mm, green or glaucous, often red-violet along the mrg, glab, smooth, awned, awns 0.8–3.2 mm; **anth** 3.5–4.5 mm. 2*n* = 42, 56, 70.

Festuca rubra subsp. *fallax* is a robust taxon that grows in damp, often disturbed places. It is native to northern and central Europe, but has been introduced widely in North America, occurring from British Columbia to eastern Quebec and south to California. It is now common in some areas, occasional in others.

Festuca rubra L. subsp. **rubra** RED FESCUE
[p. 345]
Pl rhz, usu loosely ces, with several clm arising from the same tuft, vegetative shoots 8–22(30) cm. **Clm** (20)40–90 cm. **Shth** reddish brown, scarious, pubescent, shredding into fibers; **bld** 0.5–2 mm in diameter, usu conduplicate, smt flat, abx surfaces smooth or scabrous, green or glaucous, adx surfaces scabrous or pubescent on the ribs; **abx sclerenchyma** in 5–7(9) small strands; **adx sclerenchyma** rarely present. **Infl** 7–12 cm, open, lanceolate; **br** scabrous. **Spklt** 9–14.5 mm, with 5–8 flt. **Lo glm** 3–4.5 mm; **up glm** 4–6.4 mm; **lm** (4)6–7.5(8) mm, lanceolate, usu green with red-violet borders, smt mostly red-violet, mrg smt scabrous, apc scabrous, acute to acuminate, awned, awns 0.6–3.2(4) mm; **anth** 2.4–3.5 mm. 2*n* = 42.

Festuca rubra subsp. *rubra* grows in disturbed soil. It is often planted as a soil binder, or as turf or forage grass, in mesic temperate parts of the world. Originally from Eurasia, it has been widely introduced elsewhere. Because *F. rubra* subsp. *rubra* has often been misunderstood, confounded, and lumped with other taxa of the *F. rubra* complex, statements about its distribution, including that given here, should be treated with caution.

6. **Festuca earlei** Rydb. EARLE'S FESCUE [p. 345, 498]
Pl loosely ces, often with short rhz. **Clm** (15)20–40(45) cm, glab, smooth. **Shth** closed for about ½ their length, glab, shredding into fibers, smt slowly; **col** glab; **lig** 0.1–0.5(1) mm; **bld** to 3 mm wide when flat, 0.5–1 (1.5) mm in diameter when conduplicate, veins (3)5, ribs (1)3(5), abx surfaces smooth or slightly scabrous, adx surfaces sparsely scabrous; **abx sclerenchyma** in 3–5 narrow strands less than twice as wide as high; **adx sclerenchyma** absent. **Infl** 3–5(8) cm, contracted, with 1–3 br per nd; **br** stiff, erect, scabrous, lo br with 2+ spklt. **Spklt** (4.5)5–6.5(7) mm, with 2–5 flt. **Glm** exceeded by the up flt, lanceolate to ovate-lanceolate, mostly smooth, smt scabrous distally; **lo glm** 1.5–3 mm; **up glm** 2.5–3.8 mm; **lm** 3–4.5 mm, glab or puberulent near the apc, awns (0.3)1–1.5 mm, tml; **pal** as long as or slightly shorter than the lm, intercostal region puberulent distally; **anth** 0.6–0.9(1.4) mm; **ov apc** densely and conspicuously pubescent. 2*n* = unknown.

Festuca earlei grows in rich subalpine and alpine meadows, at 2800–3800 m, in Utah, Colorado, Arizona, and New Mexico. At present, it is known from only one location in the Intermountain Region, the main portion of its range lying to the south and east. This distribution may, however, reflect taxonomic confusion with *F. rubra* and other taxa. It often grows with the non-rhizomatous species *F. brachyphylla* subsp. *coloradensis* (p. 95) and *F. minutiflora* (p. 96). It can be distinguished from the former by its pubescent ovary apices, and from the latter by its larger spikelets and lemmas. Because of its short rhizomes (which are often missing from herbarium specimens), *F. earlei* is sometimes confused with members of the *F. rubra* complex (see previous), from which it differs in having pubescent ovary apices and shorter anthers.

7. **Festuca valesiaca** Schleich. *ex* Gaudin VALAIS FESCUE [p. 345, 499]
Pl densely ces, without rhz. **Clm** 20–50(60) cm, erect, glab, smooth. **Shth** closed for about ½ their length, usu glab, smt pubescent distally, persistent; **col** glab; **lig** 0.1–0.5 mm; **bld** (0.3)0.5–0.8(1.2) mm in diameter, conduplicate, veins 5–7, ribs (1)3–5, abx surfaces glab or pubescent, not pilose, adx surfaces smooth or scabrous; **abx sclerenchyma** in 3 broad strands, smt with additional narrow strands between the midrib and mrg; **adx sclerenchyma** absent. **Infl** (3)5–10 cm, pan, contracted, with 1–2 br per nd; **br** erect

or somewhat spreading, at least at anthesis, lo br with 2+ spklt. **Spklt** (4.8)5.5–6.5(8.5) mm, with 3–5(8) flt. **Glm** exceeded by the up flt, ovate-lanceolate to lanceolate, mostly smooth or smt scabrous on the up midvein; **lo glm** 2–3 mm; **up glm** (2.3)2.5–4(4.3) mm; **lm** (3.2)3.5–4.5(5.3) mm, ovate-lanceolate to lanceolate, smooth throughout or scabrous distally, awns (0.5)1–2 mm, tml; **pal** as long as the lm, intercostal region puberulent distally; **anth** 2.2–2.6 mm; **ov apc** glab. 2*n* = 14.

Festuca valesiaca is widely distributed through central Europe and northern Asia, where it grows in steppes, dry meadows, and open rocky or sandy areas. It is sold in the North American seed trade as *F. pseudovina* Hack. *ex* Wiesb., and has been collected at a few scattered localities in western North America. It is very similar to the introduced *F. trachyphylla* (p. 95), and may be reliably distinguished only by its abaxial sclerenchyma pattern; it has three broad strands usually restricted to the margins and midvein, whereas *F. trachyphylla* usually has an irregular, more or less continuous band of abaxial sclerenchyma.

8. Festuca amethystina L. Tufted Fescue [p. 345]

Pl densely ces, without rhz. **Clm** (25)50–80(105) cm, erect, glab and smooth throughout. **Shth** closed for about ½ their length, glab, smooth, usu reddish blue towards the bases, persistent, flag lf shth tightly enclosing the clm; **col** glab; **lig** 0.1–0.5 mm; **bld** 0.6–1.2(1.5) mm in diameter, conduplicate, abx surfaces glab, smooth or scabrous, adx surfaces scabrous, veins (5)7–9, ribs 5–9; **abx sclerenchyma** in (5)7–9 broad strands, rarely some strands confluent; **adx sclerenchyma** absent. **Infl** (3)8–18 (25) cm, open or loosely contracted, with 1–2(3) br per nd; **br** somewhat lax, erect to spreading, lo br with 2+ spklt. **Spklt** (5)6–8.5(10) mm, with (3)4–6(8) flt. **Glm** exceeded by the up flt, lanceolate, mostly smooth and glab, smt scabrous distally; **lo glm** 2.5–4 mm; **up glm** 3–5(5.5) mm; **lm** (3.5)4–5.6(6.6) mm, usu smooth, smt scabrous near the apc, lanceolate, usu acute, smt obtuse, unawned, occ mucronate; **pal** about as long as the lm, intercostal region smooth or slightly scabrous distally; **anth** (2)3–4 mm; **ov apc** glab or sparsely pubescent. 2*n* = 14.

Festuca amethystina is sometimes cultivated as an ornamental species; it may occasionally escape.

9. Festuca ovina L. Sheep Fescue [p. 346]

Pl densely ces, without rhz; usu not glaucous. **Clm** (10)30–50(70) cm, glab, smooth. **Shth** closed for about ½ their length or less, glab, smooth or scabrous distally, persistent; **col** glab; **lig** shorter than 0.3 mm; **bld** 0.3–0.7(1.2) mm in diameter, conduplicate, abx surfaces smooth or scabrous,

adx surfaces scabrous, veins 5–7(9), ribs 1–3, indistinct; **abx sclerenchyma** usu a continuous band; **adx sclerenchyma** absent. **Infl** (2)5–10(12) cm, contracted, with 1–2(3) br per nd; **br** usu erect, smt spreading at anthesis, lo br with 2+ spklt. **Spklt** 4–6(7.3) mm, with 3–6(8) flt. **Glm** exceeded by the up flt, lanceolate to ovate-lanceolate, mostly smooth and glab, smt scabrous distally; **lo glm** 1–2(3) mm; **up glm** (2.2)2.6–4(4.6) mm; **lm** (2.6)3–4(5) mm, ovate-lanceolate, mostly smooth, smt scabrous or hispid near the apc, awns 0.5–2 mm, tml, smt absent; **pal** about equal to the lm, intercostal region puberulent distally; **anth** (1.4)2–2.6 mm; **ov apc** glab. 2*n* = 14, 28.

Festuca ovina was introduced from Europe as a turf grass. It is not presently used in the North American seed trade. It is included here because the name used to be interpreted very broadly in North America, including almost any fine-leaved fescue that lacked rhizomes. Consequently, much of the information reported for *F. ovina*, and the majority of the specimens identified as such, belong to other species. Species in this treatment that have frequently been included in *F. ovina* are *F. arizonica* (p. 97), *F. baffinensis* (p. 96), *F. brachyphylla* (p. 95), *F. calligera* (p. 97), *F. idahoensis* (p. 98), *F. minutiflora* (p. 96), *F. saximontana* (p. 96), and *F. trachyphylla* (p. 95). *Festuca ovina* is similar to *F. trachyphylla*, but tends to have slightly shorter glumes, lemmas, and anthers.

10. Festuca glauca Vill. Blue Fescue, Gray Fescue [p. 346]

Pl densely ces, without rhz; usu glaucous or pruinose. **Clm** (15)22–35(50) cm, erect, glab, smooth. **Shth** closed for about ½ their length, pubescent or glab, persistent; **col** glab; **lig** 0.1–0.4 mm; **bld** 0.6–1 mm in diameter, conduplicate, veins 5–7, ribs 1–3(5), indistinct; **abx sclerenchyma** forming continuous or interrupted bands; **adx sclerenchyma** absent. **Infl** 2.5–9(11) cm, compact, erect, with 1–2 br per nd; **br** stiff, erect, smooth or scabrous, lo br with 2+ spklt. **Spklt** 5.5–9(11) mm, with 3–6(7) flt. **Glm** exceeded by the up flt, ovate-lanceolate to ovate, glab or pubescent distally; **lo glm** (1.8)2–3(4) mm; **up glm** 2.8–4(5.1) mm; **lm** (3.5)4–6(6.2) mm, lanceolate to ovate-lanceolate, smooth or scabrous near the apc, smt pubescent distally, awns (0.6)1–1.5(2) mm, tml; **pal** about equal to the lm, intercostal region puberulent distally; **anth** (1.8)2–3 mm; **ov apc** glab. 2*n* = 42 [in European literature, for the horticultural forms].

Festuca glauca is widely grown as an ornamental in the Intermountain Region because of its attractive dense tufts of glaucous foliage. It is not known to have escaped cultivation. Several other Eurasian species of fescue with glaucous or bluish foliage are also sold in the horticultural trade as "*Festuca glauca*". Describing all the species involved is beyond the scope of this treatment.

11. **Festuca trachyphylla** (Hack.) Krajina
HARD FESCUE, SHEEP FESCUE [p. 346, <u>499</u>]
Pl densely ces, without rhz. **Clm** (15)20–75
(80) cm, smooth, glab or with sparse hairs.
Shth closed for less than ⅓ their length, usu
glab, rarely pubescent, persistent; **col** glab; **lig**
0.1–0.5 mm; **bld** (0.5)0.8–1.2 mm in diameter,
usu conduplicate, rarely flat, abx surfaces glab,
puberulent, or scabrous, adx surfaces scabrous
or puberulent to pubescent, veins 5–7(9), ribs
3–7, usu distinct; **abx sclerenchyma** usu in an
irregular, interrupted or continuous band, rarely
in 5–7 small strands, usu more than twice as
wide as high; **adx sclerenchyma** absent. **Infl**
(2.5)3–13(16) cm, contracted, with 1–2 br per nd;
br erect or stiffly spreading, sec br not divaricate,
lo br with 2+ spklt. **Spklt** 5–9(10.8) mm, with
3–7(8) flt. **Glm** exceeded by the up flt, ovate-
lanceolate to lanceolate, mostly smooth and
glab, smt scabrous and/or pubescent distally; **lo
glm** (1.8)2–3.5(4) mm; **up glm** 3–5(5.5) mm; **lm**
3.8–5(6.5) mm, lanceolate, usu smooth and glab
on the lo portion and scabrous or pubescent
distally, especially on the mrg, rarely entirely
pubescent, awns 0.5–2.5(3) mm, usu less than ½
as long as the lm body; **pal** about as long as the
lm, intercostal region puberulent distally; **anth**
(1.8)2.3–3.4 mm; **ov apc** glab. 2*n* = 42.

Festuca trachyphylla is native to open forests and forest
edge habitats of central Europe. It has been widely introduced
and has become naturalized in many temperate regions. In
North America, *F. trachyphylla* is generally sold under the
name 'Hard Fescue', and is popular as a durable turf grass
and soil stabilizer. It is more common in the eastern United
States and southeastern Canada, but may be grown in the
Intermountain Region. Its naturalized distribution can be
expected to expand. It is very similar to *F. ovina*, but tends to
have slightly longer glumes, lemmas, and anthers.

12. **Festuca brachyphylla** Schult. & Schult. f.
ALPINE FESCUE [p. 347, <u>499</u>]
Pl densely or loosely ces, without rhz. **Clm**
(5)8–35(55) cm, erect, usu smooth and glab, smt
sparsely scabrous or puberulent near the infl.
Shth closed for about ½ their length, smooth
or scabrous, persistent or slowly shredding
into fibers; **col** glab; **lig** 0.1–0.4 mm; **bld**
(0.3)0.5–1(1.2) mm in diameter, conduplicate,
abx surfaces smooth or sparsely scabrous, adx
surfaces scabrous, veins (3)5–7, ribs 3–5; **abx
sclerenchyma** in 3–7(9) narrow strands, usu less
than twice as wide as high; **adx sclerenchyma**
absent; **flag lf shth** not inflated, more or less
tightly enclosing the clm, bld (0.3)1–2.5(3) cm.
Infl 1.5–4(5.5) cm, contracted, usu pan, very

rarely rcm, with 1–2 br per nd; **br** usu erect, smt
spreading at anthesis, lo br with 2+ spklt. **Spklt**
3.5–7(8.5) mm, with 2–4(6) flt. **Glm** exceeded by
the up flt, ovate-lanceolate, usu glab and smooth,
smt scabrous distally; **lo glm** (1.2)1.8–3(3.5) mm;
up glm (2.4)2.6–4(4.6) mm; **lm** 2.5–4.5(6) mm,
ovate-lanceolate to lanceolate, scabrous towards
the apc, awns (0.8)1–3(3.5) mm, tml; **pal** about
as long as the lm, intercostal region scabrous or
puberulent distally; **anth** (0.5)0.7–1.1(1.3) mm;
ov apc glab. 2*n* = 28, 42, 44.

Festuca brachyphylla is a variable, circumpolar, arctic,
alpine, and boreal species of open, rocky places. It is palatable
to livestock, and is important in some areas as forage for
wildlife. The spikelets are usually tinged red to purple by
anthocyanin pigments; plants which lack anthocyanins in the
spikelets have been named *F. brachyphylla* f. *flavida* Polunin.
Festuca brachyphylla has frequently been included in *F. ovina*
(p. 94), and it is closely related to *F. saximontana* (see next) and
F. minutiflora (p. 96).

Two subspecies are known from the Intermountain
Region.

1. Culms usually less than twice as long as the
   vegetative shoot leaves; awns 1–2(2.2) mm long;
   spikelets 3.5–5(5.5) mm long; lemmas 2.5–4 mm
   long . . . . subsp. *breviculmis*
1. Culms usually twice as long as the vegetative
   shoot leaves; awns 2–3(3.2) mm long; spikelets
   4.4–5.6(7) mm long; lemmas 3–4(4.5) mm long
   . . . . subsp. *coloradensis*

Festuca brachyphylla subsp. **breviculmis** Fred.
[p. 347]
Clm usu less than twice as long as the vegetative
shoots lvs. **Spklt** 3.5–5(5.5) mm. **Lm** 2.5–4 mm;
awns 1–2(2.2) mm. 2*n* = 28(?) [chromosome count
is unknown, but has been inferred].

Festuca brachyphylla subsp. *breviculmis* is endemic to
California, where it grows in alpine habitats in the Sierra
Nevada and White Mountains.

Festuca brachyphylla subsp. **coloradensis** Fred.
[p. 347]
Clm usu about twice as long as the vegetative
shoots lvs. **Spklt** 4.4–5.6(7) mm. **Lm** 3–4(4.5) mm,
awns 2–3(3.2) mm. 2*n* = 28.

Festuca brachyphylla subsp. *coloradensis* is a common species
in alpine areas of Colorado, Utah, Wyoming, Arizona, and
New Mexico. It often grows with *F. earlei* (p. 93), from which
it can be distinguished by its lack of rhizomes, glabrous ovary
apices, and smaller spikelets and lemmas.

13. **Festuca saximontana** Rydb. ROCKY
MOUNTAIN FESCUE, MOUNTAIN FESCUE
[p. 347, <u>499</u>]
Pl usu densely, smt loosely, ces, without rhz.
Clm (5)8–50(60) cm, usu smooth and glab,
occ sparsely scabrous or puberulent below

the infl. **Shth** closed for about ½ their length, glab, smooth or scabrous, usu persistent, rarely slowly shredding into fibers; **col** glab; **lig** 0.1–0.5 mm; **bld** 0.5–1.2 mm in diameter, conduplicate, abx surfaces glab or sparsely puberulent, adx surfaces scabrous or puberulent, veins 5–7(9), ribs 1–5; **abx sclerenchyma** in 3–7 strands, smt partly confluent or forming a continuous band, usu more than twice as wide as high; **adx sclerenchyma** absent; **flag lf bld** 0.5–4 cm. **Infl** (2)3–10(13) cm, contracted, with 1–2 br per nd; **br** usu erect, spreading at anthesis, lo br with 2+ spklt. **Spklt** (3)4.5–8.8(10) mm, with (2)3–5(7) flt. **Glm** exceeded by the up flt, ovate-lanceolate to lanceolate, scabrous distally; **lo glm** 1.5–3.5 mm; **up glm** 2.5–4.8 mm; **lm** (3)3.4–4(5.6) mm, mostly smooth, often scabrous distally, awns (0.4)1–2(2.5) mm; **pal** as long as or slightly shorter than the lm, intercostal region puberulent distally; **anth** (0.8)1.2–1.7(2) mm; **ov apc** glab. $2n = 42$.

Festuca saximontana grows in grasslands, meadows, open forests, and sand dune complexes of the northern plains and boreal, montane, and subalpine regions of North America. It provides good forage for livestock and wildlife. It is closely related to *F. brachyphylla* (see previous), and is sometimes included in that species as *F. brachyphylla* subsp. *saximontana* (Rydb.) Hultén. It has also frequently been included in *F. ovina* (p. 94). It differs from alpine taxa such as *F. earlei*, *F. brachyphylla*, and *F. minutiflora* by having abaxial sclerenchyma in a continuous band or in 3–7 strands that are usually more then twice as wide as high; in the other taxa the abaxial sclerenchyma is in narrow strands that are usually less than twice as wide as high.

Three weakly differentiated taxa have been recognized at the varietal level in North America, two of which have been found in the Intermountain Region.

1. Culms 25–50(60) cm tall, usually 3–5 times the height of the vegetative shoot leaves; abaxial surfaces of the blades usually scabrous; abaxial sclerenchyma in 3–5 wide strands, usually partly confluent or forming a continuous band; plants of lowland, montane, or boreal habitats . . . . var. *saximontana*
1. Culms (5)8–37 cm tall, usually 2–3 times the height of the vegetative shoot leaves; abaxial surfaces of the blades smooth or scabrous; abaxial sclerenchyma in 5–7 narrow strands; plants of subalpine or lower alpine habitats . . . . var. *purpusiana*

Festuca saximontana var. **purpusiana** (St.-Yves) Fred. & Pavlick [p. 347]
Clm (5)8–20(25) cm, usu 2–3 times the height of the vegetative shoot lvs, usu glab below the infl. **Outer vegetative shoot shth** mostly stramineous; **bld** smooth or scabrous on the abx surfaces, ribs on the adx surfaces with hairs shorter than 0.06 mm; **abx sclerenchyma** in 5–7 narrow abx strands. **Lm** usu scabrous towards the apc and often along the mrg.

Festuca saximontana var. *purpusiana* grows in subalpine or lower alpine habitats. It is most easily distinguished from *F. brachyphylla* by its longer anthers.

Festuca saximontana Rydb. var. **saximontana** [p. 347]
Clm 25–50(60) cm, usu 3–5 times the height of the vegetative shoot lvs, usu glab below the infl, smt sparsely scabrous or pubescent; **bld** usu scabrous on the abx surfaces, scabrules to 0.1 mm; **abx sclerenchyma** in 3–5 strands, smt partly confluent or forming a continuous band. **Lm** smooth or scabrous distally.

Festuca saximontana var. *saximontana* grows throughout the range of the species.

14. **Festuca baffinensis** Polunin BAFFIN ISLAND FESCUE [p. 348, 499]
Pl densely ces, without rhz. **Clm** 5–25(30) cm, densely pubescent or shortly pilose near the infl. **Shth** closed for about ½ their length, glab, persistent or slowly shredding into fibers; **col** glab; **lig** 0.1–0.3 mm; **bld** (0.4)0.6–1(1.2) mm in diameter, conduplicate, abx surfaces smooth or sparsely scabrous, adx surfaces scabrous or puberulent, veins (3)5–7, ribs 3–5; **abx sclerenchyma** in 3–7 small strands, usu less than twice as wide as high; **adx sclerenchyma** absent; **flag lf shth** usu somewhat loosely enclosing the clm, bld 0.5–4 cm. **Infl** 1.5–4(5) cm, contracted, usu pan, rarely rcm, usu somewhat secund, with 1–2 br per nd; **br** erect, lo br with 2+ spklt. **Spklt** (4.5)5–7.5(8.5) mm, with (2)3–5(6) flt. **Glm** exceeded by the up flt, ovate-lanceolate, scabrous distally; **lo glm** 2.2–3.7(4) mm; **up glm** 3–5 mm; **lm** (3.5)4–6 mm, scabridulous near the apc, awns 0.8–2.6(3.3) mm, tml; **pal** slightly shorter than to as long as the lm, intercostal region scabrous distally; **anth** 0.3–0.7 (1.1) mm; **ov apc** usu with a few hairs, rarely glab. $2n = 28$.

Festuca baffinensis grows chiefly in damp, exposed, gravelly areas in calcareous and volcanic regions. It is circumpolar in distribution, growing in arctic and alpine habitats and extending southward in the Rocky Mountains to Colorado. It has frequently been included in *F. ovina* (p. 94).

15. **Festuca minutiflora** Rydb. LITTLE FESCUE, SMALL-FLOWERED FESCUE [p. 348, 499]
Pl loosely or densely ces, without rhz. **Clm** 4–30 cm, usu erect, smt semi-prostrate, glab, smooth. **Shth** closed for about ½ their length, glab, persistent; **col** glab; **lig** 0.1–0.3 mm; **bld** (0.2)0.3–0.4(0.6) mm in diameter, conduplicate, lax, abx surfaces glab, adx surfaces sparsely scabrous to puberulent, veins 3–5, ribs 1–3; **abx sclerenchyma** in 3–5 small strands, less than twice as wide as high; **adx sclerenchyma** absent; **flag lf bld** 0.7–3.5

cm. **Infl** 1–4(5) cm, contracted, with 1–2 br per nd; **br** erect, lo br with 2+ spklt. **Spklt** (2.5)3–5 mm, with (1)2–3(5) flt. **Glm** exceeded by the up flt, ovate to ovate-lanceolate, sparsely scabrous distally; **lo glm** 1.3–2.5 mm; **up glm** 2–3.5 mm; **lm** (2)2.2–3.5(4) mm, ovate-lanceolate, sparsely scabrous near the apc, apc abruptly acuminate, awns 0.5–1.5(1.7) mm; **pal** about as long as or slightly shorter than the lm, intercostal region scabrous distally; **anth** (0.4)0.6–1.2 mm; **ov apc** usu with a few hairs, rarely glab. 2*n* = 28.

Festuca minutiflora grows in alpine regions of the western mountains of North America. It has often been overlooked or included with *F. brachyphylla* (p. 95), from which it differs in its laxer and narrower leaves, looser panicles, smaller spikelets, more pointed lemmas, shorter awns, and scattered hairs on the ovary. In the southern Rocky Mountains, it may also grow with *F. earlei* (p. 93), which has short rhizomes and larger spikelets and lemmas. *Festuca minutiflora* has frequently been included in *F. ovina* (p. 94).

16. Festuca occidentalis Hook. WESTERN FESCUE [p. 349, 499]

Pl densely to loosely ces, without rhz. **Clm** (25)40–80(110) cm, glab, smooth. **Shth** closed for much less than ½ their length, glab, somewhat persistent or slowly shredding into fibers; **col** glab; **lig** 0.1–0.4 mm, usu longer at the sides; **bld** all alike, 0.3–0.7 mm in diameter, conduplicate, abx surfaces smooth or scabridulous, veins (3)5, ribs 1–5; **abx sclerenchyma** in 5–7 narrow strands, about as wide as the adjacent veins; **adx sclerenchyma** absent. **Infl** (5)10–20 cm, open, with 1–2 br per nd; **br** 1–15 cm, lax, widely spreading to reflexed, lo br usu reflexed at maturity, with 2+ spklt. **Spklt** 6–12 mm, with 3–6(7) flt. **Glm** exceeded by the up flt, ovate to ovate-lanceolate, glab and smooth or slightly scabrous; **lo glm** 2–5 mm; **up glm** 3–6 mm; **lm** (4)4.5–6.5(8) mm, ovate-lanceolate to attenuate, glab or finely puberulent, awns 3–12 mm, usu longer than the lm bodies; **pal** slightly shorter than the lm, intercostal region scabrous or puberulent distally; **anth** (1)1.5–2(3) mm; **ov apc** densely pubescent. 2*n* = 28 [other numbers have been reported for this species, but are probably based on misidentifications].

Festuca occidentalis grows in dry to moist, open woodlands, forest openings, and rocky slopes, up to 3100 m, primarily along the west coast. It is sometimes important as a forage grass, but is usually not sufficiently abundant. The very open inflorescence, long awns and densely pubescent ovary apices are distinctive characteristics.

17. Festuca calligera (Piper) Rydb. CALLUSED FESCUE [p. 349, 499]

Pl densely ces, without rhz. **Clm** 15–65 cm, glab, smooth. **Shth** closed for less than ½ their length, lo shth glab or retrorsely hirsute, persistent, up shth glab; **col** glab; **lig** (0.2)0.3–0.5(1) mm; **bld** all alike, 0.4–0.8 mm in diameter, conduplicate, abx surfaces sparsely scabrous, adx surfaces scabrous to pubescent, veins 5–7, ribs (1)3–5; **abx sclerenchyma** in (3)5–7 narrow to broad strands, usu wider than the adjacent veins; **adx sclerenchyma** absent. **Infl** 5–15 cm, loosely contracted, with 1–2(3) br per nd; **br** erect, lo br with 2+ spklt. **Spklt** (6)7–9(11) mm, with (2)4–6 flt. **Glm** exceeded by the up flt, lanceolate, scabrous distally; **lo glm** 2.5–4 mm; **up glm** (2.8)3–5 mm; **lm** (3.8)4–6 mm, glab, smooth or scabrous distally, awns 1–2.5 mm; **pal** slightly shorter than the lm, intercostal region scabrous or puberulent distally; **anth** 2.2–3.5 mm; **ov apc** sparsely pubescent. 2*n* = 28.

Festuca calligera is a poorly known, often overlooked species. It grows in grasslands and open montane forests, at 2500–3400 m, from southern Utah to south-central Wyoming south to Arizona and New Mexico. It is often found with *F. arizonica* (see next). *Festuca calligera* has frequently been included in *F. ovina* (p. 94).

18. Festuca arizonica Vasey ARIZONA FESCUE, PINEGRASS [p. 349, 499]

Pl densely ces, without rhz. **Clm** 35–80 (100) cm, usu densely scabrous or densely pubescent below the infl. **Shth** closed for less than ½ their length, glab, smooth or scabrous, persistent; **col** glab, smooth or scabrous; **lig** 0.5–1.5(2) mm; **bld** 0.3–0.8 mm in diameter, conduplicate, abx surfaces scabrous or puberulent, adx surfaces scabrous to pubescent, veins 5–7, ribs (1)3–5(7), distinct; **abx sclerenchyma** in 5–7 broad strands, rarely forming a complete band, forming pillars with some veins; **adx sclerenchyma** not developed. **Infl** (4)6–15(20) cm, loosely contracted or open, with 1–2 br per nd; **br** erect or spreading, lo br with 2+ spklt. **Spklt** (6)8–16 mm, with (3)4–6(8) flt. **Glm** exceeded by the up flt, lanceolate, glab, smooth or scabrous distally; **lo glm** (3)3.3–5.5 mm; **up glm** 4.5–6.6(7) mm; **lm** 5.5–9 mm, glab, smooth or scabrous towards the apc, unawned or awned, awns 0.4–2(3) mm; **pal** slightly shorter than the lm, intercostal region scabrous or puberulent distally; **anth** (2)3–4(4.2) mm; **ov apc** densely pubescent. 2*n* = 42.

Festuca arizonica grows in dry meadows and openings of montane forests, in gravelly, rocky soil, at 2100–3400 m. Its range extends from southern Nevada and southern Utah east to Colorado and south to Arizona, western Texas, and northern Mexico. It is abundant and valuable forage in some parts of its range. It is often found with *F. calligera* (see previous).

Festuca arizonica differs from *F. idahoensis* (see next), with which it is sometimes confused, in its prominently ribbed blades, shorter awns, and pubescent ovary apices. It has frequently been included in *F. ovina* (p. 94).

19. **Festuca idahoensis** Elmer IDAHO FESCUE, BLUE BUNCHGRASS, BLUEBUNCH FESCUE [p. *350*, <u>499</u>]

Pl densely ces, without rhz. **Clm** 25–85 (100) cm, usu smooth, glab, occ scabrous below the infl. **Shth** closed for less than ½ their length, smooth or scabrous, rarely pilose, persistent; **col** glab; **lig** 0.2–0.6 mm; **bld** (0.3)0.5–0.9(1.5) mm in diameter, conduplicate, abx surfaces smooth or scabrous, adx surfaces scabrous or pubescent, rarely pilose, often glaucous or bluish, veins (3)5(7), ribs (1)3–5, well defined; **abx sclerenchyma** in 5–7 wide, irregular strands; **adx sclerenchyma** absent. **Infl** (5)7–15(20) cm, loosely contracted or open, with 1–2 br per nd; **br** usu somewhat spreading at maturity, smt erect, rarely reflexed, lo br with 2+ spklt. **Spklt** (5.8) 7.5–13.5(19) mm, with (2)4–7(9) flt. **Glm** exceeded by the up flt, ovate-lanceolate to lanceolate, mostly smooth, smt scabrous distally; **lo glm** 2.4–5(6) mm; **up glm** 3–6(8) mm; **lm** 5–8.5(10) mm, scabrous at the apc, awns (1.5)2–6(7) mm, usu more than ½ as long as the lm bodies; **pal** shorter than to about as long as the lm, intercostal region scabrous or puberulent distally; **anth** 2–4.5 mm; **ov apc** glab. 2*n* = 28.

Festuca idahoensis grows in grasslands, open forests, and sagebrush meadow communities, mostly east of the Cascade Mountains, from southern British Columbia eastward to southwestern Saskatchewan and southward to central California and New Mexico. It extends up to 3000 m in the southern part of its range. It is often a dominant plant, and provides good forage. The young foliage is particularly palatable.

Festuca idahoensis differs from *F. arizonica* (see previous), with which it is sometimes confused, in its less prominently ribbed blades, longer awns, and glabrous ovary apices. It has frequently been included in *F. ovina* (p. 94).

20. **Festuca viridula** Vasey MOUNTAIN BUNCHGRASS, GREENLEAF FESCUE, GREEN FESCUE [p. *350*, <u>499</u>]

Pl loosely or densely ces, without rhz. **Clm** 35–80(100) cm, smooth, glab throughout; **nd** usu not exposed. **Shth** closed for less than ½ their length, usu glab, smt pubescent, strongly veined, persistent or slowly shredding into fibers; **col** glab; **lig** (0.2)0.3–0.8(1) mm; **bld** 0.5–1.3 mm in diameter when conduplicate, to 2.5 mm wide when flat, persistent, abx surfaces glab and smooth, adx surfaces scabrous or pubescent, veins 5–9(12), ribs 5–9, bld of the lo

cauline lvs usu rdcd to stiff horny points, bld of the up cauline lvs longer and more flexuous; **abx sclerenchyma** in strands about as wide as the adjacent veins; **adx sclerenchyma** developed; **girders** and **pillars** often present. **Infl** (4)8–15 cm, open or somewhat contracted, with 1–2 br per nd; **br** lax, spreading or loosely erect, lo br with 2+ spklt. **Spklt** 9–15 mm, with (2)3–6(7) flt. **Glm** exceeded by the up flt, ovate-lanceolate to lanceolate, glab, smooth or scabridulous distally; **lo glm** (2.4)2.8–5 mm, distinctly shorter than the adjacent lm; **up glm** 4.5–7(8.5) mm; **lm** (4.8)6–8.5 mm, lanceolate to ovate-lanceolate, glab, smooth or slightly scabrous, apc acute, unawned or awned, awns 0.2–1.5(2) mm; **pal** about as long as the lm, intercostal region scabrous or puberulent distally; **anth** (2)2.5–4(5) mm; **ov apc** densely pubescent. 2*n* = 28.

Festuca viridula grows in low alpine and subalpine meadows, forest openings, and open forests, at (900)1500–3000 m, from southern British Columbia east to Montana and south to central California and Nevada. The reduced length of the lower culm leaf blades is distinctive among the western fescues. It is highly palatable to livestock, and is an important forage species in some areas.

21. **Festuca dasyclada** Hack. *ex* Beal OPEN FESCUE, INTERMOUNTAIN FESCUE [p. *351*, <u>499</u>]

Pl loosely or densely ces, without rhz. **Clm** 20–40(50) cm, erect or somewhat geniculate at the base, densely scabrous or pubescent below the infl; **nd** usu not exposed, clm often breaking at the up nd at maturity. **Shth** closed for less than ½ their length, glab, persistent or slowly shredding into fibers; **col** glab; **lig** 0.2–0.5 mm; **bld** (1)1.2–2.5(3) mm wide, persistent, loosely conduplicate, convolute, or flat, abx surfaces glab, adx surfaces with stiff hairs, veins 7–13, ribs (6)7–13; **abx sclerenchyma** in strands opposite most of the veins, about as wide as the veins; **adx sclerenchyma** often present; **girders** or **pillars** smt present at the major veins. **Infl** 6–12 cm, open, with 2–4 br per nd; **br** stiffly divaricate, densely scabrous-ciliate on the angles, lo br with 2+ spklt; **ped** stiffly hairy. **Spklt** 5.5–8 mm, with 2(3) flt. **Glm** exceeded by the up flt, lanceolate-acuminate, sparsely scabrous to puberulent; **lo glm** 3.5–5 mm, distinctly shorter than the adjacent lm; **up glm** 5–7 mm; **lm** 5–7 mm, chartaceous, scabrous or puberulent, minutely bidentate, awned, awns 1.5–3 mm, subtml; **pal** about as long as or slightly longer than the lm, intercostal region scabrous or puberulent distally; **anth** 1.5–2.5 mm; **ov apc** pubescent. 2n = 28.

Festuca dasyclada grows on rocky slopes in open forests and shrublands of central and southern Utah and western Colorado. When the seeds are mature, the panicles break off the culms and are blown over the ground like a tumbleweed, shedding seeds as they travel. This and other unusual features, such as the divaricate branching pattern and hairy pedicels, prompted W.A. Weber to place it in the monotypic genus *Argillochloa* W.A. Weber.

8.02 LEUCOPOA Griseb.   Stephen J. Darbyshire

Pl per; unisex. **Clm** 30–120 cm. **Shth** closed only at the base; **aur** absent; **lig** memb; **bld** with sclerenchyma girders extending from the abx to adx surfaces. **Infl** open or contracted pan, usu erect to strongly ascending, not spikelike; **br** glab, smooth or somewhat scabrous, at least some br longer than 1 cm; **ped** smt longer than 3 mm, thinner than 1 mm. **Spklt** pedlt, somewhat dimorphic in unisex pl, lat compressed, with (2)3–5(6) flt; **dis** above the glm and beneath the flt. **Glm** subequal to unequal, shorter than the adjacent lm, more or less equally wide, glab, smt scabrous, mostly hyaline and thinner than the lm, memb adjacent to the midvein, unawned; **lo glm** 1-veined; **up glm** 1–3-veined; **cal** glab; **lm** memb to chartaceous, smooth or scabrous, smt hirsute, 5-veined, veins converging distally, usu extending almost to the apc, apc entire, acute, usu unawned, smt awned, awns to 2 mm; **pal** about equaling the lm, scabrous on the veins, scarious or memb distally, veins terminating at the apex; **lod** 2, memb; **anth** 3; **ov** with glab or pubescent apc. **Car** shorter than the lm, concealed at maturity, fusiform, usu adhering at least to the pal; **hila** linear. $x = 7$. Name from the Greek *leuco*, 'white', and *poa*, 'grass'.

Leucopoa is a genus of about 10 species, most of which are Asian. One species is native to the Intermountain Region. It is sometimes included in *Festuca*, but species of *Leucopoa* differ from those of *Festuca* in their dioecious habit, the hyaline glumes that are much thinner than the lemmas, and their differing ovary and caryopsis morphology. Phylogenetic studies indicate that *Leucopoa* is more closely related to *Lolium* and *Schedonorus* than to *Festuca sensu stricto* (Darbyshire and Warwick 1992; Soreng and Davis 2000; Catalán et al. 2004).

1. **Leucopoa kingii** (S. Watson) W.A. Weber
SPIKE FESCUE, SPIKEGRASS, WATSON'S FESCUE-GRASS, KING'S FESCUE [p. 351, <u>499</u>]
Pl unisex; ces, rhz (rhz usu absent from herbarium specimens). **Clm** 30–100(120) cm. **Shth** persisting, closed only at the base, glab, smooth or the lowest smt retrorsely scabrous; **lig** 0.8–2(4) mm, truncate, erose-ciliate; **bld** 14–40(50) cm long, 1.5–7(10) mm wide, bld of the flag lvs often shorter, erect, somewhat stiff, flat or loosely convolute, glab, smt glaucous, abx surfaces usu smooth, rarely scabrous, adx surfaces smt somewhat scabrous or hirsute. **Pan** 7–22 cm; **br** erect or smt spreading, smooth or somewhat scabrous, spklt-bearing to the base. **Spklt** 6–12 mm long, 2.5–4.5 mm wide, stmt spklt tending to be larger than pist spklt, with (2)3–4(6) flt. **Glm** usu unequal, smt subequal (especially in stmt spklt) ovate or ovate-lanceolate (especially in pist spklt) to lanceolate (especially in stmt spklt), smooth or somewhat scabrous; **lo glm** 3–5.5(6.5) mm, 1-veined; **up glm** 4–6.5(7.5) mm, shorter than or, particularly in stmt spklt, equal to the lowest lm, 1–3-veined; **cal** wider than long, blunt; **lm** 4.5–8(10) mm, scabrous or hirsute, apc entire, acute, unawned or subtml mucronate, mucro shorter than 1 mm; **pal** 4.4–7(9) mm, scabrous or hirsute between the veins and on the mrg; **lod** 1–2.6 mm; **anth** (2.5)3.5–5(6) mm in stmt pl, vestigial and not fnctl in pist pl; **ov** in pist pl with pubescent apc and adjacent sty bases, ov in stmt pl vestigial and non-fnctl. **Car** fusiform, 3.5–5 mm long, 1.3–1.7 mm wide. $2n = 56$.

Leucopoa kingii grows from Oregon and Montana to Nebraska, south to southern California and northern New Mexico. It occurs in habitats from dry sagebrush plains to subalpine meadows, at 1700–3600 m. Although palatable to livestock in the early part of the season, *L. kingii* is only occasionally abundant enough to be an important forage species.

8.03 SCHEDONORUS P. Beauv.   Stephen J. Darbyshire

Pl per; csp, smt rhz. **Clm** to 2 m, slender to stout, erect to decumbent. **Shth** open, rounded, smooth or scabrous; **aur** present, usu falcate and clasping, smt an undulating flange; **lig** memb, glab; **bld** flat, linear. **Infl** tml pan, erect, not spikelike; **br** glab, smooth or scabrous, most br longer than 1 cm; **ped** smt longer than 3 mm, thinner than 1 mm. **Spklt** pedlt, lat compressed, with 2–22 flt; **dis** above the glm and between the flt. **Glm** 2, shorter than the adjacent lm, more or less equally wide, lanceolate to oblong, rounded on the back, memb, 3–9-veined, apc acute, unawned; **cal** glab or sparsely hairy; **lm** lanceolate, ovate or oblong, rounded on the back, memb, chartaceous, 3–7-veined, apc acute, smt hyaline, unawned or awned, awns to 18 mm, tml or subtml, straight; **pal** narrower than the lm, memb, usu smooth, keels ciliolate, veins terminating at or beyond midlength; **lod** 2, lanceolate to ovate; **anth** 3; **ov** glab. **Car** shorter than the lm, concealed at maturity, dorsally compressed, oblong, broadly elliptic, or ovate, longitudinally sulcate, adherent to the pal; **hila** linear; **emb** ⅕–⅓ as long as the car. $x = 7$. Name from the Greek *schedon*, 'near' or 'almost', and *oros*, 'mountain' or 'summit'.

Three species of the Eurasian genus *Schedonorus* are established in North America, having been widely introduced as forage and ornamental grasses.

Schedonorus has traditionally been included in *Festuca*, despite all the evidence pointing to its close relationship to *Lolium*. This evidence includes morphological features, such as the falcate leaf auricles, flat, relatively wide leaf blades, subterminal stylar attachment, and adhesion of the mature caryopses to the paleas, none of which are found in *Festuca sensu stricto*. Fertile, natural hybrids between species of *Schedonorus* and those of *Lolium* are common in Europe, and several artificial hybrids have been registered for commercial use, primarily as forage grasses. *Schedonorus* and *Lolium* could appropriately be treated as congeneric subgenera (e.g., Darbyshire 1993). The two are treated as separate genera here for consistency with the treatments by Soreng and Terrell (1997), Holub (1998), and Edgar and Connor (2000).

1. Auricles glabrous; panicle branches at the lowest node 1 or 2, if paired the shorter with 1–2(3) spikelets, the longer with 2–6(9) spikelets; lemmas usually smooth, sometimes slightly scabrous distally, unawned or with a mucro to 0.2 mm long . . . . 1. *S. pratensis*
1. Auricles ciliate, having at least 1 or 2 hairs along the margins (check several leaves); panicle branches at the lowest node usually paired, the shorter with 1–13 spikelets, the longer with 3–19 spikelets; lemmas usually scabrous or hispidulous, at least distally, rarely smooth, unawned or with an awn up to 4 mm long . . . . 2. *S. arundinaceus*

1. **Schedonorus pratensis** (Huds.) P. Beauv.
  MEADOW FESCUE [p. 352, 499]
Pl per. **Clm** to 1.3 m. **Lvs** folded or convolute in young sht; **aur** glab; **lig** to 0.5 mm; **bld** 10–25 cm long, 2–7 mm wide. **Pan** (6)10–25 cm; **br** at the lowest nd 1 or 2, shorter br with 1–2(3) spklt, longer br with 2–6(9) spklt. **Spklt** (8.5)12–15.5(17) mm long, 2–5 mm wide, with (2)4–10(12) flt. **Lo glm** (2)2.6–4.5 mm; **up glm** 3–5 mm; **lm** 5–8 mm, usu smooth, smt slightly scabrous distally, apc unawned, smt mucronate, mucros to 0.2 mm; **pal** slightly shorter than the lm; **anth** (1.5)2–4.6 mm. **Car** 3–4 mm long, 1–1.5 mm wide. $2n = 14$.

Schedonorus pratensis is a Eurasian species that is now widely established in the Intermountain Region. It used to be a popular forage grass in the contiguous United States and southern Canada, but is rarely planted nowadays.

2. **Schedonorus arundinaceus** (Schreb.)
  Dumort. TALL FESCUE [p. 352, 499]
Pl per, smt rhz. **Clm** to 1.5(2) m. **Lvs** convolute in young sht; **aur** ciliate, having at least 1 or 2 hairs along the mrg; **lig** 1(2) mm; **bld** 11–30 cm long, 4–12 mm wide. **Pan** 10–35 cm; **br** at the lowest nd usu 2, shorter br with (1)2–9(13) spklt, longer br with (3)4–13(19) spklt. **Spklt** 8–15.5 mm long, 2–3.5 mm wide, with 3–6(9) flt. **Lo glm** 3–6 mm; **up glm** 4.5–7(9) mm; **lm** (4)5–9(11.5) mm, usu scabrous or hispidulous, at least distally, rarely smooth, awns absent or to 4 mm, tml or attached up to 0.4 mm below the apc; **pal** slightly shorter than to slightly longer than the lm; **anth** 2.5–4 mm. **Car** 2–4 mm long, 0.9–1.6 mm wide. $2n = 28$, 42, 56, 63, 70.

Schedonorus arundinaceus is a Eurasian species that has been introduced to the Intermountain Region. It is grown for forage, soil stabilization, and coarse turf. It is now grown in all but the coldest and most arid parts of North America, and often escapes. It is frequently infected with the endophytic fungus *Neotyphodium coenophialum*, which confers insect and drought resistance to the plant, among other benefits; it also produces ergot alkaloids that are toxic to livestock. Varieties with endophyte strains that do not produce toxic ergot alkaloids have been developed (Nihsen et al. 2004).

8.04 VULPIA C.C. Gmel.

Robert I. Lonard

Pl usu ann, rarely per. **Clm** 5–90 cm, erect or ascending from a decumbent base, usu glab. **Shth** open, usu glab; **aur** absent; **lig** usu shorter than 1 mm, memb, usu truncate, ciliate; **bld** flat or rolled, glab or pubescent. **Infl** pan or rcm, smt spikelike, usu with more than 1 spklt associated with each nd; **br** 1–3 per nd, appressed or spreading, usu glab, scabrous. **Spklt** pedlt, lat compressed, with 1–11(17) flt, distal flt rdcd; **dis** above the glm and beneath the flt, occ also at the base of the ped. **Glm** shorter than the adjacent lm, subulate to lanceolate, apc acute to acuminate, unawned or awn-tipped; **lo glm** much shorter than the up glm, 1-veined; **up glm** 3-veined; **rchl** terminating in a rdcd flt; **cal** blunt, glab; **lm** memb, lanceolate, 3–5-veined, veins converging distally, mrg involute over the edges of the car, apc entire, acute to acuminate, mucronate or awned; **pal** usu slightly shorter than to equaling the lm, smt longer; **anth** usu 1, rarely 3 in chasmogamous specimens. **Car** shorter than the lm, concealed at maturity, elongate, dorsally compressed, curved in cross section, falling with the lm and pal. $x = 7$. Named for J.S. Vulpius, who studied the flora of Baden, Germany.

Vulpia, a genus of 30 species, is most abundant in Europe and the Mediterranean region (Cotton and Stace 1967). There are two native and three introduced species in the Intermountain Region. Most species, including ours, are weedy, cleistogamous annuals, usually having one anther per floret. *Festuca*, in which *Vulpia* is sometimes included, consists of chasmogamous species having three anthers per floret. The two genera are closely related to each other. Sterile hybrids between *Vulpia* and *Festuca*, and *Vulpia* and *Lolium*, are known.

In the key and descriptions, the spikelet and lemma measurements exclude the awns.

1. Lower glumes less than ½ the length of the upper glumes.
  2. Lemmas 5-veined, glabrous except the margins sometimes ciliate; rachilla internodes 0.75–1.9 mm long . . . . 1. *V. myuros*
  2. Lemmas 3(5)-veined, pubescent or glabrous, the margins ciliate; rachilla internodes 0.4–0.9 mm long . . . . 5. *V. ciliata*
1. Lower glumes ½ or more the length of the upper glumes.
  3. Panicle branches 1–2 per node; spikelets with 4–17 florets; rachilla internodes 0.5–0.7 mm long; awn of the lowermost lemma in each spikelet 0.3–9 mm long; caryopses 1.7–3.7 mm long . . . . 2. *V. octoflora*
  3. Panicle branches solitary; spikelets with 1–8 florets; rachilla internodes 0.6–1.2 mm long; awn of the lowermost lemma in each spikelet 2–20 mm long; caryopses 3.5–6.5 mm long.
    4. Panicle branches appressed to erect at maturity, without axillary pulvini; paleas equal to or shorter than the lemmas. . . . . 3. *V. bromoides*
    4. Panicle branches spreading to reflexed at maturity, with axillary pulvini; paleas usually slightly longer than the lemmas . . . . 4. *V. microstachys*

1. **Vulpia myuros** (L.) C.C. Gmel. Foxtail Fescue, Rattail Fescue [p. 352, 499]
Clm 10–75(90) cm, solitary or loosely tufted, brchd or unbrchd distally. Shth usu glab; lig 0.3–0.5 mm; bld 2.4–10.5(17) cm long, 0.4–3 mm wide, usu rolled, occ flat, usu glab. Infl 3–25 cm long, 0.5–1.5(2) cm wide, dense pan or spikelike rcm, with 1 br per nd, often partially enclosed in the uppermost shth at maturity, pulvini absent; br spreading or appressed to erect. Spklt 5–12 mm, with 3–7 flt; rchl intnd 0.75–1(1.9) mm. Glm glab; lo glm 0.5–2 mm, ⅕–½ the length of the up glm; up glm 2.5–5.5 mm; lm 4.5–7 mm, 5-veined, usu scabrous distally, glab except the mrg smt ciliate, apc entire, awns 5–15(22) mm; pal 4.7–6.4 mm, minutely bifid; anth 0.5–1(2) mm. Car 3–5 mm, fusiform, glab. $2n = 14$ [f. *myuros*], 42 [f. *myuros* and f. *megalura*].

Vulpia myuros grows in well-drained, sandy soils and disturbed sites. It is native to Europe and North Africa. *Vulpia myuros* f. *megalura* (Nutt.) Stace & R. Cotton differs from *Vulpia myuros* (L.) C.C. Gmel. f. *myuros* in having ciliate lemma margins. It was once thought to be native to North America, but it occurs throughout the European and North African range of f. *myuros*, even in undisturbed areas.

2. **Vulpia octoflora** (Walter) Rydb. Sixweeks Fescue [p. 353, 499]
Clm 5–60 cm, solitary or loosely tufted, glab or pubescent. Shth glab or pubescent; lig 0.3–1 mm; bld to 10 cm long, 0.5–1 mm wide, flat or rolled, glab or pubescent. Pan 1–7(20) cm long, 0.5–1.5 cm wide, with 1–2 br per nd; br appressed to spreading. Spklt 4–10(13) mm, with (4)5–11(17) flt; rchl intnd 0.5–0.7 mm. Lo glm 1.7–4.5 mm, ½–⅔ the length of the up glm; up glm 2.5–7.2 mm; lm 2.7–6.5 mm, 5-veined, smooth, scabrous, or pubescent, apc entire, no more pubescent than the bases, awns of the lowermost lm in each spklt 0.3–9 mm; pal slightly shorter than the lm, apc entire or minutely bifid, teeth shorter than 0.2 mm; anth 0.3–1.5 mm. Car 1.7–3.7 mm. $2n = 14$.

Vulpia octoflora, a widespread native species, tends to be displaced by the introduced *Bromus tectorum* in western North America. It grows in grasslands, sagebrush, and open woodlands, as well as in disturbed habitats and areas of secondary succession, such as old fields, roadsides, and ditches. Two varieties grow in the Intermountain Region, but numerous intermediates are common.

1. Lemmas scabrous to pubescent . . . . var. *hirtella*
1. Lemmas usually smooth, sometimes scabridulous distally and on the margins . . . . var. *octoflora*

Vulpia octoflora var. **hirtella** (Piper) Henrard [p. 353]
Pan br appressed, spklt closely imbricate. Spklt usu 5.5–10 mm. Lm prominently scabrous to densely pubescent; awns of the lowermost lm in each spklt 2.5–6.5 mm.

Vulpia octoflora var. *hirtella* is most frequent from British Columbia south through the western United States and into Mexico. It is the most common variey of *V. octoflora* in the southwest.

Vulpia octoflora (Walter) Rydb. var. **octoflora**
    [p. 353]
Pan br erect to ascending, lo br smt spreading
distally. **Spklt** usu 5.5–10(13) mm, usu not or
only slightly overlapping. **Lm** usu smooth, smt
scabridulous distally and on the mrg; **awns** of
the lowermost lm in each spklt 3–9 mm.

Vulpia octoflora var. *octoflora* is widespread throughout
southern Canada, the United States, and Mexico, and has
been introduced into temperate regions of South America,
Europe, and Asia. It is more common east of the Rocky
Mountains than in the Intermountain Region.

3. **Vulpia bromoides** (L.) Gray Brome Fescue
    [p. 353, 499]
Clm 5–50 cm, solitary or loosely tufted, erect or
decumbent, smooth, scabridulous, or puberulent,
unbrchd distally. **Shth** glab or puberulent; **lig** to
0.5(1) mm; **bld** usu 2–10 cm long, 0.5–2.5 mm
wide, rolled or flat, glab or puberulent. **Pan** 1.5–15
cm long, 0.5–3 cm wide, conspicuously exserted,
with 1 br per nd; **br** usu appressed to erect at
maturity, without axillary pulvini; **ped** flattened,
smt clavate distally. **Spklt** 5–10 mm, with 4–8 flt,
not closely imbricate; **rchl intnd** 0.6–1.1 mm. **Lo
glm** 3.5–5 mm, ½–⅘ the length of the up glm; **up
glm** 4.5–9.5 mm, midveins scabrous distally; **lm**
4–8 mm, 5-veined, scabrous distally, apc entire,
awns of the lowermost lm in each spklt 2–13 mm;
pal 4–6.3 mm, equaling or shorter than the lm,
minutely bifid; **anth** 0.4–0.6(1.5) mm. **Car** 3.5–5
mm. 2*n* = 14.

Vulpia bromoides is a common European species that grows
in wet to dry, open habitats. It is adventive and naturalized
in North and South America. In North America, it is most
common on the west coast, but has been found at several
locations in the Intermountain Region.

4. **Vulpia microstachys** (Nutt.) Munro Small
    Fescue [p. 354, 499]
Clm 15–75 cm, solitary or loosely tufted, usu
glab, occ puberulent. **Shth** glab or pubescent;
lig 0.5–1 mm; **bld** usu shorter than 10 cm, 0.5–1
mm wide, usu rolled, occ flat, glab or pubescent.
Infl 2–24 cm long, 0.8–8 cm wide, usu pan, smt
spikelike rcm; **br** solitary, with axillary pulvini,
appressed to erect when immature, spreading to
reflexed at maturity. **Spklt** 4–10 mm, with 1–6 flt,
often purple-tinged; **rchl intnd** 0.6–1.2 mm. **Glm**
smooth, scabrous, or pubescent; **lo glm** 1.7–5.5
mm, ½–¾ the length of the up glm; **up glm** 3.5–7.5
mm; **lm** 3.5–9.5 mm, smooth, scabrous, or evenly
pubescent, 5-veined, awns of the lowermost lm
in each spklt (3)6–20 mm; **pal** usu slightly longer
than the lm, apc minutely bifid, teeth 0.2–0.5 mm;
anth 0.7–3 mm. **Car** 3.5–6.5 mm. 2*n* = 42.

Vulpia microstachys is native to western North America,
growing from British Columbia south through the western
United States into Baja California. Four varieties are
recognized here on the basis of spikelet indumentum, but
they frequently occur together, and intergrading forms
are known. No difference in their geographic or ecological
distribution is known.

1. Glumes and lemmas smooth or scabrous . . var. *pauciflora*
1. Glumes and/or lemmas pubescent.
    2. Glumes and lemmas pubescent . . . . var. *ciliata*
    2. Glumes or lemmas, but not both, pubescent.
        3. Glumes pubescent; lemmas glabrous . . var. *confusa*
        3. Glumes glabrous; lemmas pubescent . . . .
        . . . . var. *microstachys*

Vulpia microstachys var. **ciliata** (A. Gray)
    Lonard & Gould Eastwood Fescue [p. 354]
Spklt usu with 2–4 flt. **Glm** and **lm** sparsely or
densely pubescent.

Vulpia microstachys var. *ciliata* grows in loose, sandy soils.

Vulpia microstachys var. **confusa** (Piper)
    Lonard & Gould Confusing Fescue [p. 354]
Spklt usu with 1–3 flt. **Glm** pubescent; **lm** glab.

Vulpia microstachys var. *confusa* grows in sandy, open
sites.

Vulpia microstachys (Nutt.) Munro var.
    microstachys Desert Fescue [p. 354]
Spklt with (1)2–5 flt. **Glm** glab; **lm** sparsely to
densely pubescent.

Vulpia microstachys var. *microstachys* grows most commonly
in loose soil on open slopes and roadsides.

Vulpia microstachys var. **pauciflora** (Scribn.
    ex Beal) Lonard & Gould Pacific Fescue
    [p. 354]
Spklt with 1–6 flt. **Glm** and **lm** smooth or
scabrous.

Vulpia microstachys var. *pauciflora* grows in sandy, often
disturbed sites, and is the most common and widespread
variety of the complex. It is often intermingled with plants
of the other varieties.

5. **Vulpia ciliata** Dumort. Fringed Fescue
    [p. 354, 499]
Clm 6–45 cm, loosely tufted. **Shth** smooth, glab;
lig 0.2–0.5 mm; **bld** 3.5–10 cm long, about 0.4
mm wide, folded to involute, abx surfaces glab,
adx surfaces puberulent. **Infl** 3–20 cm long, 0.3–
1.5 cm wide, pan or spicate rcm, usu partially
enclosed in the uppermost shth at maturity, with
1 br per nd, axillary pulvini absent. **Spklt** 5–10.5
mm, with 4–10 flt; **rchl intnd** 0.4–0.9 mm. **Glm**
glab; **lo glm** 0.1–1.3 mm, less than ⅓ the length
of the up glm; **up glm** 1.5–4 mm; **lm** 4–7.7 mm,
3(5)-veined, usu pubescent on the midvein, smt
also on the body, rarely glab on both, mrg ciliate,
hairs to 1 mm, awns 6–15.3 mm; **pal** slightly

shorter than to equaling the lm, apc entire; **anth** 0.4–0.6(1.6) mm. **Car** 3.4–6.5 mm. 2n = 42.

Vulpia ciliata is native to Europe, the Mediterranean area, and southwest and central Asia. It grows in open, dry habitats. It is easily distinguished from other members of the genus because of its upper glumes with broadly membranous tips

that break off, making the glumes appear truncate or blunt. In North America, it was known until recently only from an old ballast dump record from Philadelphia. In May 2004, it was collected immediately north of the Odgen Bay Waterfowl Management Area, Weber County, Utah, in an upland area of the site. The source of the seeds is not known.

8.05 LOLIUM L.   Edward E. Terrell

Pl ann or per; csp, smt shortly rhz. **Clm** 10–150 cm, slender to stout, erect to decumbent, rarely prostrate. **Shth** open, rounded, glab, smt scabrous; **lig** to 4 mm, memb, glab; **aur** smt present; **bld** flat, linear. **Infl** distichous spikes, with solitary spklt oriented radial to the rchs, perpendicular to the rchs concavities. **Spklt** lat compressed, with 2–22 flt, distal flt rdcd; **rchl** glab; **dis** above the glm, beneath the flt. **Glm** usu 1, 2 in the tml spklt, lanceolate to oblong, rounded over the midvein, memb to indurate, 3–9-veined, unawned; **lo glm** absent from all but the tml spklt; **up glm** from shorter than to exceeding the distal flt; **cal** short, blunt, glab; **lm** lanceolate, ovate or oblong, rounded over the midvein, memb, chartaceous, 3–7-veined, apc smt hyaline, unawned or awned, awns subtml, more or less straight; **pal** memb, usu smooth, keels ciliolate; **lod** 2, free, lanceolate to ovate; **anth** 3; **ov** glab. **Car** dorsally compressed, oblong, broadly elliptic or ovate, longitudinally sulcate; **hila** linear, in the furrow; **emb** ⅕–⅓ as long as the car. x = 7. *Lolium*, first mentioned in Virgil's Georgics, is an old Latin name for darnel, *Lolium temulentum*.

As interpreted here, *Lolium* comprises five species that are native to Europe, temperate Asia, and northern Africa. Four have been introduced to the Intermountain Region, often as forage grasses.

Lolium used to be included in the *Triticeae*, but evidence from genetics, morphology, and other studies shows its closest relationship to be to the species included here in *Schedonorus*. Artificial hybrids have been produced among *L. perenne*, *L. multiflorum*, *Schedonorus pratensis*, and *S. arundinaceus*. Cultivars of these crosses have been registered for commercial use and are sometimes used for forage. Natural hybrids are not uncommon in Europe.

1. Plants either long-lived perennials with 2–10 florets per spikelet, or annuals or short-lived perennials with 10–22 florets per spikelet.
   2. Plants long-lived perennials, with 2–10 florets per spikelet; lemmas unawned or awned, awns to about 8 mm long . . . . 1. *L. perenne*
   2. Plants annuals or short-lived perennials, with 10–22 florets per spikelet; lemmas usually awned, awns to 15 mm long, rarely unawned. . . . . 2. *L. multiflorum*
1. Plants annuals, with 2–10(11) florets per spikelet.
   3. Spikelets somewhat sunken in the rachises and partly concealed by the glumes . . . . 3. *L. rigidum*
   3. Spikelets not sunken in the rachises and not concealed by the glumes. . . . . 4. *L. temulentum*

1. Lolium perenne L. PERENNIAL RYEGRASS, ENGLISH RYEGRASS [p. 355, 500]

Pl long-lived per. **Clm** to 100 cm. **Lvs** folded in the bud; **bld** usu 10–30 cm long, (1)2–4(6) mm wide. **Spikes** 3–30 cm, with 5–37 spklt; **rchs** 0.5–2.5 mm thick at the nd, often flexuous. **Spklt** 5–22 mm long, 1–7 mm wide, with (2)5–9(10) flt. **Glm** 3.5–15 mm, (⅓)½–¾ as long as to slightly exceeding the distal flt, memb to indurate; **lm** 3.5–9 mm long, 0.8–2 mm wide, unawned or awned, awns to about 8 mm, attached 0.2–0.7 mm below the apc; **pal** shorter than to slightly longer than the lm; **anth** 2–4.2 mm. **Car** 3–5.5 mm long, 0.7–1.5 mm wide, 3 or more times longer than wide. 2n = 14.

Lolium perenne, a Eurasian species, is now established in disturbed areas throughout much of North America, including the Intermountain Region. It is commercially important, being included in lawn seed mixtures as well as being used for forage and erosion prevention.

Lolium perenne intergrades and is interfertile with *L. multiflorum*. Typical *L. perenne* differs from *L. multiflorum* in being a shorter, longer-lived perennial with narrower leaves that are folded, rather than rolled, in the bud. Hybrids between the two species are called *Lolium ×hybridum* Hausskn.

2. Lolium multiflorum Lam. ANNUAL RYEGRASS, ITALIAN RYEGRASS [p. 355, 500]

Pl ann or short-lived per. **Clm** to 150 cm. **Lvs** rolled in the bud; **bld** usu 10–30 cm long, (2)3–8 (13) mm wide. **Spikes** 15–45 cm, with 5–38 spklt; **rchs** 0.8–2 mm thick at the nd, not flexuous. **Spklt** 8–31 mm long, 2–10 mm wide, with (10)11–22 flt. **Glm** 5–18 mm, ¼–½ as long as the flt, memb to indurate; **lm** 4–8.2 mm long, 1–2 mm wide, usu awned, awns to 15 mm, attached 0.2–0.7 mm below the apc, rarely unawned; **pal** shorter than to slightly longer than the lm; **anth** (2.5)3–4.5(5) mm. **Car** 2.5–4 mm long, 0.7–1.5 mm wide, 3 or more times longer than wide. 2n = 14.

Lolium multiflorum, a European species, now grows in most of North America, including the Intermountain Region. It is planted as a cover crop, as a temporary lawn grass, for roadside restoration, and for soil or forage enrichment; it often escapes from cultivation, becoming established in disturbed sites. It hybridizes with *L. perenne*.

3. Lolium rigidum Gaudin STIFF RYEGRASS
[p. *355*, <u>500</u>]

Pl ann. Clm to 70 cm. Bld to 17 cm long, 0.5–5(8) mm wide. Spikes 3–30 cm, with 2–20 spklt; rchs 0.5–3.5 mm thick at the nd, with the spklt somewhat sunken in the rchs and partly concealed by the glm. Spklt 5–18 mm long, 1–3(7) mm wide, with 2–8(11) flt. Glm 4–20(30) mm, usu from ¾ as long as to slightly exceeding the distal flt, rather indurate; lm 3–8.5(10.5) mm long, 0.9–2 mm wide; pal slightly shorter than to slightly longer than the lm, usu unawned, smt awned, awns to 10 mm; anth 1.2–3.2 mm. Car 2.7–5.5 mm long, 1–1.5 mm wide, 3 or more times longer than wide. $2n = 14$.

Lolium rigidum is native to Europe, North Africa, and western Asia. It has been found as a weed of roadsides and waste places at scattered locations in the contiguous United States and Canada.

Lolium rigidum intergrades with *L. perenne*, *L. multiflorum*, and, occasionally, *L. temulentum*. Hybrids with *L. multiflorum* are called *Lolium ×hubbardii* Jansen & Wacht. *ex* B.K. Simon.

4. Lolium temulentum L. [p. *356*, <u>500</u>]

Pl ann. Clm to 120 cm. Bld to 27 cm long, 1–12 mm wide. Spikes 2–40 cm, with 3–26 spklt; rchs 0.5–3.5 mm thick at the nd, spklt not sunken in the rchs, not concealed by the glm. Spklt 5–28 mm long, 1–8 mm wide, with 2–10 flt. Glm 5–28 mm, memb to indurate; lm 3.5–8.5 mm long, 1.2–3 mm wide, unawned or awned, awns to 23 mm, attached 0.2–2 mm below the apc; pal 1.2 mm shorter than to 0.8 mm longer than the lm, often wrinkled; anth 1.5–4 mm. Car 3.2–7 mm long, 1–3 mm wide, 2–3 times longer than wide, turgid. $2n = 14$.

Lolium temulentum is said to be the tares of the Bible. One subspecies is present in the Intermountain Region.

Lolium temulentum L. subsp. temulentum
DARNEL [p. *356*]

Bld (1.5)3–10(12) mm wide. Spikes 5–40 cm, with 5–26 spklt; rchs rather stout. Spklt 8–28 mm long, 3–8 mm wide. Glm (5.5)7–28 mm, from ¾ as long as to longer than the flt, somewhat indurate; lm 4.5–8.5 mm long, 1.5–3 mm wide, unawned or awned, awns to 23 mm, attached 0.5–2 mm below the apc. Car (3.8)4–7 mm long, (1)1.5–3 mm wide.

Lolium temulentum subsp. *temulentum* is found occasionally in disturbed sites throughout much of North America. It is native to the Eastern Hemisphere, where it is known only as a weed, especially of grain fields. Awn presence or absence and length vary, and have no taxonomic significance.

The seeds sometimes become infected with an endophytic fungus, assumed to be the source of the toxic pyrrolizidine alkaloids loline, 6-methyl loline, and lolinine, but not temuline, which is now considered an artifact of isolation (Dannhardt and Steindl 1985). Because primitive agricultural practices could not separate seeds of *Lolium temulentum* from those of wheat, infected seeds often resulted in poisonous flour.

8.06 PUCCINELLIA Parl.   Jerrold I. Davis and Laurie L. Consaul

Pl ann, bien, or per; usu csp, smt weakly or strongly stln and mat-forming. Clm 2–100 cm, erect or decumbent, smt geniculate; intnd hollow. Shth open to the base or nearly so; aur absent; lig memb, acute to truncate, entire or erose; bld flat, folded, or involute. Infl tml pan, open to contracted; br smooth or scabrous, some br longer than 1 cm; ped usu longer than 3 mm, thinner than 0.5 mm. Spklt pedlt, subterete to weakly lat compressed, with 2–10 flt; dis above the glm, beneath the flt. Glm usu unequal, smt subequal to equal, usu distinctly shorter than the lowest lm in the spklt, smt only slightly shorter, rarely longer, memb, rounded or weakly keeled, veins obscure or prominent, apc unawned; lo glm 1(3)-veined; up glm (1)3(5)-veined; cal blunt, glab or pubescent; lm memb to slightly or distinctly coriaceous, glab or pubescent, pubescence smt restricted to the bases of the veins, rounded or weakly keeled, at least distally, (3)5(7)-veined, veins obscure to prominent, more or less parallel distally, usu not extending to the apc, lat veins smt rdcd, apical mrg with or without scabrules, apc usu acute to truncate, smt acuminate, entire or serrate to erose, unawned; pal subequal to the lm, scarious or memb distally, 2-veined, veins terminating at or beyond midlength; lod 2, free, glab; anth 3; ov glab. Car shorter than the lm, concealed at maturity, oblong, terete to dorsally flattened, falling free or with the pal or both the lm and pal attached; hila oblong, about ⅓ or less the length of the car. $x = 7$. Named for Benedetto Puccinelli (1808–1850), an Italian botanist.

Puccinellia, a genus of approximately 120 species, is most abundant in the middle and high latitudes of the Northern Hemisphere. There are six species in the Intermountain Region, of which two are introduced. Most species of *Puccinellia* are halophytes, either in coastal habitats or in saline or otherwise mineralized soils of interior habitats. Polyploidy, selfing, and hybridization are widespread in the genus, and many of the species boundaries are controversial.

The angle of the panicle branches (whether erect, ascending, etc.) refers to their position when the caryopses are mature. Lemma measurements should be made on the lowest lemma in the spikelets. Principal features of the lemmas, as used in the key and descriptions, are as follows. Scabrules (short, pointed hairs, similar in form to those that occur on the pedicels and inflorescence branches of many species of *Puccinellia*, and generally requiring magnification to observe) often occur along the distal margins of the lemmas. When present, they may be either few and irregularly scattered, with gaps between them that are wider than the individual scabrules, or arranged in a continuous palisade-like row that lacks gaps (e.g., in *P. distans*, p. 106).

1. Lemmas coriaceous, including apical margins; lemmas 1.8–3 mm long; lower inflorescence branches ascending to erect, spikelet-bearing to near the base; anthers 0.6–1 mm long .... 1. *P. fasciculata*
1. Lemmas membranous or herbaceous at maturity, apical margins sometimes hyaline; lemmas 1.5–4 mm long; lower inflorescence branches erect to descending, spikelet-bearing to near the base or not; anthers 0.4–2 mm long.
  2. Plants annual.
    3. Lemma apices acute; lemmas 2.5–4 mm long, hairy between the veins on the basal ½, hairs about 0.1 mm, sparsely and evenly distributed, veins glabrous or hairy .... 2. *P. simplex*
    3. Lemma apices obtuse to truncate; lemmas 1.8–2.2 mm long, glabrous between the veins, densely hairy over the veins. .... 3. *P. parishii*
  2. Plants perennial.
    4. Lemmas 1.5–2.2 mm long; apices widely obtuse to truncate; anthers 0.4–0.8 mm long; lower panicle branches horizontal to descending .... 5. *P. distans*
    4. Lemmas (2)2.2–4 mm long; apices usually acute to obtuse, occasionally acuminate or rounded; anthers 0.6–2 mm long; lower panicle branches erect to descending.
      5. Lemma midveins often extending to the apical margins, often slightly scabrous and prominent distally; lemma apices acute, smooth to scabrous. .... 4. *P. lemmonii*
      5. Lemma midveins not extending to the margins, smooth distally; lemma apices acute to obtuse, scabrous .... 6. *P. nuttalliana*

1. **Puccinellia fasciculata** (Torr.) E.P. Bicknell
BORRER'S SALTMARSH GRASS [p. 356, <u>500</u>]
Pl short-lived per; ces, not mat-forming. **Clm** 10–65 cm, usu decumbent and geniculate, smt erect. **Lig** 1–2 mm, obtuse to truncate, entire; **bld** 2–7 mm wide, flat, folded, or involute. **Pan** 4–16 cm, compact to diffuse at maturity, linear to pyramidal, lo br ascending to erect, spklt-bearing nearly to the base; **ped** slightly to densely scabrous, often with tumid epidermal cells. **Spklt** 3–6 mm, with 2–6 flt. **Glm** rounded over the back, veins prominent to obscure, apc acute to obtuse; **lo glm** 0.8–1.6 mm; **up glm** 1.2–2.3 mm; **cal** with a few hairs; **lm** 1.8–3 mm, slightly to markedly coriaceous throughout, glab or with a few hairs near the base, backs rounded, 5-veined, veins obscure, midveins usu excurrent, smt ending at the mrg, apc acute to obtuse, entire, apical mrg smooth or with a few scattered scabrules; **pal** veins glab proximally, scabrous to shortly hispid near midlength, scabrous distally; **anth** 0.6–1 mm. 2*n* = 28.

Puccinellia fasciculata is native to Europe. In North America, it is found principally along the east coast, but it is also established at a few sites in Arizona and Utah, and has been reported from Nevada.

2. **Puccinellia simplex** Scribn. WESTERN ALKALI GRASS [p. 356, <u>500</u>]
Pl ann; not mat-forming. **Clm** usu erect, 2–25 cm. **Lig** 1–3 mm, acute to obtuse, entire; **bld** 0.7–2 mm wide, flat to involute. **Pan** 1–18 cm, compact, mostly linear at maturity, pri br usu spklt-bearing to the base, lo br erect; **ped** densely scabrous or with only a few scattered scabrules, often also with a few hairs, often with tumid epidermal cells. **Spklt** 3.5–8 mm, with 2–7 flt. **Glm** rounded to weakly keeled over the back, veins obscure to prominent, apc acute; **lo glm** 1.3–2 mm; **up glm** 2.3–3 mm; **cal** hairy; **lm** 2.5–4 mm, mostly herbaceous, usu with about 0.1 mm hairs distributed sparsely and evenly between the veins, longer hairs also usu present along the veins and near the base, bas hairs often longest, twisted and tangled, backs rounded, smt keeled distally, 5-veined, veins usu obscure, smt prominent, midveins smt reaching the mrg, other veins usu not doing so, apical mrg often hyaline, smooth or with a few scattered scabrules, apc acute, entire; **pal** veins with hairs, hairs on the proximal portion longer, twisted and somewhat tangled, hairs on the distal ⅔ usu short and straight; **anth** 0.2–0.5 mm. 2*n* = 56.

Puccinellia simplex is widespread in, and mostly confined to, saline soils of central California. The records from Utah probably reflect introductions.

3. **Puccinellia parishii** Hitchc. Parish's Alkali
    Grass [p. *357*, 500]
Pl ann; not mat-forming. **Clm** 3–22 cm, erect. **Lvs**
bas concentrated; **lig** 1–2 mm, obtuse to truncate,
entire; **bld** 0.2–1.2 mm wide, flat to involute.
Pan 1–8.5 cm, compact to diffuse at maturity,
lo br erect to descending, usu spklt-bearing to
the base; **ped** densely scabrous or with a few
scattered scabrules, often with tumid epidermal
cells. **Spklt** 3.5–5 mm, with 2–7 flt. **Glm** rounded
over the back, veins obscure to prominent, apc
acute to obtuse; **lo glm** 1–2 mm; **up glm** 1.8–2.2
mm; **cal** hairy; **lm** 1.8–2.2 mm, mostly herbaceous,
densely hairy over the proximal ½–¾ of the veins,
glab between the veins, backs rounded, 5-veined,
veins obscure to prominent, not extending to the
mrg, apical mrg hyaline, smooth or with a few
scattered scabrules, apc obtuse to truncate, entire;
pal veins glab proximally, hairy at midlength,
glab or scabrous-ciliate distally; **anth** 0.4–0.5
mm. $2n = 14$.

Puccinellia parishii grows in saline seepage areas in
California, Arizona, and New Mexico. It is not known for
sure to be in the Intermountain Region.

4. **Puccinellia lemmonii** (Vasey) Scribn.
    Lemmon's Alkali Grass [p. *357*, 500]
Pl per; ces, not mat-forming. **Clm** 5–40 cm, usu
erect. **Lvs** bas concentrated; **lig** 0.8–2.2 mm,
obtuse to acute, mostly entire, smt slightly erose;
bld involute, 1.2–1.9 mm wide when flattened.
Pan 2–18 cm, compact to diffuse at maturity, lo
br ascending to descending, usu spklt-bearing to
the base; **ped** scabrous, lacking tumid epidermal
cells. **Spklt** 3.5–8 mm, with 2–6 flt. **Glm** rounded
over the back, veins obscure, apc acute to obtuse;
lo glm 0.7–1.5 mm; **up glm** 1.4–3 mm; **cal** with
a few hairs; **lm** 2.4–4 mm, herbaceous, mostly
smooth, usu glab, smt with a few hairs near
the base, principally along the veins, backs usu
rounded, smt weakly keeled distally, 5-veined,
veins obscure, midveins often slightly scabrous
and prominent in the distal ½, often extending
to the apical mrg, lat veins not extending to
the mrg, apical mrg ranging from smooth to
scabrous, entire, apc acute, entire; **pal** veins glab
or shortly ciliate proximally, uniformly scabrous
distally; **anth** 1–2 mm. $2n = 14$.

Puccinellia lemmonii grows in non-littoral saline
environments in the western portion of the contiguous
United States.

5. **Puccinellia distans** (Jacq.) Parl. European
    Alkali Grass [p. *357*, 500]
Pl per; ces, not mat-forming. **Clm** 5–60 cm, erect
to decumbent. **Lig** 0.8–1.2 mm, obtuse to truncate,

usu entire; **bld** 1–7 mm wide, flat to involute. **Pan**
2.5–20 cm, diffuse at maturity, lo br horizontal
to descending, spklt usu confined to the distal
⅔; **ped** scabrous, lacking tumid epidermal cells.
Spklt 2.5–7 mm, with 2–7 flt. **Glm** rounded over
the back, veins obscure, apc acute to truncate; **lo
glm** 0.4–1.3 mm; **up glm** 0.9–1.8 mm; **cal** with a
few hairs; **lm** 1.5–2(2.2) mm, mostly herbaceous,
glab or sparsely hairy on the lo ½, principally
along the veins, backs rounded, 5-veined, veins
obscure, not extending to the mrg, apical mrg
hyaline and often yellowish, uniformly and
densely scabrous, apc widely obtuse to truncate,
entire; **pal** veins shortly ciliate proximally, glab,
smt scabrous distally; **anth** 0.4–0.8 mm. $2n = 14,$
$28, 42.$

Puccinellia distans is a Eurasian native, reportedly
introduced in North America. It frequently grows as a weed
in non-littoral environments, including the margins of salted
roads.

6. **Puccinellia nuttalliana** (Schult.) Hitchc.
    Nuttall's Alkali Grass [p. *358*, 500]
Pl per; ces, not mat-forming. **Clm** 10–100 cm,
usu erect. **Lvs** either concentrated at the base or
distributed along the clm; **lig** 1–3 mm, obtuse,
usu entire, smt slightly erose; **bld** 1–4 mm wide,
flat to involute. **Pan** 5–30 cm, compact to diffuse
at maturity, usu distinctly exserted from the shth,
lo br usu erect to diverging, occ descending,
spklt-bearing from the base or on the distal ⅔;
ped scabrous, lacking tumid epidermal cells.
Spklt 3.5–9 mm, with 2–7 flt. **Glm** rounded over
the back, veins obscure, apc acute to obtuse;
lo glm 0.5–1.5 mm, usu less than ½ as long as
the adjacent lm; **up glm** 1–2.8 mm; **rchl intnd**
slightly and gradually broadened to the point of
attachment with the lm; **cal** with a few hairs; **lm**
(2)2.2–3(3.5) mm, herbaceous, glab or sparsely
hairy on the proximal ½, principally along the
veins, backs rounded, 5-veined, veins obscure,
not extending to the mrg, smooth distally, lat
mrg inrolled or not, apical mrg uniformly and
densely scabrous, apc acute to obtuse, entire; **pal**
veins glab, short-ciliate, or with a few long hairs
proximally, smooth or scabrous distally; **anth**
0.6–2 mm. $2n = 28, 42, 56.$

Puccinellia nuttalliana is a widespread and variable species.
It is common in saline areas throughout the Intermountain
Region.

8.07 **SCLEROCHLOA** P. Beauv.   David M. Brandenburg

Pl ann. **Clm** 2–30 cm. **Shth** open to closed; **aur** absent; **lig** memb; **bld** flat or folded. **Infl** tml, usu rcm, smt rdcd pan, 1-sided, usu exceeded by the lvs. **Spklt** lat compressed, subsessile to pedlt; **ped** 0.5–1 mm long, 0.5–0.8 mm thick, stout, with 2–7 flt; **rchl** glab, lowest intnd thicker than those above; **dis** tardy, not strongly localized. **Glm** unequal, shorter than the lowest lm, glab, with wide hyaline mrg, apc obtuse to emgt, unawned; **lo glm** (1)3–5-veined; **up glm** (3)5–9-veined; **cal** blunt, glab; **lm** memb, with hyaline mrg, indurate at maturity, (5)7–9-veined, veins prominent, apc rounded to emgt, entire, unawned; **pal** shorter than to equaling the lm, dorsally compressed; **lod** 2, free, glab, entire to lacerate; **anth** 3. **Car** shorter than the lm, concealed at maturity, beaked from the persistent sty base, falling free; **hila** round. *x* = 7. Name from the Greek *skleros*, 'hard', and *chloa*, 'grass', alluding to the leathery glumes and lemmas.

Sclerochloa is a genus of two species, both of which are native to southern Europe and western Asia. The species found in North America, *S. dura*, is now a cosmopolitan weed.

1. **Sclerochloa dura** (L.) P. Beauv. Hardgrass, Fairground Grass [p. *358*, <u>500</u>]

Pl often matted, occ with solitary clm. **Clm** 2–15(30) cm, usu prostrate to procumbent, smt ascending, glab. **Lvs** strongly overlapping, generally exceeding the infl; **shth** completely open or closed to ½ their length; **lig** (0.3)0.7–2(3.5) mm, glab, acute; **bld** 0.5–5(7) cm long, 1–4 mm wide, glab, apc prow-tipped. **Rcm** 1–4(5) cm, often partially enclosed in the up lf shth; **ped** 0.5–0.8 mm thick. **Spklt** (3.4)5–12 mm, with (2)3–4(7) flt. **Lo glm** 1.4–3(3.7) mm; **up glm** 2.6–5.4(6.2) mm; **lowest lm** (3.4)4.5–6(7) mm, midveins scabridulous distally; **distal lm** successively smaller than the lo lm; **anth** 0.8–1.5 mm. **Car** weakly trigonous, rugulose. 2*n* = 14.

First collected in the United States in 1895, *Sclerochloa dura* is probably more widespread than indicated, because it is easily overlooked. It grows in lawns, campsites, roadsides, athletic fields, fairgrounds, and other disturbed sites. It is frequently found in severely compacted soils, because it can withstand heavy traffic by vehicles and pedestrians.

Sclerochloa dura is sometimes confused with *Poa annua*. The two species are superficially similar, occupy similar habitats, and have a similar phenology, but *S. dura* has blunt, glabrous lemmas and racemose inflorescences, whereas *P. annua* has obtuse to acute lemmas that are smooth and usually sericeous or crisply puberulent over the veins, and paniculate inflorescences. Plants of *S. dura* become stramineous in age, making them easy to locate because areas dominated by them change color.

8.08 **DACTYLIS** L.   Kelly W. Allred

Pl per; csp, smt with short rhz. **Clm** to 2.1+ m, bases lat compressed; **intnd** hollow; **nd** glab. **Lvs** mostly bas, glab; **shth** closed for at least ½ their length; **aur** absent; **lig** memb; **bld** flat to folded. **Infl** pan; **pri br** 1-sided, naked proximally, with dense clusters of subsessile spklt distally, at least some br longer than 1 cm. **Spklt** oval to elliptic in outline, lat compressed, with 2–6 flt; **rchl** glab, not prolonged beyond the distal flt; **dis** above the glm and beneath the flt. **Glm** shorter than the flt, lanceolate, 1–3-veined, ciliate-keeled, awn-tipped; **cal** short, blunt; **lm** 5-veined, scabrous to ciliate-keeled, tapering to a short awn; **pal** 2-keeled, tightly clasped by the lm, unawned, apc notched; **lod** 2, glab, toothed; **anth** 3; **ov** glab. **Car** shorter than the lm, concealed at maturity, oblong to ellipsoid, falling free or adhering to the lm and/or pal; **hila** round. *x* = 7. Name from the Greek *daktylos*, 'finger'.

Dactylis is interpreted here as a variable monotypic genus, although five species are recognized by Russian taxonomists. Numerous infraspecific taxa have been recognized in Eurasia, where *Dactylis* is native, but it does not seem feasible to identify subspecies and varieties in North America.

1. **Dactylis glomerata** L. Orchardgrass [p. *359*, <u>500</u>]

Clm to 2.1+ m, erect. **Lvs** dark green; **shth** longer than the intnd, glab, usu keeled; **lig** 3–11 mm, truncate to acuminate; **bld** (2)4–8(10) mm wide, elongate, lax, with a conspicuous midrib and white, scabridulous to scabrous mrg. **Pan** 4–20 cm, typically pyramidal, lo br spreading, up br appressed. **Spklt** 5–8 mm, subsessile. **Glm** 3–5 mm; **lm** 4–8 mm, scabridulous; **pal** slightly shorter than the lm; **anth** 2–3.5 mm. 2*n* = 14, 21, 27–31, 42.

Dactylis glomerata grows in pastures, meadows, fence rows, roadsides, and similar habitats throughout North America. Native to Eurasia and Africa, it has been introduced throughout most of the cool-temperate regions of the world, including the Intermountain Region, as a forage grass. It provides nutritious forage that is relished by all livestock, as well as by deer, geese, and rabbits. When abundant, the pollen can be a major contributor to hay fever.

8.09 **BECKMANNIA** Host

Stephan L. Hatch

Pl ann and tufted, or per and rhz. **Clm** 20–150 cm, smt tuberous at the base, erect. **Lvs** mostly cauline; **shth** open, glab, ribbed; **aur** absent; **lig** memb, acute; **bld** flat, glab. **Infl** dense, spikelike pan; **br** 1-sided, rcmly arranged, sec br few, at least some br longer than 1 cm, with closely imbricate spklt; **dis** below the glm, the spklt falling entire. **Spklt** lat compressed, circular, ovate or obovate in side view, subsessile, with 1–2 flt; **rchl** not prolonged beyond the base of the distal flt. **Glm** subequal, slightly shorter than the lm, inflated, keeled, D-shaped in side view, unawned; **cal** blunt, glab; **lm** lanceolate, inconspicuously 5-veined, unawned; **pal** subequal to the lm; **lod** 2, free; **anth** 3; **ov** glab. **Car** shorter than the lm, concealed at maturity. x = 7. Named for Johann Beckmann (1739–1811), a German botanist and author of one of the first botanical dictionaries.

Beckmannia is a genus of two species: an annual species usually with one fertile floret per spikelet that is native to North America and Asia, and a perennial species with two fertile florets per spikelet that is restricted to Eurasia.

1. **Beckmannia syzigachne** (Steud.) Fernald
    AMERICAN SLOUGHGRASS [p. *359*, <u>500</u>]
Pl ann; tufted. **Clm** 20–120 cm. **Lig** 5–11 mm, pubescent, entire or lacerate, usu folded back; **bld** 4–10(20) mm wide, flat, scabrous. **Pan** 7–30 cm; **br** spikelike, usu 1–2 cm. **Spklt** 2–3 mm, round to ovate in side view, with 1 flt, a second undeveloped or well-developed flt occ present. **Glm** appearing inflated, strongly keeled, 3-veined, apiculate; **lm** 2.4–3.5 mm, unawned, smt mucronate; **pal** subequal to the lm, acute; **anth** 0.5–1(1.5) mm, pale yellow. **Car** shorter than 2 mm, light to medium brown. $2n$ = 14.

Beckmannia syzigachne grows in damp habitats such as marshes, floodplains, the edges of ponds, lakes, streams, and ditches, and in standing water. It is a good forage grass, but frequently grows in easily damaged habitats.

8.10 **POA** L.

Robert J. Soreng

Pl ann or per; usu synoecious, smt monoecious, gynodioecious, dioecious, and/or asex; with or without rhz or stln, densely to loosely tufted or the clm solitary. **Bas brchg** invag, psdinvag, or exvag; **prophylls** of invag shoots 2-keeled and open, of psdinvag shoots not keeled and tubular, of exvag shoots scalelike. **Clm** 1–150 cm, hollow, usu unbrchd above the base. **Shth** from almost completely open to almost completely closed, terete or weakly to strongly compressed; **aur** absent; **lig** memb, truncate to acuminate; **bld** 0.4–12 mm wide, flat, folded, or involute, adx surfaces with a groove on each side of the midvein, other intercostal depressions shallow, indistinct, apc often prow-shaped. **Infl** usu tml pan, rarely rdcd and rcmlike. **Spklt** 2–12 mm, usu lat compressed, infrequently terete to subterete, usu lanceolate, smt ovate; **flt** (1)2–6(13), usu sex, smt bulb-forming; **rchl** usu terete, smt prolonged beyond the base of the distal flt; **dis** above the glm and beneath the flt. **Glm** usu shorter than the lowest lm in the spklt, usu keeled, 1–3(5)-veined, unawned; **cal** blunt, usu terete or slightly lat compressed, smt slightly dorsally compressed, glab or hairy, hairs often concentrated in 1(3) tufts or webs, smt distributed around the cal below the lm as a crown of hairs; **lm** usu keeled, infrequently weakly keeled or rounded, similar in texture to the glm, 5(7–11)-veined, lat veins smt faint, mrg scarious-hyaline distally, apc scarious-hyaline, truncate or obtuse to acuminate, unawned; **pal** from ⅔ as long as to subequal to the lm, distinctly 2-keeled, mrg and intercostal regions milky white to slightly greenish; **lod** 2, broadly lanceolate, glab, lobed; **fnctl anth** (1–2)3, 0.1–5 mm; **ov** glab. **Car** 1–4 mm, ellipsoidal, often shallowly ventrally grooved, solid, with lipid; **hila** sub-bas, round or oval, to ⅙ the length of the car. x = 7. Name from the Greek *poa*, 'grass'.

Poa includes about 500 species. It grows throughout the world, principally in temperate and boreal regions. It is taxonomically difficult because most species are polyploid, many are apomictic, and hybridization is common. A variety of sexual reproductive systems are present within the genus, although individual species are usually uniform in this regard. Apomicts derived from bisexual species usually have functional anthers; they require fertilization to stimulate endosperm (and hence seed) development. Apomicts derived from dioecious species do not require fertilization; they are normally pistillate with vestigial anthers 0.1–0.2 mm long.

Herbivores find most species of *Poa* both palatable and nutritious. *Poa fendleriana*, *P. secunda*, and *P. wheeleri* are important native forage species in western North America; *P. alpina*, *P. arctica*, and *P. glauca* are common components of alpine and arctic vegetation. Species of *Poa* sect. *Abbreviatae* are found near the limits of vegetation in both arctic and alpine regions.

Several introduced species of *Poa* are economically important. *Poa pratensis* is commonly cultivated for lawns and pasture, and is a major forage species in cooler regions of North America; *P. compressa* and *P. trivialis* are widely planted for soil stabilization and forage; *P. annua* is one of the world's most widespread weeds. *Poa bulbosa* has been cultivated; it is now widely established in the Intermountain Region.

Characteristics that may be useful for distinguishing *Poa* from other morphologically similar genera are: the two-grooved, prow-shaped blades; multiple, relatively small, unawned florets; webbed calluses; and the greenish or milky white intercostal regions of the paleas.

There is a strong correlation between the type of basal branching, prophyll structure, and blade development of the initial leaves. Extravaginal shoots have scalelike prophylls 0.5–3 mm long and initial leaves that are bladeless; intravaginal shoots have prominently keeled prophylls 10–50 mm long that are open on the abaxial side and initial leaves with well-developed blades; pseudointravaginal shoots develop intravaginally but have tubular, indistinctly keeled prophylls, and initial leaves with rudimentary blades.

In bulbiferous spikelets, the upper florets form a single tardily disarticulating offset or bulb, each lemma being thickened at the base and leaflike distally. The bulb falls as a unit, with or without the basal floret. The basal floret(s) may have pistils and stamens, and occasionally sets seed. Generally, there is a progression within an inflorescence, the earlier spikelets being bulbiferous and the later spikelets normal.

Callus hairs in *Poa* follow one of three patterns. In the most common pattern, there is an isolated dorsal tuft of crinkled or pleated hairs, the web, below the lemma keel. In a few species, additional webs may be present below the marginal veins. In the second pattern, crinkled hairs are distributed around the lemma base, but are somewhat concentrated and longer towards the back; this pattern is called a diffuse web. Webbed calluses are found only in *Poa*. In the third pattern, the hairs are straight to slightly sinuous, and more or less evenly distributed around the lemma bases; calluses with such a pattern are described as having a crown of hairs.

Two named infrasectional hybrids are included in this treatment. One, *Poa arida*, is accounted for in the key. The other, *Poa ×limosa*, is too variable to make its inclusion in the key helpful. Both are described at the end of this treatment, with comments on the probable parental taxa.

Unless stated otherwise, sheath closure is measured on the flag leaf, and ligule length on the upper 1–2 culm leaves; spikelet, floret, callus, lemma, and palea measurements are on non-bulb-forming florets; floret pubescence is evaluated on the lower florets within several spikelets; length of the callus hairs refers to their length when stretched out; anther measurements are based on functional anthers, i.e., those that produce pollen, as indicated by their being plump or, after the pollen is shed, by their open sacs. For hair lengths in the species descriptions, puberulent is to about 0.15 mm long, short-villous to about 0.3 mm long, and long-villous from 0.3–0.4+ mm long, but these are only guidelines, not discrete categories; some species are only on one end of the range, and ranges have not been confirmed for every species. In the key, no distinction is made between the different kinds of hairs. Many species key more than once, due in part to infraspecific variation.

The key below was developed by Barkworth, based on information in Soreng (2007). A multiaccess, interactive key is available at http://utc.usu.edu/keys/IMRPoa.

1. All or most spikelets bulbiferous, sometimes with a poorly developed floret below the bulbous plantlet; culm bases bulbous. . . . . 1. *P. bulbosa*
1. All or most spikelets forming only florets; culm bases not bulbous.
  2. Plants usually annual, sometimes surviving for a second season.
    3. Calluses glabrous; lemmas usually hairy over the keel and veins and glabrous between the veins, rarely glabrous throughout . . . . 3. *P. annua*
    3. Calluses webbed; lemmas either completely glabrous or with hairs over the veins and sometimes between them.
      4. Lemmas glabrous over and between the keel and veins . . . . 6. *P. bolanderi*
      4. Lemmas hairy over the keel and veins, glabrous or hairy between them. . . . . 7. *P. bigelovii*
  2. Plants perennial.
    5. Lemma backs rounded to weakly keeled over the midveins; spikelets little compressed . . . . 27. *P. secunda*
    5. Lemma backs clearly keeled over the midveins; spikelets strongly laterally compressed.
      6. Calluses glabrous, even those of the lower florets in the spikelets [opposite lead on p. 111].
        7. Plants not rhizomatous; cauline blades not strongly reduced distally [opposite lead on p. 111].
          8. Sheaths of top cauline leaves closed ¼–⅘ their length; anthers 1.3–3.5 mm long.
            9. Cauline blades 2–4.5 mm wide, flat; palea keels hairy all or most of their length, sometimes scabrous distally . . . . 2. *P. alpina* (in part)
            9. Cauline blades 0.5–3 mm wide, flat, folded, or involute; palea keels scabrous throughout.
              10. Panicle branches 2–4 cm long, ascending to widespread, bearing 1–2(3) spikelets; panicles with (1)6–17(22) spikelets; cauline leaf blades 0.5–1 mm wide, filiform, soon withering; anthers 1.3–3 mm; leaf sheaths closed ⅖–⅘ their length; plants primarily of basaltic plateaus in Washington, Oregon, Idaho, and Nevada . . . . 15. *P. leibergii*

10. Panicle branches 0.5–4(5) cm long, erect to steeply ascending, bearing 1–15 spikelets; panicles with 9–100 spikelets; cauline blades 0.5–3 mm wide, sometimes withering early; anthers 2–4 mm long; leaf sheaths closed ¼–¾ their length; plants widespread in the Intermountain Region.

    11. Lemmas hairy on the keel and marginal veins; cauline blades 0.5–1(2) mm wide, often withering early, distal blade often strongly reduced to vestigial; anthers usually 0.1–0.2 mm long, sometimes 2–3 mm long . . . . 14. *P. ×nematophylla*

    11. Lemmas usually glabrous on the keel and marginal veins, sometimes puberulent near the base, not withering early, not strongly reduced distally; anthers usually 2–4 mm long, sometimes 0.1–0.2 mm long.

        12. Sheaths closed ⅐–⅓ their length; cauline leaf blades 1.5–3 mm wide, involute; panicle branches 0.5–1(2) cm long; lemmas 2–5 mm long . . . . 16. *P. pringlei* (in part)

        12. Sheaths closed ¼–¾ their length; cauline blades 0.5–3 mm wide, flat, folded, or involute; panicle branches 0.5–4(5) cm long; lemmas (3)4–7 mm long . . . . .13. *P. cusickii* (in part)

8. Sheaths of top cauline leaves closed ¹⁄₁₀–³⁄₁₀ their length; anthers 0.2–3 mm long.

    13. Spikelets 6–10 mm long, lengths 3–6 times widths; panicle branches 3–15 cm long; lemmas 4–6 mm long . . . . 28. *P. stenantha* (in part)

    13. Spikelets 3–8(12) mm long, lengths to 3.5(3.8) times widths; panicle branches to 10 cm long; lemmas 2–4.9 mm long.

        14. Spikelets 3.9–6.2 mm long, lengths 1.5–2.5 times widths; cauline blades 2–4.5 mm wide; lemmas hairy over the intercostal regions; palea keels hairy all or most of their length, sometimes scabrous distally . . . . .2. *P. alpina* (in part)

        14. Spikelets 3–6.5(12) mm long, lengths 2–4 times widths; cauline blades 1–3 mm wide; lemmas glabrous or hairy over the intercostal regions; palea keels scabrous for at least half their length, usually throughout.

            15. Anthers 2–4 mm or 0.1–0.2 mm long; spikelets 6–8–12 mm long; lemma keels and marginal veins glabrous, scabrous, or smooth; plants unisexual; basal branching intravaginal . . . . 16. *P. pringlei* (in part)

            15. Anthers 0.2–2.5 mm long; spikelets 3–7(9) mm long; lemma keels and marginal veins usually hairy, sometimes glabrous; basal branching intra- or extravaginal.

                16. Lemma keels and marginal veins hairy; lateral veins often hairy, sometimes glabrous; panicles 1–15(20) cm long.

                    17. Basal branching intravaginal; glumes longer than or subequal to the adjacent lemmas; plants 5–12(20) cm tall; panicles 1.5–5 cm long, branches erect. . . 25. *P. abbreviata* (in part)

                    17. Basal branching extravaginal; glumes shorter than or subequal to the adjacent lemmas; plants 5–80 cm tall; panicles 1–20 cm long, branches erect to widely divergent.

                        18. Upper cauline node usually at ⅓–⅗ culm height; lemmas usually glabrous over the lateral veins and intercostal region, rarely hairy over the lateral veins; ligules 0.5–1.5(3) mm long . . . 21. *P. interior* (in part)

                        18. Upper cauline node usually at ¹⁄₁₀–⅓ culm height; lemmas usually hairy over the lateral veins, hairy or glabrous over the intercostal regions; ligules 1–4(5) mm long . . . . 22. *P. glauca* (in part)

                16. Lemma keels and marginal veins usually glabrous, sometimes sparsely hairy; lateral veins glabrous; panicles 1–8 cm long.

19. Panicles 2–8 cm long; panicle branches 1–3(4) cm long, usually ascending to weakly spreading, rarely erect, smooth or sparsely scabrous; plants of southeastern Utah . . . . 18. *P. laxa* (in part)

19. Panicles 1–4(6) cm long; panicle branches to 1.5 cm long, erect, smooth to densely scabrous; plants of California.

    20. Lower glumes usually exceeding the lower lemmas; upper florets frequently exceeded by or only slightly exceeding the glumes; anthers 0.2–0.8 mm long; spikelets 3–4 mm long . . . . 24. *P. lettermanii*

    20. Lower glumes shorter than to equaling the lower lemmas; upper glumes always exceeded by the upper florets; anthers 0.3–1.6(1.8) mm long; spikelets 3.5–6 mm long . . . . 26. *P. keckii*

7. Plants rhizomatous; cauline blades sometimes strongly reduced distally [opposite lead on p. 109].

  21. Culms, nodes, and basal leaf sheaths strongly compressed; plants long-rhizomatous . . . . 23. *P. compressa* (in part)

  21. Culms, nodes, and basal leaf sheaths not or weakly compressed; plants often only shortly rhizomatous.

    22. Palea keels hairy at least on the distal half; lemmas hairy on the keels, marginal veins, lateral veins and between the veins.

      23. Palea keels hairy most of their length . . . . 5. *P. arctica* (in part)

      23. Palea keels smooth or scabrous on the lower half . . . . 29. *P. arida* (in part)

    22. Palea keels smooth or scabrous their whole length; lemmas usually glabrous or hairy only on the keel and marginal veins.

      24. Panicle branches 3–8 cm long and strongly divergent to reflexed at maturity . . . . 9. *P. arnowiae*

      24. Panicle branches to 6.5 cm long, erect to weakly divergent at maturity.

        25. Sheaths of top cauline leaves closed for ¹⁄₁₀–¹⁄₅ their length; plants 5–15(20) cm tall; panicles 1.5–5 cm long; panicle branches to 1.5 cm long . . . . 25. *P. abbreviata* (in part)

        25. Sheaths of top cauline closed for ¼–⁹⁄₁₀ their length; plants 10–80 cm tall; panicles 2–30 cm long; panicle branches 0.5–8 cm long.

          26. Cauline blades strongly reduced distally, usually involute; blade of the top cauline leaf to 1(3) cm long; sheath of top cauline leaf usually more than 9 times as long as the blade, sometimes as little as 5 times as long as the blade . . . . 12. *P. fendleriana*

          26. Cauline leaves gradually reduced distally, flat, folded, or involute; blade of the top cauline leaf 0.5–10 cm long; sheath of the top cauline leaf 0.5–10 times as long as the blade . . . . 10. *P. wheeleri*

6. Calluses, at least those of the lowest florets in each spikelet, webbed [opposite lead on p. 109].

  27. Lemmas with hairy keels and glabrous marginal veins; spikelets 2.3–3.5 mm long . . . . 17. *P. trivialis*

  27. Lemmas usually hairy on both the keel and the marginal veins, at least on the basal third, sometimes keel and marginal veins glabrous; spikelets (2.3)3–10 mm long.

    28. Culms, nodes, and basal sheaths strongly compressed . . . . 23. *P. compressa* (in part)

    28. Culms, nodes, and basal sheaths not or only weakly compressed.

      29. Plants not rhizomatous, sometimes stoloniferous or rooting at the lower nodes.

        30. Callus hairs straight, surrounding the base of the lemma; lemma keels and marginal veins hairy . . . . 28. *P. stenantha* (in part)

30. Callus hairs wrinkled, usually in one tuft at the base of the lemma midvein, sometimes somewhat distributed around the lemma base; lemma keels and marginal veins glabrous or hairy.
  31. Sheaths of the top cauline leaves closed ⅕–¾ their length.
    32. Calluses with a diffuse web, the wrinkled hairs somewhat distributed around the base of the lemma but concentrated below the midvein; sheaths closed ¼–¾ their length . . . . 13. *P. cusickii* (in part)
    32. Calluses with a single web, the wrinkled hairs localized beneath the lemma midvein; sheaths closed ⅕–⅔ their length.
      33. Panicle branches usually spreading to reflexed at maturity; cauline blades 1–4 mm wide, flat.
        34. Palea keels usually softly puberulent at midlength; lateral veins of lemmas usually softly puberulent at least on 1 side; panicle branches smooth or sparsely scabrous . . . . 8. *P. reflexa*
        34. Palea keels glabrous or pectinately ciliate; lateral veins of lemmas glabrous; panicle branches usually densely scabrous, sometimes sparsely scabrous . . . . 19. *P. leptocoma* (in part)
      33. Panicle branches erect to weakly divergent at maturity; cauline blades 0.8–2(3) mm wide, often involute, sometimes flat or folded.
        35. Lemma keels and marginal veins hairy, rarely glabrous; plants 5–12(20) cm tall; panicle branches to 1.5 cm long . . . . 25. *P. abbreviata* (in part)
        35. Lemma keels and marginal veins usually glabrous, sometimes sparsely hairy; plants 8–35 cm tall; panicle branches 1–3(4) cm long . . . . 18. *P. laxa* (in part)
  31. Sheaths of top cauline leaves closed ⅒–⅕ their length.
    36. Culms often rooting at lower nodes; plants sometimes stoloniferous; panicles (9)13–30(41) cm long; panicle branches strongly spreading to reflexed at maturity . . . . . 20. *P. palustris*
    36. Culms not rooting at lower nodes; plants not stoloniferous; panicles 1–15(20) cm long; panicle branches erect to strongly spreading at maturity.
      37. Panicle branches smooth or slightly scabrous, often terete; culms 5–15(20) cm tall . . . . 25. *P. abbreviata* (in part)
      37. Panicle branches moderately to densely scabrous, slender to moderately stout; culms 5–80 cm tall.
        38. Lemmas usually completely glabrous between the keel and marginal veins, occasionally sparsely hairy on the lateral veins; ligules 0.5–1.5(3) mm long . . . . . . 21. *P. interior* (in part)
        38. Lemmas usually hairy on the lateral veins; ligules of cauline leaves 1–5 mm long . . . 22. *P. glauca* (in part)
29. Plants always rhizomatous, sometimes shortly so.
  39. Lemmas hairy between the keel and marginal veins.
    40. Palea keels hairy most of their length . . . . 5. *P. arctica* (in part)
    40. Palea keels smooth or scabrous ½ or more of their length . . . . 29. *P. arida* (in part)
  39. Lemmas glabrous between the keel and marginal veins.
    41. Sheaths of top culm leaves closed ⅒–¼ their length . . . . 29. *P. arida* (in part)
    41. Sheaths closed ¼–⁹⁄₁₀ their length.

42. Keel and marginal veins hairy for ⅓ the length of the
lemma . . . . 11. *P. chambersii*
42. Marginal veins hairy for more than ½ the length of the
lemma.
    43. Ligules of cauline leaves 0.8–1(3.1) mm long;
plants common in many habitats . . . . .4. *P. pratensis*
    43. Ligules of cauline leaves 1.5–6 mm long; plants
of wet areas in subalpine and alpine habitats
. . . . 19. *P. leptocoma* (in part)

Poa L. subg. Poa

Pl ann or per; smt unisex; with or without rhz or stln, densely to loosely tufted or the clm solitary. **Bas brchg** invag and/or exvag or psdinvag. **Clm** spindly to stout, terete or weakly to strongly compressed; **nd** 0–5, exserted. **Shth** terete or weakly to strongly compressed, closed only at the base or up to full length, bas shth usu glab, rarely sparsely retrorsely strigose, hairs about 0.1 mm; **lig** 0.1–18 mm, thinly memb and white to milky white or hyaline, truncate to acuminate, entire or erose to lacerate, smooth or ciliolate; **bld** flat, folded, or involute, thin to thick, smooth or sparsely to densely scabrous, adx surfaces glab or hairy, hispidulous or puberulent, apc narrowly to broadly prow-shaped. **Pan** 1–41 cm, erect to nodding or lax, tightly contracted to open, with 1–100+ spklt; **br** 0.5–20 cm, erect to reflexed, terete or angled, smooth or sparsely to densely scabrous, usu glab, rarely hispidulous, with 1 to many spklt. **Spklt** 2–12 mm, subterete to strongly lat compressed, smt bulbiferous; **flt** (1)2–8(13); **rchl intnd** smooth or scabrous, glab or pubescent. **Glm** shorter than to slightly exceeding the adjacent lm, weakly to distinctly keeled, smooth or scabrous; **cal** blunt, usu terete or slightly lat compressed, smt slightly dorsally compressed, glab, dorsally webbed, diffusely webbed, or with a crown of hairs; **lm** 1.7–11 mm, rounded to weakly or distinctly keeled, thinly memb to chartaceous, glab or hairy on the keel and veins, smt the intercostal regions also hairy, 5–7(11)-veined, mrg smooth or scabrous, glab, apc obtuse to acuminate; **pal keels** usu scabrous, infrequently smooth, glab or with hairs; **anth** (1–2)3, 0.1–4.5(5) mm.

    Poa subg. *Poa* is the largest subgenus of *Poa*. It includes all of the species in the Intermountain Region.

Poa sect. Arenariae (Hegetschw.) Stapf

Pl per; not rhz, not stln, densely tufted. **Bas brchg** invag. **Clm** 2–60 cm, terete, bases bulbous. **Shth** closed for about ¼ their length, lowest shth with swollen bases; **lig** 1–6 mm, smooth or scabrous, obtuse to acute; **bld** (0.5)1–2.5 mm wide, flat, thin, lax, soon withering. **Pan** (0.8)2–10 cm, ovoid, loosely contracted; **nd** with 2–5 br; **br** usu ascending, infrequently spreading, terete, usu smooth or sparsely scabrous, rarely moderately scabrous. **Spklt** 3–7 mm, lat compressed, some or all bulbiferous; **flt** (2)3–7, forming a bulblet, smt the bas 1–2 flt normal. **Glm** shorter than the adjacent lm, distinctly keeled, keels scabrous; **lo glm** 3-veined; **cal** terete or slightly lat compressed, glab or dorsally webbed, hairs wrinkled; **lm** normal or lflike, normal lm 2–4 mm, distinctly keeled, glab throughout or the keels and mrgl veins villous, intercostal regions glab or puberulent, lflike lm thickened at the base, bldlike distally; **pal** scabrous, keels often softly puberulent at midlength; **anth** 3, (0.6)1.2–2 mm, smt aborted late in development, smt not developed.

    Poa sect. *Arenariae* is native to Eurasia and North Africa. It includes 14 species. These are easily recognized as members of the section by the bulbous bases of their new shoots. One species is established in the Intermountain Region.

1. Poa bulbosa L. Bulbous Bluegrass [p. 360, 500]

Pl per; densely tufted, not rhz, not stln. **Bas brchg** invag. **Clm** 15–60 cm, erect or spreading, bases bulbous. **Shth** closed for about ¼ their length, terete, lowest shth with swollen bases; **lig** 1–3 mm, smooth or scabrous, apc obtuse to acute; **bld** 1–2.5 mm wide, flat, thin, lax, soon withering. **Pan** 3–12 cm, ovoid; **nd** with 2–5 br; **br** ascending to spreading, terete, usu smooth or sparsely scabrous, infrequently moderately scabrous. **Spklt** 3–5 mm, lat compressed, usu bulbiferous; **flt** 3–7, the bas flt, and smt additional flt, normal; **rchl intnd** smooth, glab. **Glm** keeled, keels scabrous; **lo glm** 3-veined; **up glm** shorter than or subequal to the lowest lm; **cal** webbed or glab; **lm** 3–4 mm, lanceolate, keeled, glab or the keels and mrgl veins short- to long-villous, intercostal regions glab or softly puberulent, apc acute; **pal** scabrous, keels often softly puberulent at midlength; **anth** 1.2–1.5 mm and fnctl, smt aborted late in development, smt not developed. $2n = 14, 21, 28, 39, 42, 45$.

    Poa bulbosa is a European species that is now established in North America. Only one subspecies is present in the Intermountain Region.

Poa bulbosa subsp. **vivipara** (Koel.) Arcang.
[p. *360*]
Clm 15–60 cm. **Spklt** bulbiferous; **flt** modified into lfy bracts, smt the bas flt within a spklt more or less normal. **Cal** usu sparsely webbed, smt glab; **lm** glab or softly puberulent over the keel and lat veins, smt between the veins; **anth** in the least deformed flt 1.2–1.5 mm or aborted late in

development, absent from modified flt. 2*n* = 21, 28, 31, 32, 33, 34, 35, 37, 39, 42+I, 44, 46, 48, 49.

Poa bulbosa subsp. *vivipara* was introduced from Europe into the Pacific Northwest as a forage grass; it has since spread across temperate areas of North America, particularly in the Pacific Northwest and northern Great Basin. It is highly tolerant of grazing and disturbance.

Poa sect. **Alpinae** (Hegetschw. *ex* Nyman) Stapf

Pl per; not rhz, not stln. **Bas brchg** invag. **Clm** 10–40 cm, terete. **Lvs** mostly bas; **shth** closed for ½–⅔ their length, terete, bas shth persistent, bases usu not swollen; **bld** flat, moderately thick, soft, straight, apc prow-shaped. **Pan** 2–6(8) cm, erect, ovoid to pyramidal, open or loosely contracted at maturity; **nd** with 1–2 br; **br** 1–3(4) cm, ascending to spreading, straight, terete, smooth or very sparsely scabrous, rarely moderately scabrous. **Spklt** ovate, lat compressed, occ bulbiferous; **flt** usu normal, bisex. **Glm** broadly lanceolate to narrowly ovate, shorter than to subequal to the adjacent lm, keeled, keels sparsely scabrous; **lo glm** 3-veined; **cal** terete, glab; **lm** broadly lanceolate, keeled, keels and mrgl veins short- to long-villous, intercostal regions glab or sparsely to moderately short-villous; **pal keels** mostly softly puberulent to short-villous, scabrous distally; **anth** 3, 1.3–2.3 mm.

Poa sect. Alpinae includes seven species. They are all cespitose perennials with intravaginal branching and broad leaves. Six species are native to Europe; one, *P. alpina*, is circumboreal.

2. **Poa alpina** L. [p. *360*, 500]
Pl per; not glaucous; densely ces, not rhz, not stln. **Bas brchg** invag. **Clm** 10–40 cm. **Lvs** mostly bas; **shth** closed for ⅛–⅔ their length, terete, bas shth persistent, overlapping, bases usu not swollen; **lig** of innovations 1–2(3) mm, those of the up cauline lvs to 4(5) mm, milky white, smooth, glab, obtuse; **bld** of innovations widely spreading, persisting through the season, bld of cauline lvs 1–5(12) cm long, 2–4.5 mm wide, flat, moderately thick, soft, straight, smooth or the mrg sparsely scabrous, apc broadly prow-shaped, bld of up cauline lvs much rdcd in length. **Pan** 2–6(8) cm, erect, ovoid to pyramidal, open or loosely contracted at maturity, fairly congested; **nd** with 1–2 br, lowest intnd 0.6–1(1.5) cm; **br** 1–3(4) cm, ascending to spreading, straight, terete, usu smooth or sparsely scabrous, rarely moderately densely scabrous; **ped** divaricate, shorter than the spklt. **Spklt** 3.9–6.2 mm, ovate, lengths 1.5–2.5 times widths, lat compressed, plump, smt bulbiferous; **flt** 3–7, usu normal; **rchl intnd** 0.5–0.8 mm, smooth, glab or sparsely softly puberulent to short-villous. **Glm** broadly lanceolate to narrowly ovate, keeled,

keels sparsely scabrous; **lo glm** 3-veined; **up glm** shorter than or subequal to the lowest lm; **cal** glab; **lm** 3–5 mm, broadly lanceolate, keeled, keels and mrgl veins short- to long-villous, lat veins moderately prominent, intercostal regions sparsely to moderately short-villous, apc acute; **pal keels** softly puberulent to short-villous over most of their length, apc scabrous; **anth** 1.3–2.3 mm. 2*n* = 22, 23, 24, 25, 26, 27, 28, 28+II, 30, 31, 32, 32+I, 33, 34, 35, 36, 37, 39, 40+I, 41, 42, ca. 43, 44, 46, ca. 48, 56.

Poa alpina is a fairly common circumboreal forest species of subalpine to arctic habitats, extending south in the Rocky Mountains to Utah and Colorado in the west, and to the northern Great Lakes region in the east. It often grows in disturbed ground and is calciphilic. The range of chromosome numbers suggests that *P. alpina* is predominantly apomictic.

Poa alpina L. subsp. **alpina** ALPINE BLUEGRASS
[p. *360*]
Spklt not bulbiferous. **Anth** 1.3–2.3 mm, well formed. 2*n* = 22, 23, 26, 27, 28, 28+I, 30, 31, 32, 32+I, 33, 34, 35, 36, 37, 39, 40+I, 41, 42, ca. 43, 44, 46, ca. 48, 56.

Poa alpina subsp. *alpina* is the only subspecies present in the Intermountain Region.

Poa sect. **Micrantherae** Stapf

Pl ann or per; green; usu neither rhz nor stln, smt stln, densely to loosely tufted. **Bas brchg** invag. **Clm** 2–20(45) cm, terete or weakly compressed; **nd** terete. **Shth** closed for ¼–⅓ their length, terete or weakly compressed, smooth, glab; **col** smooth, glab; **lig** 0.5–3(5) mm, smooth, glab, truncate to obtuse, entire; **bld** 1–3(6) mm wide, flat or weakly folded, thin, soft, smooth, mrg usu slightly scabrous, apc broadly prow-shaped. **Pan** 1–7(10) cm, erect, loosely contracted or open, ovoid to pyramidal;

nd with 1–2(5) br; **br** ascending to reflexed, straight, terete, smooth or sparsely scabrous. **Spklt** 3–6 mm, lanceolate to narrowly ovoid, lat compressed, not bulbiferous; **flt** 2–7, normal, up 1–2 flt pist in some spklt; **rchl intnd** smooth, glab. **Glm** distinctly keeled, smooth; **lo glm** distinctly shorter than the lowest lm, 1-veined; **up glm** shorter than to subequal to the lowest lm; **cal** terete, glab; **lm** 1.7–4 mm, distinctly keeled, smooth and glab or the keels, mrgl veins, and, usu, lat veins hairy, lat veins moderately prominent to prominent, intercostal regions glab, mrg smooth, glab, apc whitish, obtuse to acute; **pal keels** smooth, usu softly puberulent to long-villous, smt glab; **anth** 3, 0.1–2.5 mm, smt vestigial in the up 1–2 flt.

Poa sect. *Micrantherae* includes eight species, all of which are native to Eurasia and North Africa. They are gynomonoecious, with smooth or sparsely scabrous panicle branches. The calluses are glabrous in most species; the palea keels are usually hairy. One species grows in the Intermountain Region.

3. Poa annua L. ANNUAL BLUEGRASS [p. 361, 500]

Pl usu ann, rarely surviving for a second season; not rhz, smt stln, densely tufted. **Bas brchg** invag, innovations common, similar to the clm. **Clm** 2–20(45) cm, prostrate to erect, slender; **nd** terete, usu 1 exserted. **Shth** closed for about ⅓ their length, terete or weakly compressed, smooth; **lig** 0.5–3(5) mm, smooth, glab, decurrent, obtuse to truncate; **bld** 1–10 cm long, 1–3(6) mm wide, flat or weakly folded, thin, soft, smooth, mrg usu slightly scabrous, apc broadly prow-shaped. **Pan** 1–7(10) cm, lengths 1.2–1.6 times widths, erect; **nd** with 1–2(3) br; **br** ascending to spreading or reflexed, straight, terete, smooth, with crowded or loosely arranged spklt. **Spklt** 3–5 mm, lat compressed; **flt** 2–6; **rchl intnd** smooth, glab, concealed or exposed, distal intnd less than ½(¾) the length of the distal lm. **Glm** smooth, distinctly keeled, keels smooth; **lo glm** 1-veined; **up glm** shorter than or subequal to the lowest lm; **cal** glab; **lm** 2.5–4 mm, lanceolate, distinctly keeled, smooth throughout, the keels, mrgl veins, and, usu, lat veins crisply puberulent to long-villous, rarely glab throughout, lat veins prominent, intercostal regions glab, mrg smooth, glab, apc obtuse to acute; **pal keels** smooth, usu short- to long-villous, rarely glab; **anth** 0.6–1.1 mm, oblong prior to dehiscence, those of the up 1–2 flt usu vestigial. $2n = 28$.

Poa annua is one of the world's most widespread weeds. It thrives in anthropomorphic habitats outside of the arctic. A native of Eurasia, it is now well established throughout most of North America, including the Intermountain Region. Forms with glabrous lemmas occur sporadically within populations.

Poa L. sect. Poa

Pl per; rhz, rhz usu well developed and extensive, smt poorly developed, densely to loosely tufted or the shoots solitary. **Bas brchg** mainly exvag or equally exvag and invag. **Clm** 5–120 cm, terete or weakly compressed; **nd** terete or weakly compressed. **Shth** closed for (⅙)¼–⅗ their length, terete to slightly compressed, smooth or sparsely scabrous, usu glab, infrequently sparsely to moderately hairy, distal shth usu longer than their bld; **col** smooth, glab; **lig** 0.9–7 mm, smooth or scabrous, truncate to acute, glab or ciliolate; **innovation bld** of invag shoots involute and narrower or similar to the cauline bld and bld of exvag sht; **cauline bld** subequal or the mid bld longest, flat, folded, or weakly involute, abx surfaces smooth, glab, adx surfaces smooth or sparsely scabrous, frequently sparsely hairy, hairs 0.2–0.8 mm, apc prow-shaped, smt narrowly prow-shaped, flag lf bld 1.5–10 cm. **Pan** 2–18(20) cm, loosely contracted to open, often slightly lax to nodding, sparsely to moderately congested, with 1–7(9) br per nd; **br** 1–9 cm, ascending to widely spreading or somewhat reflexed, flexuous to straight, terete or angled, usu smooth or sparsely to moderately scabrous, infrequently densely scabrous. **Spklt** 3.5–9(12) mm, lengths to 3.5 times widths, lanceolate to broadly lanceolate, lat compressed, smt bulbiferous; **flt** 2–5(6), usu normal, bisex; **rchl intnd** smooth, glab or pubescent. **Glm** unequal to subequal, distinctly shorter than to subequal to the adjacent lm, keels weak or distinct, smooth or scabrous; **lo glm** 1- or 3-veined; **cal** terete or slightly lat compressed, usu dorsally webbed, smt with additional webs below the mrgl veins, infrequently glab; **lm** 2–8 mm, lanceolate to broadly lanceolate, distinctly keeled, keels and mrgl veins, and smt also the lat veins, hairy, all veins prominent, intercostal regions glab or hairy; **pal keels** smt with hairs at midlength, intercostal regions glab or hairy; **anth** 3, 1.2–2.5 mm, infrequently aborted late in development.

Poa section *Poa* includes 32 species. All the species are synoecious perennials; most are strongly rhizomatous.

4. **Poa pratensis** L. Kentucky Bluegrass
   [p. *362*, 500]

Pl per; green or anthocyanic, smt glaucous; extensively rhz, densely to loosely tufted or the shoots solitary. **Bas brchg** mainly exvag or evenly exvag and invag. **Clm** 5–70(100) cm, erect or the bases decumbent, not brchg above the base, terete or weakly compressed; **nd** terete or weakly compressed, 1–2(3) exposed, proximal nd(s) usu not exserted. **Shth** closed for ¼–½ their length, terete to slightly compressed, glab or infrequently sparsely to moderately hairy, bases of bas shth glab, not swollen, distal shth lengths 1.2–5(6.2) times bld lengths; **col** smooth, glab; **lig** 0.9–2(3.1) mm, smooth or scabrous, truncate to rounded, infrequently obtuse, ciliolate or glab; **bld** of exvag innovations like those of the clm, those of the invag shoots smt distinctly narrower, 0.4–1 mm wide, flat to involute; **cauline bld** 0.4–4.5 mm wide, flat, folded, or involute, soft and lax to moderately firm, abx surfaces smooth, glab, adx surfaces smooth or sparsely scabrous, frequently sparsely hairy, hairs 0.2–0.8 mm, erect to appressed, slender, curving, sinuous or straight, apc usu broadly prow-shaped, smt narrowly prow-shaped, bld subequal, the mid bld longest, the flag lf bld 1.5–10 cm. **Pan** 2–15(20) cm, narrowly ovoid to narrowly or broadly pyramidal, loosely contracted to open, sparse to moderately congested, with (25) 30–100+ spklt and (1)2–7(9) br per nd; **br** (1)2–9 cm, spreading early or late, terete or angled, smooth or sparsely to moderately densely scabrous, with 4–30(50) spklt usu fairly crowded in the distal ½. **Spklt** 3.5–6(7) mm, lengths 3.5 times widths, lat compressed, smt bulbiferous; **flt** 2–5, usu normal, smt bulb-forming; **rchl intnd** usu shorter than 1 mm, smooth, glab. **Glm** unequal to subequal, usu distinctly shorter than the adjacent lm, narrowly lanceolate to lanceolate, infrequently broadly lanceolate, distinctly keeled, keels usu sparsely to densely scabrous, infrequently smooth; **lo glm** 1.5–4(4.5) mm, usu narrowly lanceolate to lanceolate, occ sickle-shaped, 1–3-veined; **up glm** 2–4.5(5) mm, distinctly shorter than to nearly equaling the lowest lm; **cal** dorsally webbed, smt with additional webs below the mrgl veins, hairs at least ½ as long as the lm, crimped; **lm** 2–4.3(6) mm, lanceolate, green or strongly purple-tinged, distinctly keeled, keels and mrgl veins long-villous, lat veins usu glab, infrequently short-villous to softly puberulent, lat veins prominent, intercostal regions glab, lo portion smooth or finely muriculate, up portion smooth or sparsely scabrous, mrg narrowly to broadly hyaline, glab, apc acute; **pal** scabrous, keels smt softly puberulent, intercostal regions narrow, usu glab, rarely sparsely hispidulous; **anth** usu 1.2–2 mm, infrequently aborted late in development. $2n =$ 27, 28, 32, 35, 37, 41–46, 48–147.

Poa pratensis is common, widespread, and well established in many natural and anthropogenic habitats of North America. The only taxa that are clearly native to North America are the arctic and subarctic subspp. *alpigena* and *colpodea*. Outside North America, *P. pratensis* is native in temperate and arctic Eurasia. It is established in temperate regions around the world.

Poa pratensis is a highly polymorphic, facultatively apomictic species, having what is probably the most extensive series of polyploid chromosome numbers of any species in the world. It is a hybridogenic species, i.e., it comprises numerous lineages with the same basic maternal genome, but different paternal genomes. The lineages are perpetuated by agamospermic and vegetative reproduction. Some major forms are recognized as microspecies or subspecies. These have some correlated ecological and morphological differences, but the morphological boundaries between them are completely bridged; in some cases the taxa may represent environmentally induced plasticity.

Natural hybrids have been identified between *Poa pratensis* and *P. alpina*, *P. arctica*, *P. wheeleri*, and *P. secunda*. Many other artificial hybrids have been made; these involve many different, often distantly related, species. In addition, there are many cultivated forms of the species; these have been seeded widely throughout the region for lawns, soil stabilization, and forage. Most cultivated forms favor subsp. *irrigata* morphologically; others tend towards subspp. *pratensis* and *angustifolia*, the latter occurring most commonly in xeric sites.

1. Panicle branches smooth or almost smooth . . . . . .
   . . . . subsp. *alpigena*
1. Panicle branches more or less scabrous.
   2. Intravaginal innovation shoots present, intra- and extravaginal blades alike, 0.4–1 mm wide, folded to involute, somewhat firm, adaxial surfaces often sparsely and softly hairy; plants of dry meadows and forests. . . . . subsp. *angustifolia*
   2. Intravaginal innovation shoots present or absent, if present then differentiated or alike, at least some with blades 1.5–4.5 mm wide, flat or folded, adaxial surfaces rarely hairy; plants widespread, often of more mesic sites . . . .
      . . . . subsp. *pratensis*

Cultivars of **Poa pratensis** L.

Pl densely to loosely tufted, often forming turf, shoots clustered. **Bas brchg** invag and exvag or mainly exvag. **Clm** 8–50 cm. **Innovation shoot bld** usu shorter than 45 cm, (0.4)1–4 mm wide, usu flat, smt some involute, usu soft, smt somewhat firm, adx surfaces usu glab; **cauline bld** flat or folded. **Pan** 3–15 cm, broadly pyramidal, open or somewhat contracted, with 2–7(9) br per nd; **br** ascending or widely spreading, sparsely to densely scabrous, with few to many spklt per br. **Spklt** lanceolate to broadly lanceolate, not bulbiferous; **flt** normal. **Glm keels** strongly compressed, sparsely to moderately scabrous; **up glm** shorter than to nearly equaling the lowest lm; **lm** 2.8–4.3(6) mm, finely muriculate, lat veins glab; **pal keels** scabrous, glab, intercostal regions glab. 2n = 41–45, 48–59, 62, 64–74, 76, 78, 80, 81, 84–90, 95.

More than 60 cultivars of *Poa pratensis* have been released in North America. Plants grown from commercially distributed seed have generally been placed in subsp. *pratensis* by North American authors, but they appear to include genetic contributions from at least three major subspecies: subspp. *angustifolia*, *pratensis*, and *irrigata*. They are intermediate forms and are best referred to as *Poa pratensis sensu lato* or labeled as cultivated material. The chromosome counts listed here are numbers reported for the species that are probably not subspp. *alpigena*, *angustifolia*, or *colpodea*; they may represent subspp. *irrigata* or *pratensis*.

Poa pratensis subsp. alpigena (Lindm.)
Hiitonen ALPIGENE BLUEGRASS [p. 362]
Pl strongly anthocyanic; moderately to loosely tufted, shoots usu solitary. **Bas brchg** mainly exvag. **Clm** 15–70 cm. **Innovation shoot bld** shorter than 15 cm, 1–3.6 mm wide, flat or folded, soft, adx surfaces usu glab, smt sparsely pubescent; **cauline bld** flat or folded. **Pan** 3–13(20) cm, narrowly pyramidal or contracted, expanding well after emergence from the shth, with (1)2–5(7) br per nd; **br** 1–6 cm, steeply ascending to eventually spreading or somewhat reflexed, smooth or sparsely scabrous, with 5–15 spklt. **Spklt** 4–5.5 mm, narrowly lanceolate, not bulbiferous; **flt** normal. **Glm keels** distinct, smooth or sparsely scabrous near the apc; **up glm** nearly equaling the lowest lm; **lm** 2.5–3.5 mm, smooth or finely muriculate, lat veins frequently short-villous to softly puberulent; **pal keels** scabrous, often softly puberulent at midlength, intercostal regions usu glab, rarely sparsely hispidulous. 2n = 28, 32, 35, 42, 48, 50, 53, 56, 60, 63, 64, 65, 67, ca. 68, 69, 70, 72, 73, 74, 76, 77, 78, 79, 82, 84, 86, 88, 89, 92, 94.

Poa pratensis subsp. *alpigena* is a circumpolar, mesophytic to subhydrophytic, arctic and alpine subspecies that extends into boreal forests in northern parts of North America. It is infrequent south of Canada, with isolated collections being known from as far south as New Mexico in the Rocky Mountains, and New Hampshire and Maine in the east. It also grows in southern Patagonia.

Poa pratensis subsp. angustifolia (L.) Lej.
[p. 362]
Pl moderately densely to densely tufted. **Bas brchg** invag and exvag, invag shoots clustered. **Clm** 25–80 cm. **Innovation shoot bld** 10–45 cm long, 0.4–1 mm wide, all involute, smt narrower than the cauline bld, adx surfaces sparsely pubescent; **cauline bld** involute or folded, somewhat firm, adx surfaces sparsely pubescent. **Pan** 8–18 cm, narrowly pyramidal or loosely contracted, br ascending to spreading, smooth or sparsely to densely scabrous, with several to many spklt per br. **Spklt** narrowly lanceolate, not bulbiferous; **flt** normal. **Glm keels** strongly compressed, sparsely to moderately scabrous; **up glm** shorter than to nearly equaling the lowest lm; **lm** 2.5–3.5 mm, finely muriculate, lat veins glab; **pal keels** scabrous, glab, intercostal regions glab. 2n = 28, 46, 48–54, 56, 57, 59, 60, 61, 62, 63, 64, 65, 66, 68, 70, 72, 83.

Poa pratensis subsp. *angustifolia* is a western Eurasian subspecies that is also known from scattered locations throughout temperate North America. It is characterized by the predominance of fascicles of elongate, narrow, involute blades on the intravaginal vegetative shoots, and slender panicles with small spikelets. Recent research has shown that it is primarily a low polyploid.

Poa pratensis L. subsp. pratensis [p. 362]
Pl densely to loosely tufted, often forming turf, clm clustered. **Bas brchg** invag and exvag. **Clm** 8–100 cm. **Innovation shoot bld** 10–45 cm long, 0.4–4 mm wide, some distinctly narrower than the cauline bld, all flat or some involute, usu soft, adx surfaces sparsely pubescent; **cauline bld** flat or folded. **Pan** 5–18 cm, broadly pyramidal, open or somewhat contracted, with 3–5(7) br per nd; **br** spreading to somewhat reflexed, smooth or sparsely to fairly densely scabrous, with several to many spklt per br. **Spklt** lanceolate to broadly lanceolate, not bulbiferous; **flt** normal. **Glm keels** strongly compressed, sparsely to moderately scabrous; **up glm** shorter than or nearly equaling the lowest lm; **lm** 2.8–4.3 mm, finely muriculate, lat veins glab; **pal keels** scabrous, glab, intercostal regions glab. 2n = 43, 44, 48, 49, 50, 51, 52, 54, 56, 58, 59, 62, 65, 66, 67, 74, ca. 85, ca. 86, 88, 89, 95.

Poa pratensis subsp. *pratensis* grows throughout most of the range of the species, but is absent from the high arctic, and only sporadic in the low arctic. It usually has a few narrow, flat or involute, intravaginal shoot leaves, in addition to some broader, extravaginal shoot leaves.

5. Poa arctica R. Br. ARCTIC BLUEGRASS [p. 363, 500]
Pl per; usu strongly anthocyanic; rhz usu well developed, smt poorly developed, shoots usu solitary. **Bas brchg** mainly exvag. **Clm** 7.5–60 cm, slender to stout, terete or weakly compressed, bases usu decumbent, not brchg above the bases; **nd** terete, proximal nd usu not exserted, 0–2

exserted above. **Shth** closed for (⅙)⅕–⅖ their length, terete, glab, smooth or sparsely scabrous, bases of bas shth glab, distal shth lengths 1.4–4(5.3) times bld lengths; **col** smooth, glab; **lig** (1)2–7 mm, glab, smooth or sparsely to infrequently moderately scabrous, apc usu rounded to obtuse or acute, rarely truncate, entire or lacerate; **bld** 1–6 mm wide, flat or folded, somewhat involute, smooth, glab, apc broadly prow-shaped, cauline bld subequal or gradually rdcd distally, flag lf bld 0.7–9 cm. **Pan** (2)3.5–15 cm, ovoid to broadly pyramidal, usu open, sparse, with 10–40(60) spklt, proximal intnd shorter than 1.5(3) cm, with (1)2–5 br per nd; **br** 1.5–6 cm, spreading soon after emergence from the shth, thin, sinuous, and flexuous to fairly stout and straight, terete, smooth or sparsely to infrequently moderately scabrous, with (1)2–5 spklt, the spklt not crowded. **Spklt** (3.5) 4.5–8 mm, lengths to 3.5 times widths, lat compressed, smt bulbiferous; **flt** (2)3–6, infrequently bulb-forming; **rchl intnd** smooth or muriculate, proximal intnd glab or sparsely softly puberulent to long-villous. **Glm** lanceolate to broadly lanceolate, distinctly or weakly keeled, keels usu smooth, smt sparsely scabrous distally, lat veins usu moderately pronounced; **lo glm** (3)3.5–5(6) mm, 3-veined; **up glm** 3.5–5.5(6.5) mm, nearly equaling to slightly exceeding the lowest lm, or distinctly shorter; **cal** glab or webbed, hairs sparse and short to over ⅓–⅔ the lm length; **lm** (2.7)3–6(7) mm, lanceolate to broadly lanceolate, usu strongly purple, distinctly keeled, keels, mrgl veins, and lat veins long-villous, hairs on the lat veins smt shorter, lat veins prominent, intercostal regions short-villous to softly puberulent at least near the base, glab elsewhere, smooth to weakly muriculate and/or usu sparsely scabrous, infre-quently moderately scabrous, mrg broadly hyaline, glab, apc acute; **pal keels** usu short- to long-villous for most of their length, rarely nearly glab and scabrous, intercostal regions broad, usu at least sparsely softly puberulent, rarely glab, apc scabrous; **anth** 1.4–2.5 mm, smt aborted late in development. $2n$ = 36, 42, 56, 60, 62–68, 70, ca. 72, 74–76, 78–80, 82–84, 86, 88, 99, 106.

Poa arctica is a common circumboreal species of arctic and alpine regions, growing mainly in mesic to subhydric, acidic tundra and alpine meadows, and on rocky slopes. It extends south in the Rocky Mountains to New Mexico. In the southern portion of its range, *P. arctica* usually develops normal anthers. This and isozyme data for populations from alpine and low arctic regions suggest sexual reproduction is common in these habitats.

The most reliable way to distinguish *Poa arctica* from *P. pratensis* (see previous) is by the wider paleas and the presence of hairs between the palea keels. *Poa arctica* forms natural hybrids with both *P. pratensis* and *P. secunda* (p. 132).

1. Panicles erect, the branches relatively stout, fairly straight; longest branches of the lowest panicle nodes ¼–½ the length of the panicles; culms wiry, usually several together; calluses glabrous or shortly webbed; paleas sometimes glabrous; plants glaucous, growing in the southern Rocky Mountains and adjacent portions of the Intermountain Region . . . . subsp. *aperta*
1. Panicles lax to erect, the branches slender, flexuous to fairly stout and straight; longest branches of the lowest panicle nodes ⅖–⅗ the length of the panicles; culms slender to stout, varying from solitary to several together; calluses glabrous or webbed, the hairs usually more than ½ as long as the lemmas; paleas pubescent; plants sometimes glaucous, widespread in distribution.
    2. Calluses webbed, often copiously so . . subsp. *arctica*
    2. Calluses glabrous . . . . subsp. *grayana*

Poa arctica subsp. **aperta** (Scribn. & Merr.) Soreng [p. *363*]

Pl pale green, often glaucous; usu densely tufted, rhz usu short, usu well developed. **Clm** 20–60 cm, several together, wiry, bases decumbent. **Shth** closed for (⅙)⅕–⅓ their length; **lig** 3–7 mm, sparsely to moderately scabrous, acute; **bld** 1.5–2.5 mm wide, flat, folded, or somewhat involute. **Pan** 4–15 cm, erect, loosely contracted or open, with 1–3 br per nd; **br** ascending or widely spreading, fairly stout, fairly straight, smooth to very sparsely scabrous, proximal br ¼–½ the pan length. **Spklt** narrowly lanceolate to lanceolate, not bulbiferous; **flt** 2–3(4), normal; **rchl intnd** usu glab, infrequently sparsely softly puberulent; **cal** glab or webbed, hairs to ¼ the lm length; **lm** 3–4.5(6) mm; **pal keels** usu softly puberulent to long-villous at midlength, infrequently glab, intercostal regions usu softly puberulent; **anth** aborted late in development or fully developed. $2n$ = 98+I.

Poa arctica subsp. *aperta* grows in subalpine and low alpine habitats on the Wasatch Escarpment and high mountains of the Colorado Plateau in southern Utah, and the Rocky Mountains of southern Colorado and northern New Mexico. *Poa arctica* subsp. *aperta* has softer leaves, and is more densely hairy between the lemma veins and the palea keels, than subsp. *arctica*. It can be distinguished from subsp. *grayana* by its more wiry culms, and less contracted panicles with straighter branches. Many reports of *P. arida* (p. 134) growing west of the Rocky Mountains are based on misidentification of this subspecies. *Poa arctica* subsp. *aperta* may reflect introgression from *P. secunda* (p. 132).

Poa arctica R. Br. subsp. **arctica** [p. *363*]

Pl usu loosely, smt densely, tufted, rhz, rhz short or long, well developed. **Lig** (1)2–4 mm, obtuse to acute; **bld** 1.5–2.5(3) mm wide, flat or folded, thin and soon withering, flag lf bld 0.7–5.5 cm. **Pan** lax to erect, open; **br** ascending or widely spreading, sinuous and flexuous to fairly straight, smooth or sparsely scabrous, proximal br ⅖–⅗ the pan length. **Spklt** (3.5)4.5–6(7) mm, infrequently bulbiferous; **rchl intnd** usu glab, infrequently sparsely softly puberulent to long-villous; **cal** sparsely to copiously webbed; **lm** (2.7)3–4.5 mm; **pal keels** puberulent to long-villous at midlength, intercostal regions usu hairy, smt glab; **anth** usu fully developed. 2*n* = 56, 60, 62, 63, 64, 65, 68, 70, 72, 74, 75, 76, 77, 78, 79, 80, 82, ca. 83, 84, 85, 88, 106.

Poa arctica subsp. *arctica* is polymorphic and circumpolar. It grows in alpine and tundra habitats as far south as Wheeler Peak, New Mexico.

Poa arctica subsp. *arctica* has tougher leaves, and is less densely hairy between the lemma veins and palea keels, than subsp. *aperta*. Hultén (1942) recognized several variants within subsp. *arctica*; they are of ecotypic significance at best.

Poa arctica subsp. **grayana** (Vasey) Á. Löve, D. Löve & B.M. Kapoor [p. *363*]

Pl smt glaucous; densely to loosely tufted, rhz, rhz short or long, usu well developed, clm solitary or a few together. **Clm** 20–60 cm, bases decumbent, not wiry. **Shth** closed for ¼–⅖ their length; **lig** (2)3–7 mm, smooth, obtuse to acute; **bld** 1–3 mm wide, flat or folded. **Pan** lax to erect, open; **br** ascending or widely spreading, somewhat sinuous and flexuous to fairly straight, smooth to sparsely scabrous, proximal br ⅖–½ the pan length. **Spklt** (4)4.5–7 mm, not bulbiferous; **rchl intnd** usu glab, infrequently sparsely softly puberulent; **cal** glab; **lm** (2.7)3–5 mm; **pal keels** puberulent to long-villous at midlength; **anth** usu fully developed. 2*n* = 36?

Poa arctica subsp. *grayana* grows only in the alpine regions of the middle and southern Rocky Mountains of Utah, Wyoming, Colorado, and New Mexico. It is characterized by its glabrous calluses, densely hairy lemmas, and paleas that are densely hairy between the keels. It has less wiry culms, and panicles with more flexuous branches, than subsp. *aperta* and, like that subspecies, can be difficult to distinguish from *P. arida* (p. 134).

Poa sect. **Homalopoa** Dumort.

Pl ann or per; densely to loosely tufted or with solitary clm, shoots usu neither rhz nor stln, infrequently rhz. **Bas brchg** both invag and exvag or mainly exvag. **Clm** 2–120 cm, terete or somewhat compressed; **nd** terete or weakly compressed. **Shth** usu closed for ½–⅞ their length, smt only ¹⁄₂₀–¹⁄₁₀ their length, terete to distinctly compressed, smooth or scabrous; **lig** 0.7–12 mm, milky white, smooth or scabrous, truncate to acuminate; **innovation shoot bld** similar to the cauline bld; **cauline bld** 0.6–15 mm wide, flat or folded, thin or moderately thick, lax or moderately straight, abx surfaces usu smooth, smt scabrous over the midvein, adx surfaces smooth or scabrous over the veins, mrg scabrous, apc narrowly to broadly prow-shaped. **Pan** (1)2–40 cm, erect or nodding to lax, contracted or open, sparse or congested, with 1–7 br per nd; **br** erect to reflexed, terete or angled, angles smooth or scabrous, smooth or sparsely scabrous between angles. **Spklt** (2)2.4–9 mm, lat compressed, rarely bulbiferous; **flt** (1)2–7, usu normal, smt the anth aborting, rarely bulb-forming. **Glm** unequal to subequal, distinctly shorter than the adjacent lm, usu bisex, distinctly keeled; **lo glm** 1–3-veined; **cal** terete or slightly lat compressed, usu dorsally webbed, smt glab; **lm** 2–6 mm, narrowly to broadly lanceolate, distinctly keeled, glab or hairy, lat veins obscure to prominent, mrg milky white, apc obtuse to narrowly acute; **pal keels** scabrous, glab or hairy at midlength; **anth** (1, 2) 3, usu 0.1–1.1(1.8) mm, smt 1.5–3 mm and then smt aborting late in development.

Poa sect. *Homalopoa* is the largest and most heterogeneous section of the genus, having at least 170 species, including many annuals and short-lived perennials. Most species are cespitose, have sheaths closed for ¼–¾ their length and anthers up to 1 mm long. The section is widespread in its distribution, growing almost everywhere the genus is native.

6. **Poa bolanderi** Vasey BOLANDER'S BLUEGRASS [p. *364*, <u>500</u>]

Pl usu ann, rarely longer-lived; often glaucous; densely tufted, tuft bases narrow, strl shoots few, not stln, not rhz. **Bas brchg** both invag and exvag. **Clm** 20–60(70) cm, erect or geniculate at the base; **nd** terete, usu 1–3 exserted. **Shth** closed for ½–¾ their length, usu compressed and keeled, usu smooth, infrequently scabrous; **lig** 2.5–7 mm, smooth or scabrous, usu decurrent, obtuse to acute; **bld** 1.5–5 mm wide, usu flat, rarely folded, lax, soft, smooth or sparsely scabrous, mrg scabrous, apc broadly prow-shaped, cauline bld 3–15 cm, flag lf bld 1–4 cm. **Pan** (5)10–15(25) cm long, ¼–½ the pl height, usu erect, infrequently slightly nodding, usu eventually open, smt interrupted, sparse, with 1–3(5) br per nd; **br** initially erect and straight, usu some eventually spreading or reflexed, smooth or sparsely to moderately scabrous. **Spklt** (3)4–7 mm, lat compressed; **flt** 2–3(4); **rchl intnd** usu 1–1.2+ mm, smooth or sparsely scabrous, glab. **Glm** unequal, distinctly shorter than the adjacent lm, distinctly keeled, keels smooth or sparsely scabrous; **lo glm** 1–3-veined, ⅔ the length of the up glm, ½–⅔ the length of the lowest lm; **up glm** shorter than or subequal to the lowest lm; **cal** of some or all flt sparsely webbed; **lm** 2.5–4 mm, lanceolate to narrowly lanceolate, distinctly keeled, smooth or scabrous throughout, glab, lat veins obscure to moderately prominent, apc narrowly acute, usu anthocyanic near the tip; **pal keels** sparsely scabrous; **anth** 3, 0.5–1(1.8) mm. 2*n* = 28.

Poa bolanderi grows mainly in pine to fir forest openings of mountain slopes in the western United States, from Washington to California and Utah. It grows mostly at 1500–3000 m.

7. **Poa bigelovii** Vasey & Scribn. BIGELOW'S BLUEGRASS [p. *364*, 500]

Pl usu ann, rarely longer-lived; densely tufted, tuft bases narrow, usu without strl sht, not stln, not rhz. **Bas brchg** invag. **Clm** (2)5–60 (70) cm tall, 0.3–1 mm thick, usu erect, bases rarely geniculate; **nd** terete, usu 1 exserted. **Shth** closed for ¼–½ their length, usu compressed and keeled, smooth or the keels scabrous; **lig** 2–6 mm, smooth or scabrous, usu decurrent, obtuse to acute; **bld** 1.5–5 mm wide, flat, thin, soft, finely scabrous, apc broadly prow-shaped, cauline bld (1)4–15 cm, flag lf bld usu 1–4 cm. **Pan** (1)5–15 cm, erect, cylindrical, contracted, smt interrupted, congested, with 2–3(5) br per nd; **br** erect or steeply ascending, smooth or sparsely to densely scabrous. **Spklt** 4–7 mm, lat compressed; **flt** 3–7; **rchl intnd** to 1 mm, smooth, glab. **Glm** subequal, distinctly keeled, keels and smt the lat veins scabrous; **lo glm** 1(3)-veined; **up glm** shorter than or subequal to the lowest lm; **cal** webbed; **lm** 2.6–4.2 mm, lanceolate, distinctly keeled, smooth, keels, mrgl veins, and smt the lat veins short- to long-villous, keels hairy to near the apc, mrgl veins to ⅔ their length, lat veins obscure to moderately prominent, intercostal regions glab or softly puberulent, up mrg white, apc acute; **pal keels** softly puberulent to short-villous at midlength, scabrous near the apc, intercostal regions usu softly puberulent; **anth** 1–3, 0.2–1 mm. 2*n* = 28, 28+I.

Poa bigelovii grows in arid upland regions, particularly on shady, rocky slopes of the southwestern United States and northern Mexico.

8. **Poa reflexa** Vasey & Scribn. NODDING BLUEGRASS [p. *365*, 501]

Pl per, short-lived; densely tufted, tuft bases narrow or not, not stln, not rhz. **Bas brchg** mixed invag and exvag. **Clm** 10–60 cm. **Shth** closed for ⅓–⅔ their length, terete, smooth; **lig** 1.5–3.5 mm, smooth or sparsely scabrous; **bld** 1.5–4 mm wide, flat, thin, soft, apc broadly prow-shaped. **Pan** 4–15 cm, nodding, open, with numerous spklt and 1–2 br per nd; **br** (2)3–7 cm, spreading to reflexed, lo br usu reflexed, flexuous, usu terete, smooth or sparsely scabrous, with (3)6–18 spklt. **Spklt** 4–6 mm, lanceolate to broadly lanceolate, usu partly to wholly purplish, with 3–5 flt; **rchl intnd** shorter than 1 mm, smooth. **Glm** narrowly to broadly lanceolate, distinctly keeled, keels smooth or nearly so; **lo glm** 1-veined; **up glm** shorter than or subequal to the lowest lm; **cal** webbed; **lm** 2–3.5 mm, lanceolate, partly purple to fairly strongly purple, distinctly keeled, keels and mrgl veins short- to long-villous, keels hairy for ⅔–⅘ their length, lat veins usu sparsely softly puberulent at least on 1 side, lat veins obscure to moderately prominent, intercostal regions smooth, minutely bumpy, glab, apc acute, slightly bronze-colored or not; **pal keels** scabrous, usu softly puberulent at midlength; **anth** 0.6–1 mm. 2*n* = 28.

Poa reflexa grows in subalpine forests, meadows, and low alpine habitats, primarily in the central and southern Rocky Mountains.

Poa sect. **Madropoa** Soreng

Pl per; densely to loosely tufted or with solitary sht, smt stln, smt rhz. **Bas brchg** invag and/or exvag. **Clm** (5)10–125 cm, terete or weakly compressed; **nd** terete or slightly compressed. **Shth** closed from ⅐ their length to their entire length, terete to compressed, smooth or scabrous, glab or pubescent; **lig** 0.2–18 mm, milky white or colorless, usu translucent, truncate to acuminate, glab or ciliolate; **innovation bld** with the adx surfaces usu moderately to densely scabrous or hispidulous on and between the veins, smt smooth and glab; **cauline bld** flat, folded, or involute, thin or thick, lax or straight, smooth or scabrous, adx surfaces smt hairy, apc narrowly to broadly prow-shaped. **Pan** 1–29 cm, contracted to open, usu with fewer than 100 spklt; **nd** with 1–5 br; **br** 0.5–18 cm, terete or angled, smooth or scabrous, glab or hispidulous. **Spklt** 3–17 mm, lengths 3.5 times widths, lat compressed, not sex dimorphic, not bulbiferous; **flt** 2–10(13) mm, normal; **rchl intnd** smooth or scabrous, glab or hairy. **Glm** distinctly keeled, keels smooth or scabrous; **lo glm** 1, 3(5)-veined; **up glm** 3- or 5-veined; **cal** terete or slightly lat compressed, glab, webbed, or with a crown of hairs; **lm** 2.6–11 mm, lanceolate, distinctly keeled, keels, veins, and intercostal regions glab or hairy, 5–7(11)-veined; **pal** keels scabrous, glab or with hairs at midlength; **anth** 3, vestigial (0.1–0.2 mm) or 1.3–4.5(5) mm.

Poa sect. *Madropoa* is confined to North America. Its 20 species exhibit breeding systems ranging from sequential gynomonoecy to gynodioecy and dioecy. The gynomonoecious species usually grow in forests and have broad, flat leaves. The gynomonoecious and dioecious species grow mainly in more open habitats. They have normally developed anthers that are 1.3–4 mm long, and involute innovation blades that, in several species, are densely scabrous or hairy on the adaxial surfaces.

9. **Poa arnowiae** Soreng WASATCH BLUEGRASS [p. 365, 501]

Pl per; loosely tufted or with solitary sht, short-rhz. **Bas brchg** all or mostly exvag. **Clm** (15)30–80 cm, erect or the bases decumbent, terete or weakly compressed; **nd** terete, 1–3 exserted. **Shth** closed for ½–⁹⁄₁₀ their length, compressed, smooth, glab, bases of bas shth glab, distal shth lengths 1–3 times bld lengths; **col** smooth, glab; **lig** 0.5–4 mm, smooth or sparsely scabrous, truncate to obtuse; **innovation bld** similar to the cauline bld; **cauline bld** 2–5 mm wide, flat, thin, smooth or sparsely scabrous mainly over the veins, apc broadly prow-shaped, mid and up cauline bld subequal in length, flag lf bld (2.5)4–7(11) cm long. **Pan** (5)12–22 cm, usu narrowly pyramidal, open, sparse, with 20–70 spklt, proximal intnd usu (3.5)4+ cm, with 2–3(4) br per nd; **br** 3–8 cm, spreading to eventually reflexed, terete or weakly angled, sparsely to moderately scabrous, with 3–12 spklt. **Spklt** 5–9 mm, lengths to 3.5 times widths, lat compressed, not sex dimorphic; **flt** 2–6; **rchl intnd** smooth, glab, distal intnd 1+ mm. **Glm** lanceolate, distinctly keeled; **lo glm** 1–3-veined; **cal** glab; **lm** 3–6.5 mm, lanceolate, distinctly keeled, keels and mrgl veins glab or short-villous to softly puberulent to ⅓ their length, lat veins obscure, intercostal regions glab or sparsely hispidulous, rarely softly puberulent, smooth or sparsely finely scabrous, mrg glab, apc acute; **pal keels** scabrous, glab, intercostal regions glab; **anth** vestigial (0.1–0.2 mm) or (1.3)2–3.6 mm. 2*n* = unknown.

Poa arnowiae grows in openings within the coniferous forests of the mountain ranges in southeastern Idaho, northern Utah, and adjacent Wyoming. It is sequentially gynomonoecious.

10. **Poa wheeleri** Vasey WHEELER'S BLUEGRASS [p. 365, 501]

Pl per; densely to loosely tufted or with solitary sht, shortly rhz. **Bas brchg** mainly exvag. **Clm** 35–80 cm, erect or the bases decumbent, terete or weakly compressed; **nd** terete, 1–2 exserted. **Shth** closed for ⅓–¾ their length, terete to slightly compressed, at least some proximal shth densely retrorsely scabrous, hispidulous, or softly puberulent for the up ¼ of their length, bases of bas shth glab, distal shth lengths (1.4)1.7–4.6(6.2) times bld lengths; **col** of proximal lvs glab or with hairs the same length as those of their shth; **lig** 0.5–2 mm, smooth or scabrous, smt puberulent, truncate, those of the lo clm and innovation lvs 0.5–1.5 mm, abx surfaces scabrous to softly puberulent, truncate; **innovation bld** folded or involute, infrequently flat, moderately thick, soft, adx surfaces usu densely scabrous to hispidulous; **cauline bld** 2–3.5 mm wide, flat or folded, smooth or sparsely scabrous, glab or hispidulous, apc narrowly to broadly prow-shaped, bld gradually rdcd distally or the mid bld longest, flag lf bld 1–10 cm long. **Pan** 5–12(18) cm, erect or nodding, ovoid to pyramidal, loosely contracted to open, with 20–70 spklt, proximal intnd usu shorter than 3.5 cm; **nd** with 2–5 br; **br** (1)1.7–6.5 cm, ascending to spreading or reflexed, lax, terete or weakly angled, sparsely to moderately scabrous, with 2–8(12) spklt. **Spklt** 5.5–10 mm, lengths to 3.5 times widths, lat compressed, not sex dimorphic; **flt** 2–7; **rchl intnd** smooth or scabrous, glab or sparsely to densely hispidulous. **Glm** ¼–⅔(¾) as long as the adjacent lm, lanceolate, distinctly keeled; **lo glm** 1–3-veined, ¼–½ as long as the adjacent lm; **cal** glab; **lm** 3–6 mm, lanceolate, distinctly keeled, keels and mrgl veins glab or softly puberulent to short-villous, intercostal regions glab or hispidulous, infrequently puberulent, smooth or finely scabrous, lat veins obscure to moderately prominent, mrg glab, apc acute; **pal keels** scabrous, intercostal regions glab; **anth** usu vestigial (0.1–0.2 mm) or aborted late in development and up to 2 mm, rarely normal. 2*n* = 56, 61, 62, 63, 64, 66, 67, 70, ca. 74, 75, 79, 80, 81, 87, 89, 90, 91.

Poa wheeleri is common at mid- to high elevations, generally on the east side of the coastal mountains from British Columbia to California, and from Manitoba to New Mexico. It usually grows in submesic coniferous forests to subalpine habitats. Most plants have densely retrorsely pubescent or scabrous sheaths, involute innovation blades that are pubescent adaxially, and pistillate florets.

Poa wheeleri resembles *P. chambersii* (see next), but differs in having at least some proximal sheaths that are densely retrorsely scabrous or pubescent (sometimes obscurely so), and folded or involute innovation blades that are scabrous to hispidulous on the adaxial surfaces. Natural hybrids have been found between *P. wheeleri* and *P. pratensis* (p. 116).

11. **Poa chambersii** Soreng CHAMBERS'
    BLUEGRASS [p. 366, <u>501</u>]

Pl per; loosely tufted or with solitary sht, short-rhz. **Bas brchg** all or mainly exvag. **Clm** 10–50 cm, erect or the bases decumbent, terete or weakly compressed; **nd** terete, 0–1 exserted. **Shth** closed for ⅓–⅞ their length, terete to slightly compressed, smooth, glab, bases of bas shth glab, distal shth lengths (1.15)1.5–4.6(6.6) times bld lengths; **col** smooth, glab; **lig** 0.5–2(2.5) mm, smooth, truncate to obtuse; **innovation bld** similar to the cauline bld; **cauline bld** gradually rdcd in length distally, 2–5 mm wide, flat or folded, smooth or the adx surfaces sparsely scabrous, primarily over the veins, apc broadly prow-shaped, flag lf bld 0.7–6 cm. **Pan** 2–9 cm, erect, lanceoloid to ovoid, tightly to loosely contracted, with 15–35 spklt, proximal intnd shorter than 2 cm; **nd** with 1–2 br; **br** 0.9–3.2 cm, erect to ascending or slightly spreading, terete, smooth or sparsely scabrous, with 1–4 spklt. **Spklt** 6–12 mm, lengths to 3 times widths, lat compressed, not sex dimorphic; **flt** 2–7; **rchl intnd** 0.8–1.5 mm, smooth or sparsely scabrous, glab. **Glm** ⅗–⅘ as long as the adjacent lm, distinctly keeled; **lo glm** 3-veined; **cal** of at least some proximal flt sparsely webbed, with 1–2 mm hairs, others glab, rarely all glab; **lm** 5–7 mm, lanceolate, 5–7-veined, distinctly keeled, smooth or sparsely finely scabrous, glab throughout or the keels and mrgl veins sparsely softly puberulent over the proximal ¼, lat veins moderately prominent, intercostal regions glab, mrg glab, apc acute; **pal keels** sparsely scabrous, intercostal regions glab; **anth** vestigial (0.1–0.2 mm), aborted late in development, or 1.8–3.7 mm. 2*n* = unknown.

Poa chambersii is known only from upland forest openings in the Cascades of western Oregon, where it is dioecious, and from high elevations on Steens Mountain in southeastern Oregon, where it is gynodioecious. It differs from *P. wheeleri* (see previous) in having glabrous sheaths and flat or folded, glabrous innovation blades.

12. **Poa fendleriana** (Steud.) Vasey VASEY'S
    MUTTONGRASS [p. 366, <u>501</u>]

Pl per; densely to loosely tufted, rhz, often weakly so, rhz usu short and inconspicuous. **Bas brchg** mainly invag, usu some exvag. **Clm** 15–70 cm, smt stout, erect or the bases decumbent, terete or weakly compressed; **nd** terete, 0–1 exserted. **Shth** closed for about ⅓ their length, terete, smooth or scabrous, glab or occ retrorsely pubescent, bases of bas shth glab, distal shth lengths usu (5)9+ times bld lengths; **col** smooth or scabrous, glab or hispidulous; **lig** 0.2–18 mm, smooth or scabrous, decurrent or not, apc truncate to acuminate, ciliolate or glab; **innovation bld** usu moderately to densely scabrous or hispidulous on and between the veins, infrequently nearly smooth and glab; **cauline bld** strongly rdcd in length distally, (0.5)1–3(4) mm wide, usu involute, moderately thick and firm, infrequently moderately thin, abx surfaces usu smooth, infrequently scabrous, apc narrowly prow-shaped, steeply rdcd in length distally along the clm, flag lf bld often absent or very rdcd, smt to 1(3) cm. **Pan** 2–12(30) cm, erect, contracted, narrowly lanceoloid to ovoid, congested, frequently with 100+ spklt; **nd** with 1–2 br; **br** 1–8 cm, erect, terete to weakly angled, smooth or scabrous, with 3–15(25) spklt. **Spklt** (3)4–8(12) mm, lengths to 3 times widths, broadly lanceolate to ovate, lat compressed, not sex dimorphic; **flt** 2–7(13); **rchl intnd** 0.8–1.3 mm, smooth, glab or hairy, hairs to 0.3 mm. **Glm** lanceolate, distinctly keeled; **lo glm** 1–3-veined, distinctly shorter than the lowest lm; **cal** glab; **lm** 3–6 mm, lanceolate, distinctly keeled, keels, mrgl veins, and lat veins glab or short- to long-villous or softly puberulent, lat veins moderately prominent, intercostal regions softly puberulent or glab, smooth or sparsely scabrous, mrg glab, apc acute; **pal keels** scabrous, smt softly puberulent or long-villous at midlength, hairs to 0.4+ mm; **anth** vestigial (0.1–0.2 mm) or 2–3 mm. 2*n* = 28+II, 56, 56–58, 58–64.

Poa fendleriana grows on rocky to rich slopes in sagebrush-scrub, interior chaparral, and southern (rarely northern) high plains grasslands to forests, and from desert hills to low alpine habitats. Its range extends from British Columbia to Manitoba and south to Mexico. It is one of the best spring fodder grasses in the eastern Great Basin, Colorado plateaus, and southern Rocky Mountains. It is dioecious. Each of the subspecies has regions of sexual reproduction in which staminate plants are common within populations, and extensive regions where only apomictic, pistillate plants are found. The sexual populations set little seed; the apomictic populations are highly fecund.

Poa fendleriana hybridizes with *Poa cusickii* subsp. *pallida* (p. 124). The hybrids are called *P.* ×*nematophylla* (p. 124).

There are two subspecies in the Intermountain Region. They intergrade where sexual or partially sexual populations have come into contact.

1. Ligules of the middle cauline leaves 0.2–1.2 (1.5) mm long, not decurrent, usually scabrous, apices truncate to rounded, upper margins ciliolate or scabrous . . . . subsp. *fendleriana*
1. Ligules of the middle cauline leaves (1.5)1.8–18 mm long, decurrent, usually smooth to sparsely scabrous, apices obtuse to acuminate, upper margins usually smooth, glabrous . . . . subsp. *longiligula*

Poa fendleriana (Steud.) Vasey subsp. fendleriana [p. 366]

Col often scabrous or hispidulous near the throat; **lig** of mid cauline lvs 0.2–1.2(1.5) mm, scabrous, mrg not decurrent, apc truncate to rounded, usu scabrous or ciliolate; **innovation bld** usu scabrous or puberulent adx. **Rchl intnd** usu smooth and glab. **Lm** long-villous on the keels and mrgl veins, intercostal regions usu glab, infrequently softly puberulent. 2*n* = 56, 58–60, 59, 58–64.

Poa fendleriana subsp. *fendleriana* grows chiefly in the southern and middle Rocky Mountains, and in the mountains surrounding the Colorado plateaus. Sexually reproducing populations are mainly confined to Arizona, New Mexico, and Texas, are rare in California, and infrequent in Colorado and Utah. Pistillate populations are common from southern British Columbia to Manitoba and south to northern Mexico, but infrequent in the Great Basin.

Poa fendleriana subsp. longiligula (Scribn. & T. A. Williams) Soreng LONGTONGUE MUTTONGRASS [p. 366]

Col smooth to scabrous near the throat; **lig** of mid cauline lvs (1.5)1.8–18 mm, smooth or sparsely scabrous, mrg decurrent, apc obtuse to acuminate, usu smooth, glab; **innovation bld** usu scabrous, smt puberulent adx. **Rchl intnd** usu sparsely hispidulous or sparsely softly puberulent. **Lm** long-villous on the keels and mrgl veins, intercostal regions usu glab, infrequently softly puberulent. 2*n* = 56, 56–58.

Poa fendleriana subsp. *longiligula* tends to grow to the west of subsp. *fendleriana*, in areas where winter precipitation is more consistent and summer precipitation less consistent. Apomixis is far more common and widespread than sexual reproduction in this subspecies. Apomictic populations range from southwestern British Columbia to Baja California, Mexico, throughout the Great Basin and Colorado plateaus, and eastward across the Rocky Mountains. Sexual populations are mainly confined to northern Arizona, California, Nevada, and Utah.

13. Poa cusickii Vasey CUSICK'S BLUEGRASS [p. 367, 501]

Pl per; usu densely tufted, rarely moderately densely tufted, usu neither rhz nor stln, infrequently short-rhz or stln, rarely with distinct rhz. **Bas brchg** invag or invag and exvag. **Clm** 10–60(70) cm tall, 0.5–1.8 mm thick, erect or the bases decumbent, terete or weakly compressed;

nd terete, 0–2 exserted. **Shth** closed for ¼–¾ their length, terete, smooth or scabrous, glab, bases of bas shth glab, distal shth lengths 1.6–10 times bld lengths; **col** smooth or scabrous, glab; **lig** of cauline lvs 1–3(6) mm, smooth or scabrous, truncate to acute, lig of the innovation lvs 0.2–0.5(2.5) mm, scabrous, usu truncate; **innovation bld** smt distinctly different from the cauline bld, 0.5–2 mm wide, involute, moderately thick, moderately firm, adx surfaces usu densely scabrous or hispidulous to softly puberulent, infrequently nearly smooth and glab; **cauline bld** subequal or the midcauline bld longest or the bld gradually rdcd in length distally, 0.5–3 mm wide, flat, folded, or involute, usu thin, usu withering, abx surfaces smooth or scabrous, apc narrowly to broadly prow-shaped, flag lf bld 0.5–5(6) cm. **Pan** 2–10(12) cm, usu erect, contracted or loosely contracted, narrowly lanceoloid to ovoid, congested or moderately congested, with 10–100 spklt and 1–3(5) br per nd; **br** 0.5–4(5) cm, erect or steeply ascending, fairly straight, slender to stout, terete to angled, smooth or scabrous, with 1–15 spklt. **Spklt** (3)4–10 mm, lengths to 3 times widths, broadly lanceolate to narrowly ovate, lat compressed, not sex dimorphic; **flt** 2–6; **rchl intnd** 0.5–1.2 mm, smooth or scabrous. **Glm** lanceolate, distinctly keeled; **lo glm** 3-veined, distinctly shorter than the lowest lm; **cal** glab or diffusely webbed, hairs less than ¼ the lm length; **lm** (3)4–7 mm, lanceolate to broadly lanceolate, distinctly keeled, memb to thinly memb, smooth or sparsely to densely scabrous, glab or the keels and/or mrgl veins puberulent proximally, lat veins obscure to prominent, mrg glab, apc acute; **pal keels** scabrous, intercostal regions glab; **anth** vestigial (0.1–0.2 mm), aborted late in development, or 2–3.5 mm. 2*n* = 28, 28+II, 56, 56+II, 59, ca. 70.

Poa cusickii grows in rich meadows in sagebrush scrub to rocky alpine slopes, from the southwestern Yukon Territory to Manitoba and North Dakota, south to central California and eastern Colorado. It is gynodioecious or dioecious.

Sexually reproducing plants of *Poa cusickii* subspp. *cusickii* and *pallida* grow in different geographic areas, but pistillate plants of these two subspecies have overlapping ranges. Only pistillate plants are known in *Poa cusickii* subsp. *epilis*. All the alpine plants studied were pistillate.

1. Panicle branches smooth or slightly scabrous, or the basal blades more than 1.5 mm wide and flat or folded; cauline blades more than 1.5 mm wide, often flat; some basal branching extravaginal; lemmas and calluses sometimes sparsely puberulent . . . . subsp. *epilis*
1. Panicle branches moderately to strongly scabrous; basal and cauline blades usually less than 1.5 mm

wide, involute, rarely flat or folded; basal branching intravaginal; lemmas and calluses glabrous.

3. Panicle branches longer than 1.7 cm in at least some panicles; panicles open or contracted . . .
. . . . subsp. *cusickii*
3. Panicle branches up to 1.7 cm long, stout; panicles contracted . . . . subsp. *pallida*

Poa cusickii Vasey subsp. cusickii [p. 367]

Pl densely tufted. **Bas brchg** invag. **Clm** 10–60(70) cm, mostly erect, with 0–1 well-exserted nd. **Shth** closed for ¼–⅔ their length, distal shth lengths 3–10 times bld lengths; **innovation bld** 0.5–1 mm wide; **cauline bld** less than 1.5 mm wide, flat, folded, or involute, apc narrowly prow-shaped, flag lf bld (0.5)1.5–5 cm. **Pan** usu 5–10(12) cm, contracted or loosely contracted, with 20–100 spklt; **nd** with 1–5 br; **br** 1.7–4(5) cm, slender to stout, moderately to densely scabrous, with 2–15 spklt. **Spklt** 4–10 mm. **Cal** glab; **lm** 4–7 mm, glab; **anth** vestigial (0.1–0.2 mm) or 2–3.5 mm. 2*n* = 28.

Poa cusickii subsp. *cusickii* grows mainly in mesic desert upland and mountain meadows, on and around the Columbia plateaus of northern California, Oregon, southern Washington, and adjacent Idaho and Nevada. It is highly variable, with fairly open- to contracted-panicle populations, and from gynodioecious to dioecious populations. The modal and mean longest branch lengths of the narrower-panicled populations of subsp. *cusickii* serve to distinguish it from subsp. *pallida* in most cases. It appears to have hybridized with *P. pringlei* around Mount Rose, Nevada.

Poa cusickii subsp. epilis (Scribn.) W.A. Weber
SKYLINE BLUEGRASS [p. 367]

Pl densely tufted. **Bas brchg** invag and exvag. **Clm** 20–45 cm, mostly erect, with 1–2 well-exserted nd. **Shth** closed for ⅓–¾ their length, distal shth lengths 2–5 times bld lengths; **innovation bld** 0.7–1 mm wide; **cauline bld** more than 1.5 mm wide, flat or folded, apc narrowly to broadly prow-shaped, flag lf bld 1.5–5 cm, apc broadly prow-shaped. **Pan** usu 2–7 cm, usu contracted, with 20–70 spklt; **nd** with 2–5 br; **br** 1–3 cm, moderately stout, smooth to sparsely scabrous, with 1–8 spklt. **Spklt** (3)4–8 mm. **Cal** glab; **lm** 3–6 mm, glab or, rarely, the keels and mrgl veins sparsely puberulent proximally; **anth** usu aborted late in development. 2*n* = 56, ca. 70.

Poa cusickii subsp. *epilis* tends to grow around timberline. It is strictly pistillate and is usually quite distinct from subspp. *cusickii* and *pallida*. It occurs throughout most of the range of the species, and is fairly uniform even though widespread.

Poa cusickii subsp. pallida Soreng [p. 367]

Pl densely tufted. **Bas brchg** invag. **Clm** 10–40(55) cm, mostly erect, with 0(1) scarcely exserted nd. **Shth** closed for ¼–⅔ their length, distal shth

lengths 3.6–10 times bld lengths; **innovation bld** 0.5–1 mm wide, apc usu narrowly prow-shaped; **cauline bld** usu less than 1.5 mm wide, flat, folded, or involute, usu narrowly prow-shaped, infrequently broadly prow-shaped, flag lf bld 0.5–2(3) cm. **Pan** 2–6 cm, contracted, with 10–40 spklt; **nd** with 1–3 br; **br** 0.5–1.7 cm, stout, moderately to densely scabrous, with 2–5 spklt. **Spklt** 4–10 mm. **Cal** glab; **lm** 4–7 mm, glab; **anth** vestigial (0.1–0.2 mm) or 2–3.5 mm. 2*n* = 56, 56+II, 59.

Poa cusickii subsp. *pallida* grows in forb-rich mountain grasslands to alpine habitats, from the southern Yukon Territory to California, across the Great Basin and through the Rocky Mountains to central Colorado. It is found mainly east and north of subsp. *cusickii*, but pistillate plants extend into the range of that subspecies in the eastern alpine peaks of California, Nevada, and Oregon. The shorter branch length usually serves to distinguish it from narrow-panicled plants of subsp. *cusickii*. It hybridizes with *P. fendleriana* (see previous), forming *P. ×nematophylla* (see next). The hybrids may have hairy lemmas or, less often, broader leaf blades and glabrous lemmas.

14. Poa ×nematophylla Rydb. [p. 368, <u>501</u>]

Pl per; densely tufted, not stln, not rhz. **Bas brchg** invag. **Clm** 10–35 cm, erect or the bases decumbent; **nd** terete, 0–1 exserted. **Shth** closed for ¼–¾ their length, terete, apc acuminate; **innovation bld** 0.5–1(2) mm wide, involute, moderately thick, moderately firm, abx surfaces smooth or scabrous, adx surfaces usu densely scabrous or hispidulous; **cauline bld** usu gradually rdcd distally, 0.5–1(2) mm wide, flat, folded, or involute, thin, smt withering, abx surfaces smooth or scabrous, apc narrowly prow-shaped, smt the flag lf bld vestigial. **Pan** 2–8 cm, erect, narrowly lanceoloid to ovoid, contracted, congested; **nd** with 1–2 br; **br** 0.5–3 cm, erect, terete to angled, scabrous. **Spklt** 4–8 mm, lengths to 3 times widths, broadly lanceolate to narrowly ovate, lat compressed, not sex dimorphic; **flt** 2–5; **rchl intnd** 0.5–1.2 mm, smooth or scabrous. **Glm** lanceolate, distinctly keeled; **lo glm** 3-veined, distinctly shorter than the lowest lm; **cal** glab; **lm** 4–7 mm, lanceolate, distinctly keeled, memb, keels and mrgl veins usu softly puberulent, smt short-villous, intercostal regions usu glab, infrequently softly puberulent proximally, lat veins moderately prominent, mrg glab, apc acute; **pal keels** scabrous; **anth** mostly vestigial (0.1–0.2 mm), rarely 2–3 mm. 2*n* = unknown.

Poa ×nematophylla is believed to consist of hybrids between *P. cusickii* subsp. *pallida* (see previous) and *P. fendleriana* (p. 122). It is mostly pistillate and apomictic; few staminate plants have been found. It usually resembles *P. cusickii* most, but grades towards *P. fendleriana*. It tends to grow on drier slopes than either parent, mainly in and around sagebrush desert/forest interfaces.

15. Poa leibergii Scribn. LEIBERG'S BLUEGRASS [p. *368*, 501]

Pl per; densely tufted, tufts slender, not stln, not rhz. **Bas brchg** invag. **Clm** 5–35 cm tall, 0.5–0.7 mm thick, erect or the bases decumbent, with 0–1 exserted nd. **Shth** closed for ⅖–⅘ their length, terete, smooth and glab, bases of bas shth glab; **col** smooth, glab; **lig** (1)2–4 mm, colorless, transparent, smooth, mrg decurrent or not, apc truncate to acute, lig of innovation and cauline lvs alike; **innovation bld** smooth or sparsely scabrous abx; **cauline bld** 0.5–1 mm wide, flat, folded, or involute, thin, lax, filiform, usu soon withering, both surfaces smooth or sparsely scabrous, apc narrowly prow-shaped. **Pan** 1–5(8) cm, erect to lax, lanceoloid to ovoid or pyramidal, contracted to open, sparse, with (1)6–17(22) spklt; **nd** with 1–2 br; **br** 1–4 cm, erect to spreading, slender, terete, smooth or sparsely to rarely moderately densely scabrous, with 1–2(3) spklt. **Spklt** 4–8 mm, lengths to 3 times widths, broadly lanceolate to broadly ovate, lat compressed, not sex dimorphic; **flt** 2–8; **rchl intnd** glab. **Glm** thin, somewhat lustrous, distinctly keeled; **lo glm** 3-veined, distinctly shorter than the lowest lm; **cal** glab; **lm** 3.5–7 mm, lanceolate, distinctly keeled, thinly memb, smooth or scabrous, glab, lat veins moderately prominent to prominent, mrg glab, apc acute to truncate and erose; **pal keels** smooth or scabrous, glab or pectinately ciliate; **anth** vestigial (0.1–0.2 mm) or 1.3–3 mm. $2n$ = unknown.

Poa leibergii grows on mossy ledges and around vernal pools and the outer margins of *Camassia* swales, in sagebrush desert to low alpine habitats, especially where snow persists. It is found primarily on and around the basaltic Columbia plateaus, and is gynodioecious. All reports of *P. leibergii* from California, and most of those from Nevada, are based on misidentified specimens of *P. cusickii* subsp. *cusickii* (p. 124) and *P. stebbinsii*.

16. Poa pringlei Scribn. PRINGLE'S BLUEGRASS [p. *369*, 501]

Pl per; densely tufted, not stln, not rhz. **Bas brchg** invag. **Clm** 5–35 cm tall, 0.5–0.9 mm thick, erect or the bases decumbent, with 0(1) exserted nd. **Shth** closed for ⅐–⅓ their length, terete, smooth or sparsely scabrous, glab, bases of bas shth glab, distal shth lengths 2–4 times bld lengths; **col mrg** smooth or scabrous to hispidulous; **lig** of cauline lvs 1–6 mm, colorless, translucent, smooth or scabrous, truncate to acute, lig of the innovations 1–2.5 mm; **innovation bld** similar to the cauline bld, 1.5–3 mm wide, involute, thick, frequently somewhat arcuate, abx surfaces smooth, adx surfaces densely scabrous or hispidulous; **cauline bld** becoming only slightly shorter distally, 1.5–3 mm wide, involute, moderately thick, soft to moderately firm, abx surfaces smooth, apc narrowly prow-shaped. **Pan** 1–6 cm, erect, narrowly lanceoloid to ovoid, moderately congested, with 6–20(25) spklt; **nd** with 1–2 br; **br** 0.5–1.5(2) cm, erect, moderately stout, terete or weakly angled, angles smooth to fairly densely scabrous, with 1–3 spklt. **Spklt** 6–8(12) mm, lengths to 3.5 times widths, broadly lanceolate, lat compressed, not sex dimorphic, lustrous; **flt** 2–5; **rchl intnd** smooth. **Glm** subequal, isomorphic, lanceolate to broadly lanceolate, thin, lustrous, distinctly keeled, keels smooth or sparsely scabrous; **lo glm** shorter than the adjacent lm, 3-veined; **cal** glab; **lm** 5–8 mm, lanceolate, distinctly keeled, thinly memb, smooth or sparsely finely scabrous, glab, lat veins moderately prominent, mrg glab, apc acute; **pal keels** coarsely scabrous; **anth** vestigial (0.1–0.2 mm) or 2–4 mm. $2n$ = unknown.

Poa pringlei grows on rocky subalpine and alpine slopes in Oregon and California. Sierra Nevada populations are pistillate and apomictic.

Poa sect. Pandemos Asch. & Graebn.

Pl per; smt stln, smt rhz. **Bas brchg** invag and exvag. **Clm** 25–120 cm, terete or weakly compressed; **nd** terete or slightly compressed. **Shth** closed for about ¼–½ their length, compressed, distal shth lengths 0.5–4 times bld lengths; **lig** 3–10 mm, scabrous, acute to acuminate; **bld** 1–5 mm wide, flat, lax, soft, veins and mrg scabrous, apc narrowly prow-shaped. **Pan** 8–25 cm, erect or lax, pyramidal, open; **nd** with 3–7 br; **br** 2–8(10) cm, ascending to spreading, flexuous to fairly straight, angled, angles densely scabrous, crowded. **Spklt** 2.3–3.5 mm, lengths to 3 times widths, lat compressed, not bulbiferous; **flt** 2–4, bisex. **Glm** distinctly keeled, keels scabrous; **lo glm** subulate to narrowly lanceolate, usu arched to sickle-shaped, 1-veined, distinctly shorter than the lowest lm; **cal** terete or slightly lat compressed, glab or dorsally webbed; **lm** 2.3–3.5 mm, lanceolate, distinctly keeled, keels hairy, glab elsewhere or the mrgl veins pubescent, lat veins prominent; **pal keels** smooth, muriculate, tuberculate, or minutely scabrous; **anth** 3, 1.3–2 mm.

Poa sect. *Pandemos* includes two diploid species of European origin. One, *P. trivialis*, is now widespread around the world.

17. **Poa trivialis** L. ROUGH BLUEGRASS [p. *369*, 501]

Pl per, short-lived; somewhat loosely to densely tufted, usu weakly stln. **Bas brchg** invag. **Clm** 25–120 cm, decumbent to erect, smt trailing and rooting at the nd, terete or weakly compressed; **nd** terete or slightly compressed, (0)1–3 exserted. **Shth** closed for about ⅓–½ their length, compressed, usu densely scabrous, bases of bas shth glab, distal shth lengths 0.5–4 times bld lengths; **col** smooth or scabrous, glab; **lig** 3–10 mm, scabrous, acute to acuminate; **bld** 1–5 mm wide, flat, lax, soft, sparsely scabrous over the veins, mrg scabrous, apc narrowly prow-shaped. **Pan** 8–25 cm, erect or lax, pyramidal, open, with 35–100+ spklt; **nd** with 3–7 br; **br** 2–8(10) cm, ascending to spreading, flexuous to fairly straight, angled, angles densely scabrous, crowded, with 5–35 spklt in the distal ½–¾. **Spklt** 3–4.5(5) mm, lengths to 3 times widths, lat compressed; **flt** 2–4,

bisex; **rchl intnd** smooth or muriculate. **Glm** distinctly keeled, keels scabrous; **lo glm** subulate to narrowly lanceolate, usu arched to sickle-shaped, 1-veined, distinctly shorter than the lowest lm; **cal** webbed, hairs over ⅔ the lm length; **lm** 2.3–3.5 mm, lanceolate, distinctly keeled, keels usu sparsely puberulent to ⅗ their length, mrgl veins usu glab, infrequently the proximal ¼ softly puberulent, intercostal regions smooth, glab, up lm smt glab, lat veins prominent, mrg glab, apc acute; **pal keels** smooth, muriculate, tuberculate, or minutely scabrous; **anth** 1.3–2 mm. 2*n* = 14.

Poa trivialis is an introduced European species. Only *Poa trivialis* subsp. *trivialis* is present in the Intermountain Region. Several cultivars have been planted for pastures and lawns, and have often escaped cultivation. *Poa trivialis* is easily recognized by its flat blades, long ligules, sickle-shaped lower glumes, prominent callus webs, and lemmas with pubescent keels and pronounced lateral veins.

Poa sect. **Oreinos** Asch. & Graebn.

Pl per; densely to loosely tufted, smt shortly rhz or stln. **Bas brchg** mostly exvag or mixed invag and exvag. **Clm** 5–100 cm tall, 0.5–1.5 mm thick, slender, smt weak, terete; **nd** terete. **Shth** usu closed for ⅕–⅗ their length, hybrids smt closed for ¹⁄₁₀–⅕ their length, terete, smooth or sparsely scabrous; **lig** 0.5–4(6) mm, smooth or sparsely scabrous, truncate to acute, smt lacerate; **innovation bld** similar to the cauline bld; **cauline bld** 0.8–4 mm wide, flat, thin, lax, soft, adx surfaces smooth or sparsely scabrous, narrowly prow-tipped. **Pan** 1.5–15 cm, lax or slightly lax, loosely contracted to open; **nd** with 1–3(5) br; **br** 1–8 cm, steeply ascending to reflexed, capillary to slender, drooping to fairly straight, sulcate or angled, smooth or the angles scabrous, with 1–15 spklt. **Spklt** 3.2–8 mm, lengths to 3.5 times widths, narrowly lanceolate to ovate, lat compressed, not bulbiferous; **flt** 2–5, bisex; **rchl intnd** smooth, glab. **Glm** subulate to broadly lanceolate, thin, distinctly keeled, keels smooth or scabrous; **lo glm** 1–3-veined; **cal** terete or slightly lat compressed, glab or dorsally webbed; **lm** 2.5–4.6 mm, lanceolate to broadly lanceolate, distinctly keeled, thin, keels and mrgl veins short- to long-villous, lat veins usu glab, infrequently sparsely softly puberulent, lat veins obscure or moderately prominent, intercostal regions glab; **pal keels** scabrous, usu glab, infrequently pectinately ciliate; **anth** 3, 0.2–1.1(1.3) mm.

Poa sect. *Oreinos* is circumboreal. It includes seven species: four strictly Eurasian, one amphiatlantic, one primarily western North American with isolated occurrences in the Russian Far East, and one restricted to North America. The species are boreal, alpine to low arctic, and grow in bogs and on alpine slopes. They are primarily slender perennials with extravaginal tillering.

18. **Poa laxa** Haenke LAX BLUEGRASS [p. *370*, 501]

Pl per; not or only slightly glaucous; densely tufted, not stln, not rhz. **Bas brchg** mixed, mainly exvag or mainly psdinvag, smt invag. **Clm** 8–35 cm tall, 0.5–0.9 mm thick, ascending to erect, slender; **nd** terete, 0(1) exserted. **Shth** closed for ⅕–⅓ their length, terete, smooth,

glab, bases of bas shth glab; **col** smooth or scabrous, glab; **lig** 2–4 mm, smooth, apc acute, often lacerate; **innovation bld** similar to the cauline bld; **cauline bld** 1–2(3) mm wide, flat, thin, soft, smooth, narrowly prow-tipped, bld not strongly graduated or rdcd upwards. **Pan** 2–8 cm, slightly lax, usu loosely contracted and sparse, infrequently contracted and dense; **nd** with 1–3(5) br; **br** 1–3(4) cm, usu ascending or weakly spreading, infrequently erect, fairly straight or flexuous, slender, sulcate or angled, smooth or the angles sparsely scabrous, with 1–8 spklt. **Spklt** 4–6 mm, lengths to 3 times widths, lat compressed; **flt** 2–5; **rchl intnd** shorter than 1 mm, smooth, glab. **Glm** nearly equaling or slightly longer than the adjacent lm, lanceolate to broadly lanceolate, thin, distinctly keeled, keels smooth or sparsely scabrous; **lo glm** 1–3-veined; **up glm** shorter than or subequal to the lowest lm; **cal** glab or webbed, hairs usu shorter than ¼ the lm length, sparse; **lm** 3–4.6 mm, lanceolate to broadly lanceolate, thin, distinctly keeled, keels and mrgl veins short- to long-villous, lat veins glab or sparsely softly puberulent, lat veins obscure, intercostal regions glab, mrg glab, apc acute; **pal** sparsely scabrous over the keels; **anth** (0.6)0.8–1.1(1.3) mm. $2n = 28, 42, 84$.

Poa laxa is a low arctic to high alpine amphiatlantic species. Its short anthers and smoother branches usually distinguish it from *P. glauca* (p. 129), with which it can hybridize to form *P. laxa* × *glauca*.

Poa laxa has four subspecies, one of which is found in the Intermountain Region.

Poa laxa subsp. **banffiana** Soreng [p. *370*]
Bas brchg mainly exvag. **Bld** thin. **Pan** 2–8 cm, lax, loosely contracted, sparse, with 2–3(5) br per nd; **br** steeply ascending, fairly straight, usu sparsely scabrous, infrequently smooth. **Glm** lanceolate to broadly lanceolate; **lo glm** 3-veined; **cal** glab; **lm** with keels short-villous for at least ½ their length, usu the lat veins on at least 1 side of some flt sparsely softly puberulent, infrequently all the lat veins glab; **anth** 0.8–1.1 mm. $2n = 84$.

Poa laxa subsp. *banffiana* grows primarily in mesic alpine locations in the Rocky Mountains. It is sometimes difficult to distinguish from *P. glauca* (p. 129).

19. **Poa leptocoma** Trin. Western Bog Bluegrass [p. *370*, 501]
Pl per; dark to light green, often anthocyanic in part; loosely tufted, usu neither stln nor rhz, occ with short, slender rhz. **Bas brchg** mostly exvag. **Clm** 15–100 cm, slender to middling. **Shth** closed for ¼–⅗ their length, terete, smooth or sparsely scabrous, mrg not ciliate; **lig** 1.5–4(6) mm, smooth to sparsely scabrous, obtuse to acute; **bld** 1–4 mm wide, flat, thin, lax, soft, apc narrowly prow-shaped. **Pan** 5–15 cm, lax, open, sparse; **nd** with 1–3(5) br; **br** (2)3–8 cm, spreading to reflexed, capillary, usu angled, infrequently only sulcate or subterete, angles usu moderately densely scabrous, smt only sparsely so, with (3)4–15 spklt. **Spklt** 4–8 mm, lanceolate or narrowly lanceolate, green or partly purple to dark purple; **flt** 2–5; **rchl intnd** smooth, glab. **Glm** subulate to lanceolate, thin, distinctly keeled, keels usu scabrous; **lo glm** subulate to narrowly lanceolate, 1-veined; **up glm** distinctly shorter than to nearly equaling the lowest lm; **cal** sparsely webbed; **lm** 3–4 mm, lanceolate, often partly purple, distinctly keeled, thin, smooth, or with sparse hooks apically, keels and mrgl veins softly puberulent to long-villous, hairs extending ¼–⅔ the keel length, smt sparse, lat veins and intercostal regions glab, mrg glab, infolded, apc sharply acute to acuminate, usu bronze-colored; **pal keels** nearly smooth, scabrous, or pectinately ciliate; **anth** 0.2–1.1 mm. $2n = 42$.

Poa leptocoma grows around lakes and ponds and along streams, in subalpine and alpine to low arctic habitats, in western North America from Alaska to California and New Mexico, and on the Kamchatka Peninsula, Russia. It often grows with or near *P. reflexa* (p. 120), from which it differs in its more scabrous panicle branches, shorter anthers, glabrous or pectinately ciliate palea keels, and preference for wet sites. The two also differ in their ploidy level, *P. leptocoma* being hexaploid, and *P. reflexa* tetraploid.

Poa sect. **Stenopoa** Dumort.

Pl per; densely to loosely tufted, not rhz, infrequently stln. **Bas brchg** all or mostly exvag. **Clm** 5–120 cm, terete or slightly compressed; **nd** terete or slightly compressed. **Shth** closed for ¹⁄₁₀–⅕ their length, terete or slightly compressed, smooth or sparsely scabrous, distal shth shorter or longer than their bld; **lig** 0.2–6 mm, usu scabrous, smt smooth, apc truncate or obtuse and usu ciliolate, or acute and not ciliolate; **bld** 0.8–8 mm wide, mostly flat, smt folded, moderately thin, abruptly ascending to spreading, lax or straight, mrg scabrous, adx surfaces usu scabrous over the veins, apc narrowly prow-shaped. **Pan** 1–30(41) cm, erect or lax, open, narrowly lanceoloid to ovoid, sparse to moderately congested; **nd** with 2–9 br; **br** 0.4–15 cm, erect to reflexed, angled, angles scabrous. **Spklt** 3–8(9) mm, lengths 2–3.5 times widths, narrowly lanceolate to narrowly ovate, lat compressed, rarely bulbiferous; **flt** (1)2–5, bisex, rarely bulb-forming; **rchl intnd** mostly shorter than 1 mm, frequently muriculate or scabrous or pubescent. **Glm** subulate to broadly lanceolate, distinctly keeled, keels smooth or sparsely scabrous; **lo glm** 3-veined; **cal** terete or slightly lat compressed, glab or dorsally webbed; **lm** 2–4 mm, narrowly to broadly lanceolate, distinctly keeled, coriaceous-memb, usu finely muriculate, keels and mrgl veins long- to short-villous, intercostal regions glab or softly puberulent to short-villous, lat veins obscure, apc usu partially bronze-colored; **pal keels** scabrous, smt softly puberulent at midlength, intercostal regions glab or puberulent; **anth** 3, 0.8–2.5 mm.

Poa sect. *Stenopoa* includes 30 species. Most are Eurasian; three are native in, and one is restricted to, North America. The North American species are cespitose or weakly stoloniferous, and have sheaths open for much of their length, scabrous panicle branches, and faint lateral lemma veins. The new shoots for the following year are initiated late in the growing season, after flowering and fruiting; vegetative and flowering shoots are usually not present at the same time.

20. **Poa palustris** L. FOWL BLUEGRASS [p. *370*, 501]

Pl per; usu loosely, smt densely, tufted, frequently stln. **Bas brchg** exvag or mixed exvag and invag. **Clm** 25–120 cm, erect or the bases decumbent, smt brchg above the base, terete or weakly compressed, scabrous below the pan; **nd** terete or slightly compressed, proximal nd often slightly swollen, uppermost nd at or above (⅓)½ the clm length. **Shth** closed for ¹⁄₁₀–⅕ their length, slightly compressed, glab or sparsely retrorsely scabrous, bases of bas shth glab, distal shth lengths 0.7–2.2 times bld lengths; **lig** (1)1.5–6 mm, smooth or sparsely to moderately scabrous, apc obtuse to acute, frequently lacerate, usu minutely ciliolate; **bld** 1.5–8 mm wide, flat, usu several per clm, steeply ascending or spreading to 80°, often lax distally, apc narrowly prow-shaped. **Pan** (9)13–30(41) cm, lengths ⅓–½ times widths at maturity, lax, eventually open, sparsely to moderately congested, with 25–100+ spklt; **nd** with 2–9 br; **br** 4–15 cm, ³⁄₁₀–½ the pan length, initially erect, eventually widely spreading to slightly reflexed, fairly straight, slender, angles densely scabrous. **Spklt** 3–5 mm, lengths 3–3.5 times widths, narrowly to broadly lanceolate, lat compressed; **flt** (1)2–5; **rchl intnd** mostly shorter than 1 mm, usu muriculate, smt smooth, rarely sparsely hispidulous. **Glm** subulate to lanceolate, distinctly keeled, keels smooth or sparsely scabrous; **lo glm** with lengths 6.4–10 times widths, 3-veined, long-tapered to a slender point; **cal** sparsely to moderately densely webbed, hairs (½)⅔+ the lm length; **lm** 2–3 mm, narrowly lanceolate to lanceolate, distinctly keeled, keels straight or gradually arched, usu abruptly inwardly arched at the jnct of the scarious apc, keels and mrgl veins short-villous, lat veins obscure, intercostal regions muriculate, glab, mrg distinctly inrolled, glab, apc obtuse or acute, usu partially bronze-colored, frequently incurved and blunt with a short, hyaline mrg; **pal keels** scabrous, intercostal regions glab; **anth** 1.3–1.8 mm. 2*n* = 28, 30, 32, 35, 42, 56, 84.

Poa palustris is native to boreal regions of northern Eurasia and North America, and is widespread in cool-temperate and boreal riparian and upland areas. *Poa palustris* is used for soil stabilization and waterfowl feed.

Poa palustris from drier woods and meadows tends to resemble *P. interior* (see next). The best features for recognizing it include its loose growth habit, more steeply ascending leaf blades, well-developed callus webs, narrowly hyaline lemma margins, and incurving lemma keels. It also has a tendency to branch above the base.

21. **Poa interior** Rydb. INTERIOR BLUEGRASS [p. *371*, 501]

Pl per; green or less often glaucous; densely tufted, not stln, not rhz. **Bas brchg** all or mostly exvag. **Clm** 5–80 cm, usu slender, mostly erect or ascending, several to many arising together; **nd** terete or slightly compressed, (0)1–2(3) exserted, top nd usu at ⅓–⅗ the clm length. **Shth** closed for ¹⁄₁₀–⅕ their length, terete, bases of bas shth glab, distal shth lengths (0.6)0.88–1.64 times bld lengths; **lig** 0.5–1.5(3) mm, sparsely to densely scabrous, apc truncate to obtuse, ciliolate; **bld** 0.8–3 mm wide, mostly flat, thin, soft, appressed or abruptly ascending to spreading, straight or somewhat lax, apc narrowly prow-shaped. **Pan** (1.5)3–15(17) cm, lengths generally 2.5–4 times widths at maturity, usu erect, lax in shade forms, narrowly lanceloid to ovoid, sparsely to moderately congested; **nd** with 2–5 br; **br** 0.4–8(9) cm long, ¼–½ the pan length, ascending to widely spreading, fairly straight, slender to moderately stout, angled, angles moderately to densely scabrous. **Spklt** 3–6 mm, lengths 2–3 times widths, lanceolate to narrowly ovate, lat compressed, usu not glaucous; **flt** (1)2–3(5); **rchl intnd** usu shorter than 1 mm, smooth, muriculate, or scabrous, glab, hispidulous, or sparsely to densely puberulent, proximal intnd frequently curved. **Glm** lanceolate to broadly lanceolate, distinctly keeled, keels smooth or sparsely scabrous; **lo glm** 3-veined, long- or abruptly tapered to a slender point, lengths 4.5–6.3 times widths; **cal** usu webbed, infrequently glab in depauperate alpine specimens, webs usu scant, less than ½(⅔) the lm length, frequently minute; **lm** 2.4–4 mm, lanceolate, distinctly keeled, straight or gradually arched, not abruptly inwardly arched at the jnct with the scarious apc, keels and mrgl veins short-villous, hairs extending ⅔–¾ the keel length, lat veins usu glab, rarely sparsely puberulent, obscure, intercostal regions smooth, smt weakly muriculate, glab, mrg not or slightly inrolled, glab, apc acute, usu partially bronze-colored; **pal keels** scabrous, intercostal regions glab; **anth** (1.1)1.3–2.5 mm. 2*n* = 28, 42, 56.

Poa interior, a native species, grows from Alaska to western Quebec and New York, south to Arizona and New Mexico. It is restricted to North America. It is fairly common from boreal forests to low alpine habitats of the Rocky Mountains. It grows in subxeric to mesic habitats, such as mossy rocks and scree, usually in forests. It is usually tetraploid.

In alpine habitats, *Poa interior* is often quite short, and often sympatric with *P. glauca* (see next). It differs from *P. glauca* in having lemmas that are glabrous between the marginal veins and keels or, rarely, sparsely puberulent on the lateral veins. It usually also differs from *P. glauca* subsp. *rupicola* in having at least a few hairs on its calluses. It is sometimes difficult to distinguish from *P. palustris* (see previous), but differs in having lemmas with wider hyaline margins and straight or gradually arched keels, a densely tufted habit, and scantly webbed calluses.

22. Poa glauca Vahl [p. *371*, <u>501</u>]

Pl per; usu glaucous; densely tufted, not stln, not rhz. **Bas brchg** all or mostly exvag. **Clm** 5–40(80) cm, erect to spreading, straight, wiry, bases straight or slightly decumbent; **nd** terete or slightly compressed, usu 0–1 exserted, top nd at ¹⁄₁₀–⅓ the clm length. **Shth** closed for ¹⁄₁₀–⅕ their length, terete, bases of bas shth glab or sparsely minutely hairy, hairs 0.1–0.2 mm, distal shth lengths 1.1–4 times bld lengths; **lig** 1–4(5) mm, sparsely to densely scabrous, apc obtuse to acute, minutely ciliolate; **bld** 0.8–2.5 mm wide, flat or folded, thin, soft, appressed or abruptly ascending to spreading, straight, apc narrowly prow-shaped. **Pan** 1–10(20) cm, lengths 3–5 times widths at maturity, rarely racemelike with br of irregular length, erect, narrowly lanceoloid to ovoid, contracted to somewhat open, sparse, proximal intnd shorter than 1.5(4) cm; **nd** with 2–3(5) br; **br** erect, ascending or weakly spreading, fairly straight, short, stout, angled, angles moderately to densely scabrous, rarely only scabrous distally, glaucous; **ped** usu shorter than the spklt. **Spklt** 3–7(9) mm, lengths 2–3 times widths, lat compressed, rarely bulbiferous, usu glaucous; **flt** 2–5, rarely bulb-forming; **rchl intnd** to 1.2 mm, smooth, muriculate, or scabrous, glab or sparsely to densely hispidulous or puberulent. **Glm** subequal, narrowly to broadly lanceolate, distinctly keeled, keels smooth or sparsely scabrous, apc acute; **lo glm** 3-veined;

up glm 2–3.8(5.2) mm, lengths usu more than 4.1 times widths, distinctly shorter to subequal to the lowest lm; **cal** glab or webbed, webs from minute to more than ½ the lm length; **lm** 2.5–4 mm, lanceolate to broadly lanceolate, distinctly keeled, keels and mrgl veins short-villous, lat veins obscure, usu sparsely softly puberulent to short-villous, intercostal regions smooth, smt weakly muriculate, glab or puberulent, mrg glab, apc usu partially bronze-colored, obtuse or acute; **pal keels** scabrous, glab or softly puberulent at midlength, intercostal regions glab or softly puberulent; **anth** (1)1.2–2.5 mm, mature sacs 0.2 mm wide, rarely aborted late in development. 2*n* = 34, 42, 44, 47, 48, 49, 50, 56, 56, 57, 58, 60, 63, 64, 65, 70, 75, 78, ca. 100.

Poa glauca is a common, highly variable, circumboreal, boreal forest to alpine and high arctic species. It grows from Alaska to Greenland, south to California and New Mexico in the west, through Canada and the northeastern United States in the east, and at scattered locations in Patagonia. It generally favors dry habitats and tolerates disturbance well. It can be distinguished from *P. interior* (see previous) by its longer ligules, lower top culm node, and wider glumes and lemmas. It is often confused in herbaria with *P. abbreviata* subsp. *pattersonii* (p. 131), but differs in having primarily extravaginal branching and, usually, longer anthers. It is suspected to hybridize with *P. arctica* (p. 117) and *P. secunda* (p. 132). It is highly polyploid, and presumed to be highly apomictic.

1. Calluses usually webbed, sometimes glabrous; lemmas glabrous or hairy between the veins . . . . .
. . . . subsp. *glauca*
1. Calluses glabrous; lemmas hairy between the veins . . . . subsp. *rupicola*

Poa glauca Vahl subsp. **glauca** GLAUCOUS BLUEGRASS [p. *371*]

Clm 10–40(80) cm. **Pan** 3.5–10(20) cm. **Spklt** not bulbiferous; **flt** normal. **Cal** webbed or glab; **lm** usu with lat veins short-villous to softly puberulent, intercostal regions glab or short-villous to softly puberulent. 2*n* = 34, 42, 44, 47, 48, ca. 49, 50, 56, 60, 63, 64, 65, 70, 75, 78.

Poa glauca subsp. *glauca* is the widespread and common subspecies in the Northern Hemisphere. It is also disjunct in South America. It does not grow in California, and is uncommon in the Great Basin and southern Rocky Mountains. It is highly variable and is often confused in herbaria with subsp. *rupicola*, but can sometimes be distinguished by its webbed calluses and lemmas that are glabrous between the veins. In the Rocky Mountains, *P. glauca* subsp. *glauca* often grows with subsp. *rupicola* and *P. interior* (see previous).

Poa glauca subsp. **rupicola** (Nash) W.A. Weber TIMBERLINE BLUEGRASS [p. *371*]

Clm to 5–15 cm. **Pan** 1–5 cm, usu narrowly lanceoloid. **Spklt** not bulbiferous; **flt** normal.

Cal glab; **lm** at least sparsely puberulent on the intercostal regions. 2*n* = 48, 48–50, 56, 56–58, ca. 100.

Poa glauca subsp. *rupicola* is endemic to dry alpine areas of western North America. It is often confused in herbaria with

subsp. *glauca* and *P. interior* (see previous), but its calluses lack even a vestige of a web, and its lemmas have at least a few hairs between the lemma veins. It is often sympatric with both taxa outside of California.

Poa sect. **Tichopoa** Asch. & Graebn.

Pl per; rhz, usu with solitary sht, smt loosely tufted. **Bas brchg** all or mostly exvag. **Clm** 15–60 cm, distinctly compressed; **nd** distinctly compressed. **Shth** closed for ⅒–⅕ their length, distinctly compressed; **lig** 1–3 mm, moderately to densely scabrous, mrg ciliolate, apc obtuse; **bld** 1.5–4 mm wide, flat, mrg scabrous, adx surfaces smooth or scabrous mainly over the veins, apc narrowly prow-shaped. **Pan** 2–10 cm, lengths usu 3–6 times widths, erect, linear to ovoid, mostly with 1–3 br per nd; **br** 0.5–3 cm, angled, angles scabrous. **Spklt** (2.3)3.5–7 mm, lat compressed, bisex, not bulbiferous; **flt** 3–7. **Cal** terete or slightly lat compressed, glab or pubescent, webbed; **lm** coriaceous-memb, usu finely muriculate, lat veins obscure, apc usu partially bronze-colored; **anth** 3, 1.3–1.8 mm.

Poa sect. *Tichopoa* has two species, both of which are native to Europe. They are similar to species of *Poa* sect. *Stenopoa*, differing in having strongly compressed culms and nodes, and in being rhizomatous. One of the species is found in the Intermountain Region.

23. **Poa compressa** L. Canada Bluegrass
[p. 372, 501]

Pl per; usu with solitary sht, smt loosely tufted, extensively rhz. **Clm** 15–60 cm, wiry, bases usu geniculate, strongly compressed; **nd** strongly compressed, some proximal nd usu exserted. **Shth** closed for ⅒–⅕ their length, distinctly compressed, bases of bas shth glab; **lig** 1–3 mm, moderately to densely scabrous, ciliolate, apc obtuse; **bld** 1.5–4 mm wide, flat, cauline bld subequal. **Pan** 2–10 cm, generally ⅙–⅓ as wide as long, erect, linear, lanceoloid to ovoid, often interrupted, sparse to congested, with 15–80 spklt and mostly with 1–3 br per nd; **br** 0.5–3 cm, erect to ascending, or infrequently spreading, angles densely scabrous, at least in part, with

1–15 spklt. **Spklt** (2.3)3.5–7 mm, lat compressed; **flt** 3–7; **rchl intnd** usu shorter than 1 mm, smooth to muriculate. **Glm** distinctly keeled; **lo glm** 3-veined; **cal** usu webbed, smt glab; **lm** 2.3–3.5 mm, lanceolate, distinctly keeled, keels and mrgl veins short-villous, intercostal regions glab, lat veins obscure, mrg glab, apc acute; **pal** scabrous over the keels; **anth** 1.3–1.8 mm. 2*n* = 35, 42, 49, 50, 56, 84.

Poa compressa is common in much of North America. It is sometimes considered to be native, but this seems doubtful. In North America, it is often seeded for soil stabilization, and has frequently escaped. It grows mainly in riparian areas, wet meadows, and disturbed ground. Its distinctly compressed nodes and culms, exserted lower culm nodes, rhizomatous growth habit, and scabrous panicle branches make it easily identifiable.

Poa sect. **Abbreviatae** Nannf. *ex* Tzvelev

Pl per; densely tufted, not stln, not rhz. **Bas brchg** mainly invag. **Clm** usu shorter than 25(30) cm, slender, terete; **nd** terete. **Lvs** mostly bas; **shth** closed for ⅒–¼(⅓) their length, terete; **lig** 0.4–5.5 mm, milky white to hyaline, smooth or scabrous, apc truncate to acute, glab; **bld** 0.5–2 mm wide, flat, folded, or involute, thin to moderately thick, soft or moderately firm, apc narrowly prow-shaped. **Pan** 1–7 cm, erect, usu contracted, smt open; **nd** with 1–3 br; **br** 0.5–1.5(5) cm, usu erect to steeply ascending, smt ascending to spreading, sulcate to angled, smooth or the angles sparsely to densely scabrous. **Spklt** 3–7 mm, lat compressed, rarely bulbiferous; **flt** 2–5, usu bisex, smt with vestigial anth or anth that abort late in the growing season, rarely bulb-forming; **rchl intnd** usu glab, infrequently sparsely hispidulous. **Glm** usu subequal to or slightly longer than the adjacent lm, distinctly keeled, keels smooth or sparsely scabrous; **lo glm** (1)3-veined; **cal** terete or slightly lat compressed, glab or dorsally webbed; **lm** 2–5.8 mm, lanceolate to broadly lanceolate, distinctly keeled, thin, glab or the keels and mrgl veins softly puberulent to long-villous, intercostal regions glab or softly puberulent to short-villous, obscurely 5-veined; **pal keels** scabrous, glab or softly puberulent to short-villous at midlength; **anth** 3, 0.2–1.3(1.8) mm, rarely vestigial (0.1–0.2 mm) or aborted late in development.

Poa sect. *Abbreviatae* includes five North American species, three of which grow in the Intermountain Region. The species are principally high alpine to high arctic.

24. **Poa lettermanii** Vasey LETTERMAN'S
     BLUEGRASS [p. 372, 501]

Pl per; not glaucous; densely tufted, not stln, not rhz. **Bas brchg** all or mainly invag. **Clm** 1–12 cm, slender. **Shth** closed for ⅙–¼ their length, terete; **lig** 1–3 mm, milky white to hyaline, smooth; **bld** 0.5–2 mm wide, flat or folded, or slightly inrolled, thin, without papillae (at 100×), apc narrowly prow-shaped. **Pan** 1–3 cm, erect, contracted, usu exserted from the shth; **br** to 1.5 cm, erect to steeply ascending, slender, sulcate or angled, smooth or the angles sparsely scabrous; **ped** shorter than the spklt. **Spklt** 3–4 mm, lat compressed, green or anthocyanic; **flt** 2–3; **rchl intnd** shorter than 1 mm, smooth. **Glm** usu equaling or exceeding the lowest lm, smt also equaling or exceeding the up flt, lanceolate to broadly lanceolate, distinctly keeled, keels smooth; **lo glm** 3-veined; **cal** glab; **lm** 2.5–3 mm, lanceolate, distinctly keeled, thin, usu glab, keels and mrgl veins rarely sparsely puberulent proximally, apc acute; **pal keels** scabrous; **anth** 0.2–0.8 mm. $2n = 14$.

Poa lettermanii grows on rocky slopes of the highest peaks and ridges in the alpine zone, from northern British Columbia to western Alberta and south to California and Colorado, usually in the shelter of rocks or on mesic to wet, frost-scarred slopes. It is one of only three known diploid *Poa* species native to the Western Hemisphere. Its glabrous calluses and lemmas usually distinguish it from *P. abbreviata* (see next); it also differs in having flat or folded leaf blades, and shorter spikelets with glumes that are longer than the adjacent florets.

25. **Poa abbreviata** R. Br. [p. 373, 501]

Pl per; not or scarcely glaucous; densely tufted, not stln, not rhz. **Bas brchg** all or mainly invag. **Clm** 5–15(20) cm, slender, lfless above the bas tuft. **Shth** closed for ¹⁄₁₀–¼ their length, terete; **lig** 0.4–5.5 mm, milky white to hyaline, smooth or scabrous, apc truncate to acute; **bld** 0.8–1.5(2) mm wide, involute, moderately thick, soft, apc narrowly prow-shaped. **Pan** 1.5–5 cm, erect, lanceoloid to ovoid, contracted, congested; **nd** with 1–3 br; **br** to 1.5 cm, erect, slender, terete, sulcate or angled, smooth or the angles sparsely scabrous; **ped** usu shorter than the spklt. **Spklt** 4–6.5 mm, lat compressed, rarely bulbiferous, frequently strongly anthocyanic; **flt** 2–5, rarely bulb-forming; **rchl intnd** usu shorter than 1 mm, smooth or scabrous. **Glm** subequal to slightly longer than the adjacent lm, lanceolate to broadly lanceolate, distinctly keeled, keels smooth; **lo glm** (1)3-veined, lat veins often faint and short; **up glm** exceeding or exceeded by the up flt; **cal** glab or webbed; **lm** 3–4.6 mm, lanceolate to broadly lanceolate, distinctly keeled, thin, keels and mrgl veins usu short- to long-villous, hairs extending along ¾–⅚ of the keel, infrequently glab, intercostal regions glab or softly puberulent to short-villous, apc acute; **pal keels** scabrous, often short-villous to softly puberulent at midlength, smt glab; **anth** 0.2–1.2(1.8) mm. $2n = 42$.

Poa abbreviata is an alpine and circumarctic species which has two subspecies in the western cordilleras, and one in the high arctic. It grows mainly on frost scars and mesic rocky slopes, usually on open ground. In rare cases where the lemmas and calluses of *P. abbreviata* are glabrous, it can be confused with *P. lettermanii* (p. see previous), but that species has shorter spikelets and glumes that are longer than the adjacent florets.

1. Lemmas glabrous; calluses webbed . . . . . . subsp. *marshii*
1. Lemmas usually with hairs over the veins; calluses glabrous or webbed, rarely both the lemmas and calluses glabrous . . . . subsp. *pattersonii*

Poa abbreviata subsp. **marshii** Soreng MARSH'S
     BLUEGRASS [p. 373]

Lig 1–3 mm, smooth, apc obtuse to acute. **Spklt** not bulbiferous; **flt** normal. **Cal** webbed; **lm** lanceolate, glab; **pal keels** glab; **anth** 0.6–1.2 mm. $2n$ = unknown.

Poa abbreviata subsp. *marshii* is rather uncommon. It is known from scattered alpine peaks across the interior western United States: from the White Mountains of California, the Schell Creek Range of Nevada, the southern Rockies of Idaho, the Little Belt Mountains of Montana, and the Big Horn Mountains of Wyoming, mostly where subsp. *pattersonii* is absent.

Poa abbreviata subsp. **pattersonii** (Vasey) Á.
     Löve, D. Löve & B.M. Kapoor PATTERSON'S
     BLUEGRASS [p. 373]

Lig 0.8–5.5 mm, smooth or scabrous, apc obtuse to acute. **Spklt** rarely bulbiferous; **flt** rarely bulb-forming. **Cal** usu webbed, rarely glab; **lm** long-villous along ¾ of the keel and the mrgl veins, rarely glab, but then the cal also glab, intercostal regions glab or softly puberulent; **anth** 0.6–1.2(1.8) mm, rarely vestigial. $2n = 42$.

Poa abbreviata subsp. *pattersonii* is an alpine taxon that extends from the Brooks Range, Alaska, to the Sierra Nevada, California, where it is rare, and through the Rocky Mountains to southern Colorado. It also grows in the Russian Far East. It is often confused in herbaria with *P. glauca* (p. 129), but differs in having predominantly intravaginal branching, an abundance of vegetative shoots, and usually shorter anthers.

26. **Poa keckii** Soreng KECK'S BLUEGRASS
     [p. 373, 501]

Pl per; not glaucous; densely tufted, not stln, not rhz. **Bas brchg** all or mainly invag. **Clm** 2–10(18) cm, erect to spreading; **nd** terete, none exserted. **Shth** closed for ¹⁄₁₀–⅕ their length, terete, smooth, glab, bases of bas shth glab, distal shth lengths

1.5–7 times bld lengths; **col** smooth, glab; **lig** 1–3 mm, milky white, smooth or sparsely scabrous, apc obtuse to acute, lig of up innovation lvs shorter than 3 mm; **innovation bld** similar to the cauline bld; **cauline bld** 1–3.5(4.5) cm long, 0.9–1.8 mm wide, folded, moderately thick, soft, smooth, glab, adx surfaces infrequently sparsely scabrous, usu with papillae on the long cells (at 100×), apc narrowly prow-shaped, flag lf bld folded, 1–1.8 mm wide, abx surfaces with 7–15 closely spaced, slightly protruding ribs. **Pan** 1–4(6) cm, erect, ovoid to lanceoloid, contracted, congested, with 9–40 spklt; **nd** with 1–3 br; **br** 0.5–1.5 cm, erect, fairly straight, sulcate or angled, angles sparsely to densely scabrous, with 1–7 spklt; **ped** shorter than the spklt. **Spklt** 3.5–6 mm long, lengths to

3.5(3.8) times widths, lanceolate, lat compressed, fairly strongly anthocyanic, not glaucous; **flt** 2–3; **rchl intnd** terete, to 1.5 mm, smooth, smt sparsely hispidulous. **Glm** lanceolate, smooth, distinctly keeled, keels sparsely scabrous; **lo glm** shorter than to equaling the lowest lm, 3-veined; **up glm** frequently exceeding the lowest lm, exceeded by the up lm; **cal** glab; **lm** 3–4.9 mm, lanceolate, distinctly keeled, thin, smooth or finely scabrous, glab or the keels and mrgl veins sparsely puberulent proximally, lat veins obscure, mrg glab, apc acute; **pal** keels scabrous; **anth** 0.6–1.3(1.8) mm. 2*n* = unknown.

Poa keckii is endemic to high alpine frost scars and ledges, usually on open ground, in the Sierra Nevada and adjacent Sweetwater and White mountains of California.

Poa sect. **Secundae** V.L. Marsh *ex* Soreng

Pl per; usu densely, infrequently loosely, tufted, rarely weakly rhz or stln. **Bas brchg** mixed invag and exvag to completely invag. **Clm** 10–120 cm, capillary to stout, terete or weakly compressed; **nd** terete. **Shth** closed for ⅒–⅓ their length, terete, smooth or scabrous, distal shth usu longer than their bld; **lig** 0.5–7(10) mm, smooth or scabrous, apc truncate to acuminate; **bld** 0.4–3(5) mm wide, flat, folded, or involute, thin to moderately thick, soft and soon withering or moderately firm and persisting, smooth or scabrous mainly over the veins and mrg, apc narrowly prow-shaped. **Pan** 2–25(30) cm, erect or somewhat lax, narrowly lanceoloid to ovoid, usu contracted, smt open and pyramidal, sparse to congested, with 7–100(120) spklt; **nd** with 1–4(7) br; **br** 0.5–15 cm, erect to spreading, terete, sulcate or angled, smooth or the angles sparsely to densely scabrous, smt scabrous between the angles. **Spklt** (4)4.5–10 mm, lengths 3–5 times widths, terete to weakly lat compressed or distinctly compressed, smt bulbiferous; **flt** (2)3–5(10), usu normal and bisex, smt bulb-forming. **Glm** lanceolate to broadly lanceolate, shorter than to subequal to the adjacent lm, keels indistinct to distinct, smooth or scabrous; **lo glm** 3-veined; **cal** terete or slightly dorsally compressed, glab or with a crown of hairs, hairs to 2 mm; **lm** 3–7 mm, narrowly lanceolate to lanceolate or slightly oblanceolate, weakly to distinctly keeled, glab or the keels and mrgl veins and smt the lat veins with hairs, obscure, intercostal regions glab or with hairs; **anth** 3, 1.2–3.5 mm, smt aborted late in development.

Poa sect. *Secundae* includes ten species, nine of which grow in North America. All the species tend to grow in arid areas, sometimes on wetlands within such areas. They are primarily cespitose, but hybridization with members of *Poa* sect. *Poa* results in the formation of rhizomatous plants. Typically, members of sect. *Secundae* have sheaths that are closed for ⅒–¼ their length, contracted panicles, and anthers that are 1.2–3.5 mm long.

27. **Poa secunda** J. Presl SECUND BLUEGRASS [p. 374, <u>501</u>]

Pl per; frequently anthocyanic, smt glaucous; densely tufted, bas lf tufts 2–20+ cm, usu narrowly based, rarely with rhz. **Bas brchg** invag and exvag. **Clm** (10)15–120 cm, slender to stout, erect or the bases slightly decumbent, terete or weakly compressed; **nd** terete, 0–2 exserted. **Shth** closed for ⅒–¼ their length, terete, smooth or scabrous, glab, bases of bas shth glab, distal shth lengths (0.95)1.5–7(15) times bld lengths; **col** smooth or scabrous, glab; **lig** 0.5–6(10) mm, smooth or scabrous, truncate to acuminate, lig of innovation lvs similar to those of the cauline lvs or shorter and truncate; **innovation bld** similar

to the cauline bld; **cauline bld** gradually rdcd in length upwards or the mid bld longest, 0.4–3(5) mm wide, flat, folded, or involute, thin, soft, and soon withering to thick, firm, and persisting, smooth or scabrous mainly over the veins, glab, apc narrowly prow-shaped, flag lf bld 0.8–10(17) cm. **Pan** 2–25(30) cm, erect or somewhat lax, narrowly lanceoloid to ovoid, usu contracted, more or less open at anthesis, infrequently remaining open at maturity, green or anthocyanic, smt glaucous, usu moderately congested, with 10–100+ spklt; **nd** usu with 1–3 br; **br** (0.5)1–8(10) cm, usu erect or ascending, infrequently spreading at maturity, terete to weakly angled, usu sparsely to densely scabrous on and between the angles, with (1)2–20(60+) spklt in the distal ½–⅔. **Spklt** (4)5–10 mm, lengths (3.8)4–5 times widths, usu narrowly lanceolate, subterete to weakly lat compressed, drab, green or strongly anthocyanic, smt glaucous; **flt** (2)3–5(10); **rchl intnd** usu 1–2 mm, terete or slightly dorsally compressed, smooth or muriculate to scabrous. **Glm** broadly lanceolate, keels indistinct; **lo glm** 3-veined; **cal** glab or with a crown of hairs, hairs 0.1–0.5(2) mm, crisp or slightly sinuous; **lm** 3.5–6 mm, lanceolate to narrowly lanceolate or slightly oblanceolate, usu weakly keeled, glab or the keels and mrgl veins softly puberulent to short-villous, intercostal regions smooth or scabrous, glab, short-villous, crisply puberulent or softly puberulent over the bas ⅔, hairs usu 0.1–0.5 mm, hairs of the keels and veins frequently similar in length to those between the veins, usu not or only slightly denser and extending further towards the apc, lat veins obscure, mrg strongly inrolled below, broadly scarious above, glab, apc obtuse to broadly acute, blunt, or pointed; **pal keels** scabrous, glab or softly puberulent to short-villous at midlength; **anth** 1.5–3 mm. 2*n* = 42, 44+f, ca. 48, 56, ca. 62, 63, ca. 66, ca. 68, 70, ca.72, ca. 74, 78, ca. 80, 81, 82, ca. 83, 84–86, ca. 87, ca. 88, ca. 90, ca. 91, 93, ca. 94, ca. 97, ca. 98, ca. 99, 100, 104, 105–106.

Poa secunda is one of the major spring forage species of temperate western North America. It is very common in high deserts, mountain grasslands, saline wetlands, meadows, dry forests, and on lower alpine slopes, primarily from the Yukon Territory east to Manitoba and south to Baja California, Mexico.

Poa secunda is highly variable. There are two subspecies that overlap almost completely in terms of morphology, but differ ecologically and cytologically.

Poa secunda is known or suspected to hybridize with several other species, including *P. arctica* (p. 117), *P. arida* (see next), *P. glauca* (p. 129), and *P. pratensis* (p. 116). Apomixis is facultative but common.

1. Lemmas usually glabrous, the keels and marginal veins infrequently sparsely puberulent at the base; basal branching mainly extravaginal; leaves slightly lax to firm, remaining intact through the growing season; ligules of the innovations to 2 mm long . . . . subsp. *juncifolia*
1. Lemmas sparsely to densely puberulent or short-villous on the basal ⅔; basal branching mixed intra- or extravaginal or mainly intravaginal; leaves usually lax, withering with age; ligules of the innovations usually longer than 2 mm . . subsp. *secunda*

Poa secunda subsp. juncifolia (Scribn.) Soreng
ALKALI BLUEGRASS, BIG BLUEGRASS, NEVADA BLUEGRASS [p. 374]

Bas lf tufts usu medium to robust, infrequently tiny. **Bas brchg** mainly exvag. **Clm** 30–120 cm. **Lig** of clm lvs 0.5–6 mm, those of the innovations 0.5–2 mm, scabrous, apc truncate to obtuse; **bld** 1–3(5) mm, moderately thick to thick, slightly lax to firm, tending to hold their form and persist. **Pan** (4)10–25(30) cm, narrowly lanceoloid, contracted, congested; **br** erect, scabrous. **Spklt** (4)7–10 mm, lengths 4–5 times widths, narrowly lanceolate, subterete; **cal** glab; **lm** sparsely to moderately scabridulous to scabrous, usu glab, keels and mrgl veins infrequently crisply puberulent on the bas ¼, hairs usu shorter than 0.2 mm; **pal** glab. 2*n* = 42, 56, 60, 61, 62, 63, 64, ca. 65, ca. 66, 70, 78, 84, ca. 97.

Poa secunda subsp. *juncifolia* is usually more robust than subsp. *secunda*, and generally inhabits moister and sometimes saline habitats. It comprises two fairly distinct variants: a robust upland variant that is frequently used for revegetation (*P. ampla* Merr., Big Bluegrass) that grows in deep, rich, montane soils; and a riparian and wet meadow variant (*P. juncifolia* Scribn., Alkali Bluegrass). Apart from generally having glabrous lemmas, short ligules on the vegetative shoots, and leaf blades that hold their form better than those of subsp. *secunda*, subsp. *juncifolia* differs anatomically in the predominance of sinuous-walled, rectangular long cells in the blade epidermis. Smooth-walled, fusiform long cells are predominant in *P. secunda* subsp. *secunda*. Plants with glabrous lemmas and long ligules on the vegetative shoots have been called *P. nevadensis* Vasey *ex* Scribn.; they are intermediate between the subspecies. Chromosome numbers for *P. secunda* subsp. *juncifolia* center on 2*n* = 63, indicating a high degree of apomixis.

Poa secunda J. Presl subsp. secunda PACIFIC BLUEGRASS, PINE BLUEGRASS, SANDBERG BLUEGRASS, CANBY BLUEGRASS [p. 374]

Bas lf tufts usu tiny to medium, less often robust. **Bas brchg** mixed invag and exvag or mainly invag. **Clm** (10) 15–100 cm, slender to middling. **Lig** of clm lvs 2–6(10) mm, those of the innovations mostly 2–6 mm, smooth or scabrous, obtuse to acuminate; **bld** 0.4–3 mm, usu thin, lax, and soon withering, smt moderately thick, moderately firm, and somewhat persistent. **Pan** 2–15(20) cm, usu narrowly lanceoloid to ovoid, contracted at maturity and congested, or occ pyramidal, open at maturity, and sparse; **br** erect or ascending, infrequently widely spreading at maturity. **Spklt** (4)5–8 mm; **cal** glab or pubescent; **lm** with keels and mrgl veins long-villous, crisply puberulent, or softly puberulent over the bas ⅔, intercostal regions usu at least sparsely crisply or softly puberulent, hairs usu shorter than 0.5 mm; **pal keels** short-villous to softly puberulent at midlength, intercostal regions often softly puberulent. 2*n* = 42, 44+f, ca. 48, 56, ca. 62, 63, ca. 66, ca. 68, 70, ca. 72, ca. 74, ca. 78, ca. 80, 81, 82, ca. 83, 84, ca. 86, ca. 87, ca. 88, ca. 90, ca. 91, 93, ca. 94, ca. 98, ca. 99, 100, 104, 105–106.

Poa secunda subsp. *secunda* comprises several forms or ecotypes which intergrade morphologically and overlap geographically. Its chromosome numbers are centered on 2*n* = 84. It generally grows in more xeric habitats than subsp. *juncifolia*; it is also common in alpine habitats. Some of the major variants, and the names that have been applied to them, are: scabrous plants, primarily from west of the Cascade/Sierra Nevada axis (*P. scabrella* (Thurb.) Benth. *ex* Vasey, Pine Bluegrass); smoother, large plants extending eastward (*P. canbyi* (Scribn.) Howell, Canby Bluegrass); tiny, early-spring-flowering plants of stony and mossy ground (*P. sandbergii* Vasey, Sandberg Bluegrass); and slender, sparse plants, generally of mesic shady habitats, with panicles that remain open (*P. gracillima* Vasey, Pacific Bluegrass). Alpine plants have been called *P. incurva* Scribn. & T.A. Williams.

28. Poa stenantha Trin. Narrow-Flowered Bluegrass [p. 374, 502]

Pl per; glaucous or not; densely to loosely tufted, not stln, not rhz. **Bas brchg** mostly exvag, some invag. **Clm** 20–60(100) cm, bases decumbent or smt erect, terete, with 1–2 exserted nd. **Shth** closed for ¹⁄₁₀–¹⁄₅(¼) their length, terete, bases of bas shth glab; **lig** 2–5 mm, milky white, smooth or sparsely scabrous, acute to acuminate; **innovation bld** similar in texture and shape to the cauline bld; **cauline bld** not greatly rdcd upwards, 1.5–4(5) mm wide, flat or folded, thin, lax, smooth or sparsely scabrous, apc narrowly prow-shaped. **Pan** 5–18(25) cm, lax, loosely contracted to open, sparse, with 20–65 spklt and usu 2(7) br per nd; **br** 3–15 cm, ascending to spreading, angled, angles finely to coarsely, sparsely to fairly densely scabrous, infrequently smooth, with 3–10(15) spklt in the distal ½. **Spklt** 6–10 mm, lengths 3–3.6 times widths, lanceolate to narrowly ovate, lat compressed, smt bulbiferous, drab, often slightly glaucous; **flt** 3–4(7), normal or bulb-forming; **rchl intnd** 1.2–2 mm, slightly dorsally compressed, smooth or sparsely muriculate. **Glm** subequal, lanceolate to broadly lanceolate, dull, frequently glaucous, obtuse to acute; **lo glm** 3-veined; **up glm** (3.7)4.1–6.5 mm; **cal** usu crowned with 0.2–2 mm hairs, smt glab; **lm** 4–6 mm, lanceolate, distinctly compressed, distinctly keeled, keels, mrgl veins, and smt the lat veins short- to long-villous, hairs extending for ¾ of the keel, intercostal regions glab, sparsely puberulent or hispidulous proximally, usu sparsely to moderately densely scabrous distally, hairs distinctly shorter than those of the keel and veins, mrg weakly inrolled, broadly scarious, glab, apc acute; **pal keels** scabrous, often softly puberulent at midlength, intercostal regions glab or puberulent; **anth** 1.2–2 mm, smt aborted late in development or undeveloped. 2*n* = 42, [81, 84, 86?].

Poa stenantha grows in coastal meadows and on cliffs in subarctic and boreal forests; it is less common in moist, more southern subalpine and low alpine meadows and thickets. Its range extends from western Alaska to the northern Cascades and Rocky Mountains and, as a disjunct, to Patagonia.

Poa stenantha Trin. var. stenantha [p. 374]

Spklt not bulbiferous; flt normal. Anth 1.2–2 mm.

Poa stenantha var. *stenantha* can be difficult to separate from *P. secunda* subsp. *secunda* (p. 133). Its main distinguishing features are its strongly keeled lemmas with glabrous intercostal regions, and, when present, callus hairs longer than 0.2 mm. Plants with large panicles and glabrous calluses have been called *P. macroclada* Rydb. Such plants grow infrequently in the U.S. Rocky Mountain portion of the species' range. They intergrade with the more compact typical form.

Named intersectional hybrids

29. Poa arida Vasey Plains Bluegrass [p. 375, 502]

Pl per; glaucous or not; densely to loosely tufted or the clm solitary, rhz. **Bas brchg** invag and exvag. **Clm** 15–80 cm, erect or the bases decumbent, terete or weakly compressed; **nd** terete, 0–1 exserted. **Shth** closed for ¹⁄₁₀–¹⁄₅(¼) their length, terete, smooth or sparsely scabrous, glab, bases of bas shth glab, distal shth lengths (1.2)1.5–9(20) times bld lengths; **lig** (1)1.5–4(5) mm, smooth or sparsely to moderately scabrous, apc obtuse to acute; **bld** strongly to gradually rdcd in length distally, 1.5–5 mm wide, flat and moderately thin to folded and moderately thick and firm, abx surfaces smooth, adx surfaces smooth or sparsely to moderately scabrous, primarily over the veins, apc narrowly prow-shaped, flag lf bld (0.4)1–7(10) cm. **Pan** (2.5)4–12(18) cm, erect, usu narrowly lanceoloid, contracted, smt interrupted, infrequently loosely contracted, usu congested, with 25–100 spklt; **nd** with 1–5 br; **br** 1–9 cm, erect to infrequently ascending, rarely spreading, terete to weakly angled, smooth or the angles sparsely to moderately scabrous, with 3–24 spklt. **Spklt** 3.2–7 mm, lengths to 3.5(3.8) times widths, lat compressed; **flt** 2–7; **rchl intnd** smooth, smt sparsely puberulent. **Glm** lanceolate, distinctly keeled, smooth or sparsely scabrous; **lo glm** 3-veined; **cal** usu glab, infrequently webbed, hairs to ¼ the lm length; **lm** 2.5–4.5 mm, lanceolate to narrowly lanceolate, distinctly to weakly keeled, keels and mrgl veins short- to long-villous, lat veins moderately prominent, glab or puberulent, intercostal regions usu glab, infrequently hairy, hairs to 0.3 mm, mrg glab, apc acute or blunt; **pal keels** scabrous, glab or short-villous at midlength, intercostal regions usu glab, smt puberulent to short-villous; **anth** 1.3–2.2 mm. $2n = 56, 56+I, 56–58, 63, 64, 70, 76, 84$, ca. 90, 95+-5, 100, 103.

Poa arida grows mainly on the eastern slope of the Rocky Mountains and in the northern Great Plains, primarily in riparian habitats of varying salinity or alkalinity. It is spreading eastward along heavily salted highway corridors. Reports of its occurrence west of the Continental Divide and in southwestern Texas are mostly attributable to misidentifications of *P. arctica* subsp. *aperta* (p. 118), *P. arctica* subsp. *grayana* (p. 119), and rhizomatous specimens of *P. fendleriana* (p. 122). Its presence in the Intermountain Region requires confirmation. *Poa arida* may reflect past hybridization between *P. secunda* (p. 132) and a species of *Poa* sect. *Poa*.

30. Poa ×limosa Scribn. & T.A. Williams [p. 375, <u>502</u>]

Pl per; densely to loosely tufted or the clm solitary, shortly rhz. **Clm** 20–80 cm, erect or the bases decumbent. **Shth** usu closed for about ⅙ their length; **lig** 1–4 mm, smooth or sparsely scabrous, apc obtuse to acute; **innovation bld** 0.5–2 mm wide; **cauline bld** 0.5–5 mm wide, flat, folded, abx surfaces smooth or scabrous, apc narrowly prow-shaped. **Pan** 5–15 cm, erect, usu contracted, smt interrupted; **br** shorter than 4 cm, erect, angles somewhat scabrous. **Spklt** 4–7 mm, weakly lat compressed; **flt** 2–5; **rchl intnd** smooth. **Lo glm** 3-veined; **cal** glab or webbed, hairs to ¼ the lm length; **lm** 2.5–4.5 mm, narrowly lanceolate, distinctly to weakly keeled, glab throughout or the keels and mrgl veins sparsely long-villous, apc acute; **pal keels** scabrous; **anth** aborted late in development or 1.3–2.2 mm. $2n = 64$.

Poa ×limosa grows at scattered locations in western North America. It prefers wet to moist, often saline or alkaline meadows, primarily in the sagebrush zone. It is probably a hybrid between *P. pratensis* (p. 116) and *P. secunda* subsp. *juncifolia* (p. 133). Vigorous artificial hybrids with this parentage resemble *P. ×limosa*.

8.11 TORREYOCHLOA G.L. Church

Jerrold I. Davis

Pl per; rhz. **Clm** 18–145 cm, usu erect, smt decumbent and rooting at the lo nd; **intnd** hollow. **Shth** open to the base; **aur** absent; **lig** memb; **bld** flat. **Infl** tml pan; **br** scabrous, usu densely scabrid distally; **ped** less than 0.5 mm thick. **Spklt** pedlt, lat compressed to terete, with 2–8 flt; **dis** above the glm and beneath the flt. **Glm** unequal, shorter than the lowest lm, rounded to weakly keeled, memb, veins obscure to prominent, unawned; **lo glm** 1(3)-veined; **up glm** (1)3(5)-veined; **cal** blunt, glab; **lm** memb, rounded to weakly keeled, smt pubescent, particularly proximally, prominently (5)7–9-veined, veins more or less parallel, veins and interveins usu scabridulous, particularly distally, lat veins usu rdcd or absent, apc scabridulous and entire to serrate-erose, unawned; **pal** subequal to the lm, 2-veined; **lod** 2, free, glab, entire or toothed; **anth** usu 3; **ov** usu hairy, smt glab. **Car** shorter than the lm, concealed at maturity, oblong, flattened dorsally, falling free; **hila** oblong, about ⅓ the length of the car. $x = 7$. Named for John Torrey (1796–1873), an American botanist.

Torreyochloa grows in cold, wet, non-saline environments. It includes the two North American species treated below, plus two additional taxa in northeastern Asia (Koyama & Kawano 1964). Although similar to *Glyceria* and *Puccinellia*, *Torreyochloa* is not closely related to either (Church 1952; Soreng et al. 1990). It is distinguished from *Glyceria* by its open leaf sheaths, and from *Puccinellia* by the 7–9 (occasionally 5) prominent, rather than faint, lemma veins.

1. Mature inflorescences linear to narrowly elliptic, 0.3–1 cm wide, 5.5–19 times as long as wide; widest cauline blades 3.4–7.2 mm wide . . . . 1. *T. erecta*
1. Mature inflorescences conic, ovoid, or obovoid, 1–14 cm wide, 1–7.5 times as long as wide; widest cauline blades 3.6–18 mm wide. . . . . 2. *T. pallida*

1. Torreyochloa erecta (Hitchc.) G.L. Church
SPIKED FALSE MANNAGRASS [p. 376, <u>502</u>]

Clm 20–62 cm tall, 1–1.4 mm thick, usu erect. **Lig** of larger cauline lvs 2.6–6.5 mm, truncate or rounded to acute; **widest cauline bld** 3.4–6(7.2) mm wide. **Pan** 5.5–11 cm long, 0.3–1 cm wide, (5.5)6–19 times as long as wide, linear to narrowly elliptic at maturity; **lowermost br** stiff to flexuous, reflexed to erect at maturity. **Spklt** 4.2–6.4 mm, with 4–7 flt. **Lo glm** 0.8–1.6 mm; **up glm** 1–2.1 mm; **lm** 2.3–3.1 mm, obtuse to acute; **anth** 0.6–0.8 mm. $2n = 14$.

Torreyochloa erecta grows at elevations above 2000 m, in the margins of subalpine and alpine lakes and streams of the Sierra Nevada and Cascade ranges.

2. Torreyochloa pallida (Torr.) G.L. Church [p. 376, <u>502</u>]

Clm 18–145 cm tall, 0.6–4.8 mm thick, erect to decumbent, smt matted. **Lig** of larger cauline lvs 2–9 mm, truncate or acute to attenuate; **widest cauline bld** 1.5–18 mm wide. **Pan** (3)5–25 cm long, (1)1.8–16 cm wide, 1–5.75(7.5) times as long as wide, narrowly to widely conic, ovoid, or obovoid at maturity; **lowermost br** stiff to flexuous, reflexed to erect at maturity. **Spklt** 3.6–6.9 mm, with 2–8 flt. **Lo glm** 0.7–2.1 mm; **up glm** 0.9–2.7 mm; **lm** 2–3.6 mm, truncate to acute; **anth** 0.3–1.5 mm. $2n = 14$.

Only one variety of *Torreyochloa pallida* grows in the Intermountain Region. Like the other varieties, it is usually found in swamps, marshes, bogs, and the margins of lakes and streams.

Torreyochloa pallida var. **pauciflora** (J. Presl) J.I. Davis WEAK MANNAGRASS [p. *376*]
Clm 20–145 cm tall, (1.3)1.8–4.8 mm thick, usu erect. **Lig** of larger cauline lvs 3–9 mm; **widest**

cauline **bld** 3.6–18 mm wide. **Pan** (3)5–25 cm long, (1)2–14 cm wide, 1.2–5.75(7.5) times as long as wide, narrowly to widely conic, ovoid, or obovoid. **Spklt** 3.6–6.9 mm, with (3)4–8 flt. **Lo glm** 0.7–1.6 mm; **up glm** 0.9–1.8 mm; **lm** 2.2–3.3 mm, truncate or obtuse to acute; **anth** of lowest flt 0.5–0.7 mm.

Torreyochloa pallida var. *pauciflora* grows in western North America, from sea level to 3500 m. Robust plants from Pacific lowland forests are often taller than 1 m; plants from farther east and higher elevations tend to be shorter.

8.12 CATABROSA P. Beauv.   Mary E. Barkworth

Pl per; not rhz, smt stln. **Clm** 5–70 cm, usu decumbent and rooting at the lo nd; **nd** glab. **Shth** closed; **aur** absent; **lig** memb; **bld** flat. **Infl** open pan, with at least some br longer than 1 cm. **Spklt** pedlt, lat compressed to terete, with (1)2(3) flt; **rchl** glab, prolonged beyond the base of the distal ftl flt, empty or with rdcd flt; **dis** above the glm and beneath the flt. **Glm** unequal, much shorter than the lm, scarious, veinless or the up glm with 1 vein at the base, apc rounded to truncate, unawned; **cal** short, blunt, glab; **lm** glab, conspicuously 3-veined, veins raised, rounded over the midvein, apc rounded to truncate, erose and scarious, unawned; **pal** subequal to the lm, 2-veined; **lod** 2, truncate, irregularly lobed; **anth** 3; **ov** glab. **Car** shorter than the lm, concealed at maturity, fusiform; **hila** ovoid. $x = 5$. Name from the Greek *katabrosis*, 'eating up' or 'corrosion', a reference to the appearance of the lemma apices.

Catabrosa, a genus of two species, grows in marshes and shallow waters of the Northern Hemisphere and South America. It resembles members of the *Meliceae* in its closed leaf sheaths, truncate, scarious lemma apices, and chromosome base number, but lacks the distinctive lodicule morphology of that tribe. Some features support its inclusion in the *Poeae*. The lodicules are similar to those found elsewhere in the *Poeae*, and the closed leaf sheaths are not uncommon there. Chloroplast DNA data also support its placement in the *Poeae* (Soreng et al. 1990). The scarious glumes and prominently 3-veined lemmas are unusual in both the *Poeae* and *Meliceae*, but are also found in *Cutandia* of the *Poeae*. One species is native to North America.

1. **Catabrosa aquatica** (L.) P. Beauv.
BROOKGRASS, WATER WHORLGRASS [p. *376*, 502]
Pl often stln. **Clm** 10–60 cm, glab. **Shth** glab; **lig** 1–8 mm, acute to truncate, erose to subentire; **bld** (1)3–15(20) cm long, 2–13 mm wide. **Pan** 3–35 cm long, (1)2–10(12) cm wide; **nd** distant, with 3 to many, often very unequal br. **Spklt** 1.5–3.5(4) mm, terete to somewhat dorsiventrally compressed, lowest flt sessile, second flt on an

elongate intnd; **rchl** intnd 0.75–1.5 mm. **Lo glm** 0.7–1.3 mm, often not reaching the base of the distal flt; **up glm** 1.2–2.2 mm; **lm** 2–3 mm; **anth** 2–3 mm. $2n = 20, 30$.

Catabrosa aquatica grows in wet meadows and the margins of streams, ponds, and lakes in North America, Argentina, Chile, Europe, and Asia. Although palatable, it is never sufficiently abundant to be important as a forage species. The species is regarded here as being variable, but having no infraspecific taxa.

8.13 SPHENOPHOLIS Scribn.   Thomas F. Daniel

Pl usu per, rarely winter ann; usu csp, smt the clm solitary. **Clm** (5)20–130 cm, lvs evenly distributed. **Shth** open; **aur** absent; **lig** memb, erose; **bld** flat or involute, glab or pubescent. **Infl** pan, open or contracted, nodding to erect; **dis** below the glm, the distal flt smt dis first. **Spklt** pedlt, 2.1–9.5 mm, lat compressed, with 2–3 flt; **rchl** glab or pubescent, prolonged beyond the base of the distal flt as a slender bristle. **Glm** almost equaling the lowest flt, dissimilar in width, memb to subcoriaceous, mrg scarious, apc unawned; **lo glm** narrower than the up glm, 1(3)-veined, strongly keeled, apc acute; **up glm** elliptical to oblanceolate, obovate, or subcucullate, 3(5)-veined, strongly to slightly keeled, apc acuminate, acute, rounded, or truncate; **cal** glab; **lm** herbaceous, not indurate, rounded on the lo back, smooth or partly or wholly scabrous, usu keeled near the apc, 3(5)-veined, veins usu not visible, unawned or awned from just below the apc, awns straight or geniculate; **pal** hyaline, shorter than the lm; **lod** 2, free, memb, glab, toothed or entire; **anth** 3; **ov** glab. **Car** shorter than the lm, concealed at maturity, linear-ellipsoid, glab; **endosperm** liquid. $x = 7$. Name from the Greek *sphen*, 'wedge', and *pholis*, 'scale', in reference to the upper glume.

Sphenopholis includes six species, all of which are native to North America. Its greatest diversity is in the southeastern United States. Interspecific hybridization is known in the genus, but intermediate plants are not frequently encountered.

Glume widths are measured in side view, from the lateral margin to the midvein.

1. Upper glumes subcucullate, the width/length ratio 0.3–0.5; panicles usually erect, often spikelike, the spikelets usually densely arranged .... 1. *S. obtusata*
1. Upper glumes not subcucullate, the width/length ratio 0.23–0.35; panicles usually nodding, not spikelike, the spikelets usually loosely arranged........................... 2. *S. intermedia*

1. **Sphenopholis obtusata** (Michx.) Scribn.
PRAIRIE WEDGEGRASS [p. 377, <u>502</u>]

Clm (9)20–130 cm. Shth glab or hairy, smt scabridulous; lig (1)1.5–2.5 mm, erose-ciliate, more or less lacerate; bld 5–14 cm long, (1)2–8 mm wide, usu flat, rarely slightly involute, scabrous or pubescent. Pan (2)5–15(25) cm long, 0.5–2 cm wide, usu erect, often spikelike, spklt usu densely arranged. Spklt 2.2–3.6 mm. Lo glm less than ⅓ as wide as the up glm; up glm 1.5–2.5 mm, subcucullate, width/length ratio 0.3–0.5, apc rounded to truncate; lowest lm 1.9–2.8 mm, usu scabridulous distally; distal lm usu smooth on the sides, occ scabrous, unawned; anth 0.2–1 mm. 2n = 14.

Sphenopholis obtusata grows in prairies, marshes, dunes, forests, and waste places, at 0–2500 m. Its range extends from British Columbia to New Brunswick, through most of the United States, to southern Mexico and the Caribbean. The distal lemmas of *S. obtusata* are occasionally somewhat scabrous.

2. **Sphenopholis intermedia** (Rydb.) Rydb.
SLENDER WEDGEGRASS [p. 377, <u>502</u>]

Clm (5)30–120 cm. Shth smooth or scabridulous, smt pubescent; lig 1.5–2.5 mm, erose-ciliate, often lacerate; bld 8–15 cm long, (1)2–6 mm wide, flat to slightly involute. Pan (2)7–20 cm long, (0.5)1–3 cm wide, usu nodding, not spikelike, spklt usu loosely arranged. Spklt 2.1–4 mm. Lo glm less than ⅓ as wide as the up glm; up glm 1.9–2.9 mm, oblanceolate to obovate, not subcucullate, width/length ratio 0.23–0.35, apc acute, rounded, or subtruncate; lowest lm 2.1–3 mm, smooth or scabridulous; distal lm usu smooth on the sides, rarely scabridulous near the apc, unawned; anth 0.2–0.8 mm. 2n = 14.

Sphenopholis intermedia grows at 0–2500 m in wet to damp sites, sites that dry out after the growing season, and sites with clay soils that retain moisture. It is found in forests, meadows, and waste places throughout most of North America other than the high arctic. It differs from *Koeleria macrantha* (p. 163), with which it is sometimes confused, in its more open panicles and in having spikelets that disarticulate below the glumes.

8.14 **DESCHAMPSIA** P. Beauv.

Mary E. Barkworth

Pl us u per, smt ann; csp or tufted. Clm 5–140 cm, hollow, erect. Lvs usu mainly bas, often forming a dense tuft; shth open; aur absent; lig memb, decurrent, rounded to acuminate; bld often all or almost all tightly rolled or folded and some flat, smt most flat, others rolled or folded. Infl tml pan, open or contracted; dis above the glm, beneath the flt. Spklt 3–9 mm, with 2(3) flt in all or almost all spklt, flt usu bisex, smt viviparous; rchl hairy, usu prolonged more than 0.5 mm beyond the base of the distal flt, smt terminating in a highly rdcd flt. Glm subequal to unequal, usu exceeding the adjacent flt, often exceeding all flt, 1- or 3-veined, acute to acuminate; cal antrorsely strigose; lm obscurely (3)5–7-veined, rounded over the back, apc truncate-erose to 2–4-toothed, awned, awns usu attached on the lo ½ of the lm, occ subapical, straight to strongly geniculate, slightly to strongly twisted proximally, straight distally; pal shorter than the lm, 2-keeled, keels often scabrous; lod 2, lanceolate to ovate-lanceolate, usu entire; anth 3; ov glab; sty 2. Car oblong; emb about ¼ the length of the car. x = 7. Named for Louise Auguste Deschamps (1765–1842), a French naturalist.

Deschampsia includes 20–40 species. It is best represented in the Americas and Eurasia, but it grows in cool, damp habitats throughout the world. Five species are native to the Intermountain Region.

Lemma length, awn attachment, and awn length should be examined on the lower florets within the spikelets. The upper florets often have shorter lemmas, and shorter awns that are attached higher on the back than those of the lower florets.

1. Plants annual; awns strongly geniculate .... 5. *D. danthonioides*
1. Plants perennial; awns straight to strongly geniculate.
   2. Glumes mostly green, apices purple; panicles narrowly elongate, 0.5–1.5(2) cm wide, appearing greenish.. 4. *D. elongata*
   2. Glumes purplish proximally, sometimes over more than ½ their surface, whitish to golden distally; panicles usually pyramidal or ovate, sometimes narrowly elongate, 0.5–30 cm wide, appearing bronze to dark purple (*D. cespitosa* complex).

3. Spikelets strongly imbricate, often rather densely clustered on the ends of the branches, sometimes evenly distributed on the branches; glumes and lemmas dark purple proximally for over more than ½ their surface; lemmas 2.2–4 mm long
.... 3. *D. brevifolia*
3. Spikelets usually not or only moderately imbricate, not in dense clusters at the ends of the branches; glumes usually purple over less than ½ their surface, often with a green base, a distal purple band, and pale apices; lemmas 2–5(7)mm long.
    4. Basal blades with 5–11 ribs, usually most or all ribs scabridulous or scabrous, outer ribs often more strongly so, sometimes the ribs only papillose or puberulent, usually at least some blades flat and 1.5–3.5 mm wide, the majority folded or rolled and 0.5–1 mm in diameter; lower glumes often scabridulous distally over the midvein; lower panicle branches often scabridulous or scabrous, sometimes smooth.. 1. *D. cespitosa*
    4. Basal blades with 3–5 ribs, ribs usually smooth or papillose, sometimes puberulent or the outer ribs scabridulous, all blades of the current year usually strongly involute and hairlike, 0.3–0.5(0.8) in diameter; lower glumes smooth over the midvein; lower panicle branches usually smooth, sometimes sparsely scabridulous .... 2. *D. sukatschewii*

1. Deschampsia cespitosa (L.) P. Beauv. [p. 377, 502]

Pl per; loosely to tightly ces. **Clm** (7) 35–150 cm, erect, not rooting at the lo nd. **Lvs** mostly bas, smt forming a dense 10–35 cm tuft; **shth** glab; **lig** 2–13 mm, scarious, decurrent, obtuse to acute; **bld** 5–30 cm long, usu at least some flat and 1–4 mm wide, the remainder folded or rolled and 0.5–1 mm in diameter, adx surfaces with 5–11 prominent ribs, ribs usu all papillose, scabridulous, or scabrous, smt puberulent, outer ribs smt more strongly so than the inner ribs. **Pan** 8–30(40) cm, 4–30 cm wide, usu open and pyramidal, smt contracted and ovate; **br** straight to slightly flexuous, usu strongly divergent, smt strongly ascending, lo br often scabridulous or scabrous, particularly distally, with not or only moderately imbricate spklt. **Spklt** 2.5–7.6 mm, ovate to V-shaped, lat compressed, usu bisex, smt viviparous, bisex spklt usu with 2(3) flt, rarely with 1. **Glm** lanceolate, acute; **lo glm** 2.7–7 mm, entire, 1–3-veined, midvein smt scabridulous, at least distally; **up glm** 2–7.5 mm, 1–3-veined, lanceolate, midvein smooth or wholly or partly scabridulous; **cal hairs** 0.2–2.3 mm; **lm** 2–5(7) mm, smooth, shiny, glab, usu purple over less than ½ their surface, purple or green proximally, if green, often with a purple band about midlength, usu green or pale distally, usu awned, awns (0.5)1–8 mm, attached from near the base to about midlength, straight or geniculate, smt exceeding the glm; **anth** 1.5–3 mm. **Car** 0.5–1 mm. $2n$ = 18, 24, 25, 26–28, about 39, 52. The voucher specimens for these counts have not been examined.

Deschampsia cespitosa is circumboreal in the Northern Hemisphere, and also grows in New Zealand and Australia. It is an attractive taxon that grows in wet meadows and bogs, and along streams and lakes, from sea level to over 3000 m in cool-temperate, but not arctic, habitats.

There are widely varying opinions concerning the taxonomic treatment of *Deschampsia cespitosa*. Tsvelev, Aiken, Murray, and Elven (per Murray, pers. com. 2005) recommend a narrow circumscription, and consider *D. cespitosa* to be introduced and mostly ruderal in regions other than Europe and western Siberia. Chiapella and Probatova (2003) adopted a much broader interpretation of *D. cespitosa*, treating *D. sukatschewii* and *D. brevifolia*, for example, as subspecies. There have been no interdisplinary, global studies of the complex. The circumscription adopted here is narrower than has been customary in North America. Some of the distribution records shown may reflect the broad interpretation of the species. One subspecies is present in the Intermountain Region. The other, subsp. *beringensis*, is restricted to the Pacific coast.

Deschampsia cespitosa (L.) P. Beauv. subsp. cespitosa TUFTED HAIRGRASS [p. 377]

Pl densely ces, not glaucous. **Clm** (7)35–150 cm. **Lig** 2–8 mm; **bld** 5–25 cm long, 1.5–3.5 mm wide when flat. **Pan** 8–30 cm long, 4–30 cm wide, open, nodding, pyramidal; **br**, both pri and sec, usu divergent, usu sparsely to moderately scabridulous or scabrous, smt smooth. **Spklt** 2.5–7 mm, not to slightly imbricate. **Glm** subequal to the distal flt, lengths often less than 5 times widths; **lo glm** 2.5–5 mm, midveins smooth or scabridulous distally; **up glm** 2–6 mm; **lm** 2–4 mm, purple and/or green proximally, green to gold distally, the purple portion usu less than ½ the surface area, awns 1–8 mm, usu attached near the base, smt attached near midlength, straight or geniculate, exceeded by or exceeding the distal flt; **anth** 1.5–2 mm.

Deschampsia cespitosa subsp. *cespitosa* is treated here as a circumboreal taxon that is most prevalent in boreal and temperate North America, growing at 0–3000 m; many reports from arctic and alpine North America refer to what are treated here as *D. sukatschewii* or *D. brevifolia*. Even with this narrower interpretation, *D. cespitosa* is highly polymorphic.

Plants with long awns are more prevalent in western North America but, within that region, do not appear to show any geographic or ecological preference (Lawrence 1945).

Many cultivars of *Deschampsia cespitosa* subsp. *cespitosa* have been developed. At one time, the most frequently cultivated plants were distinguished by their combination of large (20–40 cm) panicles and small (2.5–4 mm) spikelets, and were called *D. cespitosa* var. *parviflora* (Thuill.) Coss. & Germ. or *D. cespitosa* subsp. *parviflora* (Thuill.) K. Richt. They are included in subsp. *cespitosa*.

2. Deschampsia sukatschewii (Popl.) Roshev.
[p. *378*, <u>502</u>]

Pl per; usu densely ces. **Clm** 5–70 cm, erect or strongly geniculate at the first nd, glab. **Lvs** mostly bas, smt forming a dense, mosslike tuft 5–20 cm in diameter; **shth** smooth, glab; **lig** 1.5–8 mm, acute; **bld** 0.5–8 cm long, usu strongly rolled and 0.5–1.3 mm in diameter, rarely flat and to 1.5(2) mm wide, abx surfaces smooth, adx surfaces with 3–5(6) ribs, ribs smooth, the outer ribs smt scabrous. **Pan** 3.5–17 cm long, 1.5–9 cm wide, usu open and pyramidal, smt closed and ovate; **br** 0.5–6 cm, spreading to reflexed, flexuous, smooth. **Spklt** 3.5–5.2 mm, shiny, purplish, with 2(3) flt. **Glm** lanceolate, smt purplish over the proximal ½, acute to acuminate; **lo glm** 2.7–4.8 mm, 0.8–0.9 times the length of the spklt, 1–3-veined, veins smooth; **up glm** 3–5 mm, equaling or exceeding the lowest flt, 1–5-veined; **cal hairs** 0.3–1 mm; **lm** 2–4 mm, smooth, shiny, glab, smt purplish distally, apc rounded or truncate, erose, awns 0.8–2.5 mm, arising at or below midlength, straight, slender, only slightly or not exserted; **anth** 0.7–2.5 mm. $2n = 26, 28, 36$, ca. 39.

Deschampsia sukatschewii is a circumboreal species that extends from northern Russia through Alaska, northern Canada, and Greenland to Svalbard, and southward in the Rocky Mountains to Nevada and Utah. It ranges from short plants that form dense, mossy tufts on the Arctic coast to larger plants in subalpine and alpine habitats of the Rocky Mountains that have frequently been included in *D. cespitosa*.

3. Deschampsia brevifolia R. Br. [p. *378*, <u>502</u>]

Pl per; ces, not glaucous. **Clm** 5–55 cm, erect, glab. **Lvs** often forming a bas tuft; **shth** glab; **lig** 1–4.5 mm, acute or acuminate, entire; **bld** 2–12 (16) cm long, usu 0.3–0.8 mm in diameter, folded or convolute, 0.5–2 mm wide when flat, abx surfaces glab, adx surfaces glab or sparsely hirtellous, smt scabrous, bld of flag lvs 0.8–3 cm. **Pan** 1.5–10(12) cm long, 0.5–2(11) cm wide, usu dense, oblong-ovate to narrowly cylindrical; **br** 1–3.6(6) cm, straight, usu stiff, erect to ascending, usu smooth or almost so, scabrules separated by 0.2+ mm, spklt-bearing to near the base. **Spklt**

2.3–6 mm, ovate to obovate, with 2(3) flt. **Glm** subequal to equal, 2.5–5.6 mm, purplish over more than ½ their surface, lanceolate, smooth, acuminate or acute; **lo glm** 1-veined, smooth; **up glm** exceeding to exceeded by the lowest flt, 3-veined; **cal hairs** 0.2–2 mm; **lm** 2.2–4 mm, oblong or lanceolate, smooth, shiny, glab, awns (0.2)0.7–4 mm, usu equaling or exceeding the lm, straight or weakly geniculate, usu attached from near the base to midlength, occ connate almost their full length; **anth** 1.2–2.5 mm. $2n = 26, 27, 28$, about 50, 52.

Deschampsia brevifolia is a circumboreal taxon that grows in wet places in the tundra, often in disturbed soils associated with riverbanks, frost-heaving, etc. It is interpreted here as extending southward through the Rocky Mountains to Colorado, where it grows at elevations up to 4300 m. It is to be expected from high elevations in British Columbia and Alberta; specimens currently identified as *D. cespitosa*, in which *D. brevifolia* is often included as a subspecies, need to be examined.

In its typical appearance, *Deschampsia brevifolia* is quite distinctive because of its dark, narrow panicles. Culm height can vary substantially from year to year, probably in response to the environment.

4. Deschampsia elongata (Hook.) Munro
SLENDER HAIRGRASS [p. *378*, <u>502</u>]

Pl per; densely ces. **Clm** (10)30–120 cm. **Lvs** smt forming a bas tuft; **shth** glab; **lig** 2.5–8(9) mm, acute to acuminate; **bld** 7–30 cm long, 0.2–2 mm wide, usu involute. **Pan** 5–30(35) cm long, 0.5–1.5(2) cm wide, erect or nodding; **br** erect to ascending. **Spklt** 3–6.7 mm, bisex, narrowly V-shaped, appressed to the br. **Glm** equaling or exceeding the flt, narrowly lanceolate, usu pale green, smt purple-tipped, 3-veined, acuminate; **lo glm** (3)3.2–5.5(6.7) mm; **up glm** (3)3.1–5.4(6) mm; **cal hairs** 0.3–1.15 mm; **lm** 1.7–4.3 mm, smooth, shiny, glab, apc weakly toothed or erose, awns 1.5–5.5(6) mm, straight to slightly geniculate, attached from slightly below to slightly above the mid of the lm, exceeding the flt by 1–2.5 mm; **anth** 0.3–0.5(0.7) mm. $2n = 26$.

Deschampsia elongata grows in moist to wet habitats, from near sea level to alpine elevations, from Alaska and the Yukon south to northern Mexico and east to Montana, Wyoming, and Arizona.

5. Deschampsia danthonioides (Trin.) Munro
ANNUAL HAIRGRASS [p. *378*, <u>502</u>]

Pl ann; tufted. **Clm** 10–40(70) cm, erect. **Lvs** not forming a bas tuft; **lig** (0.5)2–3(4.7) mm, acute to acuminate, entire; **bld** 0.3–1.5 mm wide, involute or flat. **Pan** 5–15(25) cm long, 2–8 cm wide, contracted to open, erect; **br** with the spklt confined to the distal portion. **Spklt** 4–9 mm, bisex, narrowly V-shaped, usu pale green. **Glm** exceeding the distal flt, glab to scabridulous, 3-veined; **lo glm** 4–9 mm; **up glm** 3.5–8.5 mm; **cal hairs** 0.4–1.6 mm; **lm** 1.5–3 mm, smooth, shiny, glab, pale green or purplish, apc blunt, erose to 4–toothed, ciliate, awns 4–9 mm, attached from near the base to about the mid of the lm, strongly geniculate, geniculation above the lm apc, distal segment 1.5–5 mm; **anth** 0.3–0.5 mm. $2n = 26$.

Deschampsia danthonioides grows in temperate and cool-temperate regions, usually in open, wet to dry habitats and often in disturbed ground. Its primary range extends from southern British Columbia, through Washington and Idaho, to Baja California, Mexico.

8.15 AGROSTIS L.

M.J. Harvey

Pl usu per; usu csp, smt rhz or stln. **Clm** (3)5–120 cm, usu erect. **Shth** open, usu smooth and glab, smt scabrous to scabridulous, rarely hairy; **col** not strongly developed; **aur** absent; **lig** memb, smooth or scabridulous dorsally, apc truncate, obtuse, rounded, or acute, usu erose to lacerate, the lacerations smt obscuring the shape, or entire; **bld** flat, folded, or involute, usu smooth and glab, smt scabridulous, adx surfaces somewhat ridged. **Infl** tml pan, narrowly cylindrical and dense to open and diffuse; **br** usu in whorls, usu more or less scabrous, rarely smooth, some br longer than 1 cm; **sec pan** smt present in the lf axils. **Spklt** 1.2–7 mm, pedlt, lat compressed, lanceolate to narrowly oblong or ovate, with 1(2) flt; **rchl** not prolonged beyond the base of the flt(s); **dis** above the glm, beneath the flt, smt initially at the pan base. **Glm** (1)1.3–2(4) times longer than the lm, 1(3)-veined, glab, usu mostly smooth, vein(s) often scabrous to scabridulous, backs keeled or rounded, apc acute to acuminate or awn-tipped; **lo glm** usu 0.1–0.3 mm longer than the up glm, rarely equal; **cal** poorly developed, blunt, glab or hairy, hairs to about ½ as long as the lm; **lm** thinly memb to hyaline, usu smooth and glab, smt scabridulous, occ pubescent, rarely warty-tuberculate, 3–5-veined, veins not convergent, smt excurrent as 2–5 teeth, apc acute to obtuse or truncate, smt erose, unawned or awned, smt varying within an infl, awns arising from near the lm bases to near the apc, usu geniculate, smt straight; **pal** absent, or minute to subequal to the lm, usu thin, veins not or only weakly developed; **lod** 2, free; **anth** (1)3, 0.1–2 mm, not penicillate; **sty** 2, free to the base, white; **ov** glab. **Car** with a hard, soft, or liquid endosperm, the latter resulting from the substitution of lipids for starch. $x = 7$. Name from the Greek *agros*, 'pasture' or 'green fodder'.

Agrostis in the older, broad sense is a genus comprised of species with the spikelets reduced to single florets. As such, it is found in all inhabited continents, is presumably ancient, with many of its 150–200 species only distantly related to each other. The shortage of clear-cut morphological features has hindered its subdivision into more natural units. This treatment follows Soreng (2003) in placing *A. aequivalvis* (Trin.) Trin. and *A. humilis* —together with several Central and South American species, including *A. sesquiflora* E. Desv.—in the genus *Podagrostis*.

Agrostis usually differs from *Podagrostis* in having no, or very reduced, paleas, and in rachillas that are not prolonged beyond the base of the floret. *Agrostis* is sometimes confused with *Calamagrostis* (p. 152) or *Polypogon* (p. 144). There is no single character that distinguishes all species of *Agrostis* from those of *Calamagrostis*. In general, *Agrostis* has smaller plants with smaller, less substantial lemmas and paleas than *Calamagrostis*, and tends to occupy drier habitats. It differs from *Polypogon* in having spikelets that disarticulate above the glumes.

Species of *Agrostis* growing in the temperate regions of the Northern Hemisphere and on tropical mountains are mostly perennial. The annual species predominate in warmer climates, such as the Mediterranean and the Southern Hemisphere. Of the 9 species known from the Intermountain Region, 6 are native and 3 are generally considered to be introductions.

Some species of *Agrostis* make a modest contribution to forage, a few are agricultural weeds, and some are excellent lawn grasses in cool climates. Most North American native species are narrow habitat specialists, with many being western endemics. The introduced species are all widely distributed in temperate regions of the world.

Unusual specimens of *Agrostis* with elongate or leafy spikelets are caused by infection with the nematode *Anguillina agrostis*. Other pathogens may cause stunting.

Species with awns on the lemmas frequently exhibit a developmental gradient within the inflorescence. Upper florets may possess a well-developed geniculate awn inserted at the base or on the lower half of the lemma; mid-inflorescence spikelets may have a shorter, possibly non-geniculate awn inserted high on the lemma, while basal spikelets may possess only a terminal bristle on the lemma. When using the key, it is best to examine spikelets from the upper parts of an inflorescence.

1. At least some panicle branches spikelet-bearing to near the base; lower panicle branches branching at midlength or below; panicles usually appearing somewhat dense and contracted at maturity, sometimes open.
    2. Leaves forming a dense basal tuft; panicles 1–6 cm long; plants of subalpine and alpine habitats . . . . 9. *A. variabilis*
    2. Leaves mostly cauline, not forming a dense basal tuft; panicles (3)4-30 cm long; plants usually of lower elevation habitats, sometimes subalpine.

3. Anthers 0.3–0.6 mm long; lower branches 1–2(4) cm long; plants not stoloniferous,
   sometimes rhizomatous . . . . 7. *A. exarata*
3. Anthers 0.7–1.8 mm long; lower branches 2–6 cm long; plants stoloniferous or
   rhizomatous.
   4. Plants rhizomatous, not stoloniferous; paleas absent or to 0.2 mm long, about
      ⅕ as long as the lemmas; glumes scabrous over the whole of the midvein,
      glume bodies often also sparsely scabrous . . . . 8. *A. pallens*
   4. Plants stoloniferous or rhizomatous; paleas 0.7–1.4 mm long, about ½ as long as
      the lemmas; midvein of the glumes smooth proximally, smooth, scabridulous, or
      scabrous distally.
      5. Plants stoloniferous, not rhizomatous; panicles contracted at maturity, open
         only during anthesis . . . . 3. *A. stolonifera*
      5. Plants rhizomatous, not stoloniferous; panicles open at maturity . . . . 2. *A. gigantea* (in part)
1. Panicle branches with spikelets confined to the distal half; lower panicle branches branching only at or beyond midlength; panicles appearing open or diffuse at maturity.
   6. Plants rhizomatous or stoloniferous; leaves basal and cauline or mostly cauline; paleas
      about ½ as long as the lemmas.
      7. Ligules shorter than wide, 0.3–2 mm long; panicle branches smooth or
         scabridulous; cauline nodes 2–5 . . . . 1. *A. capillaris*
      7. Ligules longer than wide, lower ligules 1–4.5 mm long, upper ligules 2–7 mm
         long; panicle branches scabrous; cauline nodes 4–7 . . . . 2. *A. gigantea* (in part)
   6. Plants neither rhizomatous nor stoloniferous; leaves mostly basal; paleas absent or up
      to ⅕ the length of the lemmas.
      8. Leaf blades 10–30 cm long, (1)2–4 mm wide, flat, usually at least some wider than
         2 mm; anthers 1–1.6 mm long . . . . 6. *A. oregonensis*
      8. Leaf blades 4–14 cm long, 0.5–2 mm wide, flat or involute; anthers 0.3–1.4 mm long.
         9. Panicle branches 4–12 cm long, scabrous; anthers 0.9–1.4 mm long . . . . 4. *A. scabra*
         9. Panicle branches 1–4 cm long, scabridulous; anthers 0.3–0.6 mm long . . . . 5. *A. idahoensis*

1. **Agrostis capillaris** L. Browntop, Rhode
Island Bent, Colonial Bent [p. 379, 502]
Pl per; **rhz** or **stln**, rhz or stln to 5 cm. **Clm** 10–75
cm, erect or geniculate, with 2–5 nd. **Lvs** bas
and cauline; **shth** smooth; **lig** 0.3–2 mm, shorter
than wide, dorsal surfaces usu scabridulous,
smt smooth, apc truncate to rounded, erose-
ciliolate, smt lacerate; **bld** 3–10 cm long, 1–5
mm wide, flat. **Pan** 3–20 cm long, less than ½
the length of the clm, (1)2–12 cm wide, stiffly
erect, widely ovate, open, exserted from the up
shth at maturity, lowest nd with (2)3–9(13) br; **br**
smooth or scabridulous, spreading during and
after anthesis, spklt usu confined to the distal
½, lo br 1.5–7 cm; **ped** 0.4–3.3 mm, adjacent ped
divergent. **Spklt** lanceolate or oblong, purplish
brown to greenish. **Glm** subequal, 1.7–3 mm,
1-veined, acute; **lo glm** scabridulous over the
midvein towards the apc; **up glm** scabridulous
or smooth over the midvein; **cal** glab, or with
a few hairs to 0.1 mm; **lm** 1.2–2.5 mm, smooth,
glab, opaque to translucent, 3(5)-veined, veins
typically prominent, apc obtuse to acute, usu
entire, smt the veins excurrent to 0.5 mm, usu
unawned, rarely awned, smt varying within
a pan, awns to 2 mm, mid-dorsal, straight or
geniculate; **pal** 0.6–1.2(1.4) mm, typically at least

½ the length of the lm, veins visible; **anth** 3,
0.8–1.3 mm. **Car** 0.8–1.5 mm; **endosperm** solid.
$2n = 28$.

Agrostis capillaris grows along roadsides and in disturbed
areas. It was introduced from Europe, and is now well
established in western and eastern North America. It is often
used for fine-leaved lawns; commercial seed sold as *Agrostis
tenuis* 'Highland' usually contains *A. capillaris*.

Agrostis capillaris differs from *A. gigantea* (see next) in its
short ligules, especially on the vegetative shoots, and the open
panicles that lack spikelets near the base of the branches.

2. **Agrostis gigantea** Roth Redtop [p. 379, 502]
Pl per; **rhz**, rhz to 25 cm, not stln. **Clm** 20–120 cm,
erect, smt geniculate at the base, smt rooting at
the lo nd, with 4–7 nd. **Lvs** mostly cauline; **shth**
smooth or sparsely scabridulous; **lig** longer than
wide, dorsal surfaces usu scabrous, smt smooth,
apc rounded to truncate, erose to lacerate, bas lig
1–4.5 mm, up lig 2–7 mm; **bld** 4–10 cm long, 3–8
mm wide, flat. **Pan** 8–25(30) cm long, less than ½
the length of the clm, (1.5)3–15 cm wide, erect,
open, broadly ovate, exserted from the up shth at
maturity, lowest nd with (1)3–8 br; **br** scabrous,
spreading during and after anthesis, usu some br
spklt-bearing to the base, lo br 4–9 cm, usu with
many shorter sec br resulting in crowding of the
spklt, spklt restricted to the distal ½ of the br and

not crowded in shade pl; **ped** 0.3–3.4(4.2) mm. **Spklt** narrowly ovate to lanceolate, green and slightly to strongly suffused with purple. **Glm** subequal, 1.7–3.2 mm, lanceolate, 1-veined, acute to apiculate; **lo glm** scabrous on the distal ½ of the midvein; **up glm** scabridulous on the distal ½ of the midvein; **cal hairs** to 0.5 mm, sparse; **lm** 1.5–2.2 mm, opaque to translucent, smooth, 3–5-veined, veins usu obscure, smt prominent throughout or distally, often excurrent to 0.2 mm, apc usu acute, smt obtuse or truncate, usu unawned, rarely with a 0.4–1.5(3) mm straight awn arising from near the apc to near the base; **pal** 0.7–1.4 mm, about ½ the length of the lm, veins visible; **anth** 3, 1–1.4 mm. **Car** 1–1.5 mm; **endosperm** solid. 2*n* = 42.

Agrostis gigantea grows in fields, roadsides, ditches, and other disturbed habitats, mostly at lower elevations. It is a serious agricultural weed, as well as a valuable soil stabilizer. It is more heat-tolerant than most species of *Agrostis*. In North America, its range extends from the subarctic to Mexico; it is considered to be native to Eurasia.

Agrostis gigantea has been confused with *A. stolonifera* (see next), from which it differs in having rhizomes and a more open panicle. *Agrostis stolonifera* has elongated leafy stolons, mainly all above the surface, that root at the nodes, and the panicles are condensed and often less strongly pigmented than in *A. gigantea*. Its distribution tends to be more northern and coastal where ditches and pond margins are common habitats, and its stolons enable it to form loose mats. *Agrostis gigantea* is also adapted to a more extreme climate—hot summers/cold winters and drought—than *A. stolonifera*. It resembles *A. capillaris* (see previous), but it differs in its longer ligules and its less open panicles with spikelets near the base of the branches

When *Agrostis gigantea* grows in damp hollows under trees it becomes more like *A. stolonifera*, particularly when the inflorescence is young, not expanded, and pale. If the rootstock is not collected, identification is a major problem.

3. **Agrostis stolonifera** L. Creeping Bent [p. *380*, <u>502</u>]

Pl per; stln, stln 5–100+ cm, rooting at the nd, often forming a dense mat, without rhz. **Clm** (8)15–60 cm, erect from a geniculate base, smt rooting at the lo nd, with (2)4–7 nd. **Lvs** mostly cauline; **shth** smooth; **lig** longer than wide, dorsal surfaces usu scabrous, rarely smooth, apc usu rounded, acute to truncate, erose to lacerate, bas lig 0.7–4 mm, up lig 3–7.5 mm; **bld** 2–10 cm long, 2–6 mm wide, flat. **Pan** (3)4–20 cm long, less than ½ the length of the clm, 0.5–3(6) cm wide, narrowly contracted, dense, oblong to lanceolate, exserted from the shth at maturity, lowest nd with 1–7 br; **br** scabrous, ascending to appressed, except briefly spreading during anthesis, usu some br at each nd spklt-bearing to the base, lo br 2–6 cm; **ped** 0.3–3.3 mm. **Spklt** lanceolate, green

and slightly to strongly suffused with purple. **Glm** subequal to unequal, 1.6–3 mm, lanceolate, 1-veined, smt scabridulous distally, at least on the midvein, acute to acuminate or apiculate; **cal hairs** to 0.5 mm, sparse; **lm** 1.4–2 mm, opaque to translucent, smooth, 5-veined, veins obscure or prominent distally, apc acute to obtuse, entire or the veins excurrent to about 0.1 mm, usu unawned, rarely with a subapical straight awn to about 1 mm; **pal** 0.7–1.4 mm, veins visible; **anth** 3, 0.9–1.4 mm. **Car** 0.9–1.3 mm; **endosperm** solid. 2*n* = 28, 35, 42.

Agrostis stolonifera grows in areas that are often temporarily flooded, such as lakesides, marshes, salt marshes, lawns, and damp fields, as well as moist meadows, forest openings, and along streams. It will also colonize disturbed sites such as ditches, clearcuts, and overgrazed pastures. Its North American range extends from the subarctic into Mexico, mostly at low to middle elevations.

Agrostis stolonifera has been confused with *A. gigantea* (see previous). It is considered to be Eurasian, but may be native in northern North America. The names *A. palustris* Huds. and *A. maritima* Lam. have been applied to plants with longer stolons; all forms intergrade. A hybrid between *A. stolonifera* and *Polypogon monspeliensis*, ×*Agropogon lutosus* (p. 146), has been reported in locations where both parents occur. It differs from *A. stolonifera* in having awned glumes and lemmas.

4. **Agrostis scabra** Willd. Ticklegrass [p. *380*, <u>502</u>]

Pl per or ann; ces, not rhz or stln. **Clm** (7.5) 15–90 cm, erect, nd usu 1–3. **Lvs** mostly bas, bas lvs usu persistent; **shth** usu smooth, smt scabridulous; **lig** 0.7–5 mm, dorsal surfaces scabrous, apc usu rounded, smt truncate or acute, erose-ciliolate, smt lacerate; **bld** 4–14 cm long, 1–2 mm wide, bas bld mostly involute, cauline bld mostly flat. **Pan** (4)8–25(50) cm long, 0.5–20 cm wide, broadly ovate, often nearly as wide as long, diffuse, exserted from the up shth, lowest nd with (1)2–7(12) br, the whole pan often detaching at the base at maturity and forming a tumbleweed; **br** scabrous, capillary, flexible, wide-spreading, readily visible, brchg beyond midlength, spklt somewhat distant, not crowded, lo br 4–12 cm; **ped** 0.4–9.6 mm. **Spklt** lanceolate, greenish purple, frequently purple at maturity. **Glm** unequal, 1.8–3.4 mm, lanceolate, 1-veined, keels scabrous at least towards the apc, apc acuminate; **cal hairs** to 0.2 mm, sparse; **lm** 1.4–2 mm, scabrous to scabridulous or smooth, translucent to opaque, 5-veined, veins prominent, apc acute to obtuse, usu entire, smt minutely toothed, unawned or awned from below midlength, awns 0.2–3 mm, exceeding the lm apc by up to 2.5 mm, geniculate or straight, persistent; **pal** absent or to

0.2 mm; **anth** 3, 0.4–0.8 mm, usu shed at anthesis. **Car** 0.9–1.4 mm; **endosperm** liquid. $2n = 42$.

Agrostis scabra grows in a wide variety of habitats, including grasslands, meadows, shrublands, woodlands, marshes, and stream and lake margins, as well as disturbed sites such as roadsides, ditches, and abandoned pastures. It occurs throughout much of North America, extending south into Mexico. It is also native to the Pacific coast from Kamchatka to Japan and Korea, and has been introduced elsewhere.

Plants in the *Agrostis scabra* aggregate are variable. Awned and unawned plants often occur together, the difference presumably being caused by a single gene. At least three groups may be distinguished within the species as treated here: widespread, lowland, rather weedy plants capable of producing very large panicles that have been introduced into the southern United States; smaller, short-leaved, slow-growing plants that are widespread in rocks and screes of the Rockies, the Appalachians, and much of Alaska, Canada, and Greenland; and luxuriant, broad-leaved plants that are characteristically found in sheltered, frost-free canyons of the southwestern United States. The second group has sometimes been called *A. scabra* var. *geminata* (Trin.) Swallen or *A. geminata* Trin.

Tercek et al. (2003) found that annual forms of *Agrostis scabra* with inflated upper sheaths and open panicles that were collected around hot springs in western North America were molecularly, and in some respects morphologically, more similar to plants identified as hot spring endemics such as *A. rossiae* Vasey and *A. pauzhetica* Prob., than they were to neighboring perennial plants of *A. scabra* that did not have inflated leaf sheaths. They differed, however, in having open, rather than contracted, panicles.

Agrostis scabra is often confused with *A. idahoensis* (see next).

5. **Agrostis idahoensis** Nash IDAHO REDTOP [p. *381*, 502]

Pl per; ces, not rhz or stln. **Clm** 8–40 cm, slender, erect, with 2–5 nd. **Lvs** mostly bas; **shth** usu smooth, smt scabridulous, not inflated; **lig** (0.7)1–3.8 mm, dorsal surfaces scabridulous, apc rounded to truncate, rarely acute, erose to lacerate; **bld** 1–7 cm long, 0.5–2 mm wide, flat, becoming involute. **Pan** 3–13 cm long, 1–6(8) cm wide, lanceolate to ovate, diffuse, exserted from the up shth at maturity, lowest nd with 1–6(10) br; **br** scabridulous, fairly stiff, more or less ascending, brchg at or above midlength, spklt not crowded, frequently solitary, lo br 1–4 cm; **ped** 0.5–6.4 mm. **Spklt** lanceolate to narrowly ovate, purplish. **Glm** subequal, 1.5–2.5 mm, 1-veined, usu scabrous to scabridulous, up glm smt smooth, apc acute to acuminate; **cal hairs** to 0.3 mm, sparse; **lm** 1.2–2.2 mm, usu smooth, smt scabridulous, translucent to opaque, 5-veined, veins usu prominent at least distally, smt obscure, apc acute to obtuse, entire, unawned; **pal** absent, or to 0.2 mm and thin; **anth** 3, 0.3–0.6 mm. **Car** 1–1.3 mm. $2n = 28$.

Agrostis idahoensis grows in western North America, from British Columbia to California and New Mexico, in alpine and subalpine meadows along wet seepage areas and bogs, and in wet openings with *Sphagnum* in coniferous forests. It is often confused with dwarf forms of *A. scabra* (p. 142), which tend to grow in better-drained habitats.

6. **Agrostis oregonensis** Vasey OREGON REDTOP [p. *381*, 502]

Pl per; ces, not rhz or stln. **Clm** 12–75 cm, erect, with up to 5 nd. **Lvs** mostly bas, bas lvs usu persistent; **shth** smooth or scabridulous, lig 1.2–6.3 mm, usu scabridulous, smt smooth, truncate to rounded, lacerate-erose; **bld** 10–30 cm long, (1)2–4 mm wide, flat, clm bld usu less substantial than the bas bld. **Pan** 8–35(60) cm long, (1.5)2.5–14 cm wide, lanceolate-ovate, usu 3 times longer than wide, open, bases exserted from the up shth at maturity, lowest nd with 1–15 br; **br** scabrous to scabridulous, ascending, mostly brchg above midlength, lo br occ brchg to near the base, spklt somewhat clustered towards the tips, lo br 2–10 cm, mostly longer than 3 cm; **ped** (0.5)1–3.5(6) mm; **sec pan** not present. **Spklt** lanceolate to narrowly ovate, yellowish green to purple. **Glm** unequal, 2–3.6 mm, 1(3)-veined, scabrous on the midvein, occ also sparsely scabridulous over the body, acute to acuminate; **cal hairs** to 0.2 mm, sparse; **lm** 1.5–2.5 mm, usu smooth, smt scabridulous or pubescent, translucent to opaque, 5-veined, veins faint or prominent distally, apc acute to obtuse, usu entire, smt erose or toothed, teeth to about 0.4 mm, usu unawned, smt awned from midlength, awns to 2 mm, straight, not exserted, smt awned and unawned spklt present on the same pl; **pal** absent, or to 0.2 mm and thin; **anth** 3, 0.5–1.2 mm. **Car** 1–1.6 mm; **endosperm** semisoft. $2n = 42$.

Agrostis oregonensis grows in wet habitats, such as stream and lake margins, damp woods, and meadows, in western North America, primarily in the Pacific Northwest from British Columbia to California and Wyoming.

7. **Agrostis exarata** Trin. SPIKE BENT [p. *381*, 502]

Pl per; usu ces, smt rhz, not stln. **Clm** 8–100 cm, erect or decumbent at the base, smt rooting at the lo nd, with (2)3–6(8) nd. **Lvs** mostly cauline; **shth** smooth or slightly scabrous; **lig** (1)1.7–8(11.2) mm, dorsal surfaces scabrous, apc truncate to obtuse, lacerate to erose; **bld** 4–15 cm long, 2–7 mm wide, flat. **Pan** (3)5–30 cm long, 0.5–4 cm wide, contracted, spikelike, oblong, or lanceolate, usu dense, rarely more open, smt interrupted near the base, bases usu exserted, rarely enclosed by the up shth at maturity, lowest nd with 1–5 br; **br** scabrous, ascending to appressed, spklt-bearing to or near the base, usu hidden by the spklt, spklt crowded, lo br 1–2(4) cm; **ped** 0.2–4.3 mm. **Spklt** lanceolate to narrowly ovate, greenish to purplish. **Glm** subequal to equal, 1.5–3.5 mm, scabrous on the midvein and smt on the back, 1(3)-veined, acute, elongate-acuminate, with an awnlike tip to 1 mm; **cal hairs** to 0.3 mm, sparse to abundant; **lm** 1.2–2.2 mm, smooth, translucent to opaque, 5-veined, veins prominent distally or obscure throughout, apc acute, entire or toothed, teeth no more than 0.12 mm, unawned or awned from above midlength, awns to 3.5 mm, straight or geniculate; **pal** absent or to 0.5 mm; **anth** 3, 0.3–0.6 mm. **Car** 0.9–1.2 mm; **endosperm** solid or soft. 2*n* = 28, 42, 56.

Agrostis exarata is common and widely distributed in western North America, usually growing in moist ground in open woodlands, river valleys, tidal marshes, and swamp and lake margins; it also grows in dry habitats such as grasslands and shrublands. It extends from Alaska into Mexico, and is also found in Kamchatka and the Kuril Islands. It readily colonizes roadsides and bare soil, and exhibits ecological and developmental flexibility. It is recognized here as a single, variable species that includes what others have treated as distinct species or varieties.

8. **Agrostis pallens** Trin. Dune Bent [p. 382, 503]

Pl per; rhz, rhz to 10 cm, not stln. **Clm** 10–70 cm, erect, smt decumbent at the base, smt rooting at the lo nd, with 3–7 nd. **Lvs** usu cauline; **shth** usu smooth, smt scabridulous; **lig** 1–6 mm, dorsal surfaces scabrous, apc truncate to rounded or acute, often lacerate to erose; **bld** 1.5–11.5 cm long, 1–6 mm wide, flat, becoming involute. **Pan** 5–20 cm long, 0.4–6(8) cm wide, lanceolate to narrowly ovate, open to contracted, exserted from the up shth at maturity, lowest nd with 1–8 br; **br** scabrous to scabridulous, usu ascending, brchg below midlength, the majority spklt-bearing to the base, lo br 2–5 cm; **ped** 0.5–7 mm. **Spklt** lanceolate to narrowly ovate, green to yellowish green or yellow, tinged with purple. **Glm** equal to subequal, 2–3.5 mm, scabrous over the midvein and smt also sparsely over the body, 1(3)-veined, acute to acuminate; **cal hairs** to 0.3(1) mm, sparse; **lm** 1.5–2.5 mm, smooth or scabridulous

or warty, 5-veined, veins prominent throughout or only distally, apc acute, entire or the veins excurrent to about 0.2 mm, usu unawned, rarely awned from below the apc, awns to 0.5(2.7) mm, straight; **pal** absent, or to 0.2 mm and thin; **anth** 3, 0.7–1.8 mm. **Car** 1–1.5 mm; **endosperm** solid. 2*n* = 42, 56.

Agrostis pallens grows on coastal sands and cliffs, in meadows, and in open, xeric woodlands to subalpine woodlands at 3500 m. It extends from British Columbia south into Baja California, Mexico, and east to western Montana and Utah. The relationship of the higher-elevation, more open-panicled plants to those of lower elevations merits further study.

9. **Agrostis variabilis** Rydb. Mountain Bent [p. 382, 503]

Pl per; ces, rarely rhz, rhz to 2 cm. **Clm** 5–30 cm, erect, smt geniculate at the base, with 2–5(7) nd. **Lvs** mostly bas, forming dense tufts; **shth** smooth; **lig** (0.7)1–2.8 mm, dorsal surfaces usu scabridulous, smt smooth, apc rounded to truncate, lacerate to erose; **bld** 3–7 cm long, 0.5–2 mm wide, flat, becoming folded or involute. **Pan** (1)2.5–6 cm long, 0.3–1.2(2) cm wide, cylindric to lanceolate, usu dense, exserted from the up shth at maturity, lowest nd with 1–5 br; **br** usu scabridulous, smt smooth, ascending to erect, brchg at or near the base and spklt-bearing to the base, to brchg in the distal ⅔, lo br 0.5–1.5 cm; **ped** 0.4–2.8(4.3) mm. **Spklt** ovate to lanceolate, greenish purple. **Glm** subequal to equal, 1.8–2.5 mm, smooth, or scabrous on the keel and smt elsewhere, 1-veined, acute to acuminate; **cal hairs** to 0.2 mm, sparse to abundant; **lm** 1.5–2 mm, smooth, translucent, (3)5-veined, veins usu prominent distally, smt obscure throughout, apc acute, entire, usu unawned, rarely awned, awns to 1(2.8) mm, arising beyond the midpoint, usu not reaching the lm apc; **pal** to 0.2 mm, thin; **anth** 3, 0.4–0.7(1) mm. **Car** 1–1.3 mm; **endosperm** soft. 2*n* = 28.

Agrostis variabilis grows in alpine and subalpine meadows and forests and on talus slopes, at elevations up to 4000 m, from British Columbia and Alberta south to California and New Mexico. It can appear similar to dwarf forms of *Podagrostis humilis* (p. 150), but differs from that species in not having paleas.

8.16 POLYPOGON Desf.

Mary E. Barkworth

Pl ann or per; not rhz. **Clm** 4–120 cm, erect to decumbent, rooting at the lo nd, sparingly brchd near the base. **Lvs** usu no more than 5 per clm, bas and cauline; **shth** open, smooth or scabridulous; **aur** absent; **lig** memb or hyaline, acute to broadly rounded, erose, ciliate; **bld** flat to convolute. **Infl** tml pan, dense, continuous or interrupted below; **br** flexible, usu some longer than 1 cm; **ped** absent and the spklt borne on a stipe, or present and terminating in a stipe; **stipes** scabrous, flaring distally; **dis**

at the base of the stipes. **Spklt** 1–5 mm, weakly lat compressed, with 1 bisex flt; **rchl** not prolonged beyond the base of the flt. **Glm** exceeding the flt, lanceolate, bases not fused, apc entire to emgt or bilobed, usu awned from the sinuses or apc, awns flexuous, glab, smt unawned; **lm** 1–3(5)-veined, often awned, awns usu tml or subtml, smt arising from just above midlength; **pal** from ⅓ as long as to equaling the lm; **lod** 2, oblong-lanceolate to lanceolate; **anth** 3; **ov** glab; **sty** separate. **Car** slightly flattened, broadly ellipsoid to oblong-ellipsoid; **hila** ⅙–¼ as long as the car, ovate. $x = 7$. Name from the Greek *poly*, 'many' or 'much', and *pogon*, 'beard', an allusion to the bristly appearance of the inflorescences.

Polypogon is a pantropical and warm-temperate genus of about 18 species. There are five species in the Intermountain Region; one species, *P. interruptus*, is native.

Polypogon is similar to *Agrostis*, and occasionally hybridizes with it. It differs from *Agrostis* in having spikelets that disarticulate below the glumes, often at the base of a stipe.

1. Glumes unawned or with awns to 3.2 mm long.
    2. Glumes unawned . . . . 1. *P. viridis*
    2. Glumes awned, the awns 1.5–3.2 mm long . . . . 2. *P. interruptus* (in part)
1. Glumes with awns 3–12 mm long.
    3. Glumes deeply lobed, the lobes more than ⅙ the length of the glume body . . . . 5. *P. maritimus*
    3. Glumes not lobed or the lobes ¹⁄₁₀ or less the length of the glume body.
        4. Plants annual; glume apices rounded, lobed, the lobes 0.1–0.2 mm long; ligules
            2.5–16 mm long . . . . 4. *P. monspeliensis*
        4. Plants perennial; glume apices acute to truncate, unlobed or the lobes shorter than
          0.1 mm; ligules 1–6 mm long.
          5. Glume awns (3)4–6 mm long; longest blades 13–17 cm long . . . . 3. *P. australis*
          5. Glume awns 1.5–3.2 mm long; longest blades 5–9 cm long . . . . 2. *P. interruptus* (in part)

1. **Polypogon viridis** (Gouan) Breistr. Water Beardgrass [p. *382*, 503]

Pl per, often flowering the first year. **Clm** 10–90 cm, smt decumbent and rooting at the lo nd. **Shth** glab, smooth; **lig** to 5 mm; **bld** 2–13 cm long, 1–6 mm wide. **Pan** 2–10 cm, ovate-oblong to pyramidal, dense but interrupted, pale green to purplish; **ped** not developed; **stipes** 0.1–0.6 mm. **Glm** 1.5–2 mm, scabrous on the back and keel, apc obtuse or truncate, unawned; **lm** about 1 mm, erose, unawned; **pal** subequal to the lm; **anth** 0.3–0.5 mm. $2n = 28, 42$.

Polypogon viridis grows in mesic habitats associated with rivers, streams, and irrigation ditches. It is native to Eurasia, but is now established in western and southwestern North America.

2. **Polypogon interruptus** Kunth Ditch Beardgrass [p. *383*, 503]

Pl per, often flowering the first year. **Clm** 20–80 (90) cm, more or less decumbent. **Shth** smooth; **lig** 2–6 mm, scabridulous-pubescent; **bld** 5–9 cm long, 3–6 mm wide. **Pan** 3–15 cm long, 0.5–3 cm wide, usu interrupted or lobed; **ped** not developed; **stipes** 0.2–0.7 mm. **Glm** 2–3 mm, subequal, scabrous, larger prickles extending up the keel beyond midlength, not tapering to the apc, apc acute to truncate, unlobed or the lobes to 0.1 mm, awned, awns 1.5–3.2 mm, those of the lo and up glm subequal; **lm** 0.8–1.5 mm, glab, smooth and shiny, apc obtuse, not emgt, awned,

awns 1–3.2 mm; **pal** about ¾ as long as the lm; **anth** 0.5–0.7 mm. $2n = 28, 42$.

Polypogon interruptus grows in moist soil at lower elevations. It is native to the Western Hemisphere, extending south from the western United States into northern Mexico, and through the American tropics to Argentina and Bolivia.

3. **Polypogon australis** Brongn. Chilean Beardgrass [p. *383*, 503]

Pl per. **Clm** 20–100 cm. **Shth** smooth to scabridulous; **lig** 1–3(4) mm, rounded to broadly acute, erose; **bld** 13–17 cm long, 5–7 mm wide, scabrous. **Pan** 8–15 cm, lobed or interrupted, usu purplish; **ped** absent or vestigial; **stipes** 0.3–0.5 mm. **Glm** 1.5–3 mm, smooth to echinate, mrg ciliate, apc acute to truncate, unlobed or lobed, lobes to 0.1 mm, awned, awns (3)4–6 mm, flexuous; **lm** 1–1.3 mm, awned, awns 2–3.5 mm, flexuous; **pal** from shorter than to subequal to the lm; **anth** 0.3–0.5 mm. $2n$ = unknown.

Polypogon australis is native to South America. It has become established in western North America, where it grows alongside ditches and streams.

4. **Polypogon monspeliensis** (L.) Desf. Rabbitsfoot Grass [p. *384*, 503]

Pl ann. **Clm** 5–65 (100) cm, erect to geniculately ascending. **Shth** glab, the uppermost shth smt inflated; **lig** 2.5–16 mm; **bld** 1–20 cm long, 1–7 mm wide. **Pan** 1–17 cm, narrowly ellipsoid, dense, smt lobed, greenish; **ped** absent or to 0.2 mm; **stipes** 0.1–0.2 mm. **Glm** 1–2.7 mm, hispidulous throughout, largest prickles restricted to the lo ½, apc rounded, lobed, lobes 0.1–0.2 mm, ⅒ or less the length of the glm body, awned from the sinus, awns 4–10 mm, yellowish; **lm** 0.5–1.5 mm, glab, awned, awns 0.5–1(4.5) mm; **pal** subequal to the lm; **anth** 0.2–1 mm. $2n = 14, 28, 35, 42$.

Polypogon monspeliensis is native to Eurasia but it is now a common weed throughout the world, including much of North America. It grows in damp to wet, often alkaline soils, particularly in disturbed areas. The English language name aptly describes the feel of the young panicles.

In Europe, *Polypogon monspeliensis* hybridizes with *Agrostis stolonifera*, producing the sterile ×*Agropogon lutosus* (p. 146); and with *P. viridis*, forming *P. ×adscendens* Guss. *ex* Bertol. Only ×*Agropogon lutosus* has been reported from the Intermountain Region. It differs from *P. monspeliensis* in having more persistent spikelets, less blunt short-awned glumes, and lemmas with subterminal rather than terminal awns.

5. Polypogon maritimus Willd.

MEDITERRANEAN BEARDGRASS [p. *384*, <u>503</u>]

Pl ann. **Clm** (5)20–40 (50) cm, geniculate. **Shth** glab, smooth, uppermost shth smt inflated; **lig** to 7 mm; **bld** (1)3–9 (14) cm long, 0.5–5 mm wide. **Pan** (1)2–8(15) cm, narrowly ellipsoid, dense, smt lobed, often purplish; **ped** to about 0.5 mm, capillary; **stipes** 0.1–1.2 mm. **Glm** 1.8–3.2 mm, hispidulous bas, hairs smt strongly inflated and obtuse, apc lobed, lobes 0.3–1.2 mm, more than ⅙ the length of the glm body, awned from the sinus, awns (4)7–12 mm; **lm** 0.5–1.5 mm, unawned or awned, awns shorter than 1 mm; **pal** subequal to the lm; **anth** 0.4–0.5 mm. $2n = 14$.

Polypogon maritimus grows in disturbed, moist places, from sea level to 700 m. It is a Mediterranean species that has been found in North America, being particularly common in, or possibly just well-reported from, California. North American plants belong to *P. maritimus* Willd. var. *maritimus*, having stipes about as long as they are wide, glumes that never become strongly indurate at the base, and uninflated, acute hairs on the glume bases.

8.17 ×AGROPOGON P. Fourn.

Mary E. Barkworth

Pl per; loosely csp to spreading, rhz. **Clm** 8–60 cm, usu brchd below, ascending from a decumbent base or the lo nd geniculate. **Shth** smooth, open; **aur** absent; **lig** memb, puberulent, acute to obtuse, often bifid or denticulate with age; **bld** usu glab, smt scabrous. **Infl** pan, moderately dense to very dense, often interrupted; **dis** tardy, below the glm. **Spklt** lat compressed, with 1 flt. **Glm** similar, narrowly oblong to elliptic, 1-veined, apc notched, midveins extending into a short awn; **cal** glab; **lm** shorter than the glm, inconspicuously 5-veined, awned from just below the minutely toothed apc; **pal** about ¾ as long as the lm; **anth** usu indehiscent. $x = 7$.

×*Agropogon* comprises hybrids between *Agrostis* and *Polypogon*; one hybrid has been reported from the Intermountain Region.

1. ×Agropogon lutosus (Poir.) P. Fourn.

PERENNIAL BEARDGRASS [p. *385*]

Clm 8–60 cm. **Shth** smt inflated; **lig** 3–7 mm; **bld** 3–20 cm long, 2–11 mm wide, usu glab, smt scabrous. **Pan** 2–18 cm long, 0.6–7 cm wide, lanceolate to narrowly ovate, green or purplish; **ped** scabrous. **Spklt** 2–3 mm. **Glm** subequal, memb, scabrous, acute, awns to about 2 mm, apical; **cal** glab; **lm** 1–1.7 mm, awns to 3 mm, subtml; **pal** about ¾ as long as the lm; **anth** about 1 mm. **Car** not produced. $2n = 28$.

×*Agropogon lutosus* is a sterile hybrid between *Agrostis stolonifera* and *Polypogon monspeliensis* that sometimes grows in locations where both parents occur, such as damp to wet, often alkaline soils on lakesides, marshes, ditches, and intermittently flooded fields. Some plants favor *A. stolonifera*, others *P. monspeliensis*. All differ from *Polypogon* in having more persistent spikelets, less blunt short-awned glumes, and lemmas with subterminal rather than terminal awns; and from *Agrostis* in having awned glumes and awned lemmas. It has been reported from Nevada, but no supporting specimens are known; it is included because both parents are present in the region.

8.18 PHLEUM L.

Mary E. Barkworth

Pl ann or per; csp, smt rhz, occ stln. **Clm** 2–150 cm, erect or decumbent; **nd** glab. **Shth** open; **aur** absent or inconspicuous; **lig** memb, not ciliate; **bld** usu flat. **Infl** dense, spikelike pan, more than 1 spklt associated with each nd; **br** often shorter than 2 mm, always shorter than 7 mm, stiff; **ped** shorter than 1 mm, smt fused; **dis** above the glm or, late in the season, beneath the glm. **Spklt** strongly lat compressed, bases usu U-shaped, smt cuneate, with 1 flt; **rchl** glab, smt prolonged beyond the base of the flt. **Glm** equal, longer and firmer than the flt, stiff, bases not connate, strongly keeled, keels usu strongly ciliate, smt glab, smt scabrous, 3-veined, apc truncate to tapered, midveins often extending into short, stiff, awnlike apc; **cal** blunt, glab; **lm** white, often translucent, not keeled, 5–7-veined, unawned, bases not connate, apc acute, entire, smt with a weak, subapical awn; **pal** subequal to the lm, 2-veined; **lod** 2, free, glab, toothed; **anth** 3; **ov** glab. **Car** elongate-ovoid; **emb** $\frac{1}{6}$–$\frac{1}{4}$ the length of the car. $x = 7$. Name from the Greek *phleos*, the name of a reedy grass.

Phleum is a genus of approximately 15 species, most of which are native to Eurasia. Two species, one native and one introduced, are present in the Intermountain Region.

Species of *Phleum* are sometimes mistaken for *Alopecurus*, but *Alopecurus* has obtuse to acute glumes that are unawned or taper into an awn, lemmas that are both awned and keeled, and paleas that are absent or greatly reduced. The species of *Phleum* in the Intermountain Region are easily recognized by their strongly ciliate, abruptly truncate, awned glumes and adnate panicle branches.

1. Sheaths of the flag leaves not inflated; panicles 2–14(17) cm long, 5–20 times as long as wide; lower internodes of the culms frequently enlarged or bulbous; widespread in the Intermountain Region. . . . . 1. *P. pratense*
1. Sheaths of the flag leaves inflated; panicles 1–6 cm long, usually 1.5–3 times as long as wide; lower internodes of the culms not enlarged or bulbous . . . . 2. *P. alpinum*

1. **Phleum pratense** L. Timothy [p. *385*, <u>503</u>]
Pl per; loosely to densely ces. **Clm** (20) 50–150 cm, usu erect, lo intnd frequently enlarged or bulbous. **Shth of the flag lvs** not inflated; **aur** occ present, inconspicuous; **lig** 2–4 mm, obtuse to acute; **bld** to 45 cm long, 4–8(10) mm wide, flat. **Pan** (3)5–10(16) cm long, 5–7.5(10) mm wide, 5–20 times as long as wide, not tapering distally; **br** adnate to the rchs. **Glm** 3–4 mm, sides usu puberulent, keels pectinate-ciliate, apc awned, awns 1–1.5(2) mm; **lm** (1.2)1.7–2 mm, about $\frac{1}{2}$ as long as the glm, usu puberulent; **anth** 1.6–2.3 mm. $2n = 42$ (21, 35, 36, 49, 56, 63, 70, 84).

Phleum pratense grows in pastures, rangelands, and disturbed sites throughout most of the mesic, cooler regions of North America. Originally introduced from Eurasia as a pasture grass, it is now well established in many parts of the world, including the Intermountain Region. North American plants belong to the polyploid *Phleum pratense* L. subsp. *pratense*. Depauperate specimens of *P. pratense* are hard to distinguish from *P. alpinum* in the herbarium.

2. **Phleum alpinum** L. Alpine Timothy [p. *385*, <u>503</u>]
Pl per; ces, smt shortly rhz. **Clm** 15–50 cm, often decumbent, lo intnd not enlarged or bulbous. **Shth of the flag lvs** inflated; **aur** not developed, lf edges smt wrinkled at the jnct of the shth and bld; **lig** 1–4 mm, truncate; **bld** to 17 cm long, 4–7 mm wide, flat. **Pan** 1–6 cm long, 5–12 mm wide, usu 1.5–3 times as long as wide, subglobose to broadly cylindric, not tapering distally; **br** adnate to the rchs. **Glm** 2.5–4.5 mm, sides scabrous, keels hispid, apc awned, awns 0.8–2.5(3.2) mm; **lm** 1.7–2.5 mm, about $\frac{3}{4}$ as long as the glm, mostly glab, keels hairy, hairs to 0.1 mm; **anth** 1–1.5(2) mm. $2n = 14, 28$.

Phleum alpinum grows along stream banks, on moist prairie hillsides, and in wet mountain meadows. It is a circumboreal species extending, in the western hemisphere, from northern North America southward through the mountains to Mexico and South America. It is also widespread in northern Eurasia. Depauperate plants of *P. pratense* may be difficult to distinguish from *P. alpinum* in the herbarium; there is never any difficulty in the field. North American plants belong to *P. alpinum* subsp. *alpinum* and are tetraploid.

8.19 SESLERIA Scop.

Mary E. Barkworth

Pl per; more or less csp, smt rhz, rarely stln. **Clm** 3–80(100) cm. **Lvs** mostly bas; **shth** closed for most of their length, often shredding when mature; **aur** absent; **lig** 0.1–1 mm, memb, truncate to obtuse, usu ciliolate; **bld** flat, plicate, or involute. **Infl** single tml pan, cylindrical to globose, usu dense, smt subtended by ovate to round, scarious or hyaline, erose bracts, smt by 1–2 pubescent scales, smt without subtending scales or bracts, more than 1 spklt associated with each nd; **br** shorter than 10 mm. **Spklt** 3–9 mm, lat compressed, with 2–5 flt, upmost flt rdcd; **rchl** usu glab, rarely sparsely pilose, smt prolonged beyond the base of the distal flt, smt terminating in a rdcd flt; **dis** above the glm, beneath the flt. **Glm** unequal, usu shorter than the lowest lm, scarious to memb, 1–3-veined, apc awned, awns glab; **cal** very short, broadly rounded, glab or with scattered hairs; **lm** memb, 5–7-veined, 3–5 veins usu extending into awnlike teeth, cent teeth longer than the lat teeth; **pal** equaling or exceeding the lm; **lod** 2, free, glab, toothed; **anth** 3; **ov** pubescent. **Car** 1.5–3 mm; **emb** $\frac{1}{4}$–$\frac{1}{3}$ the length of the car. $x = 7$. Named for Lionardo Sesler (?–1785), a Venetian naturalist and director of a botanical garden at Santa Maria di Sala.

Sesleria has approximately 30 species. Most abundant in the Balkans, its range extends from Iceland, Great Britain, and southern Sweden through central and southern Europe into northwest Asia. It is grown as an ornamental in North America.

1. Sesleria caerulea (L.) Ard. BLUE MOORGRASS
[p. *386*]

Pl densely ces, shortly rhz, glaucous. **Clm** to 40(60) cm, erect, stiff, strongly grooved, glab; **nd** 4–6. **Lvs** mostly bas; **shth** grooved, glab, keeled distally, bas shth persistent, pubescent, disintegrating at maturity; **lig** 0.3–0.5 mm, erose, often finely ciliate; **bld of innovations and lo cauline lvs** to 40 cm long, 2–4 mm wide, flat or slightly involute, divergent, apc rounded and slightly hooded, adx surfaces glaucous, particularly when young, midveins prominent, particularly bas, mrg scabrous; **bld of flag lvs** 0.5–2.5 cm long, 2–4 mm wide. **Pan** (0.9)1.2–1.4(2.4) cm long, 5–8(12) mm wide, dense, usu spherical or ovoid, usu purplish, br with 1–2 spklt; **ped** 0.5–1 mm, scabrous; **bracts** 2, 2–4 mm

long, equally broad, erose. **Spklt** 4–6 mm, with (2)3 flt, usu purplish, rarely stramineous. **Glm** 4–5 mm, ovate, hyaline, mostly glab, mrg ciliate, particularly distally, midveins with stiff hairs, apc acute, awns to 1.5 mm; **lm** 3.5–5 mm, hyaline, ovate, midveins and mrgl veins hairy at least distally, hairs about 0.1 mm, appressed hairy between the veins, midveins forming 0.2–2 mm teeth, lat veins forming 0.1–0.4 mm teeth; **pal** as long as the lm, mrg and keels hairy distally, hairs about 0.1 mm, veins forming awnlike apc to 2.3 mm; **anth** about 2.3–3.2 mm. **Car** about 2 mm, obovoid, hairy distally. $2n = 28$.

Sesleria caerulea is native to Europe, ranging from central Sweden to northwestern Russia and central Bulgaria. It usually grows in moist to wet, calcareous pastures and bogs. It is grown as an ornamental in the Intermountain Region.

8.20 **DESMAZERIA** Dumort.   Gordon C. Tucker

Pl ann. **Clm** to 60 cm, procumbent to erect, sparingly brchd at the base. **Lvs** bas and cauline; **shth** open, glab; **aur** absent; **lig** longer than wide, acute; **bld** linear, usu flat, smt convolute when dry, glab. **Infl** tml, rcm or pan, usu with 1 br per nd; **br** stiff, not secund, ped 0.5–3 mm. **Spklt** subsessile, tangential to the rchs, lanceolate to ovate, lat compressed, with 4–25 flt, distal flt rdcd; **dis** above the glm, beneath the flt; **rchl** not prolonged beyond the base of the distal flt. **Glm** unequal to subequal, shorter than or subequal to the adjacent lm, 1–5-veined, unawned; **cal** blunt, rounded, glab; **lm** narrowly elliptic, coriaceous at maturity, inconspicuously 5-veined, glab, smt scabridulous towards the apc, apc acute to obtuse, smt bifid, often mucronate, unawned; **pal** about as long as the lm, 2-veined; **lod** 2, free, lanceolate; **anth** 3, only slightly exserted at anthesis; **ov** glab. **Car** shorter than the lm, concealed at maturity, ellipsoid-oblong, dorsally flattened, falling with the pal; **hila** about ¹⁄₁₀ as long as the car, ovate. $x = 7$. Named for Jean Baptiste Henri Joseph Desmazières (1786–1862), a French merchant, amateur botanist, and horticulturalist.

Desmazeria has six or seven species, all of which are native around the Mediterranean. One species is established as a weed in the Intermountain Region. One or two genera have sometimes been segregated from *Desmazeria*; current opinion favors the treatment presented here.

1. Desmazeria rigida (L.) Tutin FERN GRASS
[p. *386*, <u>*503*</u>]

Clm to 60 cm, procumbent to erect, glab. **Shth** glab, up mrg memb, continuous with the sides of the lig; **lig** 1.5–4 mm, lacerate; **bld** 2–8(12) cm long, 1–3(4) mm wide. **Infl** usu pan, smt rcm, 1–12(18) cm long, 12–30 mm wide; **br** stiff, somewhat divaricate at maturity; **ped** 0.5–3 mm, appressed to divaricate. **Spklt** 4–10 mm, narrowly ovate, with 5–12 flt; **rchl** puberulent, hairs stiff. **Glm** usu glab, more or less keeled,

acute; **lo glm** 1.3–2 mm, (1)3–veined; **up glm** 1.5–2.3 mm, 3-veined; **lm** 2–3 mm, rounded on the back or weakly keeled distally, glab, acute to obtuse, often shortly mucronate; **anth** 0.4–0.6 mm. $2n = 14$.

Desmazeria rigida is native to Europe, and appears to have no distinctive habitat preferences. In North America, it grows as a weed in disturbed sites such as roadsides, ditches, and the edges of fields. It is now established in southern Nevada.

8.21 **VENTENATA** Koeler   William J. Crins

Pl ann; tufted. **Clm** 10–75 cm, erect, puberulent below the nd; **nd** glab. **Lvs** mostly on the lo ½ of the clm; **shth** open, glab or sparsely pubescent; **aur** absent; **lig** hyaline, acute or obtuse, usu lacerate; **bld** flat initially, involute with age. **Infl** open or contracted pan, with spklt borne near the ends of the br on clavate ped. **Spklt** lat compressed, with 2–10 flt; **dis** above the first flt and between the distal flt; **rchl** smt prolonged beyond the base of the distal flt, smt terminating in a rdcd flt. **Glm** unequal, lanceolate, hispidulous, similar in texture to the lm, mrg scarious, apc acuminate; **lo glm** 3–7-veined;

up glm 3–9-veined; **cal** of the lo flt shorter than those of the up flt, sparsely hairy, cal of the distal flt with a dense tuft of white hairs; **lm** lanceolate, chartaceous, 5-veined, mrg scarious, apc entire or bifid, awned or unawned; **lowest lm** within a spklt awned or unawned, awns straight, tml; **distal lm** within a spklt awned, awns dorsal, geniculate; **pal** shorter than the lm, memb, keels ciliate distally; **lod** 2, memb, glab, toothed or not toothed; **anth** 3; **ov** glab. **Car** shorter than the lm, concealed at maturity, glab. $x = 7$. Named for Etienne Pierre Ventenat (1757–1808), a French clergyman, librarian, and botanist.

   Ventenata is native from central and southern Europe and northern Africa to Iran. It has five species, all of which grow in dry, open habitats. Only one species is established in North America.

1. Ventenata dubia (Leers) Coss. VENTENATA, NORTH-AFRICA GRASS [p. 387, 503]

Clm 15–75 cm, puberulent below the nd; **nd** 3–4, exposed, purple-black. **Lig** 1–8 mm; **bld** 2–7(12) cm long, 0.8–2.5 mm wide. **Pan** (7)15–20 cm, open, pyramidal, lo nd with 2–5 br; **br** 1.5–7 cm, bearing 1–5 spklt distally; **ped** 2–18 mm. **Spklt** 9–15 mm, with 2–3 flt, the lowest usu stmt, the remainder bisex; **rchl** usu glab, smt pubescent abx, intnd mostly 1–1.5 mm, prolongation to 2 mm, empty or with a rdcd flt. **Lo glm** 4.5–6 mm; **up glm** 6–8 mm; **lm** 5–7.5 mm, awns of the lowest lm within a spklt to 9 mm, straight, distal lm within a spklt bifid, teeth 1–2 mm, awns 10–16 mm, geniculate; **pal** 4–5 mm; **anth** 1–2 mm. **Car** about 3 mm. $2n = 14$.

   Ventenata dubia is an African species that is now established in crop and pasture lands of eastern Washington and western Idaho (Old and Callihan 1986), and has been found, but has not necessarily become established, at scattered locations from northern California to northern Utah. Mature specimens can be confusing because the first, straight-awned floret remains after the distal, bisexual florets have disarticulated (Chambers 1985).

8.22 SCRIBNERIA Hack.   James P. Smith, Jr.

Pl ann; tufted. **Clm** 3–35 cm, ascending to erect, often brchd at the lowest nd, glab; **nd** purple. **Shth** open; **aur** absent; **lig** memb; **bld** involute, nearly filiform. **Infl** tml distichous spikes, with 1 spklt at all or most nd, occ 2 at some nd, very rarely 3 or 4 spklt per nd, lo spklt sessile, up spklt pedlt; **ped** shorter than 3 mm. **Spklt** tangential to and partially embedded in the rchs, lat compressed, with 1 flt; **rchl** prolonged beyond the base of the flt; **dis** above the glm, beneath the flt. **Glm** 2, exceeding the flt, glab, coriaceous, stiff, reddish- or purplish-tinged, 2-keeled, unawned; **lo glm** longer and narrower than the up glm, 2–3-veined; **up glm** 3–4-veined; **cal** pubescent; **lm** memb, inconspicuously 5-veined, shortly bifid, awned from the sinus, awns 2–4 mm; **pal** tightly clasped by the lm; **anth** 1; **ov** glab. **Car** about 2.5 mm, fusiform; **hila** punctiform; **emb** about ¼ the length of the car. $x = 13$. Named for Frank Lamson Scribner (1851–1938), an American agrostologist.

   Scribneria is a monospecific genus native to North America.

1. Scribneria bolanderi (Thurb.) Hack. SCRIBNER GRASS [p. 387, 503]

Clm (3)10–35 cm. **Lig** 2–4 mm; **bld** 1–3 cm long, 0.8–1.6 mm wide, abx surfaces scabrous over the midveins. **Spikes** (2)4–11 cm long, 1–2.5 mm wide. **Spklt** (3) 4–7 mm, slightly longer than the adjacent intnd. **Lm** glab or scabridulous distally and on the keels; **awns** 2–4 mm, inconspicuous; **pal** generally smaller than the lm, apc notched. $2n = 26$.

   Scribneria bolanderi grows between 500–3000 m, from southern Washington to Baja California, Mexico. It grows in diverse habitats, ranging from dry, sandy or rocky soils to seepages and vernal pools. It is often overlooked because it is relatively inconspicuous.

8.23 VAHLODEA Fr.   Jacques Cayouette and Stephen J. Darbyshire

Pl per; loosely csp. **Clm** 15–80 cm. **Shth** open nearly to the base; **aur** absent; **lig** memb; **bld** rolled in the bud, flat. **Infl** open or closed pan; **br** often flexuous, capillary, spklt distal, some br longer than 1 cm. **Spklt** pedlt, usu with 2 flt, smt with additional distal rdmt flt; **rchl** prolonged beyond the base of the distal flt about 0.5 mm or less, usu glab, smt with a few hairs; **first intnd** about 0.5 mm; **dis** above the glm, beneath the flt. **Glm** subequal to equal, equaling or exceeding the flt, memb, acute to acuminate, unawned; **lo glm** 1(3)-veined; **up glm** 3-veined; **cal** obtuse, pilose, hairs about ½ as long as the lm; **lm** ovate, with 5(7) obscure veins, awned, awns attached near the mid of the lm, twisted, geniculate, visible between the glm; **pal** subequal to the lm; **lod** 2, memb, toothed or not toothed;

anth 3; **ov** glab. **Car** shorter than the lm, concealed at maturity, ellipsoid and irregularly triangular or ovate in cross section, deeply grooved, smt adhering to the lm and pal; **hila** linear to oblong, ¼–½ the length of the car. *x* = 7. Named for the Danish botanist Jens Laurentius Moestue Vahl (1796–1854), the son of botanist Martin Vahl.

Vahlodea includes only one species. In the northern hemisphere it has a discontinuous circumboreal distribution. It also grows in southern South America.

1. **Vahlodea atropurpurea** (Wahlenb.) Fr. *ex*
   Hartm. MOUNTAIN HAIRGRASS [p. *388*, 503]
Clm 15–80 cm, erect. **Lvs** glab or pilose; lo shth usu retrorsely hirsute, smt glab; uppermost shth smooth or scabridulous; **lig** 0.8–3.5 mm, rounded to truncate, often lacerate and ciliate; **bld** flat, bld of the lo lvs to 30 cm long, 1–8.5 mm wide, bld of the flag lvs 1–10 cm long, 1–5 mm wide. **Pan** 3–20 cm; **ped** smooth or scabrous-pubescent.

Spklt 4–7 mm. **Glm** usu smooth or scabrous on the keels and mrgl veins; **lo glm** 4–5(6.5) mm; **up glm** 4–5.5(7) mm; **lm** 1.8–3 mm, apc scabrous, ciliate, awns 2–4 mm; **anth** 0.5–1.2 mm. **Car** 1–1.5 mm. 2*n* = 14.

Vahlodea atropurpurea grows in moist to wet, open woods, forest edges, streamsides, snowbeds, and meadows, in montane to alpine and subarctic habitats.

8.24 PODAGROSTIS (Griseb.) Scribn. & Merr.

M.J. Harvey and Mary E. Barkworth

Pl per; csp, smt rhz. **Clm** 5–90 cm, erect or decumbent at the base. **Lvs** bas concentrated; **shth** open to the base, smooth, glab; **aur** absent; **lig** memb, scabridulous dorsally, truncate to subacute, entire to lacerate; **bld** flat or involute. **Infl** pan, exserted at maturity, not dis; **br** ascending to erect. **Spklt** pedlt, weakly lat compressed, with 1 flt; **rchl** usu prolonged 0.1–1.9 mm beyond the base of the flt, smt absent, especially from the lo spklt within a pan, apc glab or with hairs, hairs to 0.3 mm; **dis** above the glm, beneath the flt. **Glm** equal or the lo glm longer than the up glm, flexible, acute to acuminate, smt apiculate, unawned; **cal** glab or hairy, hairs to 0.5 mm; **lm** memb, (3)5-veined, veins mostly obscure, smt prominent distally, apc truncate to rounded or acute, unawned or awned, awns to about 1.3 mm, usu subapical, occ attached near midlength; **pal** more than ½ as long as the lm, 2-veined, thinner than the lm; **anth** 3. **Car** shorter than the lm, concealed at maturity. *x* = 7. Name from the Greek *pous*, 'foot', and the genus *Agrostis*.

Podagrostis is a genus of six or more species that grow in cool, wet areas. In the past, its species have been included in *Agrostis*. Two species grow in the Intermountain Region. *Podagrostis* differs from *Agrostis* in its combination of a relatively long palea and, usually, the prolongation of the rachilla beyond the base of the floret. It differs from *Calamagrostis* in the poorly developed callus hairs and awns. This treatment differs from Harvey (2007a, b) in recognizing *P. thurberiana* as distinct from *P. humilis*, a decision based on consideration of Arnow (1987) and additional examination of specimens.

1. Glumes ovate to elliptical, exceeding the lemma by 0.2–0.4 mm; panicles 1.5–6 cm long;
   leaf blades 0.4–1.5 mm wide, highest leaf blade usually below midculm . . . . 1. *P. humilis*
1. Glumes narrowly elliptical, equalling the lemma or up to 0.1 mm longer; panicles
   (2.5)5–12 cm long; leaf blades 1–3 mm wide, highest leaf blade usually near midculm . . . . . 2. *P. thurberiana*

1. **Podagrostis humilis** (Vasey) Björkman
   ALPINE BENT [p. *389*, 503]
Pl csp, smt rhz. **Clm** 3–26 cm, erect to ascending, bases smt somewhat decumbent; **nd** 2–3. **Lvs** strongly concentrated at base; **shth** smooth; **lig** (0.2)0.5–4 mm, smooth, truncate, rounded or acute, usu entire or erose; **bld** 1.5–4(5) cm long, 0.4–1.5 mm wide, flat or folded; **flag bld** usually below, smt near midculm. **Pan** 1.5–6 cm long, 0.2–0.7(1.5) cm wide, smt wider at anthesis, linear to narrowly oblong, lowest nd with 1–3(4) br; **br** smooth, usu appressed to erect, ascending at anthesis, brchg in the distal ⅓–¾; **lo br** 0.5–1.5 cm; **ped** 0.4–1.5 mm. **Spklt** usu purplish; **rchl prolongations** 0.1–0.6 mm, glab or bristlelike, with a tuft of short hairs at the apc. **Glm**

subequal, 1.6–2 mm, exceeding the lm by 0.2–0.4 mm, ovate to elliptical, 1-veined or 3-veined at base, veins smooth or scabridulous distally, apc acute; **cal** glab or with sparse hairs to 0.5 mm; **lm** 1.5–2 mm, usually purplish, smooth, opaque, 5-veined, veins obscure or prominent distally, apc acute, entire or erose, veins occ excurrent to 0.4 mm, rarely awned, awns to about 1.3 mm, usu subapical, smt attached near midlength; **pal** 0.9–1.6 mm; **anth** 3, 0.4–1mm. **Car** 1–1.3 mm. 2*n* = 14.

Podagrostis humilis is a western North American species that grows in undisturbed alpine and subalpine meadows and screes at over 3500 m, down to meadows, fens, and open woodlands at less than 200 m. It usually differs from *P. thurberiana* in overall size and in having narrower, more basally concentrated leaves. In the field, dwarf forms of *P.*

humilis mimic *Agrostis variabilis*; they differ from that species in having paleas.

2. Podagrostis thurberiana (Hitchc.)

Hultén THURBERS's BENT [p. *389*, 503]

Pl csp, smt rhz. **Clm** 10–40(50) cm, ascending, bases smt somewhat decumbent; **nd** 2–3. **Lvs** concentrated at base; **shth** smooth; **lig** (0.2)0.5–2 mm, smooth, truncate to rounded, entire or erose; **bld** 3–10(15) cm long, 1–3 mm wide, usu flat, smt folded; **flag bld** usu near midculm. **Pan** (2.5)5–12 cm long, 0.5–0.7(2.5) cm wide, linear to narrowly oblong, lowest nd with 1–3(4) br; **br** scabridulous or smooth, erect to ascending, brchg in the distal ½–⅔; **lo br** 0.5–2.5 cm; **ped** 0.5–2 mm. **Spklt** usually greenish, smt purplish; **rchl prolongations** 0.1–0.6 mm, glab or bristlelike, with a tuft of short hairs at the apc. **Glm** subequal, 2–2.5 mm, from equalling the lm to exceeding it by 0.1 mm, narrowly elliptical, 1–3-veined, veins smooth or scabridulous distally, apc acute; **cal** glab; **lm** 2–2.5 mm, greenish or partly purplish, smooth, opaque, 5-veined, veins obscure or prominent at the apc, apc acute, entire or erose, veins occ minutely excurrent to 0.4 mm, usu unawned, rarely awned, awns to about 1.3 mm, usu subapical, smt attached near midlength; **pal** 1.2–1.6 mm; **anth** 3, 0.4–0.8 mm. **Car** 1–1.3 mm. $2n = 14$.

Podagrostis thurberiana is a western North American species that grows in undisturbed alpine and subalpine meadows and screes at over 3500 m, down to meadows, fens, and open woodlands at less than 200 m, sometimes growing with *P. humilis*. It usually differs from that species in being taller in having wide, less basally concentrated leaves.

8.25 HELICTOTRICHON Besser *ex* Schult. & Schult. f.

Gordon C. Tucker

Pl per; csp. **Clm** 5–150 cm, erect. **Shth** open nearly to the base; **aur** absent; **lig** about as long as wide, memb, truncate to rounded, ciliate-erose; **bld** convolute or involute, adx surfaces ribbed over the veins. **Infl** narrow pan or rcm, some br longer than 1 cm. **Spklt** lat compressed, with (1)2–8 flt; **rchl** pilose on all sides, terminating in rdcd flt; **dis** above the glm, beneath the flt. **Glm** equaling or exceeding the adjacent lm, exceeded by the distal flt, 1–3(5)-veined; **cal** acute, strigose; **lm** pilose or glab, 3–5-veined, apc acute, toothed, usu awned from about midlength, awns geniculate, twisted and terete below the bend, distal lm smt unawned; **pal** shorter than the lm, wings more than ½ as wide as the intercostal region; **lod** 2, lobed; **anth** 3; **ov** pubescent distally. **Car** shorter than the lm, concealed at maturity, with a solid endosperm, longitudinally grooved, with a tml tuft of hairs; **hila** more than ½ as long as the car, linear. $x = 7$. Name from the Greek *helictos*, 'twisted', and *trichon*, 'awn', referring to the lemma awn.

Helictotrichon has about 15 species. Most are native to Europe, but one is native to the Intermountain Region and another is often cultivated. The genus is sometimes interpreted as including *Avenula*, from which it differs in having truncate to rounded ligules, ribbed leaves, rachillas that are pilose on all sides, lobed lodicules, long hila, solid endosperm, and sclerenchyma rings in its roots.

1. Culms 5–20 cm tall; panicles 2–8 cm long, most branches with 1 spikelet; plants native. . . 1. *H. mortonianum*
1. Culms 30–150 cm tall; panicles 8–20 cm long, most branches with 3–10 spikelets; plants cultivated as ornamentals. . . . . 2. *H. sempervirens*

1. Helictotrichon mortonianum (Scribn.)

Henrard ALPINE OATGRASS [p. *390*, 503]

Clm 5–20 cm. **Lig** 0.5–1 mm, truncate to rounded, ciliate; **bld** 3–6 cm long, 1–2 mm wide, involute or convolute, strigose, particularly on the adx surfaces. **Pan** 2–5(8) cm; **br** erect, usu with 1 spklt each. **Spklt** 8–12 mm, with (1)2(3) flt. **Glm** equal or nearly so, 8–11 mm, acuminate or awn-tipped, awns 0.3–0.5 mm; **lowest lm** 7–10 mm, 3-veined, apc 4-toothed, awns 10–16 mm; **distal lm** unawned or awned; **anth** 1.5–2 mm. $2n = 14, 28$.

Helictotrichon mortonianum grows in alpine and subalpine meadows and summits, at 3000–4200 m. Its range extends from northeastern Utah through the central and southern Rocky Mountains.

2. Helictotrichon sempervirens (Vill.) Pilg.

BLUE OATGRASS [p. *390*]

Clm 30–100(150) cm. **Lig** 0.5–1.5 mm, truncate, ciliate; **bld** 15–60 cm long, 2–4 mm wide, usu convolute and 0.9–1.5 mm in diameter, glaucous, scabridulous, bas bld deciduous when dead. **Pan** 8–20 cm; **br** ascending, with 3–7(10) spklt. **Spklt** 10–14 mm, with 3(5) flt. **Glm** subequal to unequal; **lo glm** 7–10 mm; **up glm** 10–12 mm; **lowest lm** 7–12 mm, awns about 15 mm; **distal lm** unawned; **anth** about 5 mm. $2n = 28, 42$.

Helictotrichon sempervirens is a native of the southwestern Alps in Europe, where it grows on rocky soils and in stony pastures. It is frequently grown as an ornamental in North America, but it is not established in the region. The common cultivar is called 'Sapphire Fountain'.

8.26 **CALAMAGROSTIS** Adans.   Kendrick L. Marr, Richard J. Hebda, and Craig W. Greene†

Pl per; often csp, usu rhz. **Clm** 10–210 cm, unbrchd or brchd, more or less smooth, nd 1–8. **Shth** open, smooth or scabrous; **aur** absent; **lig** memb, usu truncate to obtuse, smt acute, entire or lacerate, lacerations often obscuring the shapes; **bld** flat to involute, smooth or scabrous, rarely with hairs. **Infl** pan, open or contracted, smt spikelike; **br** appressed to more or less drooping, some br longer than 1 cm. **Spklt** pedlt, weakly lat compressed, with 1(2) flt; **rchl** prolonged beyond the base of the distal flt(s), usu hairy; **dis** above the glm. **Glm** memb, subequal, equal to, or longer than the lm, rounded or keeled, backs smooth or scabrous, rarely long-scabrous with bent projections, veins obscure to prominent, apc acute to acuminate, rarely awn-tipped or attenuate; **lo glm** 1(3)-veined; **up glm** 3-veined; **cal** hairy, hairs 0.2–6.5 mm, sparse to abundant; **lm** 3(5)-veined, smooth or scabrous, apc usu tapering into 4 teeth, awned; **awns** arising from near the base to near the apc, straight or bent, smt delicate and indistinct from the cal hairs, smt exserted beyond the lm mrg; **pal** well developed, almost as long as to slightly longer than the lm, thin, 2-veined; **anth** 3, smt strl. **Car** shorter than the lm, concealed at maturity, oblong, usu glab. $x = 7$. Name from the Greek *calamos*, 'reed', and *agrostis*, 'grass'.

Calamagrostis grows in cool-temperate regions and is especially diverse in mountainous regions. Its species grow in both moist and xeric habitats. There are about 100 species of *Calamagrostis*, if *Deyeuxia* Clarion *ex* P. Beauv. and *Lachnagrostis* are recognized as distinct from *Calamagrostis*. The latter two genera are often considered to be restricted to the Southern Hemisphere (Edgar 1995; Jacobs 2001). According to the criteria used by Phillips and Chen (2003) to distinguish *Calamagrostis* and *Deyeuxia*, most North American species of *Calamagrostis* fit within *Deyeuxia*. The merits of their recommendation, adoption of which would require many new combinations, have not been evaluated.

Ten species of *Calamagrostis* grow in the Intermountain Region; one, *C. epigejos*, is introduced. Some species of *Calamagrostis* are rangeland forage grasses, but most occur too sparsely to be important for livestock. This treatment includes one cultivar, *Calamagrostis* ×*acutiflora* 'Karl Foerster', that is becoming increasingly popular in horticulture.

Interspecific hybridization is common in *Calamagrostis*; vivipary and agamospermy also occur in some species. Interspecific hybridization, polyploidy, and apomixis contribute to the taxonomic difficulty of the genus.

There is a high degree of misidentification of taxa within this genus (30% for some species in some herbaria), and species distributions should be taken as a guide only. Much more field collecting is needed for several of the taxa in order to verify their distributions, especially near the limits of their ranges.

Calamagrostis is sometimes confused with *Agrostis*; there is no single character that distinguishes all species of *Calamagrostis* from those of *Agrostis*, but, in general, *Calamagrostis* has larger plants with larger, more substantial lemmas and paleas than *Agrostis*, and tends to occupy wetter habitats.

Measurements of the rachilla and callus hairs reflect the longest hairs present. Panicle widths refer to pressed specimens. The following key will enable typical specimens to be identified readily, but atypical specimens are common. For this reason, most leads require observation of a combination of characters, notably awn length, length of callus hairs relative to the lemma, glume length and scabrosity, panicle size, and leaf width.

1. Callus hairs more than 1.3 times as long as the lemmas; lemmas at least 2 mm shorter than the glumes, long-acuminate . . . . .1. *C. epigejos*
1. Callus hairs usually shorter than the lemmas, if longer then the lemmas less than 2 mm shorter than the glumes and not long-acumiante.
   2. Blades usually densely hairy on the adaxial surface; glumes keeled, scabrous; awns 4.5–9 mm long. . . . . .2. *C. purpurascens*
   2. Blades glabrous or sparsely hairy on the adaxial surface; glumes keeled or rounded, scabrous or smooth; awns 0.5–17 mm long.
      3. Awns 5.4–13 mm long, always exserted and bent, if 5–6 mm long, either some blades wider than 2 mm or the abaxial blade surfaces scabrous . . . . 3. *C. tacomensis*
      3. Awns 4.5–6 mm long, exserted or not, bent or straight, if 5–6 mm long, then either all blades less than 2 mm wide or the abaxial blades surfaces smooth or nearly so.
         4. Awns attached on the distal ⅖ of the lemmas, 0.5–2 mm long, straight; blades flat; panicles contracted, 0.8–2.5(3) cm wide . . . . 4. *C. scopulorum*
         4. Awns attached on the lower ½(¾) of the lemmas, 0.9–6 mm long, straight or bent; blades flat or involute; panicles open or contracted, 0.4–9 cm wide.
            5. Blades involute, 0.2–0.4 mm in diameter; panicles (1.5)1.9–5.7(7.5) cm long; callus hairs sparse. . . . . 5. *C. muiriana*
            5. Blades flat or involute, (1)1.5–8(11) mm wide, most wider than 2 mm; panicles (2)4–30 cm long; callus hairs sparse to abundant.

6. Awns usually exserted, (2.8)3–6 mm long; callus hairs 0.1–0.7 times the length of the lemmas; leaf collars hairy or glabrous.
    7. Culms 135–210 cm tall; plants cultivated ornamentals . . . . 6. *C.* ×*acutiflora*
    7. Culms 26–120 cm tall; plants native.
        8. Awns 4–5.5 mm long; leaf collars glabrous; plants often densely cespitose; rhizomes usually 2–4 mm thick . . . . .7. *C. koelerioides*
        8. Awns 2–4.5 mm long; leaf collars sometimes hairy; plants loosely cespitose; rhizomes 0.5–2 mm thick. . . . . 8. *C. rubescens*
6. Awns usually not exserted, if exserted then barely so, 0.9-3.1(4) mm long; callus hairs (0.1)0.2–1.5 times as long as the lemmas; leaf collars glabrous or hairy, if hairy the callus hairs more than 0.7 times the length of the lemmas.
    9. Culms usually scabrous, rarely smooth; awns slightly bent; callus hairs 0.4–0.8 times as long as the lemmas; blades usually involute, 1–4 mm wide, the abaxial surfaces scabrous; nodes 1–2. . . . . 9. *C. montanensis*
    9. Culms smooth to slightly scabrous; awns straight or bent; callus hairs (0.5)0.7–1.2(1.5) times as long as the lemmas; blades flat or involute, (1)1.5–8(11) mm wide, the abaxial surfaces scabrous or smooth; nodes 1–8.
        10. Panicle branches 2.7–6(12) cm long; ligules lacerate; glumes scabrous on the keels, often throughout; blades flat, the abaxial surfaces scabridulous or scabrous; cauline nodes (2)3-7(8); panicles open . . . . 10. *C. canadensis*
        10. Panicle branches 1–5(9.5) cm long, if longer than 3.7 cm, the ligules usually entire; glumes smooth or glabrous only on the keels; blades flat or involute, the abaxial surfaces smooth or scabrous; nodes 1–3(4); panicles loosely contracted . . . . .11. *C. stricta*

1. **Calamagrostis epigejos** (L.) Roth BUSHGRASS, CHEE REEDGRASS, FEATHERTOP [p. 391, 503]
Pl with strl clm; ces, with numerous rhz 8+ cm long, 1.5–2 mm thick. **Clm** (50)100–150(160) cm, unbrchd, slightly scabrous beneath the pan; **nd** (1)2–4(6). **Shth** and **col** smooth or slightly scabrous; **lig** (1.5)3–7(13) mm, truncate to obtuse, usu entire, infrequently lacerate; **bld** (6)25–40(55) cm long, (2.5)3.5–8(13) mm wide, flat, pale green, scabrous. **Pan** (14)18–23(35) cm long, (2)2.5–4(6) cm wide, erect, contracted, greenish; **br** (3.5)5–8(11) cm, smooth or slightly scabrous, spklt usu confined to the distal ¾, infrequently confined to the distal ½. **Spklt** (4)4.5–5.5(8) mm; **rchl prolongations** about 1 mm, hairs about 3 mm. **Glm** slightly keeled, usu smooth, infrequently scabrous near the apc, lat veins prominent, apc long-acuminate; **cal hairs** (2)3.5–5(6.5) mm, (1.3)1.5–2(2.5) times as long as the lm, abundant; **lm** 2–3.5(5) mm, (1.5)2–3(4.5) mm shorter than the glm; **awns** (1.5)2–3(4) mm, attached to the lo ⅓–⅔ of the lm, not exserted, delicate, not easily distinguished from the cal hairs, usu straight, infrequently bent; **anth** about (1)1.5(2) mm. 2*n* = 28, 35, 42, 56, ±70.

Calamagrostis epigejos is an introduced Eurasian species that was first found in North America in the 1920s. It grows in waste places, along roadsides, in juniper swamps, sandy woods, and thickets, and on rehabilitated tailings and cinders of railway beds. It was collected in Idaho from reseeded rangeland plots. It is not known whether it is now established at the site.

2. **Calamagrostis purpurascens** R. Br. PURPLE REEDGRASS [p. 391, 503]
Pl apparently without strl clm; strongly ces, often with rhz 1–4 cm long, 1–2 mm thick. **Clm** (10) 30–80 cm, usu unbrchd, occ brchd, usu slightly to strongly scabrous, smt puberulent beneath the pan; **nd** (1)2(3). **Shth** scabrous; **col** usu scabrous or hairy, rarely smooth; **lig** (1.5)2–4(9) mm, usu truncate and entire, smt lacerate. **Bld** (4)5–17(30) cm long, 2–5(6) mm wide, flat or involute, stiff, abx surfaces scabrous, adx surfaces usu densely long-hairy, rarely sparsely hairy. **Pan** 4–13(15) cm long, 0.9–2(2.8) cm wide, erect, contracted, infrequently interrupted near the base, often red- or purple-tinged; **br** 1.3–3.5 cm, scabrous, prickles long, almost hairlike, spklt-bearing to the base. **Spklt** (4.5)5.5–6.5(8) mm; **rchl prolongations** about (1)2 mm, hairs about 2 mm. **Glm** keeled, usu scabrous, rarely scabrous on the keels only, lat veins obscure to prominent, apc acute; **cal hairs** (0.9)1.2–1.5(2.4) mm, 0.2–0.4(0.6) times as long as the lm, sparse; **lm** (3.5)4–4.5(5) mm, usu

1–2.5 mm shorter than, rarely equal to, the glm; **awns** (4.5)6–7(9) mm, attached to the lo ¹⁄₁₀–¹⁄₃ of the lm, usu exserted, stout, easily distinguished from the cal hairs, bent; **anth** (1.3)1.7–2.5(2.9) mm. 2*n* = 42–58, 84.

Calamagrostis purpurascens grows in alpine tundra, on subalpine slopes, in grasslands, sand dunes, meadows, coniferous and deciduous forests, and disturbed soils, usually on rocky ridgetops and slopes and, infrequently, on valley floors. It prefers well- to moderately-drained, medium- to coarse-textured substrates, including scree and talus, that are often calcareous, at elevations from 15–4000 m. Its range extends from Alaska through Canada to Greenland and Newfoundland, including the islands of the Canadian arctic, and south in the western mountains to California and northern New Mexico.

The hairy adaxial leaf surfaces are a reliable diagnostic characteristic for *Calamagrostis purpurascens*. Many specimens from Washington and Oregon currently identified as *C. purpurascens* may belong to *C. tacomensis* (see next). *Calamagrostis purpurascens* differs from *C. tacomensis* in its leaf vestiture, shorter awns and panicle branches, and more scabrous glumes. Plants of *C. purpurascens* that have short awns barely projecting beyond the lemma margins have been mistaken for *C. montanensis*, which grows mainly to the northeast of the Intermountain Region, but *C. montanensis* does not have hairy adaxial leaf surfaces.

3. **Calamagrostis tacomensis** K.L. Marr & Hebda Rainier Reedgrass [p. 391, 503]
Pl without strl clm; ces, smt densely so, usu without rhz, smt with rhz about 2 cm long, 2–3 mm thick. **Clm** (20)30–55(95) cm, unbrchd, smooth or slightly scabrous beneath the pan; **nd** (1)2(5). **Shth** and **col** smooth or slightly scabrous; **lig** (3)3.5–5.5(6) mm, usu truncate to obtuse, usu entire, smt lacerate; **bld** (6)7–14(30) cm long, (1.5)2–2.5(4) mm wide, flat, abx surfaces usu smooth, rarely slightly scabrous, adx surfaces usu slightly scabrous, rarely smooth, glab or sparsely hairy. **Pan** (5)7–10(18) cm long, (0.5)1–2(3) cm wide, loosely contracted, smt open, erect to slightly nodding, shiny green and purple; **br** (2)2.3–4(6) cm, scabrous, usu spklt-bearing on the distal ²⁄₃, smt to the base. **Spklt** (4)6–6.5(7) mm; **rchl prolongations** 1.5–2(2.5) mm, hairs (1.5)2(3) mm. **Glm** often green with a purple patch at the base, keeled, keels smooth or sparsely scabrous on the distal ½, lat veins usu prominent, apc usu acute, smt short-acuminate, not twisted; **cal hairs** (1.2)2(2.5) mm, (0.3)0.4–0.5(0.6) times as long as the lm, abundant; **lm** (3.5)4–5(5.5) mm, (0.5)1.5–2(3) mm shorter than the glm; **awns** (5.5)7–8.5(10) mm, attached to the lo ¹⁄₁₀–¹⁄₃ of the lm, exserted more than 2 mm, easily distinguished from the cal hairs, strongly bent; **anth** (1)2–3(3.5) mm. 2*n* = unknown.

Calamagrostis tacomensis grows on montane to alpine slopes in dry or wet meadows, seeps, rocky talus slopes, and cliff crevices, at 400–2200 m. It grows only in the mountains of western Washington and in the Steens Mountains of southeastern Oregon. It reaches its highest known elevations in the Steens Mountains.

This species has previously been identified as *Calamagrostis purpurascens* (see previous) (C.L. Hitchcock et al. 1969). It differs from that species in having glabrous leaves, generally longer awns and inflorescence branches, and smoother glumes.

4. **Calamagrostis scopulorum** M.E. Jones Jones' Reedgrass, Ditch Reedgrass [p. 392, 503]
Pl without strl clm; loosely ces, with rhz to 2 cm long, 2–3 mm thick. **Clm** (40)50–92 cm, smt brchd, sparsely to densely scabrous; **nd** 2–3. **Shth** and **col** smooth or scabrous; **lig** (3)4–7(9) mm, obtuse, lacerate; **bld** 10–38 cm long, (2)3–4(7) mm wide, flat, scabridulous, adx surfaces glab or sparsely hairy. **Pan** (4)7–16(18) cm long, (0.7)1.1–2(3) cm wide, nodding, contracted, pale green to purple-tinged; **br** 1–5(6.5) cm, sparsely to densely scabrous, usu spklt-bearing to the base. **Spklt** (4)4.5–6 mm; **rchl prolongations** 1–2 mm, hairy throughout, hairs 1.5–2.5 mm. **Glm** keeled, mostly smooth, keels slightly scabrous distally, lat veins obscure, apc acuminate; **cal hairs** 2–3 mm, 0.5–0.6 times as long as the lm, somewhat sparse; **lm** 3.5–5 mm, 0.5–1.5 mm shorter than the glm; **awns** (0.5)1–1.5(2) mm, attached to the up ²⁄₅ of the lm, not exserted, slender, straight, easily overlooked when short; **anth** (1.8)2–2.7(3) mm. 2*n* = 28.

Calamagrostis scopulorum grows on canyon slopes and wash bottoms, and in dry to moist montane to alpine habitats, often on rocky, sandy to silty soil, at 1000–3550 m. Its range extends from western Montana and Wyoming south to Arizona and New Mexico.

5. **Calamagrostis muiriana** B.L. Wilson & Sami Gray Muir's Reedgrass [p. 392, 504]
Pl smt with strl clm; densely ces, often with rhz 1–3 cm long, 1–2 mm thick. **Clm** (10)12–35 cm, unbrchd, smooth beneath the pan; **nd** 1–3. **Lvs** bas concentrated; **shth** and **col** smooth or scabrous; **lig** 1–2.5 mm, obtuse, entire to lacerate; **bld** (1)4–12 cm long, 0.2–0.4 mm in diameter, involute, abx surfaces scabrous, adx surfaces sparsely hairy. **Pan** (1.5)1.9–5.7(7.5) cm long, 0.4–3 cm wide, contracted to open, usu dark purple, rarely straw-colored; **br** (0.8)1.1–2(3.5) cm, smooth, spklt usu confined to the ends of the br. **Spklt** (3)3.5–4.5(5) mm; **rchl prolongations** about 2 mm, hairs 0.5–1 mm. **Glm** rounded, midvein smooth or slightly scabrous, lat veins

obscure, apc acute to acuminate, rarely awn-tipped; **cal hairs** (0.2)0.3–0.6 mm, 0.1–0.2 times as long as the lm, sparse; **lm** (2.5)3–4 mm, 0.5–1 mm shorter than the glm; **awns** 3.5–6 mm, attached to the lo ⅓ of the lm, exserted, bent, purple; **anth** 0.9–2.5 mm. $2n = 28$.

Calamagrostis muiriana grows in moist to dry, subalpine and alpine floodplain meadows, lake margins, and stream banks, at 2400–3900 m, in the Sierra Nevada south of Sonora Pass in central California.

6. Calamagrostis ×acutiflora (Schrad.) D.C. 'Karl Foerster' FOERSTER'S REEDGRASS, FEATHER REEDGRASS [p. 392]

Pl often with robust strl clm; densely ces, with rhz 1–3 cm long, about 2 mm thick. **Clm** 135–210 cm, smooth to slightly scabrous, usu unbrchd; **nd** about 3. **Shth** and **col** smooth; **lig** (1)2–7 mm, lacerate; **bld** (5)11–63(71) cm long, (1)1.5–7.5(8) mm wide, flat or involute, usu scabrous, rarely smooth, adx surfaces glab or sparsely hairy. **Pan** 15–30 cm long, 0.75–1(3) cm wide, erect, contracted, pale green to purple; **br** about 6 cm, slightly scabrous, longer br with spklt on the distal ⅔, shorter br with spklt to the base. **Spklt** 4–5 mm; **rchl prolongations** about 0.5 mm, hairs about 1.5 mm. **Glm** keeled, mostly smooth, slightly scabrous only on the keels, veins usu obscure, apc acute; **cal hairs** 2–2.5 mm, 0.4–0.6 times as long as the lm, abundant; **lm** 3–3.5 mm, 1–1.5 mm shorter than the glm; **awns** about 3.5 mm, attached to the lo ¹⁄₁₀–⅕ of the lm, exserted, slender, usu distinguishable from the cal hairs, bent; **anth** not observed.

Calamagrostis ×acutiflora is a hybrid of European origin that is now widely planted as an ornamental, especially in dry sites and gardens. The parents are *C. arundinacea* (L.) Roth and *C. epigejos* (p. 153); the hybrids are seed-sterile.

7. Calamagrostis koelerioides Vasey DENSE-PINE REEDGRASS [p. 393, 504]

Pl without strl clm; often densely ces, with rhz 2–6 cm long, 2–4 mm thick. **Clm** (26)60–85(120) cm, unbrchd, slightly scabrous; **nd** 2–3(5). **Shth** and **col** usu scabrous, rarely smooth, glab; **lig** (1.5) 2–4.5(7) mm, truncate to obtuse, entire or smt lacerate; **bld** (2)9–20(30) cm long, (2)2.5–4.5(8) mm wide, flat, slightly scabrous, adx surfaces glab or sparsely hairy. **Pan** (4)10–13(16) cm long, about 1 cm wide, contracted, erect to slightly nodding, often slightly interrupted towards the base, straw-colored or pale green to pale purple; **br** (1.1)2.8–4(6) cm, scabrous, spklt-bearing to the base. **Spklt** (4)4.5–6(7) mm; **rchl prolongations** 1.5–2.5(3) mm, hairs 1.5–2 mm. **Glm** slightly keeled, keels smooth or slightly scabrous distally,

lat veins visible but not prominent, apc acute; **cal hairs** 1.5–2 mm, 0.3–0.4 times as long as the lm, sparse; **lm** (3.5)4–5(6) mm, 0.5–1.5 mm shorter than the glm; **awns** 4–5.5 mm, attached to the lo ¹⁄₁₀–⅕ of the lm, exserted, smt barely so, stout, distinguishable from the cal hairs, bent; **anth** 2–3.5 mm. $2n = 28$.

Calamagrostis koelerioides grows in mountain meadows, chaparral, and Jeffrey pine and blue spruce forests, and on talus slopes, dry hills, and ridges, occasionally on serpentine soils, at 50–2100 m. Its range extends from Washington south to southern California and east to Montana and western Wyoming, including the nothern portion of the Intermountain Region.

Calamagrostis koelerioides is similar to *C. rubescens* (see next). The two have traditionally been distinguished by the presence of hairs on the leaf collars in *C. rubescens*, and their absence in *C. koelerioides*; more reliable differences are the longer lemmas, glumes, and awns of *C. koelerioides* compared to *C. rubescens*.

8. Calamagrostis rubescens Buckley PINEGRASS, PINE REEDGRASS [p. 393, 504]

Pl smt with strl clm; smt loosely ces, usu with rhz 15+ cm long, 1.5–2 mm thick. **Clm** (50)60–100(105) cm, unbrchd, usu smooth, rarely slightly scabrous beneath the pan; **nd** (1)2–3(4). **Shth** smooth or slightly scabrous; **col** often hairy, rarely glab; **lig** (2)3–5(6) mm, truncate to obtuse, often lacerate; **bld** (6)8–40(42) cm long, (1)2–5(8) mm wide, usu flat, abx surfaces smooth or slightly scabrous, adx surfaces smooth or scabrous, glab or sparsely hairy. **Pan** (5)6–15(25) cm long, (0.7)1.5–2(2.7) cm wide, contracted to somewhat open, erect, usu greenish, infrequently purplish; **br** (1.2)2–4(10) cm, usu slightly scabrous, rarely densely long-scabrous, spklt-bearing to the base. **Spklt** (3)4–4.5(5.5) mm; **rchl prolongations** 0.6–1.5(2) mm, hairs 1.2–2 mm. **Glm** rounded to slightly keeled, mostly smooth, keels rarely slightly scabrous, lat veins usu obscure, rarely prominent, apc acute; **cal hairs** (0.5)1–1.5(2.5) mm, 0.2–0.5(0.7) times as long as the lm, sparse; **lm** 2.5–3.5(4) mm, (0.5)1–2 mm shorter than the glm; **awns** 2.8–3.5(4.5) mm, usu attached to the lo ⅕ of the lm, rarely higher, exserted, stout and readily distinguished from the cal hairs, strongly bent; **anth** (1)1.3–2(2.6) mm. $2n = 28, 42, 56$.

Calamagrostis rubescens grows at 50–2800 m, usually in open montane pine or aspen forests and parklands, infrequently in sagebrush steppes, chaparral, and meadows. It is primarily a species of interior western North America, although it reaches the Pacific coast in southern California.

Calamagrostis rubescens is similar to *C. koelerioides* (see previous). The two have traditionally been distinguished by the presence of hairs on the leaf collars of *C. rubescens*, and their absence from *C. koelerioides*; more reliable differences are the shorter lemmas, glumes, and awns of *C. rubescens*.

9. **Calamagrostis montanensis** (Scribn.) Vasey
PLAINS REEDGRASS [p. 393, <u>504</u>]
Pl with strl clm; ces, with rhz 6+ cm long, 1–2 mm
thick. **Clm** 15–50(54) cm, unbrchd, usu scabrous,
rarely smooth; **nd** 1–2. **Shth** and **col** smooth
or slightly scabrous; **lig** (1)2–3 mm, obtuse to
acute, more or less lacerate; **bld** (5)8–19(23) cm
long, (1)2–3(4) mm wide, usu involute, seldom
reaching the pan, abx surfaces scabrous, adx
surfaces usu scabrous, rarely smooth, glab or
sparsely hairy. **Pan** 4–9(14) cm long, (0.7)1–2(3.5)
cm wide, erect, contracted and not or only slightly
interrupted, yellowish green with a light purple
tinge; **br** 1.3–3(3.7) cm, sparsely short-scabrous
to densely long-scabrous, spklt-bearing to the
base. **Spklt** (3)3.5–4.5(7) mm; **rchl prolongations**
about 1 mm, densely bearded, hairs to 2 mm.
Glm keeled, smooth or scabrous throughout, lat
veins usu somewhat obscure, rarely prominent,
apc acute to acuminate, rarely awn-tipped; **cal
hairs** (1)1.5–2(2.5) mm, 0.4–0.8 times as long as
the lm, abundant; **lm** (2.5)3–3.5(5.5) mm, 0.5–1(2)
mm shorter than the glm; **awns** (1)2–3(4) mm,
usu attached to the lo ¹⁄₁₀–²⁄₅ of the lm, rarely
above the mid, smt slightly exserted, stout,
distinguishable from the cal hairs, slightly bent;
anth (1.1)1.8–2.4(3) mm. $2n = 28$.

Calamagrostis montanensis inhabits prairie grasslands and
sagebrush flats, benchlands, valley bottoms, and occasionally
woodlands, at 200–2600 m. It grows in the continental interior
from eastern British Columbia and adjacent Alberta, south
to southern Wyoming and east to Manitoba and western
Minnesota. *Calamagrostis montanensis* may be mistaken for *C.
purpurascens* (p. 153), but the latter species has hairy adaxial
leaf surfaces and longer awns.

10. **Calamagrostis canadensis** (Michx.) P.
Beauv. BLUEJOINT [p. 394, <u>504</u>]
Pl with strl clm; ces, with rhz 2–15+ cm long, 1–3
mm thick. **Clm** (32)65–112(180) cm, often brchg
above the base, smooth or slightly scabrous
beneath the pan; **nd** (2)3–7(8). **Shth** smooth or
scabrous; **col** usu scabrous, rarely smooth or
hairy; **lig** (1)3–8(12) mm, lacerate; **bld** (10)16–
31(50) cm long, 2–8(11) mm wide, flat, lax, abx
surfaces scabrous, adx surfaces usu strongly
scabrous, rarely smooth or with scattered hairs,
often glaucous. **Pan** (6)9–17(25) cm long, (1)2–4(8)
cm wide, often contracted when young, open at
maturity, nodding, usu purplish, smt greenish to
straw-colored; **br** 2.7–6(12) cm, scabrous, spklt
sparsely to densely concentrated on the distal ²⁄₃.
Spklt 2–4.5(5.2) mm; **rchl prolongations** 0.5–1
mm, hairs 1.5–3.2 mm. **Glm** rounded or keeled,
smooth or scabrous, keels often long-scabrous,

lat veins obscure to prominent, apc acute to
acuminate; **cal hairs** (1.5)2–3.5(4.5) mm, (0.5)0.9–
1.2(1.5) times as long as the lm, abundant; **lm**
2–3.1(4) mm, 0–2.1 mm shorter than the glm;
awns 0.9–3.1 mm, attached to the lo (¹⁄₁₀)¹⁄₅–¹⁄₂(⁷⁄₁₀)
of the lm, usu not exserted, delicate, often difficult
to distinguish from the cal hairs, usu straight;
anth (0.8)1.2–1.6(2.6) mm. $2n = 42$–66.

Calamagrostis canadensis is a species of moist meadows,
thickets, bog edges, and forest openings, and grows from sea
level to 3400 m. Its range includes much of the Intermountain
Region, but it is notably absent from most of Nevada.

A high degree of pollen sterility has been documented
in some populations, suggesting that seed formation
via apomixis is common; sexual reproduction is also
documented. The many forms, varieties, and subspecies
that have been described for this species probably represent
clones. Only one of the three recognized varieties grows in
the Intermountain Region.

Calamagrostis canadensis (Michx.) P. Beauv.
var. **canadensis** [p. 394]
Clm (50)65–80(160) cm, often brchg above the
base, smooth or slightly scabrous beneath the
pan; **nd** (2)3–4(6). **Shth** smooth or scabrous; **col**
usu scabrous, rarely smooth or hairy; **lig** (1)4–
6(12) mm; **bld** (11)16–31(41) cm long, (2)2.5–5(8)
mm wide, adx surfaces usu strongly scabrous,
rarely smooth, often glaucous. **Pan** (9)11–14(19)
cm long, (1)2–3(7) cm wide, often contracted
when young, usu purplish, smt greenish to
straw-colored; **br** 2.9–4.5(5.7) cm. **Spklt** 2.5–3.5(4)
mm; **rchl prolongations** 0.5–1 mm, hairs 2.5–3.2
mm. **Glm** rounded to broadly keeled, smooth or
scabrous, often only the keels scabrous, prickles
straight, midveins raised, lat veins obscure to
prominent, apc usu acute, rarely acuminate; **cal
hairs** (1.7)2.5–2.9(3.1) mm, (0.7)0.9–1.1(1.4) times
as long as the lm; **lm** (2.2)2.5–3.1(4) mm, 0–1.6
mm shorter than the glm; **awns** 0.9–2.6 mm,
attached to the lo (¹⁄₁₀)¹⁄₅–²⁄₅(⁷⁄₁₀) of the lm, usu not
exserted, usu straight; **anth** (0.8)1.2–1.3(2) mm.
$2n = 42$–66.

Calamagrostis canadensis var. *canadensis* is widespread
throughout the range of the species.

11. **Calamagrostis stricta** (Timm) Koeler
SLIMSTEM REEDGRASS [p. 394, <u>504</u>]
Pl rarely with strl clm; ces, usu with rhz shorter
than 5 cm, 1–1.5 mm thick. **Clm** (10)35–90(120)
cm, usu unbrchd, smooth to slightly scabrous;
nd 1–3(4). **Shth** usu smooth; **col** usu smooth, smt
scabrous, rarely pubescent; **lig** (0.5)1–5.5(6) mm,
truncate to obtuse, usu entire, smt lacerate; **bld**
(5)11–25(34) cm long, (1)1.5–5(6) mm wide, flat
or involute, usu scabrous, rarely smooth, smt

puberulent. **Pan** (2)4–18(29) cm long, (0.7)1–2(2.8) cm wide, erect, contracted, smt interrupted, pale green to purple; **br** 1.4–5(9.5) mm, smooth or scabrous, usu spklt-bearing to or near the base, smt only to midlength. **Spklt** 2–4(5) mm; **rchl prolongations** 0.5–1.5 mm, hairs 1.5–3 mm. **Glm** usu less than 3 times as long as wide, rounded or keeled, usu smooth, rarely scabrous, keels smooth or scabrous, veins prominent to obscure, apc acute; **cal hairs** (1)1.5–3(4.5) mm, (0.5)0.7–0.9(1.3) times as long as the lm, abundant; **lm** 2–4(5) mm, 0.1–1.5 mm shorter than the glm; **awns** 1.5–2.5 mm, usu attached to the lo ¹⁄₁₀–¹⁄₂ of the lm, rarely beyond the midpoint, equaling or exserted slightly beyond the mrg of the glm, usu stout, rarely slender, usu distinguishable from the cal hairs, straight or bent; **anth** (0.9)1.2–1.8(2.4) mm, often strl.

Calamagrostis stricta grows throughout northern North America; it also is found in Europe and northeastern Asia. It grows in habitats ranging from meadows and grassland to wetlands, sandy shorelines, and sand dunes, from sea level to 3400 m. Primarily a species of open settings, it is frequently found in association with shrubs. It has a notable, but not exclusive, association with alkaline to saline substrates.

Calamagrostis stricta comprises both sexual and apomictic populations. Two subspecies are known from the Intermountain Region.

1. Spikelets 3–4(5) mm long; callus hairs 2–4.5 mm long; rachilla prolongations 1–1.5 mm long; panicle branches 1.5–9.5 cm long; culms usually scabrous, sometimes smooth . . . . subsp. *inexpansa*
1. Spikelets 2–2.5(3) mm long; callus hairs 1–3 mm long; rachilla prolongations 0.5–1 mm long; panicle branches 1.4–4 cm long; culms usually smooth, sometimes slightly scabrous . . . . subsp. *stricta*

Calamagrostis stricta subsp. inexpansa
(A. Gray) C.W. Greene NORTHERN REEDGRASS [p. 394]

Pl apparently without strl clm. **Clm** (29) 35–75(120) cm, usu scabrous, smt smooth. **Shth** and **col** usu smooth, col smt scabrous, rarely pubescent; **lig** (0.5)2–5.5(6) mm; **bld** (5)11–24(34) cm long, (1.5)2–5(6) mm wide, flat, usu stiff, smt puberulent. **Pan** (6)8–11(29) cm long, (0.8)1–2(2.8) cm wide, pale green, smt purple-tinged; **br** (1.5)1.6–5(9.5) cm, spklt-bearing to the base. **Spklt** 3–4(5) mm; **rchl prolongations** 1–1.5 mm. **Glm** broadly keeled or rounded; **cal hairs** (2)2.5–3(4.5) mm, (0.5)0.7–0.9(1.3) times as long as the lm; **lm** 2.5–3.5(5) mm, 0.1–1(1.4) mm shorter than the glm; **awns** 2–2.5 mm, attached to the lo ¹⁄₁₀–²⁄₅ of the lm, rarely beyond the midpoint, straight or somewhat to strongly bent; **anth** (0.9)1.5–1.8(2.4) mm, often poorly developed, strl and indehiscent. 2*n* = 28, 56, 58, 70, 84–±120.

Calamagrostis stricta subsp. *inexpansa* differs from subsp. *stricta* in its more robust growth and coarser habit. In North America, it extends from Alaska to Labrador and Newfoundland and south to California, Arizona, Minnesota, Iowa, Ohio, and New York, usually in moist meadows, sphagnum bogs, and grasslands associated with rivers and streams, and less frequently on grassy slopes, in open woods, and beside sand dunes. It grows at the edge of, rather than in, wetlands.

Calamagrostis stricta (Timm) Koeler subsp. stricta [p. 394]

Pl rarely with strl clm. **Clm** (10)35–90(100) cm, usu smooth, smt slightly scabrous beneath the pan. **Shth** and **col** smooth; **lig** (0.5)1–3.5(4) mm; **bld** (9)13–25 cm long, (1)1.5–2.5(3) mm wide, flat or often involute. **Pan** (2)8–10(13) cm long, (0.7)1–2(2.5) cm wide, smt interrupted, purple-tinged; **br** (1.4)2–2.5(4) cm, spklt-bearing to below midlength, smt to the base. **Spklt** 2–2.5(3) mm; **rchl prolongations** 0.5–1 mm. **Glm** keeled; **cal hairs** (1)1.5–2(3) mm, (0.5)0.7–0.8(1) times as long as the lm; **lm** 2–3 mm, 0.1–1.5 mm shorter than the glm; **awns** 1.5–2.5 mm, attached to the lo ¹⁄₁₀–¹⁄₂ of the lm, rarely slender, usu straight, smt bent; **anth** (1.1)1.2–1.4(1.7) mm, usu ftl and well filled with pollen, dehiscent. 2*n* = 28, 42, 56, ±70.

A circumboreal taxon, *Calamagrostis stricta* subsp. *stricta* favors moist meadows and fens, occurring less frequently in marshes and bogs, and sometimes near sand dunes. It is usually associated with fine-textured substrates. It grows throughout much of North America.

8.27 SCOLOCHLOA Link   Mary E. Barkworth

Pl per; strongly rhz, rhz succulent. **Clm** 70–200 cm, smt rooting at the lo nd. **Shth** open; **aur** absent; **lig** memb, truncate to rounded; **bld** 4–12 mm wide, flat. **Infl** tml, open pan, some br longer than 1 cm. **Spklt** pedlt, lat compressed, with 3–4 flt; **rchl** prolonged beyond the base of the distal flt, intnd glab or with a few hairs to 0.2 mm long; **dis** above the glm, between the flt. **Glm** unequal, acute to acuminate, unawned; **lo glm** shorter than the adjacent lm, 1–5-veined; **up glm** usu exceeding, smt equaling, the distal flt, 3–7-veined; **cal** short, blunt, hairy; **lm** glab, rounded, 3–9-veined, veins excurrent, apc indistinctly 3-lobed or toothed, unawned; **pal** about as long as the lm; **lod** free, glab; **anth** 3; **ov** with pubescent apc. **Car** shorter than the lm, concealed at maturity, about 2 mm, dorsiventrally compressed. *x* = 7. Name from the Greek *scolos*, 'cusp' or 'prickle', and *chloa*, 'grass', an allusion to the excurrent lemma veins.

Scolochloa is a monospecific genus that grows in shallow water and marshes of the temperate regions of North America and Eurasia.

1. **Scolochloa festucacea** (Willd.) Link COMMON RIVERGRASS, WHITETOP [p. 395, 504]

Clm 70–200 cm tall, 6–8 mm thick at the base. Lig 3–7(9) mm; bld 20–45 cm long, 4–12 mm wide. Pan 15–30 cm; pri br ascending to divergent, with spklt appressed to the br. Spklt 7–11 mm. Lo glm 5–8 mm; up glm 6.5–10 mm; cal about 0.5 mm; lm 4–9 mm; pal about as long as the lm; anth 2–4 mm. $2n = 28$.

Scolochloa festucacea grows in ponds, marshes, seasonally flooded basins, and the shallow margins of freshwater to moderately saline lakes and streams. It provides good nesting cover for some waterfowl and shorebirds, and can provide valuable forage for livestock and wildlife. It grows primarily in the northern Great Plains and prairie pothole region. Populations in the Intermountain Region probably reflect introductions.

8.28 AVENA L.

Bernard R. Baum

Pl ann or per. Clm 8–200 cm, erect or decumbent. Shth open; aur absent; lig memb; bld usu flat, smt involute, lax. Infl pan, diffuse, smt 1-sided, some br longer than 1 cm. Spklt 15–50 mm, pedlt, lat compressed, with 1–6(8) flt; rchl not prolonged beyond the base of the distal flt; dis above the glm, usu also beneath the flt, cultivated forms not dis. Glm usu exceeding the flt, memb, glab, 3–11-veined, acute, unawned; cal rounded to pointed, with or without hairs; lm usu indurate and enclosing the car at maturity, 5–9-veined, often with twisted, strigose hairs below midlength, apc dentate to bifid or biaristate, usu awned, smt unawned, awns dorsal, usu once-geniculate and strongly twisted in the bas portion; pal bifid or entire, keels ciliate; lod 2, free, glab, toothed or not toothed; anth 3; ov hairy. Car shorter than the lm, concealed at maturity, terete, ventrally grooved, pubescent; hila linear. $x = 7$. Name from the Latin *avena*, 'oats'.

Avena, a genus of 29 species, is native to temperate and cold regions of Europe, North Africa, and central Asia; it has become nearly cosmopolitan through the cultivation of cereal oats, and the inadvertent introduction of the weedy species. Three species have been introduced to the Intermountain Region.

1. Florets not disarticulating from the glumes, remaining attached to the plant even at maturity; calluses glabrous . . . . 3. *A. sativa*
1. Florets disarticulating at maturity, only the glumes remaining attached; calluses bearded.
   2. Lemma apices biaristate, 2 veins extending 2–4 mm beyond the apices . . . . 1. *A. barbata*
   2. Lemma apices erose to bifid, the veins not extending beyond the apices . . . . 2. *A. fatua*

1. **Avena barbata** Pott *ex* Link SLENDER OATS, SLENDER WILD OATS [p. 395, 504]

Pl ann. Clm 60–80 (150) cm, initially prostrate, usu becoming erect. Shth of the bas lvs pilose, up shth usu glab; lig 1–6 mm, obtuse; bld 6–30 cm long, 2–20 mm wide, glab or pilose. Pan 15–35.5 (50) cm long, 6–12 cm wide, erect or nodding. Spklt 21–30 mm, with 2–3 flt; dis beneath each flt; dis scars elliptic to triangular. Glm subequal, 15–30 mm, 7–9-veined; cal bearded, hairs 2–3 mm; lm 15–26 mm, densely strigose below midlength, apc acute, biaristate, 2 veins extending 2–4 mm beyond the apc, awns 30–45 mm, arising about midlength, geniculate; lod narrowly triangular, without lobes on the wings; anth 2.5–4 mm. $2n = 28$.

Avena barbata is native to the Mediterranean region and central Asia. It has become naturalized in western North America, particularly California, displacing native grasses. Although it has been collected in the Intermountain Region, it is not established there.

2. **Avena fatua** L. WILD OATS [p. 396, 504]

Pl ann. Clm 8–160 cm, prostrate to erect when young, becoming erect at maturity. Shth of the bas lvs with scattered hairs, up shth glab; lig 4–6 mm, acute; bld 10–45 cm long, 3–15 mm wide, scabridulous. Pan 7–40 cm long, 5–20 cm wide, nodding. Spklt 18–32 mm, with 2(3) flt; dis beneath each flt; dis scars of all flt round to ovate or triangular. Glm subequal, 18–32 mm, 9–11-veined; cal bearded, hairs to ¼ the length of the lm; lm 14–22 mm, usu densely strigose below midlength, smt sparsely strigose or glab, veins not extending beyond the apc, apc usu bifid, teeth 0.3–1.5 mm, awns 23–42 mm, arising in the mid ⅓ of the lm; lod without lobes on the wings; anth about 3 mm. $2n = 42$.

Avena fatua is native to Europe and central Asia. It is known as a weed in most temperate regions of the world; it is considered a noxious weed in some parts of Canada and the United States.

Hybrids between *Avena fatua* and *A. sativa* are common in plantings of cultivated oats. The hybrids resemble *A. sativa*, but differ in having the *fatua*-type lodicule; some also have a weak awn on the first lemma. They are easily confused with fatuoid forms of *A. sativa*.

3. Avena sativa L. Oats, Cultivated Oats, Naked Oats [p. 397, <u>504</u>]

Pl ann. **Clm** 35–180 cm, prostrate to erect when young, becoming erect at maturity. **Shth** smooth or scabridulous; **lig** 2–8 mm, truncate to acute; **bld** 8–45 cm long, 3–14 (25) mm wide, scabridulous. **Pan** (6)15–40 cm long, 5–15 cm wide, nodding. **Spklt** (18)25–32 mm, to 50 mm in 'naked oats', with 1–2 flt (to 7 in 'naked oats'); **dis** not occurring, the flt remaining attached even when mature. **Glm** subequal, (18)20–32 mm, 9–11-veined; **cal** glab; **lm** 14–18 mm, usu indurate, memb in 'naked oats', usu glab, smt sparsely strigose, apc erose to dentate, longest teeth 0.2–0.5 mm, usu unawned, smt awned,

awns 15–30 mm, arising in the mid ⅓, weakly twisted, not or only weakly geniculate; **lod** with a lobe or tooth on the wings, this smt very small; **anth** (1.7)3–4.3 mm. 2*n* = 42.

Avena sativa, a native of Eurasia, is widely cultivated in cool, temperate regions of the world, including North America. Fall-sown oats are planted in the Pacific and southern states in the United States; spring-sown oats are more important elsewhere in North America. It is sometimes planted as a fast-growing soil stabilizer along roadsides. Several forms are grown, of which the most distinctive are 'naked oats'. These differ from typical forms as indicated in the description, and in having caryopses that fall from the florets. Escapes from cultivation are common but rarely persist.

Avena sativa hybridizes readily with *A. fatua*, forming hybrids with the *fatua*-type lodicule. The hybrids are easily confused with fatuoid forms of *A. sativa*, which differ in having the *sativa*-type lodicule.

8.29 HOLCUS L.

Lisa A. Standley

Pl usu per, rarely ann; csp or rhz, rarely both csp and rhz. **Clm** (8)20–200 cm, glab or pubescent; **nd** glab or retrorsely pubescent. **Shth** open; **aur** absent; **lig** 1–5 mm, memb, entire or erose-ciliate, glab or puberulent; **bld** flat, pubescent. **Infl** tml pan, contracted to open. **Spklt** lat compressed, with 2(3) flt, lo flt bisex, up flt(s) stmt or strl; **rchl** curved below the lowest flt, smt prolonged beyond the base of the distal flt; **dis** below the glm. **Glm** equaling to exceeding the flt, strongly keeled, unawned; **lo glm** 1-veined; **up glm** 3-veined; **cal** glab or pubescent; **lm** firm, shiny, glab or pubescent, obscurely 3–5-veined, often bidentate; **lo lm** unawned; **up lm** awned from below the apc, awns hooked or geniculate; **pal** thin, subequal to the lm; **lod** 2, glab, toothed or not; **anth** 3; **ov** glab. **Car** shorter than the lm, concealed at maturity, glab. *x* = 4, 7. Name from the Greek *holkos*, a kind of grain, perhaps sorghum.

Holcus, a genus of eight species, is native to Europe, North Africa, and the Middle East. One species, *H. lanatus*, has become widely naturalized in the Americas, Japan, and Hawaii. It is the only species that has been found in the Intermountain Region.

1. Holcus lanatus L. Velvetgrass, Yorkshire Fog [p. 397, <u>504</u>]

Pl per; ces, not rhz. **Clm** 20–100 cm, erect, smt decumbent; **lo intnd** densely pilose, hairs to 1 mm; **uppermost intnd** often glab. **Shth** densely pubescent; **lig** 1–4 mm, truncate, erose-ciliolate; **bld** 2–20 cm long, (3)5–10 mm wide, densely soft-pubescent. **Pan** 3–15(20) cm long, 1–8 cm wide; **br** hairy; **ped** 0.2–1.6(4) mm, pilose, hairs to 0.3 mm. **Spklt** 3–6 mm; **rchl** 0.4–0.5 mm, glab. **Glm** exceeding and enclosing the flt, memb, ciliate on the keels and veins, usu scabrous, puberulent, or villous between the veins, especially towards the

apc, whitish green, often purple over the veins and towards the apc; **lo glm** lanceolate, narrow, acute; **up glm** ovate, wider and longer than the lo glm, midveins often prolonged as an awn to 1.5 mm, apc obtuse, somewhat bifid; **cal** sparsely hirsute; **lm** 1.7–2.5 mm, acute, erose-ciliate; **up lm** shallowly bifid, awns 1–2 mm, often purple-tipped, slightly twisted and forming a curved hook at maturity; **anth** (1.2)2–2.5 mm. 2*n* = 14.

Holcus lanatus grows in disturbed sites, moist waste places, lawns, and pastures, in a wide range of edaphic conditions and at elevations from 0–2300 m. A native of Europe, it was widely distributed in North America by 1800.

8.30 ARRHENATHERUM P. Beauv.

Stephan L. Hatch

Pl per; csp, smt rhz. **Clm** 30–200 cm, bas intnd occ globose. **Shth** open, not overlapping; **aur** absent; **lig** memb, smt ciliate; **bld** flat or convolute. **Infl** tml, narrow pan; **br** spreading until after anthesis, then becoming loosely appressed to the rchs. **Spklt** pedlt, lat compressed, with 2 flt, lo flt stmt, up flt pist or bisex, a rdmt flt occ present distally; **rchl** pubescent; **dis** above the glm, the flt usu falling together, rarely falling separately. **Glm** unequal, hyaline, unawned; **lo glm** less than ¾ the length of the up glm, 1- or 3-veined; **up glm** 3-veined; **cal** short, blunt, pubescent; **lo lm** memb, 3–7-veined, acute, awned below midlength, awns twisted and geniculate; **up lm** memb to subcoriaceous, glab or hairy, 7-veined, acute, usu unawned, smt awned from near the apc, awns short, straight, rarely awned similarly to the lo lm; **pal** subequal to the lm, 2-veined, 2-keeled, keels scabrous or hairy, apc notched; **lod** 2, free, linear, memb, glab, entire; **anth** 3, 3.4–6.5 mm; **ov** pubescent. **Car** shorter than the lm, concealed at maturity, not grooved, dorsally compressed to terete, hairy; **hila** long-linear. $x = 7$. Name from the Greek *arren*, 'masculine', and *ather*, 'awn', referring to the awned staminate florets.

Arrhenatherum is a Mediterranean and eastern Asian genus of six species; one species has become established in North America.

1. **Arrhenatherum elatius** (L.) P. Beauv. *ex* J. Presl & C. Presl Tall Oatgrass [p. 398, <u>504</u>]

Pl loosely ces, smt rhz, rhz to 3 mm thick. **Clm** 50–140 (180) cm, erect, glab, unbrchd, bas intnd swollen or not; **nd** usu glab, occ puberulent to densely hairy. **Shth** smooth; **lig** 1–3 mm, obtuse to truncate, usu ciliate; **bld** 5–32 cm long, (1)3–8(10) mm wide, flat, usu glab, rarely shortly pilose, smt scabrous. **Pan** 7–30(36) cm long, 1–6(10) cm wide, green, shiny, becoming stramineous, smt purple-tinged; **br** 15–20 mm, ascending to divergent, verticillate, usu spklt-bearing to the base; **ped** 1–10 mm. **Spklt** 7–11 mm; **rchl** stout, intnd to 0.7 mm, prolongations 1.2–2 mm, slender, apc often with a small, club-shaped rudiment. **Glm** lanceolate to elliptic; **lo glm** 4–7 mm; **up glm** 7–10 mm; **cal hairs** to 3.7 mm; **lm** (4)7–10 mm, apc bifid; **awns of lo lm** 10–20 mm, twisted below, often with alternating light and dark bands; **awns of up lm** absent or to 5 mm and arising just below the apc, rarely to 15 mm and arising from above the mid; **pal** 0.5–1 mm shorter than the lm, acute; **anth** 3.6–5(6) mm. **Car** 4–5 mm long, about 1.2 mm wide, ellipsoid, densely hairy, yellowish. $2n = 14, 28, 42$.

Arrhenatherum elatius is grown as a forage grass and yields palatable hay; it does not withstand heavy grazing. It readily escapes from cultivation, and can be found in mesic to dry meadows, the edges of woods, streamsides, rock outcrops, and disturbed areas such as fields, pastures, fence rows, and roadsides. Variegated forms, with the leaves striped green and white or yellow, are cultivated as ornamentals. There are two subspecies, both of which have been found in North America.

Arrhenatherum elatius (L.) P. Beauv. *ex* J. Presl & C. Presl subsp. *elatius* has glabrous nodes and basal internodes 2–4 mm thick. It is more common than *A. elatius* subsp. *bulbosum* (Willd.) Schübl. & G. Martens, which has densely hairy nodes and swollen basal internodes 5–10 mm thick. While both can be weedy, the latter subspecies is especially difficult to control in cultivated fields, because tilling the soil spreads the swollen internodes, which then propagate vegetatively. Plants in which both lemmas have long, geniculate awns have been called *A. elatius* var. *biaristatum* (Peterm.) Peterm., but do not merit formal taxonomic recognition.

8.31 **TRISETUM** Pers.

Mary E. Barkworth

Pl ann or per; smt rhz, smt csp. **Clm** 5–150 cm, glab or pubescent, **Bas brchg** exvag. **Shth** open the entire length or fused at the base; **aur** absent; **lig** memb, often erose to lacerate, smt ciliolate; **bld** rolled in the bud. **Infl** tml pan, open and diffuse to dense and spikelike; **br** antrorsely scabrous. **Spklt** 2.5–12 mm, usu subsessile to pedlt, rarely sessile, lat compressed, with 2–5 flt, rdcd flt smt present distally; **rchl** hairy, intnd evident, prolonged beyond the base of the distal bisex flt; **dis** usu initially above the glm and beneath the flt, subsequently below the glm, in some species initially below the glm. **Glm** subequal or unequal, keels scabrous, apc usu acute, unawned, often apiculate; **lo glm** 1(3)-veined; **up glm** 3(5)-veined, lat veins less than ½ the glm length; **cal** hairy; **lm** 3–7-veined, mrg hyaline, unawned or awned from above the mid with a single awn, apc usu bifid, smt entire; **pal** from subequal to longer than the lm, memb, 2-veined, veins usu extending as bristlelike tips; **lod** 2, shallowly and usu slenderly lobed to fimbriate; **anth** 3; **ov** glab or pubescent; **sty** 2. **Car** shorter than the lm, concealed at maturity, elongate-fusiform, compressed, brown; **emb** elliptic, to ⅓ the length of the car; **endosperm** milky. $x = 7$. Name from the Latin *tres*, 'three', and *seta*, 'bristle', alluding to the three-awned appearance of the lemmas of the type species, *Trisetum flavescens* (L.) P. Beauv.

Trisetum, a genus of approximately 75 species, occurs primarily in temperate, subarctic, and alpine regions. Six species are native to the Intermountain Region. This treatment reflects, at the species level, the taxonomy adopted by Finot et al. (2005). The descriptions and keys were revised by Barkworth and Anderton based on a combination of those provided by Finot et al. (2005) and Rumely (2007) and examination of specimens in the Intermountain Herbarium.

Trisetum usually differs from *Sphenopholis* in having longer awns that are inserted lower on the lemmas, and spikelets that disarticulate above the glumes. It differs from *Deschampsia* primarily in its more acute, bifid lemmas. In addition, all species of *Trisetum* have awns that are inserted at or above the midpoint of the lemmas; in *Deschampsia*, the awns are usually inserted at or below midlength, often near the base.

Trisetum spicatum is important as forage on native rangelands. Like other species of the genus, it is a significant component of natural food pyramids, especially in arctic and alpine regions and mountain parks.

Spikelet measurements do not include the awns.

1. Lemmas entire or slightly bilobed, unawned or with straight awns up to 2 mm long that scarcely exceed the lemma apices [*Graphephorum* Finot] . . . . 1. *T. wolfii*
1. Lemma apices bidentate, lateral veins extending into 2(4) teeth, at least some lemmas with a dorsal awn 1.3–16 mm long [*Trisetum sensu stricto*].
    2. Culms densely hairy below the panicles, hairs (0.2)0.5–1 mm; anthers 0.5–1 mm . . . . 4. *T. spicatum*
    2. Culms glabrous below the panicles or, if hairy, hairs about 0.1 mm; anthers 0.8–3 mm.
        3. Awns 3.5–4 mm; anthers 0.8–1.2 mm; ovaries glabrous . . . . 6. *T. montanum*
        3. Awns 5–14 mm; anthers 1–3 mm; ovaries glabrous or hairy near the apices.
            4. Leaf blades densely pilose on both surfaces; upper glumes equaling or exceeding the florets; panicles dense, individual branches not apparent; ligules densely pilose on dorsal surface . . . . 5. *T. projectum*
            4. Leaf blades glabrous, canescent, or sparsely pilose on the adaxial surface, glabrous or canescent on the abaxial surface; ligules glabrous or pilose, but not densely pilose, on the dorsal surface.
                5. Most panicle branches spikelet-bearing their full length, ascending to somewhat divergent; lower glumes 3–5 mm . . . . 2. *T. canescens*
                5. Most panicle branches spikelet-bearing only distally, branches at lowest 1–3 whorls strongly divergent to reflexed . . . . 3. *T. cernuum*

1. **Trisetum wolfii** Vasey WOLF'S TRISETUM
[p. *398*, <u>504</u>]

Pl per, with both ftl and strl sht; shortly rhz. **Clm** 20–100 cm, erect, glab or retrorsely pubescent below the nd. **Lvs** usu concentrated on the lo ⅓ of the clm; **shth** glab or sparsely retrorsely pilose, smt scabridulous; **lig** (1.2)2.5–4(6) mm, glab dorsally, apc truncate to rounded, dentate; **bld** to 6–20 cm long, 2–7 mm wide, flat or involute distally, ascending, lax, both surfaces smooth or scabridulous, glab or sparsely pilose, apc often prowlike. **Pan** (8)20–40(50) cm long, usu 1–1.5 cm wide, stiffly erect, green, tan, or purple-tinged, rchs scabrous; **br** 10–55 mm long, appressed-ascending, the spklt evenly distributed. **Spklt** 4–7(8) mm, usu subsessile, rarely on ped to 4 mm, ovate, with 2(3) flt; **rchl intnd** 1.5–2 mm, hairy, hairs to 0.5–1.5 mm. **Glm** subequal, usu longer than the lowest flt; **lo glm** 4–7 mm, 1-veined, keels scabrous distally; **up glm** 4–6.5 mm, slightly wider than the lo glm, 3-veined, keels scabrous distally; **cal** hairy, hairs shorter than 0.5 mm; **lm** 4–6.5 mm, lanceolate, firmer than the glm, scabridulous-puberulent, obscurely bifid, unawned or awned, awns to 2 mm, arising just below and rarely exceeding the apc; **pal** shorter than the lm; **anth** (0.6)1(1.5) mm. **Car** 2.4–3 mm, pubescent. 2*n* = 14.

Trisetum wolfii grows in moist meadows and marshes, and on stream banks in aspen groves and parks in the spruce-fir forest zone, at medium to high, but usually not alpine, elevations. It is restricted to southwestern Canada and the western United States.

Finot et al. (2005) treated *Trisetum wolfii* as one of two species in the segregate genus *Graphephorum*. Morphologically, they resemble *Sphenopholis*. In the only published phylogeny to include *T. wolfii*, it was sister to *Sphenopholis intermedia* (Quintanar et al. 2007), and the other ten species of *Trisetum* are in three other clades. It seems best to retain *T. wolfii* in *Trisetum* pending more comprehensive evaluation of its phylogenetic relationships.

2. **Trisetum canescens** Buckley TALL TRISETUM
[p. *399*, <u>504</u>]

Pl per, smt with both ftl and strl sht; ces, not rhz. **Clm** 40–120 cm, clumped, erect, usu smooth. **Lvs** 3–4 per clm; **shth** crisped-pubescent to shaggy-pilose, scabrous or smooth; **lig** (1.5)3–6 mm, rounded to truncate; **bld** 10–30 cm long, (3)7–10 mm wide, flat, erect, lax, mrg and occ the surfaces with scattered 1–3 mm hairs. **Pan** 10–25 cm long, (0.75)1–3(4) cm wide, erect or nodding at the apc, green or tan, occ purple-tinged; **br** 1–5.5 cm, ascending to somewhat divergent, most spklt-bearing for their full length, smt the lowermost br naked below. **Spklt** 7–9 mm, pedlt, with 2–4 flt; **rchl intnd** 1.5–3 mm, hairy, hairs 0.7–1 mm; **dis** above the glm, beneath the flt. **Glm** unequal to subequal; **lo glm** 3–5 mm, narrow, lanceolate to subulate, acute or long-tapered; **up glm** (3.5)5–7(9) mm long, shorter than the lowest flt, at least twice as wide as the lo glm, broadly lanceolate to obovate, widest at or below the mid, tapering to the apc, acute; **cal** hairy, hairs about 0.5 mm; **lm** 5–7 mm, glab, apc bifid, teeth to 2.5(3.2) mm, setaceous, awned, awns 7–14 mm, usu arising on the up ⅓ of the lm, exceeding the apc, geniculate;

pal as long as or slightly longer than the lm; **anth** 1–3 mm. **Car** usu to 3 mm, glab or finely hairy distally. 2*n* = 28, 42.

Trisetum canescens grows on or near stream banks and in forest margins or interiors, in moist to dry areas in western North America. It is especially abundant in ponderosa pine stands and spruce-fir forests. The vestiture of different parts varies throughout the range of the species.

3. Trisetum cernuum Trin. NODDING TRISETUM [p. 399, 504]

Pl per, with both ftl and strl sht; ces, not rhz. **Clm** (30)50–110 cm, clumped, erect, glab or pubescent. **Lvs** 2–3 per clm; **shth** scabridulous or pilose; **lig** 1.5–3 mm, truncate, erose to lacerate; **bld** (8.5)15–20+ cm long, (3)7–12 mm wide, flat, ascending, lax at maturity, often scabridulous. **Pan** 10–30 cm long, (1)2–9 cm wide, open, nodding, green or tan, occ purple-tinged; **br** 2–12+ cm, most, except smt the uppermost, spklt-bearing only towards the apc, with the bas (⅕)⅓–½ bare, filiform, flexuous, at least the lowest 1–3 whorls spreading or drooping. **Spklt** 6–12 mm, subsessile to pedlt, ped to 2 cm, usu with 2–3 fnctl flt below 1–2 rdcd flt; **rchl intnd** and **hairs** 1–2.5 mm; **dis** above the glm, beneath the flt. **Glm** unequal; **lo glm** 0.75–2(3) mm, subulate; **up glm** 3.5–5 mm long, shorter than the lowest flt, 2–3 times as wide as the lo glm, widest at or above the mid, ovate or obovate, rounded to the acuminate apc; **cal** hairy, hairs to 1 mm; **lm** 5–6 mm, broadly lanceolate, glab, bifid, teeth to 1.3 mm, awned, awns (7)9–14 mm, arising from above midlength to just below the teeth, exceeding the lm apc, arcuate to flexuous; **pal** shorter than the lm; **anth** about 1 mm. **Car** 2.5–3.2 mm, densely to sparsely pubescent. 2*n* = 42.

Trisetum cernuum grows in moist woods, on stream banks, lake and pond shores, and floodplains of western North America. The hairiness of the leaf sheaths varies, often within a plant.

4. Trisetum spicatum (L.) K. Richt. SPIKE TRISETUM [p. 400, 504]

Pl per, with both ftl and strl sht; ces, not rhz. **Clm** 10–120 cm, clumped, erect, usu glab, smt villous, smt scabridulous. **Lvs** mostly bas or evenly distributed; **shth** variously pubescent or glab; **lig** 0.5–4 mm, truncate or rounded; **bld** (3)10–20(40) cm long, 1–5 mm wide, flat, folded, or involute, erect and stiff or ascending and lax. **Pan** 2.5–7(10) cm long, (0.5)1–2.5(5) cm wide, spikelike to open, often interrupted bas, green, purplish, or tawny, usu silvery-shiny; **br** with the spklt evenly distributed. **Spklt** 5–7.5 mm, sessile, subsessile, or on ped to 1.5(3.5) mm, with 2(3) flt;

rchl intnd 0.5–1.5 mm, hariy, hairs to 1 mm. **Glm** subequal to unequal, lanceolate, usu smooth, smt sparsely scabrous, smt pilose, with wide scarious mrg, apc acute to acuminate, smt apiculate; **lo glm** 3–4(5.5) mm; **up glm** 4–7 mm long, as long as or longer than the lowest flt, less than twice as wide as the lo glm; **cal** hairy, hairs to 1 mm; **lm** 3–6(7) mm, narrowly to broadly lanceolate, glab or pilose, smt scabridulous, apc bifid, teeth usu shorter than 1 mm, awned, awns 3–8 mm, arising from the up ⅓ of the lm and exceeding the apc, geniculate, twisted bas; **anth** 0.7–1.4 mm. **Car** 1.5–3(4) mm, glab. 2*n* = 14, 28, 42.

Trisetum spicatum grows in moist meadows and forests, and on rock ledges, tundra slopes, and screes, at 0–4300 m. Its range includes both North and South America and Eurasia. Many infraspecific taxa have been based on the variation in vestiture and openness of the panicle, but none appears to be justified (see Finot et al. 2004 for a different opinion).

5. Trisetum projectum Louis-Marie SIERRAN TRISETUM [p. 400, 504]

Pl per, csp. **Clm** 35–90 cm, glab. **Shth** pilose; **lig** 0.35–1.5 mm long, oval, densely pilose dorsally, apc dentate and ciliate; **bld** 80–130 cm long, 2–3 mm wide, flat, soft, involute distally, densely pilose on both surfaces, hairs about 1.2 mm long. **Pan** 9–23 cm long, 2–3 cm wide, spikelike, interrupted, exserted, pale yellow, shiny; **rchs** scabrous; **br** ascending. **Spklt** 6–6.5 mm long, with 2 flt, apc open; **ped** scabrous; **rchl intnd** about 1.5 mm long, hairy, hairs about 0.5 mm long. **Glm** acute, translucid; **lo glm** 5–5.5 mm, as long as or shorter than spklt, 1-veined; **up glm** 6.5–8 mm, longer than spklt, 3-veined; **lm** about 5 mm long, glab, delicate and hyaline, apc bifid, teeth about 1 mm long, awns about 5.5 mm, arising from the up ⅓ of the lm, not twisted or geniculate, diversely curved; **pal** about 3.5 mm, shorter than the lm, 2-veined, veins scabrous, apc bidentate; **lod** 0.6–0.8 mm, apc bilobed; **anth** 1–1.5 mm; **ov** glab. 2*n* = unknown

Trisetum projectum is restricted to the western United States, usually in dry woods, at 1200–2900 m. Its pilose leaf sheaths, glabrous culms, and yellowish panicles distinguish it from *T. spicatum*. Unlike *T. canescens* and *T. cernuum*, it has glabrous ovaries.

6. Trisetum montanum Vasey MOUNTAIN TRISETUM [p. 400, 504]

Pl per. **Clm** 50–70 cm, glab. **Shth** shorter than intnd, glab or pilose; **lig** about 3 mm, truncate, dentate, glab dorsally; **bld** 100–150 cm long, 3–10 mm wide, flat, glab or pilose. **Pan** 10–24 cm long, lax, open to more or less contracted; **rchs** and **ped** scaberulous. **Spklt** 4.5–6 mm long, with 2–3(4) flt; **ped** to 2.5 mm long; **rchl** about 0.8 mm, pilose, hairs less than 0.5 mm. **Glm** unequal, shorter than spklt, thin, hyaline; **lo glm** 3–3.5 mm, linear-lanceolate to lanceolate, about ⅔ the length of the up glm, 1-veined; **up glm** 4–4.5 mm, oval to oval-lanceolate, 3-veined; **cal** hairy, hairs about 0.1 mm; **lm** about 4 mm, glab, apc shortly bidentate, awns 3.4–4 mm, arising from the up ⅓ or ¼ of the lm, scabrous, diversely curved, not strongly twisted or geniculate; **pal**

about 4 mm, shorter than the lm, 2-veined, veins scabrous, apc shortly bidentate; **lod** 0.5–0.8 mm, apc bilobed; **anth** 0.8–1.2 mm; **ov** glab. **Car** glab. $2n$ = unknown

Trisetum montanum grows in moist to wet, loam to rocky soils, in valleys and on mountain slopes at 1900–4500 m from Yukon Territory to Arizona and New Mexico. Its culms are usually glabrous, but occasionally canescent, beneath the panicle and it usually has a less dense panicle than *T. spicatum*.

8.32 KOELERIA Pers.

Lisa A. Standley

Pl per; usu csp, smt weakly rhz. **Clm** 5–130 cm, erect. **Shth** open; **aur** absent; **lig** memb; **bld** flat to involute, pubescent or glab. **Infl** pan, erect, usu dense and spikelike, smt lax, stiffly and narrowly pyramidal at anthesis; **main rchs** and **br** smooth, softly hairy. **Spklt** lat compressed, with 2–4 flt; **rchl** to 1 mm, glab or pubescent, usu prolonged beyond the base of the distal flt, or with a vestigial flt; **dis** above the glm, beneath the flt. **Glm** subequal to or slightly exceeding the lm, memb, scabrid to tomentose, keels smt ciliate, unawned; **lo glm** 1-veined, somewhat narrower and shorter than the up glm; **up glm** obscurely 3(5)-veined; **cal** glab or hairy; **lm** thin, memb, 5-veined, mrg shiny, scarious, apc acute, smt mucronate or awned; **pal** equaling or subequal to the lm, hyaline; **lod** 2, glab, toothed; **anth** 3; **ov** glab. **Car** glab. x = 7. Named for Georg Ludwig Koeler (1765–1807), a botanist at Mainz, Germany.

Koeleria is a cosmopolitan genus of about 35 species that grow in dry grasslands and rocky soils; one is native to the Intermountain Region. In Europe, *Koeleria* forms a series of polyploid complexes in which cytotypes are morphologically and ecologically distinct, but species boundaries are not.

1. **Koeleria macrantha** (Ledeb.) Schult.

JUNEGRASS [p. *401*, <u>504</u>]

Pl ces, smt loosely so. **Clm** 20–85(130) cm, mostly glab, pubescent below the pan and near the nd. **Lvs** primarily bas; **shth** pubescent or glab, usu breaking off with age, if disintegrating into fibers, then the fibers straight or nearly so; **lig** 0.5–2 mm; **bld** 2–20 cm long, 0.5–3(4.5) mm wide, flat, involute when dry, minutely scabrous, occ glab or densely pubescent, mrg of the bas bld glab or with hairs averaging shorter than 1 mm near the base. **Pan** 4–27 cm long, 0.5–2 cm wide, interrupted at the base, otherwise dense; **br** finely pubescent to villous. **Spklt** 2.5–6.5 mm, obovate to obelliptic, with 2(3) flt; **rchl** pubescent. **Glm** 2.5–5 mm, ovate, memb, green, scabrous except for the ciliate keels, apc acute; **cal** pubescent; **lm** 2.5–6.5 mm, memb, shiny, usu glab, smt scabrous, particularly on the keels, usu green

when young, smt purple-tinged, stramineous at maturity, acuminate, midveins prolonged into a 1 mm awn; **pal** shorter than the lm; **anth** 1–2.5(3) mm. $2n$ = 14, 28.

Koeleria macrantha is widely distributed in temperate regions of North America and Eurasia. In North America, it grows in semi-arid to mesic conditions, on dry prairies or in grassy woods, generally in sandy soil, from sea level to 3900 m. It differs from *Sphenopholis intermedia*, with which it is sometimes confused, in its less open panicles, and in having spikelets that disarticulate above the glumes.

The species is treated here as a polymorphic, polyploid complex. North American plants have sometimes been treated as a separate species, *Koeleria nitida* Nutt., but no morphological characters for distinguishing them from Eurasian members of the complex are known (Greuter 1968). Some plants from Oregon and Washington have densely pubescent culms, and high-elevation populations from western North America are often densely cespitose, with very short culms and purple leaves and inflorescences; both variants appear to intergrade with more typical plants.

8.33 ANTHOXANTHUM L.

Kelly W. Allred and Mary E. Barkworth

Pl per or ann; densely to loosely csp, smt rhz; fragrant. **Clm** 4–100 cm, erect or geniculate, smt brchd; **intnd** hollow. **Lvs** cauline or bas concentrated, glab or softly hairy; **shth** open; **aur** absent or present; **lig** memb, smt shortly ciliate or somewhat erose; **bld** flat or rolled, glab or sparsely pilose. **Infl** open or contracted pan, smt spikelike. **Spklt** pedlt or sessile, 2.5–10 mm, lat compressed, stramineous to brown at maturity, with 3 flt, lowest 2 flt stmt or rdcd to dorsally awned lm subequal to or exceeding distal flt, distal flt bisex, unawned; **rchl** not prolonged beyond the base of the distal flt; **dis** above the glm, the flt falling together. **Glm** unequal or subequal, equaling or exceeding the flt, lanceolate to ovate, glab or pilose, keeled; **cal** blunt, glab or hairy. **Lowest 2 flt: lm** strongly compressed, 3-veined, strigose, hairs brown, apc bilobed, unawned or dorsally awned. **Distal flt: lm** somewhat indurate, glab or pubescent, shiny, inconspicuously 3–7-veined, unawned; **pal** 1-veined, enclosed by the lm; **lod** 2 or absent; **anth** 2 or 3. **Car** shorter than the lm, concealed at maturity, tightly enclosed in the flt; **hila** less than ⅓ the length of the car, oval. *x* = 5. Name from the Greek *anthos*, 'flower', and *xanthos*, 'yellow', alluding to the golden color of the mature panicles.

Anthoxanthum is a cool-season genus of about 50 species that grow in temperate and arctic regions throughout the world. One species is native to the Intermountain Region. The fragrance emitted when fresh plants are crushed or burned is from coumarin. In addition to smelling pleasant, coumarin has anti-coagulant properties. It is the active ingredient in Coumadin, a prescription drug used to prevent blood clots in some patients after surgery. A disadvantage of coumarin is that it is metabolized by species of the fungal genus *Aspergillus* to dicoumarol, which induces vitamin K deficiency and a susceptibility to hemorrhaging in wounded animals. Because of this, using moldy hay containing *Anthoxanthum* as feed is dangerous.

This treatment follows the recommendation of Schouten and Veldkamp (1985) in merging what have traditionally be treated as two genera, *Anthoxanthum* and *Hierochloë*. In general, *Hierochloë* has less floral reduction, a less elaborate karyotype, and a higher basic chromosome number than *Anthoxanthum* (Weimarck 1971). The two genera appear distinct in North America but, when considered on a global level, Schouten and Veldkamp (1985) stated that the two genera overlap, with the placement of many species being arbitrary. *Phalaris* resembles *Anthoxanthum sensu lato* in its spikelet structure, differing only in the greater reduction of the lower florets. It also differs in lacking coumarin.

Anatomical studies (Pizzolato 1984) supported the close relationship of *Anthoxanthum* and *Phalaris*. Pizzolato also stated that although the bisexual florets of *Hierochloë* are described as terminal, a microscopic fourth floret is developed distal to the third (bisexual) floret.

Wherever they grow, the species that used to be treated as *Hierochloë* have been used by native peoples. Native Americans used them for incense, baskets, and decorations. In addition, they steeped them in water for a hair-, skin-, and eyewash, or for use as a cold medicine, analgesic, or insecticide. Early Europeans spread the species in churches at festivals. They can also be used to make ale (Stika 2003).

1. **Anthoxanthum hirtum** (Schrank)Y. Schouten & Veldkamp Hairy Sweetgrass [p. 401, <u>504</u>] **Pl** per; loosely ces or the clm solitary, rhz elongate, 0.7–2 mm thick. **Clm** 40–85(110) cm. **Shth** brownish or reddish; **lig** 2.5–5.5 mm; **bld** 2.5–5.5 mm wide; abx surfaces glab, shiny, adx surfaces pilose; **flag lf bld** 1–4.5(6) cm long, 3–4.5 mm wide. **Pan** (5)7.5–15 cm long, 2–10 cm wide, open, pyramidal, with 20–100+ spklt; **br** with 3+ spklt. **Spklt** 4–6.3 mm, tawny at maturity; **rchl intnd** 0.1–0.3 mm. **Glm** subequal, exceeding the flt, glab, often somewhat purplish; **lowest 2 flt** stmt; **lm** 3–5 mm, with hairs to 0.5 mm towards the apc, mrg with 16–30 hairs per mm, hairs 0.5–1 mm, apc acute, emgt, or bifid; **first lm** 3–5 mm long, 1.1–1.5 mm wide, length usu less than 4 times width, elliptic, awned, awns 0.1–1 mm;

bisex lm 2.9–3.5 mm, hairy distally, hairs 0.5–1 mm, evenly distributed around the apc, bases strongly divergent from the lm surface; **anth** of stmt flt 1.6–2.1 mm, those of bisex flt 1.2–1.3 mm. 2*n* = 56.

Anthoxanthum hirtum [≡ *Hierochloë hirta* (Scrhank) Borbás] grows throughout much of North America. It grows in wet meadows and marshes with good water, not in salt- or brackish water. Because much of its native habitat has been drained, it is becoming less common.

Weimark (1971, 1987) recognized three subspecies in *Anthoxanthum hirtum* (which he treated as *Hierochloë hirta*): subsp. *hirta*, subsp. *arctica* G. Weim., and subsp. *praetermissa* G. Weim. He stated that only *H. hirta* subsp. *arctica* grows in North America, but several North American specimens seem to fit his circumscription of *H. hirta* subsp. *hirta*. Because the variation between the two appears continuous, no subspecies of *A. hirtum* are recognized here.

8.34 PHALARIS L.   Mary E. Barkworth

Pl ann or per; smt csp, smt rhz. **Clm** 4–230 cm tall, erect or decumbent, smt swollen at the base, not brchg above the base. **Lvs** more or less evenly distributed, glab; **shth** open for most of their length, uppermost shth often somewhat inflated; **aur** absent; **lig** hyaline, glab, truncate to acuminate, entire or lacerate; **bld** usu flat, smt revolute. **Infl** tml pan, smt spikelike, ovoid to cylindrical, dense, smt interrupted, with 10–200 spklt borne singly or in clusters, spklt homogamous in species with single spklt, heterogamous in species with spklt in clusters, lo spklt in the clusters usu stmt, rarely strl, tml spklt bisex or pist. **Spklt** pedlt, lat compressed, with 1–3(4) flt, the tml or only flt usu sex, lo flt(s), if present, strl; **dis** above the glm, beneath the strl flt in species with solitary spklt, in species with clustered spklt usu at the base of the spklt clusters, smt beneath the bisex or pist spklt. **Glm** subequal, exceeding the flt, 1–5-veined, keeled, keels often conspicuously winged; **lo (strl) flt** rdcd, varying from knoblike projections on the cal of the bisex flt to linear or lanceolate lm less than ¾ as long as the bisex flt; **tml flt** usu bisex, in the lo spklt of a spklt cluster the tml flt pist or stmt, rarely strl; **lm**

of tml flt coriaceous to indurate, shiny, glab or hairy, inconspicuously 5-veined, acute to acuminate or beaked, unawned; **pal** similar to the lm in length and texture, enclosed by the lm at maturity, 1-veined, mostly glab, veins shortly hairy; **lod** absent or 2 and rdcd; **anth** 3; **ov** glab; **sty** 2, plumose. **Car** shorter than the lm, concealed at maturity, with a reticulate pericarp, falling free of the lm and pal; **hila** long-linear. $x = 6, 7$. The name of the genus is an old Greek name for a grass.

Phalaris has 22 species, most of which grow primarily in temperate regions. It is found in a wide range of habitats, although most species prefer somewhat mesic, disturbed areas. Five species are known from the Intermountain Region.

The sterile florets of *Phalaris* are frequently mistaken for tufts of hair at the base of a solitary functional floret. Close examination will reveal that the hairs are actually growing from linear to narrowly lanceolate pieces of tissue. Developmental studies have shown that these structures are reduced lemmas.

Many species of *Phalaris* are weedy. A few are cultivated for fodder, and one, *P. canariensis*, is grown for birdseed. In addition, the dense panicles of *P. paradoxa* are sometimes dyed green and used to simulate shrubs in landscape models.

1. Spikelets in clusters, heterogamous, the lower 4–7 spikelets in each cluster with a staminate (rarely sterile) terminal floret, only the terminal spikelet in the clusters with a pistillate or bisexual terminal floret; disarticulation usually at the base of the spikelet clusters, sometimes beneath the bisexual or pistillate spikelets . . . . 1. *P. paradoxa*
1. Spikelets borne singly, homogamous, all spikelets with a bisexual terminal floret; disarticulation above the glumes, beneath the sterile florets.
  2. Glume keels not winged or with wings no more than 0.2 mm wide.
    3. Plants perennial; bisexual florets with acute to somewhat acuminate apices . . . . 4. *P. arundinacea*
    3. Plants annual; bisexual florets with beaked or strongly acuminate apices . . . . 5. *P. caroliniana* (in part)
  2. Glume keels broadly winged, the wings 0.2–1 mm wide.
    4. Sterile florets usually 1, if 2, the lower floret up to 0.7 mm long and the upper floret 1–3 mm long . . . . 2. *P. minor*
    4. Sterile florets 2, equal to subequal, 0.5–4.5 mm long.
      5. Glumes 7–10 mm long, 2–2.5 mm wide; bisexual florets 4.5–6.8 mm long; anthers 2–4 mm long . . . . 3. *P. canariensis*
      5. Glumes 3.8–6(8) mm long, 0.8–1.5 mm wide; bisexual florets 2.9–4.7 mm long; anthers 1.5–2 mm long . . . . 5. *P. caroliniana* (in part)

1. **Phalaris paradoxa** L. Hooded Canarygrass [p. *401*, 505]

Pl ann; tufted. **Clm** 20–100 cm, not swollen at the base. **Lig** 3–5 mm, truncate to acute; **bld** 5–10(15) cm long, 2–5 mm wide. **Pan** 3–9 cm long, about 2 cm wide, dense, obovoid to clavate, tapering at the base, rounded to truncate at the top; **br** with groups of 5–6 usu stmt, rarely strl spklt clustered around a tml pist or bisex spklt; **ped** hispid; **dis** beneath the spklt clusters. **Spklt** heterogamous, with 3 flt, lo 2 flt strl and highly rdcd, tml flt usu stmt, pist, or bisex, rarely strl. **Glm of stmt** or **strl spklt** varying, those at the base of the pan rdcd to knobs of tissue terminating the ped, those higher up often clavate, those near the top of the pan similar to the glm of the sex spklt but somewhat narrower; **glm of pist** or **bisex spklt** 4–8 mm long, about 1 mm wide, keeled, keels winged, wings 0.2–0.4 mm wide, terminating below the apc and forming a single, prominent tooth, lat veins conspicuous, apc acuminate to awned, awns about 0.5 mm; **strl flt** of all spklt 0.2–0.4 mm, knoblike projections on the cal of the tml flt often with 1–2 hairs; **tml flt** of all spklt 2.5–3.5 mm long, 0.8–1.5 mm wide, indurate, shiny,

glab or with a few short hairs near the tip; **anth** 1.5–2.5 mm. $2n = 14$.

Phalaris paradoxa is native to the Mediterranean region; it is now found throughout the world, primarily in harbor areas and near old ballast dumps. It is an established weed in parts of Arizona and California, and has been reported in southern Nevada. Within an inflorescence, the most reduced sterile spikelets are located near the base, and the most nearly normal spikelets are near the top.

2. **Phalaris minor** Retz. Lesser Canarygrass [p. *402*, 505]

Pl ann. **Clm** 10–100 cm, not swollen at the base. **Lig** 5–12 mm, truncate to rounded, often lacerate; **bld** 3–15 cm long, 2–10 mm wide, smooth, shiny. **Pan** 1–8 cm long, 1–2 cm wide, dense, ovoid-lanceoloid, truncate to rounded at the base, rounded apically, spklt borne singly, not clustered. **Spklt** homogamous, with 2 flt, 1 bisex; **dis** above the glm, beneath the strl flt. **Glm** 3.5–6.5 mm long, 1.2–2 mm wide, keels winged distally, wings 0.3–0.5 mm wide, irregularly dentate or crenate, occ entire, varying within a pan, lat veins conspicuous, smooth; **strl flt** 1, 0.7–1.8 mm, linear, glab or almost so; **bisex flt** 2–4 mm long, 1–1.8 mm wide, hairy, dull yellow when immature, becoming shiny gray-brown at maturity, acute to somewhat acuminate; **anth** 1–2 mm. $2n = 28, 29$.

Phalaris minor is native around the Mediterranean and in northwestern Asia, but is now found throughout the world. Even where it is native, it usually grows in disturbed ground, often around harbors and near refuse dumps. It has been found at numerous locations in North America, including the southern portion of the Intermountain Region.

The compact panicle with its truncate to rounded base and the rather variable edges of the glume wings usually distinguish Phalaris minor from other species in the genus.

3. **Phalaris canariensis** L. Annual
Canarygrass [p. 402, 505]

Pl ann. **Clm** 30–100 cm. **Lig** 3–6 mm, rounded to obtuse, lacerate; **bld** 3–25 cm long, 2–10 mm wide. **Pan** 1.5–5 cm long, 1.5–2 cm wide, ovoid to oblong-ovoid, continuous, not lobed, truncate at the base; **br** not evident, spklt borne singly, not clustered. **Spklt** homogamous, with 3 flt, tml flt bisex; **dis** above the glm, beneath the strl flt. **Glm** 7–10 mm long, 2–2.5 mm wide, smooth, mostly glab, smt sparsely pilose between the veins, keels winged, wings to 0.6 mm, widening distally, lat veins inconspicuous, smooth, apc rounded to acute, smt mucronate; **strl flt** 2, equal or subequal, 2–4.5 mm, ⅓ or more the length of the bisex flt, lanceolate, sparsely hairy, acute; **bisex flt** 4.5–6.8 mm, ovate, densely hairy, shiny, stramineous to gray-brown; **anth** 2–4 mm. $2n = 12$.

Phalaris canariensis is native to southern Europe and the Canary Islands, but is now widespread in the rest of the world, frequently being grown for birdseed. The exposed ends of the glumes are almost semicircular in outline, making this one of our easier species of Phalaris to identify.

4. **Phalaris arundinacea** L. Reed Canarygrass
[p. 402, 505]

Pl per; not ces, rhz, rhz scaly. **Clm** 40–230 cm. **Lig** 4–10 (11) mm, truncate, lacerate; **bld** usu 10–30 cm long, 5–20 mm wide, flag lf bld 4–15 cm, surfaces scabrous, mrg serrate. **Pan** 5–40 cm long, 1–4 cm wide, elongate, often dense, always evidently brchd, at least near the base; **br** to 5 cm, normally appressed but spreading during anthesis, spklt borne singly, not clustered. **Spklt** homogamous, with 3 flt, tml flt bisex; **dis** above the glm, beneath the strl flt. **Glm** subequal, 4–8.1 mm long, 0.8–1 mm wide, keels smoothly curved,

usu scabrous, not or narrowly winged distally, wings to 0.2 mm wide, lat veins conspicuous, apc acute; **strl flt** 2, subequal to equal, 1.5–2 mm, less than ½ as long as the bisex flt, hairy; **bisex flt** 2.5–4.2 mm, apc acute to somewhat acuminate; **lm** glabrate proximally, hairy distally and on the mrg, dull yellow when immature, shiny gray-brown to brown at maturity, apc acute; **anth** 2.5–3 mm. $2n = 27, 28, 29, 30, 31, 35$.

Phalaris arundinacea is a circumboreal species, native to north-temperate regions. It grows in wet areas such as the edges of lakes, ponds, ditches, and creeks, often forming dense stands; in some areas it is a problematic weed. North American populations may be a mix of native strains, European strains, and agronomic cultivars (Merigliano and Lesica 1998).

A sterile form of Phalaris arundinacea with striped leaves—P. arundinacea var. picta L., also referred to as P. arundinacea forma variegata (Parn.) Druce—is known as 'Ribbon Grass' or 'Gardener's Gaiters', and is sometimes grown as an ornamental. Baldini (1995) noted that it sometimes appears to escape, but is never found far from a cultivated stand.

Phalaris arundinacea hybridizes with other species of Phalaris. One hybrid, P. ×monspeliensis Daveau [= P. arundinacea × P. aquatica] is grown for forage.

5. **Phalaris caroliniana** Walter Carolina
Canarygrass [p. 403, 505]

Pl ann. **Clm** to 150 cm. **Lig** 1.5–7 mm, truncate to broadly acute; **bld** 1.5–15 cm long, 2–11 mm wide, smooth, shiny green, apc acuminate. **Pan** 0.5–8(8.5) cm long, 0.8–2 cm wide, ovoid to subcylindrical, not lobed; **br** not evident, spklt borne singly, not clustered. **Spklt** homogamous, with 3 flt, tml flt bisex; **dis** above the glm, beneath the strl flt. **Glm** 3.8–6(8) mm long, 0.8–1.5 mm wide, keels smooth or scabridulous, narrowly to broadly winged distally, wings 0.1–0.5 mm wide, entire, smooth, lat veins prominent, usu smooth, smt scabridulous, apc acute or acuminate; **strl flt** 2, equal to subequal, 1.5–2.5 mm, ½ or more the length of the bisex flt, the bas 0.2–0.5 mm glab, the remainder hairy; **bisex flt** 2.9–4.7 mm long, 0.9–1.8 mm wide, shiny, stramineous when immature, brown when mature, apc hairy, acuminate to beaked; **anth** 1.5–2 mm. $2n = 14$.

Phalaris caroliniana grows in wet, marshy, and swampy ground. It is a common species in suitable habitats through much of the southern portion of the United States, including the southern part of the Intermountain Region.

8.35 **CINNA** L.   David M. Brandenburg

Pl per; csp, smt rhz. **Clm** 20–203 cm, solitary or clustered, often rooting at the lo nd, usu glab. **Shth** open, glab; **aur** absent; **lig** scarious; **bld** flat, mrg scabrous, surfaces scabrous or smooth. **Infl** pan; **br** spreading to ascending, some br longer than 1 cm; **ped** slightly flared, scabrous or smooth; **dis** below the glm. **Spklt** lat compressed, with 1 flt, rarely with a second rdmt or ftl flt; **rchl** usu prolonged beyond the base of the flt as a minute stub or bristle, smooth or scabridulous, smt not prolonged. **Glm**

from slightly shorter than to slightly longer than the flt, 1- or 3-veined, mrg hyaline, keeled, keels scabrous, apc acute, smt minutely awn-tipped; **lo glm** from somewhat shorter than to equaling the up glm, flt sessile or stipitate; **cal** short, glab; **lm** 3- or 5-veined, smt obscurely so, apc acute, minutely bifid, usu awned, awns subtml, smt unawned; **pal** ¾ to nearly as long as the lm, 1-veined or with 2 closely spaced veins; **anth** 1 or 2. **Car** shorter than the lm, concealed at maturity, often beaked. $x = 7$. Name of uncertain origin.

Cinna is a genus of four species, all of which generally grow in damp woods, along streams, or in wet meadows. One species, *C. latifolia*, is circumboreal. The other three species are restricted to the Western Hemisphere.

1. **Cinna latifolia** (Trevir. *ex* Göpp.) Griseb.
DROOPING WOODREED, SLENDER WOODREED
[p. *403*, 505]
Clm 20–190 cm; **nd** 4–9. **Lig** 2–8 mm; **bld** to 28 cm long, 1–20 mm wide. **Pan** 3–46 cm; **br** usu spreading, smt ascending. **Spklt** (2)2.5–4(5) mm; **rchl prolongations** 0.1–1.3 mm, smt absent. **Lo glm** (1.8)2.5–4(4.7) mm, 1-veined; **up glm** (1.9)2.5–4(5) mm, 1(3)-veined; stipes 0.1–0.45 mm; **lm** 1.8–3.8 mm, 3(5)-veined, awns 0.1–2.5 mm or absent; **pal** 2-veined, with the veins very close together, or 1-veined; **anth** 1, 0.4–1 mm. **Car** 1.8–2.8 mm. $2n = 28$.

Cinna latifolia is a circumboreal species, extending from Alaska to Newfoundland in North America, and across Eurasia from Norway to the Kamchatka Peninsula, Russia. It grows in moist to wet soil in open coniferous or mixed forests, swamps, thickets, bogs, and streamsides, at 0–2600 m, and flowers in late summer and fall. It is a variable species for which varietal names have been proposed. Because the variation is continuous, no varieties are recognized in this treatment.

8.36 ALOPECURUS L.   William J. Crins

Pl ann or per; smt csp, smt shortly rhz. **Clm** 5–110 cm, clumped or solitary, erect or decumbent, occ cormlike at the base; **nd** glab. **Lvs** inserted mostly on the lo ½ of the clm; **shth** open, up shth smt inflated; **aur** absent; **lig** 0.6–6.5 mm, truncate to acute, memb, puberulent or glab, entire to lacerate; **bld** 0.7–12 mm wide, flat or involute, glab or scabrous, bld of uppermost lvs smt short or absent. **Infl** tml pan, spikelike, capitate to cylindrical; **br** usu shorter than 5 mm, lo br smt to 2 cm; **dis** below the glm. **Spklt** 1.8–7 mm, pedlt, strongly lat compressed, oval in outline, with 1 flt; **rchl** not prolonged beyond the base of the flt. **Glm** equaling or exceeding the flt, memb or coriaceous, free or connate in at least the lo ½, narrowing from above midlength, 3-veined, keeled, keels ciliate, at least bas, apc obtuse to acute or shortly awned; **cal** blunt, glab; **lm** memb, mrg often connate in the lo ½, keeled, indistinctly 3–5-veined, apc truncate to acute, awned dorsally from just above the base to about midlength, geniculate or straight; **pal** absent or greatly rdcd; **lod** absent; **anth** 3, 0.3–4.1 mm; **ov** glab; **sty** fused, with 2 br. **Car** shorter than the lm, concealed at maturity, glab; **hila** short. $x = 7$. Name from the Greek *alopex*, 'fox', and *oura*, 'tail', referring to the cylindrical panicles.

Alopecurus is a genus of 36 species that grow primarily in open, mesic habitats, and are native to the northern temperate zone and South America. Seven species have been found in the Intermountain Region, at least two of which are introductions. Some species, including some native to North America, have been introduced as pasture grasses outside of their native ranges. Of these, only *A. pratensis* has become widely naturalized.

Species of *Alopecurus* may appear superficially similar to *Phleum*, but *Phleum* has truncate glumes that are abruptly awned or mucronate, lemmas without awns or keels, and well-developed paleas. *Alopecurus* has glumes that are obtuse to acute and gradually awned or unawned, lemmas with both awns and keels, and paleas that are absent or greatly reduced.

1. Plants annual, without rhizomes, not rooting at the lower nodes; blades 3–15 cm long, 0.9–3 mm wide; culms 5–50 cm tall.
  2. Upper sheaths conspicuously inflated; glumes 3–5 mm long; lemmas 3–5 mm long, awns exceeding the lemmas by 3–6 mm; panicles 5.5–13 mm wide, excluding the awns
    . . . . 6. *A. saccatus*
  2. Upper sheaths not or only slightly inflated; glumes 2.1–3.1 mm long; lemmas 1.9–2.7 mm long, awns exceeding the lemmas by 1.6–4 mm; panicles 3–6 mm wide, excluding the awns . . . . 7. *A. carolinianus*
1. Plants perennial, often rhizomatous, sometimes rooting at the lower nodes; blades 2–40 cm long, 1–12 mm wide; culms 5–110 cm tall.
  2. Glumes 1.8–3.7 mm long, the apices obtuse; anthers 0.5–2.2 mm long.
    3. Awns geniculate, exceeding the lemmas by 1.2–4 mm; anthers (0.9)1.4–2.2 mm long. . 4. *A. geniculatus*
    3. Awns straight, not exceeding the lemmas or exceeding them by less than 2.5 mm; anthers 0.5–1.2 mm long . . . . 5. *A. aequalis*

2. Glumes 3–6 mm long, the apices acute; anthers 2–4 mm long.
    4. Glume margins connate in the lower ⅛; glumes densely pilose throughout . . . . 3. *A. magellanicus*
    4. Glume margins connate in the lower ⅕–⅓; glumes with long hairs mainly restricted
      to the veins.
        5. Lemma apices acute; glume apices parallel or convergent . . . . 1. *A. pratensis*
        5. Lemma apices obtuse to truncate; glume apices divergent . . . . 2. *A. arundinaceus*

1. **Alopecurus pratensis** L. Meadow Foxtail [p. *403*, 505]

Pl per; shortly rhz. **Clm** 30–110 cm, erect. **Lig** 1.5–3 mm, obtuse to truncate; **bld** 6–40 cm long, 1.9–8 mm wide; **up shth** not or scarcely inflated. **Pan** 3.5–9 cm long, 6–10 mm wide. **Glm** 4–6 mm, connate in the lo ⅕–¼, memb, sides pubescent, keels not winged, finely ciliate, apc acute, parallel or convergent; **lm** 4–6 mm, connate in the lo ⅓, usu glab, keels smt ciliate distally, apc acute, awns 5–10.5 mm, geniculate, exceeding the lm by (1)2.2–5.5 mm; **anth** 2–4 mm, yellowish, orange, reddish, or purplish, smt varying within a population. **Car** 1–1.2 mm. 2*n* = 28, 42.

Alopecurus pratensis is native from temperate northern Eurasia south to North Africa. It is now widely naturalized in temperate regions throughout the world. It grows in poorly to somewhat drained soils in meadows, riverbanks, lakesides, ditches, roadsides, and fence rows. It has been widely introduced as a pasture grass; it may also have become established from ballast or imported hay.

2. **Alopecurus arundinaceus** Poir. Creeping Meadow Foxtail [p. 404, 505]

Pl short-lived per; rhz. **Clm** 30–110 cm, erect. **Lig** 1.3–5 mm, truncate; **bld** 6–40 cm long, 3–12 mm wide; **up shth** somewhat inflated. **Pan** 3–10 cm long, 7–13 mm wide. **Glm** 3.6–5 mm, connate in the lo ⅕–⅓, memb, sparsely pubescent, keels not winged, ciliate, apc acute, divergent, pale green to lead-gray; **lm** 3.1–4.5 mm, connate in at least the lo ⅓, usu glab, smt with scattered hairs near the apc, apc truncate to obtuse, awns 1.5–7.5 mm, geniculate, exceeding the lm by 0–3 mm; **anth** 2.2–3.5 mm. 2*n* = 26, 28, 30.

Alopecurus arundinaceus is native to Eurasia. It grows in wet, moderately acid to moderately alkaline soils, on flood plains, near vernal ponds, and along rivers, streams, bogs, potholes, and sloughs. It was introduced for pasture in North Dakota and is now grown more widely, having been promoted as a forage species. It is sometimes used in seed mixtures for revegetation projects. In irrigated pastures, it suppresses *Hordeum jubatum*, a troublesome, unpalatable, weedy species (Moyer and Boswall 2002).

3. **Alopecurus magellanicus** Lam. Alpine Foxtail, Boreal Foxtail [p. 404, 505]

Pl per; shortly rhz. **Clm** (6)10–80 cm, erect or decumbent. **Lig** 1–2 mm, truncate; **bld** 4–22 cm long, 2.5–7 mm wide; **up shth** inflated. **Pan** 1–5 cm long, 8–14 mm wide. **Glm** 3–5 mm, connate

in the lo ⅛, memb, densely pilose throughout, keels not winged, ciliate, apc acute and parallel; **lm** 2.5–4.5 mm, connate in the lo ½–⅔, glab proximally, finely pubescent distally, apc usu obtuse, occ truncate, awns 2–6(8) mm, geniculate, exceeding the lm by 0–5 mm; **anth** 2.3–3 mm, yellow. **Car** 0.7–2 mm. 2*n* = 98, 100, 105, 112, 117, 119, ca. 120.

Alopecurus magellanicus grows primarily in wet soils in subalpine to alpine habitats, in meadows and along streams, shorelines, gravelbars, and floodplains, and occasionally in somewhat drier forest openings, in fine or silty to stony soils or moss. The anthocyanic tint of the plant as a whole greatly increases to the north.

The morphological variability in *Alopecurus magellanicus* has prompted recognition of several segregate taxa. One of these, *Alopecurus occidentalis* Scribn. & Tweedy, refers to tall-stemmed plants found in the Rocky Mountains. Because such plants are simply an extreme in a continuum of variation, they do not merit taxonomic recognition.

4. **Alopecurus geniculatus** L. Water Foxtail [p. 404, 505]

Pl per; ces. **Clm** (5)10–60 cm, erect or decumbent, rooting at the lo nd. **Lig** 2–5 mm, obtuse; **bld** 2–12 cm long, 1–4(7) mm wide; **up shth** somewhat inflated. **Pan** 1.5–7 cm long, 4–8 mm wide. **Glm** 1.9–3.5 mm, connate at the base, memb, pubescent, keels not winged, ciliate, apc obtuse, parallel, often purplish; **lm** 2.5–3 mm, connate in the lo ½, glab or with a few scattered hairs at the apc, apc truncate to obtuse, awns 3–5(6) mm, geniculate, exceeding the lm by (1.2)2–4 mm; **anth** (0.9)1.4–2.2 mm, yellow. **Car** 1–1.5 mm. 2*n* = 28.

Alopecurus geniculatus is native to Eurasia and parts of North America, growing in shallow water, ditches, open wet meadows, shores, and streambanks, from lowland to montane zones. It has been introduced and naturalized in eastern North America. The status of populations in the west is less certain. Many occur in moist sites within native rangeland, but these areas have also been affected by European settlement, although less intensively and for a shorter period than those in eastern North America.

Alopecurus ×*haussknechtianus* Asch. & Graebn., a hybrid between *A. geniculatus* and *A. aequalis*, occurs fairly frequently in areas of sympatry, particularly in drier midcontinental areas from Alberta to Saskatchewan, south to Arizona and New Mexico. The hybrids are sterile and appear to have 2*n* = 14.

5. **Alopecurus aequalis** Sobol. [p. *405*, 505]

Pl per; ces. **Clm** 9–75 cm, erect or decumbent.
Lig 2–6.5 mm, obtuse; **bld** 2–10 cm long, 1–5(8)
mm wide; **up shth** not inflated. **Pan** 1–9 cm long,
3–9 mm wide. **Glm** 1.8–3.7 mm, connate near
the base, memb, pubescent on the sides, keels
not winged, ciliate, apc obtuse, smt erose, pale
green, occ purplish; **lm** 1.5–2.5(3.5) mm, connate
in the lo ⅓–½, glab, apc obtuse, awns 0.7–3 mm,
straight, exceeding the lm by 0–2.5 mm; **anth**
0.5–1.2 mm, usu pale to deep yellow or orange,
rarely purple. **Car** 1–1.8 mm. 2*n* = 14, 28.

Alopecurus aequalis is native to temperate zones of the
Northern Hemisphere. It generally grows in wet meadows,
forest openings, shores, springs, and along streams, as well
as in ditches, along roadsides, and in other disturbed sites,
from sea level to subalpine elevations. There are two varieties
in North America, but only one grows in the Intermountain
Region.

Alopecurus ×haussknechtianus Asch. & Graebn., a hybrid
between *A. aequalis* and *A. geniculatus*, occurs fairly frequently
in areas of sympatry, particularly in drier midcontinental
areas from Alberta to Saskatchewan, south to Arizona and
New Mexico. The hybrids are sterile and apparently have
2*n* = 14.

Alopecurus aequalis Sobol. var. **aequalis**
   Shortawn Foxtail [p. *405*]

Clm 9–75 cm. **bld** 1–5(8) mm wide. **Pan** 1–9
mm long, 3–6 mm wide. **Spklt** usu not purplish
tinged; **glm** 1.8–3 mm; **awns** not exceeding the
lm or exceeding them by less than 1 mm; **anth**
0.5–0.9 mm.

Alopecurus aequalis var. *aequalis* is the widespread variety
in most of North America.

6. **Alopecurus saccatus** Vasey Pacific Meadow
   Foxtail [p. *405*, 505]

Pl ann; tufted. **Clm** 12–45 cm, erect or decumbent.
Lig 1.5–5.5 mm, obtuse; **bld** 4–12 cm long, 1.2–4
mm wide; **up shth** conspicuously inflated. **Pan**
1.5–6.5 cm long, 5.5–13 mm wide, often dense.
Glm 3–5 mm, connate at the base, not dilated,
memb, pubescent, keels not winged, veins ciliate,
apc obtuse; **lm** 3–5 mm, connate in the lo ⅓–½,
glab, apc obtuse, awns 6–10 mm, geniculate,
exceeding the lm by 3–6 mm; **anth** 0.7–1.8 mm,
yellow to rusty brown. **Car** 1.5–2 mm. 2*n* =
unknown.

Alopecurus saccatus is a native annual that inhabits moist,
open meadows, valley plains, and vernal pools, at elevations
below 700 m, from Washington to California. Segregates have
been treated as species in the past, but the variation between
them appears to be continuous, and no habitat differentiation
is evident.

7. **Alopecurus carolinianus** Walter Tufted
   Foxtail [p. *405*, 505]

Pl ann; tufted. **Clm** 5–50 cm, erect or decumbent.
Lig 2.8–4.5 mm, obtuse; **bld** 3–15 cm long, 0.9–3
mm wide; **up shth** not or only slightly inflated.
Pan 1–7 cm long, 3–6 mm wide, always dense.
Glm 2.1–3.1 mm, connate at the base, memb
throughout, sparsely pubescent, not dilated
below, keels not winged, ciliate, apc obtuse, pale
green to pale yellow; **lm** 1.9–2.7 mm, connate
in the lo ½, glab, apc obtuse, awns 3–6.5 mm,
geniculate, exceeding the lm by 1.6–4 mm; **anth**
0.3–1 mm, yellow or orange. **Car** 1–1.5 mm.
2*n* = 14.

Alopecurus carolinianus is native to the central plains,
Mississippi valley, and southeastern United States, where it is
common in wet meadows, ditches, wetland edges, and other
moist, open habitats, as well as in arid areas of the prairies
and southwest, where it grows sporadically along sloughs
and in ditches and vernal pools. It is not clear whether the
populations in the southwest are native or naturalized.

8.37 APERA Adans.   Kelly W. Allred

Pl ann; tufted or the clm solitary. **Clm** 5–120 cm, erect or geniculate, glab. **Lvs** mostly cauline; **shth**
open, rounded to slightly keeled; **col** glab, midveins continuous; **aur** absent; **lig** memb, often lacerate
to erose; **bld** flat or weakly involute, glab. **Infl** tml pan; **br** strongly ascending to divergent. **Spklt**
pedlt, slightly lat compressed, with 1 flt, rarely more, distal flt, if present, vestigial; **rchl** prolonged
beyond the base of the flt as a bristle, rarely terminating in a vestigial flt; **dis** above the glm, beneath
the flt. **Glm** unequal, lanceolate, scabrous on the distal ½, unawned; **lo glm** 1-veined; **up glm** slightly
shorter than to slightly longer than the flt, 3-veined; **cal** blunt, glab or sparsely hairy; **lm** firmer than
the glm, folded to nearly terete, obscurely 5-veined, mrg veins not excurrent, apc entire, awned, awns
subtml; **pal** about ¾ as long as to equaling the lm, hyaline, 2-veined; **lod** 2, free, glab, usu toothed;
anth 3; **ov** glab. **Car** shorter than the lm, concealed at maturity, 1.2–2 mm, ellipsoidal, slightly sulcate;
hila broadly ovate, ⅕ the length of the car. *x* = 7. Name of uncertain origin, possibly from the Greek
a, 'not', and *peros*, 'maimed'; Adanson provided no explanation.

Apera is genus of three species, native to Europe and western Asia. It is similar to *Agrostis*, differing in its firm lemmas;
paleas that are always present and equal to the lemma, or nearly so; and prolonged rachillas. One species is present in the
Intermountain Region.

1. **Apera interrupta** (L.) P. Beauv. INTERRUPTED WINDGRASS [p. *405*, 505]
Clm (5)10–50(75) cm, weak, slender, solitary or with several shoots, smt sparingly brchd above the base; **intnd** usu longer than the shth. **Shth** often purplish; **lig** 1.5–5 mm, acute to truncate, erose, mrg decurrent; **bld** usu 4–12 cm long, 0.3–4 mm wide, flat or convolute when dry. **Pan** 3–15(20) cm long, 0.4–1.5(3) cm wide, contracted, somewhat interrupted below; **br** erect to ascending, most spklt-bearing to within 2 mm of the base; **ped** 0.5–2 mm. **Spklt** 2–2.8 mm, green or purplish; **rchl** prolonged 0.2–0.6 mm. **Lo glm** 1–2.2 mm; **up glm** 2–2.5(2.8) mm; **lm** 1.5–2.5 mm, slightly involute, awned, awns 4–10(16) mm; **anth** 0.3–0.5 mm, often purplish brown. **Car** 1–1.5 mm. 2*n* = 14, 28.

Apera interrupta grows as a weed in lawns, grain fields (especially winter wheat), sandy open ground, and roadsides. Introduced from Europe, it is now established in the northern portion of the Intermountain Region.

PACCMAD Grasses

E.A. Kellogg

The PACCMAD group of grasses consists of the *Panicoideae, Arundinoideae sensu stricto, Chloridoideae, Centothecoideae, Micrairoideae, Aristidoideae,* and *Danthonioideae*. It includes more than half the species in the *Poaceae* as a whole and, according to every molecular study, constitutes a monophyletic group. There is not, however, any obvious morphological character that distinguishes it from other taxa or groups of taxa. Two characteristics that all its members share, so far as is known, are embryos with an elongated mesocotyl internode and lodicules that lack a distal membranous portion, but neither of these features is useful for routine identification.

Despite the lack of an obvious morphological character for recognizing PACCMAD grasses, some general statements can be made. All C_4 grasses are in the PACCMAD group, but so are many C_3 grasses. Perhaps because such a high proportion of the PACCMAD grasses employ a C_4 photosynthetic pathway, the group is associated with warm climates and/or late summer blooming. Solid culms are common, but not universal, among PACCMAD species; consequently a solid culm is a good indication that a grass belongs to one of the PACCMAD subfamilies, but a hollow culm provides no information. Many PACCMAD species have punctate hila, but so do many of the non-PACCMAD grasses. It is generally easier to place a grass in its tribe or subfamily and from that determine whether it is a PACCMAD grass than to work in the other direction.

Within the PACCMAD grasses, molecular data indicate that *Panicoideae* and *Centothecoideae* are sister taxa, as are the two pairs *Arundinoideae* + *Chloridoideae* and *Aristidoideae* + *Danthonioideae*. The last four subfamilies form a group that members of the Grass Phylogeny Working Group (2000, 2001) and Bouchenak-Khelladi et al. (2008) refer to as the "Ligule of Hairs clade" because many (but not all) of its members have ligules made up of a fringe of hairs. Other, more cryptic characteristics shared by members of the four subfamilies are the presence of compound starch grains in the endosperm and embryo leaf margins that meet rather than overlap.

Despite the strong evidence supporting the monophyly of the PACCMAD grasses, the Grass Phylogeny Working Group (2000, 2001) and Bouchenak-Khelladi et al. (2008) decided not to give the group formal nomenclatural recognition. There were two reasons for this decision. One was the difficulty of identifying the group morphologically. The second was that it would lead to a drastic change in the meaning of the name *Panicoideae*, the name that would, according the International Code of Botanical Nomenclature, become the group name. "PACCMAD" is simply an acronym, a pronounceable listing of the initials of the subfamilies in the group.

4. **ARUNDINOIDEAE** Burmeist.   Grass Phylogeny Working Group and Kelly W. Allred

Pl usu per; ces or not, smt rhz, smt stln. **Clm** 15–1000 cm, ann, herbaceous to somewhat woody, intnd usu hollow. **Lvs** usu mostly cauline, often conspicuously distichous; **shth** usu open; **aur** usu absent; **abx lig** usu absent (of hairs in *Hakonechloa*); **adx lig** memb or of hairs, if memb, often ciliate; **bld** without psdpet, smt deciduous at maturity; **mesophyll** usu non-radiate (radiate in *Arundo*);

adx palisade layer absent; **fusoid cells** absent; **arm cells** usu absent (present in *Phragmites*); **Kranz anatomy** absent; **midribs** simple; **adx bulliform cells** present; **stomatal subsidiary cells** low dome-shaped or triangular; **bicellular microhairs** usu present, usu with long, narrow tml cells; **papillae** usu absent. **Infl** usu tml, ebracteate, usu paniculate, occ spicate or rcm. **Spklt** lat compressed, with 1–several bisex flt or all flt unisex and the species dioecious; **flt** 1–several, terete or lat compressed, distal flt often rdcd; **dis** above the glm. **Glm** 2, from shorter than the adjacent lm to exceeding the distal flt; **lm** (3)5–7-veined, lanceolate to elliptic, acute to acuminate, smt awned; **awns** 1 or 3, if 3 not fused into a single bas column; **pal** subequal to the lm; **lod** 2, usu free, occ joined at the base, fleshy, usu glab, not, scarcely, or heavily vascularized; **anth** (1)2–3; **ov** glab; **sty** 2, usu free, bases close together. **Car** usu punctate (long-linear in *Molinia*); **endosperm** hard, without lipid; **starch grains** compound; **haustorial synergids** absent; **emb** usu large compared to the car, waisted or not; **epiblast** absent; **scutellar cleft** present; **mesocotyl intnd** elongate; **emb lf mrg** usu meeting (overlapping in *Hakonechloa*). $x = 6, 9, 10, 12$.

The *Arundinoideae* are interpreted here as including only one tribe, the *Arundineae*. The tribe used to be interpreted more broadly (e.g., Watson et al. 1985; Clayton and Renvoize 1986; Kellogg and Campbell 1987), but the broader interpretation was generally acknowledged to be somewhat artificial. Hsiao et al. (1998) showed support for inclusion of the *Danthonieae*, *Aristideae*, and *Arundineae* in a more broadly interpreted *Arundinoideae*, but other studies (e.g., Hilu and Esen 1990; Barker et al. 1995, 1998; Grass Phylogeny Working Group 2001) have failed to support such a treatment. More recent work (Bouchenak-Khelladi et al. 2008) suggests that the *Arundineae* are basal to the *Micrairoideae*, *Aristidoideae*, *Chloridoideae*, and *Danthonioideae*.

9. ARUNDINEAE Dumort.

Kelly W. Allred

See subfamily description.

There are still questions about the circumscription of the *Arundineae*, but it clearly includes the genera in this treatment. Its morphological circumscription is also difficult. The most abundant genera in North America, *Phragmites* and *Arundo*, have tall culms bearing numerous, conspicuously distichous, broad leaves and large, plumose panicles, a habit frequently described as "reedlike", but not all members of the tribe have this habit. Linder et al. (1997) noted that *Arundo*, *Phragmites*, and *Molinia* have hollow culm internodes, punctate hila, and convex sides to the adaxial ribs in the leaf blades, but these characters have not been examined in all genera of the tribe.

Members of the *Arundineae* are found in tropical and temperate areas around the world. The reedlike species are found in marshy to damp soils, but some of the other species grow in xeric habitats.

1. Lemmas glabrous. . . . . .9.02 *Phragmites*
1. Lemmas hairy.
   2. Rachilla internodes hairy; lemmas with papillose-based hairs on the margins . . . . .9.01 *Hakonechloa*
   2. Rachilla internodes glabrous; lemmas pilose, the hairs not papillose-based . . . . 9.03 *Arundo*

9.01 HAKONECHLOA Makino *ex* Honda

Mary E. Barkworth

Pl per; loosely ces, rhz and stln. **Clm** 30–90 cm, erect or geniculate at the base. **Shth** open; **aur** absent; **abx lig** present, composed of a line of hairs across the col; **adx lig** memb and sparsely ciliate, smt lacerate, cilia subequal to the base; **bld** flat, linear-lanceolate, resupinate, in living pl the glaucous-green adx surface facing downwards and the bright green abx surface facing upwards. **Pan** not plumose. **Spklt** pedlt, somewhat lat compressed, with 5–10 flt; **rchl intnd** conspicuously pilose; **dis** at the base of the rchl segment and below each spklt. **Glm** unequal, lanceolate, unawned; **cal** 1.5–2 mm, strigose, hairs 1–1.5 mm; **lm** chartaceous, 3-veined, mrg with papillose-based hairs near the base, apc inconspicuously bidentate, awned from between the teeth; **awns** 3–5 mm, straight; **pal** 2-keeled. **Car** glab. $x = 10$. Name from Hakone, a city on the island of Honshu, Japan, and the Greek *chloa*, 'grass'.

Hakonechloa is a monospecific genus, endemic to Japan, but grown as an ornamental in North America. The resupination of the blades is not evident on herbarium specimens.

1. **Hakonechloa macra** (Munro) Makino
JAPANESE FOREST GRASS, HAKONE GRASS
[p. 406]
Rhz and **stln** covered with pale, coriaceous scales. **Clm** 30–90 cm tall, 1–1.5 mm thick, glab.

Adx lig 0.2-0.3 mm; **bld** 8–25 cm long, 0.4–1.2 cm wide, glab, abx surfaces green, adx surfaces often paler, turning orange-bronze in the fall. **Pan** 6–12 cm long, 5–7 cm wide, open, nodding, with 15–30 spklt; **br** paired, somewhat stiff, scabrous. **Spklt** 1–2 cm, yellowish green, with 5–10 flt. **Glm** broadly lanceolate; **lo glm** 3–4 mm, 1–3-veined; **up glm** 3.8–5 mm, 3-veined; **cal** 1.5–2 mm, strigose, hairs 1–1.5 mm; **lm** 6–7 mm long, 1.8–2.2 mm wide, chartaceous, 3-veined, mrg sparsely pilose with long papillose-based hairs near the base, awned; **awns** 3–5 mm; **anth** 2–3 mm. **Car** about 2 mm. 2*n* = 50.

In Japan, *Hakonechloa macra* grows on rocks along rivers. Although rhizomatous, it is not an invasive species and is recommended for mass planting. Three forms are cultivated: forma *alboaurea* Makino *ex* Ohwi, with white- and yellow-striped leaves; forma *albovariegata* Makino *ex* Ohwi, with white-striped leaves; and forma *aureola* Makino *ex* Ohwi, with yellow leaves having narrow green stripes. This last form is the one most commonly available in North America.

9.02 PHRAGMITES Adans.

Kelly W. Allred

Pl per; rhz or stln, often forming dense stands. **Clm** 1–4 m tall, 0.5–1.5 cm thick, lfy; **intnd** hollow. **Lvs** cauline, mostly glab; **shth** open; **lig** memb or of hairs, ciliate; **bld** flat or folded. **Infl** tml, plumose pan. **Spklt** with 2–10 flt, weakly lat compressed, lo 1–2 flt stmt, distal 1–2 flt rdmt, remaining flt bisex; **rchl intnd** sericeous; **dis** above the glm and below the flt. **Glm** unequal, shorter than the flt, 1–3-veined, glab; **lo glm** much shorter than the up glm; **cal** pilose, hairs 6–12 mm; **lm** 3-veined, glab, unawned; **anth** 1–3. **Car** rarely maturing. *x* = 12. Name from the Greek *phragma*, 'fence', alluding to its fencelike growth.

Phragmites is interpreted here as a monospecific genus that has a worldwide distribution. Some taxonomists (e.g., Clayton 1970; Koyama 1987; Scholz and Bohling 2000) recognize 3–4 segregate species. Recent work has identified two different genotypes in North America (Saltonstall 2002; Saltonstall et al. 2004) that preliminary data suggest may be morphologicaly distinct (see http://www.invasiveplants.net/). How these genotypes relate to the various segregate species that have been recognized is not yet known. Saltonstall (2002) demonstrated that there are three strains present in North America. Some strains approved for use in bioremediation are imported from Europe. It is not known whether they belong to the invasive strain identified here.

Plants of *Phragmites* are similar in overall appearance to *Arundo*, but the latter has subequal glumes, a glabrous rachilla, and hairy lemmas. Vegetatively, plants of *Arundo*, but not those of *Phragmites*, have a wedge-shaped, light to dark brown area at the base of the blades. They also tend to have thicker rhizomes, thicker and taller culms, and wider leaves than *Phragmites*, but there is some overlap. *Phragmites* is much more widely distributed than *Arundo* in North America.

1. **Phragmites australis** (Cav.) Trin. *ex* Steud.
COMMON REED [p. 407, <u>505</u>]

Clm 1–4 m tall, 0.5–1.5 cm thick, erect. **Lig** of hairs, about 1 mm; **bld** 15–40 cm long, 2–4 cm wide, long-acuminate, dis from the shth at maturity. **Pan** 15–35 cm long, 8–20 cm wide, ovoid to lanceloid, often purplish when young, straw-colored at maturity. **Spklt** with 3–10 flt, rchl hairs (4)6–10 mm. **Lo glm** 3–7 mm; **up glm** (4)5–10 mm; **lm** 8–15 mm, glab, linear, mrg somewhat inrolled, apc long-acuminate; **pal** 3–4 mm, memb; **anth** 1.5–2 mm, purplish; **sty** persistent. **Car** 2–3 mm, rarely maturing. 2*n* = 36, 42, 44, 46, 48, 49–54, 72, 84, 96, 120.

Phragmites australis grows in wet or muddy ground along waterways, in saline or freshwater marshes, and in sloughs throughout North America. Its tall, leafy, often persistent culms and plumose panicles make it one of our easier species to recognize. It is also one of the most widely distributed flowering plants, growing in most temperate and tropical regions of the world, spreading quickly by rhizomes. Once established, it is difficult to eradicate. Its uses include thatching, lattices, construction boards, mats, and erosion control, and it was used in the past to make arrow shafts, cigarettes and superior pen quills.

Saltonstall (2002, 2003a,b) demonstrated that there are three sets of plants in North America. All three are treated here because their distributions are not yet well established.

1. Ligules 1–1.7 mm long; lower glumes 3–6.5 mm long; upper glumes 5.5–11 mm long; leaf sheaths caducous with age; culms exposed in the winter, smooth and shiny; rarely forming a monoculture .... *P. australis* subsp. *americanus*
1. Ligules 0.4–0.9 mm long; lower glumes 2.5–5 mm long; upper glumes 4.5–7.5 mm long; leaf sheaths not caducous with age; culms not exposed in the winter, smooth and shiny or ridged and not shiny; usually forming a monoculture.
  2. Culms smooth and shiny ....
    .... *P. australis* subsp. *berlandieri*
  2. Culms ridged and not shiny .... Invasive plants

Phragmites australis subsp. **americanus**
Saltonstall, P.M. Peterson & Soreng COMMON REED [p. 407]

Phragmites australis subsp. *americanus* is the native North American strain. It grows across much of Canada and in the United States, from New England and the Mid-Atlantic states across to the Pacific coast and into the southwest. Regional structuring can be found within this lineage, with east coast, midwestern, and western populations showing different genetic profiles (Saltonstall 2003a,b).

Phragmites australis subsp. **berlandieri** (E. Fourn.) Saltonstall & Hauber [p. 407]

Phragmites australis subsp. *berlandieri* is the strain that grows from the Atlantic coast of Florida, around the Gulf of Mexico, southwestern Arizona, northern Mexico, and extends into Central America and South America. It is not yet clear whether it is native or introduced, but it probably does not grow within the Intermountain Region. Some taxonomists treat it as *P. karka* (Retz.) Trin. *ex* Steud., a species that was described from India.

Phragmites australis (Cav.) Trin. *ex* Steud.
[p. *407*] (Invasive)

The appropriate name for the introduced plants is not clear although they probably originate in Europe. The name *Phragmites australis*, and hence the name *P. australis* subsp.

australis, is based on plants collected in Australia. A number of subspecific names are available for European plants but, until their taxonomy is better understood, it is impossible to determine whether the name for the invasive strain that has been introduced to North America should be subsp. *australis* or one of the subspecies names applied to European plants. Nevertheless, because of the management issues involved, it is important to provide guidance on distinguishing the introduced plants from the two native taxa.

9.03 ARUNDO L.
Kelly W. Allred

Pl per; **rhz**, rhz short, usu more than 1 cm thick. **Clm** 2–10 m tall, 1–3.5 cm thick, usu erect, occ pendant from cliffs; **nd** glab; **intnd** hollow. **Lvs** cauline, conspicuously distichous, glab; **shth** open, longer than the intnd; **lig** memb, shortly ciliate; **bld** flat or folded, mrg scabrous. **Pan** tml, plumose, silvery to purplish. **Spklt** lat compressed, with 1–several flt; **rchl intnd** glab; **dis** above the glm and between the flt. **Glm** longer than the flt, 3–5-veined; **lm** pilose, hairs not papillose-based, 3–7-veined, apc entire or minutely awned; **pal** shorter than the lm, 2-veined; **anth** 3. $x = 12$. Name from the Latin *arundo*, 'reed'.

Arundo, a genus of three species, grows throughout the tropical and warm-temperate regions of the world. Only one species has been introduced to the Western Hemisphere.

Arundo is similar to, but usually larger than, *Phragmites*, a much more common genus in North America. In addition, *Arundo*, but not *Phragmites*, has a wedge-shaped, light to dark brown area at the base of its blades.

1. Arundo donax L. GIANT REED [p. *408*, <u>505</u>]
Clm (2)3–10 m, in large tussocks or hedges. **Lvs** distichous; **lig** 0.4–1 mm; **bld** 30–100 cm long, 2–7(9) cm wide, with a wedge-shaped, light to dark brown area at the base. **Pan** 30–60 cm long, to 30 cm wide. **Spklt** 10–15 mm, with 2–4 flt. **Glm** subequal, as long as the spklt, thin, brownish or purplish, 3-veined, long-acuminate; **lm** 8–12 mm, 3–5-veined, pilose, hairs 4–9 mm, apc bifid, midvein ending into a delicate awn; **pal** 3–5 mm, pilose at the base; **anth** 2–3 mm. **Car** 3–4 mm, oblong, light brown. $2n = 24, 100, 110$.

Arundo donax grows in the southern half of the contiguous United States, being found along ditches, culverts, and roadsides where water accumulates. It has been used extensively as a windbreak, and planted for erosion control on wet dunes. It is also grown for the ornamental value of its tall, leafy culms and large panicles, but its tendency to spread is sometimes a disadvantage. It can also persist for many years after establishment. The report of *A. donax* from Bryce Canyon, Utah, was based on a misidentification (W. Fertig, pers. comm.). Cultivars with striped or unusually wide leaves, e.g., 'Variegata' and 'Macrophylla', are of horticultural interest but do not merit taxonomic recognition.

Arundo donax has been used for thousands of years in making musical instruments, the stems being used for pipes and the tough inner rind for reeds in a wide variety of woodwind instruments. It is one of the species referred to as 'reed' in the Bible. It is still used in many parts of the world for house construction, lattice-work, mats, screens, stakes, walking sticks, and fishing poles.

5. CHLORIDOIDEAE Kunth *ex* Beilschm.
Grass Phylogeny Working Group

Pl ann or per; usu synoecious, smt monoecious or dioecious; habit varied. **Clm** usu ann, smt becoming somewhat woody, intnd solid or hollow. **Lvs** smt conspicuously distichous; **shth** usu open; **aur** absent; **abx lig** usu absent, smt present as a line of hairs; **adx lig** memb, often ciliate with cilia longer than the memb base, smt not ciliate; **bld** not psdpet; **mesophyll** usu radiate; **adx palisade layer** not present; **fusoid cells** absent; **arm cells** absent; **Kranz anatomy** present; **midrib** simple; **adx bulliform cells** present; **stomatal subsidary cells** dome-shaped or triangular; **bicellular microhairs** present, usu with a short, wide apical cell; **papillae** smt present. **Infl** ebracteate, paniculate, rcm, or spicate (occ a single spklt), if paniculate, often with spikelike br; **dis** usu beneath the flt, smt at the base of the pan br. **Spklt** usu bisex, usu lat compressed, with 1–60 flt, distal flt often rdcd. **Glm** usu 2, shorter or longer than the lm, smt exceeding the distal flt, lo or both glm occ missing; **lm** lacking uncinate hairs, smt awned, awns single or, if multiple, the bases not fused into a single column; **anth** 1–3; **ov** glab; **sty** 2, separate throughout, bases close. **Car** often with a free or loose pericarp; **hila** short; **endosperm** hard, without lipid; **starch grains** simple or compound; **haustorial synergids** absent; **emb** usu large relative to the endosperm, not waisted; **epiblasts** usu present; **scutellar cleft** present; **mesocotyl intnd** elongate; **emb lf mrg** usu meeting, rarely overlapping. $x = (7, 8,) 9, 10 (12)$.

The subfamily *Chloridoideae* is most abundant in dry, tropical and subtropical regions. In North America, it reaches its greatest diversity in the southwestern United States (Barkworth and Capels 2000). Almost all its members, and all those in North America, have C_4 photosynthesis. Most employ the NAD-ME or PCK pathways, but *Pappophorum* utilizes the NADP-ME pathway.

The subfamily has been recognized, with essentially the same limits as here, for some time, although reservations have been expressed concerning its monophyly (Campbell 1985; Jacobs 1987; Kellogg and Campbell 1987). More recent studies, both morphological (Van den Borre and Watson 1997, 2000) and molecular (Grass Phylogeny Working Group 2001; Hilu and Alice 2001, Bouchenak-Khelladi et al. 2008) support its recognition as a monophyletic unit. There is less agreement concerning the subfamily's closest relative, some studies pointing to the *Arundinoideae* (Grass Phylogeny Working Group 2001) and some to the *Danthonioideae* (Barker et al. 1995; Hilu and Esen 1993; Hilu and Alice 2001, Bouchenak-Khelladi et al. 2008).

There is considerable disagreement concerning the tribal treatment within the *Chloridoideae*, the number of tribes recognized varying from two (Prat 1936) to eight (Gould and Shaw 1983). Hilu and Wright (1982, p. 28) concluded, on the basis of their morphological study, that "... the boundaries between most of the tribes in this subfamily are not pronounced." They noted that Savile (1979) reached the same conclusion from considering the host specificity of various pathogenic fungi. Subsequent work has not resolved the issue (see, e.g., Hilu and Alice 2001 and Bouchenak-Khelladi et al. 2008).

The treatment presented here is conservative in recognizing *Pappophoreae* as a distinct tribe. It departs from most other treatments in merging all other North American taxa into a single tribe, the *Cynodonteae*. Consensus on how the *Cynodonteae sensu lato* should be broken up is unlikely to be reached until the generic limits of its members have been more thoroughly examined.

1. Lemmas 1–11-veined, unawned or with 1 or 3 awns, sometimes with hyaline lobes on either side of the central awns. . . . . 10. *Cynodonteae*
1. Lemmas 5–13-veined, all the veins extending into awns, often alternating with hyaline lobes or teeth. . . . . 11. *Pappophoreae*

10. **CYNODONTEAE** Dumort.   Mary E. Barkworth

Pl ann or per. **Clm** 1–500 cm, not woody, usu not brchd above the base. **Shth** usu open, often with coarse hairs at the top; **aur** rarely present; **lig** of hairs or memb, if memb, often ciliate, cilia smt longer than the memb base; **bld** often with stiff, coarse mrgl hairs adjacent to the lig, glab or variously pubescent elsewhere. **Infl** tml, smt also axillary, simple pan, pan of 1–many spikelike br, spikelike rcm, spikes, or, in 1 species, a solitary spklt, in dioecious taxa the stmt and pist infl smt morphologically distinct; **dis** usu beneath the ftl flt or the glm but, particularly if the pan br are short, smt at the base of the br. **Spklt** usu lat compressed, with 1–60 flt, strl or rdcd flt, if present, usu distal to the bisex flt. **Glm** from shorter than the adjacent flt to exceeding the distal flt; **lm** 1–3-veined or 7–13-veined, rarely 5-veined, if with 7–13 veins, the veins often in 3 groups; **lod** 2, or absent. $x = 7, 8, 9, 10, 12$.

Most members of the *Cynodonteae* in the Intermountain Region can be recognized by their possession of two or more of the following characteristics: 1–3- or 7–13-veined lemmas, laterally compressed spikelets, spikelike inflorescence branches, and the presence of coarse hairs near the junction of the sheath and blade. All employ the NAD-ME or PCK C_4 photosynthetic pathways, have Kranz blade anatomy, and tend to grow in hot, dry areas. Having said this, it must be acknowledged that each of these characteristics can be found in other tribes and, within the *Cynodonteae*, there are genera that lack one or more of them.

The tribe *Cynodonteae*, as interpreted here, includes genera that are normally placed in two tribes, the *Cynodonteae sensu stricto*, and the *Eragrostideae* Stapf (Clayton and Renvoize 1986; Peterson et al. 2001; Grass Phylogeny Working Group 2001, but see Campbell 1985). Genera 10.01 to 10.18 correspond to the *Eragrostideae* and genera 10.19 to 10.26 to their *Cynodonteae*. The two are treated as one here because recent morphological, anatomical, and molecular studies (Van den Borre 1994; Van den Borre and Watson 1997; Hilu and Alice 2001, Bouchenak-Khelladi et al. 2008) indicate that the distinction between the two is artificial. There is, however, no agreement on an alternative treatment. Part of the problem is that some of the genera (e.g., *Eragrostis*, *Chloris*, *Muhlenbergia*, and *Sporobolus*) appear to be polythetic (Van den Borre and Watson 1997; Hilu and Alice 2001, Bouchenak-Khelladi et al. 2008).

1. Most inflorescences clearly exceeded by the upper leaves, often completely or almost completely enclosed in the upper leaf sheaths; culms 1–30(75) cm tall.
  2. Lemmas 3-lobed, the lobes ciliate; spikelets with 4 florets. . . . . 10.09 *Blepharidachne*
  2. Lemmas not 3-lobed or the lobes not ciliate; spikelets with 1–60 florets.
    3. Spikelets (and often the plants) unisexual.
      4. Leaves strongly distichous; lemmas 9–11-veined; plants unisexual, growing in saline and alkaline soils . . . . 10.02 *Distichlis* (in part)
      4. Leaves not strongly distichous; lemmas 1–5-veined; plants unisexual or, if bisexual, with separate pistillate and staminate inflorescences, growing in a variety of soils.

5. Spikelets 5–26 mm long; pistillate and staminate inflorescences similar, simple panicles; glumes and lemmas unawned, mucronate, or 1-awned .... 10.12 *Eragrostis* (in part)

5. Spikelets 2.5–7 mm long; pistillate and staminate inflorescences strongly dimorphic; staminate inflorescences with pectinate, spikelike branches; pistillate spikelets with conspicuously 3-awned glumes or distal florets .... 10.23 *Bouteloua* (in part)

3. Spikelets bisexual, usually at least the lowest floret in each spikelet bisexual, in *Dasyochloa* the third floret in each spikelet bisexual or pistillate, if pistillate, the lowest 2 florets staminate.

6. Lemma margins with a tuft of hairs at midlength, glabrous elsewhere; blades with white, thickened margins and sharply pointed .... 10.10 *Munroa*

6. Lemma margins glabrous, or with hairs but the hairs not forming a tuft at midlength; blades without white cartilaginous margins, not sharply pointed.

7. Lemmas awned, the awns 1–11 mm long.

8. Plants stoloniferous; inflorescences 1–2.5 cm long, dense panicles; lemmas bilobed, the lobes about ½ as long as the lemmas; ligules of hairs .... 10.07 *Dasyochloa*

8. Plants not stoloniferous; inflorescences 1.5–76 cm long, not dense; lemmas entire or minutely bilobed; ligules membranous, sometimes ciliate .... 10.11 *Leptochloa* (in part)

7. Lemmas unawned, sometimes mucronate, with mucros less than 1 mm long.

9. Spikelets with 2–20 florets; inflorescences panicles of 2–120 spikelike branches .... 10.11 *Leptochloa* (in part)

9. Spikelets with 1(3) florets; inflorescences simple panicles, often highly contracted, without spikelike branches.

10. Inflorescences 0.3–7.5 cm long, dense, spikelike or capitate panicles 1–8 times longer than wide; glumes strongly keeled; plants annual .... 10.15 *Crypsis* (in part)

10. Inflorescences 1–60 cm long, sometimes dense and spikelike but, if less than 8 cm long, more than 8 times longer than wide; glumes rounded or weakly keeled; plants annual or perennial .... 10.14 *Sporobolus* (in part)

1. Most inflorescences equaling or exceeding the upper leaves; culms 1–500 cm tall.

11. Inflorescence branches disarticulating at the bases or (in *Lycurus*) at the base of the fused pedicels; branches 0.04–7 cm long, often globose or spikelike (fused to the rachis and not evident in *Lycurus*), usually with fewer than 15 spikelets per branch.

12. Upper glumes with straight or uncinate spinelike projections; spikelets crowded, the branches condensed into burs .... 10.25 *Tragus*

12. Upper glumes without spinelike projections; spikelets sometimes crowded, but not forming burlike clusters.

13. Inflorescence branches not fused or appressed to the rachises; spikelets solitary .... 10.23 *Bouteloua* (in part)

13. Inflorescences spikelike, branches fused or appressed to the rachises; spikelets in pairs or triplets.

14. Spikelets in pairs; panicle branches fused to the rachises; glumes awned, the lower glumes (1)2(3)-awned, the upper glumes 1-awned ..... 10.18 *Lycurus* (in part)

14. Spikelets in triplets; panicle branches sometimes appressed, but not fused, to the rachises; glumes unawned, 1-awned, or 3-awned. .... 10.24 *Hilaria*

11. Inflorescence branches not disarticulating; branches often longer than 4.5 cm, variously shaped, but never globose, often with more than 16 spikelets per branch.

15. Inflorescences spikes or racemes.

16. Spikelets with 1 bisexual or staminate floret and no additional florets.

17. Spikelets solitary at each node; disarticulation below the glumes or the spikelets not disarticulating .... 10.26 *Zoysia*

17. Spikelets paired, terminal on branches that are fused to the rachises; disarticulation at the base of the fused pedicel pairs .... 10.18 *Lycurus* (in part)

16. Spikelets with more than 1 floret but sometimes only 1 floret bisexual, the additional florets sterile or staminate.

18. Lemmas 9–11-veined, unawned; plants of saline habitats . . . . 10.02 *Distichlis* (in part)
18. Lemmas 3-veined, those of the pistillate florets with awns 3.4–6.8 mm long; plants of various habitats . . . . 10.05 *Scleropogon* (in part)
15. Inflorescences simple panicles (sometimes highly condensed) or panicles of 1–120 spikelike branches.
    19. Inflorescences panicles of spikelike branches, the branches digitately or racemosely arranged on the rachises.
        20. Inflorescence branches 1 or more, if more than 1, arranged in terminal, digitate clusters, sometimes with additional branches or whorls below the terminal cluster.
            21. Plants unisexual. . . . . 10.23 *Bouteloua* (in part)
            21. Plants bisexual, all spikelets with at least 1 bisexual floret.
                22. Spikelets with more than 1 bisexual floret.
                    23. Lemmas usually with hairs over the veins, at least basally, the apices often toothed, sometimes mucronate or awned . . . . 10.11 *Leptochloa* (in part)
                    23. Lemmas glabrous, the apices entire, neither mucronate nor awned . . . . 10.13 *Eleusine*
                22. Spikelets usually with only 1 bisexual floret (occasionally 2 in some genera), often with additional staminate, sterile, or modified florets.
                    24. Spikelets usually without sterile or modified florets; lemmas unawned. . . . . 10.21 *Cynodon*
                    24. Spikelets with 1 or more sterile florets distal to the bisexual floret; lemmas of the bisexual florets often awned. . . . . 10.19 *Chloris*
        20. Inflorescence branches more than 1, racemosely arranged on the rachises.
            25. All spikelets and florets unisexual . . . . 10.23 *Bouteloua* (in part)
            25. Spikelets with 1–20 bisexual florets.
                26. Spikelets with (2)3–12(20) bisexual florets. . . . . 10.11 *Leptochloa* (in part)
                26. Spikelets with only 1 bisexual floret, sometimes with sterile, rudimentary, or modified florets distal to the bisexual floret.
                    27. Spikelets with sterile, rudimentary, or modified florets distal to the bisexual floret. . . . . 10.23 *Bouteloua* (in part)
                    27. Functional spikelets with only 1 floret, lacking sterile, rudimentary, or modified florets.
                        28. Spikelets distant to slightly imbricate, appressed to the branches; branches strongly divergent . . . . 10.20 *Schedonnardus*
                        28. Spikelets clearly imbricate, appressed to strongly divergent; branches appressed to strongly divergent . . . . 10.22 *Spartina*
    19. Inflorescences simple panicles, sometimes highly contracted, even spikelike in appearance; spikelike branches not evident.
        29. Spikelets usually with only 1 floret, occasionally with 2–3 florets.
            30. Ligules membranous, hyaline, or coriaceous, sometimes ciliate; lemmas 3-veined (occasionally appearing 5-veined), usually awned, sometimes unawned or mucronate.
                31. Lemmas and paleas densely sericeous over the veins and margins, glabrous between the veins . . . . 10.08 *Blepharoneuron*
                31. Lemmas and paleas glabrous to variously hairy but not densely sericeous over the veins and margins.
                    32. Lemmas usually awned or mucronate; spikelets usually with 1 floret. . . . . 10.17 *Muhlenbergia* (in part)
                    32. Lemmas unawned or mucronate; spikelets frequently with 2–3 florets. . . . . 10.12 *Eragrostis* (in part)
            30. Ligules of hairs; lemmas 1(3)-veined, unawned, sometimes mucronate.
                33. Panicles 0.3–4(7.5) cm long, 3–15 mm wide, spikelike or capitate, 1–8 times longer than wide; plants annual . . . . 10.15 *Crypsis* (in part)

33. Panicles 1–80 cm long, 2–600 mm wide, dense to open, if less than 8 cm long, often 10 or more times longer than wide; plants annual or perennial.
  34. Calluses usually glabrous or almost so; paleas glabrous; fruits falling free of the lemma and palea . . . . 10.14 *Sporobolus* (in part)
  34. Calluses evidently hairy, the hairs ¼–⅞ as long as the lemmas; paleas hairy; fruits falling with the lemma and palea. . . . . 10.16 *Calamovilfa*
29. Spikelets with more than 1 floret.
  35. Lemmas with (5)9–11 veins.
    36. Spikelets unisexual; plants almost always unisexual, occasionally bisexual.
      37. Lemmas 9–11-veined; glumes 2–7 veined; plants rhizomatous and/or stoloniferous, found in saline or alkaline soils . . . . 10.02 *Distichlis* (in part)
      37. Lemmas 1–6-veined; lower glumes of the staminate spikelets 1-veined, those of the pistillate spikelets 1–5-veined; plants stoloniferous or rooting at the lower nodes, not rhizomatous, not found in saline or alkaline soils . . . . 10.12 *Eragrostis* (in part)
    36. All spikelets with at least 1 bisexual floret.
      38. Glumes longer than the adjacent lemmas . . . . 10.01 *Swallenia*
      38. Glumes shorter than the adjacent lemmas. . . . . 10.12 *Eragrostis* (in part)
  35. Lemmas with 1–3 veins (occasionally with scabrous lines that may be mistaken for additional veins).
    39. Florets unisexual.
      40. Plants perennial; staminate and pistillate florets strongly dimorphic; plants unisexual or bisexual, bisexual plants with unisexual or bisexual spikelets; pistillate spikelets with 3–5 functional florets and lemma awns (30)50–150 mm long; staminate spikelets with 5–10(20) florets and unawned or shortly awned (to 3 mm) lemmas. . . 10.05 *Scleropogon* (in part)
      40. Plants annual; staminate and pistillate florets similar; plants unisexual; spikelets with 4–60 florets, all or almost all functional; lemmas 1.5–10.5 mm, unawned, sometimes mucronate. . . . . 10.12 *Eragrostis* (in part)
    39. At least 1 floret in each spikelet bisexual.
      41. Lemmas, including the calluses, glabrous or inconspicuously hairy; lemma apices usually entire, sometimes minutely toothed.
        42. Spikelets with (1)2–60 florets; lemmas unawned, sometimes mucronate; ligules usually membranous and ciliate or ciliolate, sometimes of hairs . . . . . 10.12 *Eragrostis* (in part)
        42. Spikelets with 1(2–3) florets; lemmas often awned, sometimes unawned or mucronate; ligules membranous, sometimes ciliolate, not ciliate. . . . . 10.17 *Muhlenbergia* (in part)
      41. Lemma bodies conspicuously hairy over the veins and/or calluses conspicuously hairy; lemma apices usually with emarginate, bilobed, or trilobed apices, sometimes entire.
        43. Leaf margins evidently cartilaginous . . . . 10.06 *Erioneuron*
        43. Leaf margins not cartilaginous.
          44. Lemmas rounded to truncate, emarginate to bilobed; all 3 lemma veins often pilose basally . . . . 10.03 *Tridens*
          44. Lemmas acute, entire or with 3 minute teeth, glabrous or shortly pubescent on the distal ⅔, the pubescence not confined to the veins . . . . 10.04 *Redfieldia*

10.01 SWALLENIA Soderstr. & H.F. Decker   James P. Smith, Jr.

Pl per; clumped, rhz woody. **Clm** 10–60 cm, brchd above the base. **Lvs** mostly bas; **aur** absent; **lig** of hairs; **bld** flat, strongly veined, sharply pointed. **Infl** terminal, usu exceeding the up lvs, contracted pan; **br** ascending to erect. **Spklt** lat compressed, unawned, with 3–7 bisex flt, distal flt rdcd; **dis** beneath the car. **Glm** subequal, longer than the adjacent lm but exceeded by the distal flt, acuminate; **lo glm** 5–7-veined; **up glm** 7–11-veined; **cal** hairy; **lm** memb to papery, 5–7-veined, densely villous on the mrg, smt also between the veins, unawned to mucronate; **pal** equaling or exceeding the lm; **anth** 3. **Car** falling free from the lm and pal. x = 10. Named for Jason Richard Swallen (1903–1991), a U.S. Department of Agriculture botanist and a former head of the Department of Botany at the Smithsonian Institution.

  Swallenia is a monospecific genus, endemic to California. It is unusual in that only its caryopses break off the plant.

1. **Swallenia alexandrae** (Swallen) Soderstr.
   & H.F. Decker Eureka Valley Dunegrass
   [p. *408*, 505]
Clm 10–40(60) cm, stiff, erect; **nd** villous. **Shth** villous on the up mrg; **bld** 5–14 cm long, 3–8 mm wide. **Pan** 4–10 cm; **br** to 35 mm, with 1–3 spklt. **Spklt** 10–15 mm, persistent. **Glm** 9–14 mm; **lm**

7–9 mm. **Car** about 4 mm long, about 2 mm in diameter. $2n = 20$.

  Swallenia alexandrae grows on sand dunes in Inyo County, California. It is only known from four sites, all between 900–1200 m, in the Eureka Valley of northern Inyo County. At these sites, it forms dense colonies 1–2 m across. It is state-listed as rare and federally listed as endangered because of off-road vehicle activity.

10.02 DISTICHLIS Raf.   Mary E. Barkworth

Pl per; usu unisex, occ bisex; strongly rhz and/or stln. **Clm** to 60 cm, usu erect, glab. **Lvs** conspicuously distichous; **lo lvs** rdcd to scalelike shth; **up lf shth** strongly overlapping; **lig** shorter than 1 mm, memb, serrate; **up bld** stiff, glab, ascending to spreading, usu equaling or exceeding the pist pan. **Infl** tml, contracted pan or rcm, smt exceeding the up lvs. **Spklt** lat compressed, with 2–20 flt; **dis** of the pist spklt above the glm and below the flt, stmt spklt not dis. **Glm** 3–7-veined; **lm** coriaceous, stmt lm thinner than the pist lm, 9–11-veined, unawned; **pal** 2-keeled, keels narrowly to broadly winged, serrate to toothed, smt with excurrent veins; **anth** 3. **Car** glab, free from the pal at maturity, brown. x = 10. Name from the Greek *distichos*, 'two-rowed', referring to the conspicuously distichous blades.

  Distichlis, a genus of about five species, grows in saline soils of the coasts and interior deserts of the Western Hemisphere and Australia. Only *D. spicata* grows in the Intermountain Region.

1. **Distichlis spicata** (L.) Greene Saltgrass
   [p. *409*, 505]
Pl rhz and smt stln. **Clm** 10–60 cm, usu erect, smt decumbent or prostrate. **Bld** of up lvs 1–8(20) cm, rigid and divaricate to lax and ascending, usu equaling or exceeding the pist pan, varying with respect to the stmt pan. **Pist pan** 1–7 cm, often congested, with 2–20 spklt. **Pist spklt** 5–20 mm long, 4–7 mm wide, with 5–20 flt; **lo glm** 2–3 mm; **up glm** 3–4 mm; **lm** 3.5–6 mm; **pal** with serrate

keels. **Car** 2–5 mm, tapered or truncate. **Stmt pan** and **spklt** similar to the pist pan and spklt, but the lm somewhat thinner in texture and the pal not bowed-out. **Anth** 3–4 mm. $2n = 38, 40$.

  Distichlis spicata grows in saline soils of the Western Hemisphere and Australia. Plants in the Intermountain Region belong to **D. spicata** subsp. **stricta** (Torr). R.F. Thorne. They differ from coastal plants in having $2n = 38$ (not 40), earlier phenology, and chloroplast DNA (Harrington et al. 2009).

10.03 TRIDENS Roem. & Schult.   Jesús Valdés-Reyna

Pl per; usu ces, often with short, knotty rhz, occ with elongate rhz, never stln. **Clm** 5–180 cm, erect, mostly glab, lo nd smt with hairs. **Shth** shorter than the intnd, open; **lig** memb and ciliate or of hairs; **bld** 6–25 cm long, 1–8 mm wide, flat or involute, mrg not thick and cartilaginous. **Infl** tml, usu pan (smt rdcd to rcm), 5–40 cm, exceeding the up lvs, exserted. **Spklt** 4–10(13) mm, lat compressed, with 4–11(16) flt, more than 1 flt bisex; **strl flt** distal to the ftl flt; **dis** above the glm. **Glm** from shorter than to equaling the distal flt; **lo glm** 1(3)-veined; **up glm** shorter than or about equal to the lo glm, 1–3(9)-veined, unawned; **cal** usu glab, smt pilose; **lm** hyaline or memb, 3-veined, veins usu shortly hairy below, apc rounded to truncate, emgt to bilobed, midvein often excurrent to 0.5 mm, lat veins not or more shortly excurrent; **pal** glab or shortly pubescent on the lo back and mrg, veins glab or ciliolate; **lod** 2, free or adnate to the pal; **anth** 3, reddish purple. **Car** dorsiventrally compressed and reniform in cross section, dark brown; **emb** about ⅔ as long as the car. *x* = 10. Name from the Latin *tres*, 'three', and *dens*, 'tooth', referring to the three shortly excurrent veins of *Tridens flavus*, the type species.

Tridens, a genus of 14 species, is native to the Americas. Only one species grows in the Intermountain Region.

1. **Tridens muticus** (Torr.) Nash S<small>LIM</small> T<small>RIDENS</small> [p. *409, 410*, 505]

Pl ces, with knotty, shortly rhz bases. **Clm** 20–80 cm; **nd** often with soft, 1–2 mm hairs. **Shth** rounded, lo shth often strigose or pilose, up shth glab or scabrous; **lig** 0.5–1 mm, memb, ciliate; **bld** 1–4 mm wide, usu involute or loosely infolded, glab, scabrous, or sparsely pilose, attenuate distally. **Pan** 7–20(25) cm long, 0.3–0.8 cm wide; **br** erect, spklt imbricate but usu not crowded; **ped** 1–2 mm. **Spklt** 8–13 mm, with 5–11 flt. **Glm** glab, usu purple-tinged; **lo glm** 3–8(10) mm, 1–3-veined; **up glm** 4–10 mm, 1–7-veined; **lm** 3.5–7 mm, usu purple-tinged, midveins pilose on the bas ⅓–½, rarely excurrent, lat veins pilose to well above midlength, never excurrent; **pal** 1–2 mm shorter than the lm, mrg pubescent; **anth** 1–1.5 mm. **Car** 1.5–2.3 mm. 2*n* = 40.

1. Upper glumes 4–5(6) mm long, 1-veined . . . var. *muticus*
1. Upper glumes usually 5.5–10 mm long, 3–7-veined
 . . . . var. *elongatus*

Tridens muticus var. **elongatus** (Buckley) Shinners [p. *409*]

Clm usu 40–80 cm. **Bld** often 3–4 mm wide. **Up glm** usu 5.5–10 mm, 3–7-veined.

Tridens muticus var. *elongatus* grows on well-drained, clay and sandy soils from Colorado to Missouri and from Arizona to Louisiana.

Tridens muticus (Torr.) Nash var. **muticus** [p. *410*]

Clm usu 20–50 cm. **Bld** 1–2 mm wide. **Up glm** 4–5(6) mm, 1-veined.

Tridens muticus var. *muticus* is a common species on dry, sandy or clay soils in the arid southwestern United States and adjacent Mexico.

10.04 REDFIELDIA Vasey

Stephan L. Hatch

Pl per; with extensive, often deep, horizontal or vertical rhz. **Clm** 50–130 cm, erect, bases usu buried and rooting at the nd. **Lvs** cauline; **shth** shorter than the intnd, open, ribbed; **lig** memb, ciliate; **aur** absent; **bld** loosely involute, smt scabridulous, apc attenuate. **Infl** tml, conical to oblong pan, open to diffuse, exceeding the up lvs; **br** slender, widely spreading. **Ped** longer than the spklt, capillary, flexible. **Spklt** ovate to obovate, olive-green to brownish, with (1)2–6 flt; **strl flt** distal to the bisex flt; **dis** above the glm and below the flt. **Glm** unequal, usu exceeded by the flt, glab, acute; **lo glm** 1-veined; **up glm** 1- or 3-veined; **cal** with a tuft of soft hairs; **lm** lanceolate to falcate, glab or shortly pubescent, at least on the distal ⅔, 3-veined, lat veins converging distally, apc acute to awn-tipped, entire or with 3 minute teeth; **anth** 3. *x* = unknown. Named for John Howard Redfield (1815–1895), a Philadelphia businessman associated with the Philadelphia Academy of Natural Sciences.

Redfieldia is a monospecific genus that is endemic to North America.

1. **Redfieldia flexuosa** (Thurb. *ex* A. Gray) Vasey B<small>LOWOUT</small>-G<small>RASS</small> [p. *410*, 505]

Clm 50–130 cm. **Lig** to 1.5 mm; **bld** 15–45 cm long, 2–8 mm wide. **Pan** 20–50 cm long, 8–25 cm wide. **Spklt** (3)5–8 mm long, 3–5 mm wide. **Lo glm** 3–4 mm; **up glm** 3.5–4.5 mm; **cal hairs** to 1.5 mm; **lm** 4.5–6 mm, glab or shortly pubescent, veins glab,

entire or with 3 minute teeth; **pal** glab; **anth** 2–3.6 mm, yellow to reddish purple; **lod** 2, truncate. **Car** oblong, terete. 2*n* = 25.

Redfieldia flexuosa grows on sandhills and dunes. It is a common and important soil binder in blowout areas. It is only fair livestock forage but, because it grows in areas subject to blowout, this should not be of concern.

10.05 SCLEROPOGON Phil.

John R. Reeder†

Pl per; usu monoecious, less frequently dioecious, occ synoecious; bearing wiry, often arching, stln with 5–15 cm intnd, smt also weakly rhz. **Lvs** mostly bas; **shth** short, strongly veined, bas lvs commonly hispid or villous; **lig** of hairs; **bld** firm, flat or folded. **Infl** tml, usu exceeding the up lvs, spikelike rcm or contracted pan with few spklt, in bisex pl stmt and pist flt in the same spklt with the stmt flt below the pist flt or in separate spklt, bisex flt occ produced; **br** not pectinate; **dis** above the glm and below the lowest pist flt in a spklt, flt falling together, lowest flt with a bearded, sharp-pointed callus. **Stmt spklt** with 5–10(20) flt; **glm** memb, pale, 1–3-veined, acuminate; **lm** 3-veined, similar to the glm, unawned or awned, awns to 3 mm; **pal** shorter than the lm, often conspicuously so. **Pist spklt** appressed to the br axes, usu the 3–5 lo flt fnctl, up flt rdcd to awns; **glm** acuminate, strongly 3-veined, occ with a few fine accessory veins; **lm** narrow, 3-veined, veins extending into awns, awns (30)50–100(150) mm, spreading or reflexed at maturity. $x = 10$. Name from the Greek *skleros*, 'hard', and *pogon*, 'beard', in reference to the hard callus.

Scleropogon is a monospecific American genus with a disjunct distribution.

1. **Scleropogon brevifolius** Phil. BURROGRASS [p. *411*, <u>506</u>]
Stln to 50 cm, wiry, intnd 5–15 cm. **Clm** (5)10–20 cm, erect. **Lig** about 1 mm; **bld** 2–8(12) cm long, 1–2 mm wide. **Bisex spklt** 2-4 cm, stmt flt below the pist flt. **Stmt spklt** 2–3 cm. **Pist spklt** subtended by a glumelike bract; **lm bodies** 2.5–3 cm. $2n = 40$.

Scleropogon brevifolius grows on grassy plains and flats, generally being most abundant on disturbed or over-grazed land. Its North American range extends from the southwestern United States to central Mexico; its South American range is from Chile to northwestern Argentina.

10.06 ERIONEURON Nash

Jesús Valdés-Reyna

Pl per; usu ces, occ stln. **Clm** 6–65 cm, erect. **Lvs** mostly bas; **shth** smooth, glab, striate, mrg hyaline, col with tufts of 1–3 mm hairs; **bld** usu folded, pilose bas, mrg white, cartilaginous, apc acute but not sharp. **Infl** tml, simple pan (rcm in depauperate specimens), exserted well above the lvs. **Spklt** lat compressed, with 3–20 flt, distal flt stmt or strl; **dis** above the glm and between the flt. **Glm** thin, memb, 1-veined, acute to acuminate; **cal** with hairs; **lm** rounded on the back, 3-veined, veins conspicuously pilose, at least bas, apc toothed or obtusely 2-lobed, midveins often extended into awns, awns to 4 mm, lat veins smt extended as small mucros; **pal** shorter than the lm, keels ciliate, intercostal regions pilose bas; **lod** 2, adnate to the bases of the pal; **anth** 1 or 3. **Car** glossy, translucent; **emb** more than ½ as long as the car. $x = 8$. Name from the Greek *erion*, 'wool', and *neuron*, 'nerve', a reference to the hairy veins of the lemmas.

Erioneuron is an American genus of three species. Its seedlings appear to have a shaggy, white-villous indumentum, but this is composed of a myriad of small, water-soluble crystals. One species grows in the Intermountain Region.

1. **Erioneuron pilosum** (Buckley) Nash HAIRY TRIDENS, HAIRY WOOLYGRASS [p. *411*, <u>506</u>]
Clm (6)10–30(40) cm tall, (0.3)0.6–1(2.5) mm thick, glab or hispidulous. **Lig** 2–3.5 mm; **bld** (1)3–6(9) cm long, (0.5)1–1.5(2.5) mm wide, both surfaces sparsely pilose or glab, grayish green. **Pan** 1–4(6) cm; **br** with 3–9 shortly pedlt spklt. **Spklt** 6–12(15) mm, with (5)6–12(20) flt. **Glm** exceeded by the lowest flt, pale; **lo glm** 4–7 mm; **up glm** 4–7 mm; **lm** 3–6 mm, green or purplish green when young, becoming stramineous at maturity, awned, awns 0.5–2.5 mm, apc acute, entire or bidentate, teeth 0.3–0.5 mm; **anth** usu 3, 0.3–1 mm. **Car** 1–1.5 mm. $2n = 16$.

Erioneuron pilosum grows on dry, rocky hills and mesas, often in oak and pinyon-juniper woodlands. In North America, it is represented by *E. pilosum* var. *pilosum*. This variety differs from the other two varieties, both of which are restricted to Argentina, in its longer, less equal glumes and shorter awns.

10.07 DASYOCHLOA Willd. *ex* Rydb.

Jesús Valdés-Reyna

Pl per; stln, smt mat-forming. **Clm** (1)4–15 cm, initially erect, eventually bending and rooting at the base of the infl. **Lvs** not bas aggregated on the pri clm; **shth** with a tuft of hairs to 2 mm at the throat; **lig** of hairs; **bld** involute. **Infl** tml, short, dense pan of spikelike br, each subtended by lfy bracts and exceeded by the up lvs; **br** with 2–4 subsessile to shortly pedlt spklt. **Spklt** lat compressed, with 4–10 flt; **dis** above the glm. **Glm** subequal to the adjacent lm, glab, 1-veined, rounded or weakly keeled, shortly awned to mucronate; **flt** bisex; **lm** rounded or weakly keeled, densely pilose on the lo ½ and on the mrg, thinly memb, 3-veined, 2-lobed, lobes about ½ as long as the lm and obtuse, midveins extending into awns as long as or longer than the lobes, lat veins not excurrent; **pal** about as long as the lm; **anth** 3. **Car** oval in cross section, translucent; **emb** more than ½ as long as the car. $x = 8$. Name from the Greek *dasys*, 'thick with hair' and *chloë*, 'grass'.

Dasyochloa is a monospecific genus that is restricted to the United States and Mexico. It resembles *Munroa* in its leafy-bracteate inflorescence (Caro 1981). Seedlings of *Dasyochloa*, like those of *Erioneuron*, are shaggy-white-villous. This indumentum is composed of myriads of hairlike, water soluble crystals that wash off in water. They are the product of transpiration and evaporation.

1. **Dasyochloa pulchella** (Kunth) Willd. *ex*
   Rydb. Fluffgrass [p. *412*, 506]
Clm (1)4–15 cm, scabrous or puberulent;
peduncles (intnd below the pan) 3–7(11) cm.
Shth striate, mrg scarious; **lig** 3–5 mm; **bld**
(1)2–6 cm, abx surfaces scabrous, adx surfaces
scabridulous. **Pan** 1–2.5 cm long, 1–1.5 cm wide,
densely white-pubescent, light green or purple-
tinged. **Spklt** (5)6–9(10) mm, with (4)6–10 flt. **Lo**
glm 6–8.5 mm; **up glm** 6.5–9 mm, as long as or
longer than the flt; **lm** 3–5.5 mm, lobes (1)3–3.2
mm, midveins extending into straight (1.5)2.5–4
mm awns; **pal** 2–3.5 mm, keels long-pilose
proximally, ciliate distally; **anth** 0.2–0.5 mm. **Car**
1–1.5 mm, translucent. $2n = 16$.

   Dasyochloa pulchella grows in rocky soils of arid regions.
It extends from the United States to central Mexico and is
the most common grass in the *Larrea-Flourensia* scrub of the
southwestern United States and adjacent Mexico.

10.08 BLEPHARONEURON Nash

Paul M. Peterson and Carol R. Annable

Pl ann or per. **Clm** 10–70 cm. **Shth** open, glab, usu longer than the intnd; **lig** memb or hyaline,
truncate to obtuse, often decurrent; **bld** flat to involute, abx surfaces glab, smt scabrous, adx surfaces
shortly pubescent. **Infl** tml, pan, exceeding the lvs; **br** spreading to ascending; **ped** capillary, lax,
minutely glandular just below the spklt. **Spklt** with 1 flt, slightly lat compressed, grayish green; **dis**
above the glm. **Glm** subequal, ovate to obtuse, faintly 1-veined, glab; **lm** slightly longer and firmer
than the glm, 3-veined, veins and mrg densely sericeous, hairs 0.1–0.7(1) mm, apc acute to obtuse, occ
mucronate; **pal** 2-veined, densely villous between the veins; **anth** 3, purplish. $x = 8$. Name from the
Greek *blepharis*, 'eyelash', and *neuron*, 'nerve', a reference to the sericeous veins of the lemmas.

   Blepharoneuron is a genus of two species, one of which grows in the Intermountain Region.

1. **Blepharoneuron tricholepis** (Torr.) Nash
   Hairy Dropseed [p. *413*, 506]
Pl per; densely ces. **Clm** 10–70 cm, erect, glab
and smooth or scabrous just below the nd. **Lig**
(0.3)0.7–2(2.7) mm, hyaline to memb, entire; **bld**
1–15 cm long, 0.6–2.5 mm wide, scabrous. **Pan**
3–25 cm long, 1–10 cm wide; **br** ascending; **ped**
2–9 mm, straight or flexuous. **Glm** (1.5)1.8–2.6(3)
mm, often appearing 3-veined because of the
characteristic infolding of the mrg; **lm** (2)2.3–
3.5(3.9) mm; **anth** 1.2–2.1 mm, brownish. **Car**
1.2–1.4 mm. $2n = 16$.

   Blepharoneuron tricholepis grows in dry, rocky to sandy
slopes, dry meadows, and open woods in pine-oak-madrone
forests from Utah and Colorado to the state of Puebla,
Mexico, at elevations of 700–3660 m. It flowers from mid-
June through November.

10.09 BLEPHARIDACHNE Hack.

Jesús Valdés-Reyna

Pl per (rarely ann); ces, from a knotty base, often mat-forming. **Clm** 3–8(20) cm, often decumbent
and rooting at the lo nd, frequently brchd above the bases, forming short spur shoots at the ends of
long intnd; **intnd** minutely pubescent. **Lvs** clustered at the bases of the pri and spur shoots; bas shth
shorter than the intnd; **lig** of hairs or absent; **bld** linear to triangular, convolute to conduplicate, or
flat to plicate, sharp, those of the up lvs usu exceeding the infl. **Infl** tml, compact pan, exserted or
partially included in the up shth(s). **Spklt** lat compressed, subsessile or pedlt, with 4 flt per spklt;
dis above the glm but not between the flt. **Glm** subequal to each other and the lowest lm, rounded
or weakly keeled, 1-veined, awn-tipped or unawned; **lowest 2 flt** in each spklt stmt or strl; **third flt**
pist or bisex; **lm** rounded on the back, 3-veined, mostly glab but pilose across the bases and on the
mrg, strongly 3-lobed, lat lobes wider than the cent lobes, all lobes ciliate on 1 or both mrg, lo lm
with the lat lobes rounded or mucronate to awned, cent lobes awned; **third lm** 3-lobed, lobes awned;
pal from slightly shorter to slightly longer than the lm; **lod** absent; **anth** 2 or 3 (rarely 1); **sty** br 2.
Distal flt rdmt, 3-awned, plumose, or hairy. **Car** lat compressed. $x = 7$. Name from the Greek *blepharis*,
'eyelash', and *achne*, 'scale' or 'chaff', an allusion to the ciliate lemmas.

   The four species comprising *Blepharidachne* are restricted to the Americas, growing in arid and semi-arid regions of the
United States, Mexico, and Argentina. Only *B. kingii* grows in the Intermountain Region. *Blepharidachne* differs from all other
genera in the *Cynodonteae* in having four florets per spikelet, with the first two florets being sterile or staminate, the third
bisexual or pistillate, and the fourth a rudimentary 3-awned structure.

1. **Blepharidachne kingii** (S. Watson) Hack.
   King's Eyelash Grass [p. *413*, 506]
Pl ces. **Clm** 3–8(14) cm. **Shth** often with a tuft of
hairs at the throat, shth immediately below the
infl often spathelike; **lig** to 0.5 mm; **bld** 0.7–3 cm
long, less than 1 mm wide, strongly convolute to
conduplicate, stiffly arcuate, often deciduous. **Pan**
10–25 mm, subcapitate, usu partially included

in the 2 spathelike up lf shth. **Spklt** 6–9 mm. **Glm** exceeding the flt, papery and translucent, scabridulous toward the base, acuminate or awn-tipped, awns to about 1.3 mm; **lo glm** 6–7.5 mm; **up glm** 6.8–8.5 mm; **lowest 2 flt** usu strl; **lm of strl flt** 3.4–5.8 mm, lat lobes 2.2–3 mm; **pal of strl flt** linear, plumose; **third flt** bisex; **third lm** with lat lobes 0.5–1.5 mm, awned, cent awns 3–5 mm; **pal of third flt** subequal to slightly longer than the lm; **anth** 2(1), about 1.5 mm. **Car** about 2 mm, compressed. $2n = 14$.

Blepharidachne kingii grows at scattered locations in arid regions of the Great Basin, and is sometimes locally abundant.

10.10 MUNROA Torr.   Jesús Valdés-Reyna

Pl ann; stln, mat-forming; stln 2–8 cm, terminating in fascicles of lvs from which new clm arise. **Clm** 3–15(30) cm. **Lvs** mostly bas, smt with a purple tint; **shth** with a tuft of hairs at the throat; **aur** absent; **lig** of hairs; **bld** linear, usu involute, smt flat or folded, with white, thickened mrg, apc sharply pointed. **Infl** tml, capitate pan of spikelike br; **br** almost completely hidden in a subtending lfy bract, bearing 2–4 subsessile or pedlt spklt. **Spklt** lat compressed, with 2–10 flt; **lo flt** bisex or pist; tml flt strl; **dis** above the glm or beneath the lvs subtending the br. **Glm** shorter than the spklt, keeled, 1-veined, unawned; **lo glm** usu present on all spklt; **up glm** absent or rdcd on the tml spklt; **lm** with a pilose tuft of hairs along the mrg at midlength, memb or coriaceous, 3-veined, lat veins occ shortly excurrent, apc emgt or 2-lobed; **pal** glab, smooth; **lod** present or absent, truncate; **anth** 2 or 3, yellow; **sty br** elongate, 2(3), barbellate. **Car** dorsally compressed. $x = 7$ or 8. Named for Sir William Munro (1818–1880), a British botanist who collected in Barbados, the Crimea, and India.

Munroa, a genus of five species, is endemic to the Western Hemisphere. One species grows in the Intermountain Region, the remainder being confined to South America. Its closest relatives are thought to be *Blepharidachne* and *Dasyochloa*, both of which are stoloniferous, mat-forming species with leafy-bracteate panicles. *Munroa* differs from both in its annual habit.

1. **Munroa squarrosa** (Nutt.) Torr. FALSE BUFFALOGRASS [p. *414*, *506*]
Stln slender. **Clm** 3–15(30) cm, highly brchd, scabrous, often minutely puberulent. **Lig** 0.5(1) mm; **bld** 1–5 cm long, 1–2.5 mm wide. **Spklt** 6–8(10) mm, with 3–5 flt. **Glm of first 1–2 spklt** subequal, 2.5–4.2 mm, narrow, 1-veined, acute; **glm of up spklt** unequal, lo glm rdcd or even absent in the tml spklt; **lm of lo spklt** scabridulous, lanceolate, midvein excurrent, forming a stout, scabrous 0.5–2 mm awn, lat veins with a tuft of hairs on the mrg near midlength, excurrent; **anth** 1–1.5 mm. $2n = 16$.

Munroa squarrosa grows in dry, open areas, usually in sandy soil or disturbed sites, from Saskatchewan and Alberta south to Chihuahua, Mexico.

10.11 LEPTOCHLOA P. Beauv.   Neil Snow

Pl ann or per; ces. **Clm** (3)10–170 cm, usu ascending to erect, often geniculate at the lo nd, occ prostrate and rooting at the lo nd, often brchg at the aerial nd; **nd** usu glab; **intnd** usu hollow. **Lvs** usu primarily cauline, occ in bas rosettes; **shth** open; **lig** 0.2–10 mm, obtuse to attenuate, usu memb, smt ciliate; **bld** flat, involute when dry, usu ascending to erect, apc attenuate. **Pri infl** tml, pan of 2–150 non-dis, spikelike br, usu exceeding the lvs; **br** 1–22 cm, digitate, subdigitate, or rcm on the rchs, 1-sided, usu spklt-bearing throughout their length, spklt in 2 rows, axes terminating in a fnctl spklt, lo br occ with sec brchg; **sec pan** smt present, axillary to and concealed by the lo shth, their flt not dis; **dis** in the pri pan beneath the flt. **Spklt** rounded to slightly keeled, distant to tightly imbricate, not conspicuously pubescent, with (2)3–12(20) bisex flt; **rchl** rarely prolonged. **Glm** usu unequal, smt subequal, exceeded by the flt, memb, rounded to weakly keeled, 1-veined, veins scabrous, apc unawned (rarely mucronate); **lo glm** 0.5–4.9 mm; **up glm** 0.9–6 mm; **flt** usu bisex; **cal** distinct or poorly developed, glab or pubescent; **lm** memb, usu pubescent at least over the lo portion of the veins, 3(5)-veined, apc entire or minutely bilobed, unawned, mucronate, or awned; **pal** usu subequal to the lm, memb or hyaline; **anth** 1–3, 0.1–2.7 mm. **Car** obovate to elliptic, falling free of the lm and pal. $x = 10$. Name from the Greek *leptos*, 'slender', in reference to the panicle branches, and *chloa*, 'grass'.

Leptochloa is a pantropical, warm-temperate genus of 32 species. Three species grow in the Intermountain Region. Initial cladistic studies (Snow 1997) do not support recognition of the segregate genus *Diplachne*. However, preliminary molecular studies suggest *Leptochloa* is polyphyletic (Snow et al. 2008).

Leptochloa tends to grow in somewhat basic soils. Many of the species, particularly the annual species, are poor ecological competitors and grow in relatively open, seasonally inundated soils, such as are found along rivers. In disturbed areas, they are associated with roadside ditches, the margins of reservoirs, and mesic agricultural lands. A few species, primarily perennial, grow on well-drained soils. The vegetative vigor of all species is greatly influenced by the availability of soil moisture.

1. Panicle branches digitate or subdigitate; plants perennial . . . . 1. *L. dubia*
1. Panicle branches racemose; plants annual or perennial.
    2. Ligules 2–8 mm long, attenuate, becoming lacerate at maturity. . . . . 2. *L. fusca*
    2. Ligules 0.6–3.2 mm long, truncate, erose . . . . 3. *L. panicea*

1. **Leptochloa dubia** (Kunth) Nees GREEN SPRANGLETOP [p. *414*, <u>506</u>]

Pl per. **Clm** (10)30–110 cm, round or bas compressed, tillering from the bas nd, not brchg from the aerial nd, mostly glab, smt pilose bas; **intnd** solid. **Shth** smt with a pilose col; **lig** 1–2 mm, truncate, erose; **bld** (2)8–35 cm long, 2–8 mm wide, glab, strigose, or pilose. **Pan** 8–20 cm, with 2–15 subdigitate or rcm br; **sec pan** often hidden in the lowest lf shth; **br** 2–19 cm, ascending to spreading at maturity. **Spklt** 4–12 mm, light brown to dark olive green, with 4–13 flt, often widely diverging at anthesis. **Glm** narrowly triangular to ovate, acute; **lo glm** 2.3–4.8 mm; **up glm** 3.3–6 mm; **lm** 3.5–5 mm, memb, ovate to obovate, lat veins glab or sericeous, hairs often restricted to the bas portion, smt also sericeous on the midvein and between the veins, apc obtuse to truncate, usu emgt, unawned, smt mucronate; **pal** ciliate on the mrg; **anth** 3, 0.3–1.6 mm. **Car** 1.5–2.3 mm long, 0.9–1 mm wide, strongly dorsally compressed. $2n = 40, 60, 80$.

Leptochloa dubia grows from the southwestern United States and Florida through Mexico to Argentina, often in well-drained, sandy or rocky soils. It provides fair to good forage, but is seldom abundant.

2. **Leptochloa fusca** (L.) Kunth [p. *415*, <u>506</u>]

Pl ann or weakly per. **Clm** 5–170 cm, prostrate to erect; compressed, often brchg; **intnd** hollow. **Shth** glab or scabrous; **lig** 2–8 mm, memb, attenuate, becoming lacerate at maturity; **bld** 3–50 cm long, 2–7 mm wide, glab or scabrous, those of the flag lvs smt exceeding the pan. **Pan** (1.5)10–105 cm long, 0.5–22 cm wide, with 3–35 rcm br, bases of the pan smt remaining enclosed in the up lf shth at maturity; **br** 1.5–20(22) cm, ascending to reflexed. **Spklt** 5–12(14) mm, with 6–20 flt. **Lo glm** 1–3(4.9) mm; **up glm** 1.8–5.5 mm; **lm** 2–6 mm, smt with a dark spot near the base, apc acute to truncate, smt emgt to bifid, unawned, mucronate, or awned; **pal** somewhat sericeous along the veins; **anth** 1–3, 0.2–2.7 mm. **Car** 0.8–2.4 mm, elliptic to ovate or obovate. $2n = 20$.

Leptochloa fusca grows in warm areas throughout the world. Two subspecies grow in the Intermountain Region.

1. Uppermost leaf blades exceeding the panicles; panicles usually partially enclosed in the uppermost leaf sheaths; mature lemmas often smoky white with a dark spot in the basal ½ . . . . subsp. *fascicularis*
1. Uppermost leaf blades exceeded by the panicles; panicles usually completely exserted; mature lemmas usually lacking a dark spot. . . . subsp. *uninervia*

Leptochloa fusca subsp. **fascicularis** (Lam.) N. Snow BEARDED SPRANGLETOP [p. *415*]

Clm 5–110 cm, prostrate (in small circular clumps) to erect. **Bld** glab or scabrous, uppermost bld often exceeding the pan. **Pan** (1.5)10–72 cm long, 4–22 cm wide, with 3–35 br, usu partially enclosed in the uppermost lf shth; **br** 3–12(22) cm, often spreading. **Spklt** 5–12 mm. **Lo glm** 2–3 mm, lanceolate, smt asymmetric; **up glm** 2.5–5 mm, elliptic to ovate; **lm** lanceolate, smoky white at maturity, often with a dark spot on the bas ½, apc acute, mucronate, or awned, awns to 3.5 mm; **anth** 1–3, 0.2–0.5 mm. **Car** 0.8–2 mm.

Leptochloa fusca subsp. *fascicularis* extends from southern British Columbia and Ontario to Argentina.

Leptochloa fusca subsp. **uninervia** (J. Presl) N. Snow MEXICAN SPRANGLETOP [p. *415*]

Clm (15)25–110 cm, more or less erect, often brchg from the aerial nd. **Bld** usu densely scabrous on both surfaces, not exceeding the pan. **Pan** 10–57 cm long, (0.5)3–18 cm wide, often ellipsoidal, usu completely exserted from the uppermost lf shth; **br** 2–11 cm, mostly ascending. **Spklt** 5–10 mm. **Lo glm** 1–2.6 mm, narrowly triangular to ovate; **up glm** 1.8–2.8 mm, obovate to widely obovate; **lm** 2–3.6 mm, light brown, dark green, or lead-colored, usu without a bas dark spot, apc usu truncate or obtuse, rarely broadly acute, smt bifid, smt mucronate; **anth** 3, 0.2–0.6(1) mm. **Car** 1–1.5 mm.

Leptochloa fusca subsp. *uninervia* is native from the southern United States to Argentina.

3. **Leptochloa panicea** (Retz.) Ohwi [p. *415*, <u>506</u>]

Pl ann. **Clm** (5)13–150 cm, usu erect, compressed, brchg; **intnd** hollow. **Shth** sparsely or densely hairy, particularly distally, hairs papillose-based; **lig** 0.6–3.2 mm, memb, truncate, erose; **bld** 6–25 cm long, 2–21 mm wide, glab or sparsely pilose on both surfaces. **Pan** 8–30 cm, with 3–100 rcm br; **br** 1–19 cm, ascending to reflexed. **Spklt** 2–4 mm, distant to imbricate, green, magenta, or maroon, with 2–5(6) flt. **Glm** smt exceeding the flt, linear to narrowly elliptic, acute, attenuate, or aristate; **lo glm** 1.6–4 mm, linear to lanceolate; **up glm** 1.6–3.6 mm, lanceolate; **lm** 0.9–1.7 mm, glab or somewhat sericeous, acute to obtuse; **pal** glab or sericeous; **anth** 3, 0.2–0.3 mm. **Car** 0.8–1.2 mm long, 0.5–0.6 mm wide, nearly round in cross section, with or without a ventral groove, apc acute to broadly obtuse.

Leptochloa panicea is a widely ranging species that somewhat resembles *L. chinensis* (L.) Nees, an aggressive weed that has not yet been found in North America. It differs in its sparsely to densely hairy, rather than glabrous or almost glabrous, sheaths and blades. One of its three subspecies grows in the Intermountain Region.

Leptochloa panicea subsp. brachiata (Steud.)
N. Snow Red Sprangletop [p. *415*]

Clm to 150 cm. **Lig** 0.9–3.2 mm; **bld** 2–21 mm wide. **Glm** usu not exceeding the flt, lanceolate to narrowly elliptic; **lm** 1.3–1.7 mm, shortly sericeous along the veins. **Car** 0.9–1.2 mm, widely depressed, obovate or obdeltate in cross section, usu with a narrow, shallow ventral groove, apc broadly obtuse to acute. $2n = 20$.

Leptochloa panicea subsp. *brachiata* extends from the southern half of the United States to Argentina. It is common in disturbed and mesic agricultural sites, and is considered a noxious weed by the U.S. Department of Agriculture.

10.12 ERAGROSTIS Wolf
Paul M. Peterson

Pl ann or per; usu synoecious, smt dioecious; ces, stln, or rhz. **Clm** 2–160 cm, not woody, erect, decumbent, or geniculate, smt rooting at the lo nd, simple or brchd; **intnd** solid or hollow. **Lvs** not strongly distichous; **shth** open, often with tufts of hairs at the apc, hairs 0.3–8 mm; **lig** usu memb and ciliolate or ciliate, cilia smt longer than the memb base, occ of hairs or memb and non-ciliate; **bld** flat, folded, or involute. **Infl** tml, smt also axillary, simple pan, open to contracted or spikelike, tml pan usu exceeding the up lvs; **pulvini** in the axils of the pri br glab or not; **br** not spikelike, not **dis**. **Spklt** 1–27 mm long, 0.5–9 mm wide, lat compressed, with (1)2–60 flt; **dis** below the ftl flt, smt also below the glm, acropetal with deciduous glm and lm but persistent pal, or basipetal with the glm often persistent and the flt usu falling intact. **Glm** usu shorter than the adjacent lm, 1(3)-veined, not lobed, apc obtuse to acute, unawned; **cal** glab or sparsely pubescent; **lm** usu glab, obtuse to acute, (1)3(5)-veined, usu keeled, unawned or mucronate; **pal** shorter than the lm, longitudinally bowed-out by the car, 2-keeled, keels usu ciliate, intercostal region memb or hyaline; **anth** 2–3; **ov** glab; **sty** free to the bases. **Cleistogamous spklt** occ present, smt on the axillary pan, smt on the tml pan. **Car** variously shaped. $x = 10$. The origin of the name is obscure.

Eragrostis, a genus of approximately 350 species, grows in tropical and subtropical regions throughout the world. Twelve species grow in the Intermountain Region, seven of which are introduced. In most taxa native to the Western Hemisphere, disarticulation is acropetal and the lemmas fall with the caryopses, leaving the paleas attached to the rachilla.

Van den Borre and Watson (2000) and Hilu and Alice (2001) suggested that *Eragrostis* might not be monophyletic. Ingram and Doyle (2004), based on nuclear and plastid sequence data, concluded that it is, if four segregate genera are included: *Acamptocladus*, *Diandrochloa*, *Neeragrostis*, and *Pogonarthria*.

Nathaniel Wolf (1776), the person who first named *Eragrostis*, made no statement concerning the origin of its name. Clifford (1996) provides three possible derivations: from *eros*, 'love', and *Agrostis*, the Greek name for an indeterminate herb; from the Greek *er*, 'early' and *agrostis*, 'wild', referring to the fact that some species of *Eragrostis* are early invaders of arable land; or the Greek *eri-*, a prefix meaning 'very' or 'much', suggesting that the name means many-flowered *Agrostis*. Many authors have stated that the first portion of the name is derived from *eros*, but none has explained the connection between *Eragrostis* and passionate expressions of love, the kind of love to which *eros* applies.

1. Plants perennial, sometimes rhizomatous, forming innovations at the basal nodes.
  2. Caryopses somewhat laterally compressed, opaque, reddish brown . . . . 12. *E. intermedia*
  2. Caryopses dorsally compressed, translucent, mostly light brown, bases sometimes greenish.
    3. Lemmas 1.8–3 mm long; panicles 16–35(40) cm long, (4)8–24 cm wide; blades 12–50 (65) cm long; caryopses 1–1.7 mm long; ligules 0.6–1.3 mm long . . . . 2. *E. curvula*
    3. Lemmas 1.5–1.7 long; panicles 7–18 cm long, 2–8 cm wide; blades 2–12 cm long; caryopses 0.6–0.8 mm long; ligules 0.3–0.5 mm long . . . . 3. *E. lehmanniana*
1. Plants annual, tufted or mat-forming, without innovations.
  4. Plants mat-forming; panicles 1–3.5 cm long; erect portion of culms (2)5–20 cm, the basal portion prostrate and rooting at the nodes. . . . . 1. *E. hypnoides*
  4. Plants usually not forming mats; panicles 4–40 cm long; culms (2)5–130 cm tall, not prostrate or rooting at the lower nodes.

5. Caryopses with a shallow or deep ventral groove, ovoid to rectangular-prismatic, laterally compressed.. 4. *E. mexicana*
5. Caryopses without a ventral groove, usually globose, rarely flattened, pyriform, obovoid, ellipsoid, or rectangular-prismatic, the surface smooth to faintly striate.
  6. Plants without glandular pits or bands.
    7. Lemmas 1.6–3 mm long; caryopses 0.7–1.3 mm long, obovoid, smooth, light brown to white; plants cultivated, occasionally escaping .... 11. *E. tef*
    7. Lemmas 1–2.2 mm long; caryopses 0.5–1.1 mm long, pyriform or obovoid to prism-shaped, smooth or faintly striate, brownish; plants native species or established introductions, variously distributed.
      8. Lower glumes 0.5–1.5 mm long, at least ½ as long as the lowest lemmas; spikelets 1.2–2.5 mm wide; panicle branches solitary or paired at the lowest 2 nodes; lemmas with moderately conspicuous lateral veins .... 7. *E. pectinacea*
      8. Lower glumes 0.3–0.6(0.8) mm long, usually less than ½ as long as the lowest lemmas; spikelets 0.6–1.4 mm wide; panicle branches usually whorled at the lowest 2 nodes; lemmas with inconspicuous lateral veins........ 6. *E. pilosa*
  6. Plants with glandular pits or bands somewhere, the location(s) various, including any or all of the following: below the cauline nodes, on the sheaths, blades, rachises, panicle branches, or pedicels, or on the keels of the lemmas and paleas.
    9. Panicles 0.5–2 cm wide, contracted; primary panicle branches usually appressed, occasionally diverging up to 30° from the rachises; spikelets light yellowish, occasionally with reddish purple markings .... 5. *E. lutescens*
    9. Panicles 2–18 cm wide, open to somewhat contracted; primary panicle branches diverging 10–110° from the rachises; spikelets plumbeous, yellowish brown, greenish, or reddish purple.
      10. Spikelets 0.6–1.4 mm wide; pedicels 1–10 mm long, lax, appressed or divergent .... 6. *E. pilosa*
      10. Spikelets 1.1–4 mm wide; pedicels 0.2–4 mm long, stiff, straight, usually divergent.
        11. Lemmas 2–2.8 mm long, with 1–3 crateriform glands along the keels; spikelets 6–20 mm long, 2–4 mm wide, with 10–40 florets; disarticulation below the florets, the rachillas persistent; anthers yellow... 8. *E. cilianensis*
        11. Lemmas 1.4–1.8 mm long, rarely with 1 or 2 crateriform glands along the keels; spikelets 4–7(11) mm long, 1.1–2.2 mm wide, with 7–12(20) florets; disarticulation below the lemmas, both the paleas and rachillas usually persistent; anthers reddish brown.
          12. Panicles with glandular regions below the nodes, the glandular tissue forming a ring or band, often shiny or yellowish; anthers 3; blade margins without crateriform glands; pedicels without glandular bands .... 9. *E. barrelieri*
          12. Panicles sometimes with areas, but rarely rings, of glandular spots or crateriform pits below the nodes, the glands usually dull greenish gray to stramineous; anthers 2; blade margins sometimes with crateriform glands; pedicels usually with glandular bands .... 10. *E. minor*

1. **Eragrostis hypnoides** (Lam.) Britton, Sterns & Poggenb. TEEL LOVEGRASS [p. *416*, 506]
Pl ann; stln, mat-forming, without innovations, without glands. **Clm** decumbent and rooting at the lo nd, erect portion (2)5–12(20) cm, often brchd, glab or hairy on the lo intnd. **Shth** pilose on the mrg, col, and at the apc, hairs 0.1–0.6 mm; **lig** 0.3–0.6 mm; **bld** 0.5–2.5 cm long, 1–2 mm wide, flat to involute, abx surfaces glab, adx surfaces appressed pubescent, hairs about 0.2 mm. **Pan** tml and axillary, 1–3.5 cm long, 0.7–2.5 cm wide, ovate, open to somewhat congested; **pri br** 0.1–0.5 cm, appressed to strongly divergent, glab; **pulvini** sparsely pilose or glab; **ped** 0.2–1 mm, ciliate. **Spklt** 4–13 mm long, 1–1.5 mm wide, linear-oblong, often arcuate, loosely imbricate, greenish yellow to purplish, with 12–35 flt; **dis** acropetal, pal persistent. **Glm** linear-lanceolate to lanceolate, hyaline; **lo glm** 0.4–0.7 mm; **up glm** 0.8–1.2 mm; **lm** 1.4–2 mm, ovate, strongly 3-veined, veins greenish, apc acuminate; **pal** 0.7–1.2 mm, hyaline, keels scabridulous, apc acute to

obtuse; **anth** 2, 0.2–0.3 mm, brownish. **Car** 0.3–0.5 mm, ellipsoid, somewhat translucent, light brown. $2n = 20$.

Eragrostis hypnoides grows along muddy or sandy shores of lakes and rivers and in moist, disturbed sites, at 10–1600 m. It is native to the Americas, extending from southern Canada to Argentina.

2. **Eragrostis curvula** (Schrad.) Nees WEEPING LOVEGRASS [p. *416*, 506]

Pl per; ces, forming innovations at the bas nd, without glands. **Clm** (45)60–150 cm, erect, glab or glandular. **Shth** with scattered hairs, hairs to 9 mm; **lig** 0.6–1.3 mm; **bld** 12–50(65) cm long, 1–3 mm wide, flat to involute, abx surfaces glab, smt scabridulous, adx surfaces with scattered hairs bas, hairs to 7 mm. **Pan** 16–35(40) cm long, (4)8–24 cm wide, ovate to oblong, open; **pri br** 3–14 cm, diverging 10–80° from the rchs; **pulvini** glab or not; **ped** 0.5–5 mm, appressed, flexible. **Spklt** 4–8.2(10) mm long, 1.2–2 mm wide, linear-lanceolate, plumbeous to yellowish, with 3–10 flt; **dis** irregular to acropetal, proximal rchl intnd persistent. **Glm** lanceolate, hyaline; **lo glm** 1.2–2.6 mm; **up glm** 2–3 mm; **lm** 1.8–3 mm, ovate, memb, lat veins conspicuous, apc acute; **pal** 1.8–3 mm, hyaline to memb, apc obtuse; **anth** 3, 0.6–1.2 mm, reddish brown. **Car** 1–1.7 mm, ellipsoid to obovoid, dorsally compressed, adx surfaces with a shallow, broad groove or ungrooved, smooth, mostly translucent, light brown, bases often greenish. $2n = 40, 50$.

Eragrostis curvula is native to southern Africa. It is often used for reclamation because it provides good ground cover but, once introduced, it easily escapes. In North America, it grows on rocky slopes, at the margins of woods, along roadsides, and in waste ground, at 20–2400 m, usually in pine-oak woodlands, and yellow pine and mixed hardwood forests.

3. **Eragrostis lehmanniana** Nees LEHMANN'S LOVEGRASS [p. *416*, 506]

Pl per; ces, forming innovations at the bas nd, without glands. **Clm** (20)40–80 cm, erect, commonly geniculate, smt rooting at the lo nd, glab, lo portions smt scabridulous. **Shth** smt shortly silky pilose bas, hairs less than 2 mm, apc sparsely hairy, hairs to 3 mm; **lig** 0.3–0.5 mm, ciliate; **bld** 2–12 cm long, 1–3 mm wide, flat to involute, glab, abx surfaces smt scabridulous, adx surfaces scabridulous. **Pan** 7–18 cm long, 2–8 cm wide, oblong, open; **pri br** 1–8 cm, appressed or diverging to 40° from the rchs; **pulvini** glab; **ped** 0.5–4 mm, diverging or appressed, flexible. **Spklt** 5–12(14) mm long, 0.8–1.2 mm wide, linear-lanceolate, plumbeous to stramineous, with 4–12(14) flt; **dis** irregular to basipetal, pal

usu persistent. **Glm** oblong to lanceolate, memb; **lo glm** 1–1.5 mm; **up glm** 1.3–2 mm; **lm** 1.5–1.7 mm, ovate, memb, lat veins inconspicuous, apc acute to obtuse; **pal** 1.4–1.7 mm, obtuse; **anth** 3, 0.6–0.9 mm, yellowish. **Car** 0.6–0.8 mm, ellipsoid to obovoid, dorsally compressed, smt with a shallow adx groove, smooth, translucent, mostly light brown, embryo region dark brown with a greenish ring. $2n = 40, 60$.

Eragrostis lehmanniana is native to southern Africa, where it grows in sandy, savannah habitats. It was introduced for erosion control in the southern United States, where it often displaces native species. In North America, it grows in sandy flats, along roadsides, on calcareous slopes, and in disturbed areas, at 200–1830 m. It is commonly found in association with *Larrea tridentata, Opuntia, Quercus, Juniperus,* and *Bouteloua gracilis.*

4. **Eragrostis mexicana** (Hornem.) Link MEXICAN LOVEGRASS [p. *417*, 506]

Pl ann; ces, without innovations. **Clm** 10–130 cm, erect, smt geniculate, glab, smt with a ring of glandular depressions below the nd. **Shth** smt with glandular pits, pilose near the apc and on the col, hairs to 4 mm, papillose-based; **lig** 0.2–0.5 mm, ciliate; **bld** 5–25 cm long, 2–7(9) mm wide, flat, abx surfaces glab, adx surfaces scabridulous, occ pubescent near the base. **Pan** (5)10–40 cm long, (2)4–18 cm wide, ovate, rchs angled and channeled; **pri br** 3–12(15) cm, solitary to whorled, appressed or diverging to 80° from the rchs; **sec br** somewhat appressed; **pulvini** glab; **ped** 1–6(7) mm, almost appressed to narrowly divergent, stiff. **Spklt** (4)5–10(11) mm long, 0.7–2.4 mm wide, ovate to linear-lanceolate, gray-green to purplish, with 5–11(15) flt; **dis** acropetal. **Glm** subequal, 0.7–2(2.3) mm, ovate to lanceolate, memb; **lm** 1.2–2.4 mm, ovate, memb, glab or with a few hairs, gray-green, lat veins evident, often greenish, apc acute; **pal** 1–2.2 mm, hyaline, keels scabrous, apc obtuse to truncate; **anth** 3, 0.2–0.5 mm, purplish. **Car** 0.5–0.8(1) mm, ovoid to rectangular-prismatic, lat compressed, shallowly to deeply grooved on the adx surface, striate, reddish brown, distal ⅔ opaque. $2n = 60$.

Eragrostis mexicana grows along roadsides, near cultivated fields, and in disturbed open areas, at 100–3000 m. It is native to the Americas, its native range extending from the southwestern United States through Mexico, Central and northern South America, to Argentina.

1. Spikelets ovate to oblong in outline, 1.5–2.4 mm wide; lower glumes 1.2–2.3 mm long; sum of the spikelet width and lower glume length 2.7–4.7 mm; culms and sheaths sometimes with glandular depressions .... subsp. *mexicana*

1. Spikelets linear to linear-lanceolate, 0.7–1.4 wide; lower glumes 0.7–1.7 mm long; sum of the spikelet width and lower glume length 1.5–3.1 mm; culms and sheaths without glandular depressions. . . . . subsp. *virescens*

Eragrostis mexicana (Hornem.) Link subsp. mexicana [p. 417]

Clm and **shth** smt with glandular depressions. **Spklt** 1.5–2.4 mm wide, ovate to oblong in outline. **Lo glm** 1.2–2.3 mm; **sum of spklt width and lo glm length** 2.7–4.7 mm.

Eragrostis mexicana subsp. *mexicana* grows from Ontario through the midwestern United States to California, South Carolina, and Texas and southwards to Mexico.

Eragrostis mexicana subsp. virescens (J. Presl) S.D. Koch & Sánchez Vega [p. 417]

Clm and **shth** without glandular depressions. **Spklt** 0.7–1.4 mm wide, linear to linear-lanceolate. **Lo glm** 0.7–1.7 mm; **sum of spklt width and lo glm length** 1.5–3.1 mm.

Eragrostis mexicana subsp. *virescens* has a disjunct distribution, growing in California and western Nevada and, in South America, from Ecuador to Chile, southern Brazil, and northern Argentina. It has also been found, as an introduction, at various other locations in North America.

5. Eragrostis lutescens Scribn. SIXWEEKS LOVEGRASS [p. 417, 506]

Pl ann; tufted, without innovations. **Clm** (2)6–25 cm, usu erect, smt decumbent, glab, with elliptical, yellowish, glandular pits below the nd. **Shth** with elliptical glandular pits, sparsely hairy at the throat, hairs to 2 mm; **lig** 0.2–0.5 mm, ciliate; **bld** 2–12 cm long, 1–3 mm wide, flat to involute, abx surfaces scabridulous, bases with glandular pits. **Pan** tml, 4–10(15) cm long, 0.5–2 cm wide, narrowly elliptic, contracted, dense; **pri br** alternate, usu appressed, occ diverging to 30° from the rchs, rchs and br with glandular pits; **pulvini** glab; **ped** 1.4–10 mm, appressed or divergent. **Spklt** 3.6–7.5 mm long, 1.2–2 mm wide, narrowly ovate, light yellowish, occ mottled with reddish purple, with 6–11(14) flt; **dis** acropetal, pal persistent. **Glm** subequal, ovate to lanceolate, hyaline; **lo glm** (0.7)0.9–1.4 mm; **up glm** 1.2–1.8 mm; **lm** 1.5–2.2 mm, ovate, subhyaline, stramineous, veins greenish and conspicuous, apc acute; **pal** 1.2–2 mm, hyaline, keels scabridulous, apc obtuse; **anth** 3, 0.2–0.3 mm, purplish. **Car** 0.5–0.8 mm, pyriform except slightly flattened adx, smooth, light brown. 2n = unknown.

Eragrostis lutescens grows on the sandy banks of streams and lakes and in moist alkaline flats of the western United States at 300–2000 m.

6. Eragrostis pilosa (L.) P. Beauv. INDIA LOVEGRASS [p. 418, 506]

Pl ann; tufted, without innovations. **Clm** 8–45(70) cm, erect or geniculate, glab, occ with a few glandular depressions. **Shth** mostly glab, occ glandular, apc hirsute, hairs to 3 mm; **lig** 0.1–0.3 mm, ciliate; **bld** 2–15(20) cm long, 1–2.5(4) mm wide, flat, abx surfaces glab, occ with glandular pits along the midrib, adx surfaces scabridulous. **Pan** 4–20(28) cm long, 2–15(18) cm wide, ellipsoid to ovoid, diffuse; **pri br** 1–10 cm, diverging 10–80°(110°) from the rchs, capillary, whorled on the lowest 2 nd, rarely glandular; **pulvini** glab or hairy; **ped** 1–10 mm, flexible, appressed or divergent. **Spklt** (2)3.5–6(10) mm long, 0.6–1.4 mm wide, linear-oblong to narrowly ovate, plumbeous, with (3)5–17 flt; **dis** acropetal, pal tardily deciduous, rchl persisting longer than the pal. **Glm** narrowly ovate to lanceolate, hyaline; **lo glm** 0.3–0.6(0.8) mm; **up glm** 0.7–1.2(1.4) mm; **lm** 1.2–1.8(2) mm, ovate-lanceolate, memb to hyaline, grayish green proximally, reddish purple distally, lat veins inconspicuous, apc acute; **pal** 1–1.6 mm, memb to hyaline, keels scabridulous to scabrous, apc obtuse; **anth** 3, 0.2–0.3 mm, purplish. **Car** 0.5–1 mm, obovoid to prism-shaped, adx surfaces flat, smooth to faintly striate, light brown. 2n = 20, 40.

Eragrostis pilosa is native to Eurasia but has become naturalized in many parts of the world. In North America, it grows in forest margins and disturbed sites such as roadsides, railroad embankments, gardens, and cultivated fields, at 0–2500 m.

Eragrostis pilosa (L.) P. Beauv. var. pilosa [p. 418]

Clm with few or no glandular pits; **shth** and **bld** without glandular pits. **Spklt** 0.6–1.3 mm wide. **Up glm** 0.7–1.2 mm; **lm** 1.2–1.8 mm; car 0.5–0.9 mm.

Eragrostis pilosa var. *pilosa* is the only variety that grows in the Intermountain Region.

7. Eragrostis pectinacea (Michx.) Nees TUFTED LOVEGRASS [p. 418, 506]

Pl ann; tufted, without innovations, without glandular pits. **Clm** 10–80 cm, erect to geniculate or decumbent below, glab. **Shth** hirsute at the apc, hairs to 4 mm; **lig** 0.2–0.5 mm; **bld** 2–20 cm long, 1–4.5 mm wide, flat to involute, abx surfaces glab and smooth, adx surfaces scabridulous. **Pan** 5–25 cm long, 3–12(15) cm wide, ovoid to pyramidal, usu open, smt contracted; **pri br** 0.6–8.5 cm, appressed or diverging to 80° from the rchs, solitary or paired at the lowest 2 nd; **pulvini** glab or sparsely hairy; **ped** 1–7 mm, flexible, appressed to widely divergent, smt capillary. **Spklt** 3.5–11 mm long, 1.2–2.5 mm wide, linear-oblong to narrowly lanceolate, plumbeous, yellowish brown, or dark reddish purple, with 6–22 flt; **dis** acropetal, pal persistent. **Glm** subulate to ovate-lanceolate, hyaline; **lo glm** 0.5–1.5 mm, at least ½ as long as the adjacent lm; **up glm** 1–1.7 mm, usu broader than the lo glm; **lm** 1–2.2 mm, ovate-lanceolate, hyaline to memb, grayish green proximally, reddish purple distally, lat veins moderately conspicuous, apc acute; **pal** 1–2 mm, hyaline to memb, keels scabridulous, apc obtuse; **anth** 3, 0.2–0.7 mm, purplish. **Car** 0.5–1.1 mm, pyriform, slightly lat compressed, smooth, faintly striate, brownish. $2n = 60$.

Eragrostis pectinacea is native from southern Canada to Argentina. In North America, it grows in disturbed sites such as roadsides, railroad embankments, gardens, and cultivated fields, at 0–1200 m. One variety grows in the Intermountain Region.

Eragrostis pectinacea (Michx.) Nees var. **pectinacea** [p. *418*]
Ped appressed or diverging to 20° from the br axes. **Anth** 0.2–0.4 mm.

Eragrostis pectinacea var. *pectinacea* grows throughout most of the contiguous United States. It usually flowers from July–November.

8. **Eragrostis cilianensis** (All.) Vignolo *ex* Janch. Stinkgrass [p. *419*, 506]
Pl ann; tufted, without innovations. **Clm** 15–45(65) cm, erect or decumbent, smt with crateriform glands below the nd. **Shth** glab, occ glandular, apc hairy, hairs to 5 mm; **lig** 0.4–0.8 mm, ciliate; **bld** (1)5–20 cm long, (1)3–5(10) mm wide, flat to involute, abx surfaces glab, smt glandular, adx surfaces scabridulous, occ also hairy. **Pan** (3)5–16(20) cm long, 2–8.5 cm wide, oblong to ovate, condensed to open; **pri br** 0.4–5 cm, appressed or diverging 20–80° from the rchs; **pulvini** glab or hairy; **ped** 0.2–3 mm, stout, straight, stiff, usu divergent, occ appressed. **Spklt** 6–20 mm long, 2–4 mm wide, ovate-lanceolate, plumbeous, greenish, with 10–40 flt; **dis** below the flt, each flt falling as a unit, rchl persistent. **Glm** broadly ovate to lanceolate, memb, usu glandular; **lo glm** 1.2–2 mm, usu 1-veined; **up glm** 1.2–2.6 mm, often 3-veined; **lm** 2–2.8 mm, broadly ovate, memb, keels with 1–3 crateriform glands, apc obtuse to acute; **pal** 1.2–2.1 mm, hyaline, keels scabrous, smt also ciliate, cilia to 0.3 mm, apc obtuse to acute; **anth** 3, 0.2–0.5 mm, yellow. **Car** 0.5–0.7 mm, globose to broadly ellipsoid, smooth to faintly striate, not grooved, reddish brown or translucent. $2n = 20$.

Eragrostis cilianensis is an introduced European species that now grows in disturbed sites such as pastures and roadsides, at 0–2300 m, through most of the contiguous United States and southern Canada. The English name refers to the odor of fresh plants.

9. **Eragrostis barrelieri** Daveau
Mediterranean Lovegrass [p. *419*, 506]
Pl ann; tufted, without innovations. **Clm** (5)10–60 cm, erect or decumbent, much-brchd near the base, with a ring of glandular tissue below the nd, rings often shiny or yellowish. **Shth** hairy at the apc, hairs to 4 mm; **lig** 0.2–0.5 mm, ciliate; **bld** 1.5–10 cm long, 1–3(5) mm wide, flat, abx surfaces glab, adx surfaces glab, smt scabridulous, occ with white hairs to 3 mm, mrg without crateriform glands. **Pan** 4–20 cm long, 2.2–8(10) cm wide, ovate, open to contracted, rchs with shiny or yellowish glandular spots or rings below the nd; **pri br** 0.5–6 cm, diverging 20–100° from the rchs; **pulvini** glab; **ped** 1–4 mm, stout, stiff, divergent, without glandular bands. **Spklt** 4–7(11) mm long, 1.1–2.2 mm wide, narrowly ovate, reddish purple to greenish, occ grayish, with 7–12(20) flt; **dis** acropetal, pal persistent. **Glm** broadly ovate, memb, 1-veined; **lo glm** 0.9–1.4 mm; **up glm** 1.2–1.6 mm; **lm** 1.4–1.8 mm, broadly ovate, memb, apc acute to obtuse; **pal** 1.3–1.7 mm, hyaline, keels scabrous, scabridities to 0.1 mm, apc obtuse to acute; **anth** 3, 0.1–0.2 mm, reddish brown. **Car** 0.4–0.7 mm, ellipsoid, not grooved, smooth to faintly striate, light brown. $2n = 40$.

Eragrostis barrelieri is a European species that is now naturalized in North America, primarily in the southwestern United States. It grows on gravelly roadsides, in gardens, and other disturbed, sandy sites, especially near railroad yards, at 10–2000 m. The ring of glandular tissue is most conspicuous below the upper cauline nodes.

10. **Eragrostis minor** Host Little Lovegrass [p. *419*, 506]
Pl ann; tufted, without innovations. **Clm** 10–45 cm, erect to decumbent, smt with a ring of glandular tissue below the nd. **Shth** smt glandular on the midveins, hairy at the apc, hairs to 4 mm; **lig** 0.2–0.5 mm, ciliate; **bld** 1.5–10 cm long, 1–3(4) mm wide, flat, glab or sparsely white-hairy, mrg smt with crateriform glands. **Pan** 4–20 cm long, 2.2–8(10) cm wide, ovate, open to contracted, rchs smt with glandular spots or pits below the nd, rarely with a glandular ring, glands usu dull, greenish gray to stramineous; **pri br** 0.5–6 cm, diverging 20–100° from the rchs; **pulvini** glab or hairy; **ped** 1–4 mm, stiff, straight, divergent, usu with a distal ring of crateriform glands. **Spklt** 4–7(11) mm long, 1.1–2.2 mm

wide, narrowly ovate, mostly reddish purple to greenish, occ grayish, with 7–12(20) flt; **dis** acropetal, pal persistent. **Glm** broadly ovate, memb; **lo glm** 0.9–1.4 mm; **up glm** 1.2–1.6 mm; **lm** 1.4–1.8 mm, broadly ovate, memb, keels occ with 1–2 crateriform glands, apc acute to obtuse; **pal** 1.3–1.7 mm, hyaline, keels smooth or scabridulous, scabridities to 0.1 mm, apc obtuse to acute; **anth** 2, 0.2–0.3 mm, reddish brown. **Car** 0.4–0.7 mm, ellipsoid, not grooved, striate, light brown. $2n = 40$.

Eragrostis minor is a European species that now grows in gravelly roadsides and disturbed sites, especially near railroad yards, at 20–1600 m in southern Canada and the contiguous United States.

11. **Eragrostis tef** (Zucc.) Trotter TEFF [p. 420]
Pl ann; loosely tufted, without innovations, without glands. **Clm** 25–60 cm, erect, glab and shiny. **Shth** mostly glab, apc hairy, hairs to 5 mm; **lig** 0.2–0.4 mm, ciliate; **bld** 10–30 cm long, 2–5.5 mm wide, flat to involute, glab abx, scabridulous adx. **Pan** 10–45 cm long, 2.5–22 cm wide, ovate, open to contracted; **pri br** 4–17 cm, appressed or diverging up to 50° from the rchs, flexible, naked below; **pulvini** glab or hairy, hairs to 5 mm; **ped** 2.5–17 mm, appressed or divergent. **Spklt** 4–11 mm long, 1.3–2.5 mm wide, linear-lanceolate to ovate, stramineous, grayish green to purplish, with 4–16 flt; **dis** tardy, acropetal, car falling before the glm and lm, pal persistent. **Glm** lanceolate, memb to hyaline; **lo glm** 1–2 mm; **up glm** 1.5–2.8 mm; **lm** 1.6–3 mm, lanceolate, memb, apc acute; **pal** 1.4–2.2 mm, hyaline, keels scabridulous, apc obtuse; **anth** 3, 0.2–0.5 mm, purplish. **Car** 0.7–1.3 mm, obovoid, not grooved, smooth, light brown to whitish. $2n = 40$.

Eragrostis tef is native to northern Africa. In Ethiopia, it is used both as a grain and as fodder for cattle. It is also grown, but not commonly, for these purposes in North America and is occasionally found as an escape from cultivation.

12. **Eragrostis intermedia** Hitchc. PLAINS LOVEGRASS [p. 420, 506]
Pl per; ces, with innovations, without rhz, not glandular. **Clm** (30)40–90(110) cm, erect, glab below the nd. **Shth** sparsely pilose on the mrg, apc hairy, hairs to 8 mm, not papillose-based; **lig** 0.2–0.4 mm; **bld** (4)10–20(30) cm long, 1–3 mm wide, flat or involute, abx surfaces glab, adx surfaces densely hairy behind the lig, elsewhere usu glab, occ sparsely hairy. **Pan** 15–40 cm long, (8.5)15–30 cm wide, ovate, open; **pri br** 4–25 cm, diverging 20–90° from the rchs, capillary; **pulvini** hairy or glab; **ped** 2–14 mm, divergent. **Spklt** 3–7 mm long, 1–1.8 mm wide, narrowly lanceolate, olivaceous to purplish, with (3)5–11 flt; **dis** acropetal, pal persistent. **Glm** lanceolate to ovate, hyaline to memb; **lo glm** 1.1–1.7 mm, narrower than the up glm; **up glm** 1.3–2 mm, apc acuminate to acute; **lm** 1.6–2.2 mm, ovate, memb, hyaline near the mrg, lat veins inconspicuous, apc acute; **pal** 1.4–2.1 mm, hyaline, narrower than the lm, apc obtuse to acute; **anth** 3, 0.5–0.8 mm, purplish. **Car** 0.5–0.9 mm, rectangular-prismatic, somewhat lat compressed, with a well-developed adx groove, striate, opaque, reddish brown. $2n =$ ca. 54, 60, 72, ca. 74, 80, 100, 120.

Eragrostis intermedia grows in clay, sandy, and rocky soils, often in disturbed sites, at 0–1850 m. Its range extends from the United States through Mexico and Central America to South America.

10.13 ELEUSINE Gaertn.   Khidir W. Hilu

Pl ann or per; ces. **Clm** 10–150 cm, herbaceous, glab, brchg both at and above the base. **Shth** open; **lig** memb, ciliate. **Infl** tml, pan of (1)2–20 non-dis, spikelike br, exceeding the up lvs; **br** 1–17 cm, all or most in a digitate cluster, smt 1(2) br attached immediately below the tml whorl, axes flattened, terminating in a fnctl spklt. **Spklt** 3.5–11 mm, lat compressed, with 2–15 bisex flt; **dis** above the glm and between the flt (*E. coracana* not dis). **Glm** unequal, shorter than the lo lm; **lo glm** 1–3-veined; **up glm** 3–5(7)-veined; **lm** 3-veined, glab, keeled, apc entire, neither mucronate nor awned; **pal** smt with winged keels; **anth** 3, 0.5–1 mm; **ov** glab. **Fruits** modified car, pericarp thin, separating from the seed at an early stage in its development; **seeds** usu obtusely trigonous, the surfaces ornamented. $x = 8, 9,$ 10. Name from Eleusis, a Greek town where Demeter, the goddess of harvests, was worshipped.

Eight of the nine species of *Eleusine* are native to Africa, where they grow in mesic to xeric habitats; the exception, *E. tristachya*, is native to South America. Two species have become established in the Intermountain Region. When moistened, the seeds of all species are easily freed from the thin pericarp.

1. Lower glumes 1-veined; panicle branches 3–5.5 mm wide; surface of the seeds striate . . . . 1. *E. indica*
1. Lower glumes 2- or 3-veined; panicle branches 5–15 mm wide; surface of the seeds granular . . 2. *E. coracana*

1. Eleusine indica (L.) Gaertn. GOOSEGRASS [p. *420*, <u>507</u>]

Pl ann. **Clm** 30–90 cm, erect or ascending, somewhat compressed; **lo intnd** 1.5–2 mm thick. **Shth** conspicuously keeled, mrg often with long, papillose-based hairs, particularly near the throat; **lig** 0.2–1 mm, truncate, erose; **bld** 15–40 cm long, 3–7 mm wide, with prominent, white midveins, mrg and/or adx surfaces often with bas papillose-based hairs. **Pan** with 4–10(17) br, often with 1 br attached as much as 3 cm below the tml cluster; **br** (3.5)7–16 cm long, 3–5.5 mm wide, linear. **Spklt** 4–7 mm long, 2–3 mm wide, with 5–7 flt, obliquely attached to the br axes. **Lo glm** 1.1–2.3 mm, 1-veined; **up glm** 2–2.9 mm; **lm** 2.4–4 mm; **pal** with narrowly winged keels. **Seeds** ovoid, rugulose and obliquely striate, usu not exposed at maturity. $2n = 18$.

Eleusine indica is a common weed in the warmer regions of the world. In the Intermountain Region, it usually grows in disturbed areas and lawns.

2. Eleusine coracana (L.) Gaertn. [p. *421*]

Pl ann. **Clm** to 62 cm, often brchg; **lo intnd** 6–10 mm thick. **Shth** glab; **lig** 1–2 mm, ciliate, with 1–2 mm hairs; **bld** 10–60 cm long, 6–12 mm wide, smt longer than the clm, adx surfaces scabrous or pubescent. **Pan** subdigitate, with 4–20 br, 1(2) of the br attached below the tml cluster; **br** 4–17 cm long, 5–15 mm wide, spreading at maturity. **Spklt** 5–9 mm long, 3–6 mm wide, with 2–9 flt, smt not dis at maturity. **Lo glm** 1.2–3 mm, 2- or 3-veined; **up glm** 2.2–6.5 mm; **lm** 2.2–5 mm; **anth** about 1 mm. **Seeds** oblong-globose, granular, usu exposed at maturity. $2n = 36$.

Eleusine coracana is an allotetraploid, one of its genomes being derived from *E. indica*. Two subspecies are recognized; only subsp. *coracana* is known from North America.

Eleusine coracana (L.) Gaertn. subsp. coracana FINGER MILLET, RAGI [p. *421*]

Clm to 17 cm. **Bld** 30–60 cm long, 6–12 mm wide. **Br** 4–14 cm long, 7–15 mm wide, spklt closely imbricate. **Spklt** 5–9 mm long, brown, with 6–9 flt, flt not dis at maturity. **Seeds** almost globose, brownish, surfaces granular to smooth.

Eleusine coracana subsp. *coracana* is the domesticated variant of *E. coracana*. It has a long historical record dating back at least 5000 years in Africa, and 3000 years in India. Five races, based on inflorescence morphology, are recognized in East Africa where it is widely cultivated for food and drink. Biochemical data suggest that *Eleusine coracana* subsp. *coracana* evolved from a few populations of the very variable subsp. *africana*. It is cultivated at various agricultural experiment stations and occasionally escapes.

10.14 SPOROBOLUS R. Br.   Paul M. Peterson, Stephan L. Hatch, and Alan S. Weakley

Pl ann or per; usu ces, smt rhz, rarely stln. **Clm** 10–250 cm, usu erect, rarely prostrate, glab. **Shth** open, usu glab, often ciliate at the apc; **lig** of hairs; **bld** flat, folded, involute, smt terete. **Infl** tml, open or contracted pan, smt partially included in the uppermost shth. **Spklt** rounded to lat compressed, with 1(–3) flt(s) per spklt; **dis** above the glm. **Glm** 0–1-veined; **cal** poorly developed, usu glab; **lm** memb or chartaceous, 1(3)-veined, unawned; **pal** glab, 2-veined, often splitting between the veins at maturity; **anth** (2)3. **Fruits** utricles or achenes, ellipsoid, obovoid, fusiform, or quadrangular, pericarp free from the seed, becoming mucilaginous when moist in most species. **Cleistogamous spklt** occ present in the lo lf shth. $x = 9$. Name from the Greek *sporos*, 'seed', and *bolos*, 'a throw', referring to the free seeds, which are sometimes forcibly ejected when the mucilaginous pericarp dries.

Sporobolus is a cosmopolitan genus of more than 160 species that grow in tropical, subtropical, and warm-temperate regions throughout the world. Seventy-four species are native to the Western Hemisphere; 12 are present in the Intermountain Region. Two genera of the Western Hemisphere, *Calamovilfa* and *Crypsis*, resemble *Sporobolus* in having hairy ligules, spikelets with 1 floret, 1-veined lemmas, and fruits with a free pericarp (Peterson et al. 1997).

1. Plants annuals or short-lived perennials flowering in the first year.
  2. Lower panicle nodes with 7–20 branches . . . . 1. *S. pyramidatus* (in part)
  2. Lower panicle nodes with 1–3 branches.
    3. Mature panicles 10–35 cm long, 4.5–30 cm wide, open; secondary branches spreading; pedicels usually 6–25 mm long, spreading . . . . 8. *S. texanus* (in part)
    3. Mature panicles 2–5 cm long, 0.2–0.5 cm wide, contracted; secondary branches appressed; pedicels 0.1–2.5 mm long, appressed . . . . 2. *S. neglectus*
1. Plants perennial.
  4. Plants with rhizomes.
    5. Panicles (0.6)1–8 cm wide, open to somewhat contracted, narrowly pyramidal, well-exserted from the uppermost sheath; branches without spikelets on the lower ⅓
     . . . . 12. *S. interruptus* (in part)
    5. Panicles 0.4–1.6 cm wide, usually spikelike, partially included in the uppermost sheath; branches spikelet-bearing to the base . . . . 3. *S. compositus* (in part)

4. Plants without rhizomes.
　6. Spikelets 2.5–10 mm long.
　　7. Lower panicle nodes with 3 or more branches . . . . 11. *S. junceus*
　　7. Lower panicle nodes with 1–2(3) branches.
　　　8. Mature panicles 0.2–4 cm wide, spikelike; panicle branches appressed.
　　　　9. Spikelets 4–6(10) mm long, stramineous to purplish tinged; panicles
　　　　　terminal and axillary; sheaths without a conspicuous apical tuft of hairs
　　　　　. . . . . 3. *S. compositus* (in part)
　　　　9. Spikelets 1.7–3.5(4) mm long, whitish to plumbeous; panicles all terminal;
　　　　　sheaths with a conspicuous apical tuft of hairs.
　　　　　10. Culms 40–100(120) cm tall, 2–4(5) mm thick near the base; mature
　　　　　　panicles 0.2–0.8(1) cm wide; anthers 0.3–0.5 mm long . . . . 7. *S. contractus* (in part)
　　　　　10. Culms 100–200 cm tall, (3)4–10 mm thick near the base; mature panicles
　　　　　　1–4 cm wide; anthers 0.6–1 mm long . . . . 10. *S. giganteus*
　　　8. Mature panicles (0.6)1–30 cm wide, usually open, narrowly pyramidal to
　　　　pyramidal or ovate; panicle branches appressed or spreading.
　　　　11. Spikelets 2.3–3 mm long; panicles 4.5–30 cm wide, diffuse, about as long
　　　　　as wide; branches capillary; anthers 0.3–1 mm long . . . . 8. *S. texanus* (in part)
　　　　11. Spikelets 3–7.2 mm long; panicles 0.6–15 cm wide, longer than wide, not
　　　　　diffuse; branches not capillary; anthers 1.5–5 mm long . . . . 12. *S. interruptus* (in part)
　6. Spikelets 1.2–2.5(2.8) mm long.
　　12. Panicles 12–35 cm wide, open.
　　　13. Sheath apices with a conspicuous tuft of white hairs; flag blades nearly
　　　　perpendicular to the culms . . . . 6. *S. cryptandrus* (in part)
　　　13. Sheath apices glabrous or with a few scattered hairs; flag blades
　　　　ascending.
　　　　14. Secondary panicle branches spikelet-bearing to the base; pedicels
　　　　　mostly appressed, mostly 0.2–0.5 mm long; panicles 20–60 cm long . . . . 4. *S. wrightii*
　　　　14. Secondary panicle branches without spikelets on the lower ¼–½;
　　　　　pedicels mostly spreading, mostly 0.5–25 mm long; panicles 10–45 cm
　　　　　long.
　　　　　15. Pedicels 0.5–2 mm long; anthers 1.1–1.8 mm long . . . . 5. *S. airoides*
　　　　　15. Pedicels 6–25 mm long; anthers 0.3–1 mm long . . . . 8. *S. texanus* (in part)
　　12. Panicles 0.2–12(14) cm wide, contracted to open.
　　　16. Mature panicles 0.2–5 cm wide, contracted, often spikelike, the panicle
　　　　branches appressed or diverging no more than 30° from the rachises.
　　　　17. Primary panicle branches spikelet-bearing to the base . . . . 7. *S. contractus* (in part)
　　　　17. Primary panicle branches without spikelets on the lower ⅛–½ of their
　　　　　length.
　　　　　18. Lower panicle nodes with 7–12(15) branches; anthers 0.2–0.4 mm
　　　　　　long . . . . 1. *S. pyramidatus* (in part)
　　　　　18. Lower panicle nodes with 1–3 branches; anthers 0.5–1 mm long
　　　　　　. . . . . 6. *S. cryptandrus* (in part)
　　　16. Mature panicles 4.5–30 cm wide, open, pyramidal to subovate or oblong,
　　　　the panicle branches diverging more than 10° from the rachises, sometimes
　　　　reflexed.
　　　　19. Lower panicle nodes with 7–12(15) branches; anthers 0.2–0.4 mm
　　　　　long . . . . 1. *S. pyramidatus* (in part)
　　　　19. Lower panicle nodes with 1–2(3) branches; anthers 0.3–1 mm long.
　　　　　20. Pedicels 6–25 mm long, spreading; panicles 4.5–30 cm wide, about
　　　　　　as long as wide, diffuse . . . . 8. *S. texanus* (in part)
　　　　　20. Pedicels 0.1–3 mm long, appressed or spreading; panicles 2–14 cm
　　　　　　wide, longer than wide, open.
　　　　　　21. Pedicels appressed to the secondary branches; primary
　　　　　　　branches appressed, spreading, or reflexed; pulvini glabrous;
　　　　　　　rachises straight, erect; mature panicles narrowly pyramidal,
　　　　　　　lower branches longer than the middle branches . . . . 6. *S. cryptandrus* (in part)

21. Pedicels spreading from the secondary branches; primary
branches reflexed; pulvini pubescent; rachises drooping or
nodding; mature panicles subovate to oblong, lower branches
no longer than those in the middle. . . . . 9. *S. flexuosus*

1. **Sporobolus pyramidatus** (Lam.) Hitchc.
WHORLED DROPSEED [p. *421*, <u>507</u>]

Pl ann or short-lived per flowering in the first
year; ces, not rhz. **Clm** 7–35(60) cm, erect or
decumbent. **Shth** rounded below, mrg and apc
hairy, hairs to 3 mm; **lig** 0.3–1 mm; **bld** 2–12(20)
cm long, 2–6 mm wide, flat, abx surface glab, adx
surface scabridulous, smt sparsely hispid, mrg
ciliate-pectinate. **Pan** 4–15(18) cm long, 0.3–6
cm wide, open (contracted when immature),
pyramidal; **lo nd** with 7–12(15) br; **pri br** 0.5–
4.5 cm, spreading 30–90° from the rchs, with
elongated glands, without spklt on the lo ⅓–½;
sec br appressed; **ped** 0.1–0.5 mm, appressed.
Spklt 1.2–1.8 mm, plumbeous or brownish,
often secund along the br. **Glm** unequal, ovate
to obovate, memb; **lo glm** 0.3–0.7 mm, without
midveins; **up glm** 1.2–1.8 mm, at least ⅔ as long
as the flt, often longer; **lm** 1.2–1.7 mm, ovate
to elliptic, memb, glab, acute; **pal** 1.1–1.6 mm,
ovate to elliptic, memb, glab; **anth** 0.2–0.4 mm,
yellowish or purplish. **Fruits** 0.6–1 mm, obovoid,
faintly striate, light brownish. 2*n* = 24, 36, 54.

Sporobolus pyramidatus is native to the Americas,
extending from the southern United States to Argentina. It
grows in disturbed soils, roadsides, railways, coastal sands,
and alluvial slopes in many plant communities, at elevations
from 0–1500 m.

2. **Sporobolus neglectus** Nash PUFFSHEATH
DROPSEED [p. *421*, <u>507</u>]

Pl ann; tufted, delicate, slender. **Clm** 10–45 cm,
wiry, erect to decumbent. **Shth** inflated, mostly
glab but the apc with small tufts of hairs, hairs
to 3 mm; **lig** 0.1–0.3 mm; **bld** 1–12 cm long, 0.6–2
mm wide, flat to loosely involute, abx surface
glab, adx surface scabridulous, bases of both
surfaces smt with papillose-based hairs, mrg
smooth or scabridulous. **Pan** tml and axillary, 2–5
cm long, 0.2–0.5 cm wide, contracted, cylindrical,
included in the uppermost shth; **lo nd** with
1–2(3) br; **pri br** 0.4–1.8 cm, appressed, spklt-
bearing to the base; **sec br** appressed; **ped** 0.1–2.5
mm, appressed, scabridulous. **Spklt** 1.6–3 mm,
yellowish to cream-colored, smt purple-tinged.
Glm subequal, shorter than the flt, lanceolate to
ovate, memb to chartaceous, glab; **lo glm** 1.5–2.4
mm, midveins often greenish; **up glm** 1.7–2.7
mm; **lm** 1.6–2.9 mm, ovate, chartaceous, glab,
acute; **pal** 1.6–3 mm, ovate, chartaceous, glab;
anth 3, 1.1–1.6 mm, purplish. **Fruits** 1.2–1.8 mm,

obovoid, lat flattened, light brownish or orangish
brown, translucent, finely striate. 2*n* = 36.

Sporobolus neglectus is native to North America, and grows
at 0–1300 m in sandy soils, on river shores, and in dry, open
areas within many plant communities, often in disturbed
sites.

3. **Sporobolus compositus** (Poir.) Merr. ROUGH
DROPSEED [p. *421*, <u>507</u>]

Pl per; ces, smt rhz. **Clm** (20)30–130(150) cm. **Shth**
with sparsely hairy apc, hairs to 3 mm; **lig** 0.1–0.5
mm; **bld** not conspicuously distichous, 5–70 cm
long, 1.5–10 mm wide, flat, folded, or involute,
abx surface glab or pilose, adx surface glab or
scabridulous, mrg glab. **Pan** tml and axillary, 5–30
cm long, 0.4–1.6 cm wide, usu spikelike, partially
included in the uppermost shth, with 15–90
spklt per cm² (exposed portion, when pressed);
lo nd with 1–2(3) br; **pri br** 0.4–6 cm, appressed,
spklt-bearing to the base; **sec br** appressed;
pulvini glab; **ped** 0.3–3.5 mm, appressed, glab
or scabridulous. **Spklt** 4–6(10) mm, stramineous
to purplish tinged. **Glm** subequal, lanceolate,
memb to chartaceous, midveins usu greenish; **lo**
glm (1.2)2–4 mm; **up glm** (2)2.5–5(6) mm, slightly
shorter or longer than the lm; **lm** (2.2)3–6(10) mm,
lanceolate, memb to chartaceous and hyaline,
glab, smooth, occ 2- or 3-veined, acute to obtuse;
pal (2.2)3–6(10) mm, ovate to lanceolate, memb;
anth 0.2–3.2 mm, yellow to orangish. **Fruits** 1–2
mm, ellipsoid, lat flattened, often striate, reddish
brown; **pericarps** gelatinous, slipping from the
seeds when wet. 2*n* = 54, 88, 108.

Sporobolus compositus grows along roadsides and railroad
right of ways, on beaches, and in cedar glades, pine woods,
live oak–pine forests, prairies, and other partially disturbed,
semi-open sites at 0–1600 m.

The *Sporobolus compositus* complex is a difficult assemblage
of forms, perhaps affected by their primarily autogamous
breeding system (Riggins 1977). Asexual proliferation
via rhizomes adds to the species' ability to maintain local
population structure and to perpetuate unique character
combinations. One variety is present in the Intermountain
Region.

Sporobolus compositus (Poir.) Merr. var.
compositus [p. *421*]

Pl not rhz. **Clm** stout, 2–5 mm thick. **Uppermost**
shth usu 2.6–6 mm wide. **Pan** with 30–90 spklt
per cm² when pressed. **Fruits** 1–1.8 mm.

Sporobolus compositus var. *compositus* is the most
widespread of the three varieties, being found throughout
most of the range of the species.

4. Sporobolus wrightii Munro *ex* Scribn. Big
Alkali Sacaton [p. 422, 507]
Pl per; ces, not rhz. **Clm** 90–250 cm, stout. **Shth**
rounded below, shiny, glab, rarely sparsely hairy
apically, hairs to 6 mm; **lig** 1–2 mm; **bld** 20–70
cm long, 3–10 mm wide, flat (rarely involute),
glab abx, scabrous adx, mrg scabrous; **flag bld**
ascending. **Pan** 20–60 cm long, 12–26 cm wide,
open, broadly lanceolate, exserted; **pri br** 1.5–10
cm, spreading 20–70° from the rchs; **sec br**
appressed, spklt-bearing to the base; **pulvini**
glab; **ped** 0.2–0.5 mm, mostly appressed. **Spklt**
1.5–2.5 mm, crowded, purplish or greenish. **Glm**
unequal, lanceolate to ovate, memb; **lo glm** 0.5–1
mm, often appearing veinless; **up glm** 0.8–2 mm,
⅔ or more as long as the flt; **lm** 1.2–2.5 mm, ovate,
memb, glab, acute to obtuse; **pal** 1.1–2.5 mm,
ovate, memb, glab; **anth** 1.1–1.3 mm, yellowish
to purplish. **Fruits** 1–1.4 mm, ellipsoid, reddish
brown or blackish, striate. $2n = 36$.

Sporobolus wrightii grows in moist clay flats and on rocky
slopes near saline habitats, at elevations of 5–1800 m. Its
range extends to central Mexico.

5. Sporobolus airoides (Torr.) Torr. Alkali
Sacaton [p. 422, 507]
Pl per; ces, not rhz. **Clm** 35–120(150) cm,
stout. **Shth** rounded below, shiny, apc glab or
sparsely hairy, hairs to 6 mm; **lig** 0.1–0.3 mm;
bld (3)10–45(60) cm long, (1)2–5(6) mm wide,
flat to involute, glab abx, scabridulous adx, mrg
smooth or scabridulous; **flag bld** ascending.
Pan (10)15–45 cm long, 15–25 cm wide, diffuse,
subpyramidal, often included in the uppermost
shth; **pri br** 1.5–13 cm, spreading 30–90° from
the rchs; **sec br** spreading, without spklt on the
lo ¼–⅓; **pulvini** glab; **ped** 0.5–2 mm, spreading,
glab or scabrous. **Spklt** 1.3–2.8 mm, purplish
or greenish. **Glm** unequal, lanceolate to ovate,
memb; **lo glm** 0.5–1.8 mm, often without
midveins; **up glm** 1.1–2.4(2.8) mm, at least ⅔ as
long as the flt; **lm** 1.2–2.5 mm, ovate, memb, glab,
acute; **pal** 1.1–2.4 mm, ovate, memb, glab; **anth**
1.1–1.8 mm, yellowish to purplish. **Fruits** 1–1.4
mm, ellipsoid, reddish brown, striate. $2n = 80, 90,$
108, 126.

Sporobolus airoides grows on dry, sandy to gravelly flats or
slopes, at elevations from 50–2350 m. It is usually associated
with alkaline soils. Its range extends into northern Mexico.

6. Sporobolus cryptandrus (Torr.) A. Gray
Sand Dropseed [p. 423, 507]
Pl per; ces, not rhz, bases not hard and knotty.
Clm 30–100(120) cm tall, 1–3.5 mm thick, erect
to decumbent. **Shth** rounded below, glab or

scabridulous, mrg smt ciliate distally, apc with
conspicuous tufts of hairs, hairs to 4 mm; **lig** 0.5–1
mm; **bld** (2)5–26 cm long, 2–6 mm wide, flat to
involute, glab abx, scabridulous to scabrous adx,
mrg scabridulous; **flag bld** nearly perpendicular
to the clm. **Pan** 15–40 cm long, 2–12(14) cm
wide, longer than wide, initially contracted
and spikelike, ultimately open and narrowly
pyramidal; **rchs** straight, erect; **lo nd** with 1–2(3)
br; **pri br** 0.6–6 cm, appressed, spreading, or
reflexed to 130° from the rchs, without spklt
on the lo ⅛–¼; **lo br** longest, included in the
uppermost shth; **sec br** appressed; **pulvini** glab;
ped 0.1–1.3 mm, appressed, glab or scabridulous.
Spklt 1.5–2.5(2.7) mm, brownish, plumbeous, or
purplish tinged. **Glm** unequal, linear-lanceolate
to ovate, memb; **lo glm** 0.6–1.1 mm; **up glm**
1.5–2.7 mm, at least ⅔ as long as the flt; **lm**
1.4–2.5(2.7) mm, ovate to lanceolate, memb, glab,
acute; **pal** 1.2–2.4 mm, lanceolate, memb; **anth**
0.5–1 mm, yellowish to purplish. **Fruits** 0.7–1.1
mm, ellipsoid, light brownish to reddish orange.
$2n = 36, 38, 72$.

Sporobolus cryptandrus is a widespread North American
species, extending from Canada into Mexico. It grows in
sandy soils and washes, on rocky slopes and calcareous ridges,
and along roadsides in salt-desert scrub, pinyon-juniper
woodlands, yellow pine forests, and desert grasslands. Its
elevational range is 0–2900 m.

7. Sporobolus contractus Hitchc. Spike
Dropseed [p. 423, 507]
Pl per; ces, not rhz. **Clm** 40–100(120) cm tall,
2–4(5) mm thick near the base. **Shth** rounded
below, mrg hairy, particularly distally, hairs
to 3 mm, apc with conspicuous tufts of hair;
lig 0.4–1 mm; **bld** (2)4–35 cm long, 3–8 mm
wide, flat to involute, glab on both surfaces,
mrg whitish, somewhat scabridulous. **Pan** all
tml, (10)15–45(50) cm long, 0.2–0.8(1) cm wide,
contracted, spikelike, dense, usu included in
the uppermost shth; **lo nd** with 1–2(3) br; **pri
br** 0.3–1.5 cm, appressed, spklt-bearing to the
base; **sec br** appressed; **pulvini** glab; **ped** 0.2–2
mm, appressed, scabridulous. **Spklt** 1.7–3.2 mm,
whitish to plumbeous. **Glm** unequal, narrowly
lanceolate, memb, prominently keeled; **lo glm**
0.7–1.7 mm, usu 1-veined, acute to acuminate;
up glm 2–3.2 mm, at least ⅔ as long as the flt; **lm**
2–3.2 mm, linear-lanceolate, memb, glab, acute;
pal 1.8–3 mm, linear-lanceolate, memb, glab;
anth 3, 0.3–0.5 mm, light yellowish. **Fruits** 0.8–
1.2 mm, ellipsoid, lat flattened, light brownish or
translucent. $2n = 36$.

Sporobolus contractus grows in dry to moist, sandy soils, at elevations from 300–2300 m. It is found occasionally in salt-desert scrub, desert grasslands, and pinyon-juniper woodlands. Its range extends to the states of Baja California and Sonora in Mexico.

8. **Sporobolus texanus** Vasey Texas Dropseed [p. 423, <u>507</u>]

Pl per (often appearing ann); ces, with fibrous roots, not rhz. **Clm** 20–70 cm, erect to decumbent, glab or scurfy roughened below. **Shth** rounded bas, apc glab or with scattered, appressed, papillose-based hairs, hairs to 4 mm; **lig** 0.2–0.6 mm; **bld** 2.5–13(18) cm long, 1–4.2 mm wide, flat to involute, glab abx, scabrous adx, mrg scabridulous, often also with a few papillose-based hairs; **flag bld** ascending. **Pan** 10–35 cm long, 4.5–30 cm wide, open, diffuse, subpyramidal, about as long as wide, partially included in the uppermost lf shth; **lo nd** with 1–2 br; **pri br** 4–14 cm, capillary, spreading 10–80° from the rchs; **sec br** spreading, without spklt on the lo ⅓–½; **ped** 6–25 mm, spreading. **Spklt** 2.3–3 mm, purplish tinged. **Glm** unequal, linear-lanceolate to lanceolate, memb; **lo glm** 0.5–1.7 mm, often without midveins; **up glm** 1.7–3 mm, at least ⅔ as long as the flt, often longer; **lm** 1.8–3 mm, lanceolate to ovate, memb, glab, acute; **pal** 1.7–2.9 mm, ovate, memb, glab, often splitting as the fruit matures; **anth** 0.3–1 mm, yellowish. **Fruits** 1.1–1.5 mm, obovoid, light brown, translucent, occ rugulose. 2*n* = unknown.

Sporobolus texanus grows along rivers, ponds, and in wet alkaline habitats, at 100–3300 m. It is known only from the United States.

9. **Sporobolus flexuosus** (Thurb. *ex* Vasey) Rydb. Mesa Dropseed [p. 424, <u>507</u>]

Pl per (rarely appearing ann); ces, not rhz, bases not hard and knotty. **Clm** 30–100(120) cm tall, 1–3 mm thick near the base, erect to decumbent. **Shth** rounded below, smooth or scabridulous, mrg smt ciliate distally, apc with tufts of hairs, hairs to 4 mm; **lig** 0.5–1 mm; **bld** (2)5–24 cm long, 2–4(6) mm wide, ascending or strongly divergent, flat to involute, glab abx, scabridulous adx, mrg scabridulous. **Pan** 10–30 cm long, 4–12 cm wide, longer than wide, open, subovate to oblong; **rchs** drooping or nodding; **lo nd** with 1–2 br; **pri br** 1–8(12) cm, flexible, diverging at least 70° from the rchs, often strongly reflexed to 130°, tangled with each other and with br from adjacent pan; **lo br** no longer than those in the mid, usu included in the uppermost shth; **sec br** widely spreading, without spklt on the lo ⅛–½; **pulvini** pubescent; **ped** 0.3–3 mm, spreading, scabridulous. **Spklt**

1.8–2.5 mm, plumbeous. **Glm** unequal, ovate, memb; **lo glm** 0.9–1.5 mm; **up glm** 1.4–2.5 mm, subequal to the flt; **lm** 1.4–2.5 mm, lanceolate to ovate, memb, glab, acute; **pal** 1.4–2.4 mm, ovate, memb; **anth** 0.4–0.7 mm, yellow. **Fruits** 0.6–1 mm, ellipsoid, light brownish to reddish orange. 2*n* = 36, 38.

Sporobolus flexuosus grows on sandy to gravelly slopes, flats, and roadsides in the southwestern United States and northern Mexico. It is associated with desert scrub, pinyon-juniper woodlands, and yellow pine forests. Its elevational range is 800–2100 m.

10. **Sporobolus giganteus** Nash Giant Dropseed [p. 424, <u>507</u>]

Pl per; ces, not rhz. **Clm** 100–200 cm, (3)4–10 mm thick near the base. **Shth** rounded below, striate, mrg hairy distally, apc with conspicuous tufts of hairs, hairs to 2 mm; **lig** 0.5–1.5 mm; **bld** 10–50 cm long, (3)4–10(13) mm wide, flat, glab on both surfaces, mrg whitish, scabridulous. **Pan** all tml, 25–75 cm long, 1–4 cm wide, spikelike, dense, usu included in the uppermost shth; **lo nd** with 1–2(3) br; **pri br** mostly 0.5–6 cm, appressed or spreading to 30° from the rchs, spklt-bearing to the base; **sec br** appressed; **pulvini** glab; **ped** 0.5–2 mm, appressed. **Spklt** 2.5–3.5(4) mm, whitish to plumbeous. **Glm** unequal, narrowly lanceolate, memb, prominently keeled; **lo glm** 0.6–2 mm; **up glm** 2–3.5(4) mm, subequal to the lm; **lm** 2.5–3.5(4) mm, linear-lanceolate, memb, glab, acute; **pal** 2.4–3.4(3.8) mm, linear-lanceolate, memb, glab; **anth** 0.6–1 mm, yellowish. **Fruits** 0.8–1.7 mm, ellipsoid, light yellowish brown, smt translucent. 2*n* = 36.

Sporobolus giganteus grows in sand dunes and sandy areas along rivers and roadsides, at elevations from 100–1830 m. Its range extends from the southwestern United States into northern Mexico.

11. **Sporobolus junceus** (P. Beauv.) Kunth Piney Woods Dropseed [p. 425, <u>507</u>]

Pl per; ces, not rhz. **Clm** (30)40–100 cm. **Shth** rounded below, mrg and apc smt sparsely ciliate; **lig** 0.1–0.2 mm; **bld** (6)10–30 cm long, 0.8–2 mm wide, flat to tightly involute, glab abx, scabridulous adx, mrg scabrous, apc pungent. **Pan** 7–28 cm long, 2–6 cm wide, open, pyramidal; **lo nd** with 3 or more br; **pri br** 0.7–4.5 cm, spreading 20–100° from the rchs, whorled or verticellate, without spklt on the lo ⅛–½; **sec br** appressed; **ped** 0.4–2.5 mm, appressed, scabridulous. **Spklt** 2.6–3.8 mm, purplish red. **Glm** unequal, linear-lanceolate to lanceolate or ovate, hyaline to memb; **lo glm** 0.9–3 mm; **up glm** 2.6–3.8 mm, as long as or longer than the flt;

lm 2–3.6 mm, ovate, memb, glab, acute; **pal** 2–3.6 mm, ovate, memb; **anth** 1.4–2 mm, purplish. **Fruits** 1.4–1.8 mm, ellipsoid, somewhat lat flattened, somewhat rugulose, reddish brown. 2*n* = unknown.

Sporobolus junceus grows in openings in pine and hardwood forests, coastal prairies, and pine barrens, usually in sandy to loamy soils, at 2–400 m. Its range lies entirely within the southern United States.

12. Sporobolus interruptus Vasey BLACK DROPSEED [p. 425, 507]

Pl per; ces but shortly rhz, with tough, fibrous roots. **Clm** 25–60 cm. **Shth** dull and fibrous bas, with scattered, contorted hairs to 5 mm, mrg glab; **lig** 0.2–0.7 mm; **bld** (5)8–20 cm long, 1–2.5 mm wide, flat to folded, glab or scattered-pilose on both surfaces, mrg glab. **Pan** 5–20 cm long, (0.6)1–8 cm wide, longer than wide, narrowly pyramidal, open to somewhat contracted, not diffuse, well-exserted from the up lf shth; **lo nd** with 1–2(3) br; **pri br** 0.6–7 cm, appressed or spreading to 70° from the rchs, not capillary, without spklt on the lo ⅓; **ped** 0.8–5.5 mm, appressed to spreading. **Spklt** 4.5–6.6 mm, plumbeous. **Glm** unequal, lanceolate, memb; **lo glm** (2)2.5–4.2 mm; **up glm** 3.8–6.5 mm, at least ⅔ as long as the flt; **lm** 5–6.5 mm, ovate, memb, glab, acute; **pal** 4.8–6.5 mm, ovate, memb; **anth** 3–4.2 mm, yellow to purplish. **Fruits** about 3 mm long, 1.5–1.7 mm thick, pyriform-globose; **emb** dark brown to blackish; **endosperm** reddish brown. 2*n* = 30.

Sporobolus interruptus grows on rocky slopes and in dry meadows of open yellow pine and oak-pine forests and pinyon-juniper woodlands, at elevations from 1500–2300 m. It is endemic to Arizona.

10.15 CRYPSIS Aiton   Barry E. Hammel and John R. Reeder

Pl ann; synoecious. **Clm** 1–75 cm, erect to geniculately ascending, smt brchg above the base; **nd** usu exposed. **Shth** open, often becoming inflated, jnct with the bld evident; **lig** of hairs; **aur** absent; **bld** often dis. **Infl** tml or tml and axillary, spikelike or capitate pan subtended by, and often partially enclosed in, 1 or more of the uppermost lf shth, additional pan often present in the axils of the lvs below. **Spklt** 2–6 mm, strongly lat compressed, with 1 flt; **flt** bisex; **dis** above or below the glm. **Glm** 1-veined, strongly keeled; **lm** memb, glab, 1-veined, strongly keeled, not lobed, unawned, smt mucronate; **pal** hyaline, 1–2-veined; **lod** absent; **anth** 2 or 3; **ov** glab. **Fruits** oblong, pericarp loosely enclosing the seed and easily removed when wet; **hila** punctate. *x* = 8. Name from the Greek *krupsis*, 'concealment', alluding to the partially concealed inflorescence.

Crypsis, a genus of eight species, is native from the Mediterranean region to northern China. Its species tend to occur in moist soils, often in areas subject to winter flooding. The three species found in the Intermountain Region are very plastic in the lengths of their culms and leaves, e.g., the culms of *C. schoenoides* vary from 2 cm in dry sites to 75 cm under optimal conditions.

1. Spikelets 1.5–2.8 mm long; panicles 7–8 times longer than wide, usually completely exserted from the uppermost sheath at maturity . . . . 1. *C. alopecuroides*
1. Spikelets 2.5–4 mm long; panicles 1–5 times longer than wide, the bases usually enclosed in the uppermost sheath at maturity.
   2. Collars glabrous; glumes unequal, the margins glabrous; anthers 0.7–1.1 mm long . . . . . . 2. *C. schoenoides*
   2. Collars pilose; glumes subequal, at least the lower glumes pilose on the margin; anthers 0.5–0.9 mm long . . . . 3. *C. vaginiflora*

1. Crypsis alopecuroides (Piller & Mitterp.) Schrad. FOXTAIL PRICKLEGRASS [p. 426, 507]

Clm (3)5–75 cm, rarely brchd above the base. **Shth** glab; **col** glab; **lig** 0.2–1 mm; **bld** 5–12 cm long, 1.2–2.5 mm wide, not dis. **Pan** 1.5–6.5 cm long, 4–6 mm wide, 7–8 times longer than wide, often purplish, completely exserted from the uppermost shth at maturity on peduncles at least 1 cm long. **Spklt** 1.8–2.8 mm, remaining lightly attached until late in the season. **Lo glm** 1.2–2 mm; **up glm** 1.4–2.4 mm; **lm** 1.7–2.8 mm; **pal** faintly 2-veined; **anth** 3, 0.5–0.6 mm. **Car** 0.9–1.1 mm. 2*n* = 16.

Crypsis alopecuroides is common to abundant in sandy soils around drying lake margins in Oregon and southern Washington, and within the last forty years has become widespread in northern California; it is also known from several other western states. In the Eastern Hemisphere, it extends from France and northern Africa to the Urals and Iraq.

2. Crypsis schoenoides (L.) Lam. SWAMP PRICKLEGRASS [p. 426, 507]

Clm 2–75 cm, prostrate to erect, smt geniculate, usu not brchg above the base, but some pl profusely brchd. **Shth** glab or ciliate at the throat, often inflated; **col** glab; **lig** 0.5–1 mm; **bld** 2–10 cm long, 2–6 mm wide, not dis. **Pan** 0.3–4(7.5) cm long, 5–6(15) mm wide, 1–5 times longer than wide, bases usu enclosed in the uppermost lf shth at maturity. **Spklt** 2.7–3.2 mm, tardily dis. **Lo glm** 1.8–2.3 mm; **up glm** 2.2–2.7 mm; **lm** 2.4–3 mm; **pal** 2-veined; **anth** 3, 0.7–1.1 mm. **Car** about 1.3 mm. $2n = 32$.

Crypsis schoenoides is common to abundant in clay or sandy clay soils around drying lake margins and vernal pools. It is most abundant in California, but is also known from the Intermountain Region. Its native range extends from southern Europe and northern Africa through western Asia to India.

3. **Crypsis vaginiflora** (Forssk.) Opiz MODEST PRICKLEGRASS [p. 426, 507]
Clm 1–30 cm, often profusely brchg above the base, with 10–25 pan per clm. **Shth** pilose on the mrg; **col** pilose; **bld** 1–5 cm long, 1–3 mm wide, soon dis, thus many lvs on mature pl are bladeless. **Pan** 0.3–1.5(3.5) cm long, 3–6(10) mm wide, 1–5 times longer than wide, sessile or almost so, mostly included in the shth of the up 2 lvs. **Spklt** 2.5–3.2 mm, readily dis when disturbed, otherwise retained within the up shth. **Glm** about 3 mm, subequal; **lo glm** pilose on the mrg; **lm** subequal to the glm; **pal** minutely 2-veined; **anth** 3, 0.5–0.9 mm. **Car** 1.3–1.7 mm. $2n = 48$.

Crypsis vaginiflora is common to abundant in clay or sandy clay soil in California, where it was first introduced in the late 1800s. It has since been found at a few locations in Washington, Idaho, and Nevada, and will probably spread to additional sites with suitable habitat in the future. It is native to Egypt and southwestern Asia.

10.16 **CALAMOVILFA** Hack.   John W. Thieret †

Pl per; synoecious; rhz, rhz short or elongate. **Clm** 50–250 cm, solitary or few. **Lvs** cauline; **shth** open; **lig** of hairs, dense, short; **bld** elongate, long tapering. **Infl** tml, simple pan, usu exserted and exceeding the up lvs, open to contracted; **pan** 8–80 cm long, to 60 cm wide, simple, flexible, br not spikelike, not dis. **Spklt** with 1 flt, lat compressed, unawned; **rchl** not prolonged; **dis** above the glm, achenes falling with the lm and pal. **Glm** subequal to unequal, 1-veined, acute; **cal** evidently hairy, hairs ¼–⅞ as long as the lm; **lm** similar to the glm, from shorter than to longer than the up glm, 1-veined, acute, unawned; **pal** longitudinally grooved; **anth** 3, 2.4–5.5 mm; **ov** glab. **Fruit** an achene, pericarp free from the seed. $x = 10$. Name from the Greek *kalamos*, 'reed', and *Vilfa*, a genus of grasses.

Calamovilfa is a genus of five species, all of which are endemic to temperate portions of North America. One species is present in the Intermountain Region.

1. **Calamovilfa gigantea** (Nutt.) Scribn. & Merr. GIANT SANDREED [p. 427, 507]
Rhz elongate, covered with shiny, coriaceous, scalelike lvs. **Clm** to 2.5 m. **Shth** entirely glab or pubescent at the throat; **lig** 0.7–2 mm; **bld** to 90 cm long, about 12 mm wide. **Pan** 20–80 cm long, 20–60 cm wide; **br** to 35 cm, ascending to strongly divergent, lowermost br smt reflexed. **Spklt** 7–10.8 mm. **Glm** straight; **lo glm** 4.5–10.5 mm; **up glm** 6.4–10.1 mm; **cal hairs** ¼–¾ as long as the lm; **lm** 6–10 mm, straight, pubescent, smt sparsely so, very rarely glab; **pal** 6–8.3 mm, pubescent or glab; **anth** 3–5.5 mm. $2n = 60$.

Calamovilfa gigantea grows on sand dunes, prairies, river banks, and flood plains in the Rocky Mountains and central plains from Utah and Nebraska to Arizona and Texas.

10.17 **MUHLENBERGIA** Schreb.   Paul M. Peterson

Pl ann or per; usu rhz, often ces, smt mat-forming, rarely stln. **Clm** 2–300 cm, erect, geniculate, or decumbent, usu herbaceous, smt becoming woody. **Shth** open; **lig** memb or hyaline (rarely firm or coriaceous), acuminate to truncate, smt minutely ciliolate, smt with lat lobes longer than the cent portion; **bld** narrow, flat, folded, or involute, smt arcuate. **Infl** tml, smt also axillary, open to contracted or spikelike pan; **dis** usu above the glm, occ below the ped. **Spklt** with 1(2–3) flt. **Glm** usu (0)1(2–3)-veined, apc entire, erose, or toothed, truncate to acuminate, smt mucronate or awned; **lo glm** smt rdmt or absent, occ bifid; **up glm** shorter than to longer than the flt; **cal** poorly developed, glab or with hairs; **lm** glab, scabrous, or with short hairs, 3-veined (occ appearing 5-veined), apc awned, mucronate, or unawned; **awns**, if present, straight, flexuous, sinuous, or curled, smt borne between 2 minute teeth; **pal** shorter than or equal to the lm, 2-veined; **anth** (1–2)3, purple, orange, yellow, or olivaceous. **Car** elongate, fusiform or elliptic, slightly dorsally compressed. **Cleistogamous pan** smt present in the axils of the lo cauline lvs, enclosed by a tightly rolled, somewhat indurate shth. $x = 10$. Named for Gotthilf Henry Ernest Muhlenberg (1753–1815), a Lutheran minister and pioneer botanist of Pennsylvania.

Muhlenbergia is primarily a genus of the Western Hemisphere. It has approximately 155 species. All but one of the 34 species treated here are native to the Intermountain Region. Within North America, *Muhlenbergia* is represented best in the southwestern United States. *Muhlenbergia montana* is an important range grass in the southwestern United States and northern Mexico.

In the key and descriptions, "puberulent" refers to having hairs so small that they can only be seen with a 10x hand lens.

1. Plants annual.
  2. Lemmas unawned or mucronate, the mucros to 1 mm long.
    3. Glumes strigulose, at least on the margins or towards the apices, the hairs 0.1–0.3 mm long.
      4. Pedicels of most spikelets strongly curved below the spikelets, often through 90° or more; anthers olivaceous, 0.6–1.2 mm long . . . . 33. *M. sinuosa*
      4. Pedicels of most spikelets straight or curved below the spikelets, rarely curved through 90°; anthers purplish, 0.2–0.7 mm long. . . . . 34. *M. minutissima*
    3. Glumes glabrous.
      5. Panicles contracted, less than 0.5 cm wide; branches closely appressed at maturity; culms often rooting at the lower nodes. . . . . 21. *M. filiformis* (in part)
      5. Panicles open or diffuse, 0.6–11 cm wide; branches spreading at maturity; culms not rooting at the lower nodes.
        6. Primary panicle branches 0.5–3.2 cm long; pedicels stout, 1–3 mm long, 0.5–1.5 mm thick; lemmas mottled, with greenish black and greenish white areas . . . . 36. *M. ramulosa*
        6. Primary panicle branches 2.2–6.2 cm long; pedicels delicate, 6–10 mm long, about 0.02 mm thick; lemmas not mottled, purplish to light brownish . . . . 35. *M. fragilis*
  2. Lemmas awned, the awns 1.5–30 mm long.
    7. Lower glumes 2-veined, minutely to deeply bifid, the teeth aristate or with awns to 1.3 mm long; spikelets borne in subsessile-pedicellate pairs; disarticulation at the base of the pedicels. . . . . 32. *M. depauperata*
    7. Lower glumes, if present, veinless or l-veined, not bifid, unawned or with a single awn; spikelets borne singly; disarticulation above the glumes.
      8. Lemma awns 1.8–2.8 mm long . . . . 6. *M. schreberi* (in part)
      8. Lemma awns 10–30 mm long.
        9. Cleistogamous panicles not present in the axils of the lower cauline leaves; upper glumes 1.5–2.8 mm long, acute. . . . . 7. *M. tenuifolia* (in part)
        9. Cleistogamous panicles of 1–3 florets present in the axils of the lower cauline leaves; upper glumes 0.6–1.3 mm long, obtuse . . . . 8. *M. microsperma*
1. Plants perennial.
  10. Plants rhizomatous, usually not cespitose; rhizomes scaly and creeping.
    11. Panicles open, (2)4–16 cm wide; panicle branches capillary, 0.05–0.1 mm thick, diverging 30–90° from the rachises at maturity.
      12. Lemmas unawned or mucronate with mucros to 0.3 mm long . . . . 22. *M. asperifolia*
      12. Lemmas awned, awns 1–12(20) mm long.
        13. Blades stiff and pungent; lemma awns 1–1.5(2) mm long, straight; primary branches of the panicles appearing fascicled in immature plants. . . . . 17. *M. pungens*
        13. Blades not stiff and pungent; lemma awns 4–12(20) mm long, flexuous; primary branches of the panicles not appearing fascicled . . . . 12. *M. arsenei* (in part)
    11. Panicles contracted, 0.1–2(3) cm wide; panicle branches more than 0.1 mm thick, appressed or diverging up to 30(40)° from the rachises at maturity.
      14. Blades 0.2–2(2.6) mm wide, flat, involute, or folded at maturity.
        15. Lemmas awned, the awns 1–25 mm long.
          16. Lemmas pubescent for ¾ their length, the hairs about 1.5 mm long. . . 10. *M. curtifolia*
          16. Lemmas scabridulous or pubescent for no more than ½ their length, the hairs often less than 1.5 mm long.
            17. Lemmas and paleas mostly glabrous, the calluses with a few short hairs; ligules with lateral lobes, the lobes 1.5–3 mm longer than the central portion; culms erect; plants tightly cespitose at the base; sheaths and blades commonly with dark brown necrotic spots . . . . 11. *M. pauciflora* (in part)

17. Lemma midveins and margins and paleas appressed-pubescent on the lower ⅓–½; ligules with lobes less than 1.5 mm longer than the central portion; culms decumbent; plants loosely cespitose at the base; sheaths and blades usually without necrotic spots . . . . 12. *M. arsenei* (in part)

15. Lemmas unawned, mucronate, or shortly awned, the awns to 1 mm long.

18. Lemmas and paleas pubescent, the hairs to 1.2 mm long . . . . 9. *M. thurberi*

18. Lemmas and paleas scabrous, glabrous, or with hairs less than 0.3 mm long.

19. Glumes more than ½ as long as the lemmas; lemmas 2.6–4.2 mm long, attenuate . . . . 18. *M. repens*

19. Glumes ½ as long as the lemmas or less; lemmas 1.3–3.1 mm long; not attenuate.

20. Ligules 0.2–0.8 mm long; panicles usually partially included, the rachises usually visible between the branches . . . . . 19. *M. utilis*

20. Ligules 0.8–3 mm long; panicles exserted, the rachises usually hidden by the branches . . . . 20. *M. richardsonis* (in part)

14. Blades (1.5)2–15 mm wide, flat at maturity.

21. Glumes awn-tipped, 3–8 mm long (including the awns), about 1.3–2 times longer than the lemmas.

22. Internodes dull, puberulent, usually terete, rarely keeled; culms seldom branched above the base; ligules 0.2–0.6 mm long; anthers 0.8–1.5 mm long . . . . 2. *M. glomerata*

22. Internodes smooth and polished for most of their length, elliptic in cross section and strongly keeled; culms much branched above the base; ligules 0.6–1.7 mm long; anthers 0.4–0.8 mm long . . . . 1. *M. racemosa*

21. Glumes unawned or awn-tipped, 0.6–4.5 mm long (including the awns), from shorter than to about 1.2 times longer than the lemmas.

23. Lemmas glabrous . . . . 20. *M. richardsonis* (in part)

23. Lemmas with hairs, these sometimes restricted to the callus.

24. Lemma bases with hairs about as long as the florets, usually 2–3.5 mm long . . . . 4. *M. andina*

24. Lemma bases glabrous or with hairs shorter than the florets, usually shorter than 1.2 mm.

25. Ligules 0.4–1 mm long; panicles dense; pedicels up to 2 mm long; anthers 0.3–0.5 mm long . . . . 3. *M. mexicana*

25. Ligules 1–2.5 mm long; panicles not dense, pedicels 0.8–3.5 mm long; anthers 0.4–0.8 mm long . . . . 5. *M. sylvatica*

10. Rhizomes absent; plants cespitose or bushy.

26. Upper glumes usually 3-veined and 3-toothed; old sheaths flattened, ribbonlike or papery, sometimes spirally coiled.

27. Lemmas unawned, mucronate, or with awns to 5 mm long.

28. Lower glumes awned, the awns to 1.6 mm long; blades 2–6 cm long, filiform, tightly involute, sharp-tipped . . . . 23. *M. filiculmis*

28. Lower glumes unawned; blades (5)6–12 cm long, flat or loosely involute to subfiliform, not sharp-tipped . . . . 24. *M. jonesii*

27. Lemmas awned, the awns (2)6–27 mm long.

29. Upper glumes as long as or longer than the lemmas, the apices acuminate to acute, occasionally minutely 3-toothed; old sheaths conspicuously spirally coiled . . . . 25. *M. straminea*

29. Upper glumes ⅓–⅔ as long as the lemmas, the apices truncate to acute, 3-toothed; old sheaths occasionally spirally coiled . . . . 26. *M. montana*

26. Upper glumes usually 1-veined, (rarely 2- or 3-veined), rounded, obtuse, acute, or acuminate, entire or erose; old sheaths not flattened or papery, never spirally coiled.

30. Panicles loosely contracted, open, or diffuse, (1)2–40 cm wide; panicle branches usually not appressed at maturity, often naked basally.

31. Lemma awns 6–35 mm long.

32. Plants conspicuously branched, bushy in appearance; culms wiry, with geniculate, stiff, widely divergent branches.................... 13. *M. porteri*
32. Plants not conspicuously branched, not bushy in appearance, usually typical bunchgrasses; culms not wiry, when branched, the branches not both geniculate and widely divergent.
    33. Basal sheaths laterally compressed, commonly keeled; glumes (excluding any awns) exceeding the florets; culms (50)80–150 cm tall... 27. *M. emersleyi* (in part)
    33. Basal sheaths rounded; glumes (excluding any awns) exceeded by the florets; culms 40–100 cm tall........................... 29. *M. rigida*
31. Lemmas unawned or with awns to 4.2 mm long.
    34. Basal sheaths laterally compressed, usually keeled......... 27. *M. emersleyi* (in part)
    34. Basal sheaths rounded, not keeled.
        35. Culms 60–130 cm tall, erect from the base; blades (10)20–65 cm long .... 28. *M. longiligula* (in part)
        35. Culms 10–60 cm tall, somewhat decumbent; blades 1–10(16) cm long.
            36. Blades arcuate, 0.3–0.9 mm wide, 1–3(5) cm long; usually no culm nodes exposed; most leaf blades reaching no more than ⅕ of the plant height .... 15. *M. torreyi*
            36. Blades not arcuate, 1–2.2 mm wide, 4–10(16) cm long; 1 or more culm nodes exposed; leaf blades reaching ¼–½ of the plant height... 16. *M. arenicola*
30. Panicles narrow, 0.2–3(5) cm wide; branches appressed to ascending at maturity, usually spikelet-bearing their whole length.
    37. Panicles spikelike, 0.1–1.2 cm wide, sometimes interrupted near the base; branches appressed, 0.3–4 cm long.
        38. Culms 3–20(35) cm tall, often decumbent and rooting at the lower nodes; blades 1–4(6) cm long........................... 21. *M. filiformis* (in part)
        38. Culms 15–150 cm tall, stiffly erect, not rooting at the lower nodes; blades 1.4–50 cm long, at least some more than 6 cm long.
            39. Basal leaf sheaths rounded on the back; panicles 15-60 cm long .... 31. *M. rigens*
            39. Basal leaf sheaths compressed, the backs keeled; panicles 5-16 cm long .... 14. *M. wrightii*
    37. Panicles not spikelike, 0.6–5 cm wide; branches appressed, ascending, or diverging up to 70°, 0.2–13 cm long.
        40. Ligules 10–35 mm long, firm and brown basally, membranous distally; glumes as long as or longer than the florets.
            41. Basal sheaths rounded; lemmas unawned or with awns to 2 mm long; plants native in Arizona .... 28. *M. longiligula* (in part)
            41. Basal sheaths compressed-keeled, at least basally; lemmas unawned or with awns to 4 mm long; plants grown as ornamentals in southern Utah........................... 30. *M. lindheimeri*
        40. Ligules 1–3(5) mm long, usually membranous throughout, sometimes firmer basally, never brownish; glumes shorter than the florets.
            42. Lemma awns 1.5–5 mm long.
                43. Glumes 2–4 mm long; lemmas 3.5–5 mm long, the awns flexuous; ligules 1–2 mm long, with lateral lobes less than 1.5 mm longer than the central portion; anthers 1.3–3 mm long, purple .... 12. *M. arsenei* (in part)
                43. Glumes less than 2 mm long; lemmas 1.8–2.8 mm long, the awns straight; ligules 0.2–0.5 mm long, without lateral lobes; anthers 0.2–0.5 mm long, yellow .... 6. *M. schreberi* (in part)
            42. Lemma awns 4–30 mm long.
                44. Lemmas and paleas almost glabrous, with only a few short hairs on the calluses; ligules with lateral lobes, the lobes 1.5–3 mm longer than the central portion; culms erect and plants tightly cespitose at the base; sheaths and blades usually with dark brown necrotic spots.......... 11. *M. pauciflora* (in part)

44. Lemmas and paleas pubescent on the lower ⅓–½ of the midveins and margins; ligules without lateral lobes or with lobes less than 1.5 mm longer than the central portion; culms decumbent and plants loosely cespitose; sheaths and blades usually without necrotic spots.
    45. Anthers 0.9–1.5 mm long, yellowish; panicles 7–20 cm long . . . . 7. *M. tenuifolia* (in part)
    45. Anthers 1.3–3 mm long, purple; panicles 4–13 cm long . . . . 12. *M. arsenei* (in part)

1. **Muhlenbergia racemosa** (Michx.) Britton, Sterns & Poggenb. MARSH MUHLY [p. 427, 507]

Pl per; rhz, not ces. **Clm** 30–110 cm tall, 1–1.5 mm thick, stiffly erect, much brchd above the mid, intnd mostly smooth, polished, glab or puberulent immediately below the nd, elliptic in cross section and strongly keeled. **Shth** scabridulous, slightly keeled; **lig** 0.6–1.5(1.7) mm, memb, truncate, lacerate-ciliolate; **bld** 2–17 cm long, 2–5 mm wide, flat, usu scabrous or scabridulous, occ smooth. **Pan** 0.8–16 cm long, 0.3–1.8 cm wide, lobed, dense; **br** 0.2–2.5 cm, appressed; **ped** absent or to 1 mm, strigose. **Spklt** 3–8 mm. **Glm** subequal, 3–8 mm (including the awns), about 1.3–2 times longer than the lm, smooth or scabridulous distally, 1-veined, acuminate and awned, awns to 5 mm; **lm** 2.2–3.8 mm, lanceolate, pilose on the cal, lo ½ of the midveins, and mrg, hairs to 1.2 mm, apc scabridulous, acuminate, unawned or awned, awns to 1 mm; **pal** 2.2–3.8(4.5) mm, lanceolate, intercostal region loosely pilose on the lo ½, apc acuminate; **anth** 0.4–0.8 mm, yellowish. **Car** (1.2)1.4–2.3 mm, fusiform, brown. 2*n* = 40.

Muhlenbergia racemosa grows on rocky slopes, in seasonally wet meadows, in prairies, on sandstone outcrops, on stream banks, in forest ecotones, on the margins of cultivated fields, railways and roadsides, and beside irrigation ditches, at elevations of 30–3400 m. It is most common in the north-central United States, but can be found at scattered locations throughout the western United States and extends into northern Mexico.

2. **Muhlenbergia glomerata** (Willd.) Trin. SPIKE MUHLY [p. 427, 507]

Pl per; rhz, not ces. **Clm** 30–120 cm tall, 0.8–2.5 mm thick, erect, seldom brchd above the base; **intnd** dull, mostly puberulent (smt sparsely so), terete, rarely keeled, strigose immediately below the nd. **Shth** scabridulous, slightly keeled; **lig** 0.2–0.6 mm, memb, truncate, lacerate-ciliolate; **bld** 2–15 cm long, 2–6 mm wide, flat, usu scabrous or scabridulous, occ smooth. **Pan** 1.5–12 cm long, 0.3–1.8 cm wide, lobed, dense; **pri br** 0.2–2.5 cm, appressed; **ped** absent or to

1 mm, strigose. **Spklt** 3–8 mm. **Glm** subequal, 3–8 mm (including the awn), about 1.3–2 times longer than the lm, smooth or scabridulous distally, 1-veined, acuminate, awned, awns to 5 mm; **lm** 1.9–3.1 mm, lanceolate, pubescent on the cal, midveins, and mrg, hairs to 1.2 mm, apc scabridulous, acuminate, unawned or awned, awns to 1 mm; **pal** 1.9–3.1 mm, lanceolate, loosely pilose between the veins, apc acuminate; **anth** 0.8–1.5 mm, yellowish. **Car** 1–1.6 mm, fusiform, brown. 2*n* = 20.

Muhlenbergia glomerata grows in meadows, marshes, bogs, alkaline fens, lake margins, stream banks, beside hot springs and irrigation ditches, and on gravelly slopes, in many different plant communities, at elevations of 30–2300 m. It is most common in southern Canada and the northeastern United States, but grows sporadically throughout the western United States. It is not known from Mexico.

3. **Muhlenbergia mexicana** (L.) Trin. WIRESTEM MUHLY [p. 428, 507]

Pl per; rhz, not ces. **Clm** 30–90 cm tall, 0.5–2 mm thick, erect, much brchd above the base; **intnd** dull, puberulent or glab for most of their length, smt strigose immediately below the nd. **Shth** smooth or scabridulous, somewhat keeled; **lig** 0.4–1 mm, memb, truncate, lacerate-ciliolate; **bld** 2–20 cm long, 2–6 mm wide, flat, scabrous or smooth, those of the sec br similar in length and width to those of the main br. **Pan** tml and axillary, 2–21 cm long, 0.3–3 cm wide, dense, axillary pan exserted on long peduncles; **pri br** 0.3–5.5 cm, appressed or diverging up to 30° from rchs; **ped** to 2 mm, strigose. **Spklt** 1.5–3.8 mm, often purple-tinged. **Glm** subequal, 1.5–3.7 mm, equaling or slightly shorter than the lm, 1-veined, tapering from the bases to the acuminate apc, unawned or awned, awns to 2 mm; **lm** 1.5–3.8 mm, lanceolate, pubescent on the cal, lo portion of the midveins, and mrg, hairs shorter than 0.7 mm, apc scabridulous, acuminate, unawned or awned, awns to 10 mm; **pal** 1.5–3.8 mm, narrowly lanceolate, apc acuminate; **anth** 0.3–0.5 mm, yellow to purplish. **Car** 1.1–1.6 mm, fusiform, brown. 2*n* = 40.

Muhlenbergia mexicana usually grows in mesic to wet areas such as moist prairies and woodlands, stream banks, lake margins, swamps, bogs, hot springs, roadsides, and ditch banks, at elevations 50–3300 m, and is found in many different plant communities. Despite its name, *M. mexicana* grows only in Canada and the United States.

Plants with awns 3–10 mm long belong to *Muhlenbergia mexicana* var. *filiformis* (Torr.) Scribn., and those without awns or with awns less than 3 mm long to *M. mexicana* var. *mexicana*.

4. Muhlenbergia andina (Nutt.) Hitchc.

FOXTAIL MUHLY [p. *428*, 508]

Pl per; rhz, not ces. **Clm** 25–85 cm tall, 0.9–1.7 mm thick, ascending; **intnd** glab for most of their length, scabrous to strigose below the nd. **Shth** scabridulous, especially bas; **lig** 0.5–1.5 mm, memb, truncate, lacerate to ciliate; **bld** 4–16 cm long, 2–4(5) mm wide, flat, scabrous abx, pubescent adx. **Pan** 2–15 cm long, 0.5–2.8 cm wide, contracted, dense; **pri br** 0.5–5 cm, appressed to strongly ascending; **ped** 0.5–1.5 mm, appressed, strigose. **Spklt** 2–4 mm. **Glm** equal to subequal, 2–4 mm, subequal to or longer than the flt, 1-veined, veins scabridulous, apc acuminate to awn-tipped; **lm** 2–3.5 mm, lanceolate, grayish green, hairy on the cal and lm bases, hairs 2–3.5 mm, apc acuminate, awned, awns 1–10 mm; **pal** 2–3.5 mm, lanceolate, bases with silky hairs between the veins, apc acuminate; **anth** 0.4–1.5 mm, yellow. **Car** 0.9–1.1 mm, cylindrical, yellowish brown. 2*n* = 20.

Muhlenbergia andina grows in damp places such as stream banks, gravel bars, marshes, lake margins, damp meadows, around springs, and in canyons, at elevations of 700–3000 m. It grows mainly in the western contiguous United States and a few localities in southwestern Canada.

5. Muhlenbergia sylvatica (Torr.) Torr. *ex* A.

Gray WOODLAND MUHLY [p. *429*, 508]

Pl per; rhz. **Clm** 40–110 cm tall, 1–2 mm thick, erect; **intnd** puberulent for most of their length, strigose below the nd. **Shth** glab and smooth for most of their length, scabridulous distally, mrg hyaline; **lig** 1–2.5 mm, memb, truncate, lacerate-ciliolate; **bld** 5–18 cm long, 3–7 mm wide, flat, scabrous to scabridulous, occ smooth. **Pan** tml and axillary, 6–21 cm long, 0.2–1 cm wide, narrow, not dense, axillary pan usu exserted at maturity; **br** 0.8–6 cm, ascending to closely appressed; **ped** 0.8–3.5 mm, strigose. **Spklt** 2.2–3.7 mm. **Glm** subequal, 1.8–3 mm, nearly as long as the lm, 1-veined, tapering from near the base, apc scabridulous, acuminate, unawned or awned, awns to 1 mm; **lm** 2.2–3.7 mm, lanceolate to narrowly lanceolate, hairy on the cal, lo ½ of the midveins, and mrg, hairs 0.2–0.5 mm, apc

scabridulous, acuminate, awned, awns 5–18 mm, purplish; **pal** 2–3.5 mm, lanceolate, proximal ½ shortly pilose, apc scabridulous, acuminate; **anth** 0.4–0.8 mm, yellow. **Car** 1.4–2 mm, fusiform, brown. 2*n* = 40.

Muhlenbergia sylvatica grows in upland forests, along creeks and hollows, on rocky ledges derived from sandstone, shale, or calcareous parent materials, moist prairies, and swamps, at elevations from 30–1500 m. Its primary range is southeastern Canada and the midwestern and eastern United States. The record from Arizona is based on the report in Kearney and Peebles (1951) of a collection made by Toumey at Grapevine Creek in the Grand Canyon.

6. Muhlenbergia schreberi J.F. Gmel.

NIMBLEWILL [p. *429*, 508]

Pl per (appearing ann); usu ces, not rhz, smt stln. **Clm** 10–45(70) cm, geniculate, often rooting at the lo nd, glab or puberulent below the nd; **intnd** often smooth, shiny, glab. **Shth** shorter than the intnd, glab for most of their length, mrg shortly (0.3–1.2 mm) pubescent distally, not becoming spirally coiled when old; **lig** 0.2–0.5 mm, truncate, erose, ciliate; **bld** (1)3–10 cm long, 1–4.5 mm wide, flat, smooth or scabridulous. **Pan** 3–15 cm long, 1–1.6 cm wide, contracted, often interrupted below; **br** 0.4–5.5 cm, appressed or diverging up to 30° from the rchs, spklt-bearing to the base; **ped** 0.1–4 mm, scabrous to hirsute; **dis** above the glm. **Spklt** 1.8–2.8 mm, borne singly. **Glm** unequal, shorter than the flt, thin and memb throughout, unawned; **lo glm** lacking or rdmt, veinless, rounded and often erose; **up glm** 0.1–0.3 mm, veinless; **lm** 1.8–2.8 mm, oblong-elliptic, mostly scabrous, cal hairy, hairs to 0.8 mm, veins greenish, lo ¼ of the midveins with a few appressed hairs, apc acute to acuminate, awned, awns 1.5–5 mm, straight; **pal** 1.8–2.8 mm, oblong-elliptic, acute to acuminate; **anth** 0.2–0.5 mm, yellow. **Car** 1–1.4 mm, fusiform, brownish. 2*n* = 40, 42.

Muhlenbergia schreberi grows in moist to dry woods and prairies on rocky slopes, in ravines, and along sandy riverbanks, at elevations of 60–1600 m. It is also common in disturbed sites near cultivated fields, pastures, and roads at these elevations. Its geographic range includes central, but not northern, Mexico. Records from the western United States probably reflect recent introductions. The species is considered a noxious, invasive weed in California.

7. Muhlenbergia tenuifolia (Kunth) Trin.

SLENDER MUHLY [p. *430*, 508]

Pl ann or short-lived per; ces, not rhz. **Clm** 20–70 cm, erect or decumbent; **intnd** mostly scabridulous or smooth, always scabridulous below the nd. **Shth** usu shorter than the intnd, glab, smooth or scabridulous, usu without necrotic spots, not becoming spirally coiled when old; **lig** 1.2–3(5) mm, memb throughout, acute, often lacerate; **bld** 2–13 cm long, 1.2–2.5 mm wide, flat or loosely involute, scabridulous or glab abx, scabrous adx, usu without necrotic spots. **Pan** numerous, tml and axillary, 7–20 cm long, 0.3–1.4(3) cm wide, contracted, often lax, nodding, interrupted below; **pri br** 3.5–7.5 cm, ascending or diverging up to 70° from the rchs, spklt-bearing to the base; **ped** 1–3 mm, antrorsely scabrous; **dis** above the glm. **Spklt** 2–4 mm, often purplish, borne singly. **Glm** 1.2–2.8 mm, shorter than the flt, 1-veined, veins scabrous, apc often erose, unawned or awned, awns to 0.5 mm; **lo glm** 1.2–2 mm, acute to acuminate; **up glm** 1.5–2.8 mm, acute; **lm** 2–3.5(4) mm, lanceolate, mostly smooth, scabridulous distally, pubescent on the cal, lo ½ of the midveins, and mrg, hairs 0.5–1.5 mm, apc acuminate to acute, awned, awns 10–30 mm, scabrous, sinuous to flexuous; **pal** 1.8–3.4(3.8) mm, lanceolate, sparsely villous bas, apc acuminate to acute; **anth** 0.9–1.5 mm, yellowish. **Car** 1–2.2 mm, narrowly fusiform, brownish. **Cleistogamous pan** not present. $2n$ = 20, 40.

Muhlenbergia tenuifolia grows in grama grasslands and pine-oak woodlands on rocky slopes, limestone rock outcrops, gravelly roadsides, and in sandy drainages, at elevations of 1200–2200 m. Its range extends through Mexico to northern South America.

8. **Muhlenbergia microsperma** (DC.) Trin.
Littleseed Muhly [p. *430*, 508]

Pl ann, smt appearing as short-lived per; tufted. **Clm** 10–80 cm, often geniculate at the base, much brchd near the base; **intnd** mostly scabridulous or smooth, always scabridulous below the nd. **Shth** often shorter than the intnd, glab, smooth or scabridulous; **lig** 1–2 mm, memb to hyaline, truncate to obtuse; **bld** 3–8.5(10) cm long, 1–2.5 mm wide, flat or loosely involute, scabrous abx, strigulose adx. **Pan** 6.5–13.5 cm long, 1–6.5 cm wide, not dense, often purplish; **br** 1.6–4 cm, ascending or diverging up to 80° from the rchs, spklt-bearing to the base; **ped** 2–6 mm, appressed to divaricate, antrorsely scabrous; **dis** above the glm. **Spklt** 2.5–5.5 mm, borne singly. **Glm** 0.4–1.3 mm, exceeded by the flt, 1-veined, obtuse, often minutely erose; **lo glm** 0.4–1 mm; **up glm** 0.6–1.3 mm; **lm** 2.5–3.8(5.3) mm, narrowly lanceolate, mostly smooth, scabridulous distally, hairy on the cal, lo ½ of the mrg, and midveins, hairs 0.2–0.5 mm, apc acuminate, awned, awns 10–30 mm, straight to flexuous; **pal** 2.2–4.8 mm, narrowly

lanceolate, acuminate; **anth** 0.3–1.2 mm, purplish. **Car** 1.7–2.5 mm, fusiform, reddish brown. **Cleistogamous pan** with 1–3 spklt present in the axils of the lo lvs. $2n$ = 20, 40, 60.

Muhlenbergia microsperma grows on sandy slopes, drainages, cliffs, rock outcrops, and disturbed roadsides, at elevations of 0–2400 m. It is usually found in creosote scrub, thorn-scrub forest, sarcocaulescent desert, and oak-pinyon woodland associations. Its range extends from the southwestern United States through Central America to Peru and Venezuela. Morphological variation among and within its populations is marked.

9. **Muhlenbergia thurberi** (Scribn.) Rydb.
Thurber's Muhly [p. *431*, 508]

Pl per; rhz, not ces. **Clm** 12–36 cm tall, 0.3–0.5 mm thick, often clumped, usu erect, somewhat decumbent at the base; **intnd** mostly or completely glab, smt strigose below the nd. **Shth** shorter than the intnd, hirtellous near the mrg; **lig** 0.9–1.2 mm, memb, truncate to obtuse, erose; **bld** 0.2–3.7 cm long, 0.2–1 mm wide, tightly involute, straight to arcuate, smooth or scabridulous abx, hirtellous adx. **Pan** 0.7–5.5 cm long, 0.2–0.7 cm wide, contracted, not dense, smt interrupted below; **pri br** 0.4–1.8 cm, appressed; **ped** 0.1–4 mm. **Spklt** 2.6–4 mm. **Glm** subequal, 1.6–3 mm, 1-veined, scabridulous on the veins, acute, unawned; **lm** 2.6–4 mm, lanceolate, hairy on the lo ¾, hairs to 1.2 mm, tawny, apc acuminate, unawned or awned, awns to 1 mm; **pal** 2.6–4 mm, lanceolate, intercostal region tawny pubescent; **anth** 2.1–2.3 mm, yellowish purple. **Car** 2–2.2 mm, fusiform, light brownish. $2n$ = unknown.

Muhlenbergia thurberi usually grows in moist soil in seeps near canyon cliffs, sandstone slopes, and rocky ledges, at elevations of 1350–2300 m. It appears to be restricted to the southwestern United States, and flowers from July to September.

Muhlenbergia thurberi differs from *M. curtifolia* in its tightly involute blades and longer anthers and ligules. The two species have been found growing within 50 m of each other in Apache County, Arizona, but in different habitats, *M. curtifolia* growing in damp drainage areas and *M. thurberi* near a moist but dryer canyon cliff.

10. **Muhlenbergia curtifolia** Scribn. Utah
Muhly [p. *431*, 508]

Pl per; rhz, not ces. **Clm** 10–45 cm tall, 0.3–0.5 mm thick, erect or decumbent; **intnd** hirsute and strigose below the nd. **Shth** shorter than the intnd, hirsute; **lig** 0.2–0.6 mm, mrg hyaline, apc truncate to obtuse, erose, ciliate; **bld** 0.5–4(5) cm long, 0.6–1.5(2.2) mm wide, awl-shaped, flat near the base, becoming slightly involute distally, strigose to scabrous abx, hirsute adx. **Pan** 2–10 cm long, 0.2–1 cm wide, contracted, not dense; **pri br** 0.3–2.5 cm, ascending to appressed; **ped** 0.1–3 mm. **Spklt** 2.8–4 mm. **Glm** equal, 2.5–4 mm, 1-veined, veins scabridulous, apc acute; **lm** 2.8–4 mm, lanceolate, hairy on the lo ¾, hairs to 1.5 mm, tawny, midveins scabridulous distally, apc acuminate, awned, awns usu 1–6(12) mm, delicate, straight; **pal** 2.8–4 mm, lanceolate, intercostal region tawny pubescent on the bas ¾; **anth** 1.2–1.6 mm, yellowish purple. **Car** 2–2.4 mm, fusiform, brownish. $2n$ = unknown.

Muhlenbergia curtifolia grows on damp ledges and in rock crevices of vertical cliffs, and beneath large calcareous boulders above canyon floors, at elevations of 1600–2750 m, in the southwestern United States. It resembles *M. thurberi*, differing in its flatter leaf blades and shorter ligules and anthers. It also tends to grow in more mesic habitats than *M. thurberi*.

11. **Muhlenbergia pauciflora** Buckley NEW MEXICAN MUHLY [p. *431*, 508]

Pl per; tightly ces, smt rhz. **Clm** 30–70 cm tall, 0.7–1 mm thick, erect, geniculate and rooting at the lo nd; **intnd** mostly glab, smt glaucous, often striate below the nd. **Shth** usu shorter than the intnd, smooth or scabridulous, usu with dark brown necrotic spots, flat and spreading at maturity; **lig** 1–2.5(5) mm, memb, obtuse, with lat lobes 1.5–3 mm longer than the cent portion; **bld** (1)5–12(15) cm long, 0.5–1.5 mm wide, flat to involute, glab and smooth abx, scabridulous adx, often with dark brown necrotic spots. **Pan** (2)5–15 cm long, 0.5–2.8 cm wide, contracted, interrupted below; **pri br** 0.5–4(6) cm, appressed or diverging up to 30° from the rchs, spklt-bearing to the base; **ped** 0.1–3 mm, scabrous. **Spklt** 3.5–5.5 mm, occ with 2 flt. **Glm** equal, 1.5–3.5 mm, shorter than the lm, 1-veined, acuminate to acute, often erose, unawned or awned, awns to 2.2 mm; **lm** 3–5.5 mm, lanceolate, often purplish, mostly glab, cal with a few short hairs, apc acuminate, scabridulous, awned, awns (5)7–25 mm, flexuous; **pal** 3–5.5 mm, lanceolate, glab, acuminate; **anth** 1.5–2.1 mm, yellowish or purplish. **Car** 2–2.5 mm, fusiform, brownish. $2n$ = unknown.

Muhlenbergia pauciflora grows in open or closed forests on rocky slopes, cliffs, canyons, and rock outcrops of granitic or calcareous origin, at elevations of 1200–2500 m. Its range extends from the southwestern United States to central Mexico.

12. **Muhlenbergia arsenei** Hitchc. NAVAJO MUHLY [p. *432*, 508]

Pl per; rhz, rhz smt short and the pl loosely ces. **Clm** 10–50 cm tall, 0.4–1 mm thick, decumbent;

intnd glab or strigulose. **Shth** shorter than the intnd, strigulose or glab, usu without necrotic spots, not becoming spirally coiled when old; **lig** 1–2 mm, memb throughout, obtuse, strigulose or glab, erose or toothed, with lat lobes, lobes less than 1.5 mm longer than the cent portion; **bld** 1–6 cm long, 1–2 mm wide, flat to involute, smooth or scabridulous abx, hirsute adx, usu without necrotic spots. **Pan** 4–13 cm long, 1–3(5) cm wide, not dense; **pri br** 0.5–4 cm, appressed or diverging up to 30° from the rchs, spklt-bearing to the base; **ped** 0.1–3 mm. **Spklt** 3.5–5 mm. **Glm** subequal, 2–4 mm, exceeded by the flt, 1-veined, scabrous on the veins and near the apc, apc acuminate, unawned or awned, awns to 1.2 mm; **lm** 3.5–5 mm, lanceolate, mostly purplish, veins conspicuously green, pubescent on the lo ⅓–½ of the midveins and mrg, hairs to 1.5 mm, apc scabridulous, acuminate, awned, awns 4–12(20) mm, flexuous; **pal** 3.5–5 mm, narrow-lanceolate, intercostal region pubescent, apc acuminate, veins smt extending into awns to 0.5 mm; **anth** 1.3–3 mm, purple. **Car** 2–2.3 mm, fusiform, brownish. $2n$ = unknown.

Muhlenbergia arsenei grows among granitic boulders, on rocky slopes, limestone rock outcrops, and in arroyos, at elevations of 1400–2850 m. Its range extends from the southwestern United States into Baja California, Mexico. It flowers from August to September.

13. **Muhlenbergia porteri** Scribn. *ex* Beal BUSH MUHLY [p. *432*, 508]

Pl per; loosely ces from a knotty base, not rhz, distinctly bushy in appearance. **Clm** 25–100 cm tall, 0.5–1.5 mm thick, erect, geniculate, wiry, freely brchd, br stiff, geniculate, widely divergent; **intnd** scabridulous throughout. **Shth** shorter than the intnd, glab; **lig** 1–2.5(4) mm, truncate, lacerate, with lat lobes; **bld** 2–8 cm long, 0.5–2 mm wide, flat or folded, smooth or scabridulous abx, scabridulous adx. **Pan** 4–14 cm long, 6–15 cm wide, open, not dense, usu purple; **pri br** 1–7.5 cm, diverging 30–90° from the rchs, stiff, naked bas; **ped** 2–13(20) mm, scabrous. **Spklt** 3–4.5 mm. **Glm** 2–3 mm, shorter than the lm, 1-veined, veins scabrous, apc gradually acute to acuminate, occ mucronate, mucros to 0.4 mm; **lm** 3–4.5 mm, lanceolate, purplish, appressed-pubescent on the lo ½–¾ of the mrg and midveins, apc acuminate, awned, awns 2–13 mm, straight; **pal** 3–4.5 mm, lanceolate, acuminate; **anth** 1.5–2.3 mm, purple to yellow. **Car** 2–2.4 mm, oblong, compressed, yellowish brown. $2n$ = 20, 23, 24, 40.

Muhlenbergia porteri grows among boulders on rocky slopes and on cliffs, and in dry arroyos, desert flats, and grasslands, frequently in the protection of shrubs, at elevations of 600–1700 m. Its geographic range extends from the southwestern United States to northern Mexico. It is highly palatable to all classes of livestock, but is never abundant at any particular location.

14. **Muhlenbergia wrightii** Vasey *ex* J.M. Coult. SPIKE MUHLY [p. 433, 508]

Pl per; ces, not rhz. **Clm** 15–60 cm tall, 1.5–2.5 mm thick, compressed, erect, not rooting at the lo nd; **intnd** mostly hispidulous or glab, strigose to hispidulous below the nd. **Shth** usu shorter than the intnd, smooth or scabridulous, compressed-keeled, not becoming spirally coiled when old; **lig** 1–3(5) mm, memb, truncate; **bld** 1.4–12 cm long, 1–3 mm wide, flat to folded, smooth or scabridulous abx, strigose adx. **Pan** 5–16 cm long, 0.2–1.2 cm wide, spikelike, dense; **pri br** 0.3–2 cm, appressed; **ped** 0.1–1.4 mm. **Spklt** 2–3 mm, dark green or plumbeous. **Glm** equal, 0.5–1.6 mm, usu ½–¾ as long as the lm, 1-veined, scabridulous on the veins, acute or obtuse, abruptly narrowed to a short (0.5–1 mm) awn; **lm** 2–3 mm, lanceolate, appressed-pubescent on the bas ½–¾ of the midveins and mrg, hairs about 0.5 mm, apc scabridulous, acute to acuminate, mucronate, mucros 0.3–1 mm; **pal** 1.9–3 mm, lanceolate, intercostal region pubescent, apc acute to acuminate; **anth** 1.3–1.8 mm, greenish. **Car** 1.2–2 mm, fusiform, brownish. $2n$ = unknown.

Muhlenbergia wrightii grows in gravelly prairies, on rocky slopes, and in meadows on granitic, sandstone, or limestone-derived soils, at elevations of 1100–3000 m. Its range extends from the southwestern United States to northern Mexico.

15. **Muhlenbergia torreyi** (Kunth) Hitchc. *ex* Bush RING MUHLY [p. 433, 508]

Pl per; ces, not rhz. **Clm** 10–40(50) cm, decumbent, usu all the nd concealed by the shth; **intnd** mostly scabrous or smooth, hispidulous below the nd. **Lvs** strongly bas concentrated, most bld not reaching more than ⅕ of the pl height; **shth** shorter than the intnd, rounded, not keeled, scabridulous or smooth, not becoming spirally coiled when old; **lig** 2–5(7) mm, hyaline, acuminate, lacerate, often with lat lobes; **bld** 1–3(5) cm long, 0.3–0.9 mm wide, tightly involute or folded, arcuate, scabridulous, midveins and mrg not thickened, green, apc somewhat sharp-pointed. **Pan** 7–21 cm long, 3–15 cm wide, diffuse; **pri br** 1–8 cm, diverging 30–90° from the rchs, stiff, naked bas; **ped** 1–8 mm, smt appressed to the br. **Spklt** 2–3.5 mm. **Glm** equal, 1.3–2.5 mm, scabridulous,

1-veined, apc acute to acuminate, minutely erose, unawned or awned, awns to 1.1 mm; **lm** 2–3.2(3.5) mm, narrowly elliptic to lanceolate, appressed-pubescent on the bas ½–¾ of the mrg and midveins, apc scabrous, acuminate, awned, awns 0.5–4 mm; **pal** 2–3.2(3.5) mm, narrowly elliptic, intercostal region sparsely pubescent, apc acuminate; **anth** 1.2–2.1 mm, greenish. **Car** 1.7–2 mm, fusiform, brownish. $2n$ = 20, 21.

Muhlenbergia torreyi grows in desert grasslands and open woodlands on sandy mesas, calcareous rock outcrops, and rocky slopes, at elevations of 1000–2450 m. Its range extends from the southwestern United States to northern Mexico. It also grows, as a disjunct, in northwestern Argentina.

16. **Muhlenbergia arenicola** Buckley SAND MUHLY [p. 433, 508]

Pl per; ces, not rhz. **Clm** (15)20–60(70) cm, somewhat decumbent, 1 or more nd exposed; **intnd** hispidulous below the nd. **Lvs** somewhat bas concentrated, most bld not reaching more than (¼)½ of the pl height; **shth** usu a little shorter than the intnd, not keeled, scabridulous, mrg hyaline, bas shth rounded, not becoming spirally coiled when old; **lig** 2–9 mm, hyaline, acute, lacerate, often with lat lobes; **bld** 4–10(16) cm long, 1–2.2 mm wide, not arcuate, flat, folded, or involute, scabrous, often glaucous, midveins and mrg not thickened, green. **Pan** 12–30 cm long, 5–20 cm wide, diffuse; **pri br** 1–10 cm, diverging 30–80° from the rchs, naked bas; **ped** 1–4(6) mm. **Spklt** 2.5–4.2 mm. **Glm** equal, 1.4–2.5 mm, 1-veined, apc scabridulous, acute to acuminate, minutely erose, unawned or awned, awns to 1 mm; **lm** 2.5–4.2 mm, narrowly elliptic, usu purplish, scabrous distally, appressed-pubescent on the lo ½–¾ of the mrg and midveins, apc acuminate, awned, awns 0.5–4.2 mm; **pal** 2.5–3.5 mm, narrowly elliptic, intercostal region sparsely pubescent, apc acuminate, with 2 short (0.1–0.2 mm) awns; **anth** 1.5–2.1 mm, greenish. **Car** 1.9–2.3 mm, fusiform, brownish. $2n$ = 80, 82.

Muhlenbergia arenicola grows on sandy mesas, limestone benches, and in valleys and open desert grasslands, at elevations of 600–2135 m. Its range extends from the southwestern United States to central Mexico. It also grows, as a disjunct, in northwestern Argentina.

17. **Muhlenbergia pungens** Thurb. *ex* A. Gray SANDHILL MUHLY [p. 434, 508]

Pl per; rhz, not ces. **Clm** 10–70 cm, decumbent below; **intnd** cinereous-lanate, glab, or scabrous for most of their length, always cinereous-lanate below the nd. **Shth** longer than the intnd, cinereous-lanate below, glab and smooth or scabridulous distally; **lig** 0.2–1 mm, densely ciliate, obtuse, with lat lobes; **bld** 2–8 cm long, 1–2.2 mm wide, flat to tightly involute, scabrous abx, hirsute adx, stiff, pungent. **Pan** (7)8–16(19) cm long, (2)4–14 cm wide, open; **pri br** 1.5–8 cm, capillary, straight, lo br diverging 70°–90° from the rchs in mature pl, often appearing fascicled in immature pl; **ped** 10–25 mm. **Spklt** 2.6–4.5 mm. **Glm** equal, 1.2–3 mm, purplish near the base, smooth or scabridulous distally, 1-veined, acuminate or acute, unawned or awned, awns to 1 mm; **lm** 2.6–4.5 mm, lanceolate, purplish, scabridulous distally and on the mrg, apc acuminate, awned, awns 1–1.5(2) mm, straight; **pal** 2.6–4.5 mm, lanceolate, glab, acuminate, 2-awned, awns to 1 mm; **anth** 1.8–2.6 mm, purplish. **Car** 1.8–2.5 mm, fusiform, brownish. $2n = 26, 42, 60$.

Muhlenbergia pungens grows in loose sandy soils near sand dunes to sandy clay loam slopes and flats in desert shrub and open woodlands, at elevations of 600–2500 m. It is known only from the western and central contiguous United States.

18. **Muhlenbergia repens** (J. Presl) Hitchc.
CREEPING MUHLY [p. 434, 508]

Pl per; rhz, not ces. **Clm** 5–42 cm tall, 0.5–1 mm thick, decumbent near the base, forming dense mats; **intnd** glab, slightly nodulose. **Shth** shorter or longer than the intnd, glab; **lig** 0.1–1(1.8) mm, memb, truncate, occ lacerate; **bld** 0.4–6 cm long, 0.5–1.4 mm wide, involute, glab, smooth or scabridulous adx. **Pan** 1–9 cm long, 0.1–0.6 cm wide, contracted, not dense, usu partially included in the up lf shth; **pri br** 0.2–3 cm, usu closely appressed at maturity, rarely diverging up to 40° from the rchs; **ped** 0.2–3.6 mm, setulose. **Spklt** 2.6–4.2 mm, occ with 2 flt. **Glm** subequal, 1.1–3.6 mm, from ½ as long as to equaling the lm, light green, 1(2–3)-veined, acute, unawned; **lm** 2.6–3.2(4.2) mm, lanceolate, dark greenish or mottled, glab or the cal and mrg appressed-pubescent, hairs to 0.3 mm, apc scabridulous, attenuate, usu mucronate, mucros 0.1–0.3 mm; **pal** 2.1–3.3 mm, lanceolate, smooth or scabridulous, acute; **anth** 0.7–1.4 mm, yellow to purplish. **Car** 1.1–1.5 mm, ellipsoid to ovoid, brownish. $2n = 60, 70–72$.

Muhlenbergia repens grows in open, sandy meadows, canyon bottoms, calcareous rocky flats, gypsum flats, and on rolling slopes and roadsides, at elevations of 100–3120 m. Its range extends from the southwestern United States to southern Mexico.

19. **Muhlenbergia utilis** (Torr.) Hitchc.
APAREJOGRASS [p. 435, 508]

Pl per; rhz, not ces. **Clm** 7–30 cm tall, 0.5–1 mm thick, erect to decumbent; **intnd** mostly smooth to slightly nodulose, minutely pubescent or glab below the nd. **Shth** shorter or longer than the intnd, glab; **lig** 0.2–0.8 mm, memb, truncate; **bld** 0.5–4.7 cm long, 0.2–1.8 mm wide, usu involute, smt flat, often at right angles to the clm, glab abx, hirtellous adx. **Pan** 1–5 cm long, 0.1–0.4 cm wide, contracted, usu partially included in the up shth, rchs usu visible between the br; **pri br** 0.2–1.2 cm, usu closely appressed at maturity, rarely diverging up to 30° from the rchs; **ped** 0.1–1.1 mm, glab. **Spklt** 1.4–2.4 mm. **Glm** subequal, 0.5–1.4 mm, ⅓–½ as long as the lm, yellowish to light green, glab, 1(2–3)-veined, acute, unawned; **lm** 1.3–2.4 mm, lanceolate, green or purplish, glab or the cal and mrg appressed-pubescent, hairs shorter than 0.3 mm, apc acute, unawned; **pal** 1–2 mm, lanceolate, glab, acute; **anth** 0.7–1.4 mm, yellow to purplish. **Car** 0.7–1.2 mm, ellipsoid to ovoid, brown. $2n = 20$.

Muhlenbergia utilis grows in wet soils along streams, ponds, and depressions in grasslands and alkaline or gypsiferous plains, at elevations of 200–1800 m. Its range extends from the southern United States through Mexico to Costa Rica.

20. **Muhlenbergia richardsonis** (Trin.) Rydb.
MAT MUHLY [p. 435, 508]

Pl per; rhz, not ces, often mat-forming. **Clm** 5–30 cm tall, 0.4–1 mm thick, decumbent, geniculate, or erect; **intnd** usu nodulose (occ smooth) for most of their length, puberulent or nodulose below the nd. **Shth** shorter or longer than the intnd, glab; **lig** 0.8–3 mm, memb, acute to truncate, erose; **bld** 0.4–6.5 cm long, 0.5–4.2 mm wide, flat or involute, straight or arcuate-spreading, glab abx, hirtellous adx. **Pan** 1–15 cm long, 0.1–1.7 cm wide, exserted, narrow or spikelike, rchs usu concealed by the br; **pri br** 0.4–5 cm, usu closely appressed at maturity, rarely diverging up to 20° from the rchs; **ped** 0.2–2 mm, setulose. **Spklt** 1.7–3.1 mm, occ with 2 flt. **Glm** subequal, 0.6–2 mm, ⅓–½ as long as the lm, green, 1(2)-veined, acute, smt mucronate, mucros less than 0.2 mm; **lm** 1.7–2.6(3.1) mm, lanceolate, dark greenish, plumbeous, or mottled, glab, apc scabridulous, acute to acuminate, smt mucronate, mucros to 0.5 mm; **pal** 1.2–2.4(2.9) mm, lanceolate, acute; **anth** 0.9–1.6 mm, yellow to purplish. **Car** 0.9–1.6 mm, narrowly ellipsoid, brown. $2n = 40$.

Muhlenbergia richardsonis grows in open sites in alkaline meadows, prairies, sandy arroyo bottoms, talus slopes, rocky flats and the shores of rivers, at elevations of 60–3300 m. It is the most widespread species of *Muhlenbergia* in North America, extending from the Yukon Territory to Quebec in the north and to northern Baja California, Mexico, in the south.

Muhlenbergia richardsonis is sometimes confused with *M. filiformis*, which differs in being a weak annual with glabrous internodes and obtuse, erose glumes.

21. **Muhlenbergia filiformis** (Thurb. *ex* S. Watson) Rydb. PULL-UP MUHLY [p. 435, 508]
Pl ann (often appearing per), tufted. Clm (3)5–20(35) cm, erect or geniculate, often rooting at the lo nd; intnd glab. Shth shorter or longer than the intnd, glab, smooth or scabridulous; lig 1–3.5 mm, hyaline to memb, rounded to acute; bld 1–4(6) cm long, 0.6–1.6 mm wide, flat or involute, smooth or scabridulous abx, scabrous or pubescent adx. Pan 1.6–6 cm long, 0.2–0.5 cm wide, spikelike, interrupted near the base, long-exserted; pri br 0.9–1.2 cm, closely appressed at maturity; ped 1–3 mm, scabrous. Spklt 1.5–3.2 mm. Glm greenish gray, glab, 1-veined, rounded to subacute; lo glm 0.6–1.4 mm; up glm 0.7–1.7 mm; lm (1.5)1.8–2.5(3.2) mm, lanceolate, dark greenish, appressed-pubescent on the mrg and midveins, hairs shorter than 0.3 mm, apc scabridulous, acute to acuminate, unawned, smt mucronate, mucros shorter than 1 mm; pal 1.6–2.6(3.1) mm, lanceolate, scabridulous distally; anth 0.5–1.2 mm, purplish. Car 0.9–1.5 mm, fusiform, reddish brown. 2*n* = 18.

Muhlenbergia filiformis grows in open, moist meadows, on gravelly lake shores, along stream banks, and in moist humus near thermal springs, at elevations of 1060–3050 m. It is usually associated with yellow pine forests, but also grows in many other plant communities. Its range extends into northern Mexico.

Muhlenbergia filiformis resembles *M. richardsonis*, but differs in having glabrous internodes and subacute apices. Large, robust specimens have been referred to *M. simplex* Scribn. or *M. filiformis* var. *fortis* E.H. Kelso but, until there is more evidence to the contrary, it seems best to treat such plants as representing an extreme of the variation within *M. filiformis*.

22. **Muhlenbergia asperifolia** (Nees & Meyen *ex* Trin.) Parodi SCRATCHGRASS [p. 436, 508]
Pl per; rhz, not ces, occ stln. Clm 10–60(100) cm, decumbent-ascending, bases somewhat compressed-keeled; intnd glab, shiny below the nd. Shth glab, mrg hyaline; lig 0.2–1 mm, firm, truncate, ciliate, without lat lobes; bld 2–7(11) cm long, 1–2.8(4) mm wide, flat, occ conduplicate, smooth or scabridulous abx, scabridulous adx, mrg and midveins not conspicuously thickened, greenish, apc acute, not sharp. Pan 6–21 cm long, 4–16 cm wide, broadly ovoid, open; pri

br 3–12 cm, capillary, lo br spreading 30–90° from the rchs, never appearing fascicled; ped 3–14 mm, longer than the spklt. Spklt 1.2–2.1 mm, occ with 2 or 3 flt. Glm equal, 0.6–1.7 mm, purplish, scabridulous, particularly on the veins, 1-veined, apc acute; lm 1.2–2.1 mm, lanceolate to oblong-elliptic, somewhat plumbeous, glab, usu smooth, occ scabridulous near the apc, apc acute, unawned or mucronate, mucros to 0.3 mm; pal 1.2–2.1 mm, lanceolate, glab, acute; anth 1–1.3 mm, greenish yellow to purplish at maturity. Car 0.8–1 mm, fusiform, brownish. 2*n* = 20, 22, 28.

Muhlenbergia asperifolia grows in moist, often alkaline meadows, playa margins, and sandy washes, on grassy slopes, and around seeps and hot springs, at elevations of 55–3000 m. Its geographic range includes northern Mexico.

The caryopses of *Muhlenbergia asperifolia* are frequently infected by a smut, *Tilletia asperifolia* Ellis & Everhart, which produces a globose body filled with blackish brown spores.

23. **Muhlenbergia filiculmis** Vasey SLIMSTEM MUHLY [p. 436, 508]
Pl per; ces, not rhz. Clm 5–30(40) cm, erect, rounded near the base; intnd smooth or scabridulous. Shth longer than the lo intnd, glab, becoming flattened and papery or ribbonlike at maturity; lig 2–4(5) mm, memb, acute; bld 2–6 cm long, 0.4–1.6 mm wide, tightly involute, filiform, stiff, scabrous abx, sparsely hirtellous adx, apc sharp. Pan 1.5–7 cm long, 0.4–2 cm wide, contracted, smt dense; pri br 0.1–2.8 cm, mostly tightly appressed or diverging up to 30° from the rchs; ped 0.1–2 mm, scabrous. Spklt 2.2–3.5 mm. Glm subequal, 0.8–2.5 mm, scabridulous distally; lo glm 1-veined, awned, awns to 1.6 mm; up glm usu 3-veined, apc truncate to acute, 3-toothed, teeth smt shortly awned, awns to 0.6 mm; lm 2.2–3.5 mm, lanceolate, yellowish mottled with dark green, sparsely appressed-pubescent on the lo portion of the midveins and mrg, hairs to 0.4 mm, apc scabridulous, acute to acuminate, awned, awns 1–5 mm, straight or flexuous; pal 2.2–3.5 mm, lanceolate, acute to acuminate; anth 1.5–2 mm, purplish. Car 1.3–1.6 mm, fusiform, brownish. 2*n* = unknown.

Muhlenbergia filiculmis grows on rocky slopes, dry meadows, and dry gravelly flats in forest openings and grasslands, at elevations of 2500–3500 m in the southern Rocky Mountains and northern Arizona.

It is sometimes difficult to distinguish *Muhlenbergia filiculmis* from *M. montana*, but that species has longer spikelets and lemma awns, and leaf blades that are flatter and not sharply tipped.

24. **Muhlenbergia jonesii** (Vasey) Hitchc.
MODOC MUHLY [p. *436*, 509]

Pl per; tightly ces, not rhz. **Clm** 18–50 cm, erect, rounded near the base; **intnd** glab. **Shth** glab, bases becoming flattened and papery, lo shth longer than the intnd; **lig** 2–5 mm, memb, acute to acuminate; **bld** (5)6–12 cm long, 1–2.5 mm wide, flat, becoming loosely involute to subfiliform, scabrous abx, hirsute adx, apc not sharp. **Pan** 4–15 cm long, 1.5–4 cm wide, not dense; **pri br** 0.5–5 cm, appressed or diverging up to 40° from the rchs; **ped** 0.5–6 mm, flattened, scabrous. **Spklt** 2.8–3.5 mm. **Glm** subequal, 0.6–1.8 mm, scabridulous distally, truncate to obtuse, unawned; **lo glm** 1-veined; **up glm** 3-veined, 3-toothed, often erose; **lm** 2.8–3.5 mm, lanceolate, loosely pubescent on the bas ⅓ of the midveins and mrg, hairs to 0.6 mm, apc scabridulous, acute, mucronate, mucros shorter than 1 mm; **pal** 2.8–3.5 mm, lanceolate, acute; **anth** 1.4–2.3 mm, purple. **Car** 1.6–1.8 mm, fusiform, light brown. 2n = 20.

Muhlenbergia jonesii is endemic to northern California. It grows on open slopes, pumice flats, and in openings in pine forests, at elevations of 1130–2130 m.

25. **Muhlenbergia straminea** Hitchc.
SCREWLEAF MUHLY [p. *437*, 509]

Pl per; ces. **Clm** 25–70 cm, erect, rounded near the base; **intnd** glab. **Shth** glab, stiff, becoming flattened, ribbonlike or papery, and conspicuously spirally coiled when old; **lig** (6)10–20 mm, hyaline, acuminate, lacerate; **bld** 7–25 cm long, 1–4 mm wide, flat to involute, scabrous abx, spiculate adx. **Pan** 8–25 cm long, 0.5–3 cm wide, not dense; **pri br** 0.6–8 cm, appressed or diverging up to 30° from the rchs; **ped** 0.2–5 mm, scabrous. **Spklt** 3.5–7 mm, yellowish to pale greenish. **Glm** (3)3.5–6(7) mm, scabridulous, unawned or awn-tipped; **lo glm** shorter than the up glm, 1-veined; **up glm** equaling or exceeding the flt, 3-veined, acuminate to acute, occ 3-toothed, awned, awns to 1.5 mm; **lm** 3.5–5.5(6) mm, lanceolate, pubescent on the lo ½ of the midveins and mrg, hairs to 1 mm, apc scabrous, acuminate, awned, awns 12–27 mm, flexuous; **pal** 3.5–5.5 mm, lanceolate, pilose between the veins, apc scabridulous, acuminate; **anth** 2–3.5 mm, purple. **Car** 1.9–2 mm, fusiform, light brown. 2n = unknown.

Muhlenbergia straminea grows on rolling, rocky slopes, volcanic tuffs, canyon bottoms, and ridges, usually in open pine forests, at elevations of 1800–2600 m. It is known from the southwestern United States and northern Mexico.

26. **Muhlenbergia montana** (Nutt.) Hitchc.
MOUNTAIN MUHLY [p. *437*, 509]

Pl per; ces, not rhz. **Clm** 10–80 cm, erect, rounded near the base; **intnd** glab. **Shth** smooth or scabridulous, becoming flattened, papery, and occ spirally coiled when old, lo shth longer than the intnd; **lig** 4–14(20) mm, memb, acute to acuminate; **bld** 6–25 cm long, 1–2.5 mm wide, flat, becoming involute, scabrous abx, hirsute adx. **Pan** 4–25 cm long, (1)2–6 cm wide, not dense; **pri br** 0.5–10 cm, appressed or diverging up to 40° from the rchs; **ped** 0.5–6.5 mm, scabrous. **Spklt** 3–7 mm. **Glm** subequal, (1)1.5–3.2(4) mm, smooth or scabridulous distally; **lo glm** 1-veined, smt with a less than 1 mm awn; **up glm** ⅓–⅔ as long as the lm, 3-veined, truncate to acute, 3-toothed, teeth smt awned, awns to 1.6 mm; **lm** 3–4.5(7) mm, lanceolate, loosely to densely appressed-pubescent on the lo portion of the midveins and mrg, hairs to 0.8 mm, apc acute to acuminate, awned, awns (2)6–25 mm, flexuous; **pal** 3–4.5(7) mm, lanceolate, acute to acuminate; **anth** 1.5–2.3 mm, purplish. **Car** 1.8–2 mm, fusiform, light brown. 2n = 20, 40.

Muhlenbergia montana grows on rocky slopes and ridge tops and in dry meadows and open grasslands, at elevations of 1400–3500 m. Its range extends from the western United States to Guatemala. *Muhlenbergia montana* is sometimes difficult to distinguish from *M. filiculmis*, but that species has shorter spikelets and lemma awns and tightly involute or filiform, sharp blades.

27. **Muhlenbergia emersleyi** Vasey
BULLGRASS [p. *438*, 509]

Pl per; ces, not rhz. **Clm** (50)80–150 cm, stout, erect, not conspicuously brchd; **intnd** smooth for most of their length, smooth or scabridulous below the nd. **Shth** shorter or longer than the intnd, glab or puberulent, bas shth lat compressed, usu keeled; **lig** 10–25 mm, memb throughout, acuminate, lacerate; **bld** 20–50 cm long, 2–6 mm wide, flat or folded, scabrous abx, smooth or scabridulous adx. **Pan** 20–45 cm long, 3–15 cm wide, loosely contracted to open, light purplish to light brownish; **pri br** 1–17 cm, lax, loosely appressed or diverging up to 70° from the rchs, naked bas; **ped** 0.5–3 mm, smooth or scabridulous. **Spklt** 2.2–3.2 mm. **Glm** subequal, 2.2–3.2 mm, exceeding the flt, scabridulous to scabrous, faintly 1-veined, acute to obtuse, usu unawned, occ awned, awns to 0.2 mm; **lm** 2–3 mm, oblong-elliptic, shortly pubescent on the lo ½–¾, apc acute, usu awned, smt unawned, awns to 15 mm, flexuous, purplish; **pal** 1.8–2.9

mm, oblong-elliptic, acute; **anth** 1.2–1.6 mm, yellowish to purplish. **Car** 1.3–1.6 mm, fusiform, reddish brown. 2*n* = 24, 26, 28, 40, 42, 46, 60, 64.

Muhlenbergia emersleyi grows on rocky slopes, gravelly washes, canyons, cliffs, and arroyos, often in soils derived from limestone, at elevations of 1200–2500 m, and is also grown as an ornamental. Its range extends from the southwestern United States through Mexico to Panama. *Muhlenbergia emersleyi* differs from the closely related *M. longiligula* in its compressed-keeled sheaths, pubescent florets, and membranous ligules.

28. **Muhlenbergia longiligula** Hitchc.

Longtongue Muhly [p. *438*, <u>509</u>]

Pl per; ces, not rhz. **Clm** 60–130 cm, stout, erect; **intnd** mostly smooth, smt scabridulous below the nd. **Shth** shorter or longer than the intnd, smooth or scabridulous, not becoming spirally coiled, bas shth rounded; **lig** 10–30 mm, firm and brown bas, memb distally, acuminate to obtuse; **bld** (10)20–65 cm long, 3–6 mm wide, flat or inrolled at the mrg, scabrous abx, scabridulous or smooth adx. **Pan** 15–55 cm long, 1–15 cm wide, contracted to open, greenish tan to purplish; **pri br** 3–13 cm, narrowly ascending or diverging up to 60° from the rchs, stiff, spklt-bearing to the base, lo br with 30–60 spklt; **ped** 0.1–2.5 mm, usu shorter than the spklt, scabridulous or smooth, strongly divergent. **Spklt** 2–3.5 mm. **Glm** subequal, 2–3.5 mm, usu longer than the flt, scabridulous or smooth, 1(2)-veined, acute to acuminate, usu unawned, rarely awned, awns to 0.2 mm; **lm** 2–2.9 mm, oblong-elliptic, tan to purplish, smooth or scabridulous, apc acute, often bifid, teeth to 0.2 mm, unawned or awned, awns to 2 mm; **pal** 2–3 mm, oblong-elliptic, scabridulous or smooth, acute; **anth** 1–2.1 mm, yellowish to purplish. **Car** 1.1–1.5 mm, fusiform, reddish brown. 2*n* = 22, 24, 29, 30.

Muhlenbergia longiligula grows on rocky slopes, canyons, and rock outcrops derived from volcanic or calcareous parent materials, at elevations of 1220–2500 m. It is a common species in Arizona and southwestern New Mexico, and extends into northwestern Mexico. It may be confused with *M. emersleyi*, but differs from that species in its rounded basal leaf sheaths, glabrous lemmas, and panicle branches that are spikelet-bearing to the base.

29. **Muhlenbergia rigida** (Kunth) Trin.

Purple Muhly [p. *439*]

Pl per; ces, not rhz. **Clm** 40–100 cm, erect, not conspicuously brchd; **intnd** mostly smooth, smt scabridulous below the nd. **Shth** longer than the intnd, smooth or scabridulous, bas shth rounded, not becoming spirally coiled when old; **lig** (1)3–12(15) mm, firmer bas than distally, obtuse to acute, often lacerate; **bld** 12–35 cm long, 1–3 mm wide, flat or involute, smooth or

scabridulous abx, scabridulous or hirtellous adx. **Pan** 10–35 cm long, 2–5(12) cm wide, loosely contracted to open, purplish; **pri br** 0.4–10 cm, lax, capillary, usu appressed to ascending, occ diverging up to 80° from the rchs, naked bas; **ped** 1–10 mm. **Spklt** 3.5–5 mm, purplish. **Glm** equal, 1–1.7(2) mm, exceeded by the flt, usu glab, smt mostly hirtellous but glab distally, 1-veined, obtuse to subacute, unawned; **lm** 3.5–5 mm, narrowly lanceolate, purplish, cal hairy, hairs to 0.5 mm, lm bodies scabridulous to scabrous, apc acuminate, awned, awns 10–22 mm, clearly demarcated from the lm bodies, flexuous; **pal** 3.5–5 mm, narrowly lanceolate, scabridulous, acuminate; **anth** 1.7–2.3 mm, purplish. **Car** 2–3.5 mm, fusiform, brownish. 2*n* = 40, 44.

Muhlenbergia rigida grows on rocky slopes, ravines, and sandy, gravelly slopes derived from granitic and calcareous substrates, from west Texas and Arizona south to South America. Welsh et al (2008) reported that it is grown as an ornamental in Washington County, Utah.

30. **Muhlenbergia lindheimeri** Hitchc.

Lindheimer's Muhly [p. *439*]

Pl per; ces, not rhz. **Clm** 50–150 cm, stout, erect, not rooting at the lo nd; **intnd** mostly glab, smt puberulent below the nd. **Shth** shorter or longer than the intnd, glab, bas shth lat compressed, keeled, not becoming spirally coiled when old; **lig** 10–35 mm, firm and brown bas, memb distally, acuminate; **bld** 25–55 cm long, 2–5 mm wide, flat or folded, firm, scabridulous abx, scabrous and shortly pubescent adx. **Pan** 15–50 cm long, 0.6–3 cm wide, often purplish-tinged; **pri br** 0.5–7 cm, appressed or strongly ascending, rarely spreading as much as 20° from the rchs; **ped** 0.5–1.2 mm, scabrous. **Spklt** 2.4–3.5 mm, light grayish. **Glm** equal, 2–3.5 mm, shorter than or equal to the flt, scabrous or smooth, 1-veined, obtuse to acute, occ bifid and the teeth to 0.3 mm, unawned, rarely mucronate, mucros less than 0.2 mm; **lm** 2.4–3.5 mm, lanceolate, scabrous or smooth, rarely puberulent near the base, apc obtuse to acute, unawned or awned, awns to 4 mm, straight; **pal** 2.4–3.5 mm, lanceolate, obtuse; **anth** 1.1–1.5 mm, purplish. **Car** 1.2–1.6 mm, fusiform, reddish brown. 2*n* = 20, 26.

Muhlenbergia lindheimeri grows in sandy draws to rocky, calcareous soils, generally in open areas, from Texas to northern Mexico. Welsh et al. (2008) reported that it is grown as a cultivated landscaping grass in Washington County, Utah, but added that the identification of the specimens examined was tentative. *Muhlenbergia lindheimeri* differs from the closely related *M. longiligula* in its compressed-keeled basal sheaths, grayish spikelets, and, when present, bifid glume apices.

31. **Muhlenbergia rigens** (Benth.) Hitchc.
 DEERGRASS [p. *440*, 509]
Pl per; ces, not rhz. **Clm** (35)50–150 cm tall, to 5 mm thick near the base, stiffly erect, not rooting at the lo nd; **intnd** glab. **Shth** longer than the intnd, smooth or scabridulous, rounded dorsally, bases somewhat flat and chartaceous; **lig** 0.5–2(3) mm, firm, truncate, usu ciliolate; **bld** 10–50 cm long, 1.5–6 mm wide, flat or involute, stiff, glab abx, scabrous adx. **Pan** 15–60 cm long, 0.5–1.2 cm wide, spikelike, dense, grayish green, often interrupted below; **pri br** 0.2–4 cm, appressed; **ped** 0.2–3 mm, hispidulous, strongly divergent. **Spklt** 2.4–4 mm. **Glm** subequal, 1.8–3.2 mm, almost as long as the flt, scabrous to scabridulous, 1-veined, usu acute or obtuse, occ acuminate or notched, occ mucronate, mucros to 0.6 mm; **lm** 2.4–4 mm, lanceolate, pubescent on the cal, lo ⅙ of the mrg, and midveins, hairs to 0.4 mm, apc scabrous, acute or obtuse, unawned, rarely mucronate, mucros to 1 mm; **pal** 2.3–3.8 mm, lanceolate, acute; **anth** 1.3–1.8 mm, yellow to purple. **Car** 1.8–2.2 mm, fusiform, brownish. 2*n* = 40.

Muhlenbergia rigens grows in sandy washes, gravelly canyon bottoms, rocky drainages, and moist, sandy slopes, often along small streams, at elevations of 90–2500 m. Its geographic range extends to central Mexico. It is available commercially as an ornamental.

32. **Muhlenbergia depauperata** Scribn.
 SIXWEEKS MUHLY [p. *440*, 509]
Pl ann; tufted. **Clm** 3–15 cm; **intnd** mostly scabridulous or pubescent, pubescent or strigose below the nd. **Shth** often longer than the intnd, somewhat inflated, smooth or scabrous, keeled, mrg scarious, not flattened, papery, or spirally coiled when old; **lig** 1.4–2.5 mm, memb, acute, with lat lobes; **bld** 1–3 cm long, 0.6–1.5 mm wide, flat or involute, scabrous to strigose, midveins and mrg thickened, whitish. **Pan** 2.5–8.5 cm long, 0.5–0.7 cm wide, contracted; **pri br** 1–2.2 cm, appressed, spklt-bearing to the base, spklt borne in subsessile-pedlt pairs; **longer ped** 3–6 mm, scabrous; **dis** beneath the spklt pairs. **Spklt** 2.5–5.1 mm, appressed. **Glm** 2.3–5.1 mm, equaling or exceeding the flt; **lo glm** 2.3–4 mm, subulate, 2-veined, minutely to deeply bifid, teeth aristate or with awns to 1.3 mm; **up glm** 3–5.1 mm, lanceolate, 1-veined, entire, acuminate; **lm** 2.5–4.5 mm, narrowly lanceolate, light greenish brown to purplish, scabrous, appressed-pubescent on the mrg and midveins, apc acuminate, awned, awns 6–15 mm, stiff;

pal 2.4–3.6 mm, lanceolate, intercostal region appressed-pubescent, apc acuminate; **anth** 0.5–0.8 mm, purplish to yellowish. **Car** 1.5–2.3 mm, narrowly fusiform, brownish. 2*n* = 20.

Muhlenbergia depauperata grows in gravelly flats, rock outcrops, exposed bedrock, and sandy banks, in grama grassland associations, usually on soils derived from calcareous parent materials, at elevations of 1530–2400 m. Its range extends from the southwestern United States to southern Mexico.

Muhlenbergia depauperata shares several features with *Lycurus*: spikelets borne in pairs, 2-veined and 2-awned lower glumes, 1-veined and 1-awned upper glumes, acuminate, awned lemmas with short pubescence along the margins, and pubescent paleas. See comment under *Lycurus setosus* (p. 211).

33. **Muhlenbergia sinuosa** Swallen
 MARSHLAND MUHLY [p. *441*, 509]
Pl ann; delicate. **Clm** 12–50 cm, erect to geniculate; **intnd** mostly glab and smooth or scabridulous, scabridulous or strigulose below the nd. **Shth** usu longer than the intnd, glab, smooth or scabridulous; **lig** 1.5–3.1 mm, hyaline, truncate to obtuse, irregularly toothed to lacerate, with lat lobes that exceed the cent portion; **bld** 2–8.5 cm long, 0.8–2 mm wide, flat, smt involute, scabridulous abx, shortly pubescent to minutely villous adx, midveins prominent abx. **Pan** 10–26 cm long, 2.8–8 cm wide; **pri br** 2.6–7 cm, often capillary, diverging 25–80° from the rchs; **ped** 4–7 mm, usu curved, often through 90° or more. **Spklt** 1.4–2 mm. **Glm** equal, 0.7–1.2 mm, usu conspicuously strigulose, particularly near the mrg and apc, 1-veined, acute to obtuse, unawned; **lm** 1.4–2 mm, oblong-elliptic, greenish, smt purplish tinged, shortly appressed-pubescent on the midveins and mrg, apc acute or obtuse, unawned; **pal** 1.3–1.8 mm, oblong-elliptic, intercostal region sparsely short-pilose or glab; **anth** 0.6–1.2 mm, olivaceous. **Car** 0.8–1.2 mm, fusiform, brownish. 2*n* = 20, 24.

Muhlenbergia sinuosa grows in sandy soil along washes, on open slopes and rocky ledges, and in roadside ditches, at elevations of 1650–2300 m. It is usually found in oak-pine forests, pinyon-juniper woodlands, oak-grama savannahs, and riverine woodlands. Its range extends from the southwestern United States into northern Mexico.

34. **Muhlenbergia minutissima** (Steud.)
 Swallen ANNUAL MUHLY [p. *441*, 509]
Pl ann. **Clm** 5–40 cm, slender, erect; **intnd** mostly glab, scabridulous or smooth, scabridulous or strigulose below the nd. **Shth** shorter or longer than the intnd, smooth or scabridulous; **lig** 1–2.6 mm, hyaline, truncate to obtuse, smt with lat lobes; **bld** 0.5–4(10) cm long, 0.8–2 mm wide, flat

or involute, scabrous abx, shortly pubescent adx. **Pan** 5–16.2(21) cm long, 1.5–6.5 cm wide, open; **pri br** 8–42 mm, often capillary, diverging 25–80° from the rchs; **ped** 2–7 mm, straight or curved, but rarely curved as much as 90°. **Spklt** 0.8–1.5 mm. **Glm** sparsely strigulose, at least near the apc, 1-veined; **lo glm** 0.5–0.8 mm, obtuse to acute; **up glm** 0.6–0.9 mm, broader than the lo glm, obtuse; **lm** 0.8–1.5 mm, lanceolate, brownish to purplish, glab or the midveins and mrg appressed-pubescent, apc obtuse to subacute, unawned; **pal** 0.8–1.4 mm, shortly pubescent or glab; **anth** 0.2–0.7 mm, purplish. **Car** 0.6–0.9 mm, fusiform to elliptic, brownish. $2n = 60, 80$.

Muhlenbergia minutissima grows in sandy and gravelly drainages, rocky slopes, flats, road cuts, and open sites. It is usually found in yellow pine and oak-pine forests, pinyon-juniper woodlands, thorn-scrub forests, and oak-grama savannahs, at elevations of 1200–3000 m. Its range extends from the western United States to southern Mexico.

35. **Muhlenbergia fragilis** Swallen Delicate Muhly [p. *442*, 509]

Pl ann. **Clm** 10–38 cm, erect or spreading; **intnd** mostly glab, smooth or scabridulous, scabrous or strigulose below the nd. **Shth** often longer than the intnd, scabridulous, mrg hyaline; **lig** 1–3 mm, hyaline, obtuse, irregularly toothed to lacerate, with lat lobes; **bld** 1–10 cm long, 0.4–2 mm wide, flat, scabrous abx, strigulose adx, mrg and midveins thickened bas, whitish. **Pan** 10–24 cm long, 3.5–11 cm wide, diffuse; **pri br** 2.2–6.2 cm long, about 0.1 mm thick, diverging 80–100° from the rchs, straight; **ped** 6–10 mm long, about 0.02 mm thick, delicate; **dis** above the glm. **Spklt** 1–1.2 mm, appressed to slightly divergent. **Glm** equal to subequal, 0.5–1 mm, glab throughout or obscurely puberulent, hairs about 0.06 mm, 1-veined, obtuse or subacute; **lm** 1–1.2 mm, oblong-elliptic, purplish to light brownish, not mottled, glab or densely appressed-puberulent on the mrg and midveins, apc obtuse, unawned;

pal 0.9–1.2 mm, oblong-elliptic; **anth** 0.3–0.5 mm, purplish. **Car** 0.7–0.9 mm, elliptic, reddish brown. $2n = 20$.

Muhlenbergia fragilis grows on rocky talus slopes, cliffs, canyon walls, road cuts, and sandy slopes, often over calcareous parent materials, at elevations of 480–2200 m. It is usually found in oak-grama savannahs, thorn-scrub forests, oak-yellow pine forests, and pinyon-juniper woodlands. Its range extends from the southwestern United States to southern Mexico.

Populations may have individual plants with completely glabrous lemmas or may consist entirely of such plants. This morphological variation is not correlated with any distributional or habitat characteristics.

36. **Muhlenbergia ramulosa** (Kunth) Swallen Green Muhly [p. *442*, 509]

Pl ann. **Clm** (3)5–25 cm, erect, brchg, but not rooting, bas; **intnd** glab, mostly smooth, smt scabridulous below the nd. **Shth** usu shorter than the intnd, glab, smooth or scabridulous; **lig** 0.2–0.5 mm, hyaline, truncate, ciliate, without lat lobes; **bld** 0.5–3 cm long, 0.8–1.2 mm wide, involute or flat, glab abx, puberulent adx. **Pan** (1)2–9 cm long, 0.6–2.7 cm wide, open, ovoid or deltoid, spklt sparse; **pri br** (0.5)1–3.2 cm long, about 0.1 mm thick, ascending, spreading 20–100° from the rchs; **ped** 1–3 mm long, 0.5–1.5 mm thick, stiff. **Spklt** 0.8–1.3 mm, appressed or divaricate. **Glm** equal, 0.4–0.7 mm, glab, 1-veined, obtuse or subacute, unawned; **lm** 0.8–1.3 mm, oval, plump, mottled with greenish black and greenish white or ochroleucous patches, glab or shortly appressed-pubescent on the mrg and midveins, apc acute, unawned; **pal** 0.7–1.3 mm, oval; **anth** 0.2–0.3 mm, purplish. **Car** 0.5–1 mm, elliptic, brownish. $2n = 20$.

Muhlenbergia ramulosa grows in open, well-drained areas including slopes, sandy meadows, washes, gravelly road cuts, and rock outcrops in yellow pine-oak forests and in open meadows of pine-fir forests, at elevations of 2100–3500 m. Its range includes the southwestern United States, Mexico, Guatemala, Costa Rica, and Argentina.

10.18 LYCURUS Kunth   Charlotte G. Reeder

Pl per; ces. **Clm** 10–60 cm, erect to somewhat decumbent, usu brchd. **Shth** open, compressed-keeled, glab, smooth or scabridulous, mostly shorter than the intnd, a 2-veined prophyllum often present; **lig** hyaline, strongly decurrent, truncate or rounded to elongate and acuminate, smt with narrow triangular lobes extending from the edges of the shth on either side; **bld** folded or flat, rather stiff, with prominent, firm, scabrous mrg, midveins smt extending as short mucros or fragile, scabrous, awnlike apc. **Infl** tml and axillary, dense, bristly, spikelike pan; **br** short, fused to the rchs, terminating in a pair of unequally pedlt spklt or a pedlt spklt and a short sec br bearing two spklt, occ in a solitary spklt, usu the lo spklt in a pair stmt or strl and the up spklt bisex, smt vice versa, or both spklt bisex; **dis** at the fused base of the ped or ped and br, paired spklt falling as a unit, leaving a cuplike tip. **Spklt** with 1 flt. **Glm** subequal, awned; **lo glm** with (1)2(3) awns, usu unequal, awns commonly longer than the body; **up glm** 1-veined, with a single flexuous awn that is usu longer than the glm body, rarely a finer second awn present; **lm** lanceolate, 3-veined, pubescent on the mrg, mostly glab over the back, tapering to a scabrous awn that is usu shorter than the lm body; **pal** about equal to the lm, acute or occ the 2 veins extending as very short mucros, pubescent between the veins and on the sides, except for the narrow, glab, hyaline mrg; **anth** 3. **Car** fusiform, brownish. $x = 10$. Name from the Greek *lykos*, 'wolf', and *oura*, 'tail', an allusion to the spikelike inflorescence.

Lycurus is a genus of three species of open rocky slopes and mesas. It is native to two disjunct regions, one extending from Colorado and southern Utah to southern Mexico and Guatemala, the other from Colombia through western South America to west-central Argentina. One species is native to the Intermountain Region.

1. **Lycurus setosus** (Nutt.) C. Reeder BRISTLY WOLFSTAIL [p. 443, <u>509</u>]
Pl densely ces. **Clm** 30–50(60) cm, erect, with several nd, sparingly brchd, scabrous to puberulent at or near the nd. **Lig** (2)3–10(12) mm, hyaline, acuminate, smt shortly cleft on the sides (in dried specimens the fragile lig may appear shorter because of the folded tip); **bld** 4–9(13) cm long, 1–2 mm wide, glab, smooth or scabridulous abx, scabridulous or hispidulous adx, with prominent whitish midribs and scabrous mrg, midribs extending as fragile, easily broken, awnlike, (3)4–7(12) mm apc. **Pan** 4–8(10) cm long, (5)7–8 mm wide; shorter ped 0.8–1(1.5) mm; longer ped 1–2 mm. **Spklt** 3–4 mm. **Glm** 1–1.5(2) mm, scabrous apically; **lo glm** 2-veined, with (1)2(3) unequal scabridulous awns, shorter

awns 1–1.5 mm, longer awns (1)1.5–3(3.5) mm; **up glm** 1-veined, with a single, flexuous 2.5–4(5) mm awn; **lm** 3–4 mm, tapering to a scabrous 1.5–2(3) mm awn; **anth** 1.5–2 mm, yellowish. **Car** about 2 mm, brownish. $2n = 40$.

Lycurus setosus grows on rocky slopes and open mesas, at elevations of 570–3400 m. Its range extends from the southwestern United States to northern Mexico, and, as a disjunct, in Argentina and Bolivia. Its flowering time is July–October.

Lycurus setosus is sometimes confused with *Muhlenbergia wrightii*, which has a somewhat similar aspect but is normally found in moist habitats. Also, in *M. wrightii* the first glume is 1-veined with a very short awn, and the lemma is acuminate and unawned or with an awn no more than 1 mm long.

Peterson and Columbus (2008) transferred *Lycurus setosus* to *Muhlenbergia*, as *M. alopecuroides* (Griseb.) P.M. Peterson & Columbus. It is retained in *Lycurus* until the treatment of the other two species in *Lycurus* has been clarified.

10.19 **CHLORIS** Sw.
Mary E. Barkworth

Pl ann or per; habit various, rhz, stln, or ces. **Clm** 10–300 cm; **intnd** pith-filled. **Shth** strongly keeled, glab, scabrous, or pubescent; **lig** memb, erose to lacerate or ciliate, occ absent; **bld**, particularly those of the bas lvs, often with long, coarse hairs near the base of the adx surface and mrg. **Infl** tml, pan with (1)5–30 spikelike br, these usu borne digitately, occ in 2–several whorls, smt with a few isolated br below the pri whorl(s), all br usu exceeding the up lvs; **br** with spklt in 2 rows on 1 side of the br axes. **Spklt** solitary, sessile to pedlt, lat compressed, with 2–3(5) flt, usu only the lowest flt bisex, rarely the lo 2 flt bisex, remaining flt(s) strl or stmt; **flt** lat compressed or terete, cylindrical to obovoid, awned or unawned, strl and stmt flt progressively rdcd distally if more than 1 present; **dis** usu beneath the lowest flt in the spklt, all flt falling as a unit, smt at the uppermost cauline nd, pan falling intact. **Glm** unequal, exceeded by the flt, lanceolate, acute to acuminate, usu unawned, occ awned, awns to 0.3 mm; **cal** bearded; **lm of bisex flt** 3-veined, mrgl veins pubescent, midveins usu glab, smt scabrous, usu extending into an awn, smt merely mucronate, awns to 37 mm, lm apc truncate or obtuse, entire or bilobed, lobes, when present, smt awn-tipped, awns to 0.6 mm; **pal** shorter than the lm, 2-veined, veins scabrous; **anth** 3; **lod** 2. **Car** ovoid, elliptic, or obovoid. $x = (9)10$. Named for Chloris, the Greek goddess of flowers.

As interpreted here, *Chloris* is a tropical to subtropical genus of 55–60 species. It is most abundant in the Southern Hemisphere. Two of the three species treated here are native to North America, but may be introduced to the Intermountain Region. Many species of *Chloris*, including the native species, provide good forage.

Anderson (1974) interpreted *Chloris* as including *Enteropogon*, *Eustachys*, and *Trichloris*. Support for treating the genus more narrowly can be found in Van den Borre (1994), Van den Borre and Watson (2000), and Hilu and Alice (2001), but the limits of many genera in the *Cynodonteae*, including *Chloris*, are not clear.

1. Lowest lemmas with a conspicuous glabrous or pubescent groove on each side; plants annual or short-lived perennials . . . . 1. *C. pilosa*
1. Lowest lemmas not conspicuously grooved on the sides; plants perennial or annual.
   2. Panicle branches digitate, in a single terminal cluster . . . . 2. *C. virgata*
   2. Panicle branches borne in 2 or more, clearly distinct whorls . . . . 3. *C. verticillata*

1. **Chloris pilosa** Schumach. [p. 443, 509]

Pl ann or short-lived per; smt shortly stln. **Clm** 30–70(200) cm, erect or somewhat decumbent. **Shth** glab or sparsely to densely pilose; **bld** to 30 cm long, 2–10 mm wide, with coarse hairs behind the lig and on the lo portion of the mrg. **Pan** digitate, with 5–9 clearly distinct or easily separable br; **br** 3–5 cm, with 5–7 spklt per cm. **Spklt** barely imbricate, pale to dark gray, often mottled when mature, with 1 bisex and (1)2 strl flt. **Lo glm** 1.1–1.6 mm; **up glm** 1.9–2.3 mm, awned, awns to 0.3 mm; **lowest lm** 2.3–3.5 mm, broadly ovate or elliptic, keels gibbous, sides with a conspicuous glab or pubescent groove, mrg glab or appressed pubescent, apc awned, awns to 6 mm; **second flt** 1.5–2.2 mm, widened and inflated distally, mucronate or awned, awns to 3 mm; **distal flt** less than 1 mm, turbinate; **anth** 0.4–0.5 mm. **Car** 1.3–1.5 mm long, 0.5–0.6 mm wide, trigonous. 2*n* = 20, 30.

Chloris pilosa is native to equatorial Africa, but it is sometimes planted for forage. It has been collected, possibly from an experimental forage planting, in Grand County, Utah.

2. **Chloris virgata** Sw. FEATHER WINDMILL-GRASS, FEATHER FINGERGRASS [p. 443, 509]

Pl ann; usu tufted, occ stln. **Clm** 10–100+ cm. **Shth** usu glab; **lig** to 4 mm, erose or ciliate; **bld** to 30 cm long, to 15 mm wide, bas hairs to 4 mm, otherwise usu glab, occ pilose. **Pan** digitate, with 4–20, evidently distinct br; **br** 5–10 cm, erect to ascending, averaging 10 spklt per cm. **Spklt** strongly imbricate, with 1 bisex and 1(2) strl flt(s). **Lo glm** 1.5–2.5 mm; **up glm** 2.5–4.3 mm; **lowest lm** 2.5–4.2 mm, keels usu prominently gibbous, glab, or conspicuously pilose, sides not grooved, mrg glab, scabrous or pilose bas, with conspicuously longer hairs distally, hairs longer than 1.5 mm, lm apc not conspicuously bilobed, awned, awns 2.5–15 mm; **second flt** 1.4–2.9 mm long, 0.4–0.8 mm wide, somewhat widened distally, not inflated, bilobed, lobes less than ⅓ as long as the lm, awned from the sinuses, awns 3–9.5 mm; **third flt** greatly rdcd, unawned and shorter than the subtending rchl segment or absent but the rchl segment present. **Car** 1.5–2 mm long, about 0.5 mm wide, elliptic. 2*n* = 20, 26, 30, 40.

Chloris virgata is a widespread species that grows in many habitats, from tropical to temperate areas with hot summers, including much of the United States. It is a common weed in alfalfa fields of the southwestern United States.

3. **Chloris verticillata** Nutt. TUMBLE WINDMILL-GRASS [p. 444, 509]

Pl per; ces. **Clm** 14–40 cm, erect or decumbent, smt rooting at the lo nd. **Shth** mostly glab, with hairs to 3 mm adjacent to the lig; **lig** 0.7–1.3 mm, shortly ciliate; **bld** to 15 cm long, 2–3 mm wide, with bas hairs, otherwise glab or scabrous. **Pan** with 10–16, evidently distinct br in several well-separated whorls, and a solitary, vertical tml br; **br** 5–15 cm, spklt-bearing to the base, with 4–7 spklt per cm; **dis** at the uppermost cauline nd, pan falling intact. **Spklt** with 1 bisex and 1 strl flt. **Lo glm** 2–3 mm; **up glm** 2.8–3.5 mm; **lowest lm** 2–3.5 mm long, 1.5–1.9 mm wide, elliptic to lanceolate, keels glab or appressed pubescent, sides not conspicuously grooved, mrg glab or appressed pubescent, acute to obtuse, awned, awns 4.8–9 mm; **second flt** 1.1–2.3 mm, oblong, somewhat inflated, truncate, not or inconspicuously bilobed, lobes less than ⅓ as long as the lm, midveins excurrent, forming 3.2–7 mm awns. **Car** 1.3–1.5 mm long, about 0.5 mm wide, elliptic. 2*n* = ca. 28, 40, 63.

Chloris verticillata is a common weed of roadsides, lawns, and waste areas in the central United States that has been introduced to the Intermountain Region. Prior to disruption of the native vegetation, it grew in low areas of the central prairies.

10.20 **SCHEDONNARDUS** Steud.   Neil Snow

Pl per; ces. **Clm** 8–55 cm, smt geniculate and brchd bas, usu curving distally; **intnd** minutely retrorsely pubescent, mostly solid. **Lvs** mostly bas; **shth** compressed-keeled, closed, glab, mrg scarious; **lig** 1–3.5 mm, memb, lanceolate; **bld** 1–12 cm long, 0.7–2 mm wide, stiff, usu folded, often spirally twisted, midrib well-developed, mrg thick and whitish. **Infl** tml, pan of widely spaced, rcmly arranged, spikelike br, exceeding the up lvs; **br** strongly divergent, with distant to slightly imbricate, closely appressed spklt. **Spklt** 3–5.5 mm, mostly sessile, compressed lat, with 1 flt; **flt** bisex; **dis** at the base of the pan and above the glm. **Glm** lanceolate, unequal, 1-veined; **lm** usu exceeding the glm, 3-veined, unawned or shortly awned; **pal** subequal to the lm; **anth** 3; **sty** 2. **Car** fusiform. *x* = 10. Name from the Greek *schedon*, 'near'; Steudel considered *Schedonnardus* to be closely related to *Nardus*.

Schedonnardus is a monotypic North American genus that grows in the prairies and central plains of Canada, the United States, and northwestern Mexico. It is only distantly related to *Nardus*, which is a basal member of the *Pooideae* (Bouchenak-Khelladi et al., 2008).

1. Schedonnardus paniculatus (Nutt.) Trel.

TUMBLEGRASS [p. *444*, <u>509</u>]

Pan 5–50 cm, rchs becoming curved; **br** 2–8(16) cm. **Lo glm** 1.5–3 mm; **up glm** 1.5–4(5.5) mm; **lm** 3–5 mm; **anth** 0.7–1.4 mm. **Car** 2.5–3.5 mm. 2n = 20, 30.

Schedonnardus paniculatus is frequently found in disturbed areas. At maturity, the panicle breaks at the base and functions as a tumbleweed for seed dispersal.

10.21 CYNODON Rich.

Mary E. Barkworth

Pl per; smt stln, smt also rhz, often forming dense turf. **Clm** 4–100 cm. **Shth** open; **aur** absent; **lig** of hairs or memb; **bld** flat, conduplicate, convolute, or involute, smt dis. **Infl** tml, digitate or subdigitate pan of spikelike br; **br** (1)2–20, 1-sided, with 2 rows of solitary, subsessile, appressed, imbricate spklt. **Spklt** lat compressed, with 1(–3) flt, only the lowest flt fnctl; **rchl extension** usu present, smt terminating in a rdcd flt; **dis** above the glm. **Glm** usu shorter than the lm, memb, keeled, usu muticous; **lo glm** 1-veined; **up glm** 1–3-veined, occ shortly awned; **lm** memb to cartilaginous, 3-veined, keeled, keels with hairs, occ winged, apc mucronate or muticous; **pal** about as long as the lm, 2-keeled; **anth** 3; **sty br** 2, plumose; **lod** 2. x = 9. Name from the Greek *kyon*, 'dog', and *odous*, 'tooth', a reference to the sharp, hard scales of the rhizomes.

Cynodon is a genus of nine species, all of which are native to tropical regions of the Eastern Hemisphere. Several species are used as lawn and forage grasses in tropical and warm-temperate regions. The most widespread species, *C. dactylon*, is also the most frequently encountered species in North America. It is used for lawns, putting greens, and pastures in southern portions of the region, but is generally considered a weed in other parts.

Many cultivars of *Cynodon* have been developed. These cultivars may exhibit combinations of features that are not found in the wild species, making it difficult to accommodate them in a key. Here we treat only the commonly encountered species. Those interested in a more complete treatment should consult Barkworth et al. (2003).

1. Cynodon dactylon (L.) Pers BERMUDAGRASS

[p. *445*, <u>509</u>]

Pl stln, usu also rhz. **Clm** 5–40(50) cm, not becoming woody. **Shth** glab or with scattered hairs; **col** usu with long hairs, particularly at the mrg; **lig** about 0.5 mm, of hairs; **bld** 1–6(16) cm long, (1)2–4(5) mm wide, flat at maturity, conduplicate or convolute in bud, glab or the adx surfaces pilose. **Pan** with (2)4–6(9) br; **br** 2–6 cm, in a single whorl, axes triquetrous. **Spklt** 2–3.2 mm. **Lo glm** 1.5–2 mm; **up glm** 1.4–2.3 mm; **lm** 1.9–3.1 mm, keels not winged, pubescent, mrg usu less densely pubescent; **anth** dehiscent at maturity; **pal** glab. 2n = 18, 36.

Cynodon dactylon is a variable species, but taxonomists disagree on just how variable. Caro and Sánchez (1969) limited *C. dactylon* to plants with conduplicate leaves, placing those with convolute leaves in a number of other species, such as *C. affinis* Caro & Sánchez and *C. aristiglumis* Caro & Sánchez; de Wet and Harlan (1970) did not mention this character in their study of *Cynodon*. Caro and Sánchez also employed several other characters in the key separating *C. dactylon* from the species with convolute immature leaves, but the overlap between the two sides of the lead is substantial. Pending further study, the broader interpretation, in which *C. dactylon*

includes plants with both convolute and conduplicate leaves, has been adopted.

Several varieties of *C. dactylon* have been described, in addition to which numerous cultivars have been developed, some as turf grasses for lawns or putting greens, others as pasture or forage grasses. Their useful range is limited because *C. dactylon* is not cold hardy, going dormant and turning brown when nighttime temperatures fall below freezing or average daytime temperatures are below 10°C.

Cynodon dactylon is considered a weed in many countries and it is true that, once established, it is difficult to eradicate. It does, however, have some redeeming values. It is rich in vitamin C, and its leaves are sometimes used for an herbal tea. It is claimed to have various medicinal properties, but these have not been verified. It is considered a good pasture grass, in addition to which it is sometimes grown as an ornamental and for erosion control on exposed soils.

The most commonly encountered variety, both in North America and in other parts of the world, is *C. dactylon* var. *dactylon*, largely because it thrives in severely disturbed, exposed sites; it does not invade natural grasslands or forests. Determining how many other varieties are established in North America is almost impossible, because there has been no global study of variation in the species. The presence of numerous cultivars complicates an already difficult problem. For most purposes, it is probably neither necessary nor feasible to identify the variety of *C. dactylon* encountered.

10.22 SPARTINA Schreb.

Mary E. Barkworth

Pl per; ces from knotty bases or rhz. **Clm** 10–350 cm, erect, terete, solitary or in small to large clumps. **Lvs** mostly cauline; **shth** open, smooth, smt striate; **lig** memb, ciliate, cilia longer than the memb bases; **bld** flat or involute. **Infl** tml, usu exceeding the up lvs, 3–70 cm, pan of 1–75 spikelike br attached to an elongate rchs; **br** rcmly arranged, alternate, opposite, or whorled, appressed to strongly divergent, axes 3-sided, spklt usu sessile on the 2 lo sides, usu divergent to strongly divergent; **dis** beneath the glm. **Spklt** lat compressed, with 1 flt. **Glm** unequal, strongly keeled; **lo glm** shorter than the flt, 1-veined; **up glm** usu longer than the flt, 1–6-veined; **lm** shorter than the pal, 1–3-veined, midveins keeled, lat veins usu obscure; **pal** thin, papery, 2-veined, obscurely keeled; **anth** 3; **lod** smt present, truncate, vascularized; **sty** 2, plumose. **Car** rarely produced. $x = 10$. Name from the Greek *spartine*, a cord made from *Spartium junceum*, probably applied to *Spartina* because of the tough leaves.

Spartina is a genus of 15–17 species, most of which grow in moist to wet, saline habitats, both coastal and interior. Reproduction of all the species is almost entirely vegetative. There are two species in the Intermountain Region.

Dr. Loran Anderson (in a letter to Barkworth, 2008) reported that inflorescences of *Spartina* may be unisexual or have unisexual inflorescence branches. It is not known how general this phenomenon is.

1. Rhizomes whitish; upper glumes with all lateral veins on the same side of the keels;
   panicles with 3–12 branches, the branches 1.5–8 cm long . . . . 1. *S. gracilis*
1. Rhizomes light brown to brownish purple; upper glumes with lateral veins on either side
   of the keels; panicles with 5–50 branches, the branches 1.5–15 cm long . . . . 2. *S. pectinata*

1. **Spartina gracilis** Trin. ALKALI CORDGRASS [p. 445, 509]

Pl strongly rhz; rhz elongate, 1.5–5 mm thick, whitish, scales not inflated, closely imbricate. **Clm** 40–100 cm tall, 2–3.5 mm thick, usu solitary, erect, terete, indurate, glab. **Shth** smooth or striate, mostly or completely glab, throats occ ciliate; **lig** 0.5–1 mm; **bld** 6–30 cm long, 2.5–8 mm wide, flat, becoming involute, abx surfaces glab, adx surfaces scabrous, mrg scabrous. **Pan** 8–25 cm, not smooth in outline, with 3–12 br; **br** 1.5–8 cm, alternate, differing only slightly in length and spacing within a pan, usu appressed, rarely spreading, with 10–30 spklt. **Spklt** 6–11 mm, ovate to lanceolate. **Glm** with glab or sparingly hispidulous mrg, apc acute or mucronate; **lo glm** 3–7 mm, sides narrow, glab or sparsely pubescent, keels glab or strigose; **up glm** 6–10 mm, usu equaling the flt, keels strigose, hairs 0.2–0.5 mm, lat veins 2, inconspicuous, both on the same side of the keel; **lm** glab or sparsely hirsute, keels hirsute, at least distally, hairs 0.3–1 mm, mrg sparsely hairy, apc obtuse to rounded, smt obscurely lobed; **pal** sparsely hispid distally, obtuse to slightly rounded; **anth** 2.5–5 mm, well-filled, dehiscent at maturity. $2n = 40, 42$.

Spartina gracilis is found on the margins of alkaline lakes and along stream margins and river bottoms. Its range extends from the southern portion of the Northwest Territories, Canada, to central Mexico.

2. **Spartina pectinata** Link PRAIRIE CORDGRASS [p. 446, 509]

Pl strongly rhz; rhz elongate, (2)3–8 mm thick, purplish brown or light brown (drying white), scales closely imbricate. **Clm** to 250 cm tall, 2.5–11 mm thick, solitary or in small clumps, indurate. **Shth** mostly glab, throats often pilose; **lig** 1–3 mm; **bld** 20–96 cm long, 5–15 mm wide, flat when fresh, becoming involute when dry, glab on both surfaces, mrg strongly scabrous, bld of the second lf below the pan 32–96 cm long, 5–14 mm wide, usu involute. **Pan** 10–50 cm, not smooth in outline, with 5–50 br; **br** 1.5–15 cm, appressed to somewhat spreading, with 10–80 spklt. **Spklt** 10–25 mm. **Glm** shortly awned, glab or sparsely hispidulous; **lo glm** 5–10 mm, from ¾ as long as to equaling the adjacent lm, keels hispid, apc awned; **up glm** 10–25 mm (including the awn), exceeding the flt, glab or sparsely hispid, keels scabrous to hispid, trichomes about 0.3 mm, lat veins usu glab (rarely hispid), on either side of, and close to, the keels, apc awned, awns 3–8 mm; **lm** glab, keels pectinate distally, apc bilobed, lobes 0.2–0.9 mm; **anth** 4–6 mm, well-filled, dehiscent. $2n = 40, 40+1, 42, 70, 80, 84$.

Spartina pectinata is native to Canada and the United States, but it has been introduced at scattered locations on other continents. In western North America, it grows in both wet and dry soils, including dry prairie habitats and along roads and railroads. L. Anderson (pers. comm.) observed pistillate inflorescences on plants in Cache Valley, Utah.

10.23 **BOUTELOUA** Lag.

J.K. Wipff

Pl ann or per; synoecious or dioecious; habit various, ces, stln, or rhz. **Clm** 1–80 cm. **Lvs** usu mostly bas; **shth** open; **lig** of hairs, memb, or memb and ciliate. **Infl** tml, pan of 1–80 solitary, spikelike br, usu exceeding the up lvs, if pist, smt partially hidden in a leafy shth; **br** 4–50(75) mm, not woody, usu 1-sided and elongate, usu rcm on elongate rchs, smt pist and burlike, smt digitate or subdigitate, with 1–130+ sessile to subsessile spklt in 2 rows, axes terminating in a spklt or extending beyond the base of the distal spklt. **Spklt** closely imbricate, appressed to pectinate, lat compressed or terete, with 1–2(3) flt, lowest flt in each spklt usu bisex, distal flt stmt or strl, smt unisex; **dis** at the base of the br or above the glm. **Glm** unequal or subequal, 1 or both glm equaled or exceeded by the distal flt,

1-veined, acute or acuminate, smt shortly awned; **lo glm** usu shorter than the lowest flt; **up glm** of pist spklt smt globose, white, indurate; **lm of lowest flt** entire, bilobed, trilobed, or 4-lobed, 3-veined, veins usu extended into 3 teeth or awns; **pal of lowest flt** 2-veined, veins smt excurrent; **distal flt(s)** stmt or strl, varying from similar to the lowest flt in shape, size, and venation to strl and rdcd to an awn column with well-developed awns or to a flabellate scale. $x = 10$. Named for the brothers Claudio (1774–1842) and Esteban (1776–1813) Boutelou Agraz, Spanish botanists.

Bouteloua, a genus of the Western Hemisphere, has about 40 species; all 11 species treated here are native to the Intermountain Region. Two taxa that are particularly important in North America are *B. curtipendula* and *B. gracilis*. These were major constituents of the shortgrass prairie that once covered the drier portions of the Great Plains. Both are excellent forage species. Irrigation has converted much of the area they once occupied to agricultural use, but large areas of *Bouteloua* grasslands remain.

An editorial decision was made to include *Buchloë* Engelm. in *Bouteloua* (see Columbus 1999).

1. Plants unisexual; pistillate panicles with burlike branches. . . . . 12. *B. dactyloides*
1. Plants bisexual; panicle branches not burlike.
  2. Panicle branches deciduous, disarticulation occurring at their bases; spikelets usually 1–15 per branch, usually appressed rather than pectinate.
    3. All or most panicle branches with 1 spikelet . . . . 2. *B. uniflora*
    3. All or most panicle branches with 2–15 spikelets.
      4. First (proximal) spikelet on each branch with 1 floret, the remaining spikelets with 2 florets; plants annual; panicles with 1–15 branches. . . . . 3. *B. aristidoides*
      4. Spikelets all alike or with 2 or more florets; plants annual or perennial; panicles with 1–80 branches.
        5. Second florets sterile, usually rudimentary, usually without paleas; central awns 1.5–7 mm long; panicles with 5–40 branches. . . . . 1. *B. curtipendula*
        5. Second florets bisexual, pistillate or staminate, with well-developed paleas; central awns 4–10 mm long; panicles with 7–12 branches. . . . . 4. *B. repens*
  2. Panicle branches persistent; disarticulation above the glumes; spikelets 6–130 or more per branch, pectinate.
    6. Upper glumes of at least some spikelets with papillose-based hairs.
      7. Panicle branches extending beyond the base of the terminal spikelets. . . . . 6. *B. hirsuta*
      7. Panicle branches terminating in a spikelet.
        8. Plants tufted annuals or short-lived stoloniferous perennials; panicle branches 4–8, the axes with papillose-based hairs; lowest lemmas 3–4 mm long. . . . . 11. *B. parryi*
        8. Plants perennial, often shortly rhizomatous; panicle branches 1–3(6), the axes scabrous, never with papillose-based hairs; lowest lemmas 3.5–6 mm long . . .
        . . . . 5. *B. gracilis* (in part)
    6. Upper glumes glabrous, scabrous, or hairy, but the hairs not papillose-based.
      9. Lower cauline internodes woolly-pubescent. . . . . 7. *B. eriopoda*
      9. Lower cauline internodes glabrous or mostly so, sometimes pubescent immediately below the nodes.
        10. Central awns of lemmas not flanked by membranous lobes . . . . 8. *B. trifida*
        10. Central awns of lemmas flanked by 2 membranous lobes.
          11. Lowest paleas in the spikelets awned, awns 1–2 mm long; panicles with 2–11 branches . . . . 9. *B. barbata*
          11. Lowest paleas in the spikelets unawned, but the veins sometimes excurrent for less than 1 mm; panicles with 1–6 branches.
            12. Plants annual . . . . 10. *B. simplex*
            12. Plants perennial . . . . 5. *B. gracilis* (in part)

1. Bouteloua curtipendula (Michx.) Torr.

Sideoats Grama [p. 446, <u>509</u>]

Pl per; ces or not, with or without rhz. **Clm** 8–80 cm, erect or decumbent, solitary or in small to large groups. **Lvs** evenly distributed; **shth** mostly glab, smt with hairs distally; **lig** 0.3–0.5 mm, memb, ciliate; **bld** 2–30 cm long, (1.4)2.5–7 mm wide, at least some over 2.5 mm wide, flat or folded when dry, usu smooth abx and scabrous adx, occ pubescent, bases usu with papillose-based hairs on the mrg. **Pan** 13–30 cm, secund, with (12)30–80 reflexed br; **br** (5)10–30(40) mm, deciduous, with (1)2–7(15) spklt, axes terminating 3–5 mm beyond the base of the tml spklt, apc entire; **dis** at the base of the br. **Spklt** appressed, all alike, with 1 bisex and 1–2 strl, rdmt flt. **Glm**

unequal, glab or scabrous; **lo glm** 2.5–6 mm, ½ or more as long as the up glm; **up glm** 5.5–8 mm; **lowest lm** 3–6.5 mm, glab or scabrous-strigose, often minutely rugose, acute or inconspicuously 3-lobed, 3-veined, veins usu extending as short mucros or awns to 6 mm; **cent mucros** or **awns** not flanked by memb lobes; **lowest pal** acute, unawned; **anth** 1.5–3.5 mm, yellow, orange, red, or purple; **distal flt(s)** 0.4–3.5 mm, strl, variable, usu a glab lm having a short memb base, no pal, and 3 unequally developed awns, cent awns 1.5–7 mm. 2*n* = (20), 40, 41–103.

Bouteloua curtipendula is a common, often dominant or co-dominant species in open grasslands and wetlands of the drier portions of the central grasslands of North America. It is highly regarded as a forage species and is also an attractive ornamental. Its range extends from southern Canada through Mexico and Central America to western South America.

Two of its three varieties grow in the Intermountain Region; the third, *B. curtipendula* var. *tenuis* Gould & Kapadia, is endemic to Mexico.

1. Plants not long-rhizomatous, bases sometimes knotty with short rhizomes; culms in large or small clumps . . . . var. *caespitosa*
1. Plants long-rhizomatous; culms solitary or in small clumps . . . . var. *curtipendula*

Bouteloua curtipendula var. **caespitosa** Gould & Kapadia [p. 446]

Pl ces, often with a knotty base, not or shortly rhz. **Clm** in large or small clumps, stiffly erect. **Bld** usu narrow, but at least some over 2.5 mm wide. **Pan** with 12–80 br, averaging 2–7 spklt per br. **Glm** and **lm** bronze or stramineous to green, or various shades of purple; **anth** usu yellow or orange, occ red or purple. 2*n* = 58–103.

Bouteloua curtipendula var. *caespitosa* grows on loose, sandy or rocky, well-drained limestone soils at 200–2500 m in the southwestern United States, Mexico, and South America.

Bouteloua curtipendula (Michx.) Torr. var. **curtipendula** [p. 446]

Pl not ces, with long rhz. **Clm** solitary or in small clumps. **Bld** 3–7 mm, flat. **Pan** with 40–70 br, averaging 3–7 spklt per br. **Glm** and **lm** typically purple or purple-tinged; **anth** red or red-orange, infrequently yellow, orange, or purple. 2*n* = 40, 41–66.

Bouteloua curtipendula var. *curtipendula* is the common variety of *B. curtipendula* in most of the Intermountain Region. It grows on rich, loamy, well-drained prairie soils, at 100–2500 m.

2. Bouteloua uniflora Vasey NEALLY'S GRAMA [p. 447, <u>510</u>]

Pl per; ces, without rhz or stln. **Clm** 20–60 cm, stiffly erect, glab. **Shth** mostly glab, a few long hairs present near the lig; **lig** 0.2–0.5 mm, of hairs; **bld** 6–16 cm long, 1–2 mm wide, involute when dry, glab, bases usu with papillose-based hairs on the mrg. **Pan** 5–10(14) cm, with 15–70 br; **br** 5–9 mm, deciduous, scabrous, with 1 spklt (lo br occ with 2 spklt), axes extending 3–4 mm beyond the tml spklt, apc entire; **dis** at the base of the br. **Spklt** appressed, with 1 bisex and 0–1 rdmt flt. **Glm** acute to slightly cleft and minutely apiculate, midveins usu scabrous; **lo glm** 2.5–4 mm; **up glm** 6.2–8 mm, mostly smooth, midveins usu scabrous; **lowest lm** 6–7.5 mm, acute or minutely cleft, glab, unawned, smt mucronate; **lowest pal** unawned, glab; **anth** 2.5–3 mm, bright yellow; **second flt** absent or rdcd to 1 or 3 short awns, glab. **Car** about 3 mm. 2*n* = 20.

Bouteloua uniflora grows primarily in fertile, rocky, limestone soils of Texas and adjacent Coahuila, Mexico at 300–1000 m. A disjunct collection has been reported from Zion National Park, Utah. Plants in the United States belong to *B. uniflora* Vasey var. *uniflora*.

3. Bouteloua aristidoides (Kunth) Griseb. [p. 447, <u>510</u>]

Pl ann; tufted. **Clm** 4–60 cm, outer clm of a tuft decumbent, smt geniculate, brchd at the lo nd. **Lig** 0.2–0.5 mm, memb, lacerate or ciliate; **bld** 2–5(9) cm long, 0.7–2 mm wide, flat or folded, adx surfaces smt with papillose-based hairs, mrg usu with papillose-based hairs near the lig. **Pan** 2.5–10.5 cm, with (1)4–15 br; **br** 5–45 mm, deciduous, densely pubescent (at least bas), with 2–10 spklt per br, axes extending 2–10 mm beyond the base of the tml spklt, apc entire; **dis** at the base of the br, the break forming a sharp tip. **Spklt** appressed. **Proximal spklt on each br** with 1 flt; **lo glm** 1.5–3.5 mm, glab, narrow to subulate; **up glm** 5.5–6.2 mm, densely pubescent, at least on the bas ½; **lm** 5.8–6 mm, acuminate, unawned; **lowest pal** almost as long as the lm, bifid, glab; **rchl** prolonged beyond the flt for about 0.5 mm. **Distal spklt** with 1 bisex and 1 rdmt flt; **glm** unequal, glab, minutely scabrous on the keels; narrowly acute or acuminate; **lo glm** 1.5–2 mm; **up glm** 5–6 mm, glab or sparsely pubescent bas, often divergent; **lowest lm** 6–8 mm, veins pubescent, lat veins excurrent as short (to 1 mm) awns, acuminate, midvein extended into a setaceous tip or a short awn; **lowest pal** 5–7 mm, bifid, veins often excurrent as short awns; **anth** about 2.5 mm, yellow or yellow and red; **distal flt** rdcd to a pubescent, 3-awned, awn column, awns 2–7 mm, exserted. **Car** 2.5–3 mm. 2*n* = 40.

There are two varieties, one of which grows in the Intermountain Region.

Bouteloua aristidoides (Kunth) Griseb. var. **aristidoides** NEEDLE GRAMA [p. *447*]
Pan br 5–16 mm to the base of the tml spklt and extending an additional 6–10 mm, with 2–5 spklt.

Bouteloua aristidoides var. *aristidoides* grows in dry mesas, plains, and washes from near sea level to about 2000 m. It matures rapidly following summer rains, and can be abundant over large areas within its range, which extends from California to western Texas and Mexico.

4. **Bouteloua repens** (Kunth) Scribn. & Merr.
SLENDER GRAMA [p. *448*, 510]
Pl per; ces, usu not dense, hard, or knotty, without rhz or stln. Clm 15–65 cm, erect, geniculate, or decumbent, smt rooting at the lo nd, usu brchg from the aerial nd. Shth glab or pubescent; lig 0.2–0.3 mm, memb, ciliate; bld 5–20 cm long, 1–5 mm wide, bases with papillose-based hairs on the mrg, both surfaces glab or pubescent. Pan 4–14 cm, with (3)7–12 br; br 10–20 mm, with 2–8 spklt, extending 4–6 mm beyond the base of the tml spklt, apc entire; dis at the base of the br. Spklt appressed, all alike, with 1 bisex and 1 stmt (rarely rdmt) flt. Glm glab, veins scabrous or strigose; lo glm 4–7 mm; up glm 4–9 mm, mostly glab, smt scabrous or strigose over the veins, apc acute, unawned or awn-tipped, awns about 1 mm; lowest lm 4.5–8 mm, usu glab, rarely pubescent bas, 3-awned, awns wide bas, cent awns slightly longer than the lat awns, often flanked by 2 memb 0.5–1.5 mm lobes; lowest pal 6–8 mm, bilobed, often shortly 2-awned; anth 3–5.5 mm, usu orange or yellow, occ red or purple; second lm 5.5–7 mm, glab, 3-awned, cent awns 4–10 mm, often flanked by memb lobes, lat awns 2–10 mm; second pal 4–7 mm; anth smaller than those of the lowest flt; rchl prolonged beyond the second flt as a short bristle. Car 3–4 mm. 2*n* = 20, 40, 60.

Bouteloua repens grows in open, usually hilly terrain on many soil types, from sandy ocean shores to montane slopes, reaching elevations of 2500 m. Its native range extends from the southwestern United States through the Caribbean islands, Mexico, and Central America to Colombia and Venezuela.

5. **Bouteloua gracilis** (Kunth) Lag. *ex* Griffiths
BLUE GRAMA, EYELASH GRASS [p. *448*, 510]
Pl per; usu densely ces, often with short, stout rhz. Clm 24–70 cm, not woody bas, erect, geniculate, or decumbent and rooting at the lo nd, not brchd from the aerial nd; nd usu 2–3, glab or puberulent; lo intnd glab. Lvs mainly bas; shth glab or sparsely hirsute; lig 0.1–0.4 mm, of hairs, often with mrgl tufts of long hairs; bld 2–12(19) cm long, 0.5–2.5 mm wide, flat to involute at maturity, hairs usu present bas. Pan with 1–3(6) br, these rcm on 2–8.5(12.5) cm rchs or digitate; br 13–50(75) mm, persistent, arcuate, scabrous, without papillose-based hairs, with 40–130 spklt, terminating in a spklt; dis above the glm. Spklt pectinate, with 1 bisex and 1 rdmt flt. Glm mostly glab or scabrous, midveins smt with papillose-based hairs; lo glm 1.5–3.5 mm; up glm 3.5–6 mm; lowest lm 3.5–6 mm, pubescent at least bas, 5-lobed, cent and lat lobes veined and awned, awns 1–3 mm, cent awns flanked by 2 memb lobes; lo pal about 5 mm, shallowly bilobed, veins excurrent for less than 1 mm; rchl intnd subtending second flt with a distal tuft of hairs; anth 1.7–2.9 mm, yellow or purple; up flt strl, 0.9–3 mm, lobed almost to the base, lobes rounded, 3-awned, awns equal, 1–3 mm. Car 2.5–3 mm long, about 0.5 mm wide. 2*n* = 20, 28, 35, 40, 42, 60, 61, 77, 84.

Bouteloua gracilis grows in pure stands in mixed prairie associations and disturbed habitats from Canada to Mexico, usually on rocky or clay soils and mainly at elevations of 300–3000 m. It is an important native forage species and also an attractive ornamental.

6. **Bouteloua hirsuta** Lag. [p. *448*, 510]
Pl per; densely or loosely ces, occ stln. Clm 15–75 cm, erect or decumbent, smt brchd bas, smt brchd aerially; nd 3–6; intnd glab or sparsely to densely pubescent with papillose-based hairs. Lvs bas or mainly cauline; shth mostly glab, finely scabrous, or pubescent, pilose near the lig; lig 0.2–0.5 mm, of hairs; bld 1–30 cm long, 1–2.5 mm wide, flat to involute, papillose-based hairs often present on both surfaces, usu present on the bases of the mrg. Pan usu with 0.7–18 cm rchs bearing 1–6 br, the br smt digitate; br 10–40 mm, persistent, straight, with 20–50 spklt, axes extending 5–10 mm beyond base of the tml spklt; dis above the glm. Spklt pectinate, green to dark purple, with 1 bisex flt and 1–2 rdmt flt. Glm acuminate or awn-tipped; lo glm 1.4–3.5 mm; up glm 3–6 mm, midveins with papillose-based hairs; lowest lm 2–4.5 mm, pubescent, 1–3-awned, cent (or only) awns 0.2–2.5 mm, not flanked by memb lobes, lat lobes acuminate, unawned or with awns no longer than the cent awn; lo pal ovate, unawned; anth 2–3.4 mm, cream or yellow; rchl intnd subtending second flt glab or pubescent, smt with a distal tuft of hairs; second lm 0.5–2 mm, bilobed, 3-awned, awns 2–4(6) mm; third lm, if present, minute, memb scales, glab. Car 1.5–2.6 mm. 2*n* = 20, 40, 50, 60; numerous dysploid numbers also reported.

Bouteloua hirsuta is a widespread species, one subspecies of which grows in the Intermountain Region.

Bouteloua hirsuta Lag. subsp. hirsuta HAIRY GRAMA [p. *448*]

Pl loosely or densely ces, smt stln. Clm 15–60 cm, usu decumbent and brchd bas, smt erect, brchd or unbrchd from the aerial nd; nd usu 4–6; intnd glab or sparsely to densely pubescent with papillose-based hairs. Lvs bas clustered, smt not strongly so; shth glab or pubescent, hairs not papillose-based, smt scabrous. Pan with 1–4 br on 0.7–7.5(9.2) cm rchs or digitate; br 1–4. Anth 2–2.5 mm; rchl intnd subtending second flt without a distal tuft of hairs. Car 1.4–2 mm. $2n$ = 20, 40, 50, 60; numerous dysploid numbers also reported.

Bouteloua hirsuta subsp. *hirsuta* grows from open plains to slightly shaded openings in woods and brush on well-drained, often rocky, soils at 50–300 m. Its range extends from North Dakota and Minnesota to central Mexico. In the northern portion of its range, it is not densely tufted and the culms are decumbent and branched; in the southwestern United States and northern Mexico, it grows in isolated, dense clumps, with erect, stout, unbranched culms and mostly basal leaves.

7. Bouteloua eriopoda (Torr.) Torr. BLACK GRAMA [p. *449*, <u>*510*</u>]

Pl per; often shortly rhz, stln, stln long, densely woolly-pubescent. Clm 20–60(75) cm, wiry, decumbent, rooting at the lo nd; lo intnd densely woolly-pubescent. Shth mostly glab or sparsely pilose, usu pilose near the lig; lig 0.1–0.4 mm, of hairs; bld 2.5–6 cm long, 0.5–2 mm wide, scabrous adx, mrg with papillose-based hairs bas. Pan (1)2–16 cm, with (1)2–8 br; br 14–50 mm, persistent, densely woolly-pubescent bas, with 8–18 spklt, axes terminating in entire, smt scarious apc; dis above the glm. Spklt pectinate, with 1 bisex flt and 1 rdmt flt. Glm unequal, smooth or scabrous; lo glm 2–4.5 mm; up glm 4.5–8(9) mm, glab, scabrous, or with hairs, hairs to 0.5 mm, not papillose-based; lo lm 4–7 mm, pubescent bas, glab or sparsely puberulent distally, acuminate, cent awns 0.5–4 mm, lat awns absent or shorter than 1 mm; lo pal acuminate, unawned; anth 1.5–3 mm, yellow to orange; rchl segment to second flt about 2 mm, with a distal tuft of hairs; up flt rdmt, an awn column terminating in 3 awns of 4–9 mm. Car 2.5–3 mm. $2n$ = 20, 21, 28.

Bouteloua eriopoda grows on dry plains, foothills, and open forested slopes, often in shrubby habitats, and also in waste ground. It is usually found between 1000–1800 m, but extends to 2500 m. Once a dominant in much of its range, under heavy grazing *B. eriopoda* persists only where protected by shrubs or cacti because it is highly palatable. Its range extends from the southwestern United States to northern Mexico.

8. Bouteloua trifida Thurb. *ex* S. Watson RED GRAMA [p. *449*, <u>*510*</u>]

Pl per; ces, older pl occ shortly rhz. Clm 5–40 cm, slender, wiry, erect or slightly geniculate at the lo nd; lo intnd glab, shorter than those above. Lvs mostly bas; shth glab, smt scabridulous, becoming flattened, persistent; lig 0.2–0.5 mm, of hairs; bld 0.7–8 cm long, 0.5–1.5(2) mm wide, scabridulous, mrg often with papillose-based hairs bas. Pan 3–9 cm, with 2–7 br; br 7–25 mm, persistent, spreading, ascending, or appressed, straight to slightly arcuate, with 8–24(32) spklt, axes terminating in a spklt; dis above the glm. Spklt appressed to pectinate, reddish purple, with 1 bisex flt and 1 rdmt flt. Glm bilobed; lo glm 1.7–3.4 mm, slightly shorter than the up glm, veins excurrent to 0.6 mm; up glm 1.9–4 mm, glab or pubescent, hairs not papillose-based, veins excurrent to 1 mm; lo lm 1.2–2.2 mm, glab, sparsely appressed-pubescent along the veins or densely appressed-pubescent for much of their length and on the mrg, trilobed, lobes veined, tapering into 3 awns, awns 2.2–6.6 mm, cent awns not flanked by memb lobes; anth 0.2–0.4 mm, yellow; rchl intnd glab; up flt glab, of 3 equal awns, awns 2–7 mm. Car 0.8–1.5 mm long, 0.3–0.6 mm wide. $2n$ = 20.

Bouteloua trifida grows on dry open plains, shrubby hills, and rocky slopes, at 2200–2500 m. Its range extends from the southwestern United States to central Mexico. It is a drought-resistant species that is sometimes mistaken for *Aristida* because of its delicate, cespitose growth habit and purplish, 3-awned spikelets. Juvenile plants may also be confused with *B. barbata* but that species is annual, has a central awn flanked by two membranous lobes, and its lowest paleas are 4-lobed and 2-awned. One variety of *Bouteloua trifida* grows in the Intermountain Region.

Bouteloua trifida Thurb. *ex* S. Watson var. trifida [p. *449*]

Pan br appressed to ascending, occ divergent. Lo glm 2.2–3.4 mm, midveins excurrent for 0.1–0.6 mm; up glm 2.7–4 mm, midveins excurrent for 0.2–1 mm; lo lm glab or sparsely appressed pubescent along both sides of the veins, awns (3.2)4–6.6 mm; anth 0.3–0.4 mm. Car 1.3–1.5 mm long, 0.4–0.5 mm wide, grooved adx. $2n$ = unknown.

Bouteloua trifida var. *trifida* grows in dry plains and rocky slopes, mostly at 300–1500 m, from southern California, Nevada, and Utah to Texas and Mexico.

9. Bouteloua barbata Lag. [p. *449*, <u>*510*</u>]

Pl ann or short-lived per; tufted, smt with stln. Clm 1–75 cm, prostrate, decumbent, or erect, smt rooting at the lo nd; lo intnd glab. Lvs bas or cauline; shth usu glab, except for tufts of long hairs on either side of the col; lig 0.1–1 mm, memb, ciliate; **bld** 0.5–10 cm long, 0.7–4 mm wide, adx surfaces usu sparsely pubescent with a few papillose-based hairs bas. **Pan** 0.7–25 cm, with (2)4–9(11) br; **br** 10–30 mm, persistent, straight to arcuate, glab, scabridulous, or with papillose-based hairs, with 20–55 spklt, axes terminating in a well-developed spklt; **dis** above the glm. **Spklt** 2.5–5 mm, pectinate, with 1 bisex and 2 rdmt flt. **Glm** unequal, glab, smt scabridulous, apc smt shortly bilobed, acuminate or mucronate; **lo glm** 0.7–1.5 mm; **up glm** 1.5–2.5 mm, glab, scabrous, or strigose, hairs not papillose-based; **lowest lm** 1.7–4 mm, densely pilose, at least on the mrg, 3-awned, awns 0.5–3 mm, cent awns flanked by 2 memb lobes; **lowest pal** 1.5–4 mm, pubescent on the mrg, 4-lobed, 2-awned, awns 1–2 mm; **anth** 0.4–0.7 mm; **rchl intnd subtending second flt** terminating in a dense tuft of hairs; **second flt** rdmt, 1.5–4 mm, 2-lobed, lobes rounded, 3-awned, awns 0.5–4 mm; **rchl intnd subtending third flt** with glab or puberulent apc; **third flt** rdmt, flabellate, unawned. **Car** to 1 mm. 2*n* = 20.

The range of *Bouteloua barbata* extends from the southwestern United States to southern Mexico. There are three varieties recognized. The two that grow in the Intermountain Region are often sympatric, but are usually easily distinguished in the field by their growth habit. According to Gould (1979), in the southern portion of their range the differences between the two varieties are less evident, particularly on herbarium specimens.

Bouteloua barbata is often confused with juvenile plants of the perennial *B. trifida*, but in *B. barbata* the central awn is flanked by two membranous lobes and the lowest paleas are 4-lobed and 2-awned.

1. Plants annual; culms usually decumbent and geniculate, occasionally rooting at the lower nodes
   . . . . var. *barbata*
1. Plants short-lived perennials; culms erect from the base . . . . var. *rothrockii*

Bouteloua barbata Lag. var. **barbata** Sixweeks Grama [p. *449*]

Pl ann; not stln. **Clm** 1–35 cm, usu decumbent and geniculate, occ rooting at the lo nd. **Lig** 0.4–1 mm; **bld** 0.5–5(9) cm long, 0.7–3 mm wide. **Pan** 0.7–9 cm, with (2)4–11 br; **br** 10–27 mm, scabrous, with 20–40 spklt. 2*n* = 20.

Bouteloua barbata var. *barbata* grows in loose sands, rocky slopes, and washes, often on disturbed soils, usually at elevations below 2000 m. Its range extends from the southwestern United States to northwestern Mexico.

Bouteloua barbata var. **rothrockii** (Vasey) Gould Rothrock's Grama [p. *449*]

Pl short-lived per; not stln. **Clm** 25–60 (75) cm, stiffly erect or slightly geniculate-spreading bas; **lig** 0.1–0.5 mm; **bld** 6–10 cm long, 1–4 mm wide.

Pan (3.5)5–25 cm, with 3–8 br; **br** 15–30 mm, scabrous, with 35–50 (55) spklt. 2*n* = 40.

Bouteloua barbata var. *rothrockii* grows on dry slopes and sandy flats, mostly at 750–1700 m. It grows throughout the southwestern United States and Mexico, sometimes covering large areas. It used to be the most important forage grass in southern Arizona and neighboring regions.

Bouteloua barbata var. *rothrockii* resembles *B. parryi* var. *parryi*, but can be easily distinguished from that taxon by the lack of papillose-based hairs on the keels of its upper glumes.

10. **Bouteloua simplex** Lag. Mat Grama [p. *450*, *510*]

Pl ann. **Clm** 3–35 cm, usu decumbent, occ erect, rarely brchg; **intnd** glab. **Shth** smooth, deeply striate; **lig** 0.1–0.2 mm, of short hairs, smt with a few papillose-based hairs on either side; **bld** 2–8 cm long, 0.5–1.5 mm wide, flat to involute, adx surfaces mostly glab, often pilose bas. **Pan** usu with only 1 br (terminating the clm), or with 2–4 br and subdigitate; **br** 10–25(40) mm, persistent, straight, arcuate, or circular, with 30–80 spklt, axes terminating in a rdcd spklt; **dis** above the glm. **Spklt** pectinate, with 1 bisex flt and 1–2 rdmt flt. **Glm** glab, smt scabrous distally, acute or acuminate; **lo glm** 1.5–2.5 mm; **up glm** 3.5–5 mm; **lowest lm** 2.5–3.5 mm, pilose over the veins, 3-awned, awns stout and flattened, cent awns 1–2 mm, flanked by 2 memb lobes, lat awns shorter than the cent awns; **lowest pal** obovate, unawned; **rchl intnd subtending second flt** with densely pubescent apc; **second flt** rdcd to an awn column with 3 awns of 5–6 mm; **third flt**, if present, flabellate scales. 2*n* = 20.

Bouteloua simplex grows on rocky, open slopes in grassy and open shrub vegetation at 1200–2500 m. Its native range extends from the southwestern United States through Mexico and Central America to western South America.

11. **Bouteloua parryi** (E. Fourn.) Griffiths Parry's Grama [p. *450*, *510*]

Pl ann or short-lived per; tufted, smt stln. **Clm** 20–60 cm, erect or somewhat geniculate at the base. **Lvs** mostly bas; **shth** pubescent, usu with tufts of long hairs on either side of the col; **lig** 0.1–0.5 mm, of hairs; **bld** 1–3 cm long, 1–2.5 mm wide, mrg and usu both surfaces with papillose-based hairs. **Pan** 2.5–10 cm, with 4–8 br; **br** 20–35 mm, persistent, with papillose-based hairs, with 40–65 spklt, br terminating in a spklt; **dis** above the glm. **Spklt** pectinate, with 1 bisex flt and 2 rdmt flt. **Glm** unequal; **lo glm** about 2 mm, glab or sparsely pubescent at the base, mucronate; **up glm** 3–4 mm, keels with papillose-based hairs, apc bilobed, awned from between the teeth, awns to 0.7 mm; **lowest lm** 3–4 mm, pilose or villous proximally, 3-awned, awns 2–3 mm, cent awns flanked by 2 memb lobes; **lowest pal** about 2.5 mm, 4-lobed, 2-awned; **anth** 1.8–2 mm, yellow; **rchl intnd subtending second flt** with densely pubescent apc; **second flt** lobed nearly to the base, lobes ovate, awns 2–4 mm, exceeding those of the lowest lm; **third flt** rdcd to minute scales, glab, unawned or with a single awn. **Car** 1.3–1.5 mm. $2n = 20$.

Bouteloua parryi grows on sandy slopes and flats at elevations from near sea level to 2000 m. Its range extends from the southwestern United States to central Mexico. Plants in the Intermountain Region belong to *B. parryi* (E. Fourn.) Griffiths var. *parryi,* which differs from *B. parryi* var. *gentryi* (Gould) Gould in comprising tufted annuals rather than stoloniferous perennials. *Bouteloua parryi* var. *parryi* resembles *B. barbata* var. *rothrockii,* but differs in the papillose-based hairs on the keels of its upper glumes.

12. **Bouteloua dactyloides** (Nutt.) Engelm.
    BUFFALOGRASS [p. 450, 510]

Pl per; usu dioecious; strongly stln, smt mat-forming. **Clm** 1–30 cm, erect, solid, mostly unbrchd, those of the pist infl much shorter than those of the stmt infl; **nd** mostly glab. **Lvs** bas tufted, not clustered or strongly distichous; **shth** open, rounded, often sparsely pilose near the col;

lig 0.5–1 mm; **bld** 2–15 cm long, 1-2.5 mm wide, usu flat bas, curling when dry, glab or sparsely pilose, apc involute. **Stmt infl** usu exceeding the up lvs, pan of 1–3(4) rcmly arranged, unilateral, pectinate br; **br** not enclosed at maturity, spklt densely crowded in 2 rows. **Stmt spklt** 4–6 mm long, 1.3–1.8 mm wide, with 2 flt; **glm** unequal, glab, 1- or 2-veined; **lm** 3-veined, glab, unawned; **anth** 2.5–3 mm, brownish to red or orange. **Pist infl** partially hidden within bracteate lf shth; **br** 2–3(4), 2.5–4.5 mm, burlike, with 3–5(7) spklt; **dis** at the base of the pan br. **Pist spklt** to 7 mm long, about 2.5 mm wide, with 1 flt, almost completely enclosed by the up glm; **lo glm** irregular and rdcd; **br axes** and **lo portion of up glm** globose, white, indurate, terminating in 3 awnlike teeth; **lm** firmly memb, glab, 3-veined, unawned or shortly 3-awned. **Car** 2–2.5 mm. $2n = 20, 40, 56, 60.$

Bouteloua dactyloides is a frequent dominant on upland portions of the semi-arid, shortgrass component of the Great Plains, ranging from the southern prairie provinces of Canada through the desert southwest of the United States to much of northern Mexico.

Bouteloua dactyloides provides valuable forage for livestock and wildlife, and withstands heavy grazing. It may be confused in the southern portion of its range with *Hilaria belangeri,* which consistently has pilose nodes.

10.24 HILARIA Kunth   Mary E. Barkworth

Pl per or ann; tufted or ces, smt stln, per species smt rhz. **Clm** 5–250 cm, erect or decumbent; **nd** usu villous or pilose, particularly the up nd. **Shth** open, glab or pilose, lo shth often glab bas and pilose distally, mrg smt villous or pilose, up shth often glab even if the lo shth are pilose; **lig** 0.5–5 mm, memb, lacerate or ciliate. **Infl** tml, spikelike pan of rdcd, dis br, exceeding the up lvs; **br** with 3 spklt, appressed to the rchs, bases straight, seated in a ciliate, cuplike structure, smt with a 0.5–2 mm cal, cal pilose, axes not extending past the distal flt; **dis** at the base of the br, leaving the zig-zag rchs. **Lat spklt of each br** shortly pedlt, with 1–4(5) strl or stmt flt; **glm** almost as long as the flt, deeply cleft into 2 or more lobes, with 1 or more dorsal awns; **lm** memb, hyaline. **Cent spklt** sessile, with 1 pist or bisex flt; **glm** shorter than the flt, rigid, indurate and fused bas, apc with 2 or more lobes; **lm** memb, awned or unawned. $x = 9$. Named for Auguste François César Prouvençal de St.-Hilaire (1779–1853), a French explorer, botanist, and entomologist.

Hilaria is a genus of 10 species that ranges from the southwestern United States to northern Guatemala, growing primarily in dry grasslands and desert areas. Most of the species are important forage species. The stoloniferous species are important soil binders.

Hilaria and *Pleuraphis* Torr. are sometimes treated as separate genera but, although they differ consistently in some morphological characters, their overall similarity is striking. One molecular study (Columbus et al. 1998) included representatives of both groups. It showed them to be sister taxa; there seems little value in promoting each to generic rank.

In the descriptions below, the term "fascicle" refers to a branch and its spikelets. Actual branch lengths are much shorter and harder to measure.

1. Glumes thickened, indurate, and conspicuously fused at the base; central spikelets with 1 pistillate floret . . . . 4. *H. belangeri*
1. Glumes papery or membranous throughout, not conspicuously fused at the base; central spikelets with 1 bisexual floret.

2. Glumes of the lateral spikelets flabellate, the awns not exceeding the apical lobes; cauline nodes usually only shortly pubescent, sometimes glabrous . . . . 1. *H. mutica*
2. Glumes of the lateral spikelets lanceolate or parallel-sided, the awns exceeding the apical lobes; cauline nodes pilose, villous, or glabrous.
   3. Lower cauline internodes tomentose . . . . 2. *H. rigida*
   3. Lower cauline internodes glabrous . . . . 3. *H. jamesii*

1. **Hilaria mutica** (Buckley) Benth. Tobosagrass [p. 451, 510]

Pl per; ces, rhz. **Clm** 30–60 cm, erect, geniculate at the mid nd; **nd** glab or pubescent, hairs to 0.3 mm. **Shth** glab or sparsely pilose on the mrg; **lig** 0.5–2 mm, lacerate; **bld** 2–15 cm long, 2–4 mm wide, mostly scabrous on both surfaces, with papillose-based hairs behind the lig. **Pan** 4–8 cm; fascicles 5–8 mm. **Lat spklt** with 1 or 2(4) stmt flt; **glm** not conspicuously fused bas, thin, papery, flabellate, dorsally awned, awns not exceeding the apc, apical lobes rounded, ciliate to finely laciniate, veins not or scarcely excurrent; **anth** 3, 2.5–3.5 mm. **Cent spklt** with 1 bisex flt; **glm** with 1 or more divergent, dorsal awns, apical lobes, ciliate to finely laciniate, veins excurrent; **lm** exceeding the glm, bilobed, mucronate. 2*n* = 36, 54.

Hilaria mutica grows in level upland areas and desert valleys subject to occasional flooding but lacking permanent streams. Its range extends into northern Mexico. Although *H. mutica* has moderate forage value, its palatability is low and it is frequently infected with ergot.

2. **Hilaria rigida** (Thurb.) Benth. *ex* Scribn. Big Galleta [p. 451, 510]

Pl per; ces, smt rhz. **Clm** 35–250 cm, decumbent, much brchd above the base, becoming almost woody; **up nd** glab or villous, hairs to 1.5 mm; lo **intnd** tomentose. **Lig** 1–2 mm, densely ciliate; **bld** 2–10(16) cm long, 2–5 mm wide, flat bas, involute distally. **Pan** 4–12 cm; fascicles 6–12 mm. **Lat spklt** with 2–4 flt, lo 2 flt stmt, other flt (if present) usu strl; **glm** thin, memb, not fused at the base, lanceolate or parallel-sided, 7-veined, awned, awns exceeding the glm apc, apc 2–4-lobed, lobes acute to rounded, long-ciliate, smt with 1–3 excurrent veins that form additional slender awns to 1.8 mm; **lo glm** with dorsal, divergent awns; **up glm** with subapical awns; **anth** 3, 4–4.5 mm. **Cent spklt** equaling or exceeding the lat spklt, with 1 stipitate, bisex flt; **glm** thin, memb, narrow, deeply cleft into few–several acuminate, ciliate lobes and slender awns; **lm** often exceeding the glm, thin, ciliate, 2-lobed, midveins excurrent. 2*n* = 18, 36, 54.

Hilaria rigida grows in deserts and open juniper stands, at low elevations, from the southwestern United States to central Mexico. Although almost shrubby, it is very popular with pack horses.

3. **Hilaria jamesii** (Torr.) Benth. Galleta [p. 451, 510]

Pl per; strongly rhz or stln. **Clm** 20–65 cm, erect, bases much brchd; **nd** usu pilose or villous, smt glab; **lo intnd** glab. **Shth** glab, smt slightly scabrous; **col** pilose at the edges; **lig** 1.5–5 mm, often laciniate; **bld** 2–20 cm long, 2–4 mm wide, involute and curled when dry, sparsely villous behind the lig, abx surfaces scabridulous, adx surfaces scabrous. **Pan** 2–6 cm; fascicles 6–8 mm. **Lat spklt** with 3 stmt flt; **glm** thin, memb, lanceolate or parallel-sided, not conspicuously fused at the base, apc acute to rounded, often ciliate, veins rarely excurrent; **lo glm** dorsally awned, awns exceeding the apc; **anth** 3, about 5 mm. **Cent spklt** with 1 bisex flt; **glm** with excurrent veins forming distinct awns; **lm** exceeding the glm, ciliate, the midveins smt excurrent. 2*n* = 18, 36.

Hilaria jamesii is endemic to the southwestern United States, where it grows in deserts, canyons, and dry plains. It has medium grazing value but low palatability. It is usually less pubescent than *H. rigida*, the difference being most marked on the lower cauline nodes.

4. **Hilaria belangeri** (Steud.) Nash Curly Mesquite [p. 452, 510]

Pl per; ces, usu stln. **Clm** 5–35 cm, erect; **nd** villous. **Shth** striate, glab; **lig** 1–3 mm, often lacerate; **bld** 3–15 cm long, 1–3.5 mm wide, adx surfaces sparsely pilose, hairs papillose-based, mrg sparsely pilose bas, with similar hairs. **Pan** 2–4 cm; fascicles 5–8 mm. **Lat spklt** with 2(3) stmt flt, or 1 strl flt; **glm** unequal, thick, indurate, and conspicuously fused bas, thinner distally, asymmetrically lobed, scabrous, pale to purplish, bases smt spotted with a few dark glands, mrg wide, hyaline, awns 1 or more, attached below midlength, equaling or exceeding the cent spklt, antrorsely scabrous; **lo glm** wider, more deeply lobed, with longer awn(s) than the up glm; **anth** 3, 3–3.7 mm. **Cent spklt** as long as or longer than the lat spklt, with 1 pist flt; **glm** terminating in 1 or more antrorsely scabrous awns. 2*n* = 36, 72, 74.

Both varieties of *Hilaria belangeri* are found on mesas and plains within the Intermountain Region.

1. Plants stoloniferous; blades 3–10 cm long, 1–2 mm wide; ligules about 1–1.5 mm long . . . . var. *belangeri*
1. Plants not stoloniferous; blades 3–15 cm long, to 3.5 mm wide; ligules 2.5–3 mm long . . . . . . var. *longifolia*

Hilaria belangeri (Steud.) Nash var. belangeri [p. 452]

Pl stln. **Lig** 1–1.5 mm; **bld** 3–10 cm long, 1–2 mm wide.

Hilaria belangeri var. *belangeri* grows from Arizona to Texas, and south through Mexico. It was the dominant grass on Texas' shortgrass prairies.

Hilaria belangeri var. longifolia (Vasey) Hitchc. [p. 452]

Pl not stln. **Lig** 2.5–3 mm; **bld** 3–15 cm long, to 3.5 mm wide.

Hilaria belangeri var. *longifolia* is more restricted than var. *belangeri* in its distribution, growing from Arizona to Texas, and south to northwestern Mexico.

10.25 TRAGUS Haller

J.K. Wipff

Pl ann or per; ces. **Clm** (2)5–65 cm, herbaceous, usu rooting at the lo nd; **nd** and **intnd** glab. **Lvs** cauline; **shth** open, usu shorter than the intnd, mostly glab but long-ciliate at the edges of the col; **lig** memb, truncate, ciliate; **bld** usu flat, mrg ciliate. **Infl** tml, exceeding the up lvs, narrow, cylindrical pan; **br** 0.5–5 mm, resembling burs, with 2–5 spklt; **dis** at the base of the br. **Spklt** crowded, attached individually to the br, with 1 flt; **proximal spklt(s)** bisex, larger than the distal spklt(s); **tml spklt** often strl. **Glm** unequal; **lo glm** absent or minute, veinless, memb; **up glm** usu exceeding the flt, 5–7-veined, with 5–7 longitudinal rows of straight or uncinate spinelike projections; **lm** 3-veined; **pal** 2-veined, hyaline, memb. $x = 10$. Name from the Greek *tragos*, 'he-goat'.

Tragus has seven species, all of which are native to the tropics and subtropics of the Eastern Hemisphere; one has been introduced into the Intermountain Region. The genus is easily recognized by the spinelike projections on the upper glumes.

1. **Tragus berteronianus** Schult. SPIKE BURGRASS [p. 452, 510]
Pl ann. **Clm** (2)3.5–45 cm. **Lig** 0.5–1 mm; **bld** (0.5)0.7–8.5 cm long, 1.2–5 mm wide, glab. **Pan** (1)2–13 cm long, (3)4–8 mm wide; **rchs** pubescent; **br** (0.5)0.7–2.7 mm, pubescent, with 2(3) spklt, axes occ extending past the distal spklt; **proximal intnd** 0.2–0.6(0.7) mm, shorter than the second intnd. **Proximal spklt** (1.8)2–4.3 mm; **second spklt** (0.8)1–3.9 mm, smt strl. **Lo glm** 0.1–0.6 mm, memb, minutely pubescent; **up glm** 1.8–4.3 mm, minutely pubescent, 5-veined, rarely with 1–2 additional veins adjacent to the midvein; **glm projections** (4)6–14, in 5 rows, (0.2)0.3–1 mm, uncinate; **lm** (1.5)1.8–3.1 mm, sparsely pubescent on the back, midveins occ excurrent to 0.6 mm; **pal** (1.3)1.5–2.4 mm; **anth** 3, 0.4–0.6 mm, yellow, occ purple- or green-tinged. **Car** (0.9)1.2–2 mm long, 0.4–0.8 mm wide. $2n = 20$.

Tragus berteronianus is native to Africa and Asia, and is now established in Arizona, New Mexico, and Texas.

10.26 ZOYSIA Willd.

Sharon J. Anderson

Pl per, rhz, mat-forming. **Clm** 5–40 cm tall. **Lig** to 0.3 mm, of hairs, often with longer hairs at the base of each bld immediately behind the lig; **bld** usu glab abx, smt with a ciliate callus at the base of each col, adx surfaces glab, scabrous, or sparsely pilose, apc often sharply pointed. **Infl** tml, exceeding the lvs, solitary spikelike rcm (a single spklt in *Z. minima*), spklt solitary, shortly pedlt, lat appressed to the rchs; **dis** beneath the glm or not occurring. **Spklt** lat compressed, with 1 flt; **flt** bisex. **Lo glm** usu absent; **up glm** enclosing the flt, chartaceous to coriaceous, awned, awns to 2.5 mm; **lm** thin, lanceolate or linear, acute to emgt, 1-veined; **pal** thin, rarely present. $x = 10$. Named for Carl von Zois (1756–1800), a German botanist.

Zoysia has 11 species. They are native to coastal sands between 42°N and 42°S and from Mauritius to Polynesia. Some species also grow in disturbed inland areas. Three species are used as lawn grasses in mesic tropical and subtropical areas, including parts of the Intermountain Region. For a more complete account of *Zoysia* in North America, see Barkworth et al. (2003).

1. **Zoysia japonica** Steud. JAPANESE LAWNGRASS, KOREAN LAWNGRASS [p. 453]
Pl rhz. **Clm** intnd 2–10 mm. **Shth** glab; **lig** 0.05–0.25 mm; **bld** to 6.5 cm long, 0.5–5 mm wide, ascending, flat to loosely involute when fully hydrated, involute when stressed, surfaces glab or pilose. **Peduncles** exserted, extending 0.8–6.5 cm beyond the shth of the flag lvs. **Rcm** 2.5–4.5 cm, with 25–50 spklt; **ped** 1.6–3.5 mm. **Spklt** 2.5–3.4 mm long, 1–1.4 mm wide, ovate, awned, awns 0.1–1.1 mm. $2n = 40$.

Zoysia japonica was the first species of *Zoysia* introduced into cultivation in the United States, with the introduction of the cultivar 'Meyer' in the 1950s. It is the most cold-tolerant and coarsely textured of the three species that have been introduced to North America, and is the only species that is currently available as seed in the United States.

11. PAPPOPHOREAE Kunth   John R. Reeder†

Pl per; rarely ann. **Clm** herbaceous. **Lig** of hairs; **bld epdm** with bicellular microhairs. **Infl** tml, dense, narrow to somewhat open pan; **dis** above the glm but not between the flt (except in *Cottea*). **Spklt** with 3–10 flt, smt only the lowest flt bisex. **Lm** 5–13-veined, veins extending into awns, often with intermixed hyaline lobes. **Car** with punctate hila; **emb** ½ or more as long as the car. *x* = 10.

The tribe *Pappophoreae* includes five genera and approximately 40 species. It is represented in tropical and warm regions around the world. Recent work (e.g. Bouchenak-Khelladi et al. 2008) does not support its recognition as a tribe.

11.01 ENNEAPOGON Desv. *ex* P. Beauv.   John R. Reeder†

Pl per or ann; ces, more or less hairy throughout. **Clm** 3–100 cm; **nd** hairy; **intnd** hollow. **Shth** open; **lig** of hairs; **microhairs of bld** each with an elongated bas cell and an inflated tml cell. **Infl** tml, spikelike to somewhat open pan, bases often included within the uppermost lf shth; **dis** above the glm but not between the flt, flt falling as a unit. **Spklt** with 3–6 flt, frequently only the lowest flt bisex, distal flt progressively rdcd. **Glm** subequal, as long as or slightly shorter than the flt (including the awns), more or less pubescent; **lo glm** 5–7-veined; **lm** firm, rounded on the backs, villous below the mid, strongly 9-veined, veins extending into equal, plumose awns 3–5 times as long as the lm bodies and forming a pappuslike crown; **pal** longer than the lm, entire, thinly memb, 2-veined, 2-keeled, keels hairy; **anth** 3, 0.2–1.5 mm; **sty** 2, free to the base, white. *x* = 10. Name from the Greek *ennea*, 'nine', and *pogon*, 'beard', a reference to the nine hairy awns of the lemmas.

Enneapogon includes about 28 species. It is found in tropical and warm regions of the world, especially in Africa and Australia. One species grows in the Intermountain Region.

1. **Enneapogon desvauxii** P. Beauv. NINEAWN PAPPUSGRASS [p. 453, <u>510</u>]
Pl per. **Clm** 20–45 cm, about 1 mm thick, ascending to erect from a hard knotty base, often brchg; **nd** pubescent. **Shth** usu shorter than the intnd, more or less pubescent; **lig** about 0.5 mm; **bld** mostly 2–12 cm long, 1–2 mm wide, more or less hairy, soon involute. **Pan** 2–10 cm, spikelike, grayish green or lead-colored. **Spklt** mostly 5–7 mm, usu only the lowest flt bisex. **Glm** 3–5 mm, subequal, thin, puberulent; **up glm** often 3- or 4-veined; **lowest lm** 1.5–2 mm, firm, rounded on the back; **awns** 3–4 mm; **anth** 0.3–0.5 mm. **Car** 1–1.2 mm, oval, plump; **emb** subequal to the car. **Cleistogamous spklt** commonly present in the lo shth, their lm larger than those of the flt in the aerial pan, unawned or with awns that are much rdcd. 2*n* = 20.

Enneapogon desvauxii grows in open areas of the southwestern United States and in much of Mexico, as well as in the Eastern Hemisphere.

6. DANTHONIOIDEAE N.P. Barker & H.P. Linder   Grass Phylogeny Working Group

Pl usu per, smt ann; when per, ces, rhz, or stln. **Clm** usu solid, rarely hollow. **Lvs** distichous; **shth** usu open; **abx lig** usu absent; **aur** usu absent; **adx lig** of hairs or memb and ciliate; **bld** not psdpet; **mesophyll** non-radiate; **adx palisade layer** absent; **fusoid cells** absent; **arm cells** absent; **Kranz anatomy** absent; **midrib** simple, usu with 1 vascular bundle (an arc of bundles in *Cortaderia*); **adx bulliform cells** present or not; **stomata** usu with dome-shaped or parallel-sided subsidiary cells (rarely slightly triangular or high dome-shaped); **bicellular microhairs** usu present, distal cell long, narrow; **papillae** usu absent. **Infl** ebracteate (subtending lf somewhat spatheate in *Urochlaena* Nees), usu paniculate, smt rcm or spicate, occ a single spklt; **dis** usu above the glm and between the flt, smt below the glm or in the clm. **Spklt** bisex (smt with unisex flt) or unisex, with 1–7(20) bisex or pist flt, distal flt in the bisex spklt often strl or stmt; **rchl extension** present. **Glm** 2, usu equal, (1)3–7-veined, usu exceeding the distal flt; **flt** lat compressed; **lm** firmly memb to coriaceous, 3–9-veined, rounded across the back, glab or with non-uncinate hairs, these smt in tufts or fringes, lm apc shortly to deeply bilobed, lobes often setaceous, midveins often extended as awns, awns usu geniculate, bas segment often flat and twisted; **pal** well developed, smt short relative to the lm; **lod** 2, usu free, usu fleshy, rarely with a memb apical flap, glab or ciliate, often with microhairs, smt heavily vascularized; **anth** 3; **ov** usu glab, rarely with apical hairs; **haustorial synergids** present, smt weakly developed; **sty** 2, bases usu widely separated. **Car** separate from the lm and pal; **hila** punctate or long-linear; **emb** large or small relative to the car; **endosperm** hard; **starch grains** usu compound; **epiblasts** absent; **scutellar cleft** present; **mesocotyl intnd** elongated; **emb lf mrg** usu meeting, smt overlapping. x = 6, 7, 9.

The *Danthonioideae* include only one tribe, the *Danthonieae*, which used to be included in the *Arundinoideae*. Conert (1987) placed *Cortaderia* in a tribe of its own, but its traditional inclusion in the *Danthonieae* is supported by more recent work (Barker et al. 2000; Grass Phylogeny Working Group 2001, Bouchenak-Khelladi et al. 2008). The combination of haustorial synergids, ciliate ligules, elongated embryo mesocotyls, and C_3 photosynthesis distinguishes the *Danthonioideae* from other subfamilies of the *Poaceae*.

12. DANTHONIEAE Zotov

Mary E. Barkworth

See subfamily description.

The *Danthonieae*, the only tribe in the *Danthonioideae*, include approximately 13 genera and 290 species, most of which grow in mesic to xeric, open habitats such as grasslands, heaths, and open woods. It is most abundant in the Southern Hemisphere. Only *Danthonia* is native in the Northern Hemisphere.

1. Culms 200–700 cm tall; inflorescences plumose, 30–130 cm long . . . . 12.01 *Cortaderia*
1. Culms 2–100 cm tall; inflorescences not plumose, 0.5–12 cm long.
   2. Plants perennial . . . . 12.02 *Danthonia*
   2. Plants annual. . . . . 12.03 *Schismus*

12.01 CORTADERIA Stapf

Kelly W. Allred

Pl per; often dioecious or monoecious; ces. **Clm** 2–7 m, erect, densely clumped. **Lvs** primarily bas; **shth** open, often overlapping, glab or hairy; **aur** absent; **lig** of hairs; **bld** to 2 m, flat to folded, arching, edges usu sharply serrate. **Infl** tml, plumose pan, 30–130 cm, subtended by a long, ciliate bract; **br** stiff to flexible. **Spklt** somewhat lat compressed, usu unisex, smt bisex, with 2–9 unisex flt; **dis** above the glm and below the flt. **Glm** unequal, nearly as long as the spklt, hyaline, 1-veined; **cal** pilose; **lm** 3–5(7)-veined, long-acuminate, bifid and awned or entire and mucronate; **lm of pist and bisex flt** usu long-sericeous; **lm of stmt flt** less hairy or glab; **lod** 2, cuneate and irregularly lobed, ciliate; **pal** about ½ as long as the lm, 2-veined; **anth** of bisex flt 3, 1.5–6 mm, those of the pist flt smaller or absent. **Car** 1.5–3 mm; **hila** linear, about ½ as long as the car; **emb** usu shorter than 1 mm. x = 9. Name from the Spanish *cortada*, 'cutting', referring to the sharply serrate blades.

Cortaderia, a genus of about 25 species, is native to South America and New Zealand, with the majority of species being South American. Recent evidence suggests that the species in the two regions represent different lineages, each of which merits generic recognition (Barker et al. 2003). The species treated here would remain in *Cortaderia* if this change were made.

One species is grown as an ornamental in the Intermountain Region.

1. **Cortaderia selloana** (Schult. & Schult. f.) Asch. & Graebn. PAMPAS GRASS [p. 453, 510] **Pl** usu dioecious, smt monoecious. **Clm** 2–4 m, usu 2–4 times as long as the pan. **Lvs** primarily bas; **shth** mostly glab, with a dense tuft of hairs at the col; **lig** 1–2 mm; **bld** to 2 m long, 3–8 cm wide, mostly flat, cauline, ascending, arching, bluish green, abx surfaces glab bas. **Pan** 30–130 cm, only slightly, if at all, elevated above the foliage, whitish or pinkish when young. **Spklt** 15–17 mm; **cal** to 1 mm, with hairs to 2 mm; **lm** long-attenuate to an awn, awns 2.5–5 mm; **pal** to 4 mm; **stigmas** exserted. **Car** and **flt** not separating easily from the rchl. $2n$ = 72.

Cortaderia selloana is native to central South America. It is cultivated as an ornamental in the warmer parts of North America. It was thought that it would not become a weed problem because most plants sold as ornamentals are unisexual, but it is now considered an aggressive weed in California and Bendigo, Australia. The weedy Australian plants are bisexual (Walsh 1994).

12.02 DANTHONIA DC.

Stephen J. Darbyshire

Pl per; ces, smt shortly rhz. **Clm** 7–130 cm, erect. **Shth** open to the base, with tufts of hairs at the aur position, smt with a line of hairs around the col; **aur** absent; **lig** of hairs; **bld** rolled in the bud, flat or involute when dry. **Infl** tml; pan, rcm, or a solitary spklt, to 12 cm; **rchs**, **br**, and **ped** scabrous or hirsute. **Spklt** terete or lat compressed, with 3–12 flt, tml flt rdcd; **dis** beneath the flt, also at the cauline nd in some species. **Glm** subequal or the lo glm a little longer than the up glm, usu exceeding the flt (excluding the awns and lm teeth), lanceolate, chartaceous, 1–7-veined, keels glab or sparsely scabrous; **rchl** glab; **cal** densely strigose on the sides; **lm bodies** obscurely (5)7–11-veined, backs glab or pilose, mrg usu densely pilose proximally, apc with 2 acute to aristate lobes, mucronate or awned between the lobes; **awns**, when present, geniculate and twisted below the geniculation; **pal** about as long as the lm bodies, 2-veined, veins scabrous, apc obtuse, smt bifid; **lod** 2, glab or with a few hairs; **anth** 3, their size depending on whether the flt are chasmogamous or cleistogamous; **ov** glab. **Car** 1.5–5.5 mm, ovate to obovate, dorsally flattened, brown; **hila** linear, ⅓–¾ as long as the car. **Clstgn** usu present in the lo shth, with 1(–10) flt, not dis; **rchl intnd** about as long as or longer than the adjacent flt; **lm** coriaceous, glab or scabrous near the apex, entire, unawned; **pal** smt slightly longer than the lm; **anth** 3, minute; **ov** glab; **car** more linear than in the aerial flt. $x = 12$. Named for the French botanist Étienne Danthoine, who worked in the early nineteenth century.

Danthonia is interpreted here as a genus of about 20 species that are native in Europe, North Africa, and the Americas. All five of the species found in the Intermountain Region are native.

In the key and descriptions, lemma lengths do not include the apical teeth. Callus characteristics are best seen on the middle to upper florets in the spikelet.

1. Calluses of the middle florets from shorter to slightly longer than wide, convex abaxially; lemma bodies 2.5–6 mm long, the backs usually pilose, occasionally glabrous or sparsely pilose . . . . 1. *D. spicata*
1. Calluses of the middle florets longer than wide, concave abaxially; lemma bodies 3–11 mm long, the backs usually glabrous or sparsely pilose.
  2. Lower inflorescence branches (pedicels if the inflorescence racemose) flexible, slightly to strongly divergent; pedicels usually as long as or longer than the spikelets.
    3. Uppermost cauline blades usually erect to ascending; inflorescences usually paniculate; pedicels shorter than to as long as the spikelets; lemmas pilose over the back, at least basally; mature culms not disarticulating at the nodes . . . . 3. *D. parryi*
    3. Uppermost cauline blades usually strongly divergent or reflexed; inflorescences usually racemose; pedicels usually much longer than the spikelets and usually strongly divergent; lemmas glabrous or sparsely hairy over the back; mature culms disarticulating at the nodes . . . . 4. *D. californica*
  2. Lower inflorescence branches (pedicels if the inflorescence racemose) stiff, erect; pedicels from shorter than to as long as the spikelets.
    4. Spikelets (4)5–10; lower inflorescence branches usually with 2–3 spikelets; lemma bodies 3–6 mm; mature culms not disarticulating at the nodes. . . . . 2. *D. intermedia*
    4. Spikelets 1(–3), if 2–3, the inflorescence a raceme; lemma bodies 5.5–11 mm; mature culms disarticulating at the nodes . . . . 5. *D. unispicata*

1. **Danthonia spicata** (L.) P. Beauv. *ex* Roem. & Schult. POVERTY OATGRASS [p. *454*, <u>510</u>]
Clm (7)10–70(100) cm, dis at the nd when mature. **Shth** pilose or glab; **bld** 6–15(20) cm long, 0.8–3(4) mm wide, usu becoming curled at maturity, glab or pilose, uppermost cauline bld erect to ascending. **Infl** with 5–10(18) spklt; **br** stiff, appressed to strongly ascending after anthesis; **lo br** with 1–3 spklt; **ped** on the lowest br from shorter than to equaling the spklt. **Spklt** 7–15 mm. **Cal of mid flt** about as long as wide, convex abx; **lm bodies** 2.5–5 mm, usu pilose (smt glab) over the back, mrg pilose to about midlength, longest hairs 0.5-2 mm, apical teeth 0.5–2 mm, acute to aristate, less than ⅔ as long as the lm bodies; **awns** 5–8 mm; **anth** to 2.5 mm. **Car** 1.5–2(2.3) mm long, 0.7–1 mm wide. $2n = 31, 36$.

Danthonia spicata grows in dry rocky, sandy, or mineral soils, generally in open sunny places. Its range includes most of boreal and temperate North America and extends south into northeastern Mexico.

Phenotypically, *Danthonia spicata* is quite variable, expressing different growth forms under different conditions (Dore and McNeill 1980; Darbyshire and Cayouette 1989). Slow clonal growth, extensive cleistogamy, and limited dispersal contribute to the establishment of morphologically uniform populations, some of which have been given scientific names. For instance, *D. spicata* var. *pinetorum* Piper is sometimes applied to depauperate plants and *D. allenii* Austin misapplied to more robust plants (Dore and McNeill 1980). Plants of shady or moist habitats often lack the distinctive curled or twisted blades usually found on plants growing in open habits. Such plants, which tend to have smaller spikelets and pilose foliage, have been called *D. spicata* var. *longipila* Scribn. & Merr. The terminal inflorescence is usually primarily cleistogamous, but plants with chasmogamous inflorescences are found throughout the range of the species. Chasmogamous plants differ in having divergent inflorescence branches at anthesis, larger anthers, and well-developed lodicules.

2. **Danthonia intermedia** Vasey TIMBER OATGRASS [p. *454*, 510]

Clm 10–50(70) cm, not dis at maturity. **Shth** usu glab; **bld** 5–10 cm long, 1–3.5 mm wide, glab or slightly pilose. **Infl** with (4)5–10 spklt; **br** stiff, appressed or strongly ascending; **lo br** with (1)2–3(5) spklt; **ped** on the lowest br shorter than the spklt. **Spklt** 11–15(19) mm. **Cal of mid flt** longer than wide, concave abx; **lm bodies** 3–6 mm, glab over the back, densely pilose along the mrg, teeth 1.5–2.5 mm, acute to acuminate or aristate; **awns** 6.5–8 mm; **anth** usu tiny, smt to 4 mm. **Car** (2)2.3–3 mm long, 0.7–1.1 mm wide. **Clstgn** rarely produced. 2*n* = 18, 36, 98.

Danthonia intermedia grows in boreal and alpine meadows, open woods, and on rocky slopes and northern plains. Its range extends from Kamchatka, Russia, to northern North America, extending south along the cordillera. Its primarily cleistogamous reproduction has probably facilitated its establishment and spread through more boreal and alpine habitats than other members of the genus.

Tsvelev (1976) treats the American plants as *D. intermedia* (Vasey) subsp. *intermedia* and the Russian plants, which have 2*n* = 18, as *D. intermedia* subsp. *riabuschinskii* (Kom.) Tzvelev.

3. **Danthonia parryi** Scribn. PARRY'S OATGRASS [p. *455*, 511]

Clm 30–80(100) cm, not dis at maturity. **Shth** glab or sparsely pubescent; **bld** 15–25 cm long, to 4 mm wide, glab or scabrous (rarely pilose), uppermost cauline bld erect or diverging less than 20° from the clm at maturity. **Infl** usu pan, smt rcm, with (3)4–11 spklt; **br** appressed to ascending, somewhat flexible; **ped** on the lowest br from shorter than to as long as the spklt. **Spklt** 16–24 mm. **Cal of mid flt** longer than wide, concave abx; **lm bodies** 5.5–10 mm, backs usu pilose, especially near the base (rarely glab), mrg pilose, teeth 2.5–8 mm, aristate; **awns** 12–15 mm; **anth** to 6.5 mm. **Car** rarely produced, 3.5–5.2 mm long, 0.9–1.8 mm wide. 2*n* = 36.

Danthonia parryi is endemic to western North America and is often a major component of grasslands on the eastern foothills of the Rocky Mountains. It grows in open grassland, open woods, and rocky slopes, at elevations up to 4000 m. It rarely produces caryopses in the terminal inflorescences. This and its somewhat intermediate morphology have led to speculation that it is derived from hybridization between *D. californica* and *D. intermedia*.

4. **Danthonia californica** Bol. CALIFORNIA OATGRASS [p. *455*, 511]

Clm (10)30–130 cm, dis at the nd at maturity. **Shth** glab or pilose, up shth usu glab or unevenly pilose; **bld** 10–30 cm long, (1)2–5(6) mm wide, flat to rolled or involute, glab or pilose, uppermost cauline bld strongly divergent to reflexed at maturity. **Infl** usu rcm, with (2)3–6(10) widely spreading spklt; **br** flexible, strongly divergent to reflexed at maturity, pulvini usu present at the base; **ped** on the lowest br longer than the spklt, often crinkled. **Spklt** (10)14–26(30) mm. **Cal of mid flt** usu longer than wide, concave abx; **lm bodies** 5–10 mm, glab or sparsely pilose over the back, mrg pubescent (rarely glab), apical teeth (2)4–6(7) mm, aristate; **awns** (7)8–12 mm; **anth** to 4 mm. **Car** 2.5–4.2 mm long, 1.3–1.6 mm wide. 2*n* = 36.

Danthonia californica grows in prairies, meadows, and open woods. It has a disjunct distribution, growing in western North America and in Chile.

Plants with pilose foliage have been called *Danthonia californica* var. *americana* (Scribn.) Hitchc. and plants with sparsely pilose lemma backs *D. californica* var. *macounii* Hitchc., but the variation is not taxonomically significant.

5. **Danthonia unispicata** (Thurb.) Munro *ex* Vasey ONE-SPIKE OATGRASS [p. *455*, 511]

Clm (10)15–30(42) cm, dis at the nd at maturity. **Shth** usu densely pilose, hairs smt papillose-based (up shth smt glab); **bld** 3–8(20) cm long, 1–3 mm wide, both surfaces sparsely to densely pilose, smt also scabrous or hirsute (rarely glab). **Infl** with 1–2(3) spklt, if more than 1, rcm; **ped** stiff, appressed, shorter than the spklt. **Spklt** (8)12–26 mm. **Cal of mid flt** longer than wide, concave abx; **lm bodies** 5.5–11 mm, glab over the back (rarely with a few scattered hairs), mrg pilose (rarely glab), apical teeth 1.5–7 mm, acute to aristate; **awns** 5.5–13 mm; **anth** to 3.5 mm. **Car** 2.2–4 mm long, about 1 mm wide. 2*n* = 36.

Danthonia unispicata is restricted to western North America, where it grows in prairies and meadows, on rocky slopes, and in dry openings up to timberline in the mountains. It differs from *D. californica*, to which it is closely related, in its shorter stature, usually densely pilose foliage, short, erect pedicels, and the usually erect cauline leaf blades. Some authors prefer to treat it as *D. californica* var. *unispicata* Thurb.

12.03 **SCHISMUS** P. Beauv.   Elizabeth A. Kellogg

Pl ann or short-lived per; tufted. **Clm** 2–30 cm, smt decumbent, glab. **Shth** open, usu shorter than the intnd, with tufts of 1.5–4 mm hairs on the mrg of the col; **aur** absent; **lig** memb, ciliate; **bld** flat or folded, becoming involute on drying. **Infl** tml, dense pan, 1–7 cm long, 0.5–2(3) cm wide, br 1–2 per nd; **dis** initially above the glm, glm and ped smt falling together later. **Spklt** with (4)5–7(10) flt. **Glm** subequal, exceeding or exceeded by the distal flt, 3–7-veined, mrg hyaline; **lm** 7–9-veined, mrg and intercostal regions usu pubescent, varying to glab, mrg hyaline, apc bifid or merely notched, sinuses smt mucronate, mucros to 1.5 mm; **pal** spatulate, memb, 2-veined, 2-keeled; **anth** 3, 0.2–0.5 mm. **Car** ovoid. $x = 6$. Name from the Greek *schizo*, 'split' referring to the bidentate lemmas.

Schismus is a genus of five species that is native to Africa and Asia. Two species are established in the Intermountain Region. In using the key and descriptions, the lowest floret in a spikelet should be examined. Succeeding florets tend to have shorter, less acute or acuminate lobes, and a shallower sinus.

1. Lower glumes equaling or exceeding the distal florets; lemma lobes longer than wide,
   acute to acuminate; paleas always shorter than the lemmas .... 1. *S. arabicus*
1. Lower glumes exceeded by the distal florets; lemma lobes as wide as or wider than
   long, acute to obtuse; paleas of the lower florets in the spikelets as long as or longer than
   the lemmas .... 2. *S. barbatus*

1. Schismus arabicus Nees ARABIAN SCHISMUS [p. 456, <u>511</u>]

Pl ann. **Clm** (2)6–16 cm. **Lig** 0.5–1.5 mm, of hairs; **bld** 4–6 cm long, 0.5–2 mm wide, abx surfaces glab or sparsely pubescent, adx surfaces sparsely to densely pubescent. **Pan** (1)2–3.5 cm. **Spklt** 5–7 mm. **Lo glm** 4.2–6.2 mm, equaling or exceeding the distal flt; **up glm** 4–6 mm; **lm** 1.8–2.6 mm, with dense, spreading pubescence between the veins, lobes longer than wide, acute to acuminate; **pal** 1.5–2.2 mm, shorter than the lm; **anth** 0.2–0.5 mm. **Car** 0.5–0.8 mm. $2n = 12$.

Schismus arabicus is native to southwestern Asia, but it is now established in the southwestern United States, growing in open and disturbed sites.

2. Schismus barbatus (Loefl. *ex* L.) Thell.

COMMON MEDITERRANEAN GRASS [p. 456, <u>511</u>]

Pl ann. **Clm** 6–27 cm. **Lig** 0.3–1.1 mm, of hairs; **bld** 3–15 cm long, 0.3–1.5 mm wide, abx surfaces glab or scabrous, adx surfaces scabrous, sparsely long-pubescent near the lig. **Pan** 1–6(7.5) cm. **Spklt** 4.5–7 mm. **Lo glm** 4–5.2 mm, exceeded by the distal flt; **up glm** 4–5.3 mm; **lm** 1.5–2(2.5) mm, with sparse, appressed pubescence between the veins, or glab and with spreading hairs on the mrg, lobes as wide as or wider than long, acute to obtuse; **pal** 1.7–2.2(2.6) mm, those of the lo flt in the spklt as long as or longer than the lm; **anth** 0.2–0.4 mm. **Car** 0.6–0.8 mm. $2n = 12$.

Schismus barbatus is native to Eurasia, but it is now established in the southwestern United States. It grows in sandy, disturbed sites along roadsides and fields and in dry riverbeds.

7. ARISTIDOIDEAE Caro   Grass Phylogeny Working Group and Kelly W. Allred

Pl ann or per; usu ces. **Clm** ann, erect, solid or hollow, usu unbrchd. **Lvs** distichous; **shth** usu open; **aur** absent; **abx lig** absent or of hairs; **adx lig** memb and ciliate or of hairs; **bld** without psdpet; **mesophyll cells** radiate or non-radiate; **adx palisade layer** absent; **fusoid cells** absent; **arm cells** absent; **Kranz anatomy** absent or present, when present, with 1 or 2 parenchyma shth; **midribs** simple; **adx bulliform cells** present; **stomatal subsidiary cells** dome-shaped or triangular; **bicellular microhairs** present, with long, slender, thin-walled tml cells. **Infl** tml, not lfy, usu pan, smt spikes or rcm. **Spklt** bisex, with 1 flt; **rchl extension** absent; **dis** above the glm. **Glm** 2, usu longer than the flt, usu acute or acuminate; **flt** terete or lat compressed, with well-developed cal; **lm** 1- or 3-veined, more or less coriaceous, with a germination flap, lm mrg overlapping at maturity and concealing the pal, apc evidently 3-awned, awn bases often forming a column, lat awns occ rdcd or absent; **pal** less than ½ as long as the lm; **lod** usu present, 2, free, memb, glab, heavily vascularized; **anth** 1–3; **ov** glab; **haustorial synergids** absent; **sty** 2, free to the base but close. **Car** usu fusiform, falling with the lm and pal attached; **hila** short or long, linear; **endosperm** hard, without lipid; **starch grains** compound; **emb** small or large relative to the car; **epiblasts** absent; **scutellar cleft** present or absent; **mesocotyl intnd** elongated; **emb lf mrg** meeting. $x = 11, 12$.

The subfamily *Aristidoideae* includes only one tribe, the *Aristideae*. Earlier taxonomists have generally included the *Aristidoideae*, with the *Danthonieae* and *Arundineae*, in the *Arundinoideae* (e.g., Watson et al. 1985; Clayton and Renvoize 1986; Kellogg and Campbell 1987), but Esen and Hilu (1991) demonstrated that *Aristida* is clearly distinct from the *Danthonieae* and *Arundineae* in terms of its prolamins. Subsequent work has provided further support for the monophyly of the three tribes, but their position relative to each other and other members of the PACCMAD clade is more equivocal (Bouchenak-Khelladi et al. 2008).

13. ARISTIDEAE C.E. Hubb.   Kelly W. Allred

See subfamily description.

The tribe *Aristideae* has three genera and 300–350 species, and is primarily pantropical in its distribution. Its members are usually readily recognized by their terete, 3-awned lemmas with overlapping margins. *Aristida*, which has many more species than the other two genera combined, is the only genus found in the Americas.

13.01 ARISTIDA L.   Kelly W. Allred

Pl usu per; herbaceous, usu ces, occ rhz. **Clm** 10–150 cm, not woody, smt brchd above the base; **intnd** usu pith-filled, smt hollow. **Lvs** smt predominantly bas, smt predominantly cauline; **shth** open; **aur** lacking; **lig** of hairs or very shortly memb and long-ciliate, the 2 types generally indistinguishable. **Infl** tml, usu pan, smt rcm, occ spikes; **pri br** without axillary pulvini and usu appressed to ascending, or with axillary pulvini and ascending to strongly divergent or divaricate. **Spklt** with 1 flt; **rchl** not prolonged beyond the flt; **dis** above the glm. **Glm** often longer than the flt, thin, usu 1–3-veined, acute to acuminate; **flt** terete or weakly lat compressed; **cal** well developed, hirsute; **lm** fusiform, 3-veined, convolute, usu glab or scabridulous, usu enclosing the pal at maturity, usu with 3 tml awns, lat awns rdcd or obsolete in some species, lm apc smt narrowed to a straight or twisted beak below the awns; **awns** ascending to spreading, usu straight, bases smt twisted together into a column or the bases of the individual awns coiled, twisted, or otherwise contorted, occ dis at maturity; **pal** shorter than the lm, 2-veined, occ absent; **anth** 1 or 3. **Car** fusiform; **hila** linear. x = 11, 12. Name from the Latin *arista*, 'awn'.

Aristida is a tropical to warm-temperate genus of 250–300 species. It grows throughout the world in dry grasslands and savannahs, sandy woodlands, arid deserts, and open, weedy habitats and on rocky slopes and mesas. All 10 species in this treatment are native to the Intermountain Region.

The divergent awns aid in wind and animal transportation of the florets and, by holding the florets and the caryopses they contain at an angle to the ground, in establishment. The presence of *Aristida* frequently indicates soil disturbance or abuse. Although generally poor forage grasses and, because of the calluses, potentially harmful to grazing animals, some species of *Aristida* are an important source of spring forage on western rangelands. Quail and small mammals eat small amounts of the seed.

Lemma lengths are measured from the base of the callus to the divergence of the awns.

1. Lower glumes 3–7-veined.
    2. Awns nearly equal, the lateral awns 8–66 mm long and at least ¾ as long as the central
       awns . . . . 7. *A. oligantha*
    2. Awns markedly unequal, the lateral awns 1–4 mm long, no more than ½ as long as the
       central awns, sometimes absent . . . . 3. *A. schiedeana* (in part)
1. Lower glumes 1–2(3)-veined.
    3. Lateral awns markedly reduced, usually ⅓ or less as long as the central awns.
        4. Panicles 0.5–3 cm wide, the branches erect-appressed to strongly ascending, without
           axillary pulvini or the pulvini only weakly developed.
            5. Plants annual; culms often highly branched above the base. . . . . 8. *A. adscensionis* (in part)
            5. Plants perennial; culms rarely branched above the base.
                6. Primary panicle branches 3–6 cm long; lateral awns (1)8–140 mm long . . 9. *A. purpurea* (in part)
                6. Primary panicle branches 6–16 cm long; lateral awns absent or to 1(3) mm
                   long . . . . 3. *A. schiedeana* (in part)
        4. Panicles 3–45 cm wide, at least the lower branches spreading and having well-
           developed axillary pulvini.
            7. Lateral awns absent or no more than 3 mm long.
                8. Central awns often deflexed at a sharp angle when mature; lemma apices
                   often twisted at maturity . . . . 3. *A. schiedeana* (in part)
                8. Central awns usually straight or arcuate; lemma apices not twisted . . . . . . 2. *A. ternipes* (in part)
            7. Lateral awns 3–23 mm long.
                9. Anthers 1.2–2.4 mm long. . . . . 2. *A. ternipes* (in part)
                9. Anthers 0.8–1 mm long.
                    10. Spikelets usually divergent and the pedicels with axillary pulvini;
                        secondary branches usually absent; primary branches 2–6 cm long; lemma
                        apices with 0–2 twists when mature . . . . 5. *A. havardii* (in part)

10. Spikelets usually appressed and the pedicels without axillary pulvini; secondary branches usually well developed; primary branches 5–13 cm long; lemma apices with 4 or more twists when mature . . . . 4. *A. divaricata* (in part)
3. Lateral awns well developed, usually at least ½ as long as the central awns.
  11. Junction of the lemma and awns evident; awns disarticulating at maturity . . . . 1. *A. californica*
  11. Junction of the lemma and awns not evident; awns not disarticulating at maturity.
    12. Lower primary panicle branches (pedicels in racemose species) appressed, without axillary pulvini.
      13. Lower glumes usually ⅓–¾ as long as the upper glumes.
        14. Plants annual. . . . . 8. *A. adscensionis* (in part)
        14. Plants perennial . . . . 9. *A. purpurea* (in part)
      13. Lower glumes usually more than ¾ as long as the upper glumes.
        15. Plants annual. . . . . 8. *A. adscensionis* (in part)
        15. Plants perennial . . . . 10. *A. arizonica*
    12. At least the lower primary panicle branches divergent and with axillary pulvini.
      16. Panicles narrow and contracted above, usually only the lower 1–2 branches spreading and with a pulvinus; lemma apices 0.2–0.3 mm wide . . . . 9. *A. purpurea* (in part)
      16. Almost all panicle branches spreading and with axillary pulvini; lemma apices 0.1–0.2 mm wide.
        17. Anthers 0.8–1 mm long.
          18. Spikelets usually divergent and the pedicels with axillary pulvini; secondary branches absent or nearly so; primary branches 2–6 cm long; lemma apices straight or with 1 or 2 twists . . . . 5. *A. havardii* (in part)
          18. Spikelets usually appressed and the pedicels without axillary pulvini; secondary branches usually well developed; primary branches 5–13 cm long; lemma apices with 4 or more twists at maturity . . . . 4. *A. divaricata* (in part)
        17. Anthers 1–3 mm long.
          19. Base of the blades with scattered hairs 1.5–3 mm long on the adaxial surfaces. . . . . 2. *A. ternipes* (in part)
          19. Base of the blades glabrous or puberulent on the adaxial surface, the hairs, if present, less than 0.5 mm long.
            20. Glumes reddish, the lower glumes often shorter than the upper glumes; awns ascending to divaricate, (8)13–140 mm long; terminal spikelets usually appressed and without axillary pulvini . . . . 9. *A. purpurea* (in part)
            20. Glumes brownish, equal or unequal; awns spreading to horizontal, 5–15 mm long; terminal spikelets often spreading from axillary pulvini . . . . 6. *A. pansa*

1. Aristida californica Thurb. MOJAVE THREEAWN [p. *456*, <u>511</u>]

Pl per; smt flowering the first year. **Clm** 10–40 cm, highly brchd above the base in age; **intnd** glab or pubescent, smt nearly lanose. **Lvs** cauline; **shth** shorter than the intnd, glab or puberulent; **col** glab or pubescent at the sides; **lig** 0.5–1 mm; **bld** usu less than 6 cm long, 0.5–1 mm wide, pale green, involute, glab or puberulent abx. **Infl** paniculate or rcm, 5–10 cm long, 1–2 cm wide, with few spklt; **rchs nd** glab or with straight hairs; **pri br** 1–2 cm, appressed, without axillary pulvini. **Spklt** appressed. **Glm** unequal, 1–2-veined; **lo glm** 4–10 mm; **up glm** 7–15 mm; **cal** about 1 mm; **lm** 5–7 mm, purple or mottled, jnct of the lm and awns evident; **awns** twisted together bas into a

4–26 mm column, free portions 12–50 mm, those of the cent and lat awns similar in length, curved to arcuate bas, straight and divergent distally, dis at the base of the column at maturity; **anth** 3, about 2 mm long. 2*n* = 22.

The range of both varieties of *Aristida californica* extends from the southwestern United States into northwestern Mexico.

1. Cauline internodes puberulent to nearly lanose . . . . var. *californica*
1. Cauline internodes glabrous . . . . var. *glabrata*

Aristida californica Thurb. var. **californica** [p. *456*]

Cauline intnd puberulent to nearly lanose. **Lo glm** 6–10 mm; **up glm** 11–13 mm; **awns** forming a 7–26 mm column, free portions 25–50 mm.

Aristida californica var. *californica* grows in dry, sandy plains, dunes, and flats of the Sonoran and Mojave deserts at elevations of 0–700 m.

Aristida californica var. **glabrata** Vasey [p. 456]
Cauline intnd glab. Lo glm 4–8 mm; up glm 7–15 mm; awns forming a 4–16 mm column, free portions 12–40 mm.

Aristida californica var. *glabrata* grows in sandy to rocky soils of desert grassland and desert thorn-scrub communities in the Sonoran and Mojave deserts at elevations of 500–1400 m, generally to the east of var. *californica*.

2. **Aristida ternipes** Cav. [p. 457, 511]
Pl per; ces. Clm 25–120 cm, wiry, erect to sprawling, unbrchd. Lvs bas and cauline; shth usu longer than the intnd, glab; col glab or strigillose; lig less than 0.5 mm; bld 5–40 cm long, 1–2.5 mm wide, flat to folded, straight to lax at maturity, adx surfaces with scattered, 1.5–3 mm hairs near the lig. Infl paniculate, 15–40 cm long, (8)10–35(45) mm wide; rchs nd glab or strigillose; pri br 5–25 cm, remote, stiffly ascending to divaricate, with axillary pulvini, usu naked near the base; sec br and ped usu appressed. Spklt usu congested. Glm 9–15 mm, subequal, 1-veined, acuminate; cal 1–1.2 mm; lm 9–15 mm long, smooth to tuberculate-scabrous, narrowing to slightly keeled, usu not twisted, 0.1–0.2 mm wide apc, jnct with the awns not evident; awns not dis at maturity, unequal or almost equal; cent awns 8–25(30) mm, straight to arcuate at the base; lat awns absent or to 0–23 mm; anth 3, 1.2–2.4 mm. Car 6–8 mm, light brownish. 2n = 22, 24.

One variety grows in the Intermountain Region.

Aristida ternipes var. **gentilis** (Henrard) Allred
Hook Threeawn [p. 457]
Awns subequal to unequal, ascending to spreading; cent awns 10–25(30) mm; lat awns (2)6–23 mm. 2n = 44.

Aristida ternipes var. *gentilis* grows on dry slopes and plains and along roadsides from Californica to Texas and south through Mexico to Guatemala.

3. **Aristida schiedeana** Trin. & Rupr. Single
Threeawn [p. 457, 511]
Pl per; ces. Clm 30–120 cm, erect, unbrchd. Lvs bas and cauline, pale green, smt glaucous; shth longer or shorter than the intnd, glab except at the summit; col densely to sparsely pilose or glab; lig less than 0.5 mm; bld 8–30 cm long, 1–2 mm wide, usu flat, often curled at maturity. Infl paniculate, 10–30 cm long, (4)8–26 cm wide; rchs nd with straight hairs, hairs to 0.8 mm; pri br 6–16 cm, abruptly spreading to divaricate, stiff to

lax, with axillary pulvini, usu not spklt-bearing below midlength. Spklt appressed, rarely spreading. Glm 1(3)-veined, brown or purple at maturity, acuminate; lo glm 6–13 mm; up glm equaling or to 4 mm shorter than the lo glm; cal 0.8–1.2 mm; lm 10–15(17) mm, terminating in a strongly twisted, 2–4 mm awnlike beak, jnct with the awns not conspicuous; awns not dis at maturity; cent awns 5–12 mm, markedly bent near the base; lat awns absent or to 1(3) mm, erect; anth 1.2–2.2 mm, brownish. Car 6–8 mm. 2n = 22, 44.

Aristida schiedeana grows on rocky slopes and plains, generally in pinyon-juniper, oak, or ponderosa pine communities. Plants from the southwestern United States and northern Mexico belong to *A. schiedeana* var. *orcuttiana* (Vasey) Allred & Valdés-Reyna, in which the lower glumes are usually glabrous and longer than the upper glumes, and the collar and throat are usually glabrous.

4. **Aristida divaricata** Humb. & Bonpl. *ex* Willd.
Poverty Grass [p. 457, 511]
Pl per; ces. Clm 25–70 cm, erect or prostrate, unbrchd or sparingly brchd. Lvs tending to be bas; shth longer than the intnd, glab except at the summit; col densely pilose; lig 0.5–1 mm; bld 5–20 cm long, 1–2 mm wide, flat to loosely involute, glab. Infl paniculate, 10–30 cm long, 6–25 cm wide, peduncles flattened and easily broken; rchs nd glab or with hairs, hairs to 0.5 mm; pri br 5–13 cm, stiffly divaricate to reflexed, with axillary pulvini, usu naked on the bas ½; sec br usu well developed. Spklt overlapping, usu appressed, smt divergent and the ped with axillary pulvini. Glm 8–12 mm, 1-veined, acuminate or shortly awned, awns to 4 mm; cal about 0.5 mm; lm 8–13 mm long, the tml 2–3 mm with 4 or more twists when mature, narrowing to 0.1–0.2 mm wide just below the awns, jnct with the awns not evident; awns (7)10–20 mm, not dis at maturity; cent awns almost straight to curved at the base, ascending to somewhat divergent distally; lat awns slightly thinner and from much to slightly shorter than the cent awns, ascending to divergent; anth 3, 0.8–1 mm. Car 8–10 mm, light brown. 2n = 22.

Aristida divaricata grows on dry hills and plains, especially in pinyon-juniper-grassland zones, from the southwestern United States through Mexico to Guatemala. It occasionally intergrades with *A. havardii*, but that species has lemma beaks that are straight or have only 1–2 twists, shorter primary branches, usually no secondary branches, and pedicels that more frequently have axillary pulvini so the spikelets are more frequently divergent than in *A. divaricata*.

5. Aristida havardii Vasey HAVARD'S THREEAWN [p. *457*, 511]

Pl per; ces. **Clm** 15–40 cm, slender, usu erect, occ decumbent, often tightly clustered into hemispheric clumps, unbrchd. **Lvs** mostly bas; **shth** longer than the intnd, glab except at the summit; **col** densely pilose; **lig** 0.5–1 mm; **bld** 5–20 cm long, 1–2 mm wide, flat to loosely involute, glab. **Infl** paniculate, 8–18 cm long, 4–12 cm wide, peduncles often flattened and easily broken; **rchs nd** glab or with straight, less than 0.3 mm hairs; **pri br** 2–6 cm, stiffly divaricate to reflexed, with axillary pulvini, usu naked on the lo ½; **sec br** usu absent. **Spklt** usu divergent, ped usu with axillary pulvini. **Glm** 8–12 mm, 1-veined, acuminate or awned, awns to 4 mm; **cal** about 0.5 mm; **lm** 8–13 mm long, glab, smooth or scabrous, tml 2–3 mm straight or with 1–2 twists, narrowing to 0.1–0.2 mm wide, jnct with the awns not evident; **awns** (7)10–22 mm, not dis at maturity, from almost straight to somewhat curved bas, ascending to divergent distally; **lat awns** slightly shorter and thinner than the cent awns; **anth** 3, 0.8–1 mm. **Car** 8–10 mm, light brown. $2n = 22$.

Aristida havardii grows on dry hills and plains in desert grassland to pinyon-juniper zones, and in sandy to rocky ground from the southwestern United States to northern Mexico. It occasionally intergrades with *A. divaricata*, but that species differs in having more twisted lemma beaks, longer primary branches, well-developed secondary branches, and, usually, appressed spikelets.

6. Aristida pansa Wooton & Standl. WOOTON'S THREEAWN [p. *458*, 511]

Pl per; ces. **Clm** 20–60(75) cm, erect, unbrchd. **Lvs** bas and cauline; **shth** usu longer than the intnd, glab except at the summit; **col** densely pilose, hairs 1–3 mm, cobwebby and tangled, often deflexed; **lig** less than 0.5 mm; **bld** 4–28 cm long, less than 1 mm wide, usu involute, infrequently flat, usu arcuate, abx surfaces glab, adx surfaces glab or puberulent near the base, scabrous or puberulent distally. **Infl** paniculate, 10–20 cm long, 3–10(12) cm wide; **rchs nd** usu glab, smt with straight, less than 0.3 mm hairs; **pri br** 2–11 cm, stiffly ascending to spreading, with axillary pulvini; **sec br** and **ped** with or without pulvini; tml spklt often divergent. **Spklt** clustered on the distal ½ of the br. **Glm** equal or subequal, 1-veined, acuminate or awned, awns to 6 mm, brownish; **lo glm** 5–10 mm; **up glm** 6–12 mm; **cal** 0.5–1 mm; **lm** 7–13 mm, terminating in an obscure, narrow beak 1–4 mm long, 0.1–0.2 mm wide, jnct with the base of the awns not evident;

awns 6–15 mm, not dis at maturity, cent and lat awns similar in length and thickness, spreading to horizontal; **anth** 3, 1–3 mm, brown. **Car** 6–8 mm, tan. $2n$ = unknown.

Aristida pansa grows in desert scrub, commonly in the Chihuahuan Desert of the southwestern United States and Mexico, but its ecological range extends into the lower juniper zones and its geographic range to southern Mexico. It prefers cobbly to sandy, often gypsiferous soil. It has been confused with *A. purpurea* var. *perplexa*, which differs in having reddish glumes of unequal length and longer ascending awns.

7. Aristida oligantha Michx. OLDFIELD THREEAWN [p. *458*, 511]

Pl ann. **Clm** 25–55 cm, erect or geniculate at the base, highly brchd. **Lvs** cauline; **shth** usu shorter than the intnd, lowermost shth appressed-pilose bas; **col** glab; **lig** less than 0.5 mm; **bld** usu 4–12 cm long, 0.5–1.5 mm wide, flat or loosely involute, somewhat lax, glab or scabridulous, pale green. **Infl** spicate or rcm, (5)7–20 cm long, 2–4 cm wide; **pri br** rarely developed. **Spklt** divergent, ped with axillary pulvini. **Glm** unequal, glab, brownish green with a purple tinge; **lo glm** (9)12–22(28) mm, 3–7-veined, midvein extended into a 1–13 mm awn between 2 delicate setae; **up glm** (7)11–20(24) mm, 1-veined; **cal** 0.5–2 mm; **lm** (9)12–22(23) mm, glab, light-colored, often mottled; **awns** (8)12–65(70) mm, subequal, spreading; **anth** usu 1 and less than 0.5 mm, rarely 3 and 3–4 mm. **Car** 8–14 mm, brown. $2n = 22$.

Aristida oligantha grows in waste places, dry fields, roadsides, along railroads, and in burned areas, usually in sandy soil.

8. Aristida adscensionis L. SIXWEEKS THREEAWN [p. *458*, 511]

Pl short- to long-lived ann. **Clm** (3)10–50(80) cm, often highly brchd above the base. **Lvs** cauline, glab; **shth** shorter than the intnd, not disintegrating into threadlike fibers; **lig** 0.4–1 mm; **bld** 2–14 cm long, 1–2.5 mm wide, flat to involute. **Infl** pan, 5–15(20) cm long, 0.5–3 cm wide, often interrupted below; **nd** glab or with straight, less than 0.5 mm hairs; **pri br** 1–4 cm, erect to ascending, without axillary pulvini, with 3–8 spklt. **Spklt** crowded. **Glm** unequal, 1-veined, acuminate; **lo glm** 4–8 mm; **up glm** 6–11 mm; **cal** 0.5–0.8 mm; **lm** 6–9 mm, slightly keeled, midveins scabrous, jnct with the awns not evident; **awns** not dis at maturity, flattened and straight to somewhat curved at the base, cent rib flanked by equally wide pale wings; **cent awns** 7–15(20) mm; **lat awns** somewhat shorter, occ only 1–2 mm; **anth** 3, 0.3–0.7 mm. $2n = 22$.

Aristida adscensionis grows in waste ground, along roadsides, and on degraded rangelands and dry hillsides, often in sandy soils. It is associated with woodland, prairie, and desert shrub communities. Its range extends from the United States south through Mexico and Central America to South America.

Because *Aristida adscensionis* is highly variable in height, panicle size, and awn development, several varieties have been described. None are recognized here because most of the variation appears to be environmentally induced.

9. **Aristida purpurea** Nutt. [p. *459*, <u>511</u>]

Pl per; densely ces, without rhz. **Clm** 10–100 cm, erect to ascending, usu unbrchd. **Lvs** mostly bas or mostly cauline; **shth** shorter or longer than the intnd, glab, not disintegrating into threadlike fibers at maturity; **col** glab, or sparsely pilose at the sides with straight hairs; **lig** less than 0.5 mm; **bld** 4–25 cm long, 1–1.5 mm wide, tightly involute to flat, usu glab, smt scabridulous abx, gray-green, lax to curled at maturity. **Infl** usu sparingly brchd pan, occ rcm, 3–30 cm long, 2–12 cm wide, with 2 or more spklt per nd; **nd** glab or with straight, about 0.5 mm hairs; **pri br** 3–6 cm, appressed to divaricate, varying smt within a pan, stiff to flexible, bases appressed or abruptly spreading, usu without axillary pulvini. **Spklt** divergent or appressed, with or without axillary pulvini. **Glm** usu unequal, lo glm shorter than the up glm, smt subequal, light to dark brown or purplish, glab, smooth or scabridulous, 1(2)-veined, acuminate, unawned or awned, awns to 1 mm; **lo glm** 4–12 mm; **up glm** 7–25 mm; **cal** 0.5–1.8 mm; **lm** 6–16 mm, glab, scabridulous, or tuberculate, whitish to purplish, apc 0.1–0.8 mm wide, not beaked or the beak less than 3 mm, jnct with the awns not conspicuous; **awns** (8)13–140 mm, ascending to divaricate, not dis at maturity; **cent awns** thicker than the lat awns; **lat awns** (8)13–140 mm, usu subequal to the cent awns, occ less than ⅓ as long as the cent awns; **anth** 3, 0.7–2 mm. **Car** 6–14 mm, tan to chestnut. $2n = 22, 44, 66, 88$.

Aristida purpurea is composed of several intergrading varieties.

1. Lower or all primary panicle branches stiff, divergent to divaricate from the base, with axillary pulvini; awns 13–30 mm.
    2. Lower glumes ½–⅔ as long as the upper glumes. . . . . .var. *perplexa*
    2. Lower glumes from ¾ as long as to equaling the upper glumes . . . . var. *parishii* (in part)
1. Primary panicle branches appressed or ascending at the base, sometimes drooping distally, without axillary pulvini; awns 8–140 mm.
    3. Awns 35–140 mm long.
        4. Lemma apices 0.1–0.3 mm wide; awns 0.1–0.2(0.3) mm wide at the base, 35–60 mm

long; upper glumes usually shorter than 16 mm . . . . .var. *purpurea* (in part)
    4. Lemma apices 0.3–0.8 mm wide; awns 0.2–0.5 mm wide at the base, 40–140 cm long; upper glumes 14–25 mm long . . . . . . .var. *longiseta*
    3. Awns 8–35 mm long.
        5. Lemma apices 0.1–0.3 mm wide distally; awns 0.1–0.3 mm wide at the base.
            6. All or most of the panicle branches straight (lower branches sometimes lax); pedicels straight, appressed to ascending . . . . .var. *nealleyi*
            6. All or most of the panicle branches and pedicels drooping to sinuous distally . . . . .var. *purpurea* (in part)
        5. Lemma apices 0.2–0.3 mm wide; awns stout, 0.2–0.3 mm wide at the base.
            7. Mature panicle branches and pedicels flexible, lax or drooping distally . . . . .var. *purpurea* (in part)
            7. Mature panicle branches and pedicels usually stiff, straight.
                8. Panicles usually 3–15 cm long; blades 4–10 cm long . . . . var. *fendleriana*
                8. Panicles usually 15–30 cm long; blades 10–25 cm long.
                    9. Glumes and lemmas reddish or dark-colored at anthesis or earlier (fading to stramineous), usually in marked contrast with the current foliage; panicles dense, the lower nodes with 8–18 spikelets; flowering March to May, after winter rains . . . . . . var. *parishii* (in part)
                    9. Glumes and lemmas tan to brown (also fading to stramineous), giving the panicle a brownish appearance; old growth gray-green, not in marked contrast with the current foliage; panicles less dense, the lower nodes with 2–10 spikelets . . . . var. *wrightii*

Aristida purpurea var. **fendleriana** (Steud.) Vasey FENDLER'S THREEAWN [p. *459*]

Clm 10–40 cm. **Lvs** mostly cauline; **bld** 4–10 cm, involute. **Pan** 3–14(15) cm; **pri br** mostly appressed, stiff, straight, without axillary pulvini, with few spklt. **Lo glm** 5–8 mm; **up glm** 10–15 mm; **lm** 8–14 mm long, apc 0.2–0.3 wide; **awns** subequal, 18–40 mm long, occ slightly longer, 0.2–0.3 mm wide at the base. $2n = 22, 44$.

Aristida purpurea var. *fendleriana* grows on open slopes, hills, and sandy flats, at low to medium elevations, from the western United States into northern Mexico. It is often confused with var. *longiseta*, having short basal leaves and short panicles, but plants of var. *fendleriana* have narrower lemma apices and thinner, shorter awns than those of var. *longiseta*.

Aristida purpurea var. **longiseta** (Steud.) Vasey RED THREEAWN [p. *459*]

Clm 10–40(50) cm. **Lvs** smt mostly bas, smt mostly cauline; **bld** 4–16 cm, usu involute. **Pan**

5–15 cm; **pri br** appressed or ascending at the base, without axillary pulvini, stout and straight to delicate and drooping distally, usu neither flexible nor tangled. **Lo glm** 8–12 mm; **up glm** (14)16–25 mm; **lm** 12–16 mm long, apc 0.3–0.8 mm wide; **awns** subequal, 40–100(140) mm long, 0.2–0.5 mm wide at the base. $2n = 22, 44, 66, 88$.

Aristida purpurea var. *longiseta* grows on sandy or rocky slopes and plains, and in barren soils of disturbed ground from western Canada to northern Mexico. It is the most variable variety of *Aristida purpurea*, ranging from short plants with basal leaves and short panicles suggestive of var. *fendleriana*, to tall plants with long cauline leaves and long, drooping panicles resembling var. *purpurea*. The length of its glumes, width of its lemma apices, and the length and thickness of its awns distinguish it from all the other varieties. The calluses and long, stiff awns are especially troublesome to sheep and cattle.

Aristida purpurea var. **nealleyi** (Vasey) Allred NEALLEY'S THREEAWN [p. *459*]

Clm 20–45 cm. **Bld** 5–15 cm, mostly bas, involute. **Pan** 8–18(20) cm; **pri br** and ped mostly appressed to narrowly ascending, without axillary pulvini, stiff, straight, lo br occ flexible. **Glm** usu unequal; **lo glm** 4–7 mm; **up glm** (7)8–14 mm; **lm** 7–13 mm long, narrowing to about 0.1 mm wide, up portion smt twisted; **awns** 15–22(30) mm long, subequal, about 0.1 mm wide at the base. $2n = 22, 44$.

Aristida purpurea var. *nealleyi* grows on dry slopes and plains at lower elevations than the other varieties, frequently in desert grassland vegetation. Its range extends from the southwestern United States into Mexico. Although var. *nealleyi* is more distinct than the other varieties, having tight tufts of foliage exceeded by narrow, straw-colored panicles, it grades into var. *purpurea*, and the panicles resemble those of var. *wrightii*. It may also be confused with *A. arizonica*, but differs in having involute, generally straight leaf blades and shorter awns.

Aristida purpurea var. **parishii** (Hitchc.) Allred PARISH'S THREEAWN [p. *459*]

Clm 20–50 cm. **Lvs** mostly cauline; **bld** more than 10 cm, loosely involute to flat. **Pan** 15–24 cm; **pri br** stiff, lo br strongly divergent to divaricate, with axillary pulvini, up br appressed to ascending, without axillary pulvini, lo nd associated with 8–18 spklt. **Glm** red or dark at anthesis, fading to stramineous; **lo glm** 7–11 mm, ¾ as long as to equaling the up glm; **up glm** 10–15 mm; **lm** 10–13 mm long, narrowing to 0.2–0.3 mm wide near the apex; **awns** subequal, 20–30 mm long, 0.2–0.3 mm wide at the base. $2n =$ unknown.

Aristida purpurea var. *parishii* grows on sandy plains and hills of the southwestern United States and Baja California, Mexico. In many respects it is intermediate between *A. purpurea* and other species of *Aristida* with spreading panicle branches, especially *A. ternipes* var. *gentilis*. Its spikelets are indistinguishable from those of var. *wrightii*, but var. *parishii* frequently has axillary pulvini associated with the lower branches. The two also differ in their phenology: var. *parishii* flowers from March through May in response to winter rains, whereas var. *wrightii* flowers from May through October in response to summer rains.

Aristida purpurea var. **perplexa** Allred & Valdés-Reyna JORNADA THREEAWN [p. *459*]

Clm 30–65 cm. **Bld** 8–20 cm, involute. **Pan** 8–20 cm; **pri br** stiff, lo br diverging or divaricate, with axillary pulvini, up br usu strongly divergent, smt ascending; **ped** often with axillary pulvini; **tml spklt** usu appressed. **Glm** reddish; **lo glm** (4.5)5–7(7.5) mm, ½–⅔ as long as the up glm; **up glm** 8–11(12) mm; **lm** (8)10–12(13) mm long, narrowing to 0.1–0.2 mm wide; **awns** subequal, (13)18–30 mm long, 0.1–0.2 mm wide. $2n =$ unknown.

Aristida purpurea var. *perplexa* grows in sandy to rocky plains and on mesas in desert grassland and scrub communities, often in calcareous soils, in both the United States and Mexico. It is sometimes confused with *A. pansa*, which differs in having cobwebby hairs at the collar, equal glumes, and shorter awns.

Aristida purpurea Nutt. var. **purpurea** PURPLE THREEAWN [p. *459*]

Clm 26–60 cm. **Bld** 3–17 cm, bas and cauline, involute. **Pan** 10–25 cm; **pri br** appressed at the base, without axillary pulvini, capillary, drooping to sinuous distally; **ped** capillary, usu lax to sinuous. **Lo glm** 4–9 mm; **up glm** 7–16 mm; **lm** 6–12 mm long, narrowing to 0.1–0.3 mm wide; **awns** subequal, (15)20–60 mm long, 0.1–0.3 mm wide at the base. $2n = 22, 44, 66, 88$.

Aristida purpurea var. *purpurea* grows in sandy to clay soils, along right of ways, and on dry slopes and mesas. Its range extends to Mexico and Cuba. As treated here, var. *purpurea* is, admittedly, a broadly defined taxon, incorporating slender plants with small spikelets that used to be referred to *A. roemeriana* Scheele, but also occasional plants with somewhat flexible branches that are intermediate to var. *wrightii* and var. *nealleyi*.

Aristida purpurea var. **wrightii** (Nash) Allred WRIGHT'S THREEAWN [p. *459*]

Clm 45–100 cm. **Bld** 10–25 cm, involute or flat. **Pan** (12)14–30 cm; **pri br** usu erect, without axillary pulvini, stiff, straight, lo nd associated with 2–10 spklt. **Glm** tan to brown, fading to stramineous. **Lo glm** 5–10 mm; **up glm** 9–16 mm; **lm** 8–14 mm long, narrowing to 0.2–0.3 mm wide; **awns** (8)20–35 mm long, 0.2–0.3 mm wide at the base, lat awns usu subequal to the cent awn, rarely 1–3 mm. $2n = 22, 44, 66$.

Aristida purpurea var. *wrightii* grows on sandy to gravelly hills and flats from the southwestern United States to southern Mexico. It is the most robust variety of *A. purpurea*, and has dark, stout awns and long panicles. It may be confused with var. *nealleyi*, which has narrower lemmas and awns and a light-colored panicle, but it also intergrades with var. *purpurea* and var. *parishii*. *Aristida purpurea* forma *brownii* (Warnock) Allred & Valdés-Reyna refers to plants with short central awns and lateral awns that are only 1–3 mm long.

10. **Aristida arizonica** Vasey   ARIZONA
      THREEAWN [p. 460, <u>511</u>]

Pl per; usu ces, occ with rhz. **Clm** 30–80(100) cm, erect, unbrchd. **Lvs** mostly bas; **shth** usu longer than the intnd, mostly glab, throat smt with hairs, not disintegrating into threadlike fibers; **col** glab or with hairs at the sides; **lig** 0.2–0.4 mm; **bld** 10–25(30) cm long, 1–3 mm wide, usu flat, often curling like wood shavings when mature, glab. **Infl** spikelike pan, 10–25 cm long, 1–3 cm wide; **nd** glab or with straight, about 0.5 mm hairs; **pri br** 2–6 cm, appressed, without axillary pulvini, with 2–8 spklt. **Glm** 10–15(18) mm, brownish, acuminate to awned, awns to 3 mm; **lo glm** slightly shorter than to equaling the up glm, 1–2-veined; **cal** 1–1.8 mm; **lm** 12–18 mm, glab, rarely sparsely pilose, terminating in a 3–6 mm twisted column, jnct with the awns not conspicuous; **awns** 20–35 mm, straight to curved bas, ascending distally, not dis at maturity; **cent awns** 20–35 mm; **lat awns** slightly shorter than the cent awns; **anth** 3, 1.3–1.9 mm. $2n = 22$.

Aristida arizonica grows in pine, pine-oak, and pinyon-juniper woodlands from the southwestern United States to southern Mexico. It may be confused with *A. purpurea* var. *nealleyi*, but differs in having flat, curly leaf blades and longer awns.

8. **PANICOIDEAE** Link   Grass Phylogeny Working Group

Pl ann or per; synoecious, monoecious, or dioecious; primarily herbaceous, habit varied. **Clm** ann, usu solid, smt somewhat woody, smt decumbent, often brchd above the base. **Lvs** distichous; **shth** usu open; **aur** usu absent; **abx lig** usu absent, occ present as a line of hairs; **adx lig** memb, smt also ciliate, or of hairs, smt absent; **bld** smt psdpet; **mesophyll** radiate or non-radiate; **adx palisade layer** absent; **fusoid cells** usu absent; **arm cells** usu absent; **Kranz anatomy** absent or present; **midribs** usu simple, rarely complex; **adx bulliform cells** present; stomata with triangular or dome-shaped subsidiary cells; **bicellular microhairs** usu present, with a long, narrow distal cell; **papillae** absent or present. **Infl** ebracteate (*Paniceae*) or bracteate (most *Andropogoneae*) pan, rcm, spikes, or complex arrangements of rames (in the *Andropogoneae*), usu bisex, smt unisex; **dis** usu below the glm, frequently in the sec and higher order axes of the infl. **Spklt** bisex or unisex, frequently paired or in triplets, the members of each unit usu with ped of different lengths or 1 spklt sessile. **Glm** usu 2, equal or unequal, shorter or longer than the adjacent flt, smt exceeding the distal flt; **flt** 2(–4), usu dorsally compressed, smt terete or lat compressed; **lo flt** strl or stmt, frequently rdcd to a lm; **up flt** usu bisex; **lm** hyaline to coriaceous, lacking uncinate hairs, often tml awned; **awns** single; **pal of bisex flt** well developed, rdcd, or absent; **lod** usu 2, smt absent, cuneate, free, fleshy, usu glab; **anth** 1–3; **ov** usu glab; **haustorial synergids** absent; **sty** br 2, free and close or fused at the base. **Car:** hila usu punctate; **endosperm** hard, without lipid; **starch grains** simple; **emb** large in relation to the car, usu waisted; **epiblasts** usu absent; **scutellar cleft** present; **mesocotyl intnd** elongated; **emb lf mrg** usu overlapping, rarely just meeting. $x = 5$, (7), 9, 10, (12), (14).

The subfamily *Panicoideae* is most abundant in tropical and subtropical regions, particularly mesic portions of such regions, but several species grow in temperate regions of the world. In North America, the *Panicoideae* are most abundant in the eastern United States (Barkworth and Capels 2000). Within the Intermountain Region, the *Panicoideae* are represented by 27 genera and 69 species. Photosynthesis may be either C_3 or C_4. All three pathways are found in the subfamily, but the PCK and NAD-ME variants appear to have evolved only once, while the NADP-ME pathways seems to have evolved several different times (Giussani et al. 2001).

The *Panicoideae* were first recognized as a distinct unit by Brown (1814), earlier than any of the other subfamilial taxa of the *Poaceae*. Its early recognition is undoubtedly attributable to its distinctive spikelets. Spikelets with two florets are found in many other subfamilies, but rarely do they follow the pattern of the lower floret being sterile or staminate and the upper floret bisexual.

The *Paniceae* and *Andropogoneae* have their conventional interpretation in this treatment, so far as the North American taxa are concerned. Molecular studies, however, while strongly supporting the monophyly of the *Andropogoneae*, show the *Paniceae* to be paraphyletic, with two distinct clades. In one of these clades, most taxa have a chromosome base number of $x = 9$, but some have $x = 10$, and the taxa are pantropical in origin. The taxa in the other clade, with one exception, have a chromosome base number of $x = 10$ and are American in origin. This latter clade is sister to the *Andropogoneae*, which also have a chromosome base number of $x = 10$ (Gómez-Martínez and Culham 2000; Giussani et al. 2001; Barber et al. 2002, Bouchenak-Khelladi et al. 2008).

1. Glumes usually conspicuously unequal; lower glumes usually greatly exceeded by the upper floret; upper glume from subequal to longer than the distal floret; lemma of the upper floret usually coriaceous to indurate; disarticulation usually beneath the glumes, not in the axes of the inflorescence branches. . . . . 14. *Paniceae*
1. Glumes usually subequal, usually exceeding and concealing the florets; lemmas of the upper florets hyaline to membranous; disarticulation frequently in the axes of the inflorescence branches . . . . 15. *Andropogoneae*

14. PANICEAE R. Br.   Mary E. Barkworth

Pl ann or per; habit various. **Clm** 3–800 cm, ann, usu not woody. **Lvs** bas and/or cauline; **shth** usu open; **lig** of hairs or memb, memb lig often ciliate, cilia smt longer than the memb base; **bld** occ psdpet, seldom dis at maturity. **Infl** tml, smt also axillary, occ subterranean pan; **br** smt spikelike and secund, smt less than 1 cm; **dis** usu below the glm, smt at the base of the pan br, occ below the flt. **Spklt** usu dorsally compressed, varying to terete or lat compressed, with 2(3) flt, lo flt stmt, strl, or rdcd, up flt usu bisex; **cal** not developed. **Glm** usu memb; **lo glm** usu less than ½ as long as the spklt, smt absent; **up glm** usu subequal to the up flt, occ absent; **lo lm** similar to the up glm in length and texture; **up lm** indurate, coriaceous, or cartilaginous, with a germination flap at the base, mrg usu widely separated and involute at maturity, smt flat and hyaline; **up pal** similar to the up lm in length and texture; **lod** short; **anth** usu 3; **stigmas** usu red. **Car** usu dorsally compressed or terete; **emb** ½ or more the length of the car. *x* = 9, 10.

The tribe *Paniceae* includes about 100 genera and 2000 species and is primarily tropical in distribution. Within the Intermountain Region, it is represented by 13 genera and 46 species.

The tribe is so morphologically distinct that it was first recognized, in essentially its current sense, by Robert Brown in 1814. Its primary distinguishing features are the unusual spikelet structure combined with the indurate to coriaceous upper florets. The germination flap is a small soft tissue at the base of the upper lemma through which the primary root of the seedling grows.

1. Inflorescences spikelike panicles, with the branches partially embedded in the flattened rachises; plants perennial, stoloniferous . . . . 14.11 *Stenotaphrum*
1. Inflorescences panicles, sometimes spikelike, but the branches not embedded in the rachises or the rachises not flattened; plants annual or perennial, sometimes stoloniferous.
    2. Most spikelets or groups of 2–11 spikelets subtended by 1–many, distinct to more or less connate, stiff bristles or bracts.
        3. Bristles persistent; disarticulation below the spikelets . . . . 14.09 *Setaria* (in part)
        3. Bristles falling with the spikelets at maturity; disarticulation at the base of the reduced panicle branches (fascicles).
            4. Bristles plumose or antrorsely scabrous, free or fused no more than ½ their length . . . . .14.07 *Pennisetum*
            4. Bristles glabrous, smooth, retrorsely scabrous, or strigose, usually at least some bristles fused for more than ½ their length . . . . 14.08 *Cenchrus*
    2. All or most spikelets not subtended by stiff bristles, sometimes the terminal spikelet on each branch subtended by a single bristle, and occasionally other spikelets with a single subtending bristle.
        5. Terminal spikelet on each branch subtended by a single bristle; other spikelets occasionally with a single stiff subtending bristle . . . . 14.09 *Setaria* (in part)
        5. None of the spikelets subtended by a stiff bristle.
            6. Lower glumes or lower lemmas awned, sometimes shortly so (the awn reduced to a point in *Echinochloa colona*) . . . . 14.02 *Echinochloa* (in part)
            6. Lower glumes and lower lemmas unawned.
                7. Upper lemmas and paleas cartilaginous and flexible at maturity; lemma margins flat, hyaline; lower glumes absent or to ¼ the length of the spikelets . . . . 14.01 *Digitaria*
                7. Upper lemmas and paleas chartaceous to indurate and rigid at maturity; lemma margins not hyaline, frequently involute; lower glumes varying from absent to subequal to the spikelets or extending beyond the distal floret.
                    8. Spikelets subtended by a cuplike callus . . . . 14.06 *Eriochloa*
                    8. Spikelets not subtended by a cuplike callus.

9. At least the upper leaves, often all leaves, without ligules; ligules, when present, of hairs . . . . 14.02 *Echinochloa* (in part)
9. All leaves with ligules, ligules membranous or of hairs.
　10. Panicle branches not 1-sided, usually with well-developed secondary branches; inflorescences sometimes spikelike.
　　11. Lemmas of upper florets transversely rugose . . . . 14.13 *Zuloagaea*
　　11. Lemmas of upper florets usually smooth, sometimes verrucose, not rugose.
　　　12. Blades of the basal leaves clearly distinct from the cauline leaves; basal leaves ovate to lanceolate, cauline leaves with longer and narrower blades; basal leaves forming a distinct winter rosette . . . . 14.03 *Dichanthelium* (in part)
　　　12. Blades of the basal and cauline leaves similar, usually linear to lanceolate, varying from filiform to ovate; basal leaves not forming a distinct winter rosette.
　　　　13. Panicles terminating the culms usually appearing in late spring; branches usually developing from the lower and middle cauline nodes in summer, the branches rebranching 1 or more times by fall; upper florets not disarticulating at maturity, plump . . . . 14.03 *Dichanthelium* (in part)
　　　　13. Panicles terminating the culms usually appearing after midsummer; branches usually not developing from the lower and middle cauline nodes, when present, rarely rebranched; upper florets disarticulating or not very plump at maturity. . . . . 14.04 *Panicum* (in part)
　10. Panicle branches 1-sided; secondary branches usually absent or few and clearly shorter than the primary branches, sometimes many and almost as long as the primary branches; panicles never spikelike.
　　14. Lower glumes, sometimes both glumes, absent from all or almost all spikelets. . . . . 14.12 *Paspalum* (in part)
　　14. Lower glumes present on all or most spikelets.
　　　15. Lower glumes from ¾ as long as to equaling the spikelets; upper glumes 7–11-veined; upper florets mostly glabrous but hairy at the apices . . . . 14.10 *Hopia*
　　　15. Lower glumes usually less than ¾ the length of the spikelets; upper florets usually glabrous, sometimes hairy all over, never just at the apices; upper glumes 0–13-veined.
　　　　16. Upper lemmas rugose at maturity; upper glumes 5–13-veined . . . . 14.05 *Urochloa*
　　　　16. Upper lemmas smooth at maturity; upper glumes usually 0–5-veined, sometimes 7-veined.
　　　　　17. Spikelets at least twice as long as wide; lower glumes and upper lemmas glabrous . . . . . . 14.04 *Panicum* (in part)
　　　　　17. Spikelets less than twice as long as wide and/or lower glumes and upper lemmas hairy . . . 14.12 *Paspalum* (in part)

14.01 **DIGITARIA** Haller   J.K. Wipff

Pl ann, per, or of indefinite duration. **Clm** 5–250 cm, erect or decumbent, brchg bas or at aerial clm nd, when ann or of indefinite duration usu decumbent and rooting at the lo nd. **Shth** open; **lig** memb, smt ciliate; **bld** usu flat. **Infl** tml, smt also axillary, usu pan of 1-sided spikelike br (smt only 1 br) attached digitately or rcmly to a rchs, smt simple pan of solitary, pedlt spklt; **spikelike br**, if present, smt with sec br, pri br axes triquetrous, bearing spklt abx, in 2 rows, usu in unequally pedlt groups of 2–5, occ borne singly. **Spklt** 1.2–8.2 mm, lanceoloid to ellipsoid, dorsally compressed, apc obtuse to acuminate, unawned, with 2 flt; **dis** beneath the glm. **Lo glm** absent or to ¼ as long as the spklt; **up glm** usu from ⅙ as long as to equaling the spklt, occ absent, 0–5-veined, usu pubescent; **lo flt** strl; **lo lm** memb, usu as long as the up lm, usu pubescent, (3)5–7(13)-veined; **lo pal** absent or rdcd; **up**

lm mostly stiffly chartaceous to cartilaginous, obscurely veined, with 0.5–1 mm hyaline mrg that embrace the up pal; **up pal** similar to the up lm in texture and size; **lod** 3, cuneate; **anth** 3. **Car** plano-convex; **emb** ⅕–½ as long as the car; **hila** punctiform to ellipsoid. $x = 9$. Name from the Latin *digitus*, 'finger', a reference to the digitate inflorescence of some species.

Digitaria has approximately 200 species that grow primarily in tropical and warm-temperate regions, often in disturbed, open sites. Some species are grown as cereals; others for forage or as lawn grasses. In North America, the genus is best known for two of its weedy species, *D. sanguinalis* and *D. ciliaris*. There are four species known from the Intermountain Region.

1. Spikelets in groups of 3 on the middle portions of the primary branches, the lower portions of the longer pedicels in each group often adnate to the branch axes . . . . .2. *D. ischaemum*
1. Spikelets paired on the middle portions of the primary branches; pedicels not adnate to the branch axes.
    2. Upper lemmas brown when immature, almost always dark brown when mature; primary branches not wing-margined . . . . 1. *D. californica*
    2. Upper lemmas pale yellow, tan, or gray, sometimes purple-tinged, when immature, becoming brown, often purple-tinged, at maturity; primary branches wing-margined, the wings at least ½ as wide as the midribs.
        3. Lateral veins of the lower lemmas scabrous for the distal ⅔ of their length, sometimes scabrous throughout (use 20× magnification); leaf blades usually with papillose-based hairs on both surfaces . . . . 3. *D. sanguinalis*
        3. Lateral veins of the lower lemmas smooth throughout or scabrous only on the distal ⅓; leaf blades with a few papillose-based hairs at the base of the adaxial surfaces, occasionally over the whole adaxial surface . . . . 4. *D. ciliaris*

1. **Digitaria californica** (Benth.) Henrard
[p. 460, 511]

Pl per; ces, neither rhz nor stln. **Clm** 40–100 cm, erect, smt geniculate, not rooting, at the lo nd. **Bas shth** villous; **up shth** glab, densely villous or densely tomentose, or sparsely to densely hairy, with papillose-based hairs; **lig** (1)1.5–6 mm, entire or lacerate, not ciliate; **bld** 2–12(18) cm long, 2–5(7) mm wide, glab or the adx surfaces sparsely to densely villous or tomentose. **Pan** with 4–10 spikelike pri br on 5–10 cm rchs, rarely with sec br; **pri br** 3–6 cm, appressed to ascending, axes not wing-margined; **intnd** 2–5.5 mm (midbr), bearing spklt in unequally pedlt pairs; **sec br** rarely present; **ped** not adnate to the br axes; **shorter ped** 0.1–0.3 mm; **longer ped** 1–2 mm; **tml ped of br** 1.7–6(7) mm. **Spklt** homomorphic, (3.7)4–7.5 mm (including pubescence), 3–5.4 mm (excluding pubescence). **Lo glm** 0.4–0.6 mm; **up glm** 2.5–5.1 mm (excluding pubescence), narrower than the up flt, 3-veined, densely villous, hairs 1.5–5 mm, silvery white to purple, widely divergent at maturity; **lo lm** 2.7–5 mm (excluding pubescence), pubescence exceeding the up flt by 2.2–4 mm, 7-veined, veins unequally spaced, only the 3 or 5 cent veins visible, mrg and outer lat veins densely pubescent, hairs 1.5–5 mm, silvery white to purple, widely divergent at maturity, intercostal regions glab, apc attenuate (acuminate); **up lm** 2.5–3.4 mm, ovate-lanceolate, brown to dark brown, acuminate. **Car** 1.3–2 mm. $2n = 36, 54, 70, 72$.

Digitaria californica grows on plains and open ground from Arizona, southern Colorado, and Oklahoma through Mexico and Central America to South America. The name reflects the fact that the first collection was made in Baja California, Mexico. Plants in the Intermountain Region belong to *D. californica* (Benth.) Henrard var. *californica*. They differ from those of *D. californica* var. *villosissima* Henrard in having densely villous, rather than densely tomentose, leaves.

2. **Digitaria ischaemum** (Schreb.) Muhl.
SMOOTH CRABGRASS [p. 460, 511]

Pl ann or of indefinite duration. **Clm** 20–55(70) cm, decumbent, brchg and rooting at the lo nd; **nd** 3–4. **Shth** glab or sparsely pubescent; **lig** 0.6–2.5 mm; **bld** 1.5–9 cm long, 3–5 mm wide, glab, with a few papillose-based hairs bas. **Pan** tml and axillary; **tml pan** with 2–7 spikelike pri br, subdigitate or on 0.5–2 cm rchs; **pri br** 6–15.5 cm, axes wing-margined, wings at least ½ as wide as the midribs, bearing spklt in groups of 3, lo portions of the longer ped adnate to the axes; **sec br** rarely present; **axillary infl** always present in some of the lo shth, entirely or partially concealed. **Spklt** 1.7–2.3 mm, homomorphic, narrowly elliptic. **Lo glm** absent or a veinless, memb rim; **up glm** 1.3–2.3 mm, from ¾ as long as to equaling the up lm, appressed-pubescent; **lo lm** 1.7–2.3 mm, 7-veined, veins unequally spaced, smooth, pubescent; **up lm** dark brown at maturity; **anth** 0.4–0.6 mm. $2n = 36$.

Digitaria ischaemum is a Eurasian weed that is now common in lawns, gardens, fields, and waste ground of warm-temperate regions throughout the world, including the Intermountain Region.

3. Digitaria sanguinalis (L.) Scop. Hairy
   Crabgrass [p. *461*, <u>511</u>]
Pl ann. **Clm** 20–70(112), often decumbent and
rooting at the lo nd. **Shth** keeled, usu sparsely
pubescent with papillose-based hairs; **lig** 0.5–2.6
mm; **bld** 2–11(14) cm long, 3–8(12) mm wide,
usu with papillose-based hairs on both surfaces,
smt glab. **Pan** with 4–13 spikelike pri br, these
subdigitate or on rchs to 6 cm; **pri br** 3–30 cm
long, 0.7–1.5 mm wide, flattened and winged,
wings more than ½ as wide as the midribs, lo and
mid portion of the br bearing spklt in unequally
pedlt pairs, ped not adnate to the br; **sec br**
rarely present. **Spklt** homomorphic, 1.7–3.4 mm
long, 0.7–1.1 mm wide. **Lo glm** 0.2–0.4 mm long,
veinless; **up glm** 0.9–2 mm, ⅓–½ as long as the
spklt, 3-veined, pubescent on the mrg; **lo lm** usu
exceeded or equaled by the up flt, smt exceeding
them but by no more than 0.2 mm, glab, 7-veined,
lat (or all) veins scabrous throughout or smooth
on the lo ⅓(½) and scabrous distally, 3 mid veins
usu widely spaced, remaining veins on each side
close together and near the mrg; **up lm** 1.7–3 mm,
yellow or gray, frequently purple-tinged when
immature, often becoming brown at maturity;
anth 0.5–0.9 mm. 2*n* = 36, 28, 34, 54.

Digitaria sanguinalis is a weedy Eurasian species that is
now found in waste ground of fields, gardens, and lawns
throughout much of the world, including the Intermountain
Region.

4. Digitaria ciliaris (Retz.) Koeler Southern
   Crabgrass [p. *461*, <u>511</u>]
Pl ann or of indefinite duration. **Clm** 10–100 cm
long, erect portion 30–60 cm, long-decumbent,

rooting and brchg at the decumbent nd, sparingly
brchd or unbrchd from the up nd; **nd** 2–5, glab.
Shth with papillose-based hairs; **lig** 2–3.5 mm,
erose; **bld** 1.5–14.4(18.9) cm long, 3–9 mm wide,
flat, glab, a few scattered papillose-based hairs at
the base of the adx surfaces (occ over the whole
adx surface), usu also scabrous on both surfaces.
Pan with 2–10 spikelike pri br, these digitate or in
1–3 whorls on rchs to 2 cm; **lowest pan nd** with
hairs more than 0.4 mm; **pri br** 3–24 cm long,
0.6–1.2(2) mm wide, glab or with less than 1 mm
hairs, axes wing-margined, wings at least ½ as
wide as the midribs, lo and mid portions of the
br bearing spklt in unequally pedlt pairs; **sec br**
absent; **shorter ped** 0.5–1 mm; **longer ped** 1.5–4
mm. **Spklt** (2.7)2.8–4.1 mm long, homomorphic.
Lo glm 0.2–0.8 mm, acute; **up glm** (1.2)1.5–2.7
mm, about ⅔ to almost as long as the spklt,
3-veined, mrg and apc pilose; **lo lm** 2.7–4.1 mm,
7-veined, veins unequally spaced, outer 3 veins
crowded together near each mrg, well separated
from the midvein, usu smooth, occ the lat veins
scabridulous on the distal ⅓, mrg and regions
between the 2 inner lat veins hairy, hairs 0.5–1
mm (rarely glab), smt also with glassy yellow
hairs between the 2 inner lat veins, these more
common on the up spklt; **up lm** 2.5–4 mm, glab,
yellow, tan, or gray when immature, becoming
brown, often purple-tinged (occ completely
purple) at maturity; **anth** 0.6–1 mm. 2*n* = 36, 54.

Digitaria ciliaris is a weedy species, found in open,
disturbed areas in most warm-temperate to tropical regions,
primarily in the eastern United States. Only var. *ciliaris* has
been found in the Intermountain Region.

14.02 ECHINOCHLOA P. Beauv.

P.W. Michael

Pl ann or per; with or without rhz. **Clm** 10–460 cm, prostrate, decumbent or erect, distal portions smt
floating, smt rooting at the lo nd; **nd** usu glab; **intnd** hollow or solid. **Shth** open, compressed; **aur**
absent; **lig** usu absent but, if present, of hairs; **bld** linear to linear-lanceolate, usu more than 10 times
longer than wide, flat, with a prominent midrib. **Infl** tml, pan of simple or compound spikelike br
attached to elongate rchs, axes not terminating in a bristle, spklt subsessile, densely packed on the
angular br; **dis** below the glm (cultivated taxa not or tardily dis). **Spklt** plano-convex, with 2(3) flt;
lo flt strl or stmt; **up flt** bisex, dorsally compressed. **Glm** memb; **lo glm** usu ¼–⅖ as long as the spklt
(varying to more than ½ as long), unawned to minutely awn-tipped; **up glm** unawned or shortly
awned; **lo lm** similar to the up glm in length and texture, unawned or awned, awns to 60 mm; **lo**
pal vestigial to well developed; **up lm** coriaceous, dorsally rounded, mostly smooth, apc short or
elongate, firm or memb, unawned; **up pal** free from the lm at the apc; **lod** absent or minute; **anth** 3.
Car ellipsoid, broadly ovoid or spheroid; **emb** usu 0.7–0.9 times as long as the car. *x* = 9. Name from
the Greek *echinos*, 'hedgehog', and *chloa*, 'grass', in reference to the bristly or often awned spikelets.

Echinochloa is a tropical to warm-temperate genus of 40–50 species that are usually associated with wet or damp places.
Many of the species are difficult to distinguish because they tend to intergrade. Some of the characters traditionally used for
distinguishing taxa, e.g., awn length, are affected by the amount of moisture available; others reflect selection by cultivation,
e.g., non-disarticulation in grain taxa, mimicry of rice as weeds of rice fields. There are five species in the Intermountain
Region.

The most abundant species appears to be the introduced, weedy *Echinochloa crus-galli*, which closely resembles the native *E. muricata*. The confusion between the two species has caused them to be treated as the same species. This confusion is probably reflected in the mapping of both *E. crus-galli* and *E. muricata*.

1. Lower lemmas usually unawned; spikelets, particularly those near the base of the panicles, not disarticulating at maturity; upper lemmas wider and longer than the upper glumes at maturity and, hence, exposed at maturity.
   2. Spikelets always green and pale at maturity, their apices usually obtuse, varying to acute; rachis nodes not or only sparsely hispid with papillose-based hairs; caryopses whitish . . . . 3. *E. frumentacea*
   2. Spikelets purplish to blackish brown at maturity, their apices obtuse to shortly acute; rachis nodes densely hispid with papillose-based hairs; caryopses brownish . . . . 5. *E. esculenta*
1. Lower lemmas often awned; spikelets disarticulating at maturity; upper lemmas not or scarcely exceeding the upper glumes in length and width at maturity.
   3. Panicle branches 0.7–2(4) cm long, without secondary branches; spikelets 2–3 mm long, unawned; lower glumes ½ as long . . . . 2. *E. colona*
   3. Panicle branches 1.5–10 cm long, usually rebranched, the secondary branches often short and inconspicuous; spikelets 2.5–5 mm long, awned or unawned; lower glumes of variable length.
      4. Upper lemmas with rounded or broadly acute coriaceous apices that pass abruptly into a membranous tip, a line of minute hairs present at the base of the tip; upper paleas abuptly narrowed . . . . 4. *E. crus-galli*
      4. Upper lemmas with acute or acuminate coriaceous apices that extend into thesomewhat stiff membranous tip, without hairs at the base of the tip; upper paleas prominently acuminate . . . . 1. *E. muricata*

1. **Echinochloa muricata** (P. Beauv.) Fernald
AMERICAN BARNYARD GRASS [p. 462, <u>511</u>]
Pl ann. **Clm** 80–160 cm, erect or spreading, smt rooting at the lowest nd, often developing short axillary flowering shoots at most up nd when mature; **lo nd** glab or puberulent; **up nd** glab. **Shth** glab; **lig** absent; **bld** 1–27 cm long, 0.8–30 mm wide. **Pan of pri clm** 7–35 cm, rchs and br glab or hispid, hairs to 3 mm, papillose-based; **pri br** 2–8 cm, usu spreading and rather distant, often with sec br. **Spklt** 2.5–5 mm, dis at maturity, usu purple or streaked with purple, usu hispid, hairs papillose-based. **Up glm** about as long as the spklt; **lo flt** strl; **lo lm** unawned or awned, awns to 16 mm; **lo pal** well developed; **up lm** broadly obovoid or orbicular, narrowing to an acute or acuminate coriaceous portion that extends into the memb tip, boundary between the coriaceous and memb portions not marked by minute hairs; **anth** 0.4–1.1 mm. **Car** 1.2–2.5 mm, broadly obovoid or spheroid, yellowish; **emb** 1.4–2 mm, 80–91% as long as the car. $2n = 36$.

Echinochloa muricata is native to North America, growing from southern Canada to northern Mexico in moist, often disturbed sites (but not rice fields). It resembles *E. crus-galli* in gross morphology and ecology, but differs consistently by the characters used in the key. Its two varieties tend to be distinct, but there is some overlap in both morphology and geography. One of its two varieties grows in the Intermountain Region.

Echinochloa muricata var. **microstachya** Wiegand [p. 462]
Spklt 2.5–3.8 mm. **Lo glm** 0.9–1.6 mm; **up glm** 2.8–3.8 mm; **lo lm** unawned or awned, awns to 10 mm; **anth** 0.4–0.7 mm.

Echinochloa muricata var. *microstachya* is the common variety in the western part of North America, extending east to the Missouri River and the Texas panhandle.

2. **Echinochloa colona** (L.) Link AWNLESS BARNYARD GRASS [p. 462, <u>512</u>]
Pl ann; erect or decumbent, ces or spreading, rooting from the lo cauline nd. **Clm** 10–70 cm; **lo nd** glab or hispid, hairs appressed; **up nd** glab. **Shth** glab; **lig** absent, lig region frequently brown-purple; **bld** 8–22 cm long, 3–6(10) mm wide, mostly glab, smt hispid, hairs papillose-based on or near the mrg. **Pan** 2–12 cm, erect, rchs glab or sparsely hispid; **pri br** 5–10, 0.7–2(4) cm, erect to ascending, spikelike, somewhat distant, without sec br, axes glab or sparsely hispid, hairs 1.5–2.5 mm, papillose-based. **Spklt** 2–3 mm, dis at maturity, pubescent to hispid, hairs usu not papillose-based, tips acute to cuspidate. **Lo glm** about ½ as long as the spklt; **up glm** about as long as the spklt; **lo flt** usu strl, occ stmt; **lo lm** unawned, similar to the up glm; **lo pal** subequal to the lm; **up lm** 2.6–2.9 mm, not or scarcely exceeding the up glm, elliptic, coriaceous portion rounded distally, passing abruptly into a sharply differentiated, memb, soon-withering

tip; **anth** 0.7–0.8 mm. **Car** 1.2–1.6 mm, whitish; **emb** 63–83% as long as the car. $2n = 54$.

Echinochloa colona is widespread in tropical and subtropical regions. It is adventive and weedy in North America, growing in low-lying, damp to wet, disturbed areas, including rice fields. The unbranched, rather widely spaced panicle branches make this one of the easier species of *Echinochloa* to recognize.

3. **Echinochloa frumentacea** Link Siberian Millet, White Panic [p. *462*]

Pl ann. **Clm** 70–150 cm, erect, glab. **Shth** glab; **lig** absent; **bld** 8–35 cm long, 3–20(30) mm wide, glab. **Pan** 7–18 cm, erect to slightly drooping at maturity, rchs not or only sparsely hispid, nd with papillose-based hairs; **br** numerous, appressed or ascending, spikelike, not or only sparsely hispid, hairs papillose-based; **pri br** 1.5–4 cm, glab or sparsely hispid, hairs to 3 mm, papillose-based; **sec br**, if present, usu concealed by the densely packed spklt; **longer ped** 0.2–0.5 mm. **Spklt** 3–3.5 mm, often with 1 strl and 2 bisex flt, not dis at maturity (particularly those near the bases of the pan), scabrous or short-hispid but without papillose-based hairs, green and pale at maturity, apc usu obtuse, varying to acute. **Up glm** narrower and shorter than the up lm; **lo flt** strl; **lo lm** unawned; **lo pal** subequal to the lo lm; **up lm** 2.5–3 mm, ovate to elliptic, coriaceous portion terminating abruptly at the base of the memb tip; **anth** 0.8–1 mm. **Car** 1.7–2.2 mm long, 1.6–1.8 mm wide, whitish; **emb** 66–86% as long as the car. $2n = 36, 54$.

Echinochloa frumentacea originated in India, and possibly also in Africa. It is grown for grain, fodder, and beer, but not as extensively as in the past. It is found occasionally in the contiguous United States and southern Canada, the primary source being birdseed mixes. It used to be confused with *E. esculenta*, from which it differs in its whitish caryopses and proportionately smaller embryos. It appears to be a domesticated derivative of *E. colona* (Yabuno 1962). Hybrids between *E. frumentacea* and *E. colona* are partially fertile; those with *E. esculenta* are sterile.

4. **Echinochloa crus-galli** (L.) P. Beauv. Barnyard Grass [p. *463*, <u>*512*</u>]

Pl ann. **Clm** 30–200 cm, spreading, decumbent or stiffly erect; **nd** usu glab or the lo nd puberulent. **Shth** glab; **lig** absent, lig region smt pubescent; **bld** to 65 cm long, 5–30 mm wide, usu glab, occ sparsely hirsute. **Pan** 5–25 cm, with few to many papillose-based hairs at or below the nd of the pri axes, hairs smt longer than the spklt; **pri br** 1.5–10 cm, erect to spreading, longer br with

short, inconspicuous sec br, axes scabrous, smt also sparsely hispid, hairs to 5 mm, papillose-based. **Spklt** 2.5–4 mm long, 1.1–2.3 mm wide, dis at maturity. **Up glm** about as long as the spklt; **lo flt** strl; **lo lm** unawned to awned, smt varying within a br, awns to 50 mm; **lo pal** subequal to the lm; **up lm** broadly ovate to elliptical, coriaceous portion rounded distally, passing abruptly into an early-withering, acuminate, memb tip that is further demarcated from the coriaceous portion by a line of minute hairs (use 25× magnification); **anth** 0.5–1 mm. **Car** 1.3–2.2 mm long, 1–1.8 mm wide, ovoid or oblong, brownish; **emb** 59–86% as long as the car. $2n = 54$.

Echinochloa crus-galli is a Eurasian species that is now widely established in North America, where it grows in moist, disturbed sites, including rice fields. Some North American taxonomists, including Welsh et al. (2008), interpret *E. crus-galli* much more widely; others treat it as here, but recognize several infraspecific taxa based on such characters as trichome length and abundance, and awn length. There are several ecological and physiological ecotypes within the species, but the correlation between these and the species' morphological variation has not been established, so no infraspecific taxa are recognized here.

5. **Echinochloa esculenta** (A. Braun) H. Scholz Japanese Millet [p. *463*]

Pl ann. **Clm** 80–150 cm tall, 4–10 mm thick, glab. **Shth** glab; **lig** absent, lig region smt pubescent; **bld** 10–50 cm long, 5–25 mm wide. **Pan** 7–30 cm, dense, rchs nd densely hispid, hairs papillose-based, intnd scabrous; **pri br** 2–5 cm, erect or spreading, simple or brchd, often incurved at maturity, nd hispid, hairs papillose-based, intnd usu scabrous; **longer ped** 0.5–1 mm. **Spklt** 3–4 mm long, 2–2.5 mm wide, not or only tardily dis at maturity, obtuse to shortly acute, purplish to blackish brown at maturity. **Up glm** narrower and shorter than the up lm; **lo flt** strl; **lo lm** usu unawned; **lo pal** shorter and narrower than the lm; **up lm** longer and wider than the up glm, broadly ovate to ovate-orbicular, shortly apiculate, exposed distally at maturity; **anth** 1–1.2 mm. **Car** 1.2–2.3 mm, brownish; **emb** 84–96% as long as the car. $2n = 54$.

Echinochloa esculenta was derived from *E. crus-galli* in Japan, Korea, and China. It is cultivated for fodder, grain, or birdseed. It has sometimes been included in *E. frumentacea*, from which it differs in its brownish caryopses and longer pedicels. It appears to be a domesticated derivative of *E. crus-galli* (Yabuno 1962). Hybrids between *E. crus-galli* and *E. esculenta* are fully fertile, but those with *E. frumentacea* are sterile.

14.03 **DICHANTHELIUM** (Hitchc. & Chase) Gould　　Robert W. Freckmann and Michel G. Lelong

Pl per; ces, smt rhz, smt with hard, cormlike bases, often with bas winter rosettes of lvs having shortly ovate to lanceolate bld, these often sharply distinct from the bld of the cauline lvs. **Clm** 5–150 cm, herbaceous, hollow, usu erect or ascending, rarely sprawling, in the spring often spreading, smt decumbent in the fall, usu brchg from the mid- or lo clm nd in summer and fall; **br** rebrchg 1–4 times, terminating in small sec pan that are usu partly included in the shth. **Cauline lvs** 3–14, usu distinctly longer and narrower than the rosette bld; **lig** of hairs, memb, or memb and ciliate, smt absent; **psdlig** of 1–5 mm hairs often present at the bases of the bld immediately behind the true lig; **bld** usu distinctly longer and narrower than those of the bas rosette, cross sections with non-Kranz anatomy; photosynthesis C_3. **Infl** pan, tml on the clm and br; **strl br** and bristles absent; **dis** below the glm. **Pri pan** terminating the clm, developing April–June(July), smt also in late fall, usu at least partially chasmogamous, often with a lo seed set than the sec pan; **sec pan** terminating the br, produced from (May)June to fall, usu partially or totally cleistogamous. **Spklt** 0.8–5.2 mm, not subtended by bristles, dorsally compressed, surfaces unequally convex, apc unawned. **Glm** apc not or only slightly gaping at maturity; **lo glm** ⅕–¾ as long as the spklt, 1–5-veined, truncate, acute, or acuminate; **up glm** slightly shorter than the spklt or exceeding the up flt by up to 1 mm, 5–11-veined, not saccate, apc rounded to attenuate. **Lo flt** strl or stmt; **lo lm** similar to the up glm; **lo pal** smt present, thin, shorter than the lo lm; **up flt** bisex, sessile, plump, usu apiculate to mucronate, smt minutely so, or subacute to (rarely) acute; **up lm** striate, chartaceous-indurate, shiny, usu glab, mrg involute; **up pal** striate; **lod** 2; **anth** 3. **Car** smooth; **pericarp** thin; **endosperm** hard; **hila** round or oval. $x = 9$. Name from the Greek *di*, 'twice' and *anth*, 'flowering', a reference to the two flowering periods.

　　Dichanthelium is a genus of approximately 72 species, two of which grow in the Intermountain Region. It is sometimes included in *Panicum*, the two taxa being similar in gross morphology. Molecular data reinforce the morphological arguments for recognizing *Dichanthelium* as a distinct genus.

　　When the branches of *Dichanthelium* develop, in late summer or fall, the culms acquire a very different aspect; comments about the 'fall phase' refer to the appearance of the plant or its culms following this branching. Unless stated otherwise, descriptions and measurements refer to structures of the culms and primary panicles, not those of the branches and secondary panicles. Ligule measurements usually include the hairs of the pseudoligule, if present, because the two are often difficult to distinguish with less than 30× magnification.

1. Spikelets 2.7–3.5 mm long, broadly obovoid-ellipsoid, turgid; upper glumes with a prominent orange to purplish spot at the base, the veins prominent; ligules 1–1.5 mm long . . . . 1. *D. oligosanthes*
1. Spikelets 1.5–2.1 mm long, ellipsoid or obovoid, not turgid; upper glumes lacking an orange or purple spot at the base and the veins not prominent; ligules and pseudoligules 2–5 mm long . . . . 2. *D. acuminatum*

1. **Dichanthelium oligosanthes** (Schult.) Gould
Few-Flowered Panicgrass [p. 463, <u>512</u>]

Pl ces, with caudices. **Bas rosettes** well differentiated; **bld** 2–6 cm, few, ovate to lanceolate. **Clm** 20–75 cm, geniculate bas, stiffly erect distally; **nd** glab or sparsely pubescent; **intnd** often purplish, glab, puberulent, or papillose-hirsute; **fall phase** brchg from the midclm nd, br initially ascending to erect, smt developing simultaneously with and overtopping the pri pan, later rebrchg to form short, bushy clumps of bld and small, included sec pan. **Cauline lvs** 5–7; **shth** not overlapping, glab, puberulent, or ascending papillose-hispid, mrg ciliate, col loose, puberulent; **lig** 1–3 mm, of hairs; **bld** 5–12 cm long, 4–15 mm wide, flat or partly involute, glab or pubescent abx, with 7–9 major veins only slightly more prominent

than the minor veins, bases ciliate, rounded to truncate, mrg cartilaginous. **Pri pan** 5–9 cm long, 3–6 cm wide, partly enclosed to long-exserted, with 6–60 spklt; **br** stiff or wiry, puberulent or scabridulous. **Spklt** 2.7–4.2 mm long, 1.7–2.4 mm wide, ellipsoid to broadly obovoid, turgid, glab or sparsely pubescent. **Lo glm** 1–1.6 mm, acute, similar in texture and vein prominence to the up glm; **up glm** strongly veined, often orange to purplish at the base; **lo flt** strl; **up flt** with minutely umbonate apc. $2n = 18$.

　　Dichanthelium oligosanthes grows in Canada and the United States and extends into northern Mexico. The primary panicles are briefly open-pollinated, then cleistogamous, from late May to early June; the secondary panicles, which are produced from June to November, are cleistogamous. One subspecies grows in the Intermountain Region.

Dichanthelium oligosanthes subsp.
  scribnerianum (Nash) Freckmann & Lelong
  SCRIBNER'S PANICGRASS [p. 463]

Clm 20–50 cm; **intnd** often lustrous, glab or sparsely papillose-hispid (rarely puberulent). **Cauline shth** often lustrous, glab or sparsely papillose-hispid (rarely puberulent); **lig** 1–1.5 mm; **bld** usu 6–15 mm wide, less than 10 times longer than wide, flat, ascending to spreading, glab or sparsely pubescent abx, acute. **Pri pan** denser than in subsp. *oligosanthes*, br more flexible; **ped** mostly shorter than 5 mm. **Spklt** usu 2.7–3.5 mm long, 2–2.4 mm wide, broadly obovoid-ellipsoid, usu glab. **Up glm** with a prominent orange to purplish spot at the base.

Dichanthelium oligosanthes subsp. *scribnerianum* grows in sandy or clayey banks and prairies. Its range extends from southern British Columbia to the east coast of the United States, and south into northern Mexico.

2. **Dichanthelium acuminatum** (Sw.) Gould &
  C.A. Clark HAIRY PANICGRASS [p. 464, 512]

Pl more or less densely ces. **Bas rosettes** usu well differentiated; **bld** ovate to lanceolate. **Clm** 15–100 cm (rarely taller), usu thicker than 1 mm, weak and wiry or relatively stout and rigid, erect, ascending or decumbent; **nd** occ swollen, glab or densely pubescent, often with a glab or viscid ring below; **intnd** purplish or olive green or grayish green, to yellowish green, variously pubescent, with hairs of 2 lengths or glab; **fall phase** erect, spreading, or decumbent, usu brchg extensively at all but the uppermost nd, ultimately forming dense fascicles of branchlets with rdcd, flat or involute bld and rdcd sec pan with few spklt. **Cauline lvs** 4–7; **shth** usu shorter than the intnd, glab or densely and variously pubescent with hairs shorter than 3 mm, mrg ciliate or glab; **lig** and **psdlig** 1–5 mm, of hairs; **bld** 2–12 cm long (rarely longer), 2–12 mm wide (rarely wider), firm or lax, spreading to reflexed or stiffly ascending, yellowish green or grayish green to olivaceous, densely to sparsely and variously pubescent, mrg similar or occ whitish scabridulous, mrg often with papillose-based

cilia, at least bas, bases rounded or subcordate. **Pri pan** 3–12 cm, ¼–¾ as wide as long, usu open, well exserted, rather dense; **rchs** glab, puberulent, or more or less densely pilose, at least bas. **Spklt** 1.1–2.1 mm, obovoid to ellipsoid, yellowish green to olivaceous or purplish, variously pubescent, obtuse or subacute. **Lo glm** usu ¼–½ as long as the spklt, obtuse to acute; **up glm** and lo lm subequal, equaling the up flt at maturity, or occ the up glm slightly shorter, not strongly veined; **lo flt** strl; **up flt** 1.1–1.7 mm long, 0.6–1 mm wide, ellipsoid, obtuse to acute or minutely umbonate or apiculate. $2n = 18$.

Dichanthelium acuminatum is common and ubiquitous in dry to wet, open, sandy or clayey woods, clearings, bogs, and swamps, or in saline soil near hot springs, growing in much of North America and extending into northern South America. It is probably the most polymorphic and troublesome species in the genus. Only one subspecies grows in the Intermountain Region.

Dichanthelium acuminatum subsp.
  fasciculatum (Torr.) Freckmann & Lelong
  [p. 464]

Pl yellowish green to olivaceous or purplish. **Clm** 15–75 cm, suberect, ascending or spreading; **nd** often with spreading hairs, occ with a glab ring below. **Cauline shth** with ascending to spreading, papillose-based hairs, occ with shorter hairs underneath; **midclm shth** about ½ as long as the intnd; **lig** and **psdlig** 2–5 mm; **bld** 5–12 cm long, 6–12 mm wide, spreading to ascending, bases with papillose-based cilia, abx surfaces usu pubescent, adx surfaces pilose or glab, hairs shorter than 3 mm. **Spklt** 1.5–2 mm (tending to be longer in the western part of its range), obovoid to ellipsoid.

Dichanthelium acuminatum subsp. *fasciculatum* grows primarily in disturbed areas, open or cut-over woods, thickets, and grasslands, in dry to moist soils, including river banks, lake margins, and marshy areas. It is widespread in temperate North America, growing from Canada to Mexico, but it is somewhat less common in the western part of the species range, where it often grows in moister areas.

Dichanthelium acuminatum subsp. *fasciculatum* includes probably the most widespread, ubiquitous, and variable assemblages of forms in the species.

14.04 **PANICUM** L.   Robert W. Freckmann and Michel G. Lelong

Pl ann or per; their habit variable. **Clm** 2–300 cm, herbaceous, smt hard and almost woody, or woody, simple or brchd, bases smt cormlike; **intnd** solid, spongy, or hollow. **Lvs** cauline, bas, or both, bas lvs not forming a winter rosette; **lig** memb, usu ciliate; **bld** filiform to ovate, flat to involute, glab or pubescent, cross sections with Kranz anatomy and 1 or 2 bundle shth or with non-Kranz anatomy; **photosynthesis** C_4 with NAD-ME or NADP-ME pathways, or, in pl with non-Krantz anatomy, C_3. **Infl** tml on the clm and br, often also axillary, tml pan typically appearing after midsummer; **strl br** and **bristles** absent; **dis** usu below the glm, smt at the base of the up flt, if at the base of the up flt, then the flt not very plump at maturity. **Spklt** 1–8 mm, usu dorsally compressed, smt subterete or lat compressed, unawned. **Glm** usu unequal, herbaceous, glab or pubescent, rarely tuberculate or glandular, apc not or only slightly gaping at maturity; **lo glm** minute to almost equaling the spklt, 1–9-veined, truncate, acute, or acuminate; **up glm** slightly shorter to much longer than the spklt, 3–13(15)-veined, bases rarely slightly sulcate, apc rounded to attenuate; **lo flt** strl or stmt; **lo lm** similar to the up glm; **lo pal** absent, or shorter than the lo lm and hyaline; **up flt** bisex, sessile or stipitate, apc acute, puberulent, or with a tuft of hairs; **up lm** usu more or less rigid and chartaceous-indurate, usu shiny, glab or (rarely) pubescent, usu smooth, smt verrucose or transversely rugose, mrg involute, usu clasping the pal, rarely with bas wings or lunate scars, apc obtuse, acute, apiculate, or with small green crests; **up pal** striate, rarely transversely rugose; **lod** 2; **anth** usu 3. **Car** smooth; **pericarp** thin; **endosperm** hard, without lipid, starch grains simple or compound, or both; **hila** round or oval. $x = 9$ (usu), smt 10, with polyploid and dysploid derivatives. Name from the Latin *panis*, 'bread', or *panus*, an ear of millet.

   Panicum is a large genus, but just how large is difficult to estimate because its limits are not yet clear. Recent work supports some aspects of the treatment presented here, but not all of them. For instance, Aliscioni et al. (2003) and Giussani et al. (2001) suggest that *Panicum* subg. *Panicum* is a monophyletic group that should have a rank equivalent to *Dichanthelium* and *Steinchisma*. This treatment excludes *P. plenum* and *P. obtusum* (now placed in *Hopia*), and *P. bulbosum* (now in *Zuloagaea*).

   Most species of *Panicum* are tropical, but many grow in warm, temperate regions. Within North America, *Panicum* is most abundant in the southeastern United States. Ten species grow in the Intermountain Region. Many species grow in early seral stages or weedy areas; some grow at forest edges, in prairies, savannahs, deserts, forests, beaches, and in shallow water.

   Panicum miliaceum has been grown since prehistory in China and India as a cereal grain, and is a common component of bird seed. Seeds of *P. hirticaule* subsp. *sonorum* have been used for food by the Cocopa tribe of the southwest. *Panicum virgatum* is an important hay and range species.

   Apomixis, polyploidy, and autogamy have produced numerous microspecies in some groups; hybridization and introgression has resulted in a reticulum of intergrading forms in some complexes. The number of taxa recognized has varied widely over the past century.

1. Plants with rhizomes about 1 cm thick and with large, pubescent, scalelike leaves; culms hard, almost woody; margins of upper lemmas nearly flat . . . . 10. *P. antidotale*
1. Plants without rhizomes or with rhizomes less than 0.5 cm thick and with small, glabrous, scalelike leaves; culms clearly not woody; margins of upper lemmas inrolled.
   2. Glumes, lower lemmas, and upper lemma margins villous, with whitish hairs . . . . 9. *P. urvilleanum*
   2. Glumes and lemmas usually glabrous, sometimes the lower lemmas sparsely pilose on the margins and near the apices.
      3. Plants perennial, usually with vigorous scaly rhizomes; lower florets staminate . . . . 8. *P. virgatum*
      3. Plants annual, or perennials usually without rhizomes, sometimes rooting at the lower nodes; lower florets sterile.
         4. Lower glumes obtuse to acute, ¼–⅓ as long as the spikelets; sheaths compressed, glabrous or sparsely pubescent; culms usually succulent. . . . . 7. *P. dichotomiflorum*
         4. Lower glumes acute to attenuate, usually ⅓–¾ as long as the spikelets; sheaths rounded, usually hirsute or hispid; culms not succulent.
            5. Spikelets 4–6 mm long. . . . . 1. *P. miliaceum*
            5. Spikelets 1.9–4.2 mm long.
               6. Plants perennial; panicle branches usually with all or most secondary branches confined to the distal ⅓ . . . . 6. *P. hallii*
               6. Plants annual; panicle branches usually with secondary branches and pedicels attached to the distal ⅔.
                  7. Panicles more than 2 times longer than wide at maturity; branches ascending to somewhat divergent; spikelets narrowly ovoid, usually about 3 times longer than wide . . . . 3. *P. flexile*
                  7. Panicles less than 1.5 times longer than wide at maturity; branches diverging; spikelets variously shaped, less than 3 times longer than wide.
                     8. Spikelets 1.9–4 mm long; lower glumes usually ⅓–½ as long as the spikelets; lower paleas usually small or absent; ligules 0.5–1.5 mm long. . . . . 2. *P. capillare*
                     8. Spikelets 1.9–3.3 mm long; lower glumes ½–¾ as long as the spikelets; lower paleas 0.2–0.9 mm long, from ⅓ as long as the lower lemmas to equaling them; ligules 0.2–0.4 mm or 1.5–3.5 mm long.

9. Primary panicle branches appressed to the main axis; culms 2–8 cm long; spikelets 2–2.2 mm long . . . . 5. *P. mohavense*

9. Primary panicle branches divergent; culms 11–70 cm long; spikelets 1.9–3.3 mm long . . . . 4. *P. hirticaule*

1. **Panicum miliaceum** L. BROOMCORN, PROSO MILLET, HOG MILLET, PANIC MILLET [p. 464, 512]

Pl ann; smt brchg from the lo nd. **Clm** 20–210 cm, stout, not woody; **nd** puberulent; **intnd** usu with papillose-based hairs, smt nearly glab, not succulent. **Lvs** numerous; **shth** terete, densely pilose, with papillose-based and caducous hairs; **lig** memb, ciliate, cilia 1–3 mm; **bld** 15–40 cm long, 7–25 mm wide. **Pan** 6–20 cm long, 4–11 cm wide, included or shortly exserted at maturity, dense; **br** stiff, appressed to spreading, spklt solitary, confined to the distal portions; **ped** 1–9 mm, scabrous and sparsely pilose. **Spklt** 4–6 mm, ovoid, usu glab. **Lo glm** 2.8–3.6 mm, ½–¾ as long as the spklt, 5–7-veined, veins scabridulous distally, apc attenuate; **up glm** 4–5.1 mm, slightly exceeding the up flt, 11–13(15)-veined, veins scabridulous distally; **lo flt** strl; **lo lm** 4–4.8 mm, slightly exceeding the up flt, 9–13-veined, veins scabridulous distally; **lo pal** 1.2–1.6 mm, ½ or less the length of the up flt, truncate to bilobed; **up flt** 3–3.8 mm long, 2–2.5 mm wide, smooth or striate, more or less shiny, stramineous to orange, red-brown, or blackish, persisting in the spklt or dis at maturity. 2*n* = 36, 40, 42, 49, 54, 72.

Panicum miliaceum is native to Asia, where it has been cultivated for thousands of years. In North America, it is grown for bird seed and is occasionally planted for game birds. It is also found in corn fields and along roadsides. In Asia, *P. miliaceum* is still grown for fodder and as a cereal, its fast germination and short growth period enabling it to be sown following a spring crop. It also has one of the lowest water requirements of any cereal grain.

1. Mature upper florets stramineous to orange, not disarticulating; culms 20–120 cm tall; panicles usually nodding, not fully exserted, more than twice as long as wide; panicle branches ascending to appressed; pulvini almost absent . . . subsp. *miliaceum*

1. Mature upper florets blackish, disarticulating at maturity; culms 70–210 cm tall; panicles erect, exserted at maturity, about twice as long as wide; panicle branches ascending to spreading; pulvini well developed . . . . subsp. *ruderale*

Panicum miliaceum L. subsp. **miliaceum** [p. 464]

Clm 20–120 cm. **Pan** more than twice as long as wide, relatively contracted, usu nodding, not fully exserted; **br** ascending to appressed; **pulvini** almost absent. **Up flt** stramineous to orange, not dis at maturity.

Panicum miliaceum subsp. *miliaceum* is the subspecies used in bird seed. It probably rarely persists because of the retention of the upper florets on the plant and, in northern states, poor seed survival over winter.

Panicum miliaceum subsp. **ruderale** (Kitag.) Tzvelev [p. 464]

Clm 70–210 cm. **Pan** about twice as long as wide, open, erect, exserted; **br** ascending to spreading; **pulvini** well developed. **Up flt** blackish, shiny, dis at maturity.

Panicum miliaceum subsp. *ruderale* is now naturalized over much of North America. It is sometimes a major weed, especially in corn fields.

2. **Panicum capillare** L. WITCHGRASS [p. 465, 512]

Pl ann; hirsute or hispid, hairs papillose-based, often bluish or purplish. **Clm** 15–130 cm, slender to stout, not woody, erect to decumbent, straight to zigzag, simple to profusely brchd; **nd** sparsely to densely pilose. **Shth** rounded, hirsute or hispid, hairs papillose-based; **lig** memb, ciliate, cilia 0.5–1.5 mm; **bld** 5–40 cm long, 3–18 mm wide, linear, spreading. **Pan** 13–50 cm long, 7–24 cm wide, usu more than ½ as long as the pl, included at the base or exserted at maturity, dis at the base of the peduncles at maturity and becoming a tumbleweed; **br** spreading; **ped** 0.5–2.8 mm, scabrous, pilose. **Spklt** 1.9–4 mm, ellipsoid to lanceoloid, often red-purple, glab. **Lo flt** strl; **lo glm** ⅓–½ as long as the spklt, 1–3-veined; **up glm** 1.8–3.1 mm, 7–9-veined, midveins scabridulous; **lo lm** 1.9–3 mm, extending 0.4–1.1 mm beyond the up flt, often stiff, straight, prominently veined distally; **up flt** stramineous or nigrescent, smt with a prominent lunate scar at the base, often dis before the glm, leaving the empty glm and lo lm temporarily persisting on the pan. 2*n* = 18.

Panicum capillare grows in open areas, particularly in disturbed sites such as fields, pastures, roadsides, waste places, ditches, sand, rock crevices, etc. It grows throughout temperate North America, including northern Mexico.

Panicum capillare L. subsp. **capillare** [p. 465]

Clm medium to robust, ascending to erect, rarely delicate or spreading, usu green or red-purple, rarely bluish green, often brchg at the base. **Pan** br spreading; **sec br** and ped strongly divergent. **Spklt** 1.9–4 mm. **Lo pal** absent; **mature up flt** about ½ as wide as long, stramineous or tan, smt blackish, without a lunate scar at the base.

Panicum capillare subsp. *capillare* is the common subspecies, growing in weedy and dry habitats throughout the range of the species. Plants in the western United States and Canada have spikelets over 2.6 mm long more often than those in the east. Robust plants germinating early in the season and growing on better soils tend to spread more, and have wider, shorter blades and more exserted panicles than plants in the eastern United States and Canada growing under comparable conditions. They are sometimes included in *P. capillare* var. *occidentale* Rydb., but these traits are not well correlated, and several environmental factors apparently affect their expression.

3. **Panicum flexile** (Gatt.) Scribn. WIRY
   WITCHGRASS [p. 465, 512]

Pl ann; delicate, green or yellow-green. **Clm** 10–75 cm, about 1 mm thick, simple or with erect bas br; **nd** densely pilose, hairs ascending; **intnd** glab or shortly pubescent distally. **Shth** longer than the intnd, green to purplish, hispid, mrg sparsely ciliate; **lig** 0.5–1.5 mm; **bld** 3–32 cm long, 1–7 mm wide, ascending to erect, linear, narrowing bas, flat or the mrg involute, surfaces sparsely hirsute or pilose (rarely glab), hairs near the base papillose-based, mrg prominent, apc acute. **Pan** 5–45 cm long, 1–6 cm wide, at least ½ as long as the pl and 3 times longer than wide, open; **rchs** glab; **pri br** usu alternate or subopposite, ascending to slightly divergent, sec br and ped attached to the distal ⅔; **sec br** diverging; **ped** 0.5–17 mm, ascending to appressed. **Spklt** 2.5–3.7 mm long, 0.6–1.1 mm wide, narrowly ovoid, glab, acute; **lo glm** 0.8–1.3 mm, ⅓–½ as long as the spklt, acuminate; **up glm** 2.3–3.3 mm, 7–9-veined, exceeding the up flt by about 0.6 mm; **lo flt** strl; **lo lm** 2.2–2.7 mm, exceeding the up flt by about 0.6 mm, 7- or 9-veined, apc scabridulous, pointed; **lo pal** absent; **up flt** 1.6–1.7 mm long, about 0.6 mm wide, usu smooth, usu pale, smt becoming dark at maturity. $2n = 18$.

Panicum flexile grows in fens and other calcareous wetlands, in dry, calcareous or mafic rock barrens, and in open woodlands, especially on limestone derived soils.

4. **Panicum hirticaule** J. Presl ROUGHSTALKED
   WITCHGRASS [p. 466, 512]

Pl ann; glab or hispid, hairs papillose-based. **Clm** 11–110 cm, erect to decumbent; **nd** shortly hirsute or glab. **Shth** shorter than the intnd, greenish to purplish, glab or with papillose-based hairs, ciliate on 1 mrg, glab on the other; **col** hirsute; **lig** 1.5–3.5 mm, of hairs; **bld** 3–30 cm long, 3–30 mm wide, flat, usu hirsute or sparsely pubescent, hairs papillose-based, smt glab, bases rounded to cordate-clasping, mrg ciliate, cilia papillose-based, apc acute. **Pan** 9–30 cm long, 5–8 cm wide, erect or nodding, partially included to well exserted, rchs glab or sparsely hispid bas;

pri br usu alternate to opposite, divergent, sec br and ped confined to the distal ⅔; **pulvini** inconspicuous; **sec br** appressed; **ped** 9–27 mm, appressed. **Spklt** 1.9–4 mm long, 0.8–1 mm wide, ovoid to almost spherical, often reddish brown, glab, veins prominent, scabridulous, apc abruptly acuminate. **Lo glm** 1.3–2.4 mm, ½–¾ as long as the spklt, 3–5-veined; **up glm** 1.8–3.3 mm, 7–11-veined; **lo flt** strl; **lo lm** similar to the up glm, 9-veined; **lo pal** 0.4–0.9 mm; **up flt** 1.5–2.4 mm long, 0.4–0.8 mm wide, ellipsoid, smooth or conspicuously papillate, shiny, stramineous, often with a lunate scar at the base.

Panicum hirticaule grows in rocky or sandy soils in waste places, roadsides, ravines, and wet meadows along streams. Its range extends from southeastern California and southwestern Texas southward through Mexico, Central America, Cuba, and Hispaniola to western South America and Argentina. One subspecies grows in the Intermountain Region.

Panicum hirticaule J. Presl subsp. **hirticaule**
   [p. 466]

Clm 11–70 cm tall, usu simple; **nd** usu hirsute. **Shth** hirsute, hairs papillose-based; **bld** 3–16 mm wide, rounded bas. **Pan** erect. **Spklt** 1.9–3.3 mm. **Lo pal** less than ½ as long as the up flt. $2n = 18, 36$.

Panicum hirticaule subsp. *hirticaule* is the most common of the subspecies, growing throughout the range of the species but occurring more often in arid habitats. It includes *P. alatum* Zuloaga and Morrone, a recently described species that differs from *P. hirticaule* subsp. *hirticaule* by the presence of paired elaiosomes at the base of a slightly stipitate upper floret.

5. **Panicum mohavense** Reeder MOHAVE
   WITCHGRASS [p. 466, 512]

Pl ann. **Clm** 2–8 cm, erect-spreading; **nd** 1–2, hispid; **intnd** pilose, hairs papillose-based. **Shth** rounded, much longer than the intnd, with prominent veins, hispid, hairs papillose-based; **lig** 0.2–0.4 mm, memb, ciliate; **bld** 1–4 cm long, 1–3 mm wide, flat or involute apically, glab bas, mrg ciliate, cilia papillose-based. **Pan** congested, partially included in the shth, less than 1.5 times longer than wide; **br** ascending, narrow; **pri br** appressed to the main axes, sec br and ped attached to the distal ⅔; **ped** appressed, 1–2 mm. **Spklt** 2–2.2 mm long, 1–1.3 mm wide, plump-ellipsoid, glab. **Lo glm** 1.2–1.3 mm, acute to attenuate; **up glm** and lo lm 2–2.2 mm, 7–9-veined, apc purplish, acute; **lo flt** strl; **lo pal** 0.2–0.4 mm; **up flt** 1.4–1.8 mm long, about 1 mm wide, broadly ovoid.

Panicum mohavense is known only from arid limestone terraces in Arizona and New Mexico.

6. **Panicum hallii** Vasey HALL's WITCHGRASS
[p. *466*, 512]

Pl per; ces. **Clm** 10–100 cm, 2–10 mm thick, erect, simple or sparingly brchd bas; **nd** sericeous, pilose or glab; **intnd** usu glaucous. **Lvs** often crowded bas; **shth** rounded, glab or hirsute, hairs fragile, papillose-based, mrg smt ciliate distally; **lig** 0.6–2 mm; **bld** 4–23 cm long, 1–10 mm wide, erect to spreading, flat or smt involute (on strl br), often curling at maturity, glaucous, abx surfaces smt with prominent papillae along the midribs, bases rounded or narrowing to the shth, mrg cartilaginous, ciliate bas, scabridulous elsewhere, apc acute. **Tml pan** 7–31 cm long, 3–15 cm wide; **rchs** glab, tending to break at maturity; **br** usu alternate, slender, stiff, ascending to divergent; **ped** 1–15 mm, appressed. **Spklt** 2.1–4.2 mm long, 0.8–1 mm wide, usu ovoid, glab. **Lo glm** 1.2–2.4 mm, ½–¾ as long as the spklt, attenuate; **up glm** and **lo lm** similar, 7–11-veined, acuminate, extending 0.3–1.2 mm beyond the up flt; **lo flt** strl; **lo pal** 0.8–2 mm; **up flt** 1.5–2.4 mm long, 0.7–1.2 mm wide, ovoid to ellipsoid, smooth, nigrescent. $2n = 18$.

Panicum hallii grows on sandy, gravelly, or rocky land, including roadsides, pastures, rangeland, oak and pine savannahs, chaparral, and moist areas in deserts and on mesas. Its range extends from the southwestern United States to southern Mexico.

Panicum hallii subsp. **filipes** (Scribn.) Freckmann & Lelong [p. *466*]

Pl often taller than subsp. *hallii*, sparsely pubescent to almost glab. **Bld** relatively lax, ascending to spreading, not strongly clustered bas or curling at maturity. **Pan** scarcely exceeding the bld, with more closely spaced spklt than in subsp. *hallii*; **main br** rarely whorled, more crowded. **Spklt** 2.1–3 mm.

Panicum hallii subsp. *filipes* often grows in moist soil. Its range extends from Arizona, Texas, and Louisiana to southern Mexico.

7. **Panicum dichotomiflorum** Michx. FALL PANICUM [p. *467*, 512]

Pl ann or short-lived per in North America, per in the tropics; usu terrestrial, smt aquatic but not floating. **Clm** 5–200 cm tall, 0.4–3 mm thick, decumbent to erect, commonly geniculate to ascending, rooting at the lo nd when in water, simple to divergently brchd from the lo and mid nd, usu succulent, slightly compressed, glab; **nd** usu swollen, smt constricted on robust pl, glab; **intnd** glab, shiny, pale green to purplish. **Shth** compressed, inflated, sparsely pubescent near the base, elsewhere mostly glab, sparsely pilose, or hispid, hairs smt papillose-based, mrg or throat ciliate, with papillose-based hairs; **lig** 0.5–2 mm; **bld** 10–65 cm long, 3–25 mm wide, glab or sparsely pilose, often scabrous near the mrg, midribs stout, whitish. **Pan** 4–40 cm, diffuse, lax, with a few spklt; **br** to 15 cm, alternate or opposite, occ verticillate, ascending to spreading, stiff, scabrous; **ped** 1–6 mm, sharply 3-angled, scabrous, expanded to cuplike apc, appressed mostly to the abx side of the br. **Spklt** 1.8–3.8 mm long, 0.7–1.2 mm wide, ellipsoid to narrowly ovoid, light green to red-purple, glab, acute to acuminate. **Lo glm** 0.6–1.2 mm, ¼–⅓ as long as the spklt, 0–3-veined, obtuse to acute; **up glm** and lo lm similar, exceeding the up flt by 0.3–0.6 mm, 7–9-veined; **lo pal** vestigial to almost as long as the lo lm; **lo flt** strl; **up flt** 1.4–2.5 mm long, 0.7–1.1 mm wide, narrowly ellipsoid, smooth, shiny, stramineous to nigrescent, with pale veins. $2n = 36, 54$.

Panicum dichotomiflorum grows in open, often wet, disturbed areas such as cultivated and fallow fields, roadsides, ditches, open stream banks, receding shores, clearings in flood plain woods, and sometimes in shallow water. It is probably native throughout the eastern United States and adjacent Canada, but introduced elsewhere, including in the western United States. Its size and habit may be partly under genetic control, but these features also seem to be strongly affected by moisture levels, soil richness, competition, and the time of germination.

Panicum dichotomiflorum Michx. subsp. **dichotomiflorum** [p. *467*]

Clm 5–200 cm. **Shth** glab or sparsely pilose, not hispid with papillose-based hairs. **Ped** usu less than 3 mm and shorter than the spklt. **Spklt** 2.3–3.8 mm, tapered from below the mid to the acuminate apc; **up glm** and **lo lm** subcoriaceous.

Panicum dichotomiflorum subsp. *dichotomiflorum* is the most common of the three subspecies and is the only one found in the Intermountain Region. In the past, members of this subspecies have been treated as two different taxa, var. *geniculatum* (Alph. Wood) Fernald and var. *dichotomiflorum*, with more erect, slender plants having fewer long-exserted panicles with slender, ascending branches and less crowded spikelets being placed in var. *dichotomiflorum*. Such plants are more common in the southern part of the subspecies' range, but the traits are poorly correlated and the differences are at least in part affected by photoperiod, nighttime temperatures, and the time of seed germination.

8. **Panicum virgatum** L. SWITCHGRASS [p. *467*, 512]

Pl per; rhz, rhz often loosely interwoven, hard, with closely overlapping scales, smt short or forming a knotty crown. **Clm** 40–300 cm tall, 3–5 mm thick, solitary or forming dense clumps, erect or decumbent, usu simple; **nd** glab; **intnd** hard, glab or glaucous, green or purplish. **Shth** longer than the lo intnd, shorter than those above, glab or pilose, especially on the throat, mrg usu ciliate; **lig** 2–6 mm; **bld** 10–60 cm long, 2–15 mm wide, flat, erect, ascending or spreading, glab or pubescent, adx surfaces smt densely pubescent, particularly bas, bases rounded to slightly narrowed, mrg scabrous. **Pan** 10–55 cm long, 4–20 cm wide, exserted, open; **pri br** thin, straight, solitary to whorled or fascicled, ascending to spreading, scabrous, usu rebrchg once; **ped** 0.5–20 mm, appressed to spreading. **Spklt** 2.5–8 mm long, 1.2–2.5 mm wide, narrowly lanceoloid, turgid to slightly lat compressed, glab, acuminate. **Lo glm** 1.8–3.2 mm, ½–⅘ as long as the spklt, glab, 5–9-veined, acuminate; **up glm** and **lo lm** extending 0.4–3 mm beyond the up flt, 7–11-veined, strongly gaping at the apc; **lo flt** stmt; **lo pal** 3–3.5 mm, ovate-hastate, lat lobes folded over the anth before anthesis; **up flt** 2.3–3 mm long, 0.8–1.1 mm wide, narrowly ovoid, smooth, glab, shiny; **up lm** clasping the pal only at the base. $2n = 18, 21, 25, 30, 32, 35, 36, 54–60, 67–72, 74, 77, 90, 108$.

Panicum virgatum grows in tallgrass prairies, especially mesic to wet types where it is a major component of the vegetation, and on dry slopes, sand, open oak or pine woodlands, shores, river banks, and brackish marshes. Its range is primarily on the eastern side of the Rocky Mountains. It has been introduced as a forage grass outside its native range, and is probably introduced in the Intermountain Region.

Panicum virgatum is an important and palatable forage grass, but its abundance in native grasslands decreases with grazing. Several types are planted for range and wildlife habitat improvement.

9. Panicum urvilleanum Kunth Silky Panicgrass [p. 468, 512]

Pl per. **Clm** 50–100 cm, erect, solitary or in small tufts from stout, scaly, creeping to vertical rhz or stolons, simple or brchg at the base; **nd** densely villous. **Shth** densely villous; **lig** memb, ciliate, hairs 1.5–2 mm; **bld** 20–60 cm long, 4–10 mm wide, ascending to spreading, strigose to subglab, flat bas, tapering to a long, involute point. **Pan** 20–30 cm long, 3–9 cm wide, narrow, shortly exserted; **br** slender, ascending; **sec br** and ped 1–4 mm, crowded, ascending to appressed. **Spklt** 5–7 mm, densely villous, hairs silvery or tawny-white. **Lo glm** about ¾ the length of the spklt, 7–11-veined; **up glm** and **lo lm** 7–15-veined; **lo flt** stmt; **lo pal** about as long as the lo lm; **up flt** striate, mrg of the up lm villous, hairs white; **lod** very large. $2n = 36$.

Panicum urvilleanum grows on desert sand dunes and in creosote bush scrubland in the Mojave and Colorado desert regions of southern California, southern Nevada, and western Arizona. It also grows in Peru, Chile, and Argentina.

10. Panicum antidotale Retz. Blue Panicgrass [p. 468, 512]

Pl per; ces, rhz, rhz about 1 cm thick, knotted, pubescent, with large, scalelike lvs. **Clm** 50–300 cm tall, 2–4 mm thick, often compressed, erect or ascending, hard, becoming almost woody; **nd** swollen, glab or pubescent; **intnd** glab, glaucous. **Shth** not keeled, shorter than or equal to the intnd, glab or the lo shth at least partially pubescent, hairs papillose-based; **lig** 0.3–1.5 mm; **bld** 10–60 cm long, 3–20 mm wide, elongate, flat, abx surfaces and mrg scabrous, adx surfaces occ pubescent near the base, with prominent, white midveins, bases rounded to narrowed. **Pan** 10–45 cm, to ½ as wide as long, open or somewhat contracted, with many spklt; **br** 4–12 cm, opposite or alternate, ascending to spreading; **ped** 0.3–2.5 mm, scabridulous to scabrous, appressed to diverging less than 45° from the br axes. **Spklt** 2.4–3.4 mm long, 1–1.3 mm wide, ellipsoid-lanceoloid to narrowly ovoid, often purplish, glab, acute. **Lo glm** 1.4–2.2 mm, ⅓–½ as long as the spklt, 3–5-veined, obtuse; **up glm** and **lo lm** subequal, glab, 5–9-veined, mrg scarious, acute; **lo flt** stmt; **up flt** 1.8–2.8 mm long, 0.9–1.1 mm wide, smooth, lustrous, acute, mrg of up lm nearly flat. $2n = 18, 36$.

Panicum antidotale is native to India. It is grown in North America as a forage grass, primarily in the southwestern United States. It is now established in the region, being found in open, disturbed areas and fields. *Panicum antidotale* is considered to be somewhat unusual in the genus. Freckmann and Lelong (2002) placed it in its own section, sect. *Antidotalia*.

14.05 UROCHLOA P. Beauv.   J.K. Wipff and Rahmona A. Thompson

Pl ann or per; usu ces, smt mat-forming, smt stln. **Clm** 5–500 cm, herbaceous, erect, geniculate, or decumbent and rooting at the lo nd. **Shth** open; **aur** rarely present; **lig** apparently of hairs, the bas memb portion inconspicuous; **bld** ovate-lanceolate to lanceolate, flat. **Infl** tml or tml and axillary, usu pan of spikelike pri br in 2 or more ranks, rchs not concealed by the spklt; **pri br** usu alternate or subopposite, spikelike, and 1-sided, less frequently verticillate, axes flat or triquetrous, usu terminating in a well-developed, rdmt spklt; **sec br** present or absent, axes flat or triquetrous; **dis** beneath the spklt. **Spklt** solitary, paired, or in triplets, subsessile or pedlt, divergent or appressed, ovoid to ellipsoid, dorsally compressed, in 1–2(4) rows, with 2 flt, lo or up glm adjacent to the br axes. **Glm** not saccate bas; **lo glm** usu ⅕–⅔ as long as the spklt, occ equaling the up flt, (0)1–11-veined; **up glm** 5–13-veined; **lo flt** strl or stmt; **lo lm** similar to the up glm, 5–9-veined; **lo pal** if present, usu hyaline, 2-veined; **up flt** bisex, sessile, ovoid to ellipsoid, usu plano-convex, usu glab, not dis, mucronate or acuminate; **up lm** indurate, transversely rugose and verrucose, 5-veined, mrg involute, apc round to mucronate, or aristate; **up pal** rugose, shiny or lustrous; **lod** 2, cuneate, truncate; **anth** 3. **Car** ovoid to elliptic, dorsally compressed; **emb** ½–¾ as long as the car; **hila** punctate to linear. x = 7, 8, 9, or 10. Name from the Greek *ouros*, 'tail' and *chloa*, 'grass', a reference to the abruptly awned lemmas of some species.

Urochloa is a genus of approximately 100 tropical and subtropical species. One species grows in the Intermountain Region. The rugose, often mucronate or aristate, distal florets of *Urochloa* distinguish it from most species of *Panicum*.

1. Urochloa arizonica (Scribn. & Merr.) Morrone & Zuloaga ARIZONA SIGNALGRASS [p. 469, <u>512</u>]

Pl ann. **Clm** 15–65 cm, erect or geniculate, brchg from the lo nd; **nd** glab or hispid. **Shth** glab or with papillose-based hairs, mrg ciliate distally; **lig** 1–1.6 mm; **bld** 5–15 cm long, 5–12 mm wide, glab. **Pan** 6–20 cm long, 2–5 cm wide, ovoid, with 6–12 spikelike pri br in more than 2 ranks; **pri br** 3–7 cm, divergent, axes about 0.4 mm wide, triquetrous, densely pubescent with papillose-based hairs; **sec br** short, divergent; **ped** shorter than the spklt, with papillose-based hairs. **Spklt** 3.2–4 mm long, 1.2–1.6 mm wide, mostly paired, in 2 rows, appressed to the br. **Glm** scarcely separate, rchl intnd short, not pronounced; **lo glm** 1.5–2 mm, to ½ as long as the spklt, glab, 5-veined, smt with evident cross venation near the apc; **up glm** 2.5–3.2 mm, glab or shortly hirsute, 7-veined, with evident cross venation distally; **lo flt** stmt or strl; **lo lm** 2.5–3.2 mm, glab or shortly hirsute, 5-veined, about as long as the spklt; **lo pal** present; **up lm** 2.8–3 mm long, 1.2–1.6 mm wide, acute, beaked or mucronate; **anth** 0.8–1 mm. **Car** 1.5–2 mm; **hila** punctiform. $2n = 36$.

Urochloa arizonica is native to the southwestern United States and northern Mexico. It grows in open, dry areas with rocky or sandy soils.

14.06 ERIOCHLOA Kunth   Robert B. Shaw, Robert D. Webster, and Christine M. Bern

Pl ann or per; ces, smt with short rhz or stln, not producing subterranean spklt. **Clm** 20–250 cm, erect or decumbent, usu with 2–5 nd. **Shth** open; **aur** absent; **lig** memb, ciliate. **Infl** tml, pan of spikelike br on elongate rchs; **br** with many pedlt, loosely appressed spklt, terminating in a spklt, without stiff bristles or flat bracts, spklt in pairs, triplets, or solitary, often solitary distally when in pairs or triplets at the mid of the br; **ped** terminating in a well-developed disk; **dis** below the glm(s). **Spklt** with 2 flt, lo flt usu strl, up flt bisex. **Lo glm** typically rdcd (smt absent) and fused with the glab callus to form a cuplike structure; **up glm** lanceolate to ovate, glab or variously pubescent, 3–9-veined, unawned or awned; **lo lm** similar to the up glm in length, shape, venation, and pubescence, unawned; **lo pal** absent to fully developed; **up lm** lanceolate to ovate, indurate, rugose, dull, glab, rounded on the back, veins not pronounced, mrg involute; **anth** 3; **lod** 2, papery; **sty** with 2 br, purple, plumose. **Car** not longitudinally grooved; **endosperm** solid. $x = 9$. Name from the Greek *erion*, 'wool', and *chloe*, 'grass', a reference to the usually pubescent pedicels and rachises.

Eriochloa, a genus of 20–30 species, grows in tropical, subtropical, and warm-temperate areas of the world. Two species grow in the Intermountain Region.

1. Spikelets solitary at the middle of the branches, sometimes in unequally pedicellate pairs near the base. . . . . .1. *E. contracta*
1. Spikelets in unequally pedicellate pairs or triplets at the middle of the branches, sometimes solitary distally. . . . . 2. *E. acuminata*

1. Eriochloa contracta Hitchc. PRAIRIE CUPGRASS [p. 469, <u>512</u>]

Pl ann; ces. **Clm** 20–100 cm, erect or decumbent, smt rooting at the lo nd; **intnd** pilose or pubescent; nd pubescent to puberulent. **Shth** sparsely to densely pubescent; **lig** 0.4–1.1 mm; **bld** 6–12(22) cm long, 2–8 mm wide, linear, flat to conduplicate, straight, appressed to divergent, both surfaces sparsely to densely pubescent with short, evenly spaced hairs. **Pan** 6–20 cm long, 0.3–1.2 cm wide; **rchs** pilose, longer hairs 0.1–0.8 mm; **br** 10–20(28), 15–45(60) mm long, 0.2–0.4 mm wide, appressed, pubescent to setose, not winged, with 8–16 mostly solitary spklt, occ paired at the base of the br; **ped** 0.2–1 mm, variously hirsute below, apc with fewer than 10 hairs more than 0.5 mm long. **Spklt** (3.1)3.5–4.5(5) mm long, 1.2–1.7 mm wide, lanceolate. **Up glm** as long as the lo lm, with sparsely appressed pubescence on the lo ⅔, scabrous or glab distally, 3–9-veined, acuminate and awned, awns 0.4–1 mm; **lo flt** strl; **lo lm** 3–4.3 mm long, 1.2–1.7 mm wide, lanceolate, setose, 3–7-veined, acuminate, unawned or mucronate; **lo pal** absent; **up lm** 2–2.5 mm, indurate, elliptic, 5–7-veined, acute to rounded and awned, awns 0.4–1.1 mm; **up pal** indurate, faintly rugose, blunt. $2n = 36$.

Eriochloa contracta grows in fields, ditches, and other disturbed areas. It is known only from the United States, being native and common in the central United States, and adventive to the east and southwest. It differs from *E. acuminata* in its tightly contracted, almost cylindrical panicles and longer lemma awns, but intermediate forms can be found.

2. Eriochloa acuminata (J. Presl) Kunth

SOUTHWESTERN CUPGRASS [p. *469*, <u>512</u>]

Pl ann; ces. **Clm** 30–120 cm, erect or decumbent, smt rooting at the lo nd; **intnd** glab or with scattered hairs; **nd** glab or pilose. **Shth** smt conspicuously inflated, glab or pubescent; **lig** 0.2–1.2 mm; **bld** 5–12(18) cm long, (2)5–12(16) mm wide, linear, flat or folded, straight or lax, glab or sparsely pubescent adx. **Pan** 7–16 cm long, 1–6 cm wide, loosely contracted; **rchs** scabrous or hairy; **br** 5–20, 1–5 cm long, 0.4–0.6 mm wide, appressed to divergent, pubescent, smt setose, not winged, with 20–36 spklt, spklt mostly in unequally pedlt pairs, solitary distally; **ped** 0.1–1 mm, hairy. **Spklt** 3.8–5(6) mm long, 1.1–1.4 mm wide, lanceolate to ovate. **Lo glm** absent; **up glm** equaling the lo lm, lanceolate to ovate, hairy, 5(7)-veined, acuminate to acute, unawned or awned, awns to 1.2 mm; **lo lm** 3.6–5 mm long, 1.1–1.4 mm wide, lanceolate to ovate, setose, 5(7)-veined, acuminate to acute, unawned; **lo pal** absent; **anth** absent; **up lm** 2.3–3.3 mm, 0.7–0.9 times as long as the lo lm, indurate, elliptic, rounded, 5-veined, awned, the awns 0.1–0.3 mm; **up pal** indurate, blunt, rugose. $2n = 36$.

Eriochloa acuminata is native to the southern United States and northern Mexico, but has become established outside this region. There are two varieties, differing as shown in the key below. Both grow in Mexico as well as the United States.

1. Spikelets 4–6 mm long, long-acuminate or tapering to a short awn . . . . var. *acuminata*
1. Spikelets 3.8–4 mm long, acute . . . . var. *minor*

Eriochloa acuminata (J. Presl) Kunth var. acuminata [p. *469*]

Spklt 4–6 mm long, long-acuminate or tapering to a short awn.

Eriochloa acuminata var. *acuminata* generally grows in ditches, fields, right of ways, and other disturbed areas of the southern United States.

Eriochloa acuminata var. minor (Vasey) R.B. Shaw [p. *469*]

Spklt 3.8–4 mm long, acute.

Eriochloa acuminata var. *minor* is common in irrigated fields, orchards and disturbed areas of the southwestern United States.

14.07 PENNISETUM Rich.

J.K. Wipff

Pl ann or per; habit various. **Clm** 3–800 cm, not woody, smt brchg above the base; **intnd** solid or hollow. **Lig** memb and ciliate, or of hairs, rarely completely memb; **bld** smt psdpet. **Infl** spicate pan with highly rdcd br termed fascicles; **pan** 1–many per pl, tml on the clm or on both the clm and the sec br, or tml and axillary, or only axillary, usu completely exposed at maturity; **rchs** usu terete, with (1)5–many fascicles; **fascicle axes** 0.2–7.5(28) mm, with (1)3–130+ bristles and 1–12 spklt. **Bristles** free or fused at the base, dis with the spklt at maturity; of 3 kinds, outer, inner, and pri, in some species all 3 kinds present below each spklt, in others 1 or more kinds missing from some or all of the spklt; **outer (lo) bristles** antrorsely scabrous, terete; **inner (up) bristles** antrorsely scabrous or long-ciliate, usu flatter and wider than the outer bristles; **pri (tml) bristles** located immediately below the spklt, solitary, antrorsely scabrous or long-ciliate, often longer than the other bristles associated with the spklt; **dis** usu at the base of the fascicles, smt also beneath the up flt. **Spklt** with 2 flt; **lo glm** absent or present, 0–5-veined; **up glm** longer, 0–11-veined; **lo flt** strl or stmt; **lo lm** usu as long as the spklt, memb, 3–15-veined, mrg usu glab; **lo pal** present or absent; **up lm** memb to coriaceous, 5–12-veined; **up pal** shorter than the lm but similar in texture; **lod** 0 or 2, glab; **anth** 3, if present. $x = 5, 7, 8, 9$ (usu 9). Name from the Latin *penna*, 'feather', and *seta*, 'bristle', an allusion to the plumose bristles of some species.

Pennisetum has 80–130 species, most of which grow in the tropics and subtropics, and occupy a wide range of habitats. Twenty-five species are native to the Western Hemisphere, but none to North America. Most of the species treated here are cultivated for food, forage, or as ornamental plants. Many species, including several cultivated species, are weedy. Four are classified as noxious weeds by the U.S. Department of Agriculture. Records known to be based on cultivated plants are not included in the distribution maps but, in many cases, it is not possible to determine whether a record is based on a cultivated plant or an escape.

The placement of the boundary between *Pennisetum* and *Cenchrus* is contentious. As treated here, *Pennisetum* has antrorsely scabrous bristles that are not spiny, fascicle axes that terminate in a bristle, and chromosome base numbers of 5, 7, 8, and 9. *Cenchrus* has retrorsely (rarely antrorsely) scabrous, spiny bristles, fascicle axes that are terminated by a spikelet, and a chromosome base number of 17 (Wipff 2001). In both genera, the bristles are reduced branches (Goebel 1882; Sohns 1955).

Pedicel length is the distance from the base of the primary bristles to the base of the terminal spikelets. Fascicle axis lengths and fascicle densities are measured in the middle of the panicle; spikelet measurements refer to the largest spikelets in the fascicles.

1. Most or all bristles scabrous, the primary bristles sometimes sparsely and inconspicuously long-ciliate.
   2. Terminal panicle drooping; fascicles subsessile, the bases 0.5–0.7 mm long. . . . . 1. *P. macrostachys*
   2. Terminal panicle erect; fascicles with a stipelike base 1.5–5.6 mm long . . . . 2. *P. alopecuroides*
1. Bristles, at least the primary bristles, conspicuously long-ciliate.
   3. Spikelets 9–12 mm long. . . . . 4. *P. villosum*
   3. Spikelets 3–7 mm long.
      4. Upper florets readily disarticulating at maturity; upper lemmas smooth and shiny, conspicuously different in texture from the lower lemmas . . . . 3. *P. polystachion*
      4. Upper florets not disarticulating at maturity; lower and upper lemmas similar in texture.
         5. Midculm leaves 2–3.5 mm wide, convolute or folded, green, the midvein noticeably thickened; lower florets of the spikelets usually sterile, sometimes staminate
         . . . . 5. *P. setaceum*
         5. Midculm leaves 6–11 mm wide, flat, usually burgundy, rarely green, the midvein not noticeably thickened; lower florets of the spikelets staminate. . . . . 6. *P. advena*

1. **Pennisetum macrostachys** (Brongn.) Trin.
PACIFIC FOUNTAINGRASS [p. 470]

Pl per, or ann in temperate climates; ces. **Clm** 100–300 cm, erect, brchg; **nd** glab. **Lvs** burgundy; **shth** glab; **lig** 0.1–0.3 mm; **bld** 30–53.5 cm long, (15)18–35 mm wide, flat, glab. **Pan** tml, (15.5)18–40 cm long, 32–50 mm wide, fully exserted from the shth, flexible, drooping, burgundy; **rchs** terete, shortly pubescent. **Fascicles** 17–22 per cm; **fascicle axes** 0.5–0.7 mm, with 1 spklt; **outer bristles** 21–40, 1.2–22.3 mm, scabrous; **inner bristles** absent; **pri bristles** 20–23 mm, not noticeably longer than the other bristles, scabrous. **Spklt** 4.4–4.9 mm, sessile or pedlt, glab; **ped** to 0.1 mm; **lo glm** 1.1–1.3 mm, veinless; **up glm** 2.1–2.8 mm, usu about ½ as long as the spklt, 1–3-veined; **lo flt** stmt (strl); **lo lm** 4–4.5 mm, 5-veined; **lo pal** absent or to 2.6 mm; **anth** absent or 1.4–1.6 mm; **up lm** 4.3–4.8 mm, 5-veined; **anth** 1.6–1.8 mm. 2n = 68.

Pennisetum macrostachys is native to the South Pacific. It is grown in North America as an ornamental species, being sold as 'Burgundy Giant'.

2. **Pennisetum alopecuroides** (L.) Spreng.
FOXTAIL FOUNTAINGRASS [p. 470]

Pl per; ces. **Clm** 30–100 cm, erect; **nd** glab. **Shth** glab, mrg ciliate; **lig** 0.2–0.5 mm, memb, ciliate; **bld** (10)30–60 cm long, 2–8(12) mm wide, flat to folded, glab, mrg ciliate bas. **Pan** all tml, 6–20 cm long, 20–53 mm wide, fully exserted from the shth, erect, green to brown, deep purple, or stramineous to creamy white; **rchs** terete, with pubescent hairs. **Fascicles** 9–16 per cm; **fascicle axes** 1.5–5.6 mm, with a stipelike base of 1–5.6 mm and 1(2) spklt(s); **outer bristles** 13–19, 0.8–15.6 mm; **inner bristles** 7–10, 11.2–30 mm, scabrous; **pri bristles** 26.7–35 mm, scabrous, usu not noticeably longer than the other bristles. **Spklt** 5.5–8.4 mm, sessile or subsessile, glab; **ped** to 0.1 mm; **lo glm** 0.2–1.4 mm, veinless; **up glm** 2–4.9 mm, to ½ as long as the spklt, 1–5-veined, acute to broadly acute; **lo flt** strl; **lo lm** 4.9–8.1 mm, 7–9(10)-veined; **lo pal** absent; **up lm** 5.2–7.6 mm, 5–7-veined, acuminate; **anth** 3, 3–4.5 mm. 2n = 18.

Pennisetum alopecuroides is native to southeast Asia. It is frequently grown as an ornamental in North America.

3. **Pennisetum polystachion** (L.) Schult.
MISSION GRASS [p. 471, 512]

Pl ann or per; ces from a hard, knotty base. **Clm** 30–200 cm, erect, brchg; **nd** glab. **Shth** glab, mrg ciliate; **lig** 1.5–2.7 mm; **bld** 15–55 cm long, 4–18 mm wide, flat, glab or pubescent. **Pan** tml,

10–25 cm long, 15–30 mm wide, fully exserted from the shth, erect to drooping, white, yellow, light brown, or pink to deep purple; **rchs** terete, scabrous. **Fascicles** 33–45 per cm, dis at maturity; **fascicle axes** 0.2–0.5 mm, with 1 spklt; **outer bristles** 13–30, 1.3–5 mm, scabrous; **inner bristles** 6–14, 4.3–11.5 mm, long ciliate; **pri bristles** 14–25 mm, long-ciliate, noticeably longer than the other bristles. **Spklt** 3–4.5 mm, sessile; **lo glm** absent or to 2 mm, veinless; **up glm** 3–4.5 mm, glab, 5–7-veined, 3-lobed; **lo flt** strl or stmt; **lo lm** 3–3.9 mm, 5–7-veined, apc lobed; **lo pal** 2.9–3.7 mm; **anth** absent or 1.7–2 mm; **up flt** dis at maturity; **up lm** 1.7–3 mm, coriaceous, shiny, 5-veined, apc ciliate; **anth** 1.3–2.1 mm. **Car** about 1.7 mm, concealed by the lm and pal at maturity. $2n = 18$, 36, 45, 48, 52, 53, 54, 56, 78.

Pennisetum polystachion is a polymorphic, weedy African species that has become established in the tropics and subtropics. The U.S. Department of Agriculture considers it a noxious weed. Only *Pennisetum polystachion* subsp. *setosum* (Sw.) Brunken has been found in North America.

4. Pennisetum villosum R. Br. *ex* Fresen.
FEATHERTOP [p. 471]

Pl per; rhz. **Clm** 16–75 cm, erect; **nd** glab. **Shth** glab, mrg ciliate; **lig** 1–1.3 mm; **bld** 5–40 cm long, 2–4.5 mm wide, flat to folded, glab, pubescent, or scabrous, mrg ciliate or glab bas. **Pan** tml, 4–11.5 cm long, 50–75 mm wide, fully exserted from the shth, erect, white; **rchs** terete, pubescent (bas). **Fascicles** 7–11 per cm; **fascicle axes** 1.5–2.5 mm, with 1–4 spklt; **outer bristles** (0)1–8, 1–13.5 mm; **inner bristles** 23–41, 13–50.5 mm, densely plumose; **pri bristles** 40–50 mm, ciliate, usu not longer than the other bristles. **Spklt** 9–12 mm, glab; **ped** 0.1–0.4 mm; **lo glm** 0.3–1.3 mm, veinless; **up glm** 2.5–5.2 mm, 1(3)-veined; **lo flt** stmt or strl; **lo lm** 7.5–10.5 mm, 7–9(10)-veined; **lo pal** absent or 5.5–8.5 mm; **anth** absent or 3.8–4.5 mm; **up lm** 9–11 mm, 7-veined, apc scabridulous; **anth** 3.5–5 mm. **Car** concealed by the lm and pal at maturity. $2n = 45$.

Pennisetum villosum is native to Ethiopia, northern Somalia, and the Arabian Peninsula. It is grown as an ornamental in North America.

5. Pennisetum setaceum (Forssk.) Chiov.
TENDER FOUNTAINGRASS [p. 472, 512]

Pl per, or ann in temperate climates; ces. **Clm** 40–150 cm, erect, pubescent beneath the pan;

nd glab. **Lvs** green, smt glaucous; **shth** glab, mrg ciliate; **lig** 0.5–1.1 mm; **bld** 20–65 cm long, 2–3.5 mm wide, convolute or folded, scabrous, midvein noticeably thickened. **Pan** (6)8–32 cm long, 40–52 mm wide, erect or arching, pink to dark burgundy; **rchs** pubescent. **Fascicles** 8–10 per cm; **fascicle axes** 2.3–4.5 mm, with 1–4 spklt; **outer bristles** 28–65, 0.9–19 mm; **inner bristles** 8–16, 8–27 mm, ciliate; **pri bristles** 26.5–34.3 mm, ciliate, noticeably longer than the other bristles. **Spklt** 4.5–7 mm, sessile or pedlt; **ped** to 0.1 mm; **lo glm** absent or to 0.3 mm, veinless; **up glm** 1.2–3.6 mm, (0)1-veined; **lo flt** usu strl, smt stmt; **lo lm** 4–6 mm, 3-veined, acuminate, midvein excurrent to 0.7 mm; **lo pal** usu absent, if present, to 4.4 mm; **anth** absent or 2.3–2.4 mm; **up lm** 4.5–6.7 mm, attenuate, 5-veined, midvein excurrent to 0.7 mm, mrg glab; **anth** 2.1–2.7 mm. $2n = 27$.

Pennisetum setaceum is a desert grass native to the eastern Mediterranean region. It is a popular ornamental throughout the southern United States, but it is also an invasive weed.

6. Pennisetum advena Wipff & Veldkamp
PURPLE FOUNTAINGRASS [p. 472]

Pl per, or ann in temperate climates; ces. **Clm** 1–1.5 m, erect, smt brchg above, pubescent beneath the pan; **nd** glab. **Lvs** burgundy (rarely green); **shth** glab, mrg ciliate; **lig** 0.5–0.8 mm; **bld** 33–52 cm long, 6–11 mm wide, flat, antrorsely scabridulous, mrg ciliate bas, midvein not noticeably thickened. **Pan** 23–32 cm long, 30–58 mm wide, fully exserted from the shth, flexible, drooping, burgundy (rarely pale or whitish green); **rchs** terete, pubescent. **Fascicles** 10–17 per cm, **dis** at maturity; **fascicle axes** 1–2 mm, with 1–3 spklt; **outer bristles** 43–68, 1.2–18.5 mm, terete, scabrous; **inner bristles** 4–10, 11.7–25 mm, long-ciliate; **pri bristles** 21.3–33.6 mm, ciliate, noticeably longer than the other bristles. **Spklt** 5.3–6.5 mm; **ped** 0.1–0.3 mm; **lo glm** 0.5–1 mm, veinless; **up glm** 1.9–3.6 mm, 0–1-veined; **lo flt** stmt; **lo lm** 4.7–6.1 mm, 5(6)-veined; **lo pal** 4.5–5 mm; **anth** 2–2.5 mm; **up flt** not dis at maturity; **up lm** 5.2–6.1 mm, 5-veined; **anth** 2.5–2.7 mm. **Car** concealed by the lm and pal at maturity. $2n = 54$.

The origin of *Pennisetum advena* is uncertain. It is frequently cultivated as an ornamental, usually being sold as *P. setaceum* 'Rubrum'.

14.08 CENCHRUS L.
Michael T. Stieber and J.K. Wipff

Pl ann or per. **Clm** 5–200 cm, erect or decumbent, usu geniculate; **nd** and **intnd** usu glab. **Shth** open, usu glab; **lig** memb, ciliate, cilia as long as or longer than the bas membrane; **bld** flat or folded, mrg cartilaginous, scabridulous. **Infl** tml, spikelike pan of highly rdcd br termed fascicles ("burs");

fascicles consisting of 1–2 series of many, stiff, partially fused, usu retrorsely scabridulous to strigose, sharp bristles surrounding, smt almost concealing, 1–4 spklt; **outer (lo) bristles**, if present, in 1 or more whorls, terete or flattened; **inner (up) bristles** usu strongly flattened, fused at least at the base and forming a disk, frequently to more than ½ their length and forming a cupule; **dis** at the base of the fascicles. **Spklt** sessile, with 2 flt; **lo flt** usu strl; **up flt** bisex. **Lo glm** ovate, scarious, glab, 1-veined, acute to acuminate; **up glm** and **lo lm** ovate, 3–9-veined; **lo pal** equaling the lm, tawny or purplish; **up lm** and **pal** subequal, indurate, ovate, obscurely veined, acuminate. **Car** obtrulloid. $x = 17$. Name from the Greek *kengchros*, 'millet'.

   Cenchrus has about 16, primarily tropical species, most of which are readily (and painfully) recognized by their spiny fascicles. Most of its species differ from those of *Pennisetum* in having retrorsely scabrous or strigose inner bristles that are fused to well above their bases. The species are usually considered to be undesirable weeds. Two species grow in the Intermountain Region.

1. Inner bristles 1–2 mm wide at the base; fascicles with 8–40 bristles . . . . 1. *C. spinifex*
1. Inner bristles 0.5–0.9(1.4) mm wide at the base; fascicles with 45–75 bristles . . . . 2. *C. longispinus*

1. **Cenchrus spinifex** Cav. Coastal Sandbur, Common Sandbur [p. 473, <u>512</u>]

Pl ann or per but short-lived; tufted. **Clm** 30–100 cm, geniculate. **Shth** compressed, glab or sparsely pilose; **lig** 0.5–1.4 mm; **bld** 3–28 cm long, (1)3–7.2 mm wide, glab or sparsely long-pilose adx. **Pan** 3–5(8.5) cm; **fascicles** 5.5–10.2 mm long, 2.5–5 mm wide, imbricate, ovoid to globose, glab or sparsely to moderately pubescent; **outer bristles**, when present, mostly flattened; **inner bristles** 8–40 (rarely more), 2–5.8 mm long, 1–2 mm wide, fused at least ½ their length, forming a distinct cupule, the distal portions usu diverging from the cupule at multiple, irregular intervals, smt diverging at more or less the same level, ciliate at the base, pubescent, stramineous to mauve or purple, flattened. **Spklt** 2–4 per fascicle, 3.5–5.9 mm, glab. **Lo glm** 1–3.3 mm; **up glm** (2.8)3.5–5 mm, 5–7-veined; **lo flt** smt stmt; **lo lm** 3–5(5.9) mm, 5–7-veined; **lo pal** smt rdcd or absent; **anth** 1.3–1.6 mm; **up lm** 3.5–5(5.8) mm; **anth** 0.5–1.2 mm. **Car** about 2.5 mm long, 1–2 mm wide, ovoid. $2n = 34$ (32).

   Cenchrus spinifex is common in sandy woods, fields, and waste places throughout the southern United States and southwards into South America. It has been confused with *C. longispinus*, but differs in having shorter spikelets, fewer bristles overall, wider inner bristles, and outer bristles that are usually flattened rather than usually terete.

2. **Cenchrus longispinus** (Hack.) Fernald Mat Sandbur, Longspine Sandbur [p. 473, <u>513</u>]

Pl ann; tufted. **Clm** 20–90 cm, smt decumbent, often with many br arising from the base. **Shth** strongly compressed-keeled; **lig** 0.6–1.8 mm; **bld** 4–27 cm long, 1.5–5(7.5) mm wide, adx surfaces scabrous or sparsely pilose. **Pan** 1.5–8(10) cm; **fascicles** 8.3–11.9 mm long, 3.5–6 mm wide, somewhat globose, medium- to short-pubescent; **bristles** 45–75; **outer bristles** numerous, shorter and thinner than the inner bristles, imbricate, mostly terete, reflexed; **inner bristles** 3.5–7 mm long, 0.5–0.9(1.4) mm wide at the base, irregularly placed, fused for ½ their length or more, forming a distinct cupule, the distal portions diverging at irregular intervals from the cupule, often grooved along the mrg, purple-tinged. **Spklt** 2–3(4) per fascicle, (4)5.8–7.8 mm. **Lo glm** 0.8–3 mm; **up glm** 4–6 mm, 3–5-veined; **lo flt** often stmt; **lo lm** 4–6.5 mm, 3–7-veined; **anth** 1.5–2 mm; **up lm** 4–7(7.6) mm; **anth** 0.7–1 mm, seemingly not well developed at anthesis. **Car** 2–3.8 mm long, 1.5–2.6 mm wide, ovoid. $2n = 34$ (38).

   Cenchrus longispinus grows in sandy woods, fields, and waste ground from southern Canada and the contiguous United States southwards to Venezuela. It is often confused with *C. spinifex*; see comment under that species.

14.09 SETARIA P. Beauv.   James M. Rominger

Pl ann or per; ces, rarely rhz. **Clm** 10–600 cm, erect or decumbent. **Lig** memb and ciliate or of hairs; **bld** flat, folded, or involute, or plicate and petiolate (subg. *Ptychophyllum*). **Infl** tml, pan, usu dense and spikelike, occ loose and open; **dis** usu below the glm, spklt falling intact, bristles persistent. **Spklt** 1–5 mm, usu lanceoloid-ellipsoid, rarely globose, turgid, subsessile to short pedlt, in fascicles on short br or single on a short br, some or all subtended by 1–several, terete bristles (strl branchlets). **Lo glm** memb, not saccate, less than ½ as long as the spklt, 1–7-veined; **up glm** memb to herbaceous at maturity, ½ as long as to nearly equaling the up lm, 3–9-veined; **lo flt** stmt or strl; **lo lm** memb, equaling or rarely exceeding the up lm, rarely absent, not constricted or indurate bas, 5–7-veined; **lo pal** usu hyaline to memb at maturity, rarely absent or rdcd, veins not keeled; **up flt** bisex; **up lm** and

pal indurate, transversely rugose, rarely smooth; **anth** 3, not penicillate; **sty** 2, free or fused bas, white or red. **Car** small, ellipsoid to subglobose, compressed dorsiventrally. $x = 9$. Name from the Latin *seta*, 'bristle' and *aria*, 'possessing'.

Setaria, a genus of about 140 species, grows predominantly in tropical and warm-temperate regions, but it is particularly well represented in Africa, Asia, and South America. Species from the Intermountain Region fall into one of three categories: native to North America, native to South America, or native to the Eastern Hemisphere. There are nine species in the Intermountain Region; four are native, and five are established introductions. *Setaria macrostachya* and *S. leucopila* provide valuable forage in the southwestern United States. *Setaria italica* has been cultivated for centuries in Asia and Europe, providing food for humans and their livestock. The majority of species in temperate North America are aggressive, exotic annuals which collectively are a major nuisance, particularly in the corn and bean fields of the midwestern states.

1. Bristles 4–12 below each spikelet.
  2. Plants perennial . . . . 8. *S. parviflora*
  2. Plants annual.
    3. Panicles erect; bristles 3–8 mm long; spikelets 2–3.4 mm long; blades 4–10 mm wide . . . . . 9. *S. pumila*
    3. Panicles arching and drooping from near the base; bristles about 10 mm long;
      spikelets 2.5–3 mm long; blades 10–20 mm wide. . . . . 7. *S. faberi*
1. Bristles 1–3 (rarely 6) below each spikelet.
  4. Bristles retrorsely scabrous . . . . 4. *S. verticillata*
  4. Bristles antrorsely scabrous.
    5. Plants perennial.
      6. Lower paleas narrow, ½–¾ as long as the lemmas; spikelets elliptical . . . . 2. *S. leucopila*
      6. Lower paleas broad, subequal to the lemmas in length; spikelets subspherical . . . 1. *S. macrostachya*
    5. Plants annual.
      7. Upper lemmas smooth and shiny, occasionally obscurely transversely rugose . . . . 6. *S. italica*
      7. Upper lemmas distinctly transversely rugose, dull.
        8. Panicles loosely spicate; rachises visible, hispid . . . . 3. *S. grisebachii*
        8. Panicles densely spicate; rachises not visible, villous.
          9. Blades softly pilose on the upper surface; spikelets 2.5–3 mm long; panicles
            nodding from the base . . . . 7. *S. faberi*
          9. Blades scabrous; spikelets 1.8–2.2 mm long; panicles nodding only from
            near the apex . . . . 5. *S. viridis*

1. Setaria macrostachya Kunth Plains Bristlegrass [p. 473, 513]

Pl per; densely ces. **Clm** 60–120 cm, rarely brchd distally, scabrous below the nd and pan. **Shth** keeled, glab, usu with a few white hairs at the throat; **lig** 2–4 mm, densely ciliate; **bld** 15–20 cm long, 7–15 mm wide, flat, adx surface scabrous. **Pan** 10–30 cm long, 1–2 cm wide, uniformly thick from the base to the apex, dense, rarely lobed bas; **rchs** scabrous and loosely pilose; **bristles** usu solitary, 10–20 mm, soft, antrorsely scabrous. **Spklt** 2–2.3 mm, subspherical. **Lo glm** ⅓–½ as long as the spklt, 3–5-veined; **up glm** about ¾ as long as the spklt, 5–7-veined; **lo lm** equaling the up lm, 5-veined; **lo pal** nearly equaling the up pal in length and width; **up lm** transversely rugose; **up pal** convex, ovate. $2n = 54$.

Setaria macrostachya is an abundant and valuable forage grass in the desert grasslands of the southwestern United States, particularly in southern Arizona and Texas. It extends south through the highlands of central Mexico. It also grows in the West Indies, but is not common there.

2. Setaria leucopila (Scribn. & Merr.) K. Schum. Streambed Bristlegrass [p. 474, 513]

Pl per; ces. **Clm** 20–100 cm. **Shth** compressed, glab, mrg villous distally; **lig** 1–2.5 mm, ciliate; **bld** 8–25 cm long, 2–5 mm wide, flat or folded, scabrous on both surfaces. **Pan** 6–15 cm, tightly spikelike, pale green; **rchs** scabrous or villous; **bristles** usu solitary, 4–15 mm, ascending. **Spklt** 2.2–2.8(3) mm, elliptical. **Lo glm** about ½ as long as the spklt, 3-veined; **up glm** from ¾ as long as to equaling the flt, 5-veined; **lo lm** equaling the up lm, 5-veined; **lo pal** ½–¾ as long as the up pal, lanceolate; **up lm** apiculate, finely and transversely rugose; **up pal** similar. $2n = 54$, 68, 72.

Setaria leucopila grows in the southwestern United States and northern Mexico. It is the most common of the perennial "Plains bristlegrasses."

3. Setaria grisebachii E. Fourn. Grisebach's Bristlegrass [p. 474, 513]

Pl ann. **Clm** 30–100 cm; **nd** pubescent, hairs appressed. **Shth** with ciliate mrg; **lig** ciliate; **bld** to 12(25) cm long, to 10(20) mm wide, flat, hispid on both surfaces. **Pan** 3–18 cm, loosely spicate, interrupted, often purple; **rchs** hispid; **bristles** 1–3, 5–15 mm, flexible, antrorsely scabrous. **Spklt** 1.5–2.2 mm. **Lo glm** about ⅓ as long as the spklt, distinctly 3-veined, lat veins coalescing with the cent veins below the apc; **up glm** nearly equaling the up lm, obtuse, 5-veined; **lo lm** equaling the up lm; **lo pal** about ⅓ as long as the lo lm, narrow; **up lm** finely and transversely rugose; **up pal** similar to the up lm. 2*n* = unknown.

Setaria grisebachii is the most widespread and abundant native annual species of *Setaria* in the southwestern United States. It grows in open ground and extends along the central highlands of Mexico to Guatemala, usually at elevations of 750–2500 m.

4. Setaria verticillata (L.) P. Beauv. Hooked Bristlegrass [p. 474, 513]

Pl ann. **Clm** 30–100 cm; **nd** glab. **Shth** glab, mrg ciliate distally; **lig** to 1 mm, densely ciliate; **bld** 5–15 mm wide, flat, abx surfaces scabrous. **Pan** 5–15 cm, tapering to the apc; **rchs** retrorsely rough hispid; **bristles** solitary, 4–7 mm, retrorsely scabrous. **Spklt** 2–2.3 mm. **Lo glm** about ⅓ as long as the spklt, obtuse, 1(3)-veined; **up glm** nearly as long as the spklt; **lo pal** about ½ as long as the spklt, broad; **up lm** finely and transversely rugose; **up pal** similar to the up lm. 2*n* = 18, 36, 54, 72, 108.

Setaria verticillata is a European adventive that is now common throughout the cooler regions of the contiguous United States and in southern Canada.

5. Setaria viridis (L.) P. Beauv. Green Bristlegrass [p. 475, 513]

Pl ann. **Clm** 20–250 cm; **nd** glab. **Shth** glab, smt scabridulous, mrg ciliate distally; **lig** 1–2 mm, ciliate; **bld** to 20 cm long, 4–25 mm wide, flat, scabrous or smooth, glab. **Pan** 3–20 cm, densely spicate, nodding only from near the apc; **rchs** hispid and villous; **bristles** 1–3, 5–10 mm, antrorsely scabrous, usu green, rarely purple. **Spklt** 1.8–2.2 mm. **Lo glm** about ⅓ as long as the spklt, triangular-ovate, 3-veined; **up glm** nearly equaling the up lm, elliptical, 5–6-veined; **lo lm** slightly exceeding the up lm, 5-veined; **lo pal** about ⅓ as long as the lo lm, hyaline; **up lm** very finely and transversely rugose, pale green, 5–6-veined; **up pal** similar to the up lm. 2*n* = 18.

Setaria viridis resembles *S. italica* but differs in its shorter spikelets and rugose upper florets, and mode of disarticulation. It is also a more aggressive weed. It is native to Eurasia but is now widespread in warm temperate regions of the world. One variety grows in the Intermountain Region.

Setaria viridis (L.) P. Beauv. var. **viridis** Green Foxtail [p. 475]

Clm 20–100 cm; **nd** 6–7. **Bld** 4–12 mm wide. **Pan** usu 3–8 cm, producing around 600–800 car. 2*n* = 18.

Setaria viridis var. *viridis* is an aggressive adventive weed throughout temperate North America.

6. Setaria italica (L.) P. Beauv. Foxtail Millet [p. 475, 513]

Pl ann. **Clm** 10–100 cm. **Shth** mostly glab, mrg sparsely ciliate; **lig** 1–2 mm; **bld** to 20 cm long, 1–3 cm wide, flat, scabrous. **Pan** 8–30 cm, dense, spikelike, occ lobed below; **rchs** hispid to villous; **bristles** 1–3, to 12 mm, tawny or purple. **Spklt** about 3 mm, dis between the lo and up flt. **Lo glm** 3-veined; **up glm** 5–7-veined; **lo pal** absent or ½ as long as the lo lm; **up lm** very finely and transversely rugose to smooth and shiny, exposed at maturity. 2*n* = 18.

Setaria italica was cultivated in China as early as 2700 B.C. and during the Stone Age in Europe. Nowadays it is grown mostly for hay or as a pasture grass, but it has been used as a substitute for rice in northern China. It is sometimes cultivated in North America, but it is better known as a weed in moist ditches. It is closely related to *S. viridis*, differing in the longer (3 mm) spikelets and smooth, shiny upper florets which readily disarticulate above the lower florets. It exhibits considerable variation in seed and bristle color, bristle length, and panicle shape. Using these characters, Hubbard (1915) recognized several infraspecific taxa; they are not treated here.

7. Setaria faberi R.A.W. Herrm. Chinese Foxtail [p. 475, 513]

Pl ann. **Clm** 50–200 cm. **Shth** glab, fringed with white hairs; **lig** about 2 mm; **bld** 15–30 cm long, 10–20 mm wide, usu with soft hairs on the adx surface. **Pan** 6–20 cm, densely spicate, arching and drooping from near the base; **rchs** densely villous; **bristles** (1)3(6), about 10 mm. **Spklt** 2.5–3 mm. **Lo glm** about 1 mm, acute, 3-veined; **up glm** about 2.2 mm, obtuse, 5-veined; **lo lm** about 2.8 mm, obtuse; **lo pal** about ⅔ as long as the lo lm; **up lm** pale, finely and distinctly transversely rugose; **up pal** similar to the up lm. 2*n* = 36.

Setaria faberi spread rapidly throughout the North American corn belt after being accidentally introduced from China in the 1920s. It has become a major nuisance in corn and bean fields of the midwestern United States.

8. Setaria parviflora (Poir.) Kerguélen Knotroot Bristlegrass [p. 476, 513]

Pl per; rhz, rhz short, knotty. **Clm** 30–120 cm; **nd** glab. **Shth** glab; **lig** shorter than 1 mm, of hairs; **bld** to 25 cm long, 2–8 mm wide, flat, scabrous above. **Pan** 3–8(10) cm, of uniform width throughout their length, densely spikelike;

rchs scabrous-hispid; **bristles** 4–12, 2–12 mm, antrorsely barbed, yellow to purple. **Spklt** 2–2.8 mm, elliptical and turgid. **Lo glm** about ⅓ as long as the spklt, 3-veined; **up glm** ½–⅔ as long as the spklt, 5-veined; **lo flt** often stmt; **lo lm** occ indurate and faintly transversely rugose; **lo pal** equaling the lo lm; **up lm** distinctly transversely rugose, often purple-tipped. $2n = 36, 72$.

Setaria parviflora is a common, native species of moist ground. It is most frequent along the Atlantic and Gulf coasts, but also grows from the Central Valley of California east through the central United States and southward through Mexico to Central America, as well as in the West Indies. It is the most morphologically diverse and widely distributed of the indigenous perennial species of *Setaria*.

9. Setaria pumila (Poir.) Roem. & Schult.

YELLOW FOXTAIL, PIGEON GRASS [p. 476, 513]

Pl ann. **Clm** 30–130 cm. **Shth** glab; **lig** ciliate; **bld** 4–10 mm wide, loosely twisted, adx surfaces with papillose-based hairs bas. **Pan** 3–15 cm, uniformly thick, erect, densely spicate; **rchs** hispid; **bristles** 4–12, 3–8 mm, antrorsely scabrous. **Spklt** 2–3.4 mm, strongly turgid. **Lo glm** about ⅓ as long as the spklt, 3-veined, acute; **up glm** about ½ as long as the spklt, 5-veined, ovate; **up flt** often stmt; **lo lm** equaling the up lm; **lo pal** equaling the lo lm, broad; **up lm** conspicuously exposed, strongly transversely rugose. $2n = 36, 72$.

Setaria pumila (Poir.) Roem. & Schult. subsp. pumila [p. 476]

Bld yellowish green. **Bristles** yellowish. **Spklt** 3–3.4 mm. $2n = 36, 72$.

Setaria pumila subsp. *pumila* is a European adventive that has become a common weed in lawns and cultivated fields throughout temperate North America.

14.10 HOPIA Zuloaga & Morrone   Mary E. Barkworth

Pl per; shortly rhz and long stln. **Clm** not cormous at the base, not brchd above the base; **nd** swollen, villous, particularly on the stln. **Lig** membranous, papery; **bld** linear-lanceolate, cross-sections with Kranz anatomy with a single mestome shth surrounding the vascular bundles and in contact with the metaxylem vessels. **Pan** contracted; **br** spikelike, appressed, 1-sided, spklt borne in pairs. **Spklt** ellipsoid to obovoid. **Lo glm** ¾–⅘ as long as the spklt, 5–7-veined; **up glm** subequal to the lo lm, 7–11-veined, blunt; **lo flt** stmt; **lo pal** well developed; **up flt** obovoid, indurate, smooth, with microhairs and simple papillae near the base and tip. $x = 10$. Name from the Native American Hopi tribe of the southwestern United States where the genus is native.

Hopia is a monospecific genus that extends from the southwestern United States into northern Mexico. Its only species used to be included in *Panicum*. Zuloaga et al. (2007) recognized it as a distinct genus based on its unique combination of characteristics: microhairs and papillae on the upper floret, possession of the XYMS- subtype of C_4 photosynthesis, chromosome base number of 10, and molecular sequence data.

1. Hopia obtusa (Kunth) Zuloaga & Morrone [p. 477, 513]

Pl per; shortly rhz, stln to 2 m, with long intnd. **Clm** 15–80 cm, geniculate or not; **nd** glab or hairy, hairs whitish. **Shth** usu shorter than intnd, lo shth with papillose-based hairs, up shth glab; **lig** 1–2 mm; **bld** 3.5–20 cm long, 2–4 mm wide, flat or mrg slightly involute, bases narrow, with papillose-based hairs, apc attenuate, surfaces glab or adx surfaces with papillose-based hairs. **Infl** 3–20 cm long, 1–2 cm wide, exserted, contracted, rchs flat, glab, br ascending, triquetrous, scabrous, pulvini shortly hairy. **Spklt** in pairs, 1.4–1.7 mm long, ellipsoid to obovoid, biconvex, pale brown or tinged with purple, lo spklt usu aborted. **Lo glm** ¾–⅘ spklt length, glab, 3–7-veined, veins anastomosing distally, apc acute; **up glm** subequal to lo lm, 7–11-veined, acute; **lo lm** 7–9-veined, obtuse to acute; **lo flt** stmt, pal elliptic; **up flt** 2.9–3.7 mm long, 1.3–1.6 mm wide, obovoid, apc with hairs; **anth** 3, 1.4–1.5 mm. **Car** not seen, $2n = 20, 40$. **Anthesis** between April and October.

Hopia obtusa grows in the southwestern United States through central Mexico, in seasonally wet sand or gravel, especially on stream banks, ditches, roadsides, wet pastures, and rangeland, at 0–2850 m elevation. It is a good source of forage, remaining green throughout the winter.

14.11 STENOTAPHRUM Trin.   Kelly W. Allred

Pl ann or per; smt rhz or stln. **Clm** 10–60 cm, usu compressed; **intnd** solid. **Lvs** cauline; **shth** shorter than the intnd, compressed; **lig** memb and ciliate or of hairs; **bld** flat or folded. **Infl** spikelike pan; **br** very short, with fewer than 10 spklt, appressed to and partially embedded in the flattened, corky rchs; **dis** below the glm, often with a segment of the br. **Spklt** lanceolate to ovate, unawned, lo glm oriented away from the br axes. **Glm** memb; **lo glm** scalelike, usu without veins; **up glm** 5–7-veined;

lo flt stmt or strl, **lm** 3–9-veined; **up flt** bisex; **up lm** longer than the glm, papery to subcoriaceous, 3–5-veined; **up pal** generally indurate, 2-veined; **anth** 3. **Car** lanceolate to ovate, often failing to develop. $x = 9$. Name from the Greek *stenos*, 'narrow', and *taphros*, 'trench', referring to the cavities in the rachises.

Stenotaphrum is a genus of seven species that usually grow on the seashore or near the coast, primarily along the Indian Ocean rim. Three species are endemic to Madagascar, and one species is thought to be native to North America.

1. **Stenotaphrum secundatum** (Walter) Kuntze
   St. Augustine Grass [p. 477, 513]
Pl stln. **Clm** 10–30 cm, decumbent, rooting at the lo nd, brchd above the base, with prominent prophylls. **Shth** sparsely pilose, constricted at the summit; **lig** about 0.5 mm, memb, ciliate; **bld** 3–15(18) cm long, 4–10 mm wide, thick, flat, glab, apc blunt. **Pan** 4.5–10 cm long, less than 1 cm wide; **rchs** flattened, winged; **br** 12–20, with 1–5 spklt. **Spklt** 3.5–5 mm, partially embedded in 1 side of the br axes; **lo glm** about 1 mm, rounded, irregularly toothed; **up glm** and **lo lm** 3–4 mm, about equal; **up lm** papery, 5-veined, mrg weakly

clasping the pal; **anth** 2–2.5 mm, tan or purple. **Car** about 2 mm, oblong to obovate. $2n = 18$.

Stenotaphrum secundatum grows on sandy beaches, at the edges of swamps and lagoons, and along inland streams and lakes. It may be native to the southeastern United States, being known from the Carolinas prior to 1800, but it has become naturalized in most tropical and subtropical regions of the world.

Stenotaphrum secundatum is planted for turf in the southern United States and is now established from California to North Carolina and Florida. Numerous cultivars have been developed. Specimens with variegated foliage (often called *S. secundatum* var. *variegatum* Hitchc.) are sometimes used as ornamentals in hanging baskets and greenhouses.

14.12 PASPALUM L.   Charles M. Allen and David W. Hall

Pl ann or per; ces, rhz, or stln. **Clm** 3–400 cm, erect, spreading or prostrate, smt trailing for 200+ cm. **Shth** open; **aur** smt present; **lig** memb. **Infl** tml, smt also axillary, pan of 1–many spikelike br, these digitate or rcm on the rchs, spreading to erect, 1 or more br completely or partially hidden in the shth in some species; **br axes** flattened, usu narrowly to broadly winged, usu terminating in a spklt, smt extending beyond the distal spklt but never forming a distinct bristle; **dis** below the glm. **Spklt** subsessile to shortly pedlt, plano-convex, rounded to acuminate, dorsally compressed, not subtended by bristles or a ringlike callus, solitary or paired (1 spklt of the pair rdcd in some species), in 2 rows along 1 side of the br, with 2 flt, first rchl segment not swollen, up glm and up lm adjacent to the br axes; **lo flt** strl; **up flt** sessile or stipitate, bisex, acute or rounded. **Lo glm** absent or present only on some spklt of each br, without veins or 1-veined, unawned; **up glm** and **lo lm** subequal, memb, apc rounded, unawned; **lo pal** absent or rdmt; **up lm** convex, indurate, smooth to slightly rugose, stramineous to dark brown, mrg scarious, involute, clasping the pal; **up pal** indurate, smooth to slightly rugose, stramineous to dark brown. **Car** orbicular to elliptical, plano-convex or flattened, white, yellow, or brown. $x = 10, 12$. Name from the Greek *paspalos*, a kind of millet.

Paspalum includes 300–400 species, most of which are native to the Western Hemisphere. Two species grow in the Intermountain Region, one of which is native. *Paspalum scrobiculatum* is grown as a grain in India, and several species are grown as forage plants. There are also many weedy species in the genus.

1. Spikelets solitary, not associated with a naked pedicel or rudimentary spikelets; edge of
   spikelet not ciliate . . . . 1. *P. distichum*
1. Spikelets paired, if only 1 spikelet functional, a naked pedicel or rudimentary, non-
   functional spikelet present; edge of spikelet distinctly ciliate . . . . 2. *P. dilatatum*

1. **Paspalum distichum** L. Knotgrass,
   Thompsongrass [p. 478, 513]
Pl per; rhz or ces. **Clm** 5–65 cm, erect; **nd** glab. **Shth** glab, sparsely long pubescent distally; **lig** 1–2 mm; **bld** to 14 cm long, 1.8–11.5 mm wide, flat or conduplicate, glab or pubescent, apc involute. **Pan** tml, usu composed of a digitate pair of br, a third br smt present below; **br** 1.4–7 cm, diverging, often arcuate; **br axes** 1.2–2.2 mm wide, winged, glab, mrg scabrous, terminating

in a spklt. **Spklt** 2.4–3.2 mm long, 1.1–1.6 mm wide, solitary (rarely paired), appressed to the br axes, broadly elliptic, stramineous, smt partially purple. **Lo glm** absent or, if present, to 1 mm and triangular; **up glm** sparsely and shortly pubescent on the back, 3-veined; **lo lm** glab, 3-veined; **up flt** stramineous. **Car** 1.9–2.1 mm, yellow. $2n = 20, 30, 40, 48, 60, 61$.

Paspalum distichum grows on the edges of lakes, ponds, rice fields, and wet roadside ditches. It is native in warm regions throughout the world, being most abundant in humid areas. In the Western Hemisphere, it grows from the United States to Argentina and Chile.

2. Paspalum dilatatum Poir. Dallisgrass [p. 478, 513]

Pl per; ces, rhz, rhz short (less than 1 cm), forming a knotty base. **Clm** 50–175 cm, erect; **nd** glab. **Shth** glab or pubescent, lo shth more frequently pubescent than the up shth; **lig** 1.5–3.8 mm; **bld** to 35 cm long, 2–16.5 mm wide, flat, mostly glab, adx surfaces with a few long hairs near the base. **Pan** tml, with 2–7 rcmly arranged br; **br** 1.5–12

cm, rcm, divergent; **br axes** 0.7–1.4 mm wide, winged, glab, mrg scabrous, terminating in a spklt. **Spklt** 2.3–4 mm long, 1.7–2.5 mm wide, paired, appressed to the br axes, ovate, tapering to an acute apex, stramineous (rarely purple). **Lo glm** absent; **up glm** and **lo lm** 5–7-veined, mrg pilose; **up flt** stramineous. **Car** 2–2.3 mm, white to brown. $2n = 20, 40, 50–63$.

Paspalum dilatatum is native to Brazil and Argentina. It is now well established in North America, generally as a weed in waste places. It is also used as a turf grass.

14.13 ZULOAGAEA Bess   Mary E. Barkworth

Pl per; ces, rhz, rhz short, thin. **Clm** clumped or solitary, often with hard, cormlike bases, slightly compressed, erect or geniculate at the lo nd. **Shth** shorter than the intnd, keeled, often pilose, hairs papillose-based near the throat; **lig** memb-based, ciliate; **bld** (6)20–75 cm long, 1.5–15 mm wide, flat, adx surfaces scabrous. **Pan** pyramidal, open, bearing 4–40 br, bas nd with 1 br; **pri br** to 30 cm, with 3–6 orders of brchg, straight or flexible, ascending to reflexed; **ped** scabridulous, divergent. **Spklt** 2.5–4.2(5.5) mm long, ellipsoid or lanceoloid, purplish or greenish, glab, acute or obtuse. **Lo glm** 1–3.7 mm, about ⅔ as long as the up glm, 3–5-veined; **up glm** slightly shorter to subequal to the lo lm, glab, 5–7-veined; **lo flt** strl or stmt; **lo lm** 2.9–3.3 mm, glab, 5-veined, acute; **lo pal** 3–4 mm, hyaline; **up flt** 2.1–5 mm long, exceeding the up glm, mrg embracing the the lo lm, dull, pale, finely transversely rugose, apc acute, puberulent; **anth** about 2 mm, yellow-brown; **stigmas** pale purple, plumose. **Car** oblong, compressed; **emb** about ⅓ the length of the car.

Zuloagaea is a monospecific genus that is native to western North America. It used be included in Panicum, but molecular phylogenetic studies consistently place it in the bristle clade of the Paniceae, although it never develops bristles. Bess et al. (2006) concluded that it is best treated as a monospecific genus. They noted that it is recognizable by its "open, loosely flowered pyramidal panicle with small spikelets that are purple in color if the plant is growing in the sun, or green in color if it is growing in the shade." Its distinctive vegetative characters include the thickened, often cormous, culm bases, elongate blades, and rather short, often pilose sheaths.

1. Zuloagaea bulbosa (Kunth) Bess Bulbous Panicgrass [p. 479, 513]

Pl per, ces or single-stalked, erect. **Clm** 20–200 cm tall, 1–8 mm thick, glab, lowest intnd often thickened into a hard, cormlike base. **Shth** pilose to glab; **lig** memb, membrane 0.1–1.8 cm long, fringed with hairs 0.1–3 mm long; **bld** flat, 8–75 cm long, 1.5–12 mm wide, abx surface glab and sometimes pilose, adx surface scabrous, mrg scabrous. **Pan** 10–75 cm long, bearing 4–40 br, brchg in 3 to 6 orders, pri br minutely scabrous, sec br glab. **Br** flexuose to straight, ascending to divergent, loosely flowered, to 30 cm long, spklt pedlt; **ped** < 1 mm long, apc cuplike, slightly scabrous to smooth. **Spklt** dorsally compressed, 3–5.5 mm long; **flt** 2, the lo stmt or strl, the up perfect; **dis** below the glm. **Lo glm** triangular,

3-veined, 1–3.7 mm long, about ⅔ as long as up glm; **up glm** acute, 5-veined, 1.4–4 mm long, slightly shorter than lo lm, purple or green; **lo lm** acute, 5-veined, 2.9–3.3 mm long; **lo pal** hyaline, about as long as up glm; **up lm** indurate, scabrous at the apc, 2.1–5 mm long, extending beyond the up glm, apc acute, mrg wrapping around up pal mrg; **up pal** flattened, indurate.

Zuloagaea bulbosa grows in roadside ditches, on gravelly river banks and moist mountain slopes, often in ponderosa pine and oak woodlands, from southern Nevada and Arizona to western Texas and central Mexico. It is considered an important forage grass and is sometimes cut for hay but is not known to be cultivated. Flowering is from July to mid-October. Plants growing in sunlight tend to have purple spikelets; those growing in the shade tend to have green spikelets.

15. ANDROPOGONEAE Dumort.   Mary E. Barkworth

Pl usu per. **Clm** 7–600 cm, ann, not woody, often reddish or purple, particularly at the nd, often brchd above the base. **Shth** open; **lig** usu scarious to memb, ciliate or not; **bld** mostly well developed, lvs subtending an infl or an infl unit often with rdcd bld. **Photosynthetic pathway** NADP-ME; **bundle**

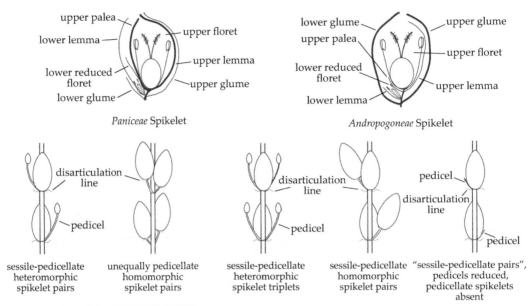

Paniceae Spikelet   Andropogoneae Spikelet

| sessile-pedicellate heteromorphic spikelet pairs | unequally pedicellate homomorphic spikelet pairs | sessile-pedicellate heteromorphic spikelet triplets | sessile-pedicellate homomorphic spikelet pairs | "sessile-pedicellate pairs", pedicels reduced, pedicellate spikelets absent |

SPIKELETS and SPIKELET UNITS

shth single. **Infl** tml, frequently on both the clm and their br, smt also axillary, usu of 1–many spikelike br, these in digitate clusters of 1–13+ on a peduncle or attached, directly or indirectly, to elongate rchs, often partially to almost completely enclosed by the subtending lf shth at maturity, in some taxa axillary infl composed of multiple-stalked pedunculate clusters of infl br subtended by a modified lf; **dis** usu in the br axes beneath the sessile flt, the dispersal unit being a sessile flt, the intnd to the next sessile flt, the pedicel, and the pedlt spklt (br with dis axes are termed *rames* in the following accounts), smt beneath the glm, the br axes remaining intact. **Spklt** in unequally pedlt pairs, sessile-pedlt pairs, or triplets, or apparently solitary and sessile, pedlt spklt and smt the ped rdcd or absent, triplets usu with 1 sessile and 2 pedlt spklt, tml spklt units on the br often with 2 pedlt spklt even if the others have only 1 (all spklt units with 2 sessile and 1 pedlt spklt in *Polytrias*). **Spklt** pairs or triplets homogamous (spklt in the unit sexually alike) or heterogamous (spklt in the unit sexually dissimilar); spklt of unequally pedlt pairs usu homogamous and homomorphic; **spklt in sessile-pedlt pairs** or **triplets** usu heterogamous and heteromorphic; **sessile spklt** usu bisex; **pedlt spklt** usu smaller than the sessile spklt, often stmt or strl, smt absent. **Spklt** usu with 2 flt (1 in *Polytrias*). **Glm** exceeding and usu concealing the flt (excluding the awns), rounded or dorsally compressed, usu tougher than the lm; **lo flt** in bisex or pist spklt strl or stmt, often rdcd to a hyaline scale; **up flt** bisex or pist, lm often hyaline, smt with an awn that exceeds the glm; **lod** cuneate; **anth** usu 3. **Ped** free or fused to the rchs intnd. **Pedlt spklt** variable, smt similar to the sessile spklt, smt differing in sexuality and shape, smt missing. x = usu 9 or 10, or possibly 5 with 9 and 10 reflecting ancient polyploidy.

The tribe *Andropogoneae* includes about 87 genera and 1060 species, of which 14 genera and 24 species have been found in the Intermountain Region; some of these have not become established. The tribe is common in tropical and subtropical regions, particularly in areas with significant summer rains, such as the central plains of North America. Two of the grasses that used to dominate the prairies of central North America, *Andropogon gerardii* and *Schizachyrium scoparium* (Big and Little Bluestem, respectively), are members of the *Andropogoneae*. The reddish purplish coloration that characterizes the culms and leaves of many *Andropogoneae* gives a striking aspect to grasslands (and lawns) dominated by its members.

Members of the *Andropogoneae* differ from those of *Paniceae* in the reduced lemmas and paleas of their florets and, usually, in their paired, unequally pedicellate spikelets, disarticulating inflorescence branches (*rames*), and the manner in which these branches are aggregated into inflorescences. Unequally pedicellate spikelet pairs are found in many other tribes, but they are more common, and the pedicels more strikingly unequal in length, in the *Andropogoneae*. Recent molecular work supports recognition of the tribe with one modification of its traditional limits, the incorporation of *Arundinella* and *Tristachya* (Kellogg 2000). There is less agreement on the tribe's internal structure and its relationship to the *Paniceae* (Kellogg 2000; Spangler 2000; Giussani et al. 2001, Bouchenak-Khelladi 2008).

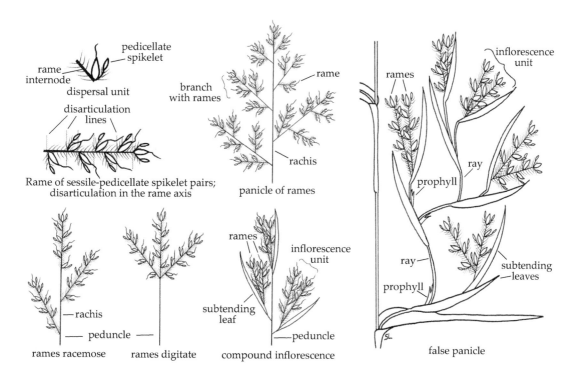

pedicellate spikelet

rame internode

dispersal unit

disarticulation lines

Rame of sessile-pedicellate spikelet pairs; disarticulation in the rame axis

rame

branch with rames

rachis

panicle of rames

rames

inflorescence unit

ray

prophyll

rachis

peduncle

rames racemose

rames digitate

rames

inflorescence unit

subtending leaf

peduncle

compound inflorescence

ray

prophyll

subtending leaves

false panicle

INFLORESCENCE STRUCTURES

Inflorescence Structures

Describing inflorescence structures in the *Andropogoneae* is not simple. There is a basic pattern, but its many modifications have resulted in great structural diversity. The following paragraphs provide an overview of this diversity and explain the words and phrases used in describing it. Diagrammatic representations of many of the structures mentioned are presented on above and on p. 258.

Spikelets

Members of the *Andropogoneae*, like those of the *Paniceae*, generally have two florets per spikelet, the lower floret usually being reduced in size and sterile or staminate, and the upper floret bisexual (p. 258). Despite this similarity, spikelets of the two tribes are easy to distinguish. In the *Paniceae*, the lowest glume is usually much shorter than the floret, and the upper florets usually have lemmas that are thicker and tougher than the glumes and lower lemmas. In the *Andropogoneae*, the glumes usually exceed and enclose both florets, and are thicker and tougher than the lemmas. The florets of the *Andropogoneae* contrast strongly with the glumes, having hyaline or thinly membranous lemma bodies and hyaline paleas, or, in many cases, no palea. They are almost always completely concealed by the glumes, except that the upper floret often has an awn that projects beyond the glumes.

In some *Andropogoneae*, the glumes are merely thickly membranous, but most genera have coriaceous or indurate glumes. The lower glumes are sometimes tougher and larger than the upper glumes, and may even conceal the upper glumes as, for example, in *Heteropogon* (p. 271). In such genera, the lower glumes may be mistaken for lemmas. In dioecious species, or monoecious species with strongly differentiated staminate and pistillate spikelets, the staminate spikelets usually have softer glumes than the pistillate spikelets.

Spikelet Units

The basic element of the inflorescence structure in the *Andropogoneae* is the spikelet unit. These units usually consist of pairs of spikelets, one sessile and one pedicellate, but they may consist of a pair of unequally pedicellate spikelets or of three spikelets. If there are three spikelets in the unit, one is usually sessile and the other two pedicellate, but a few genera, such as *Polytrias*, have two sessile spikelets and one pedicellate spikelet.

Unequally pedicellate spikelet pairs or triplets are found in other tribes, but in the *Andropogoneae* the spikelets usually differ in size, shape, and sexuality. Spikelet units with spikelets that differ in their sexuality are described as heterogamous; those with sexually similar spikelets are said to be homogamous. Spikelet units with morphologically dissimilar spikelets are heteromorphic; those with morphologically similar spikelets are homomorphic. In most *Andropogoneae*, the spikelet units are heterogamous and heteromorphic. The sessile spikelets usually contain a bisexual or pistillate floret, and often exhibit features such as awns and calluses that are related to seed dispersal and establishment (Peart 1984); the pedicellate spikelets are usually staminate, sterile, vestigial, or even absent. In some genera the situation is reversed, the pedicellate spikelets being bisexual or pistillate, and the sessile spikelets staminate or sterile. Sterile and staminate spikelets are sometimes morphologically similar to the pistillate or bisexual spikelets, but usually lack the features associated with seed dispersal and establishment.

A few genera have no staminate or sterile spikelets, but merely empty pedicels associated with the bisexual sessile spikelets, as in *Sorghastrum* (p. 265), or with only a stump where the pedicel and its spikelet would be.

Inflorescence Structure

Further complexity is introduced to the *Andropogoneae* inflorescence structure by the manner in which the spikelet units are aggregated and the mode of disarticulation. Three patterns can be identified. The simplest pattern consists of inflorescences similar to those common in other tribes, in which neither the rachis nor the inflorescence branches break up at maturity. Genera with such inflorescences [e.g., *Miscanthus* (p. 262) and *Imperata* (p. 263)] have unequally pedicellate spikelets, and disarticulation is below the glumes. Such inflorescences are, however, in the minority within the *Andropogoneae*.

A more common situation is for the spikelets to be in sessile-pedicellate pairs and disarticulation to be in the branch axes, immediately below the attachment of the sessile spikelets. The resulting dispersal unit consists of the spikelet pair plus the internode that extends from the sessile spikelet to the next most distal sessile spikelet. These disarticulating inflorescence branches, termed rames in this treatment, form the basic unit of the typical *Andropogoneae* inflorescence. In other publications, the rames are often called racemes, a word that is restricted in this treatment to an entire inflorescence, not just an inflorescence branch.

Rames are usually composed of several spikelet units, but sometimes of only one. The spikelets may be evenly distributed, or the base of the rame axis may be naked. Individual plants may bear few to many rames, and the rames themselves may be aggregated in a wide array of primary and secondary arrangements; they may also be branched.

One or more rames may be borne on a single stalk. If this stalk is attached to a rachis, the unit formed by the stalk and its rame(s) constitutes an inflorescence branch. Such a pattern is seen, for example, in *Sorghum halepense* (p. 264). A more common situation is for one or more rames to be attached digitately to a common stalk, the peduncle. This peduncle may terminate a culm, or be axillary to a subtending leaf, as in *Andropogon hallii* (p. 268). Each peduncle and its associated rame(s) constitutes an inflorescence unit.

False panicles represent a further level of complexity. In these, the inflorescence units terminate rays, each of which has a prophyll, a 2-veined structure, in its axil. Several rays may develop within the axil of a single leaf sheath, and rays may themselves give rise to subtending leaves with multiple rays in their axils. The result is a complex, tiered inflorescence in which only the ultimate units are easily described. Such inflorescences are found, for example, in *Andropogon glomeratus* (p. 268). Fortunately, identification of the *Andropogoneae* does not require analyzing false panicles, merely their ultimate inflorescence units.

In another inflorescence pattern, the rame axes are thick and the pedicels are either closely appressed or even fused to the rame axes. In these genera, the pedicellate spikelets are often highly reduced or absent. Pistillate rames of wild taxa of *Zea* (p. 271) represent an extreme example of this pattern. In these genera, the sessile spikelets are completely embedded in the rame axes, the lower glumes being indurate and completely concealing the florets.

1. Leaves smelling of lemon oil or citronella, the sheaths without glandular depressions on the keel; plants known only in cultivation . . . . 15.10 *Cymbopogon*
1. Leaves not aromatic (or, if so and smelling of citronella, the sheaths with glandular depressions along the keel and plants annual); plants cultivated or not.
    2. All spikelets unisexual, the pistillate and staminate spikelets in separate inflorescences or the pistillate spikelets below the staminate spikelets in the same inflorescence.
        3. Pistillate spikelets completely concealed within a hard, globose, beadlike structure (a modified leaf sheath) from which the staminate rames protrude . . . . 15.14 *Coix*
        3. Pistillate spikelets enclosed by 1 or more subtending leaf sheaths and a hyaline prophyll; staminate spikelets in a separate inflorescence on the same plant. . . . . 15.13 *Zea*
    2. Some spikelets bisexual (usually the sessile or more shortly pedicellate spikelet of each spikelet pair or triplet).
        4. Spikelets apparently solitary and sessile, the pedicellate spikelets absent; pedicels absent or present.
            5. Inflorescences terminal and axillary, composed of digitate clusters of 1–13 rames on a common peduncle; peduncles subtended by, and often partially included in, a modified leaf . . . . 15.09 *Andropogon* (in part)
            5. Inflorescences terminal, with elongate rachises and branches with several to many rames; peduncles and branches not subtended by a modified leaf . . . . 15.07 *Sorghastrum*
        4. Spikelets in sessile-pedicellate or unequally pedicellate pairs or triplets, the pedicellate spikelets often smaller than the sessile spikelets, sometimes rudimentary.
            6. All spikelet units homogamous, frequently also homomorphic.
                7. Spikelets in sessile-pedicellate pairs or triplets; disarticulation in the rames, below the sessile spikelets.
                    8. Panicle branches alternate, with multiple rames; rames with more than 5 spikelet units. . . . . 15.02 *Saccharum*
                    8. Panicle branches subverticillate, with 1–3 rames; rames with 2–5 spikelet units. . . . . 15.01 *Spodiopogon*

7. Spikelets in unequally pedicellate pairs; disarticulation below the glumes, the branches remaining intact at maturity.
    9. Spikelets usually awned; inflorescence branches usually 7–35 cm long .... 15.03 *Miscanthus*
    9. Spikelets unawned; inflorescence branches 1–7 cm long .... 15.04 *Imperata*
6. All or most spikelet units heterogamous, usually also heteromorphic, sometimes the proximal units on the rames or racemes homomorphic and homogamous.
    10. Terminal inflorescences with elongated rachises.
        11. Rame internodes and pedicels with a translucent median line ..15.08 *Bothriochloa* (in part)
        11. Rame internodes and pedicels without a translucent median line .... 15.06 *Sorghum*
    10. Terminal inflorescences and individual inflorescence units without elongated rachises, composed of 1–13 rames, or a raceme in which disarticulation occurs below the pedicellate spikelets.
        12. Disarticulation occurring below the pedicellate spikelets, not in the inflorescence axes; pedicellate spikelets bisexual .... 15.05 *Trachypogon*
        12. Disarticulation occurring below the sessile spikelets, in the inflorescence axes; pedicellate spikelets staminate, sterile, or absent.
            13. Rame internodes and pedicels with a translucent median groove ..
            .... 15.08 *Bothriochloa* (in part)
            13. Rame internodes and pedicels without a translucent median groove.
                14. Distal spikelet units on each rame awned, awns 6–10 cm long; basal spikelet units unawned........................... 15.12 *Heteropogon*
                14. All spikelet units on each rame alike, unawned or with awns to 2.5 cm long.
                    15. Rames usually solitary on the peduncles, occasionally 2; rame internodes cupulate or fimbriate distally; lower glumes of the sessile spikelets veined between the keels ..
                    .... 15.11 *Schizachyrium*
                    15. Rames usually 2–13 on the peduncles, occasionally solitary; rame internodes neither fimbriate nor cupulate distally; lower glumes of the sessile spikelets usually without veins between the keels........................... 15.09 *Andropogon* (in part)

15.01 **SPODIOPOGON** Trin.

Mary E. Barkworth

Pl usu per; smt rhz. **Clm** 40–150 cm, erect, simple or brchg. **Lvs** not aromatic; **lig** memb; **bld** lanceolate to broadly linear, smt psdpet. **Infl** tml, open or contracted pan, with evident rchs with numerous subverticellate br that terminate in 1–3 short rames; **rames** with slender intnd and 2–5 sessile-pedlt homogamous spklt pairs; **dis** in the rames, below the sessile spklt. **Spklt** usu lanceolate. **Glm** equal, chartaceous, often pilose, scarcely keeled, with several raised veins, acute; **cal** glab or densely hairy; **lo flt** usu stmt, unawned; **up flt** bisex; **up lm** bilobed, with a geniculate awn; **anth** 3. $x = 10$. **Ped** slender, not fused to the rame axes. Name from the Greek *spodios*, 'ash-colored' or 'gray', and *pogon*, 'beard', a reference to the spikelet hairs.

Spodiopogon is a genus of 10–15 species, most of which grow in subtropical regions of the Eastern Hemisphere, although *Spodiopogon sibiricus* extends north to Irkutsk, Russia. One species is cultivated as an ornamental in North America.

1. **Spodiopogon sibiricus** Trin. Silver Spike [p. 479, 513]

Pl rhz. **Clm** 90–150 cm tall, 2–4 mm thick. **Bas lvs** bladeless or with rdcd bld; **cauline shth** mostly glab, but pilose at the col; **lig of cauline lvs** 2–3 mm, ciliate on the erose mrg; **cauline bld** to 35 cm long, 8–20 mm wide, pilose on both surfaces, mrg ciliate near the base, cilia papillose-based. **Pan** 12–20 cm long, 2–4 cm wide, shortly exserted; **rchs** glab, smooth; **pri br** 2–6 cm; **rames** 2–3.5 cm. **Spklt** 4.5–5.5 mm. **Lo glm** pilose throughout, 5–9-veined; **up glm** of the sessile spklt pilose on the mrg, those of the pedlt spklt pilose throughout; **cal hairs** about ¼ as long as the spklt; **awns** 0.7–1.2 cm; **anth** of the sessile spklt about 2 mm, those of the pedlt spklt about 3 mm. $2n = 40, 42$.

Spodiopogon sibiricus is native to the grasslands of the montane regions that extend from central China to northeastern Siberia. It is grown as an ornamental in North America.

15.02 SACCHARUM L.   Robert D. Webster

Pl per; **ces**, often with a knotty crown, smt rhz, rhz usu short but elongate in some species, rarely stln. **Clm** 0.8–6 m, erect. **Lvs** cauline, not aromatic; **shth** usu glab, smt ciliate at the throats; **lig** memb, ciliate; **bld** flat, lax, smooth, usually glab. **Infl** tml, large, often plumose, fully exserted pan with evident rchs and numerous, ascending to appressed br terminating in multiple rames, br alternate, smt naked below; **rames** with numerous sessile-pedlt spklt pairs and a tml triad of 1 sessile and 2 pedlt spklt, **intnd** slender, without a translucent median groove; **dis** beneath the pedlt spklt and in the rames beneath the sessile spklt, sessile spklt falling with the adjacent intnd and ped. **Spklt pairs** homogamous and homomorphic, or almost so, not embedded in the rame axes, dorsally compressed. **Sessile spklt: cal** truncate, usu with silky hairs; **glm** subequal, chartaceous to coriaceous, glab or villous, 2-keeled, veins not raised; **lo flt** strl; **lo lm** hyaline or memb; **lo pal** absent or vestigial, entire; **up flt** bisex; **up lm** entire or bidentate, muticous or awned; **lod** 2, truncate; **anth** 2 or 3. **Ped** neither appressed nor fused to the rame axes. **Pedlt spklt** well developed, from slightly shorter than to equaling the sessile spklt. $x = 10$. Name from the Latin *saccharum*, 'sugar', a reference to the sweet juice.

Saccharum is a genus of 35–40 species that grow throughout the tropics and subtropics. One species may be found as an ornamental in the Intermountain Region. Some species of *Saccharum* hybridize naturally with other, presumably closely related, genera such as *Miscanthus*, *Imperata*, and *Sorghum*. The most familiar species of *Saccharum* is *S. officinarum*, sugar cane.

1. **Saccharum ravennae** (L.) L. Ravennagrass [p. *480*]

Pl ces. **Clm** 2–4 m, glab; **nd** glab. **Shth** glab; **aur** absent; **lig** 0.6–1.1 mm; **bld** 50–100 cm long, 5–14 mm wide, glab. **Peduncles** 40–80 cm, glab; **pan** lanceolate; **rchs** 30–70 cm, glab; **pri br** 6–20 cm, appressed or spreading; **rame intnd** 1–2 mm, with hairs. **Sessile spklt** 4–6 mm long, 0.7–0.9 mm wide, straw-colored. **Cal hairs** 4–6 mm, subequal to the spklt, white; **lo glm** smooth, 4–5-veined; **up glm** 3-veined; **lo lm** 3–5 mm, 1-veined; **up lm** subequal to the lo lm, without veins, entire; **awns** 2–5 mm, flat, straight or curved at the base; **lod** veins not extending into hairlike projections; **anth** 3. **Ped** 1–3 mm, pubescent. **Pedlt spklt** similar to the sessile spklt. $2n = 20$.

Saccharum ravennae is native to southern Europe and western Asia. It is grown as an ornamental in North America, occasionally escaping and persisting.

15.03 MISCANTHUS Andersson   Mary E. Barkworth

Pl per; ces, smt rhz. **Clm** 40–400 cm, erect. **Lvs** not aromatic; **shth** open; **lig** memb, truncate, ciliate; **bld** flat. **Infl** tml, ovoid or corymbose pan, with elongate rchs and numerous ascending, spikelike br; **br** usu more than 10 cm long, with unequally pedlt spklt pairs, spklt homogamous and homomorphic; **dis** below the glm. **Cal** short, blunt, pilose, with fine hairs, hairs often exceeding the spklt. **Glm** memb to coriaceous; **lo glm** broadly convex to weakly 2-keeled, without raised veins; **lo flt** strl; **up flt** bisex; **up lm** entire and unawned or bidentate and awned from the sinuses; **anth** 2 or 3. **Ped** free. $x = 19$. Name from the Greek *mischos*, 'pedicel', and *anthos*, 'flower', both spikelets ("flowers") being pedicellate.

Miscanthus is a genus of approximately 25 species. Most of the species are native to southeast Asia; a few extend into Africa. Some species hybridize with *Saccharum*, from which *Miscanthus* differs in its non-disarticulating branches and unequally pedicellate, rather than sessile-pedicellate, spikelets.

The two species described below are grown as ornamentals because of their large, plumose panicles and striking growth habit. They flower in late summer to fall.

1. Spikelets 3–3.5 mm long; blades 15–40 mm wide; rachises ¾–⅘ as long as the panicles . . . . . . 1. *M. floridulus*
1. Spikelets 3.5–7 mm long; blades 6–20 mm wide; rachises ⅓–⅔ as long as the panicles . . . . 2. *M. sinensis*

1. **Miscanthus floridulus** (Labill.) Warb. *ex* K. Schum. & Lauterb. Giant Chinese Silvergrass [p. *481*]

Pl ces, forming large clumps. **Clm** 1.5–4 m tall, 8–16 mm thick below. **Lvs** crowded at the base; **shth** glab or sparsely pubescent, mrg glab or ciliate; **lig** 1–3 mm; **bld** 30–80 cm long, 15–40 mm wide, adx surfaces pubescent near the bases, glab elsewhere, midveins whitish, conspicuous both abx and adx. **Pan** 30–50 cm long, 10–20 cm wide, exserted, dense, ovoid-ellipsoid, white, usu with more than 15 br; **rchs** 25–40 cm, hispid-pubescent, ¾–⅘ as long as the pan; **br** 10–25 cm long, 8–10 mm wide, often brchd at the base; **intnd** 3–5 mm, glab. **Shorter ped** 1–1.5 mm; **longer ped** 2.5–3.5 mm, becoming somewhat recurved. **Spklt** 3–3.5 mm, lanceolate to lance-ovate; **cal hairs** 4–6 mm, to twice as long as the spklt, white. **Lo glm** glab or puberulent distally; **awns of up lm** 5–15 mm, weakly geniculate. $2n = 36, 38, 57$.

Miscanthus floridulus is the most widespread species of *Miscanthus* in southeast Asia. The culms are used for arrow-shafts in Papua New Guinea and as support and drying racks for climbing vegetables and tobacco in the Philippines. In North America it is grown as an ornamental. The blades of the lower leaves tend to fall off in late summer, leaving the culms naked at the base. It is tolerant of wind and salt spray.

2. Miscanthus sinensis Andersson EULALIA [p. 481]

Pl ces, forming large clumps, with short, thick rhz. **Clm** 60–200 cm tall, 3–7 mm thick below. **Lvs** predominantly bas; **shth** mostly glab, throats pilose; **lig** 1–2 mm; **bld** 20–70 cm long, 6–20 mm wide, midveins conspicuous abx, 1–2 mm wide, whitish. **Pan** 15–25 cm long, 8–28 cm wide, dense to loose, usu with more than 15 br; **rchs** 6–15 cm, ⅓–⅔ as long as the infl; **br** 8–15(30) cm long, about 10 mm wide, smt brchd at the base; **intnd** 4–8 mm, glab. **Shorter ped** 1.5–2.5 mm; **longer ped** 3.5–6 mm, slightly recurved at maturity. **Spklt** 3.5–7 mm, lanceolate to lance-ovate; **cal hairs** 6–12 mm, to twice as long as the spklt, white, stramineous to reddish. **Glm** subequal; **lo glm** 3-veined, ciliolate on the mrg; **up glm** 1-veined; **awns of up lm** 6–12 mm, geniculate below. $2n = 38, 40$, and dysploids from 35–42.

Miscanthus sinensis is native to southeastern Asia. It is frequently cultivated in the United States and southern Canada, and is now established in some parts of the United States. Approximately 40 forms and cultivars are available, some having white-striped leaves, others differently colored callus hairs and, consequently, differently colored panicles.

15.04 IMPERATA Cirillo

Mark L. Gabel

Pl per; strongly rhz. **Clm** 10–150(217) cm, mostly erect and unbrchd, usu with 3–4 nd. **Lvs** not aromatic; **shth** open, ciliate at the mrg of the col; **lig** memb; **bld** of the bas lvs linear to lanceolate, smt ciliate bas, those of the cauline lvs rdcd. **Infl** tml, cylindrical to conical pan with an evident rchs; **rchs** often with numerous long hairs; **infl br** 1–7 cm, usu shorter than the rchs, with spklt in unequally pedlt pairs; **dis** below the glm. **Spklt** homogamous and homomorphic, unawned; **cal** very short, hairy, hairs 7–16 mm. **Glm** equal to subequal, memb, 3–9-veined, with hairs longer than the flt over at least the lo ½; **lo flt** rdcd to hyaline or memb lm; **up flt** bisex, lm, if present, hyaline, unawned; **anth** 1–2, yellow to brown; **stigmas** elongate, purple to brown; **sty** connate or free. **Ped** not fused to the br axes, terminating in cuplike tips. **Car** ovate to obovate, light to dark brown. $x = 10$. Named after Ferrante Imperato (1550–1625) of Naples, an apothecary and author of a folio work on natural history.

Imperata has nine species and is widely distributed in warm regions of both hemispheres. Its economic importance is primarily negative, as both *I. cylindrica* and *I. brasiliensis* are weedy (Gabel 1989), but new shoots of both species are used for hay or grazing. One species grows in the Intermountain Region.

1. Imperata brevifolia Vasey SATINTAIL [p. 482, 513]

Clm 51–129 cm. **Lig** 0.7–2.9 mm; **bld** 7–14 mm wide, linear to lanceolate, abx surfaces smooth, adx surfaces smt densely pilose bas, otherwise scabrous. **Pan** 16–34 cm, dense; **lo br** 2–5 cm, divergent. **Cal hairs** 8–12 mm; **glm** 2.7–4.1 mm; **lo lm** 2.5–3.9 mm, memb, glmlike; **up lm** 1.4–2.4 mm, completely surrounding the ov; **sta** 1, filaments dilated at the base; **anth** 1.3–2.3 mm, yellow to orange; **sty** 0.9–2.4 mm; **stigmas** 2.1–4 mm, purple to brown. $2n = 20$.

Imperata brevifolia is native to wet or moist sites in the southwestern deserts from California, Nevada, and Utah to western Texas. Most collections were made before 1945, in sites that are now used for housing or agriculture, but there have been several post-1990 collections in California. It also continues to grow in Grand Canyon National Park but has not been collected in Nevada or Utah since the 1970s.

15.05 TRACHYPOGON Nees

Kelly W. Allred

Pl ann or per; ces or shortly rhz. **Clm** 30–200 cm, unbrchd; **intnd** semi-solid. **Lvs** cauline, not aromatic; **shth** shorter than the intnd, rounded; **lig** memb; **bld** flat to involute. **Infl** tml, solitary rcm of heterogamous subsessile-pedlt spklt pairs (rarely of 2 digitate spikelike br), axes slender, without a translucent median groove; **dis** beneath the pedlt spklt. **Subsessile spklt** stmt or strl, without a callus and unawned, otherwise similar to the pedlt spklt. **Ped** slender, not fused to the rames axes. **Pedlt spklt** bisex; **cal** sharp, strigose; **glm** firm, enclosing the flt; **lo glm** several-veined, encircling the up glm; **up glm** 3-veined; **lo flt** strl; **up flt** bisex, lm firm but hyaline at the base, tapering to an awn; **awns**

(4)6–15 cm, twisted, pubescent to plumose; **pal** absent; **anth** 3. $x = 10$. Name from the Greek *trachys*, 'rough', and *pogon*, 'beard', referring to the plumose awn of the bisexual florets.

Trachypogon is a tropical or warm-temperate genus that is native to Africa and tropical to subtropical America. Estimates of the number of species included range from one to ten. One species, *Trachypogon secundus*, is native to North America, but some taxonomists (e.g., Dávila 1994) include it in *T. plumosus* (Humb. & Bonpl. *ex* Willd.) Nees and others (e.g., Judziewicz 1990) include it, *T. plumosus*, and various other taxa in *T. spicatus* (L. f.) Kuntze. The traditional treatment and nomenclature for North American plants is retained here, pending formal study of the taxa involved.

1. Trachypogon secundus (J. Presl) Scribn.
CRINKLE-AWN [p. *482*, <u>513</u>]

Pl per. **Clm** 60–120 cm, erect; **nd** appressed-hirsute. **Shth** sparsely appressed-pilose; **lig** 2–5 mm, stiff, acute; **bld** usu 12–35 cm long, 3–8 mm wide, with a broad midrib. **Rcm** 10–18 cm, the intnd glab. **Pedlt spklt** 6–8 mm; **glm** pilose; **awns** 4–6 cm, pilose below, with 1–2 mm hairs, nearly glab distally; **anth** 4–5 mm, orange. $2n = 20$.

Trachypogon secundus is found in sandy prairies, woodlands, rocky hills, and canyons, in well-drained soils at 500–2000 m. Statements about its range are difficult to make because of disagreement as to whether northern plants, such as those found in the United States, belong to the same species as those found elsewhere.

Trachypogon secundus resembles *Heteropogon*, but differs in the longer, non-disarticulating inflorescence and shorter, pale awns. It rates as fairly good fodder when green, but is seldom abundant enough to be an important forage grass.

15.06 SORGHUM Moench   Mary E. Barkworth

Pl ann or per. **Clm** 50–500+ cm; **intnd** solid. **Lvs** not aromatic, bas and cauline; **aur** absent; **lig** memb and ciliate or of hairs; **bld** usu flat. **Infl** tml, pan with evident rchs; **pri br** whorled, compound, the ultimate units rames; **rames** with most spklt in heterogamous sessile-pedlt spklt pairs, tml spklt unit on each rame usu a triplet of 1 sessile and 2 pedlt spklt, rame axes without a translucent median line; **dis** in the rames below the sessile spklt, smt also below the pedlt spklt (cultivated taxa not or only tardily dis). **Sessile spklt** dorsally compressed, cal blunt or pointed; **lo glm** dorsally compressed and rounded bas, 2-keeled or winged distally, 5–15-veined, usu unawned; **up glm** 2-keeled, smt awned; **lo flt** rdcd to hyaline lm; **up flt** pist or bisex, lm hyaline, smt awned. **Ped** slender, neither appressed nor fused to the rame axes. **Pedlt spklt** stmt or strl, well developed, often subequal to the sessile spklt in size. $x = 10$. Name from the Italian word for the plant, *sorgho*.

Most of the approximately 25 species of *Sorghum* are native to tropical and subtropical regions of the Eastern Hemisphere, but one is native to Mexico. Some species are grown as forage, although they produce cyanogenic compounds. *Sorghum bicolor* is widely cultivated, being used as a grain, for syrup, and as a flavoring for beer.

Spangler (2000) found, using ndhF data, that *Sorghum* is polyphyletic, forming two distinct clades. The two species treated here were in the same clade.

1. Plants perennial, rhizomatous; spikelets disarticulating at maturity; caryopses not exposed at maturity. . . . . 1. *S. halepense*
1. Plants usually annual, sometimes short-lived perennials; spikelets either not disarticulating or doing so tardily; caryopses often exposed at maturity. . . . . 2. *S. bicolor*

1. Sorghum halepense (L.) Pers. JOHNSON
GRASS [p. *483*, <u>513</u>]

Pl per; rhz. **Clm** 50–200 cm tall, 0.4–2 cm thick; **nd** appressed pubescent; **intnd** glab. **Lig** 2–6 mm, memb, conspicuously ciliate; **bld** 10–90 cm long, 8–40 mm wide. **Pan** 10–50 cm long, 5–25 cm wide, pri br compound, terminating in rames of 1–5 spklt pairs; **dis** usu beneath the sessile spklt, smt also beneath the pedlt spklt. **Sessile spklt** bisex, 3.8–6.5 mm long, 1.5–2.3 mm wide; **cal** blunt; **glm** indurate, shiny, appressed pubescent; **up lm** unawned, or with a geniculate, twisted awn to 13 mm; **anth** 1.9–2.7 mm. **Ped** 1.8–3.3 mm. **Pedlt spklt** stmt, 3.6–5.6 mm; **glm** memb to coriaceous, unawned. **Car** not exposed at maturity. $2n = 20$, 40; several dysploid counts also reported.

Sorghum halepense is native to the Mediterranean region. It is sometimes grown for forage in North America, but it is considered a serious weed in warmer parts of the United States. It hybridizes readily with *S. bicolor*, and derivatives of such hybrids are widespread. The annual *Sorghum ×almum* Parodi, which has wider (2–2.8 mm) sessile spikelets with more veins in the lower glumes (13–15 versus 10–13) than *S. halepense*, is one such derivative.

2. Sorghum bicolor (L.) Moench SORGHUM
[p. *483*, <u>513</u>]

Pl ann or short-lived per; often tillering, without rhz. **Clm** 50–500+ cm tall, 1–5 cm thick, smt brchg above the base; **nd** glab or appressed pubescent; **intnd** glab. **Lig** 1–4 mm; **bld** 5–100 cm long, 5–100 mm wide, smt glab. **Pan** 5–60 cm long, 3–30 cm wide, open or contracted, pri br compound, terminating in rames with 2–7 spklt pairs; **dis**

usu not occurring or tardy. **Sessile spklt** bisex, 3–9 mm, lanceolate to ovate; **cal** blunt; **glm** coriaceous to memb, glab, densely hirsute, or pubescent, keels usu winged; **up lm** unawned or with a geniculate, twisted, 5–30 mm awn; **anth** 2–2.8 mm. **Ped** 1–2.6 mm. **Pedlt spklt** 3–6 mm, usu shorter than the sessile spklt, stmt or strl. **Car** often exposed at maturity. $2n = 20, 40$.

Sorghum bicolor was domesticated in Africa 3000 years ago, reached northwestern India before 2500 B.C., and became an important crop in China after the Mongolian conquest. It was introduced to the Western Hemisphere in the early sixteenth century, and is now an important crop in the United States and Mexico. Numerous cultivated strains exist. They are all interfertile with each other and with other wild species of *Sorghum*.

The treatment presented here is based on de Wet (1978) and is artificial. *Sorghum bicolor* subsp. *arundinaceum* is the wild progenitor of the cultivated strains, all of which are treated as *S. bicolor* subsp. *bicolor*. These strains tend to lose their distinguishing characteristics if left to themselves. They will also hybridize with subsp. *arundinaceum*, and these hybrids can backcross to either parent, resulting in plants that may strongly resemble one parent while having some characteristics of the other. All such hybrids and backcrosses are treated here as *S. bicolor* subsp. ×*drummondii*.

1. Inflorescence branches remaining intact at maturity; caryopses exposed at maturity; sessile spikelets 3–9 mm long, elliptic to oblong . . . . subsp. *bicolor*
1. Inflorescence branches rames, disarticulating at maturity, sometimes tardily; caryopses not exposed at maturity; sessile spikelets 5–8 mm long, lanceolate to elliptic.
   2. Rames readily disarticulating . . subsp. *arundinaceum*
   2. Rames disarticulating tardily . . . subsp. ×*drummondii*

15.07 **SORGHASTRUM** Nash

Sorghum bicolor subsp. **arundinaceum** (Desv.) de Wet & J.R. Harlan *ex* Davidse [p. 483]

Pl ann or weakly biennial. **Clm** to 4 m, slender to stout. **Rames** readily dis at maturity, with 1–5 nd. **Sessile spklt** 5–8 mm, lanceolate to elliptic. **Car** not exposed at maturity.

Sorghum bicolor subsp. *arundinaceum* is native to, and most common, in Africa, but some strains have been introduced into the Western Hemisphere.

Sorghum bicolor (L.) Moench subsp. **bicolor** Sorghum, Broomcorn, Sorgo [p. 483]

Pl ann. **Clm** to 5 m or more, stout, frequently tillering. **Infl** br remaining intact at maturity, with 1–5 nd. **Sessile spklt** 3–9 mm long, 2–5 mm wide, elliptic to oblong. **Car** exposed at maturity.

All the cultivated sorghums are placed in *Sorghum bicolor* subsp. *bicolor*. 'Grain sorghums' have short panicles and panicle branches, 'broomcorns' have elongate panicles and panicle branches, and 'sweet sorghums' or 'sorgo' produce an abundance of sweet juice in their stems. For a more detailed treatment, see Harlan and de Wet (1972).

Sorghum bicolor subsp. ×**drummondii** (Steud.) de Wet *ex* Davidse Chicken Corn, Sudangrass [p. 483]

Pl ann. **Clm** to 4 m, relatively stout. **Rames** usu tardily **dis**, mostly with 3–5 nd. **Sessile spklt** 5–6 mm, lanceolate to elliptic. **Car** not exposed at maturity.

The hybrids treated here as *Sorgum bicolor* subsp. ×*drummondii* are most common in the Eastern Hemisphere, but a few are cultivated in the United States. Among these are the plants known as 'chicken corn' and 'Sudangrass' [= *S. sudanense* (Piper) Stapf] (de Wet 1978).

Patricia D. Dávila Aranda and Stephan L. Hatch

Pl ann or per; ces, smt rhz. **Clm** 50–300+ cm, erect, nodding or clambering, unbrchd; **nd** densely pubescent, particularly in young pl. **Lvs** not aromatic; **lig** memb, glab or pubescent; **bld** flat, involute, or folded. **Infl** tml, secund or equilateral pan with evident rchs and numerous br, not subtended by modified lvs; **br** capillary, rebrchg, with many rames, not subtended by modified lvs; **dis** in the rames, beneath the sessile spklt. **Spklt** sessile, subtending a hairy ped (2 ped in the tml spklt units), dorsally compressed. **Cal** blunt or sharp; **glm** coriaceous; **lo glm** pubescent, 5–9-veined, acute; **up glm** slightly longer, usu glab, 5-veined, truncate; **lo flt** rdcd to hyaline lm; **up flt** bisex, lm hyaline, bifid, awned from the sinuses; **awns** usu once- or twice-geniculate, often spirally twisted, shortly strigose, brownish; **anth** 3; **ov** glab. **Car** flattened. **Ped** 3–6.5 mm, slender, not fused to the rame axes; **pedlt spklt** absent. $x = 10$. Name from *Sorghum* and the Latin suffix *astrum*, 'a poor imitation of', alluding to its similarity to *Sorghum*.

Sorghastrum includes about 18 species. Most are native to tropical or subtropical America, two are African, and four are native to North America, including one in the Intermountain Region. Absence of the pedicellate spikelet, while confusing at first, makes *Sorghastrum* a readily recognizable genus. Its species range from sea level to approximately 3000 m, and can be found in a wide range of habitats.

1. **Sorghastrum nutans** (L.) Nash Indiangrass [p. 484, 514]

Pl rhz, rhz short, stout, scaly. **Clm** 50–240 cm tall, 1.5–4.5 mm thick, erect; **intnd** glab. **Shth** glab or sparsely hispid; **lig** 2–6 mm, usu with thick, pointed aur; **bld** 10–70 cm long, 1–4 mm wide, usu glab. **Pan** 20–75 cm, loosely contracted, yellowish to brownish; **br** often flexible. **Spklt** 5–8.7 mm. **Cal** blunt, villous; **lo glm** 5–8 mm, pubescent, 7–9-veined; **up glm** 5–8 mm, 5-veined;

awns 10–22(30) mm, about 2–3 times longer than the spklt, once-geniculate; **anth** (2)3–5 mm. **Car** 2–3 mm. **Ped** 3–6 mm, flexible. $2n = 20, 40, 80$.

Sorghastrum nutans grows in a wide range of habitats, from prairies to woodlands, savannahs, and scrubland vegetation. It is native from Canada to Mexico and was one of the four

principal grasses of the tallgrass prairie that occupied the central United States prior to agricultural development of the region. It is frequently used for forage, for erosion control on slopes and along highways, and in restoration work. It is an attractive plant and can be used to advantage in flower arrangements. It grows readily from seed if adequate moisture is available. There are several cultivars on the market.

15.08 **BOTHRIOCHLOA** Kuntze

Kelly W. Allred

Pl per; ces or stln. **Clm** 30–250 cm, with pithy intnd. **Lvs** bas or cauline, not aromatic; **shth** open; **aur** absent; **lig** memb, smt also ciliate; **bld** usu flat, convolute in the bud. **Infl** tml, pan of subdigitate to rcmly arranged br, each br with (1)2–many rames, br not subtended by modified lvs; **rames** with spklt in heterogamous sessile-pedlt pairs, intnd with a translucent, longitudinal groove, often villous on the mrg; **dis** in the rames, beneath the sessile spklt. **Spklt** dorsally compressed; **sessile spklt** with 2 flt; **lo glm** rounded, several-veined, smt with a dorsal pit, mrg clasping the up glm; **up glm** somewhat keeled, 3-veined; **lo flt** hyaline scales, unawned; **up flt** bisex; **up lm** with a midvein that usu extends into a twisted, geniculate awn, occ unawned; **anth** 3. **Ped** similar to the intnd. **Pedlt spklt** rdcd or well developed, strl or stmt, unawned. **Car** lanceolate to oblong, somewhat flattened; **hila** punctate, bas; **emb** about ½ as long as the car. $x = 10$. Name from the Greek *bothros*, 'trench' or 'pit', and *chloë*, 'grass', alluding either to the groove in the pedicels or to the pit in the lower glumes of some species.

Bothriochloa is a genus of about 35 species that grow in tropical to warm-temperate regions. Three are native to the Intermountain Region; one has been introduced. Most species provide fair forage in summer and fall. Polyploidy has been an important mechanism of speciation in the genus.

1. Pedicellate spikelets about as long as the sessile spikelets . . . . 4. *B. ischaemum*
1. Pedicellate spikelets much shorter than the sessile spikelets.
    2. Sessile spikelets 2.5–4.5 mm long; awns 8–17 mm long . . . . 1. *B. laguroides*
    2. Sessile spikelets 4.5–8.5 mm long; awns 18–35 mm long.
        3. Rachises 5–10 cm long, with numerous branches . . . . 2. *B. barbinodis*
        3. Rachises 1–5 cm long, with 2–9 branches . . . . 3. *B. springfieldii*

1. **Bothriochloa laguroides** (DC.) Herter SILVER BLUESTEM [p. *484*, <u>514</u>]

Clm 35–115(130) cm tall, usu less than 2 mm thick, erect or geniculate at the base, brchd at maturity; **nd** shortly hirsute, pilose with erect hairs, or glab. **Lvs** usu bas (smt cauline on robust pl), usu glaucous; **lig** 1–3 mm; **bld** 5–25 cm long, 2–7 mm wide, flat to folded, mostly glab. **Pan** 4–12(14) cm, narrowly oblong or lanceolate, silvery white or light tan; **rchs** 4–8 cm, with more than 10 br; **br** 1–5.5 cm, erect-appressed, rarely with axillary pulvini, lo br shorter than the rchs, usu with more than 1 rame; **rame intnd** with a groove wider than the mrg, mrg copiously hairy, hairs 3–9 mm, at least somewhat obscuring the spklt. **Sessile spklt** 2.5–4.5 mm, ovate, somewhat glaucous, apc blunt; **lo glm** glab or hirtellous, rarely with a dorsal pit; **awns** 8–16 mm; **anth** 0.6–1.4 mm. **Pedlt spklt** 1.5–2.5(3.5) mm, shorter than the sessile spklt, strl. $2n = 60$.

Bothriochloa laguroides grows in well-drained soils of grasslands, prairies, roadsides, river bottoms, and woodlands, often on limestone, usually at 20–2100 m. Plants from the United States and northern Mexico belong to *B. laguroides* subsp. *torreyana* (Steud.) Allred & Gould. Occasional plants are found with spreading branches and axillary pulvini; they do not merit formal recognition. *Bothriochloa laguroides* subsp.

torreyana is sometimes used in landscaping. It does well on rocky slopes and sandy banks.

2. **Bothriochloa barbinodis** (Lag.) Herter CANE BLUESTEM [p. *485*, <u>514</u>]

Clm 60–120 cm tall, rarely more than 2 mm thick, erect, geniculate at the base, often brchd at maturity, not glaucous below the nd; **nd** hirsute, hairs 3–4 mm, mostly erect to ascending, tan or off-white. **Lvs** cauline; **lig** 1–2 mm, often erose; **bld** 20–30 cm long, 2–7 mm wide, not glaucous, glab or sparingly pilose near the throat. **Pan** 5–14(20) cm on the larger sht, oblong to somewhat fan-shaped, silvery white; **rchs** 5–10 cm, straight, exserted or partially included in the shth, with numerous br; **br** 4–9 cm, erect, with several rames; **rame intnd** with a memb groove wider than the mrg, mrg densely pilose, longest hairs 3–7 mm, concentrated distally. **Sessile spklt** 4.5–7.3 mm; **lo glm** short pilose, with or without a dorsal pit; **awns** 20–35 mm; **anth** 0.5–1 mm, often remaining within the spklt. **Pedlt spklt** 3–4 mm, narrowly lanceolate, strl. $2n = 180$.

Bothriochloa barbinodis is a common species, at 500–1200 m, along roadsides, drainage ways, and gravelly slopes in desert grasslands, from the southwestern United States through Mexico and Central America to Bolivia and

Argentina. Plants with a pit on the back of their lower glumes occur sporadically; they do not differ in any other respect from those without pits. The species is sometimes used as an ornamental.

Bothriochloa barbinodis has been confused with *B. springfieldii*, but differs in having taller culms, wider leaves, shorter nodal hairs, and more numerous, less hairy panicle branches.

3. **Bothriochloa springfieldii** (Gould) Parodi
SPRINGFIELD BLUESTEM [p. *485*, <u>514</u>]

Clm 30–80 cm, erect, unbrchd; **nd** prominently bearded, hairs 3–7 mm, spreading, silvery white. **Lvs** mostly bas; **lig** 1–2.5 mm; **bld** 5–30 cm long, 2–3(5) mm wide, flat to folded, glab or sparsely hispid adx, pilose near the throat. **Pan** 4–9 cm, oblong to fan-shaped; **rchs** 1–5 cm, with 2–9 br; **br** 4–8 cm, longer than the rchs, with 1(2) rames; **rame intnd** with a memb groove wider than the mrg, mrg densely white-villous, hairs 5–10 mm, obscuring the sessile spklt. **Sessile spklt** 5.5–8.5 mm, lanceolate; **lo glm** densely short-pilose on the lo ½, smt with a dorsal pit; **awns** 18–26 mm; **anth** 1–1.5 mm. **Pedlt spklt** 3.5–5.5 mm, strl. $2n = 120$.

Bothriochloa springfieldii grows in rocky uplands, ravines, plains, sandy areas, and roadsides, from southern Utah to western Texas and Mexico at 900–2500 m, and as a disjunct in northwest Louisiana. It differs from *B. barbinodis* in its less robust habit, narrower blades, longer nodal hairs, and fewer, more hairy panicle branches.

4. **Bothriochloa ischaemum** (L.) Keng [p. *486*, <u>514</u>]

Pl usu ces, occ stln or almost rhz under close grazing or cutting. **Clm** 30–80(95) cm, stiffly erect; **nd** glab or short hirsute. **Lvs** tending to be bas; **lig** 0.5–1.5 mm; **bld** 5–25 cm long, 2–4.5 mm wide, flat to folded, glab or with long, scattered hairs at the base of the bld. **Pan** 5–10 cm, fan-shaped, silvery reddish purple; **rchs** 0.5–2 cm, with (1)2–8 br; **br** 3–9 cm, longer than the rchs, erect to somewhat spreading from the axillary pulvini, usu with only 1 rame; **rame intnd** with a cent groove narrower than the mrg, mrg ciliate, with 1–3 mm hairs. **Sessile spklt** 3–4.5 mm, narrowly ovate; **lo glm** hirsute below, with about 1 mm hairs, lacking a dorsal pit; **awns** 9–17 mm, twisted, geniculate; **anth** 1–2 mm. **Pedlt spklt** about as long as the sessile spklt, but usu narrower, strl or stmt. $2n = 40, 50, 60$.

Bothriochloa ischaemum grows along roadsides and in waste ground and rangeland pastures, at 50–1200 m. It is native to southern Europe and Asia. It was introduced to the United States for erosion control along right of ways and for livestock forage in the southwest. It is now established in the region and has spread along roadsides into other central and southern states. There are two variants that are sometimes recognized as varieties, plants with glabrous nodes being called *B. ischaemum* var. *ischaemum* and plants with pubescent nodes being called *B. ischaemum* var. *songarica* (Rupr. *ex* Fisch. & C.A. Mey.) Celarier & J.R. Harlan. The varieties are not recognized here.

15.09 **ANDROPOGON** L.

Christopher S. Campbell

Pl per; usu ces, smt rhz. **Clm** 20–310 cm, erect, much-brchd distally. **Lvs** not aromatic; **lig** memb, smt ciliate; **bld** linear, flat, folded, or convolute. **Infl** tml and axillary or a false pan; **infl units** 1–600+ per clm; **peduncles** initially concealed by the subtending lf shth, smt exserted beyond the shth at maturity, with (1)2–5(13) rames; **rames** not reflexed at maturity, axes slender, terete to flattened, not longitudinally grooved, usu conspicuously pubescent, with spklt in heterogamous sessile-pedlt pairs (the tml spklt smt in triplets of 1 sessile and 2 pedlt spklt), apc of the intnd neither cupulate nor fimbriate; **dis** in the rames, below the sessile spklt. **Sessile spklt** bisex, awned, with short, blunt cal; **lo glm** 2-keeled, flat or concave, usu not veined between the keels, smt 2–9-veined; **anth** 1, 3(2). **Ped** usu longer than 3 mm, similar to the rame intnd in shape, length, and pubescence color, not fused to the rame axes. **Pedlt spklt** usu vestigial or absent, smt well developed and stmt. $x = 10$. Name from the Greek *andro*, 'man', and *pogon*, 'beard', referring to the pubescent pedicels of the staminate spikelets.

Andropogon is a cosmopolitan genus of tropical and temperate zones, comprising approximately 120 species. Several taxa are ecologically important in North America. *Andropogon gerardii* is one of the most important native grasses in North America, being one of the dominant species in the tallgrass prairies that used to cover the center of the continent. Many varieties of *A. glomeratus* aggressively colonize abandoned fields, cutover timberlands, and roadsides. Some species are used in restoration and landscaping. Three species have been found in the Intermountain Region.

1. Pedicellate spikelets vestigial or absent, sterile; sessile spikelets 3–5 mm long . . . . 3. *A. glomeratus*
1. Pedicellate spikelets usually well developed, (3.5)6–12 mm long, usually staminate; sessile spikelets 5–12 mm long.
    2. Sessile spikelets with awns 8–25 mm long; ligules 0.4–2.5 mm long; hairs of the rame internodes 2.2–4.2 mm long, sparse to dense; rhizomes sometimes present, the internodes usually less than 2 cm long . . . . 1. *A. gerardii*

2. Sessile spikelets unawned or with awns less than 11 mm long; ligules (0.9)2.5–4.5 mm long; hairs of the rame internodes 3.7–6.6 mm long, usually dense; rhizomes always present, the internodes often more than 2 cm long . . . . 2. *A. hallii*

1. **Andropogon gerardii** Vitman Big Bluestem [p. 486, 514]

Pl often forming large clumps, rhz, if present, with intnd shorter than 2 cm. **Clm** 1–3 m, often glaucous. **Shth** glab or pilose; **lig** 0.4–2.5 mm; **bld** 5–50 cm long, (2)5–10 mm wide, usu pilose adx, at least near the col. **Infl units** usu only tml; **peduncles** with 2–6(10) rames; **rames** 5–11 cm, exserted at maturity, usu purplish, smt yellowish; **intnd** sparsely to densely pubescent, hairs 2.2–4.2 mm, usu white, rarely yellowish. **Sessile spklt** 5–11 mm, scabrous; **awns** 8–25 mm; **anth** 3, 2.5–4.5 mm. **Pedlt spklt** 3.5–12 mm, usu well developed and stmt. $2n = 20, 40, 60$ (usu), 70, 80.

Andropogon gerardii grows in prairies, meadows, and generally dry soils. It is a widespread species, extending from southern Canada to Mexico, and was once dominant over much of its range. It is frequently planted for erosion control, restoration, or as an ornamental. It hybridizes with *A. hallii*, the two sometimes being treated as conspecific subspecies.

2. **Andropogon hallii** Hack. Sand Bluestem [p. 487, 514]

Pl strongly rhz, rhz intnd often longer than 2 cm. **Clm** (40)60–150(200) cm, strongly glaucous. **Lig** (0.9)2.5–4.5 mm, ciliate; **bld** 3–40(51) cm long, (1.5)2–10 mm wide, often pilose, at least near the col. **Infl units** usu only tml; **peduncles** with 2–7 rames; **rames** 4–7(9) cm, exserted at maturity; **intnd** usu densely pubescent, hairs 3.7–6.6 mm, often strongly yellowish. **Sessile spklt** (5)6.5–12 mm; **lo glm** often ciliate; **awns** absent or to 11 mm; **anth** 3, (2.3)4–6 mm. **Pedlt spklt** 3.5–12 mm, usu well developed and stmt. $2n = 60$ (usu), 70, 100.

Andropogon hallii grows on sandhills and in sandy soil. Its range extends through the central plains into northern Mexico. It is similar to *A. gerardii*, differing primarily in its rhizomatous habit, more densely pubescent rames and pedicels, and greater drought tolerance. *Andropogon hallii* and *A. gerardii* are sympatric in some locations. The two species can hybridize and are sometimes treated as conspecific subspecies.

3. **Andropogon glomeratus** (Walter) Britton, Sterns & Poggenb. Bushy Bluestem, Bushy Beardgrass [p. 487, 514]

Pl ces, up portion dense, oblong to oblanceolate or obpyramidal. **Clm** 20–250 cm; **intnd** green, smt glaucous; **br** mostly erect, straight. **Shth** usu scabrous, smt smooth; **lig** 0.6–2.2 mm, smt ciliate, cilia to 0.9 mm; **bld** 13–109 cm long, 2.9–9.5 mm wide, glab or sparsely to densely pubescent,

hairs usu spreading, rarely appressed. **Infl units** 10–600 per clm; **subtending shth** (2.0)2.9–4.4(6.5) cm long, (1.5)2.3–3.4(4.4) mm wide; **peduncles** (1)6–14(60) mm, with 2(4) rames; **rames** (1)1.7–2.5(3.5) cm, exserted or not at maturity, pubescence sparse bas and increasing in density distally within each intnd. **Sessile spklt** 3–5 mm; **cal hairs** 1–2.5 mm; **keels of lo glm** smt scabrous below midlength, usu scabrous distally; **awns** 6–19 mm; **anth** 1(3), 0.5–1.5 mm, yellow, red, or purple. **Pedlt spklt** vestigial or absent, strl. $2n = 20$.

Two varieties are found in the Intermountain Region.

1. Sheaths subtending the inflorescence units 1.5–3 mm wide; leaf sheaths usually smooth; cilia of ligules 0.2–0.9 mm long . . . . var. *pumilus*
1. Sheaths subtending the inflorescence units (2.3)2.9–4.5(6.3) mm wide; leaf sheaths usually scabrous; cilia of ligules 0.2–0.5 mm long . . . . var. *scabriglumis*

Andropogon glomeratus var. **pumilus** (Vasey) L.H. Dewey [p. 487]

Pl oblanceolate to obpyramidal in the up portion. **Clm** to 2.5 m, but as short as 20 cm in poor soils; **intnd** not glaucous. **Shth** usu smooth; **lig** ciliate, cilia 0.2–0.9 mm; **bld** green, smooth or pubescent. **Subtending shth of infl units** (2)2.9–4.3(5.2) cm long, 1.5–3 mm wide; **peduncles** (2)8–15(40) mm; **rames** 1.3–3 cm, exserted. **Keels of lo glm** usu smooth below midlength, scabrous distally; **anth** often retained within the spklt.

Andropogon glomeratus var. *pumilus* is weedy and grows in disturbed, wet or moist sites. It is abundant and widespread, extending from the southern United States through Central America to northern South America.

Andropogon glomeratus var. **scabriglumis** C.S. Campb. [p. 487]

Pl oblanceolate to obpyramidal in the up portion. **Clm** 80–150 cm; **intnd** not glaucous. **Shth** usu scabrous; **lig** ciliate, cilia 0.2–0.5 mm; **bld** green, smooth or pubescent. **Subtending shth of infl units** (2.3)2.9–4.5(6.3) cm long, (1.5)2.3–3.3(4.4) mm wide; **peduncles** (2)5–10(16) mm; **rames** (1.7)1.9–2.3(2.8) cm, exserted. **Keels of lo glm** usu scabrous below and above the midpoint.

Andropogon glomeratus var. *scabriglumis* grows in moist soils of seepage slopes and the edges of springs, from California to New Mexico and southward into Mexico.

15.10 CYMBOPOGON Spreng.

Mary E. Barkworth

Pl usu per; ces. **Clm** 15–300 cm. **Lvs** aromatic, smelling of lemon oil or citronella; **shth** open, not strongly keeled except near the summit; **lig** memb; **bld** usu glab or mostly so, with long filiform apc. **Infl** tml and axillary, false pan; **peduncles** often enclosed in the subtending lf shth at maturity, with 2 rames; **rames** with 4–7 heterogamous spklt pairs, axes slender, without a median groove, lo rame of each pair with 1 homogamous spklt pair at the base, its ped swollen and more or less fused to the adjacent intnd, up rames with short, strl, flattened bases that are usu deflexed at maturity, without homogamous spklt units. **Heterogamous spklt units: sessile spklt** dorsally compressed, with 2 flt; **lo glm** chartaceous, concave or flat, 2-keeled, with or without intercostal veins, often streaked with oil glands; **up flt** with a short, glab awn (rarely unawned); **ped** linear, free from the rame axes; **pedlt spklt** well developed. *x* = 10. Name from the Greek *kymbe*, 'boat', and *pogon*, 'beard', referring to the boat-shaped leaf sheaths subtending the usually hairy rames (Clifford 1996).

 Cymbopogon comprises 55 species, and is native to the tropics and subtropics of the Eastern Hemisphere. It is cultivated in southern Florida and California, sometimes persisting for a considerable period. Plants grown outdoors in North America generally remain vegetative, but can usually be identified to genus by their lemony aroma.

1. Cymbopogon citratus (DC.) Stapf LEMON GRASS [p. 488]

Pl per. **Clm** to 200 cm, flexuous; **nd** not swollen. **Bas shth** closely overlapping, gaping at maturity, forming somewhat flattened fans, glab, strongly glaucous; **lig** 0.5–2 mm, truncate; **bld** to 90 cm long, 6.5–15 mm wide. **Infl** to 60 cm, nodding; **rames** 10–25 mm; **intnd** and **ped** pilose on the mrg and dorsal surface. **Sessile spklt of heterogamous pairs** 5–6 mm; **lo glm** shallowly concave below, flat distally, keels narrowly winged; **up lm** entire or bidentate, unawned or with a 1–2 mm awn. **Pedlt spklt** 4–4.5 mm, unawned. 2*n* = 40, 60.

 Cymbopogon citratus is now known only in cultivation, even in Asia. Young shoots are used as a spice, and the oils are extracted for lemon oil. It has been grown in Florida. In the Intermountain Region, it needs to be overwintered in a greenhouse.

15.11 SCHIZACHYRIUM Nees

J.K. Wipff

Pl ann or per; ces or rhz, smt both ces and shortly rhz. **Clm** 7–210 cm, brchd above the bases, often purplish near the nd. **Lvs** not aromatic, shth open; **aur** usu absent; **lig** memb; **bld** flat, folded, or involute, those of the uppermost lvs often greatly rdcd. **Infl** axillary and tml, of 1, rarely 2, rames, peduncles subtended by a modified lf; **rames** not reflexed, with spklt in heterogamous sessile-pedlt spklt pairs, intnd more or less flattened, filiform to clavate, without a median groove, apc cupulate or fimbriate; **dis** in the rame axes, below the sessile spklt. **Spklt** somewhat dorsiventrally compressed. **Sessile spklt** with 2 flt; **glm** exceeding the flt, lanceolate to linear, memb; **lo glm** enclosing the up glm, convex, weakly 2-keeled, with several (smt inconspicuous) intercostal veins; **lo flt** rdcd to hyaline lm; **up flt** bisex, lm hyaline, bilobed or bifid to ⅞ of their length (rarely entire), awned from the sinuses; **anth** 3. **Ped** free of the rame axes, usu pubescent. **Pedlt spklt** usu shorter than to as long as the sessile spklt, occ longer, strl or stmt, with 1 flt, often dis as the rame matures; **lm** present in stmt spklt, hyaline, unawned or with a straight awn of less than 10 mm. *x* = 10. Name from the Greek *schizo*, 'split', and *achyron*, 'chaff', referring to the divided lemmas.

 Schizachyrium is a genus of approximately 60 species that are native to tropical and subtropical regions of the world; three grow in the Intermountain Region. In North America, the best known species is *S. scoparium*, which was one of the major constituents of the grasslands that used to cover the central plains. Hitchcock (1951) included both *Schizachyrium* and *Bothriochloa* in *Andropogon*. Most species of *Schizachyrium* differ from species of the other two genera in having only one rame per peduncle. More reliable, but less conspicuous distinguishing features of *Schizachyrium* are the cupulate tips of the rame internodes, the convex lower glumes, and the presence of veins between the keels of the lower glumes. A few species of *Andropogon* have solitary rames, but they do not have these other features.

1. Pedicellate spikelets 6–8 mm long, about as long as the sessile spikelets, usually staminate, sometimes sterile, unawned . . . . 3. *S. cirratum*
1. Pedicellate spikelets 0.7–10 mm long, usually shorter than the sessile spikelets, sterile, unawned or awned, the awns up to 6 mm long.
 2. Upper lemmas cleft for ⅔–⅞ of their length; lower glumes glabrous or pubescent . . . . . . 2. *S. sanguineum*
 2. Upper lemmas cleft for up to ½ of their length; lower glumes glabrous . . . . 1. *S. scoparium*

1. **Schizachyrium scoparium** (Michx.) Nash [p. *489*, <u>514</u>]

Pl ces or rhz, green to purplish, smt glaucous. **Clm** 7–210 cm tall, usu 1–3 mm thick, not rooting or brchg at the lo nd. **Shth** rounded or keeled, glab or pubescent, smt glaucous; **lig** 0.5–2 mm, col neither elongate nor narrowed; **bld** 7–105 cm long, 1.5–9 mm wide, without a longitudinal stripe of white, spongy tissue. **Peduncles** 0.8–10 cm; **rames** 2.5–8 cm, partially to completely exserted, usu somewhat open; **intnd** 3–7 mm, usu arcuate at maturity, ciliate on at least the distal ½ (smt throughout), hairs 1.5–6 mm. **Sessile spklt** 3–11 mm; **cal** 0.5–1(2) mm, hairs 0.3–4 mm; **lo glm** glab; **up lm** memb throughout, cleft to ½ their length; **awns** 2.5–17 mm. **Ped** 3–7.5 mm long, 0.1–0.2 mm wide at the base, flaring above midlength to 0.3–0.5 mm, straight or curving outwards. **Pedlt spklt** 0.7–10 mm, smt shorter than the sessile spklt, strl or stmt, unawned or awned, awns to 4 mm, when strl, the lm usu absent. 2*n* = 40.

Schizachyrium scoparium is a widespread grassland species extending from Canada to Mexico. It is one of the principal grasses in the tallgrass prairies that used to dominate the central plains of North America. It exhibits considerable variation, much of it clinal. One variety grows in the Intermountain Region.

Schizachyrium scoparium (Michx.) Nash var. **scoparium** Little Bluestem [p. *489*]

Pl usu ces, smt producing short rhz. **Clm** 30–210 cm. **Shth** usu glab, keeled; **bld** 9–45 cm long, 1.5–9 mm wide, flat, usu glab, occ pubescent. **Peduncles** to 10 cm; **rames** 2–8 cm, with 6–13 spklt, exserted. **Sessile spklt** 6–11 mm; **cal** about 0.5 mm, hairs to 2.5 mm, awns 2.5–17 mm. **Ped** 3–7.5 mm, straight or curving out at maturity. **Pedlt spklt** usu 1–6 mm, strl, without lm, occ stmt and with a lm, unawned or awned, awns to 4 mm.

Schizachyrium scoparium var. *scoparium* grows in a variety of soils and in open habitats. It was once a dominant component of the prairie grasslands that extended through the central plains of North America and into Mexico, but it has largely been replaced by fields of maize, wheat, sorghum, sunflowers, and field mustard. It is the most variable of the varieties recognized within *S. scoparium*, with morphological features that vary independently and continuously across its range, coming together in distinctive combinations in some regions. Some of these phases have been named as varieties, or even species, but they have proven to be untenable taxonomic entities when plants from throughout the range of the species are considered.

2. **Schizachyrium sanguineum** (Retz.) Alston [p. *489*, <u>514</u>]

Pl ces. **Clm** 40–120 cm, erect, not rooting or brchg at the lo nd, glab. **Shth** glab, rounded; **lig** 0.7–2 mm; **bld** 7–20 cm long, 1–6 mm wide, usu with long, papillose-based hairs bas, glab elsewhere, smt scabrous, without a longitudinal stripe of white, spongy tissue. **Peduncles** 4–6 cm; **rames** 4–15 cm, not open, usu almost fully exserted at maturity; **intnd** 4–6 mm, straight, from mostly glab with a tuft of hairs at the base to densely hirsute all over. **Sessile spklt** 5–9 mm; **cal** 0.5–1 mm, hairs to 2 mm; **lo glm** glab or densely pubescent; **up lm** cleft for (⅔)¾–⅞ of their length; **awns** 15–25 mm. **Ped** 3–6 mm long, 0.3–0.5 mm wide at the base, gradually widening to about 0.6–0.8 mm at the top, straight. **Pedlt spklt** 3–5 mm, usu evidently shorter than the sessile spklt, strl or stmt, awned, awns 0.3–6 mm.

Schizachyrium sanguineum extends from the southern United States to Chile, Paraguay, and Uruguay. One variety grows in the Intermountain Region.

Schizachyrium sanguineum var. **hirtiflorum** (Nees) S.L. Hatch Hairy Crimson Bluestem [p. *489*]

Clm 40–120 cm, brchg at the up nd, glaucous. **Lig** 1–2 mm; **bld** 10–20 cm long, 1.5–5 mm wide. **Rames** 4–10(12) cm; **intnd** scabrous, glab or hirsute. **Sessile spklt** 5–9 mm; **lo glm** sparsely to densely hirsute on the back; **awns** 15–25 mm. **Ped** arcuate at maturity, ciliate on both edges distally. **Pedlt spklt** 3–5 mm, stmt or strl, awns 0.3–5 mm. 2*n* = 40, 60, 70, 100.

Schizachyrium sanguineum var. *hirtiflorum* grows on rocky slopes and well-drained soils from Arizona to southwestern Texas and Florida It is considered a good forage species.

3. **Schizachyrium cirratum** (Hack.) Wooton & Standl. Texas Schizachyrium, Texas Beardgrass [p. *489*, <u>514</u>]

Pl ces or shortly rhz. **Clm** 31–75 cm, often decumbent, not rooting or brchg at the lo nd, glab, glaucous, smt purplish. **Lig** 1–2.5 mm; **bld** 6–17 cm long, 2–4 mm wide, glab, without a longitudinal stripe of white, spongy tissue. **Rames** 4–6 cm, usu exserted, straight, often somewhat stiff, not flexuous, appearing linear; **intnd** straight, with a tuft of hairs near the base, elsewhere glab or ciliate on the mrg. **Sessile spklt** 8–10 mm; **cal** 0.3–0.6 mm, hairs 0.5–1.2 mm; **glm** glab or scabrous; **awns** 13–24 mm. **Ped** 3.5–5 mm long, 0.2–0.5 mm wide at the base, widening to 0.5–1 mm, straight, with a tuft of hairs at the base, distal ½ usu ciliate on 1 side, smt on both sides. **Pedlt spklt** 6–8 mm, about as long as the sessile spklt, usu stmt, smt strl, unawned. $2n = 20$ (for var. *cirratum*).

Schizachyrium cirratum grows on rocky slopes, mostly at elevations of 5000 feet or higher, from southern California to western Texas into Mexico. It is an excellent forage grass. Plants in the United States differ from those in central Mexico in being essentially non-rhizomatous and in having glabrous rame axes and pedicels that are ciliate only on the distal half.

15.12 HETEROPOGON Pers.

Mary E. Barkworth

Pl ann or per; ces. **Clm** 20–200 cm, simple or brchd. **Lvs** smt aromatic and smelling of lemon oil or citronela; **shth** keeled, smt with a row of glandular depressions on the keel; **lig** memb, glab or ciliate. **Infl** tml and axillary; **peduncles** usu with 1 rame, smt with several in a digitate cluster; **rames** with 3–10 homogamous, unawned, sessile-pedlt spklt pairs on the lo ¼–⅔ and heterogamous, awned, sessile-pedlt spklt pairs distally, axes slender, without a translucent median groove; **dis** in the rames, beneath the sessile spklt of the heterogamous spklt pairs, smt also below their pedlt spklt. **Homogamous spklt units** strl or stmt; **cal** poorly developed; **glm** memb, many-veined, keels winged above. **Heterogamous spklt units: sessile spklt** bisex, terete; **cal** 1.5–3 mm, sharp, antrorsely strigose, hairs golden brown; **glm** coriaceous, pubescent, concealing the flt; **lo glm** enclosing the up glm, obscurely 5–9-veined; **up glm** sulcate, 3-veined; **lo flt** strl, rdcd to a hyaline lm; **up flt** bisex, lm with conspicuous, geniculate awns; **awns** 5–15 cm, with hairs. **Car** lanceolate, sulcate on 1 side. **Ped** short, free of the rame axes, not grooved; **pedlt spklt** strl or stmt, larger than the sessile spklt; **cal** long, glab, functioning as ped; **glm** memb, many-veined, keels winged above. $x = 10, 11$. Name from the Greek *heteros*, 'different', and *pogon*, 'beard', alluding to the difference between the calluses of the spikelets in the heterogamous pairs.

Heteropogon is a pantropical genus of eight to ten species. Many grow well on poor soils. One species grows in the Intermountain Region.

1. **Heteropogon contortus** (L.) P. Beauv. *ex* Roem. & Schult. Tanglehead [p. 490, <u>514</u>] **Pl** per. **Clm** 20–150 cm, erect. **Shth** smooth, reddish; **lig** 0.5–0.8 mm, cilia 0.2–0.5 mm; **bld** 10–15 cm long, 2–7 mm wide, flat or folded, glab or pubescent. **Rames** 3–7 cm, secund, with 12–22, brown to reddish brown, sessile-pedlt spklt pairs. **Homogamous spklt** 6–10 mm. **Heterogamous spklt: sessile spklt** 5–10 mm, brown, awned; **cal** 1.8–2 mm, strigose; **awns** 6–10 cm; **pedlt spklt** 6–10 mm, unawned; **glm** ovate-lanceolate, glab or with papillose-based hairs distally, without glandular pits, greenish to purplish brown, becoming stramineous when dry. $2n = 40, 50, 60$.

Heteropogon contortus grows on rocky hills and canyons in the southern United States into Mexico, and worldwide in subtropical and tropical areas, occupying a variety of different habitats, including disturbed habitats. It is probably native to the Eastern Hemisphere but is now found in tropical and subtropical areas throughout the world.

Heteropogon contortus is a valuable forage grass if continuously grazed so as to prevent the calluses from developing. It is also considered a weed, being able to establish itself in newly disturbed and poor soils.

15.13 ZEA L.

Hugh H. Iltis

Pl ann or per; monoecious, infl unisex or bisex with the pist spklt bas and the stmt spklt distal. **Clm** (0.2)0.5–6 m tall, 1–5 cm thick, solitary or several to many together, monopodial, often brchg (br frequently highly rdcd and hidden within the subtending lf shth), usu succulent when young, becoming woody with age; **lo nd** with prop roots; **intnd** pith-filled. **Lvs** not aromatic, cauline, distichous; **shth** open; **aur** smt present; **lig** memb, shortly ciliate; **bld** 2–12 cm wide, flat. **Pist or partially pist infl** tml on axillary br; **stmt infl** (*tassels*) pan, of 1-many br or rames, smt with sec and tertiary brchg. **Wild taxa: Pist infl** solitary, distichous rames (*ears*), these often tightly clustered in false pan, each usu wholly or partially enclosed by a thin prophyll and an equally thin bladeless lf shth; **rames** composed of 5–15 spklt in 2 ranks; **dis** in the rame axes, dispersal units (*fruitcases*) consisting of an indurate, shiny rame segment and its embedded spklt. **Pist spklt** solitary, sessile, with 1 flt; **ped** and **pedlt spklt** suppressed; **lo glm** exceeding the flt, indurate on the cent, exposed portion, hyaline on the mrg, concealing the car at maturity. **Domesticated taxon: Pist infl** solitary, polystichous spikes (*ears*) terminating rdcd br, each spike surrounded by several to many, often bladeless lf shth and a prophyll (*husks*), with 60–1000+ spklt in 8–24 rows, neither spikes nor spklt **dis** at maturity. **Pist spklt** in subsessile pairs, each spklt with 1 fnctl flt; **glm** shorter than the spklt, indurate bas, hyaline distally; **lo flt** suppressed. **ALL TAXA: lm** and **pal** hyaline, unawned; **lod** absent; **ov** glab; **sty** (silks) 2, appearing solitary by being fused except at the very tip, filamentous, sides stigmatic. **Car** subspherical to dorsally compressed; **hila** round; **emb** about ⅔ as long as the car. **WILD TAXA: Stmt pan** tml on the clm and pri br, smt also on the sec br and pist infl; **rames** distichous, similar in thickness and structure, axes dis below the sessile spklt after pollination, abscission layers evident. **DOMESTICATED TAXON: Stmt pan** tml on the clm, cent axes always much thicker than the lat br and irregularly polystichous, lat br distichous to more or less polystichous, not dis, without abscission layers below the sessile spklt. **ALL TAXA: Stmt spklt** in sessile-pedlt pairs, each with 2 stmt flt; **glm** memb to chartaceous, stiff to flexible, smt with a pair of winged keels, 5–14(28)-veined, acute; **lm** and **pal** hyaline; **lod** 2; **anth** 3. $x = 10$. Name from the Greek *zea* or *zeia*, a kind of grain.

Zea is an American genus of five species, four of which are native to montane Mexico and Central America. The fifth species, *Z. nicaraguensis* H.H. Iltis & B.F. Benz, is said to have been ubiquitous at one time in coastal Pacific Nicaragua, but is now known from only four or five small populations near sea level in seasonally flooded savannahs and riverine forests inland from the Bay of Fonseca, Nicaragua.

Zea mays subsp. *mays*, the most widespread taxon in the genus, was first domesticated about 7,000 years ago and soon became widely planted in the Americas. It is now grown in all warmer parts of the world and is the world's third most important crop plant. No other American grass has such agricultural importance.

1. **Zea mays** L. [p. *490*]

Pl ann. **Clm** (0.5)1–3(6) m tall, (0.5)1–5 cm thick. **Bld** mostly 30–90 cm long, 2.5–12 cm wide. **Pist infl** rames or spikes, usu shortly pedunculate (smt sessile), solitary, 4–30(40) cm long, (0.5)1–10 cm thick, with 2 or more rows of paired spklt, hence the spklt 4 or more ranked, rarely terminating in an unbrchd stmt infl. **Car** concealed in fruitcases (wild taxa) or exposed (domesticated taxon); **fruitcases of wild taxa** distichous, triangular in side view; **domesticated taxon** without fruitcases, glm rdcd and shallow or collapsed and embedded in the rchs. **Stmt pan** 10–25+ cm, with 1–60(235) br, intnd 1.5–8.2 mm; **spklt** 9–14 mm long, 2.5–5 mm wide; **lo glm** rounded dorsally, flexible, translucent, papery, loosely enclosing the up glm, the 2 lat veins subequal to the others, not winged. $2n = 20$.

Of the five subspecies of *Zea mays*, only the domesticated subspecies, *Z. mays* subsp. *mays*, is widely grown outside of research programs.

Zea mays L. subsp. **mays** CORN, INDIAN CORN, MAIZE [p. *490*]

Clm (1)2–4(6) m tall, (1)2–5 cm thick. **Bld** 50–90 cm long, 3–12 cm wide. **Pist infl** spikes, 15–25(40) cm long, 2–5(10) cm thick, cylindrical, tightly and permanently enclosed in several to many lf shth and a large prophyll, with 8–24 or more rows of paired spklt on a thickened, strongly vascularized, tough rchs (*cob*), not dis at maturity; **fruitcases** not developed, rchs intnd fused into the extra-vascular cylinder, glm rdcd, shallow. **Car** 60–1000+, exposed and naked. **Stmt pan** with a polystichous cent axis and non-dis br; **cent axes** usu much denser and thicker than the usu distichous lat br, these lacking abscission layers. $2n = 20$.

Zea mays subsp. *mays* is the familiar domesticated corn (or maize), from which around 400 indigenous races and many different kinds of cultivars have been developed. It is an obligate cultigen, unable to persist outside of cultivation because the caryopses are permanently attached to the rachis and enclosed by the subtending leaf sheaths. Supersweet cultivars have a double recessive gene that delays the conversion of sugar to starch; flint corns have unusually hard endosperm; and waxy cultivars have endosperm with an unusually high level of proteins and oils. Popcorns have a core of soft, relatively moist endosperm surrounded by hard endosperm. The grains "pop" when heat causes the moisture of the inner endosperm to vaporize.

15.14 **COIX** L.

John W. Thieret †

Pl ann or per; monoecious, pist and stmt spklt on separate rames in the same infl. **Clm** to 3 m, erect, creeping, or floating, brchd; **intnd** solid. **Lvs** not aromatic; **lig** memb. **Infl** axillary, of 2(3) rames, 1 pist, the other(s) stmt, pist rames completely enclosed in indurate, globose to cylindric, modified lf shth, termed *involucres*, from which the stmt rames protrude. **Pist rames** each with 3 spklt, 1 sessile and pist, the other 2 pedlt and rdmt; **sessile spklt** somewhat dorsally compressed; **glm** coriaceous, beaked; **stigmas** protruding from the involucres. **Car** more or less globose. **Stmt rames** flexible, exserted from the involucre; **spklt** in pairs or triplets, 1 sessile, the other(s) pedlt, rdcd, or absent; **lo glm** chartaceous, with 15 or more veins, 2-keeled, keels winged above; **up glm** similar, with 1 keel; **lo**

flt smt strl; **up flt** stmt; **sta** 0 or 3; **lod** 2. **Ped** not fused to the rame axes. $x = 5$. Name from the Greek *koix*, a palm.

Coix is a genus of about five species, one of which has been introduced to North America. All the species are native to tropical Asia, where *C. lacryma-jobi* is harvested for food.

1. **Coix lacryma-jobi** L. Job's-Tears [p. 490]
Pl ann or per. **Clm** to 3 m. **Lvs** mostly cauline, evidently distichous; **bld** to 75 cm long, 1.5–6 cm wide. **Involucres** usu 8–12 mm, varying in color. **Lo glm** of fnctl pist spklt 6–10 mm, hyaline below, 5–7-veined, with a 1–3 mm coriaceous beak. **Stmt rames** 10–35 mm, with 3–25 spklt pairs, dis at maturity; **spklt** 5–9 mm, dorsally compressed; **glm** exceeding the flt, with 15+ veins; **lo glm** elliptic to obovate, somewhat asymmetrical, mrg folded inward, apc obtuse; **up glm** lanceolate to narrowly elliptic, keels often winged, apc acute; **up lm** 5–8 mm, hyaline, elliptic to ovate, 3-veined; **up pal** similar but 2-veined; **anth** 3–6 mm. $2n = 20$.

Coix lacryma-jobi is a tall, maize-like plant. In North America, it is usually grown as an ornamental, but it has become established at scattered locations. The involucres, which can be used as beads, may be white, blue, pink, straw, gray, brown, or black, with the color being distributed evenly, irregularly, or in stripes. Cultivars with easily removed involucres are grown for food and beverage, especially in Asia.

Illustrations

The illustrators of the plates are indicated by the intials placed on them. In many instances, two sets of initials occur together. The first set of initials belongs to the illustrator who prepared the rough illustrations; the second set belongs to the illustrator who converted the rough pencil illustration into a finished inked illustration. If only one set of initials appears on a plate, all the species on that plate are the work of the individual(s) indicated; otherwise initials appear next to each species. Because of the number of illustrators involved, only those who served as the primary illustrator for one or more plates are listed on the cover page. The initials of all illustrators are listed below in alphabetical order.

BFG	Bee F. Gunn
K	Karen Klitz
SL	Sandy Long
AM	Annaliese Miller
BM	Linda Bea Miller
HP	Hana Pazdírková
CR	Christine Roberts
CTR	Cindy Roché
AS	Andy Sudkamp
LAV	Linda Ann Vorobik

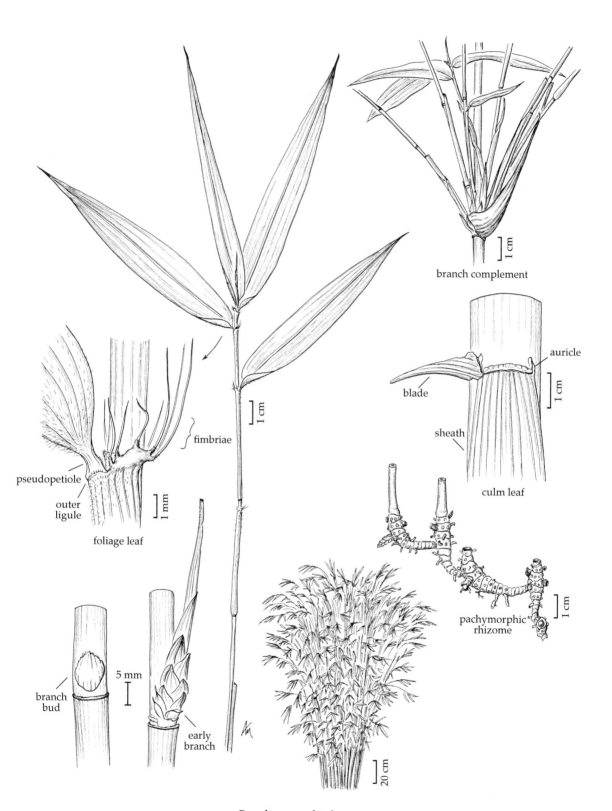

branch complement

auricle

blade

sheath

culm leaf

fimbriae

pseudopetiole

outer
ligule

foliage leaf

pachymorphic
rhizome

branch
bud

5 mm

early
branch

20 cm

Bambusa multiplex

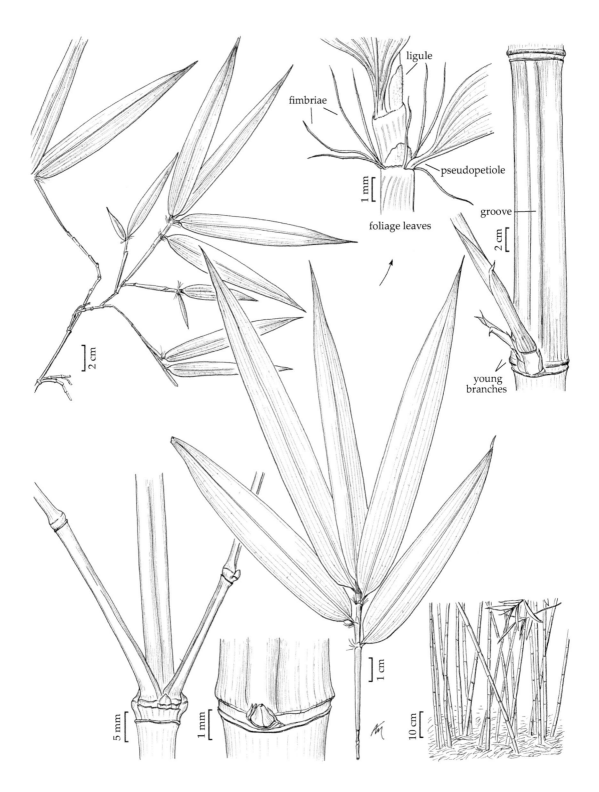

fimbriae

ligule

pseudopetiole

groove

foliage leaves

young
branches

1 mm

2 cm

2 cm

5 mm

1 mm

1 cm

10 cm

Phyllostachys bambusoides

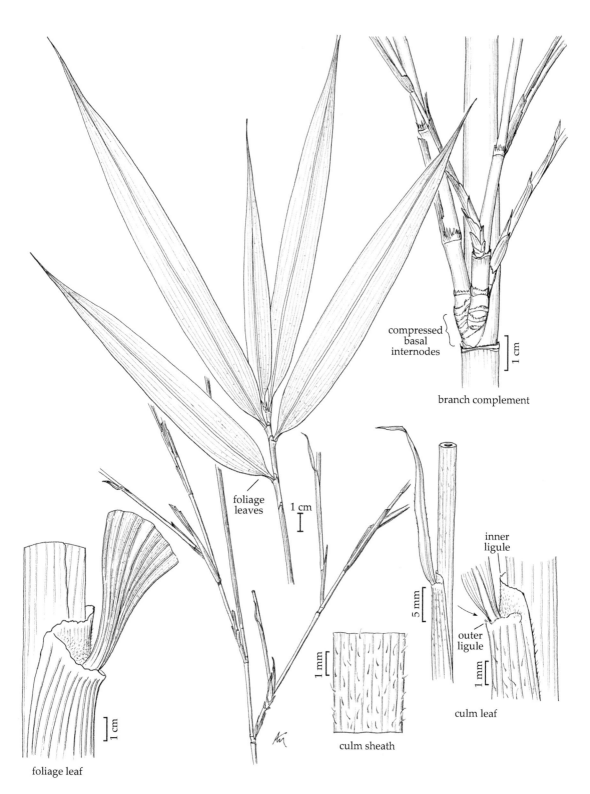

compressed
basal
internodes

branch complement

foliage
leaves

1 cm

5 mm

inner
ligule

outer
ligule

culm leaf

1 mm

culm sheath

1 mm

foliage leaf

1 cm

Pseudosasa japonica

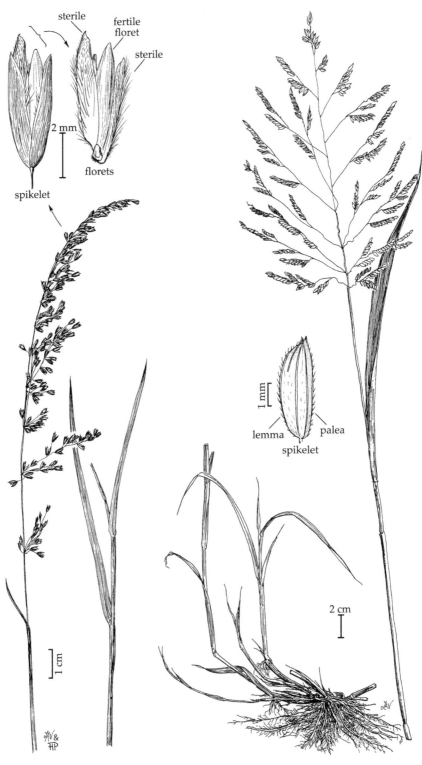

sterile   fertile
floret

sterile

2 mm

florets

spikelet

lemma   palea
spikelet

1 mm

2 cm

1 cm

Ehrharta calycina

Leersia oryzoides

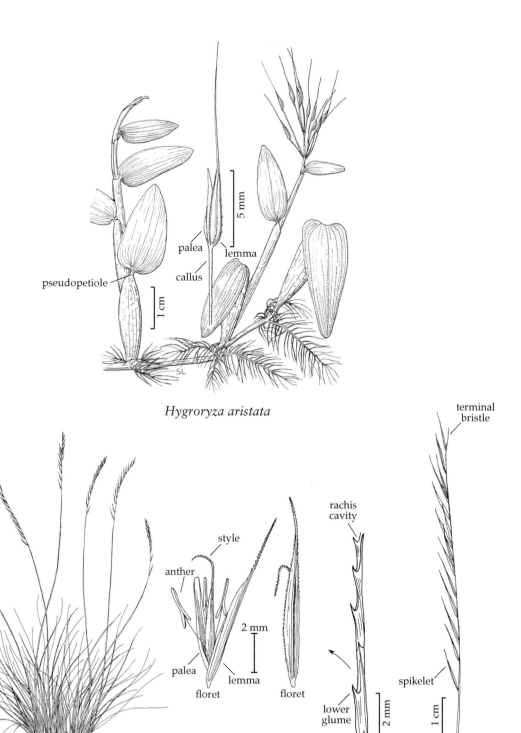

Hygroryza aristata

Nardus stricta

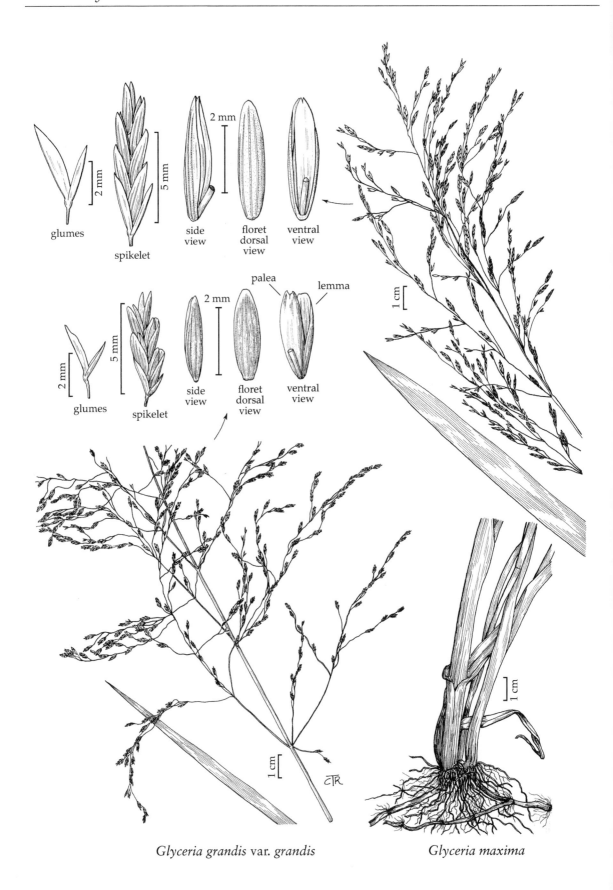

glumes

spikelet

2 mm

side
view

floret
dorsal
view

ventral
view

glumes

spikelet

palea

lemma

side
view

floret
dorsal
view

ventral
view

1 cm

1 cm

1 cm

CTR

Glyceria grandis var. *grandis*

Glyceria maxima

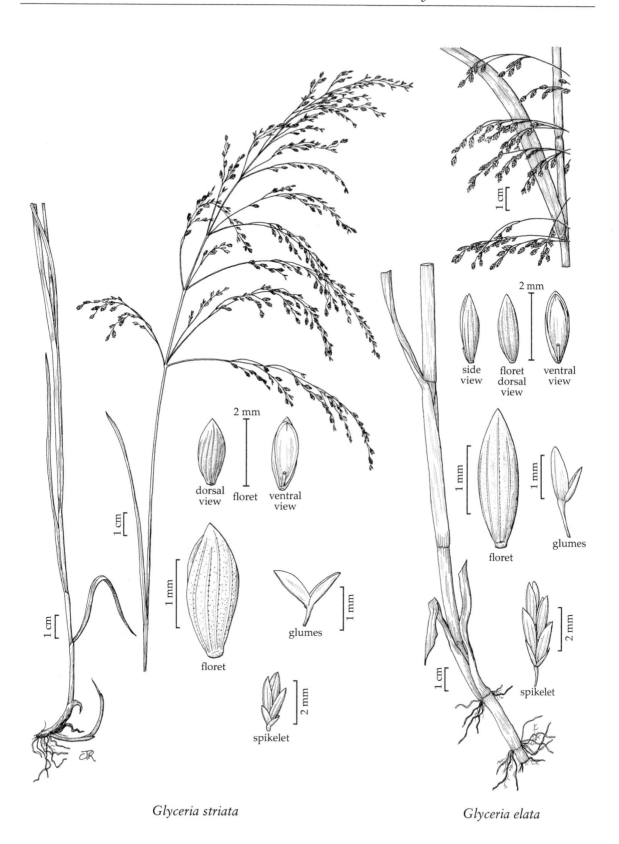

side
view

floret
dorsal
view

ventral
view

2 mm

dorsal
view

floret

ventral
view

2 mm

floret

glumes

1 mm

1 mm

floret

glumes

spikelet

2 mm

spikelet

2 mm

Glyceria striata

Glyceria elata

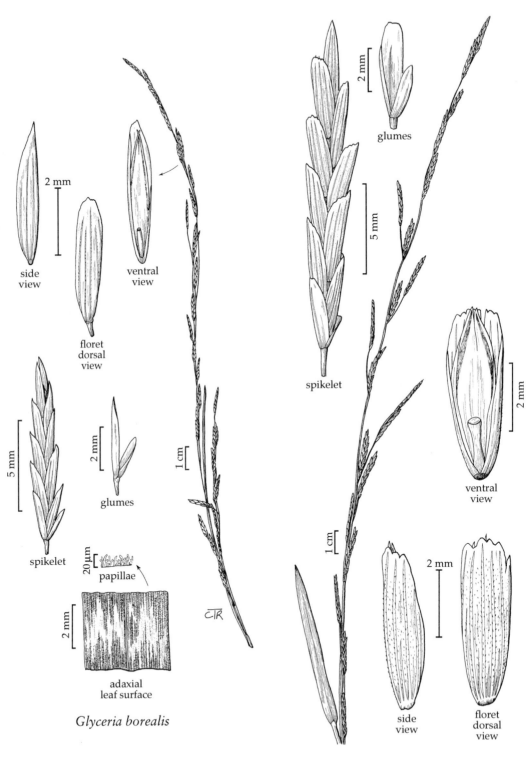

side
view

floret
dorsal
view

ventral
view

spikelet

glumes

papillae

adaxial
leaf surface

Glyceria borealis

glumes

spikelet

ventral
view

side
view

floret
dorsal
view

Glyceria declinata

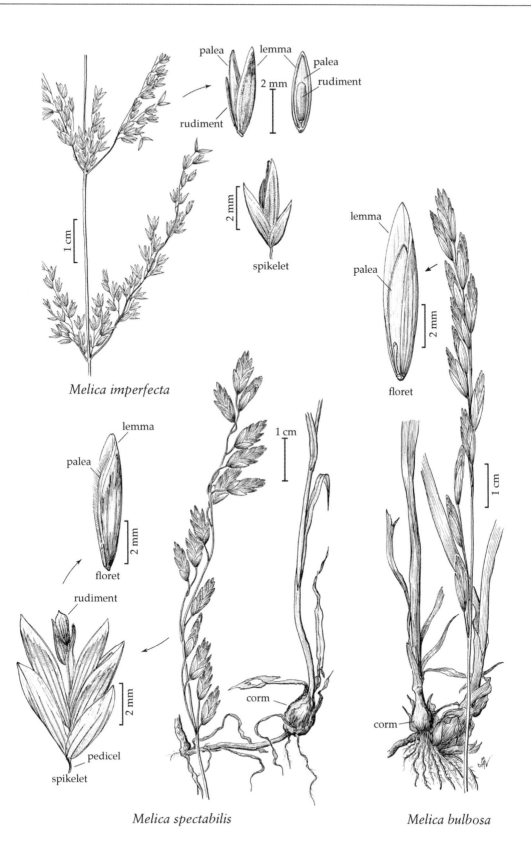

palea   lemma   palea   rudiment

2 mm

rudiment

1 cm

2 mm

spikelet

Melica imperfecta

lemma

palea

floret

2 mm

lemma

palea

floret

2 mm

rudiment

1 cm

2 mm

pedicel

spikelet

corm

Melica spectabilis

1 cm

corm

Melica bulbosa

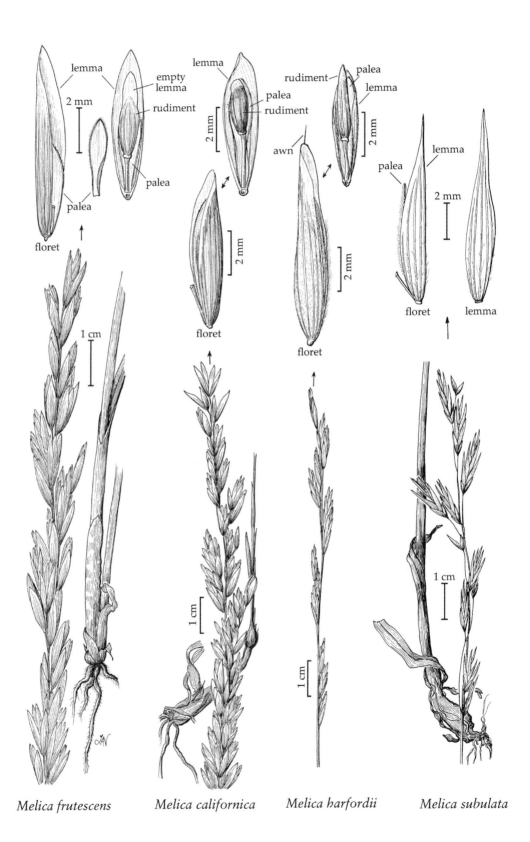

Melica frutescens   *Melica californica*   *Melica harfordii*   *Melica subulata*

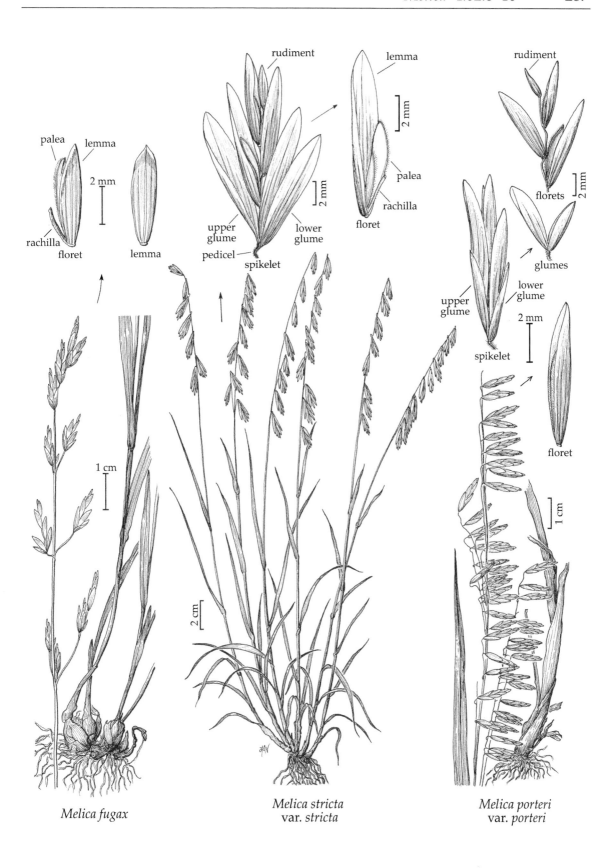

palea lemma

2 mm

rachilla
floret

lemma

rudiment

upper
glume
pedicel
lower
glume
spikelet

lemma

2 mm

palea

rachilla

floret

rudiment

florets

2 mm

glumes

upper
glume

lower
glume

2 mm

spikelet

floret

1 cm

2 cm

1 cm

Melica fugax

Melica stricta
var. *stricta*

Melica porteri
var. *porteri*

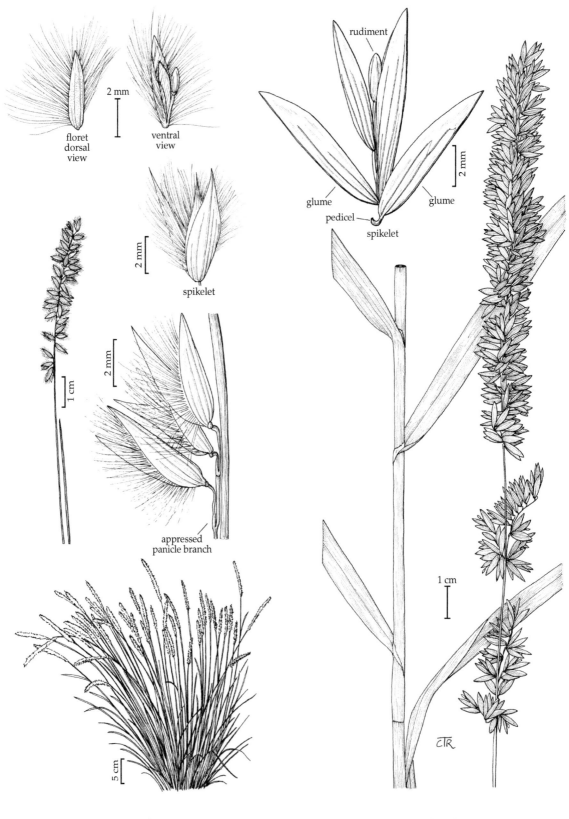

floret
dorsal
view

2 mm

ventral
view

rudiment

2 mm

glume   glume

pedicel

spikelet

spikelet

2 mm

2 mm

appressed
panicle branch

1 cm

1 cm

5 cm

CTR

Melica ciliata   *Melica altissima*

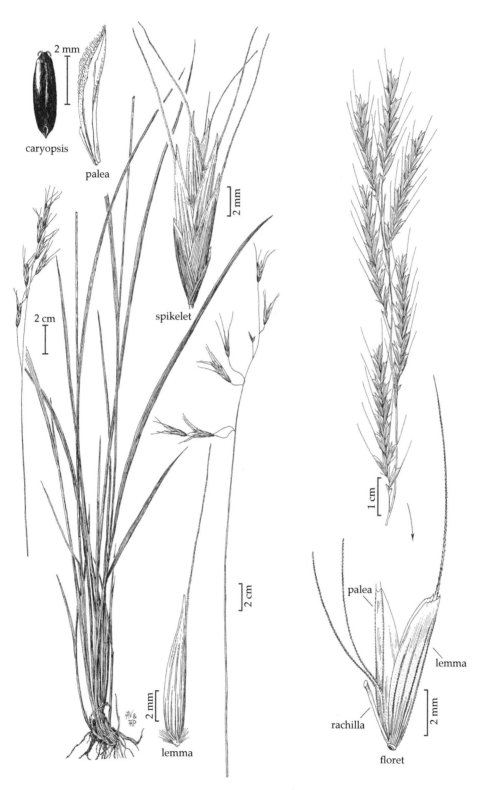

caryopsis

palea

spikelet

2 cm

2 cm

lemma

Schizachne purpurascens

palea

rachilla

lemma

floret

Pleuropogon oregonus

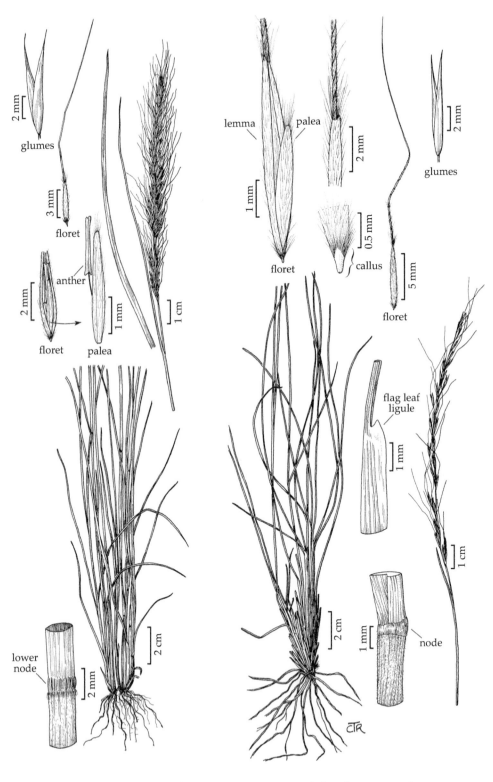

glumes

3 mm

floret

anther

floret   palea

lemma   palea

floret   callus

glumes

floret

flag leaf ligule

node

lower node

Achnatherum lettermanii

Achnatherum nevadense

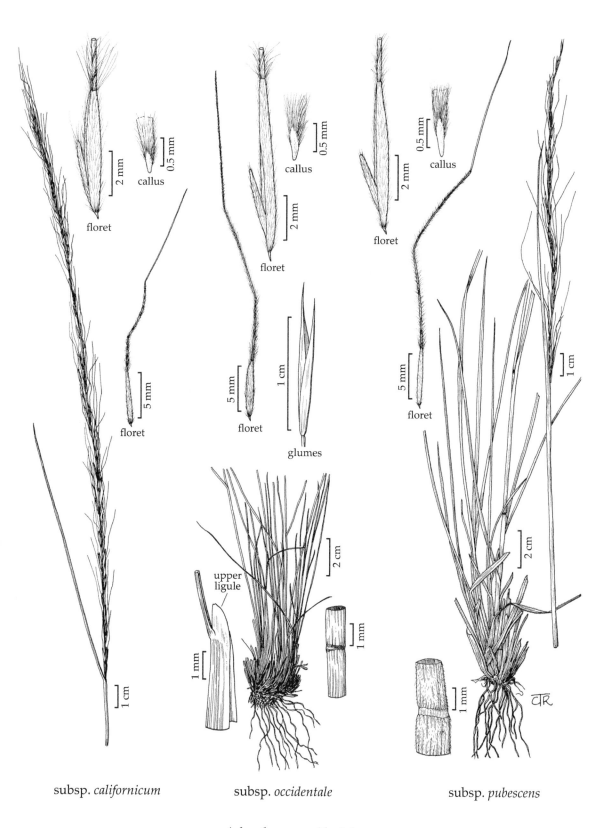

subsp. *californicum*   subsp. *occidentale*   subsp. *pubescens*

Achnatherum occidentale

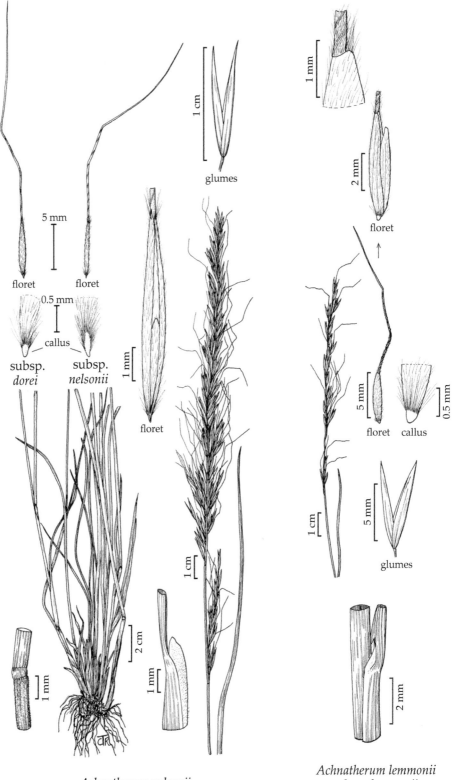

5 mm

floret   floret

0.5 mm

callus

subsp.
dorei

subsp.
nelsonii

glumes

1 cm

floret

1 mm

1 cm

1 mm

2 cm

1 mm

Achnatherum nelsonii

1 mm

2 mm

floret

5 mm

floret   callus

0.5 mm

glumes

5 mm

1 cm

2 mm

Achnatherum lemmonii
subsp. *lemmonii*

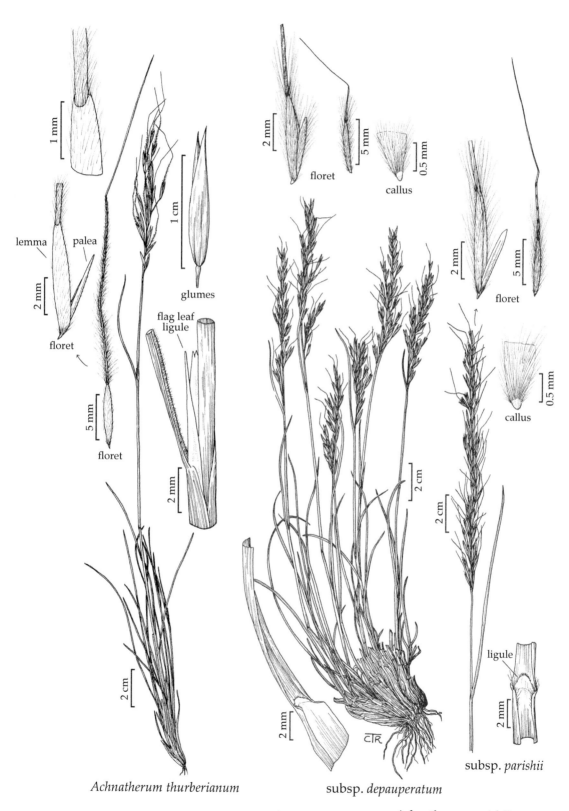

lemma
palea
floret
2 mm
5 mm
floret

1 mm

glumes
1 cm

flag leaf
ligule
2 mm

2 cm

Achnatherum thurberianum

floret
2 mm
5 mm
callus
0.5 mm

2 cm

2 mm
CTR

subsp. *depauperatum*

floret
2 mm
5 mm
callus
0.5 mm

2 cm

ligule
2 mm

subsp. *parishii*

Achnatherum parishii

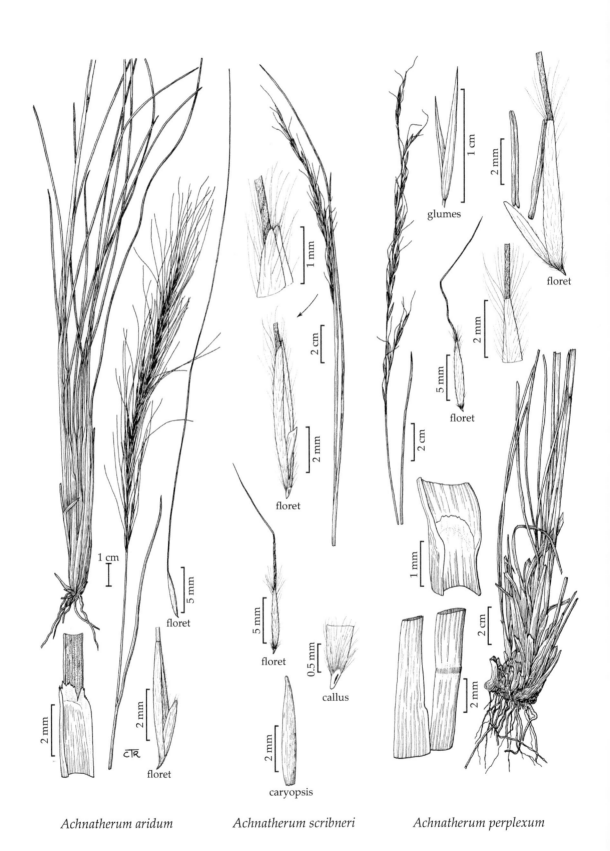

Achnatherum aridum   *Achnatherum scribneri*   *Achnatherum perplexum*

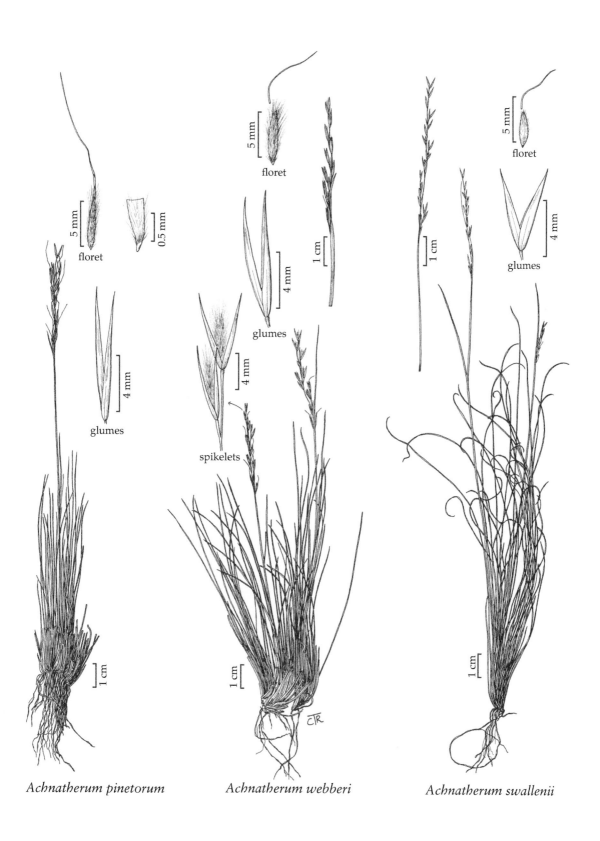

Achnatherum pinetorum   *Achnatherum webberi*   *Achnatherum swallenii*

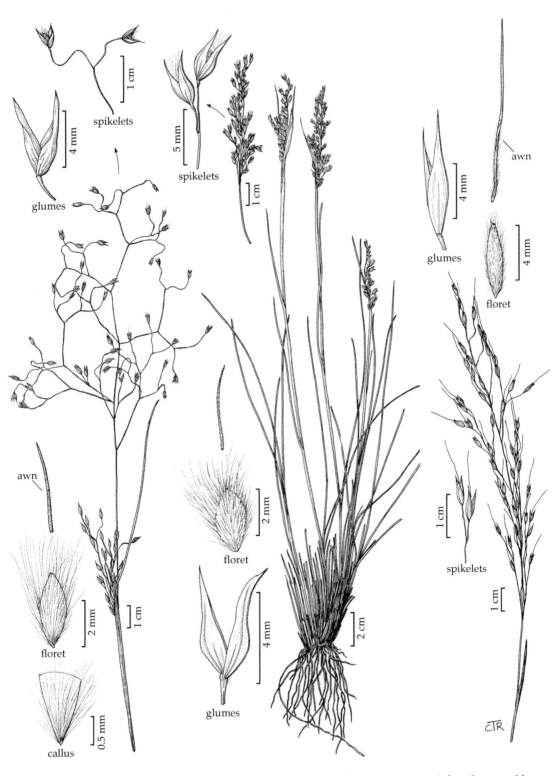

spikelets

glumes

spikelets

awn

glumes

floret

spikelets

awn

floret

floret

floret

callus

glumes

spikelets

CTR

Achnatherum hymenoides   *Achnatherum arnowiae*   *Achnatherum* ×*bloomeri*

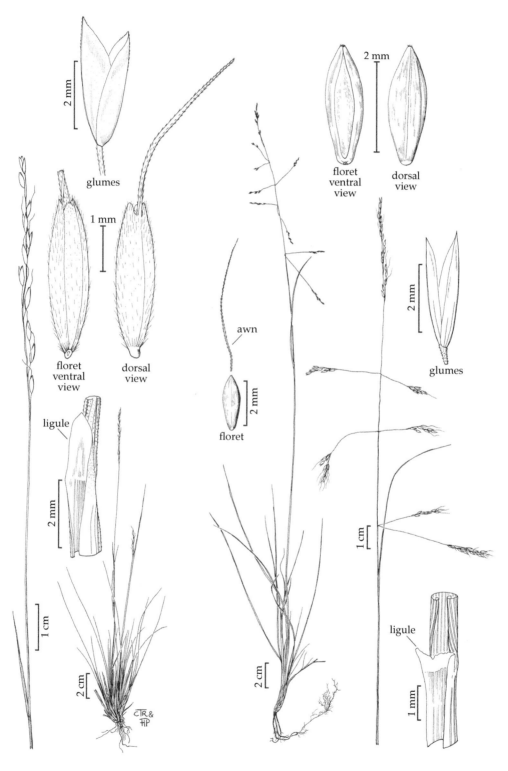

glumes

floret
ventral
view

dorsal
view

2 mm

1 mm

ligule

2 mm

1 cm

2 cm

awn

floret

2 mm

Piptatherum exiguum

floret
ventral
view

dorsal
view

2 mm

glumes

2 mm

1 cm

ligule

1 mm

2 cm

Piptatherum micranthum

CTR &
HP

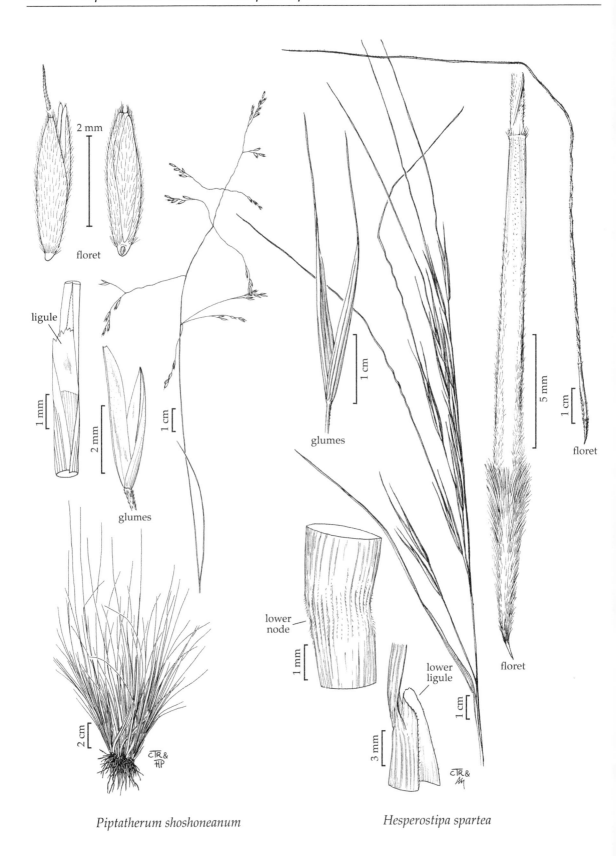

Piptatherum shoshoneanum   *Hesperostipa spartea*

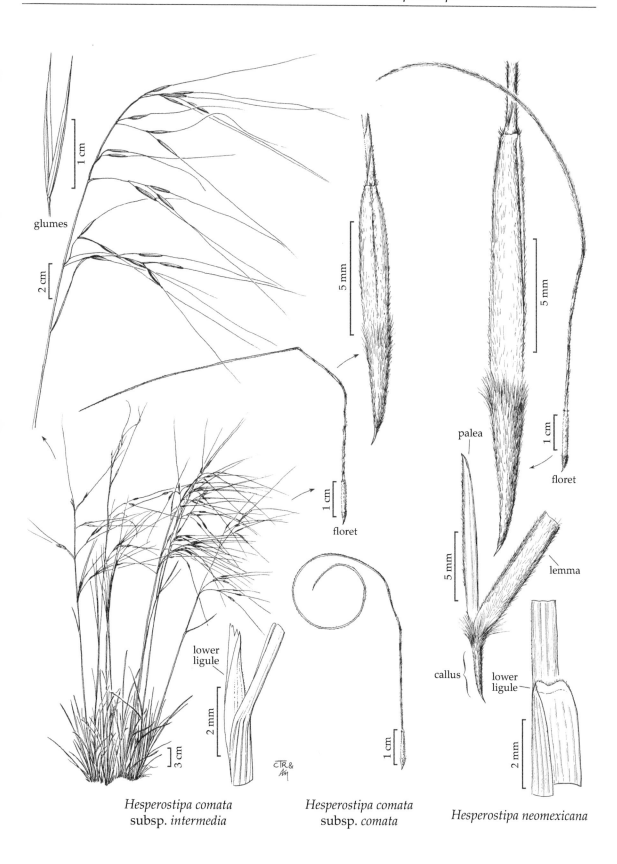

glumes

floret

palea

lemma

callus

lower ligule

lower ligule

Hesperostipa comata
subsp. *intermedia*

Hesperostipa comata
subsp. *comata*

Hesperostipa neomexicana

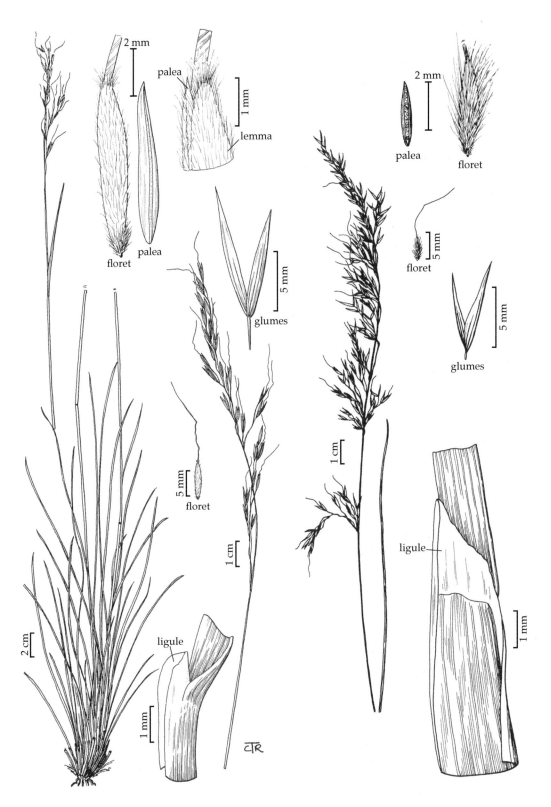

Piptochaetium pringlei

Piptochaetium lasianthum

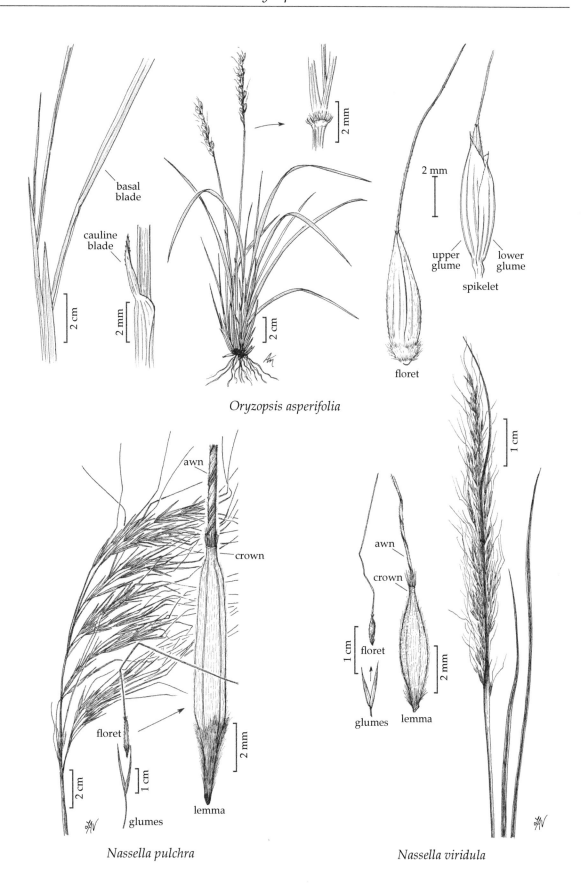

basal blade

cauline blade

2 cm

2 mm

2 mm

Oryzopsis asperifolia

2 mm

upper glume

lower glume

spikelet

floret

2 cm

awn

crown

floret

glumes

lemma

2 cm

1 cm

2 mm

Nassella pulchra

awn

crown

floret

glumes

lemma

1 cm

2 mm

1 cm

Nassella viridula

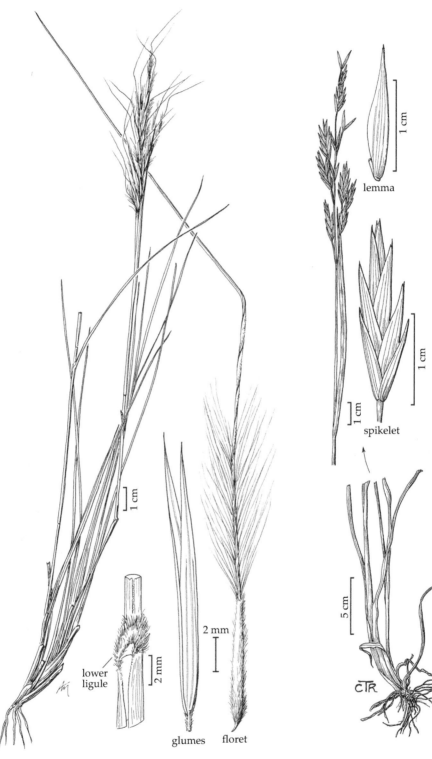

lemma

spikelet

lower
ligule

glumes   floret

Pappostipa speciosa

Bromus catharticus
var. *catharticus*

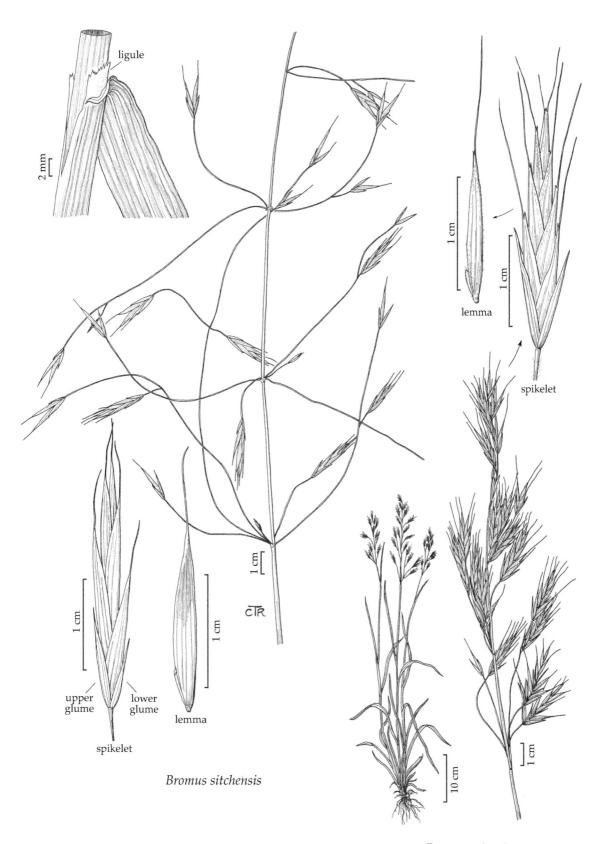

ligule

2 mm

1 cm

lemma

1 cm

spikelet

1 cm

upper
glume

lower
glume

spikelet

1 cm

lemma

1 cm

CTR

Bromus sitchensis

10 cm

1 cm

Bromus arizonicus

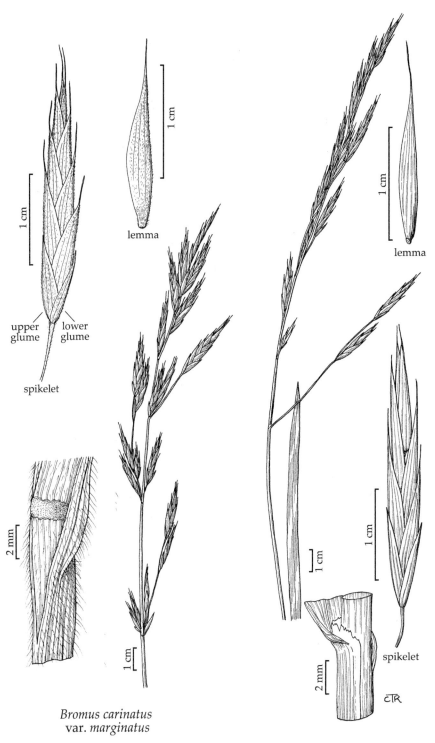

upper   lower
glume   glume

spikelet

lemma

2 mm

1 cm

Bromus carinatus
var. *marginatus*

lemma

1 cm

spikelet

2 mm

CTR

Bromus polyanthus

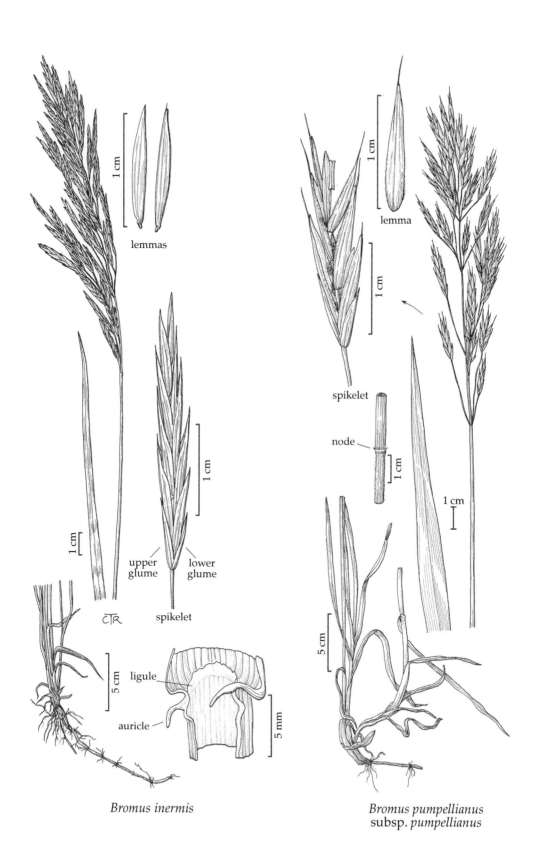

Bromus inermis

Bromus pumpellianus
subsp. *pumpellianus*

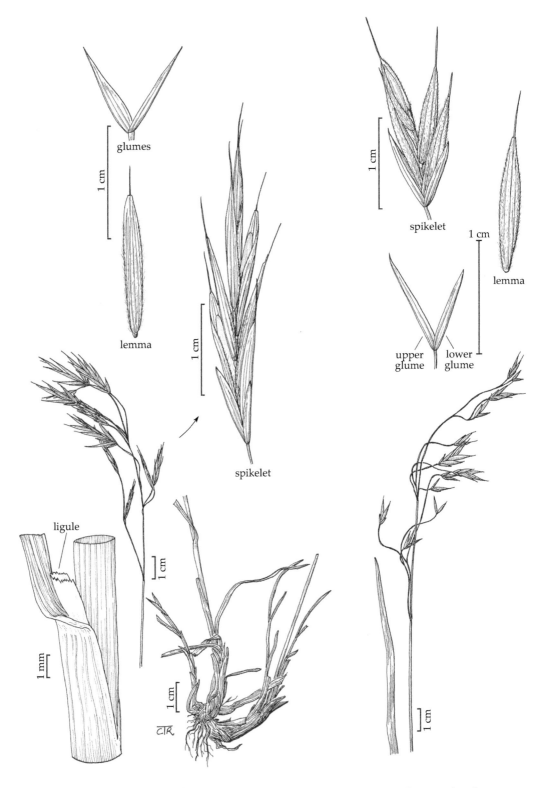

glumes

1 cm

lemma

spikelet

1 cm

ligule

1 mm

1 cm

1 cm

CTR

spikelet

1 cm

lemma

upper
glume

lower
glume

1 cm

1 cm

Bromus laevipes

Bromus frondosus

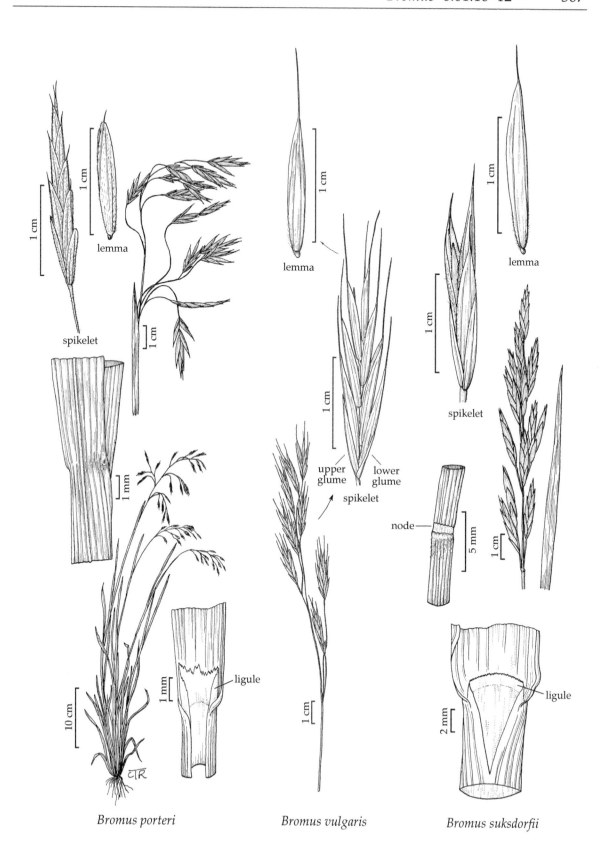

spikelet

lemma

1 cm

1 cm

1 cm

1 mm

ligule

1 mm

10 cm

Bromus porteri

lemma

1 cm

1 cm

upper
glume

lower
glume

spikelet

1 cm

node

5 mm

Bromus vulgaris

lemma

1 cm

1 cm

spikelet

1 cm

ligule

2 mm

Bromus suksdorfii

lemma

glumes

upper
glume

spikelet

Bromus mucroglumis

glumes

lemma

spikelet

lemma

upper
glume

lower
glume

spikelet

CTR

Bromus lanatipes

Bromus ciliatus

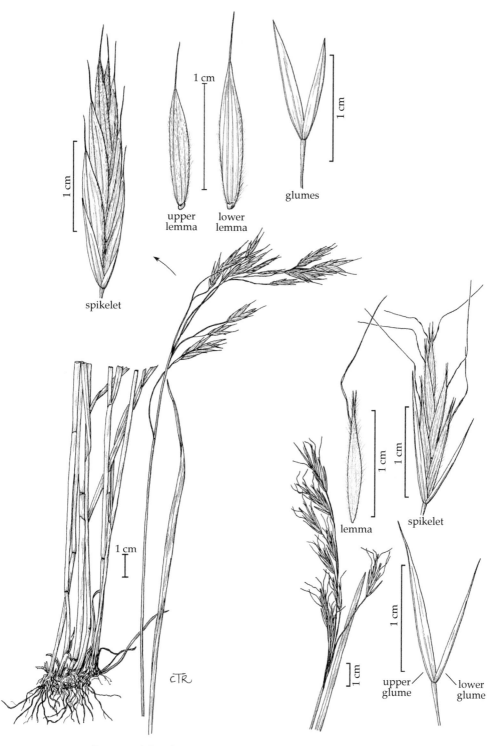

upper
lemma

lower
lemma

glumes

spikelet

1 cm

1 cm

1 cm

1 cm

lemma

spikelet

1 cm

1 cm

upper
glume

lower
glume

1 cm

1 cm

1 cm

CTR

Bromus richardsonii

Bromus berteroanus

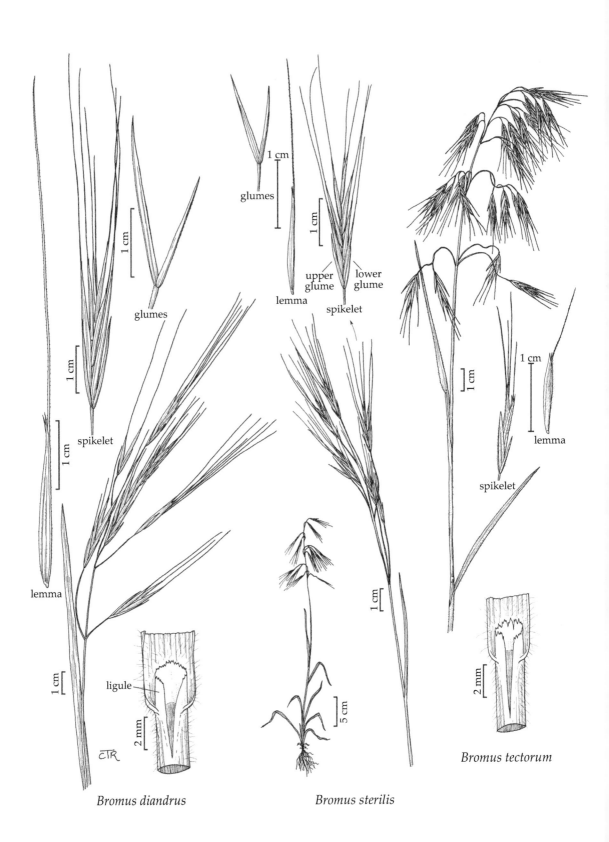

glumes

1 cm

glumes

1 cm

spikelet

1 cm

lemma

1 cm

ligule

2 mm

CTR

Bromus diandrus

glumes

1 cm

1 cm

upper
glume

lower
glume

lemma   spikelet

1 cm

5 cm

1 cm

Bromus sterilis

1 cm

1 cm

lemma

spikelet

2 mm

Bromus tectorum

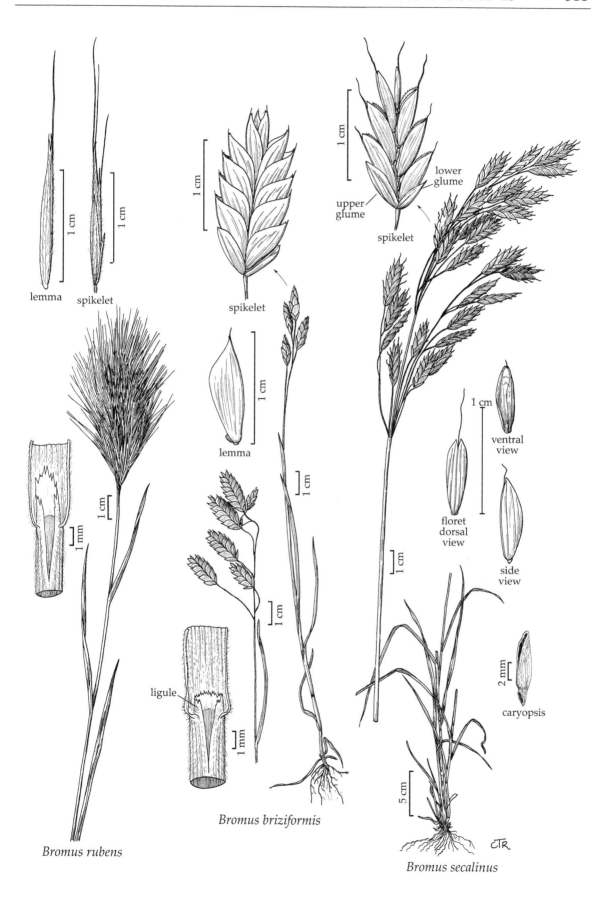

lemma

spikelet

spikelet

lemma

upper
glume

lower
glume

spikelet

ligule

ventral
view

floret
dorsal
view

side
view

caryopsis

Bromus rubens

Bromus briziformis

Bromus secalinus

CTR

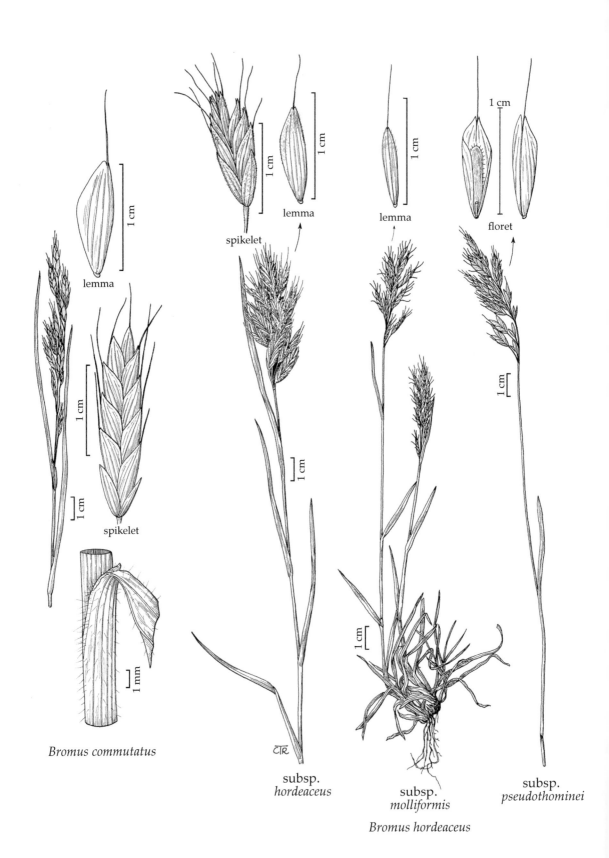

lemma

spikelet

lemma

1 cm

lemma

floret

spikelet

1 cm

Bromus commutatus

subsp.
hordeaceus

subsp.
molliformis

subsp.
pseudothominei

Bromus hordeaceus

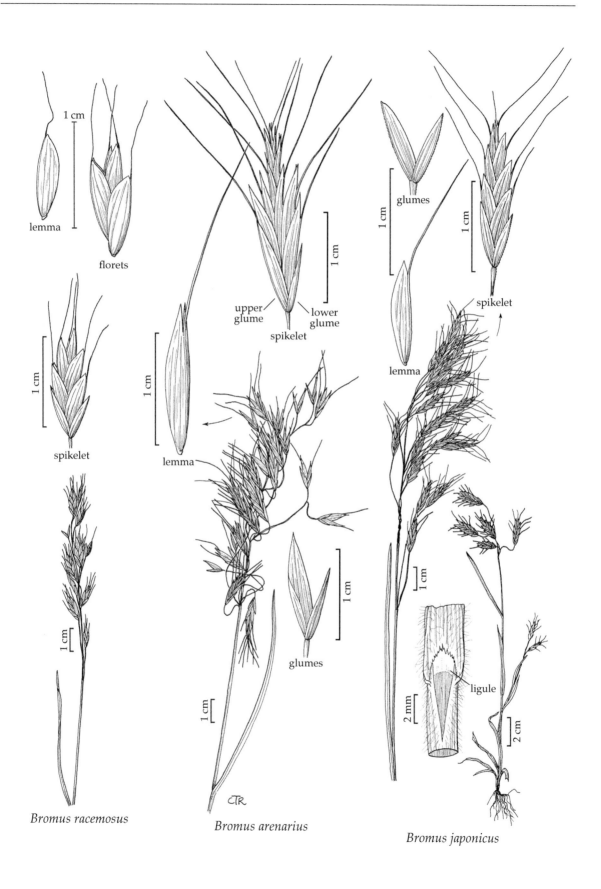

lemma

florets

spikelet

upper glume

lower glume

spikelet

spikelet

lemma

glumes

lemma

glumes

spikelet

ligule

Bromus racemosus

Bromus arenarius

Bromus japonicus

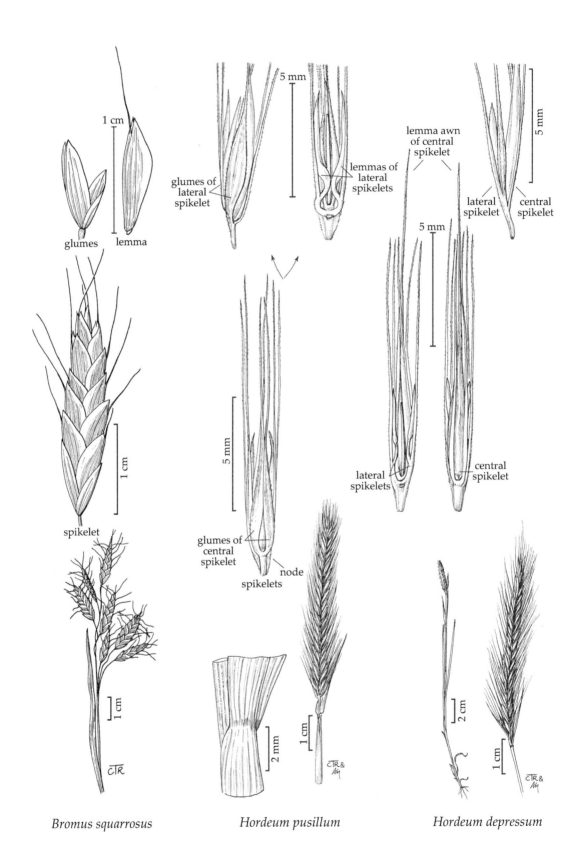

1 cm

glumes   lemma

spikelet

1 cm

Bromus squarrosus

5 mm

glumes of
lateral
spikelet

lemmas of
lateral
spikelets

5 mm

glumes of
central
spikelet

node
spikelets

Hordeum pusillum

lemma awn
of central
spikelet

5 mm

lateral
spikelet

central
spikelet

5 mm

lateral
spikelets

central
spikelet

Hordeum depressum

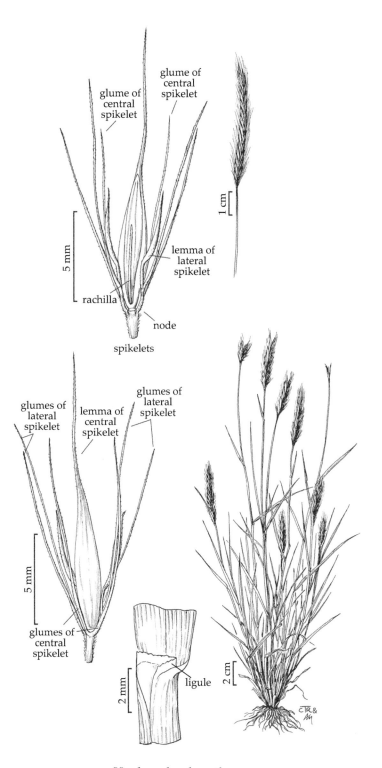

glume of
central
spikelet

glume of
central
spikelet

lemma of
lateral
spikelet

5 mm

rachilla

node

spikelets

1 cm

glumes of
lateral
spikelet

lemma of
central
spikelet

glumes of
lateral
spikelet

5 mm

glumes of
central
spikelet

ligule

2 mm

2 cm

CTR &
MM

Hordeum brachyantherum
subsp. *brachyantherum*

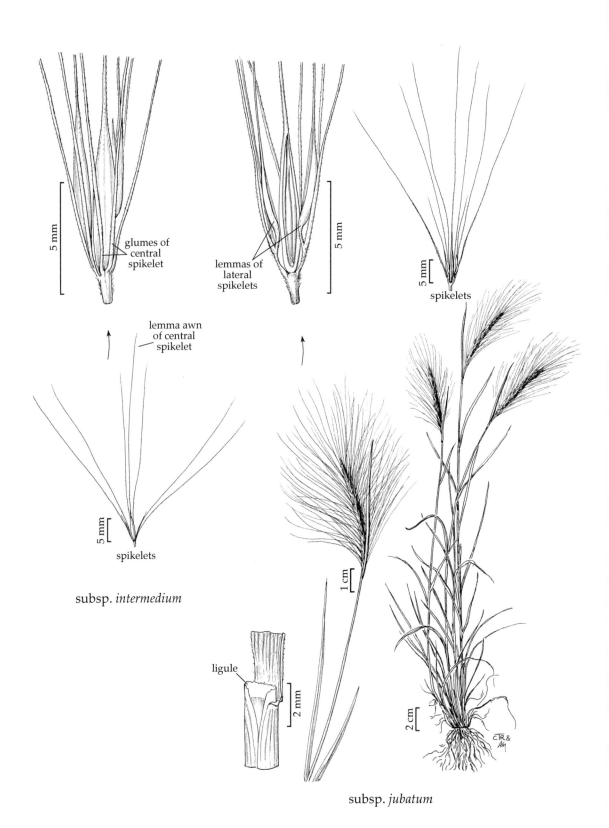

glumes of central spikelet

lemmas of lateral spikelets

5 mm

5 mm

spikelets

lemma awn of central spikelet

5 mm

spikelets

subsp. *intermedium*

ligule

2 mm

1 cm

2 cm

subsp. *jubatum*

Hordeum jubatum

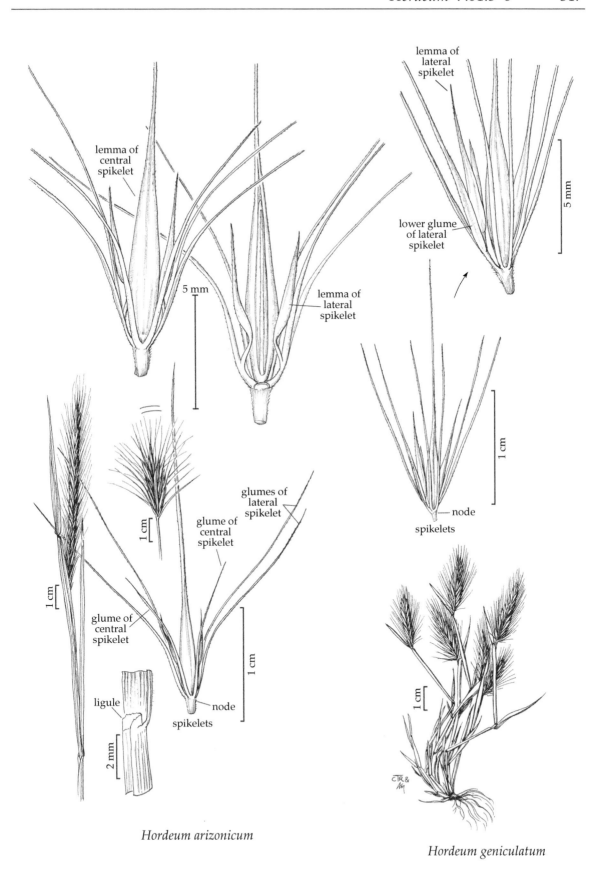

lemma of
central
spikelet

5 mm

lemma of
lateral
spikelet

lemma of
lateral
spikelet

5 mm

lower glume
of lateral
spikelet

glumes of
lateral
spikelet

glume of
central
spikelet

1 cm

1 cm

1 cm

glume of
central
spikelet

node

spikelets

node

spikelets

1 cm

ligule

2 mm

1 cm

Hordeum arizonicum

Hordeum geniculatum

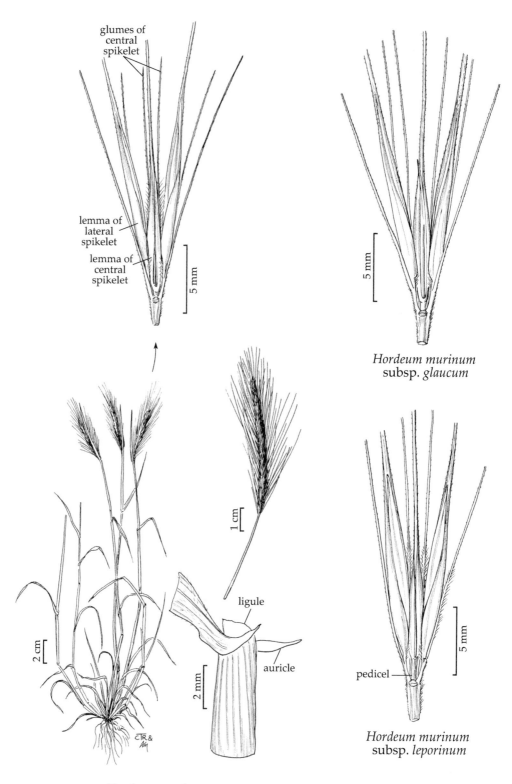

glumes of
central
spikelet

lemma of
lateral
spikelet

lemma of
central
spikelet

5 mm

Hordeum murinum
subsp. *glaucum*

5 mm

1 cm

ligule

auricle

2 mm

pedicel

5 mm

Hordeum murinum
subsp. *leporinum*

2 cm

CTR &
M

Hordeum murinum
subsp. *murinum*

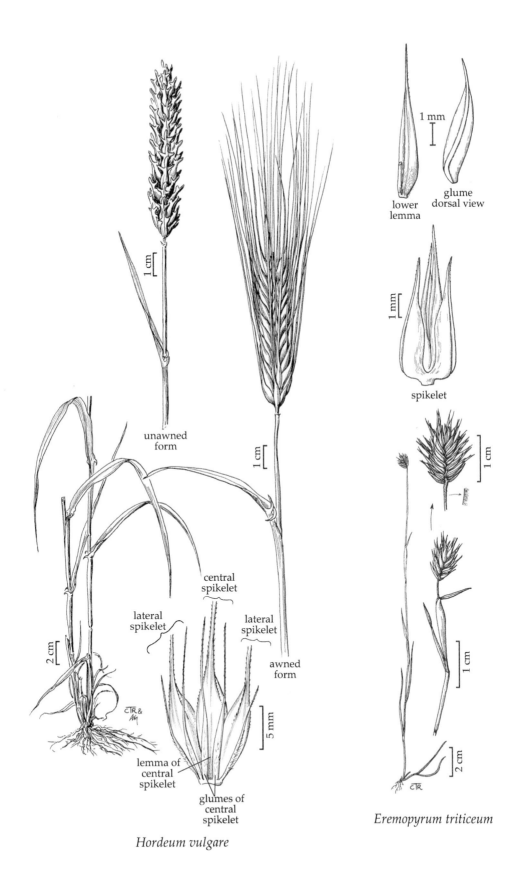

1 mm

lower
lemma

glume
dorsal view

1 mm

spikelet

1 cm

1 cm

1 cm

2 cm

1 cm

unawned
form

central
spikelet

lateral
spikelet

lateral
spikelet

awned
form

5 mm

2 cm

lemma of
central
spikelet

glumes of
central
spikelet

Eremopyrum triticeum

Hordeum vulgare

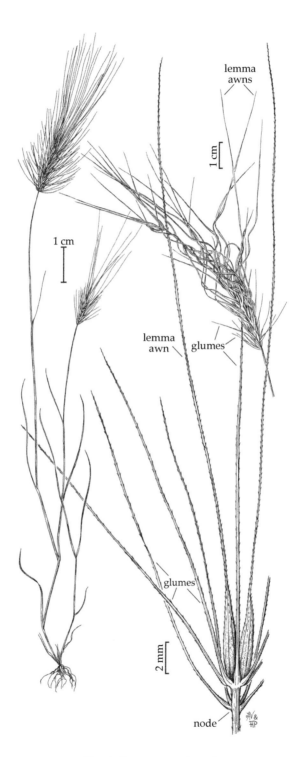

Taeniatherum caput-medusae

lemma

spikelet

glume   glume

spikelet

Secale cereale   ×*Triticosecale*

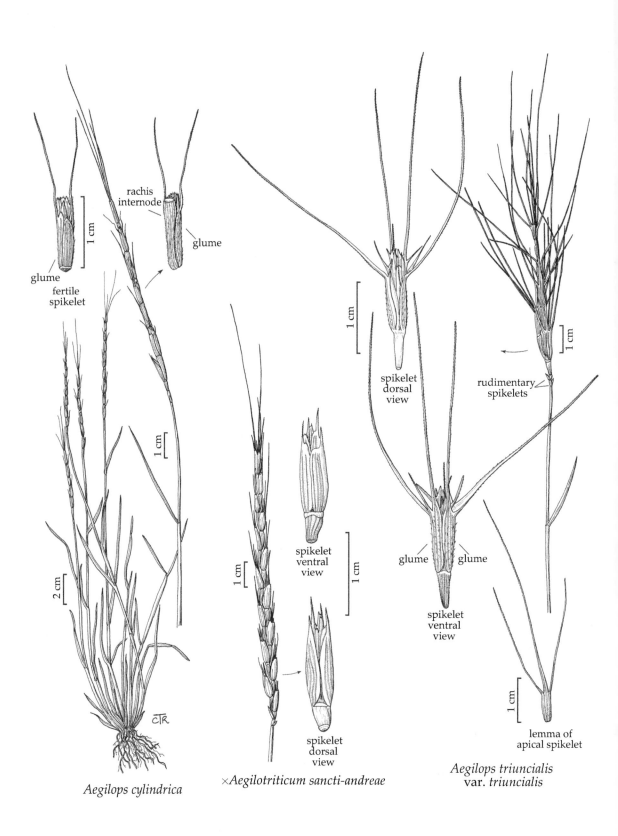

glume

fertile
spikelet

rachis
internode

glume

1 cm

Aegilops cylindrica

2 cm

spikelet
ventral
view

spikelet
dorsal
view

×*Aegilotriticum sancti-andreae*

spikelet
dorsal
view

rudimentary
spikelets

glume   glume

spikelet
ventral
view

lemma of
apical spikelet

Aegilops triuncialis
var. *triuncialis*

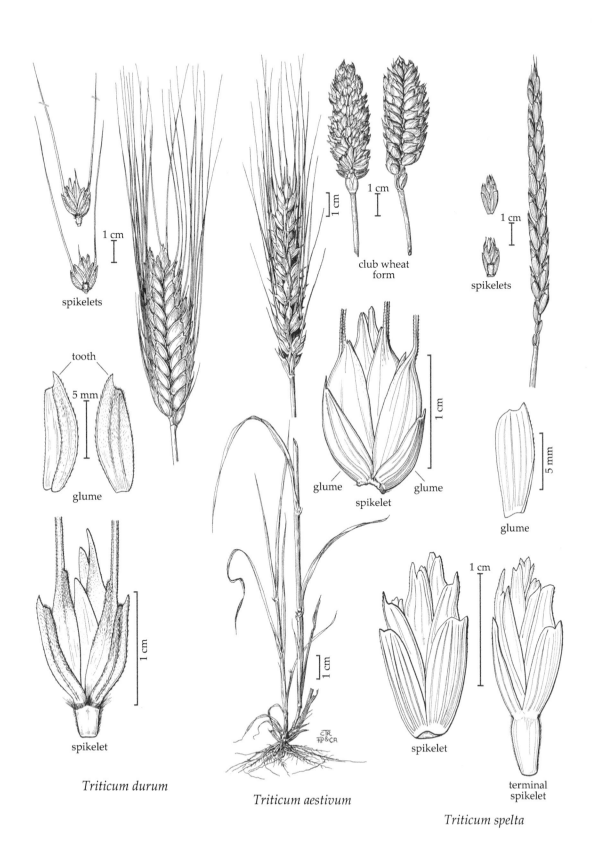

spikelets

tooth

5 mm

glume

1 cm

spikelet

Triticum durum

club wheat
form

1 cm

1 cm

glume   glume

spikelet

1 cm

Triticum aestivum

spikelets

1 cm

glume

5 mm

spikelet

1 cm

terminal
spikelet

Triticum spelta

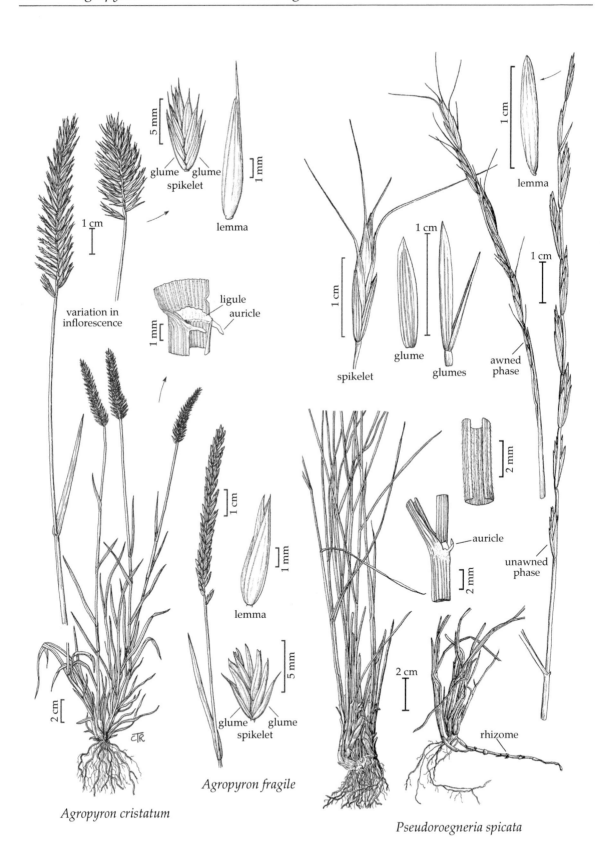

variation in inflorescence

glume   glume
spikelet

lemma

ligule
auricle

spikelet

glume

glumes

awned phase

lemma

unawned phase

auricle

rhizome

lemma

glume   glume
spikelet

Agropyron fragile

Agropyron cristatum

Pseudoroegneria spicata

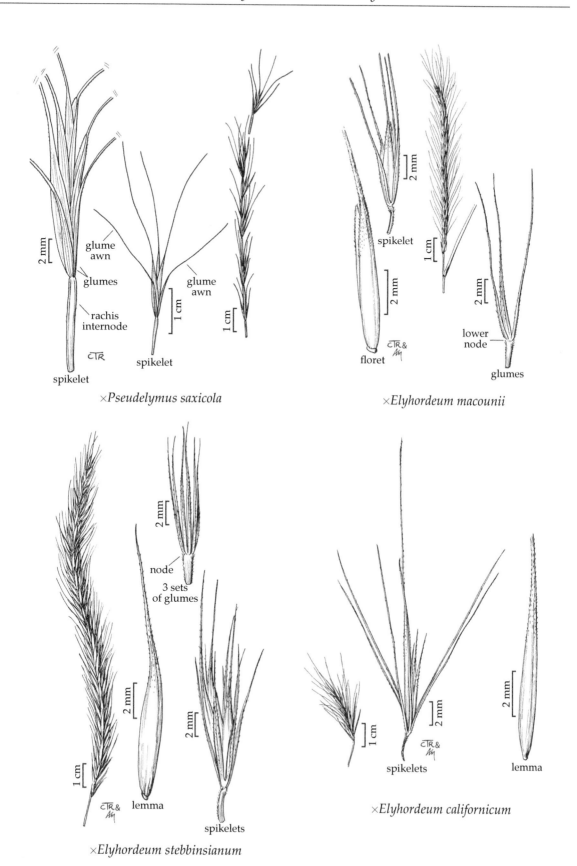

×*Pseudelymus saxicola*

×*Elyhordeum macounii*

×*Elyhordeum stebbinsianum*

×*Elyhordeum californicum*

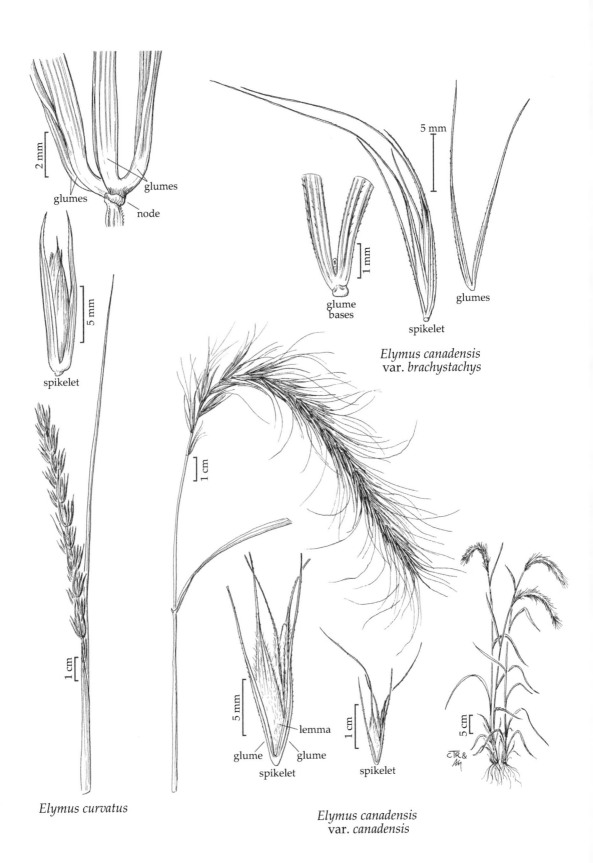

Elymus canadensis
var. *brachystachys*

Elymus curvatus

Elymus canadensis
var. *canadensis*

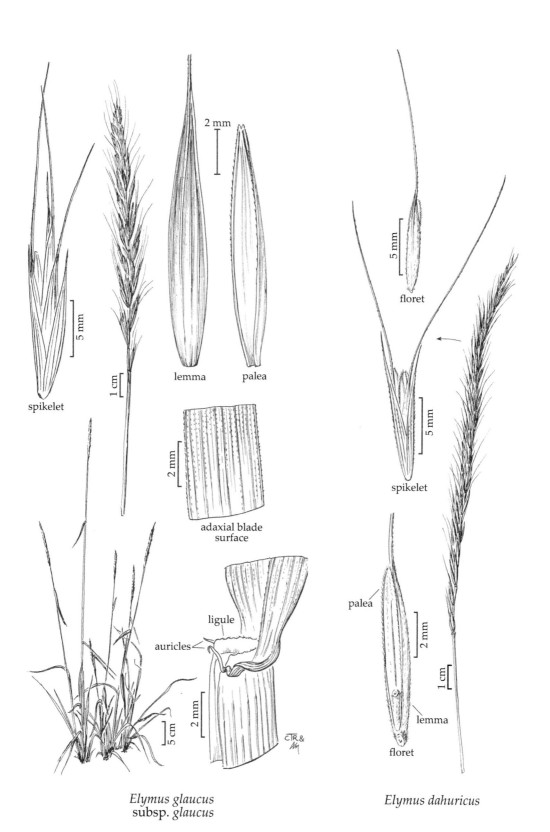

spikelet

lemma   palea

2 mm

1 cm

adaxial blade
surface

2 mm

ligule

auricles

2 mm

5 cm

Elymus glaucus
subsp. *glaucus*

floret

5 mm

spikelet

5 mm

palea

2 mm

lemma

floret

1 cm

Elymus dahuricus

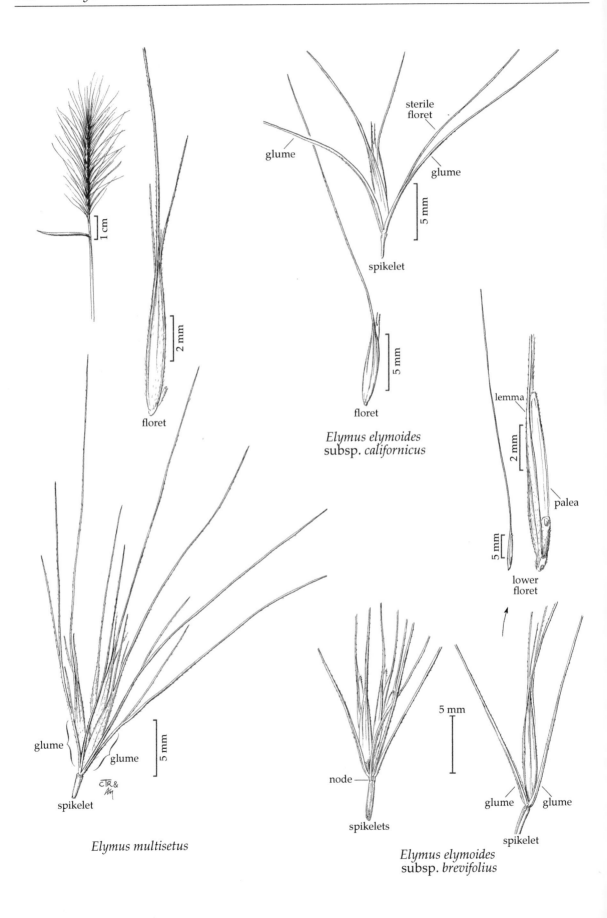

sterile floret

glume

glume

5 mm

spikelet

5 mm

floret

Elymus elymoides
subsp. *californicus*

lemma

2 mm

palea

5 mm

lower floret

floret

2 mm

glume

glume

spikelet

CTR &
AM

5 mm

Elymus multisetus

node

spikelets

5 mm

glume   glume

spikelet

Elymus elymoides
subsp. *brevifolius*

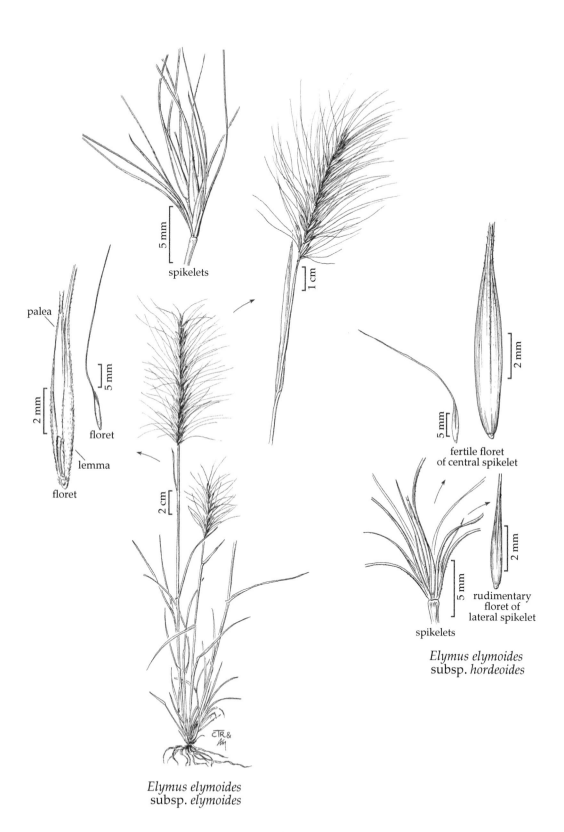

spikelets

5 mm

1 cm

palea

2 mm

5 mm

floret

lemma

floret

floret

2 cm

fertile floret
of central spikelet

5 mm

2 mm

2 mm

5 mm

rudimentary
floret of
lateral spikelet

spikelets

Elymus elymoides
subsp. *hordeoides*

CTR &
AM

Elymus elymoides
subsp. *elymoides*

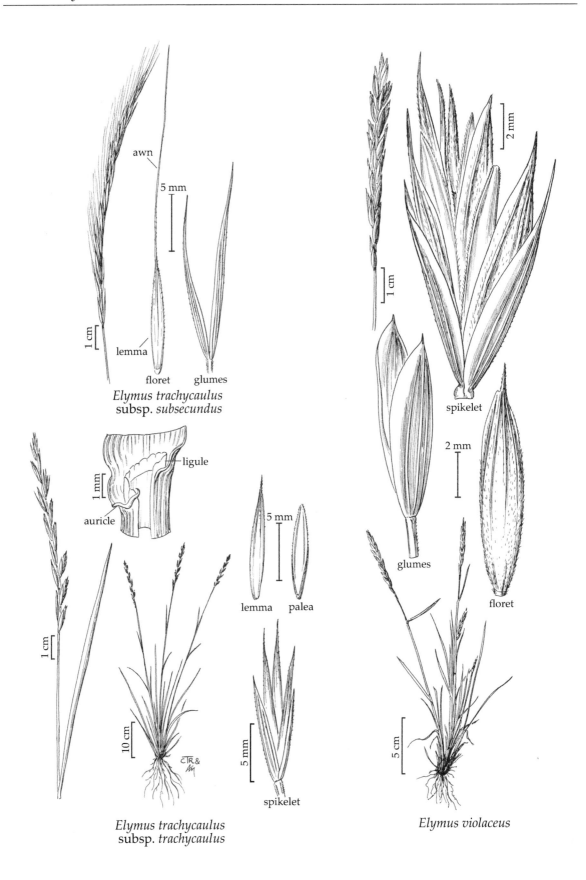

Elymus trachycaulus
subsp. *subsecundus*

Elymus trachycaulus
subsp. *trachycaulus*

Elymus violaceus

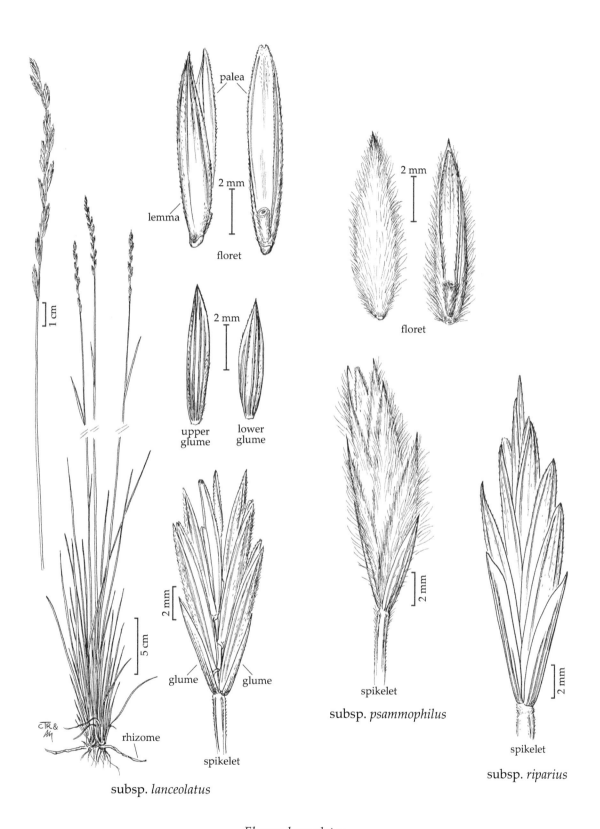

palea

lemma

floret

2 mm

upper glume

lower glume

glume   glume

spikelet

2 mm

subsp. *lanceolatus*

rhizome

1 cm

5 cm

floret

2 mm

subsp. *psammophilus*

spikelet

2 mm

subsp. *riparius*

spikelet

2 mm

Elymus lanceolatus

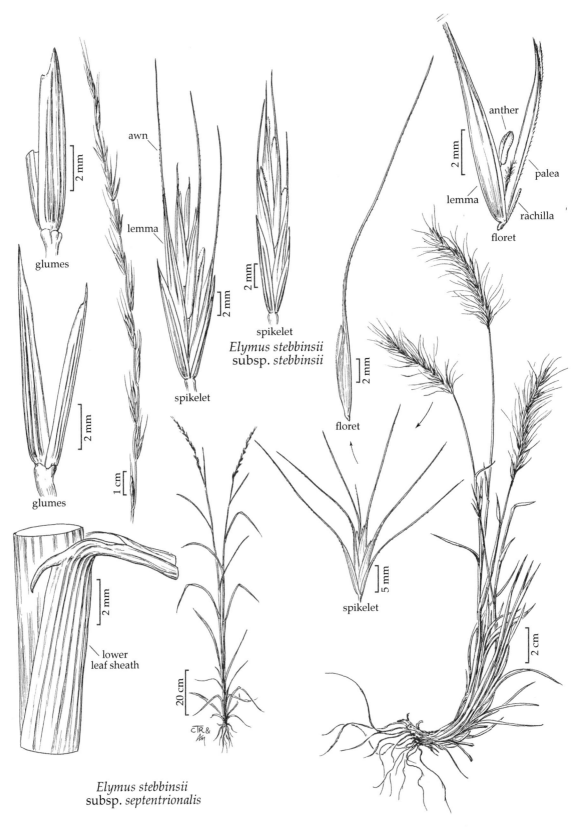

glumes

awn

lemma

glumes

spikelet

spikelet

Elymus stebbinsii
subsp. *stebbinsii*

anther

lemma

palea

rachilla

floret

floret

spikelet

lower
leaf sheath

Elymus stebbinsii
subsp. *septentrionalis*

Elymus scribneri

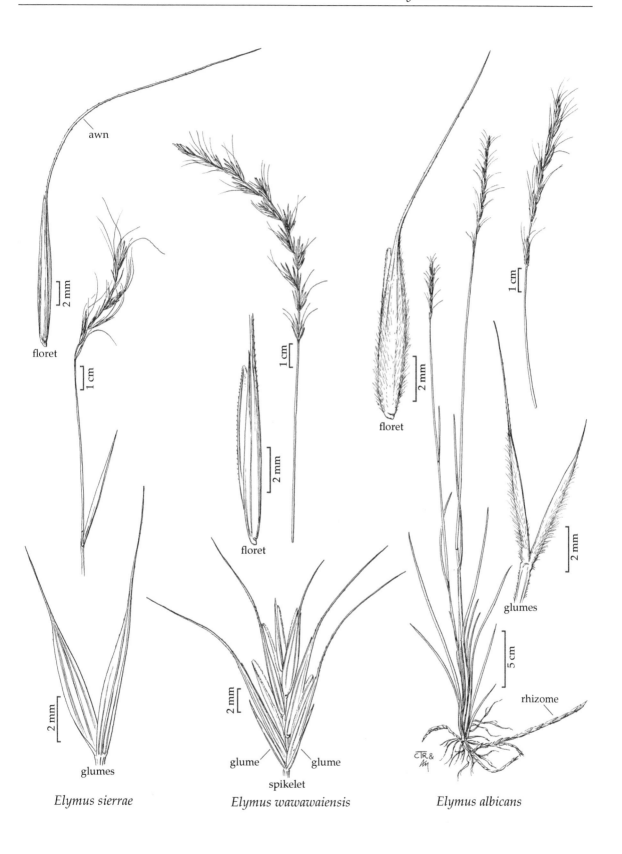

Elymus sierrae   *Elymus wawawaiensis*   *Elymus albicans*

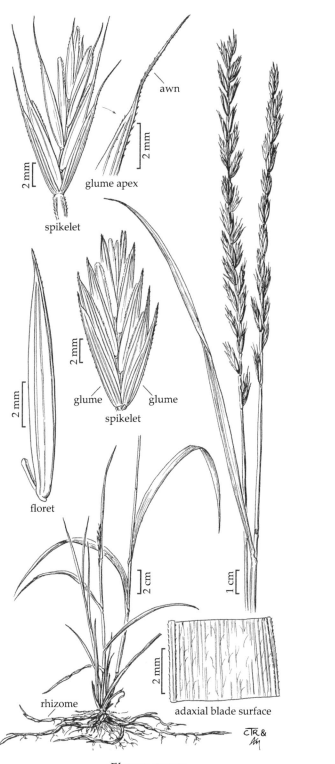

awn

2 mm

glume apex

2 mm

spikelet

2 mm

floret

2 mm

glume   glume

spikelet

2 cm

1 cm

2 mm

adaxial blade surface

rhizome

Elymus repens

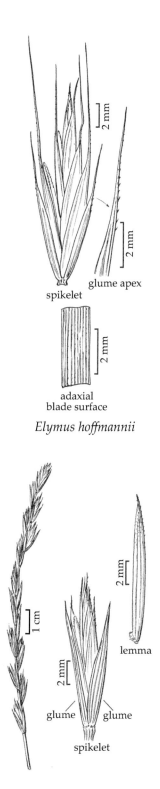

2 mm

glume apex

2 mm

spikelet

2 mm

adaxial
blade surface

Elymus hoffmannii

1 cm

2 mm

2 mm

lemma

glume   glume

spikelet

Elymus ×pseudorepens

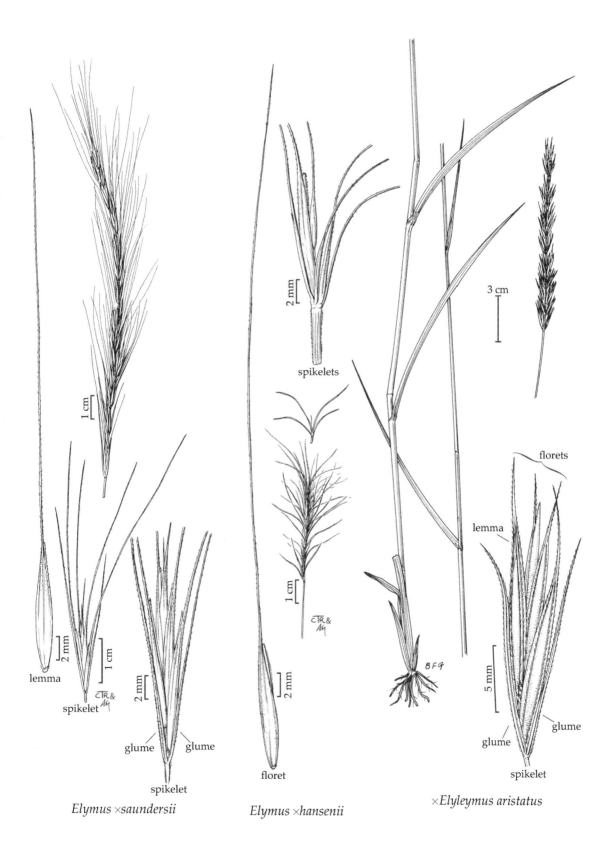

lemma

spikelet

glume   glume

spikelet

Elymus ×*saundersii*

spikelets

floret

Elymus ×*hansenii*

3 cm

florets

lemma

glume   glume

spikelet

×*Elyleymus aristatus*

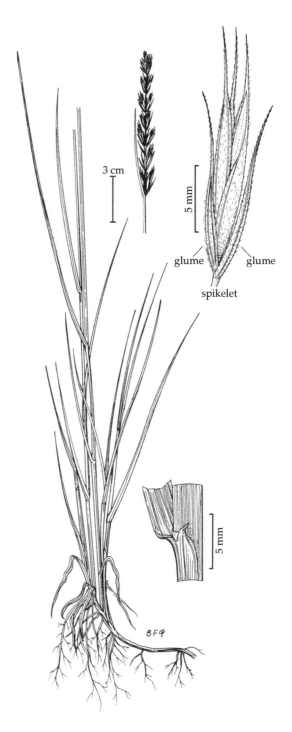

3 cm

5 mm

glume   glume

spikelet

5 mm

BFG

×*Elyleymus hirtiflorus*

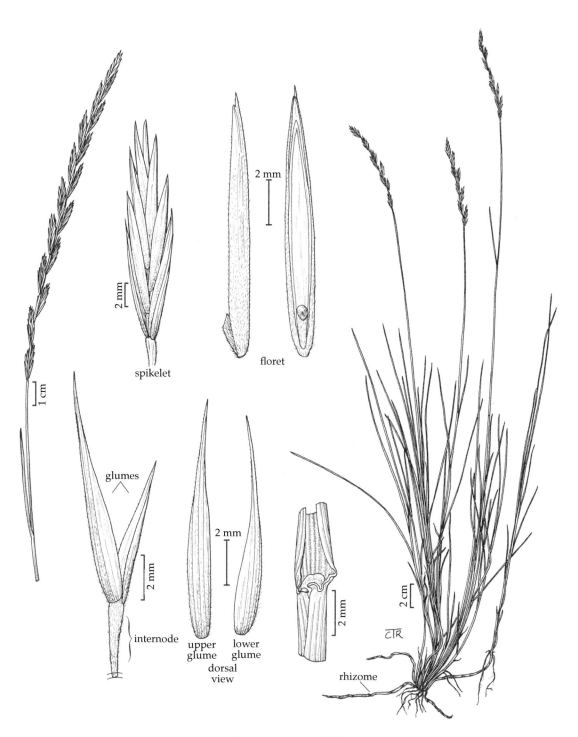

1 cm

2 mm

spikelet

2 mm

floret

glumes

2 mm

internode

2 mm

upper
glume

lower
glume

dorsal
view

2 mm

2 cm

CTR

rhizome

Pascopyrum smithii

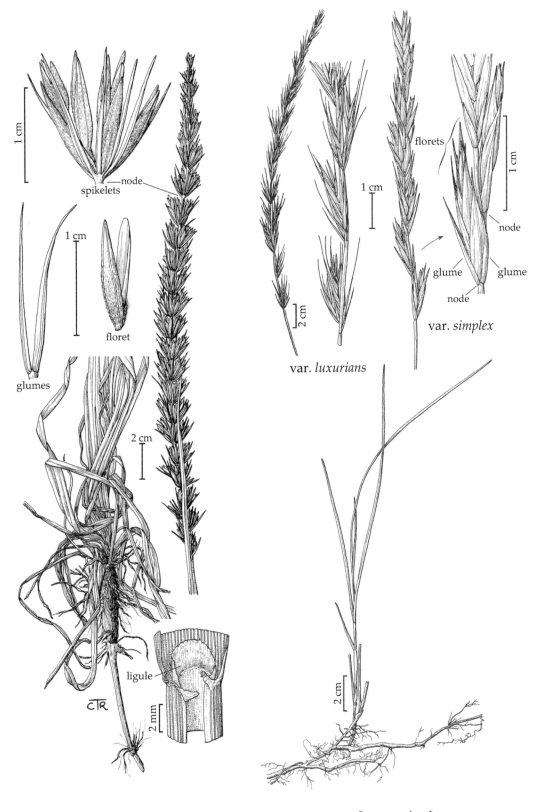

spikelets

node

1 cm

1 cm

floret

glumes

2 cm

ligule

2 mm

CTR

florets

1 cm

node

glume

glume

node

var. *simplex*

var. *luxurians*

2 cm

2 cm

Leymus racemosus

Leymus simplex

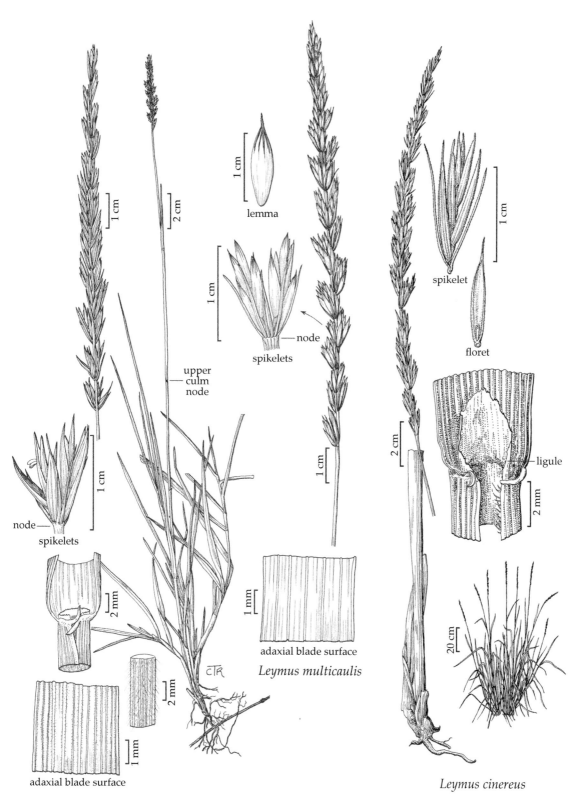

lemma

node
spikelets

node
spikelets

upper
culm
node

adaxial blade surface

Leymus triticoides

adaxial blade surface

Leymus multicaulis

spikelet

floret

ligule

Leymus cinereus

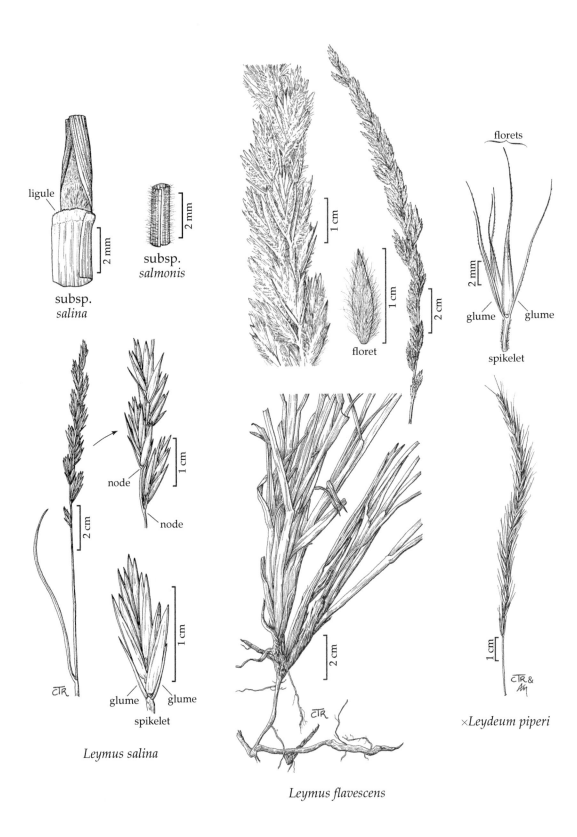

ligule

subsp.
salina

subsp.
salmonis

florets

glume   glume

spikelet

floret

node

node

glume   glume

spikelet

Leymus salina

Leymus flavescens

×*Leydeum piperi*

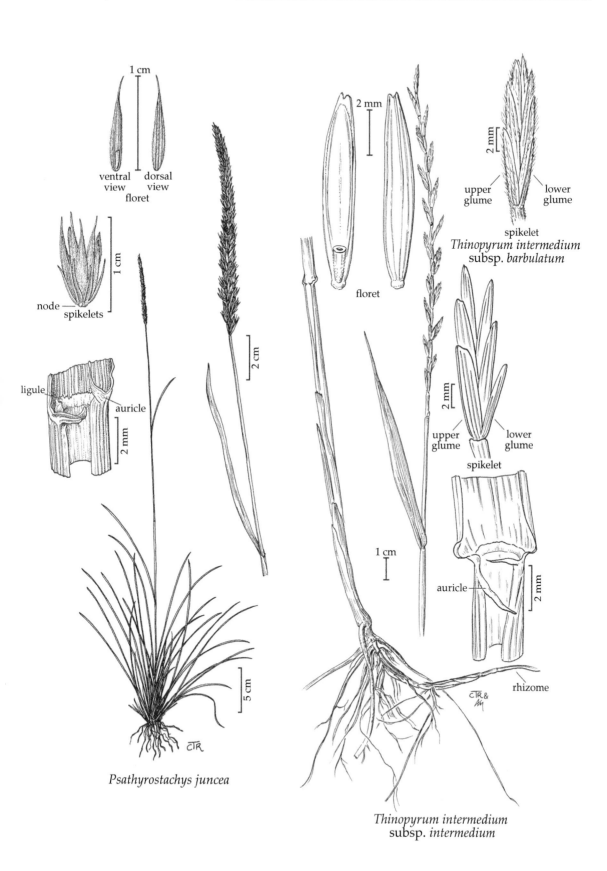

1 cm

ventral
view
dorsal
view
floret

node
spikelets

ligule
auricle
2 mm

1 cm

2 cm

floret

2 mm

Thinopyrum intermedium
subsp. *barbulatum*

upper
glume
lower
glume
spikelet

2 mm

upper
glume
lower
glume
spikelet

auricle
2 mm

rhizome

5 cm

Psathyrostachys juncea

Thinopyrum intermedium
subsp. *intermedium*

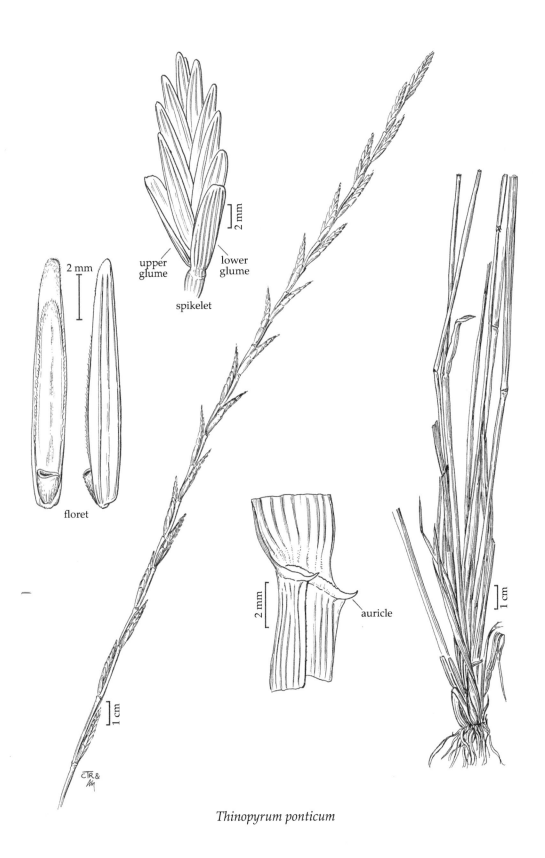

upper
glume

lower
glume

spikelet

floret

auricle

Thinopyrum ponticum

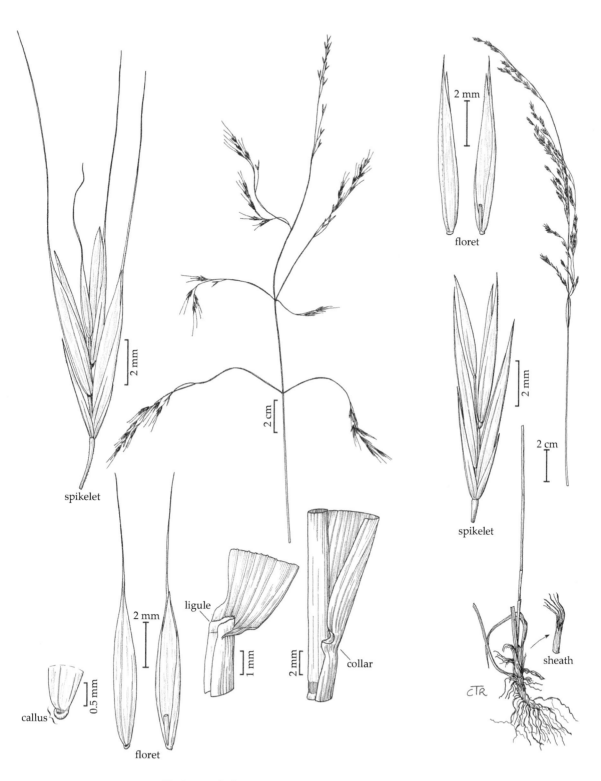

spikelet

callus

2 mm

0.5 mm

floret

ligule

1 mm

2 mm

collar

2 mm

2 cm

Festuca subulata

2 mm

floret

2 mm

spikelet

2 cm

sheath

Festuca sororia

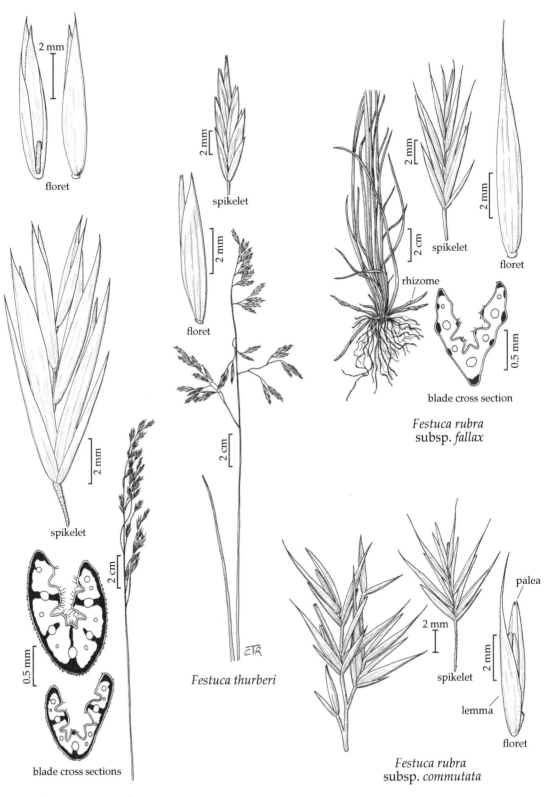

floret

spikelet

floret

floret

spikelet

spikelet

rhizome

blade cross section

Festuca rubra
subsp. *fallax*

spikelet

blade cross sections

Festuca campestris

Festuca thurberi

palea

spikelet

lemma

floret

Festuca rubra
subsp. *commutata*

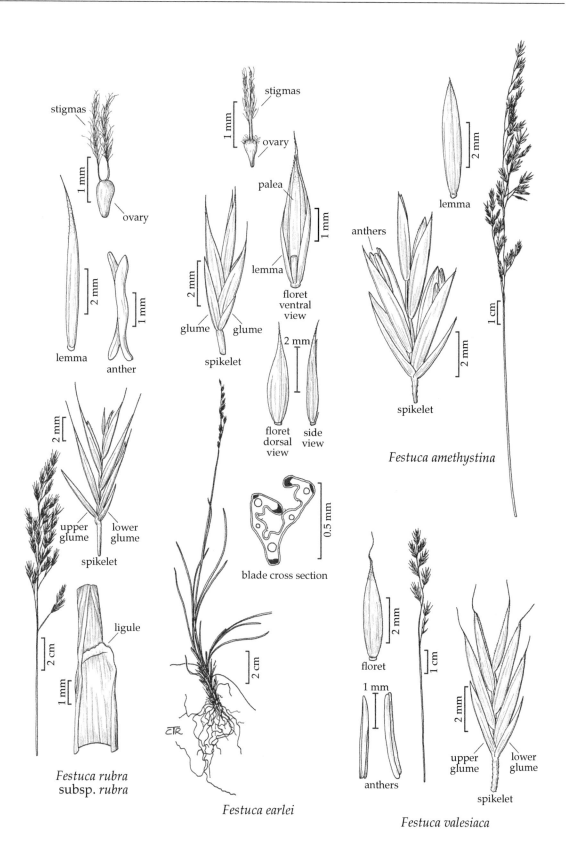

stigmas

ovary

1 mm

lemma

2 mm

anther

1 mm

stigmas

ovary

1 mm

palea

lemma

1 mm

floret
ventral
view

glume   glume

2 mm

spikelet

2 mm

floret
dorsal
view

side
view

0.5 mm

blade cross section

lemma

2 mm

anthers

spikelet

2 mm

Festuca amethystina

1 cm

2 mm

upper
glume   lower
glume

spikelet

2 cm

ligule

1 mm

Festuca rubra
subsp. *rubra*

ETR

2 cm

Festuca earlei

floret

2 mm

1 mm

anthers

1 cm

upper
glume   lower
glume

2 mm

spikelet

Festuca valesiaca

floret

spikelet

sheath

Festuca ovina

spikelet

floret

Festuca glauca

florets

spikelet

ligule

blade
cross section

Festuca trachyphylla

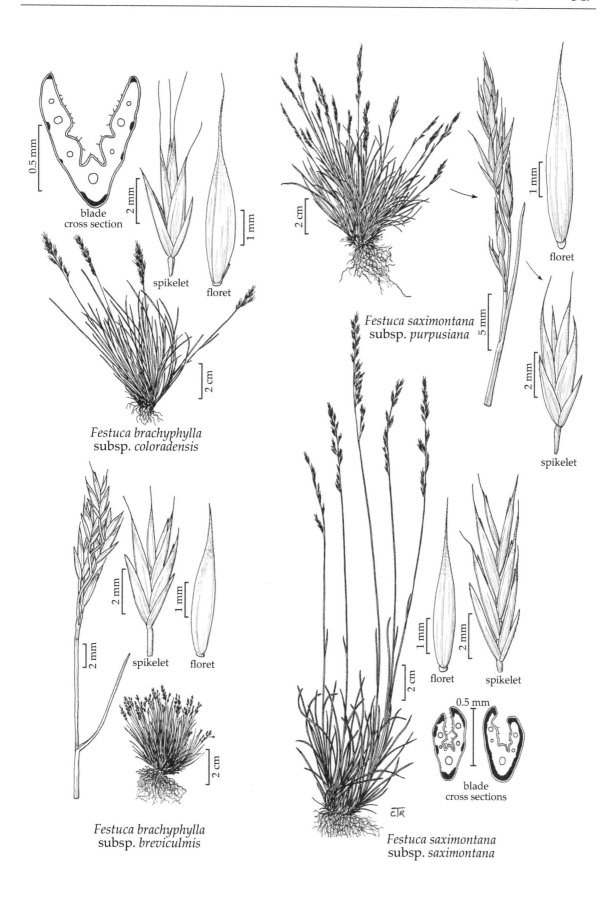

blade
cross section

0.5 mm

2 mm

spikelet

floret

1 mm

Festuca brachyphylla
subsp. *coloradensis*

2 cm

Festuca saximontana
subsp. *purpusiana*

5 mm

floret

1 mm

spikelet

2 mm

2 mm

spikelet

floret

1 mm

Festuca brachyphylla
subsp. *breviculmis*

2 cm

2 cm

floret

1 mm

spikelet

2 mm

0.5 mm

blade
cross sections

Festuca saximontana
subsp. *saximontana*

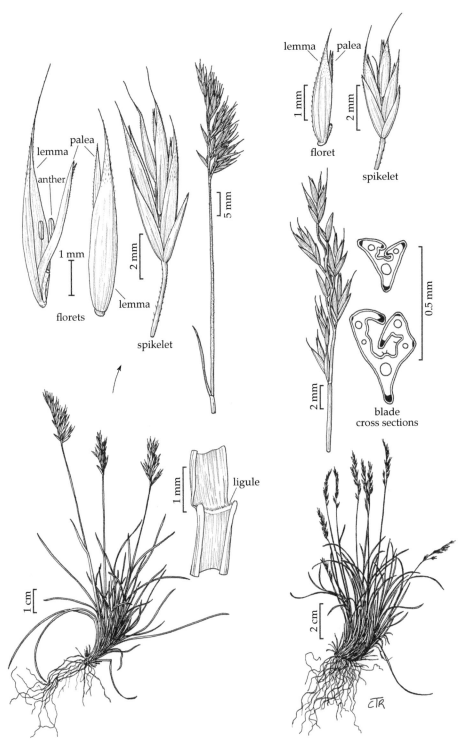

Festuca baffinensis   *Festuca minutiflora*

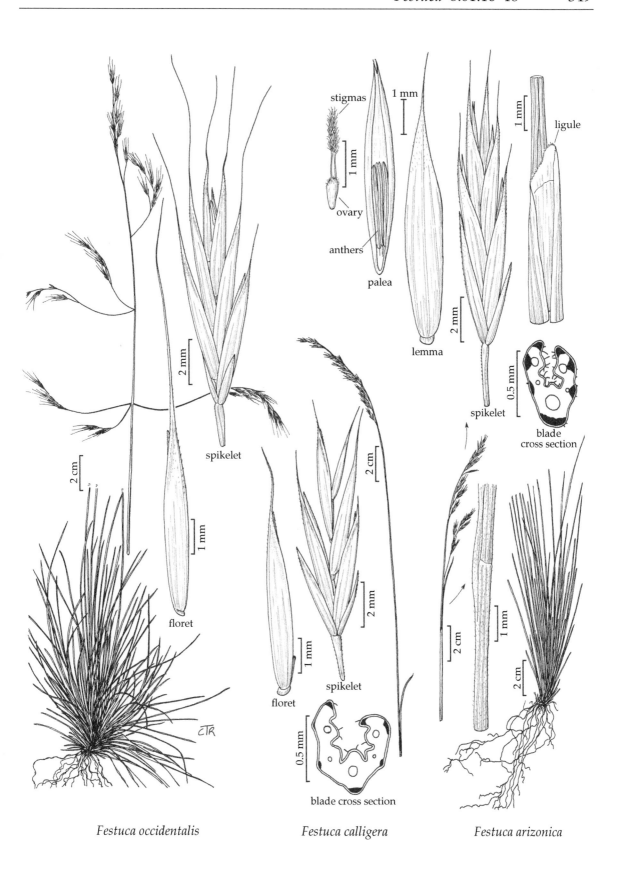

Festuca occidentalis

Festuca calligera

Festuca arizonica

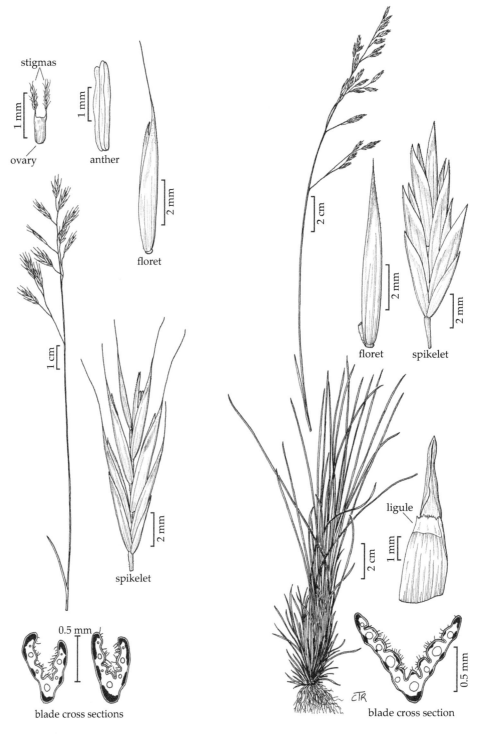

stigmas

ovary

anther

floret

spikelet

blade cross sections

Festuca idahoensis

floret   spikelet

ligule

blade cross section

Festuca viridula

CTR

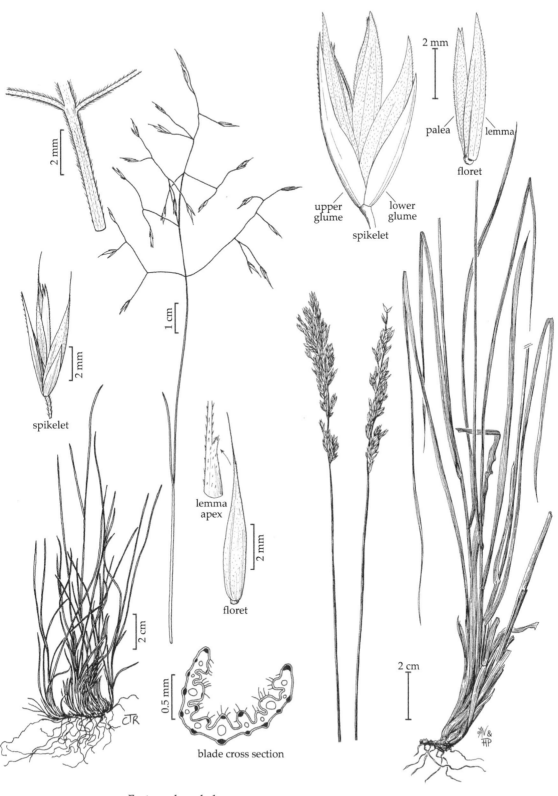

upper glume
lower glume
spikelet

palea lemma
floret

2 mm

2 mm

2 mm

1 cm

spikelet

2 mm

lemma apex

floret

2 mm

2 cm

0.5 mm

blade cross section

2 cm

Festuca dasyclada

Leucopoa kingii

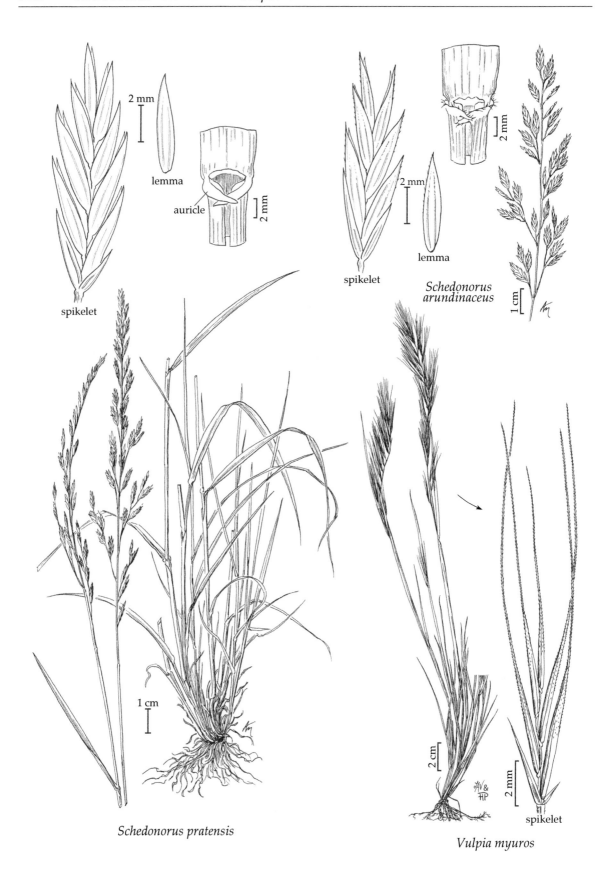

spikelet

2 mm

lemma

auricle

2 mm

spikelet

Schedonorus arundinaceus

2 mm

lemma

1 cm

1 cm

Schedonorus pratensis

2 cm

2 mm

spikelet

Vulpia myuros

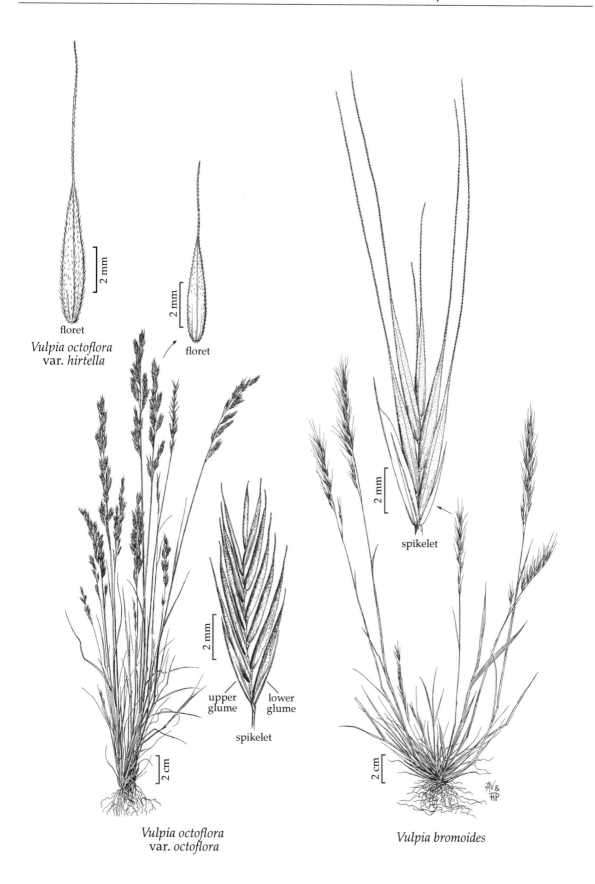

floret

Vulpia octoflora
var. *hirtella*

floret

Vulpia octoflora
var. *octoflora*

upper
glume

lower
glume

spikelet

spikelet

Vulpia bromoides

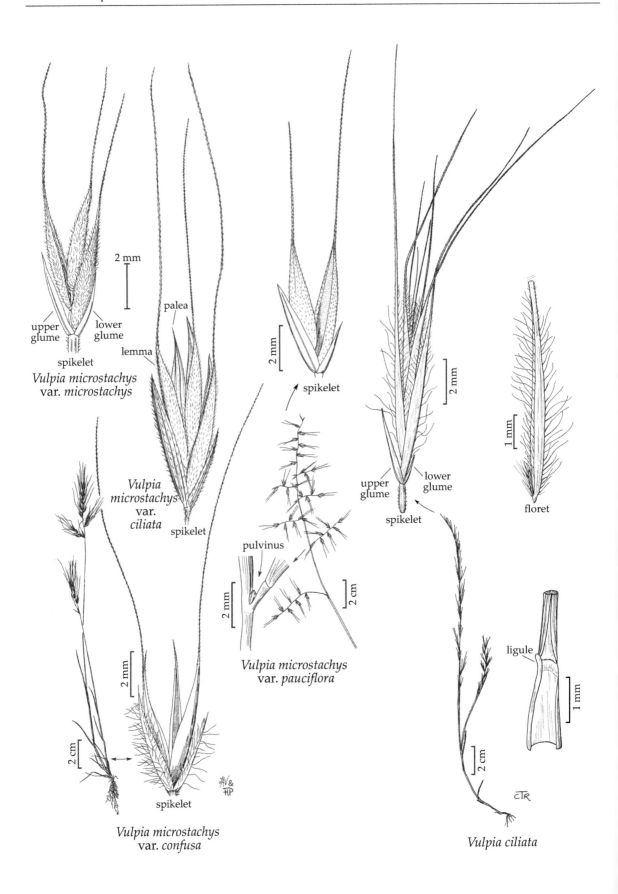

2 mm

upper
glume

lower
glume

spikelet

Vulpia microstachys
var. *microstachys*

palea

lemma

spikelet

*Vulpia
microstachys
var.
ciliata*

2 mm

spikelet

2 mm

spikelet

pulvinus

2 mm

2 cm

Vulpia microstachys
var. *pauciflora*

2 mm

upper
glume

lower
glume

spikelet

2 mm

floret

1 mm

2 mm

2 cm

spikelet

Vulpia microstachys
var. *confusa*

ligule

1 mm

2 cm

Vulpia ciliata

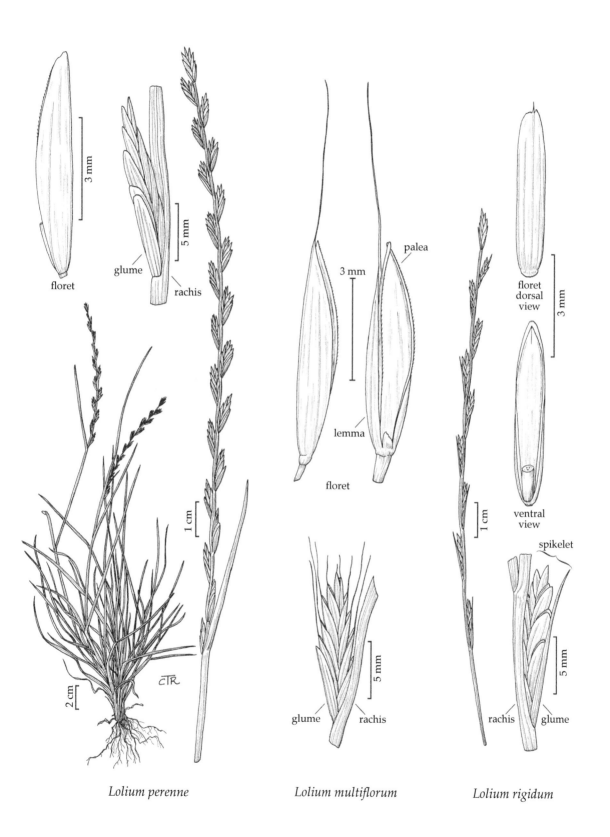

floret

glume
rachis

palea

lemma

floret

glume   rachis

floret
dorsal
view

ventral
view

spikelet

rachis   glume

Lolium perenne   *Lolium multiflorum*   *Lolium rigidum*

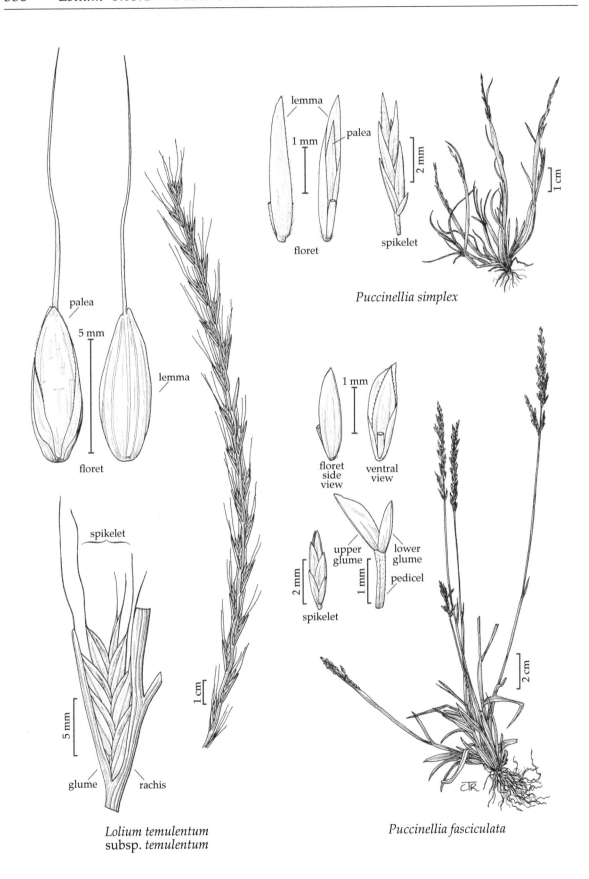

palea

5 mm

lemma

floret

spikelet

glume   rachis

1 cm

Lolium temulentum
subsp. *temulentum*

lemma

palea

1 mm

2 mm

floret

spikelet

Puccinellia simplex

1 cm

1 mm

floret
side
view

ventral
view

upper
glume

lower
glume

pedicel

2 mm

1 mm

spikelet

2 cm

Puccinellia fasciculata

floret

spikelet

lemma

pedicel

spikelet

lemma

palea

floret

palea

lemma

spikelet

Puccinellia parishii

Puccinellia lemmonii

Puccinellia distans

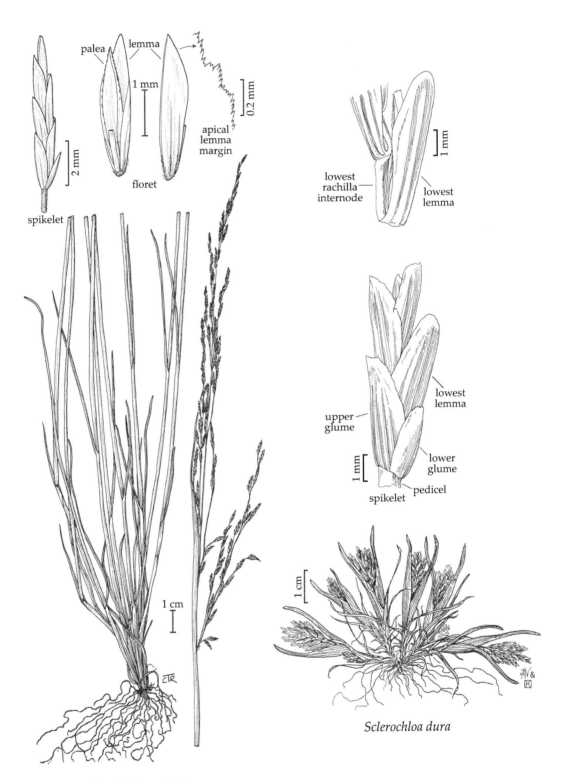

palea lemma

1 mm

apical
lemma
margin

0.2 mm

floret

2 mm

spikelet

lowest
rachilla
internode

lowest
lemma

1 mm

upper
glume

lowest
lemma

lower
glume

1 mm

pedicel

spikelet

1 cm

Puccinellia nuttalliana

1 cm

Sclerochloa dura

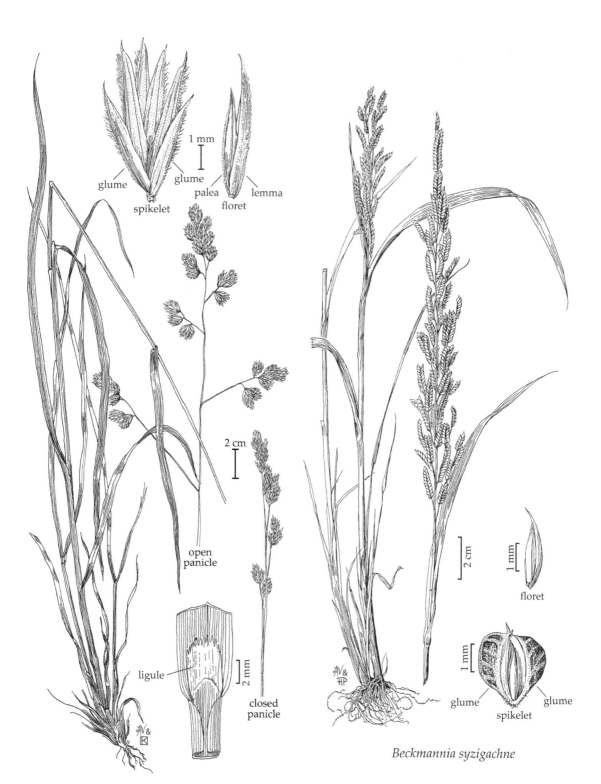

glume

glume

spikelet

palea   lemma

floret

open
panicle

2 cm

ligule

closed
panicle

Dactylis glomerata

1 mm

2 cm

floret

1 mm

glume   glume

spikelet

Beckmannia syzigachne

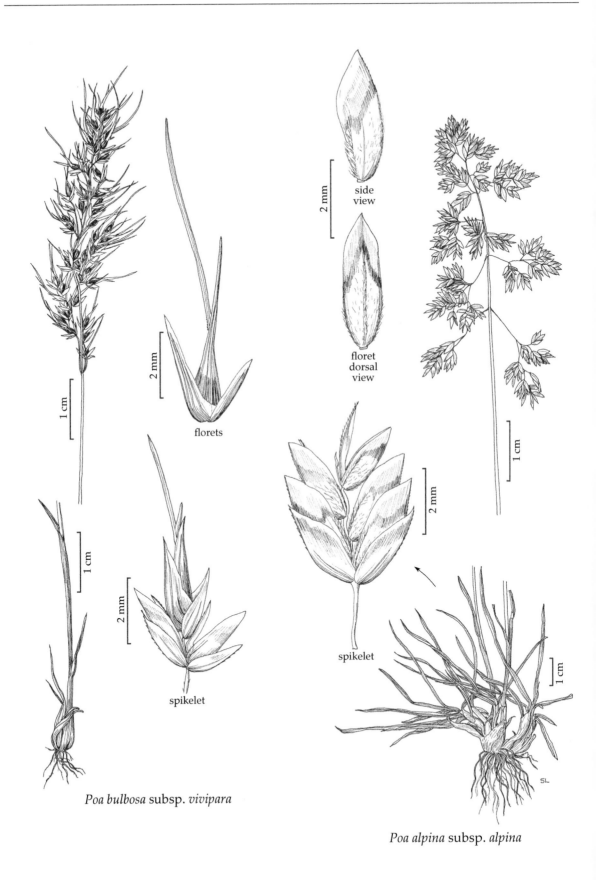

side
view

floret
dorsal
view

florets

spikelet

spikelet

Poa bulbosa subsp. *vivipara*

Poa alpina subsp. *alpina*

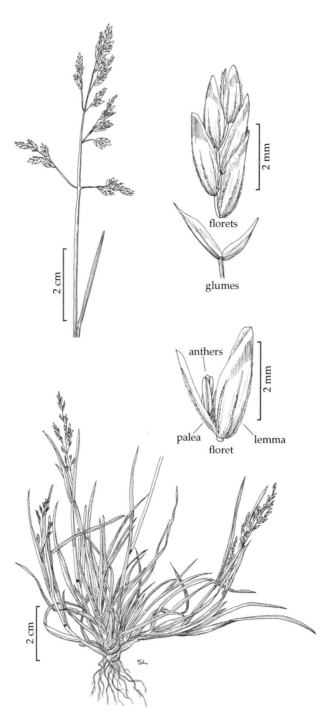

Poa annua

lemma

spikelet

cauline
blade

palea

ligule

rhizomes

subsp. *alpigena*

subsp. *angustifolia*

subsp. *pratensis*

Poa pratensis

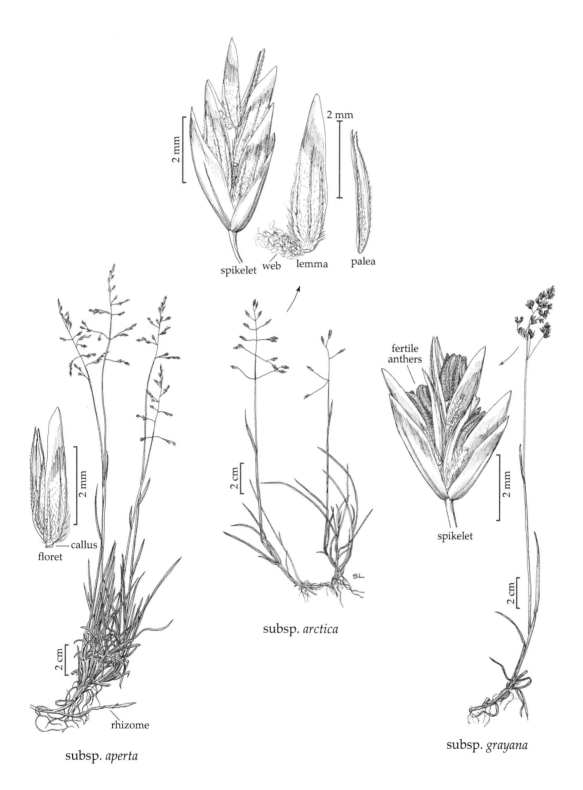

2 mm

2 mm

spikelet   web   lemma   palea

2 mm

callus
floret

2 cm

rhizome

subsp. *aperta*

2 cm

subsp. *arctica*

fertile
anthers

2 mm

spikelet

2 cm

subsp. *grayana*

Poa arctica

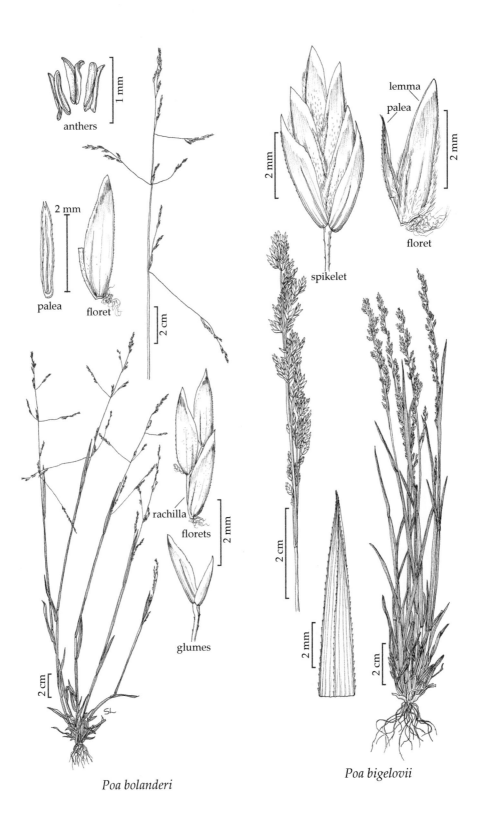

anthers

1 mm

2 mm

palea

floret

2 mm

2 cm

rachilla

florets

2 mm

glumes

Poa bolanderi

lemma

palea

2 mm

floret

2 mm

spikelet

2 cm

2 mm

2 cm

Poa bigelovii

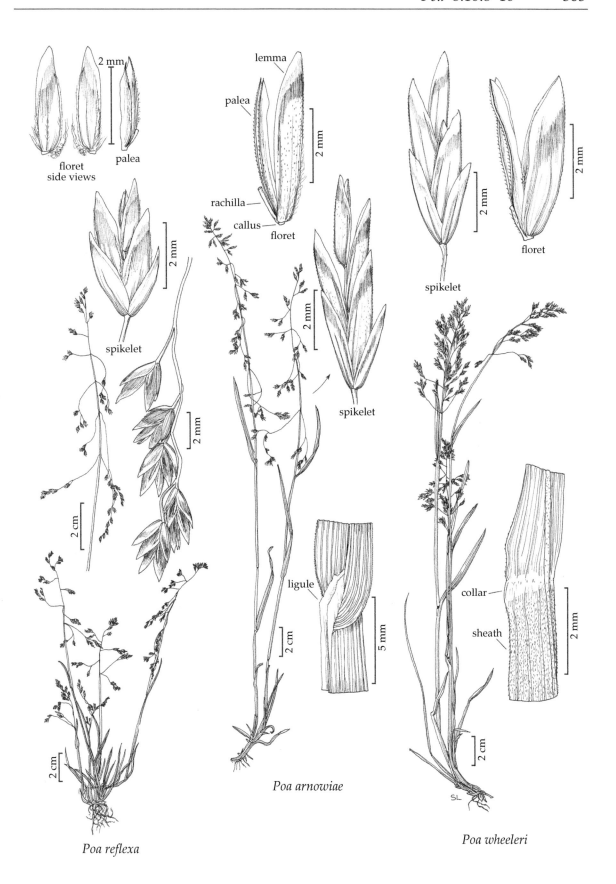

floret
side views

palea

Poa reflexa

lemma

palea

rachilla

callus

floret

spikelet

ligule

Poa arnowiae

floret

spikelet

collar

sheath

Poa wheeleri

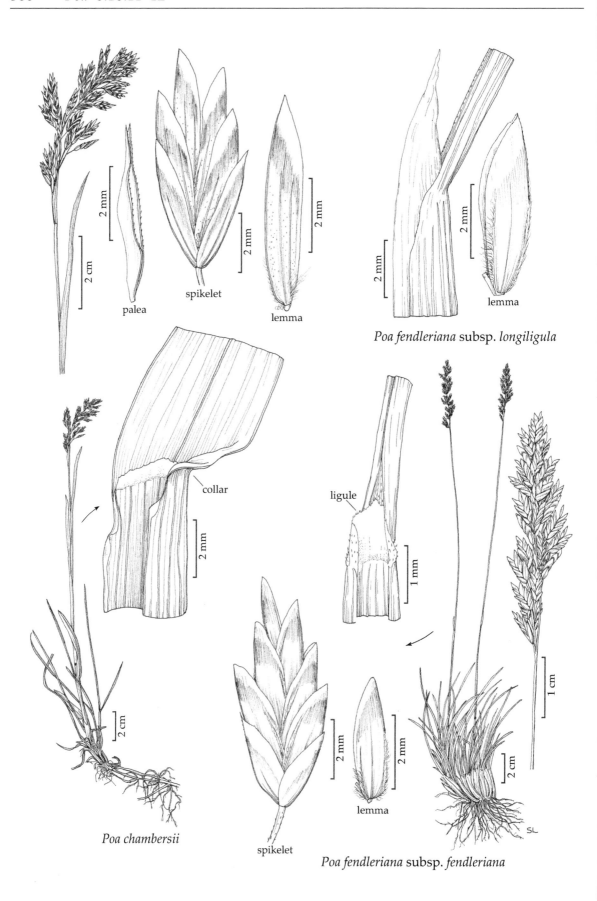

palea

spikelet

lemma

2 mm

2 mm

2 mm

2 cm

lemma

2 mm

Poa fendleriana subsp. *longiligula*

collar

2 mm

ligule

1 mm

Poa chambersii

2 cm

spikelet

lemma

2 mm

2 mm

1 cm

2 cm

Poa fendleriana subsp. *fendleriana*

SL

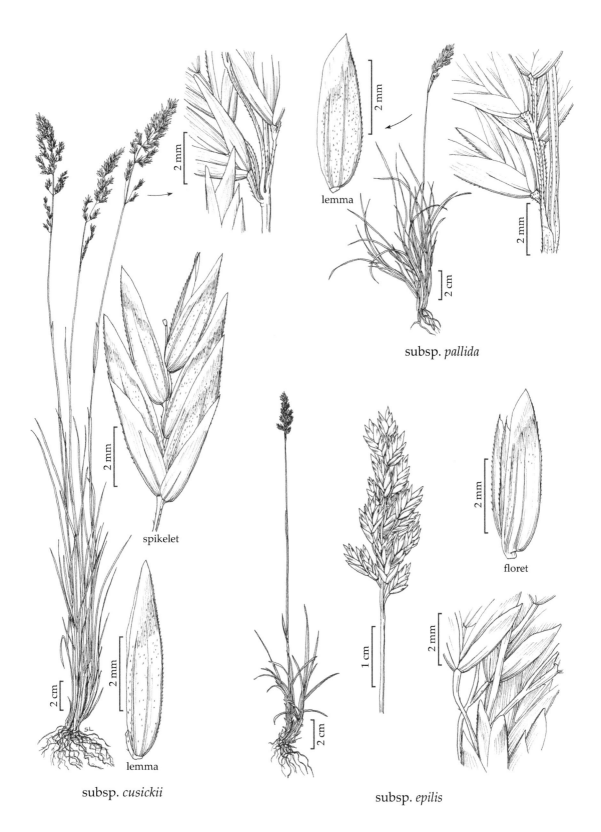

subsp. *pallida*

spikelet

lemma

subsp. *cusickii*

floret

subsp. *epilis*

Poa cusickii

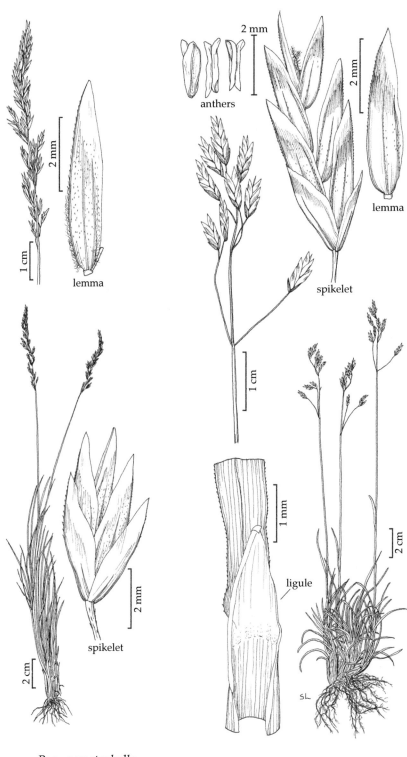

2 mm

anthers

2 mm

2 mm

lemma

lemma

1 cm

spikelet

1 cm

spikelet

2 mm

1 mm

ligule

2 cm

2 cm

spikelet

SL

Poa ×*nematophylla*

Poa leibergii

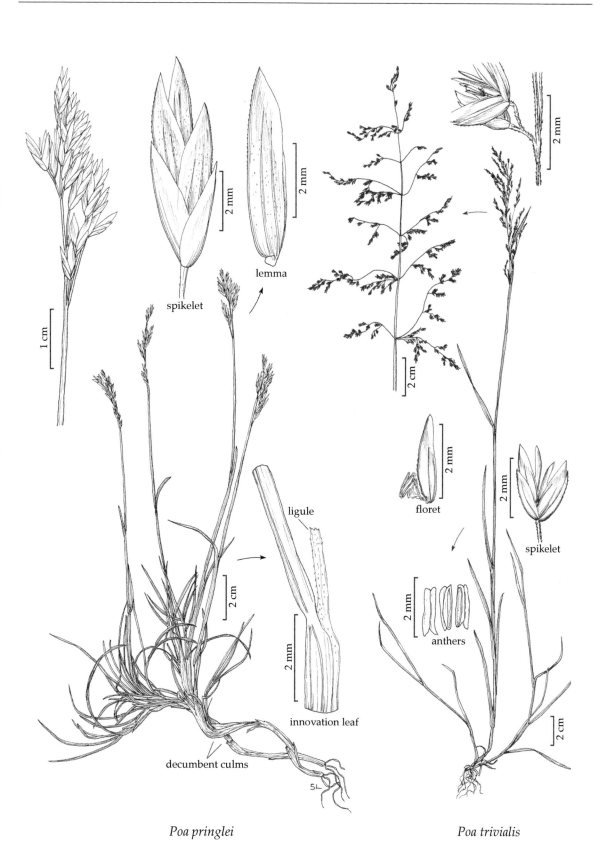

spikelet

lemma

2 mm

2 mm

1 cm

ligule

innovation leaf

2 mm

2 cm

2 mm

decumbent culms

SL

Poa pringlei

2 cm

2 mm

floret

2 mm

anthers

2 mm

spikelet

2 cm

Poa trivialis

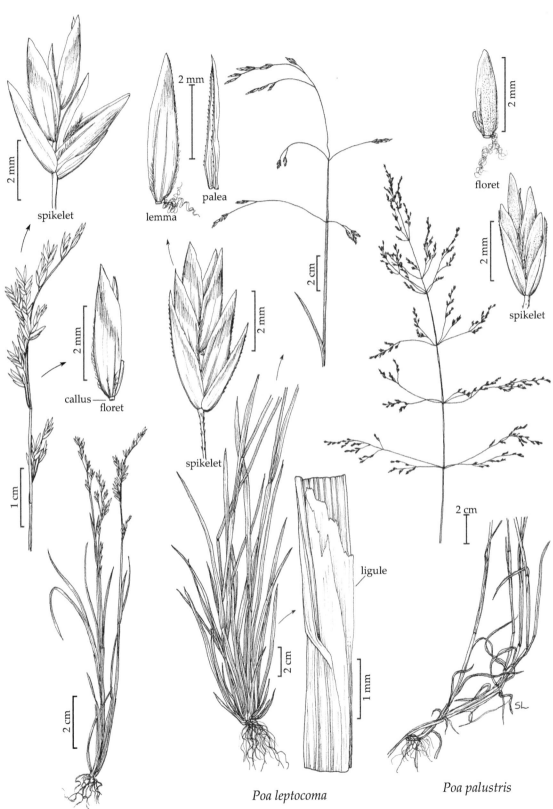

spikelet

lemma

palea

floret

spikelet

callus — floret

2 mm

2 cm

2 cm

2 mm

2 mm

2 mm

spikelet

1 cm

ligule

spikelet

2 cm

2 mm

2 cm

1 mm

2 cm

SL

Poa laxa subsp. *banffiana*

Poa leptocoma

Poa palustris

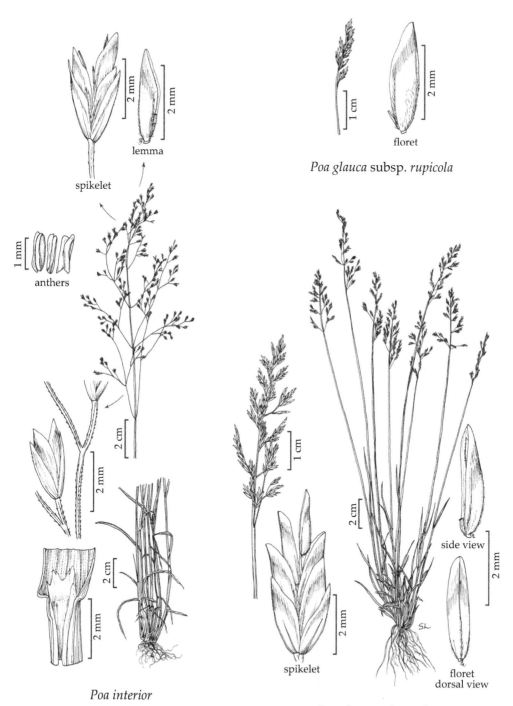

spikelet

lemma

2 mm

2 mm

anthers

1 mm

2 cm

2 mm

2 mm

2 cm

Poa interior

floret

1 cm

2 mm

Poa glauca subsp. *rupicola*

spikelet

1 cm

2 mm

2 cm

side view

2 mm

floret
dorsal view

Poa glauca subsp. *glauca*

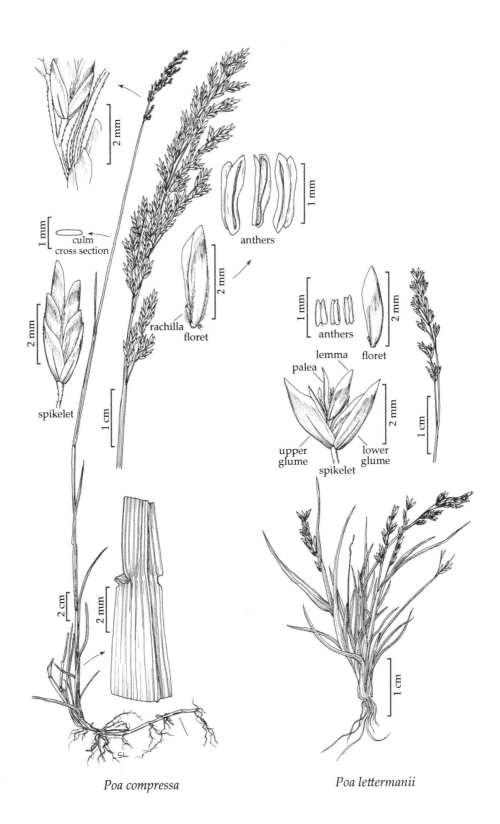

Poa compressa

Poa lettermanii

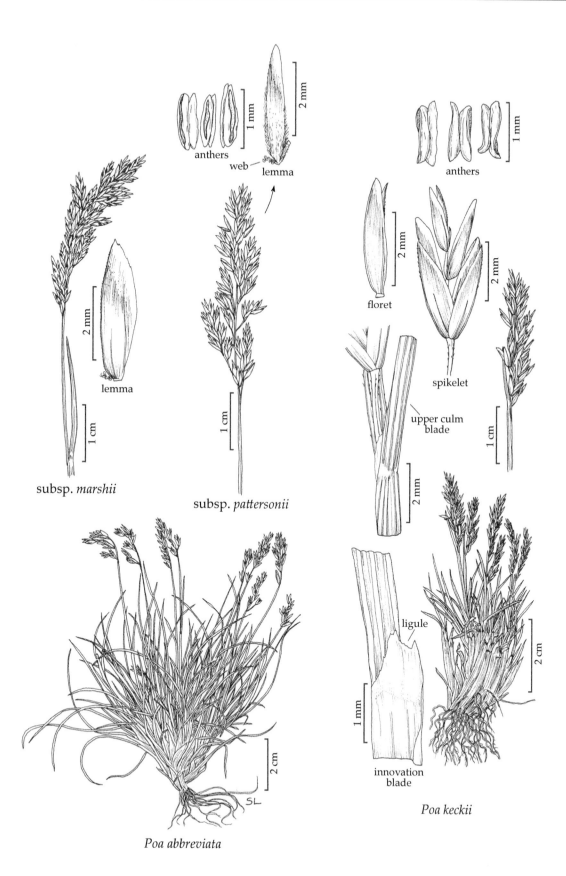

anthers

web lemma

2 mm

1 mm

lemma

subsp. *marshii*

subsp. *pattersonii*

anthers

floret

spikelet

upper culm
blade

ligule

innovation
blade

Poa abbreviata

SL

Poa keckii

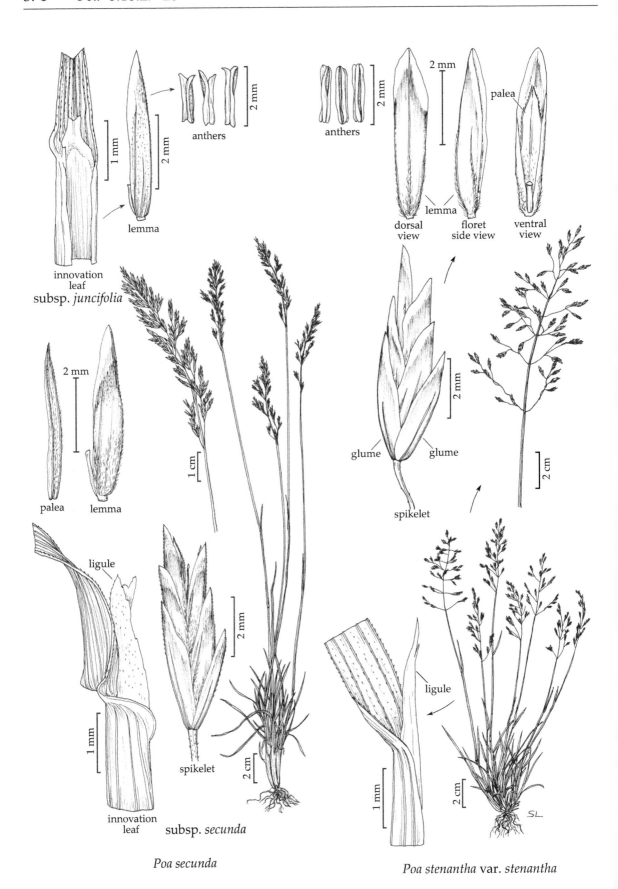

innovation
leaf
subsp. *juncifolia*

anthers

lemma

anthers

palea

lemma

dorsal
view

floret
side view

ventral
view

palea   lemma

glume   glume

spikelet

ligule

innovation
leaf

spikelet

subsp. *secunda*

ligule

Poa secunda

Poa stenantha var. *stenantha*

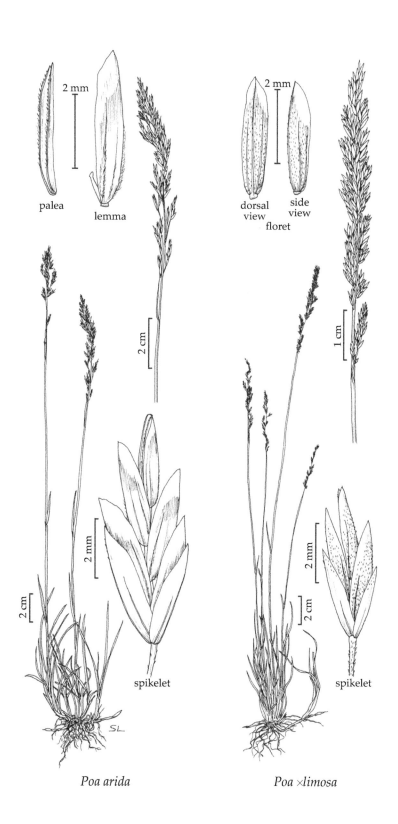

palea

lemma

2 mm

dorsal view

side view

floret

2 mm

2 cm

2 mm

1 cm

2 mm

2 cm

2 cm

spikelet

spikelet

SL

Poa arida

Poa ×limosa

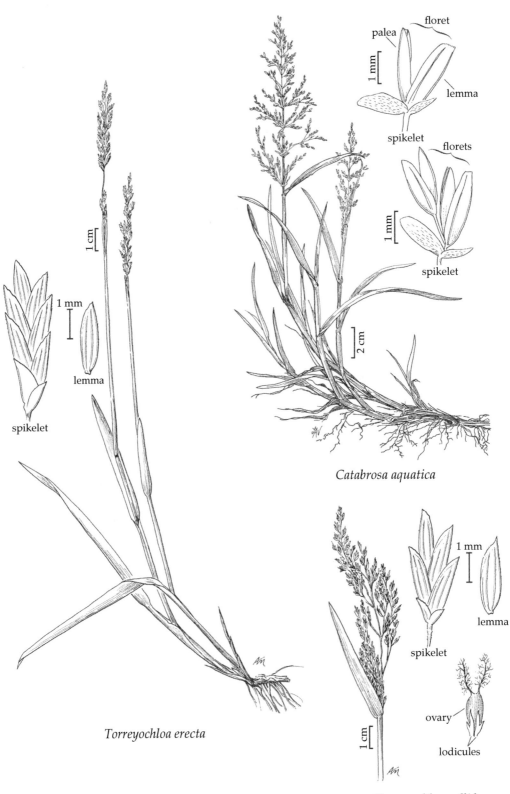

floret
palea
lemma
spikelet

florets
lemma
spikelet

Catabrosa aquatica

spikelet
lemma

Torreyochloa erecta

spikelet
lemma

ovary
lodicules

Torreyochloa pallida
var. *pauciflora*

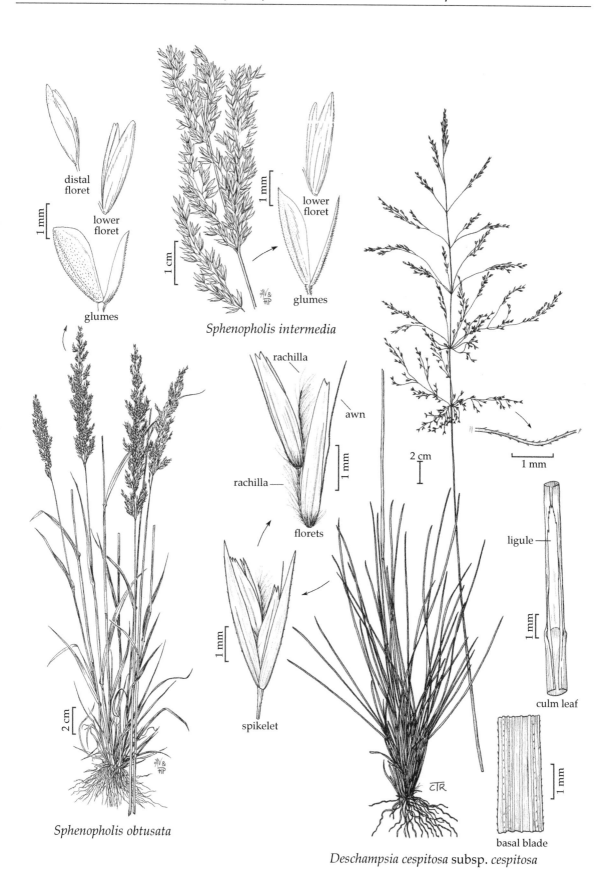

distal
floret

1 mm

lower
floret

glumes

lower
floret

1 mm

lower
floret

glumes

Sphenopholis intermedia

rachilla

awn

1 mm

rachilla

florets

2 cm

1 mm

1 mm

spikelet

ligule

1 mm

culm leaf

Sphenopholis obtusata

2 cm

basal blade

1 mm

Deschampsia cespitosa subsp. *cespitosa*

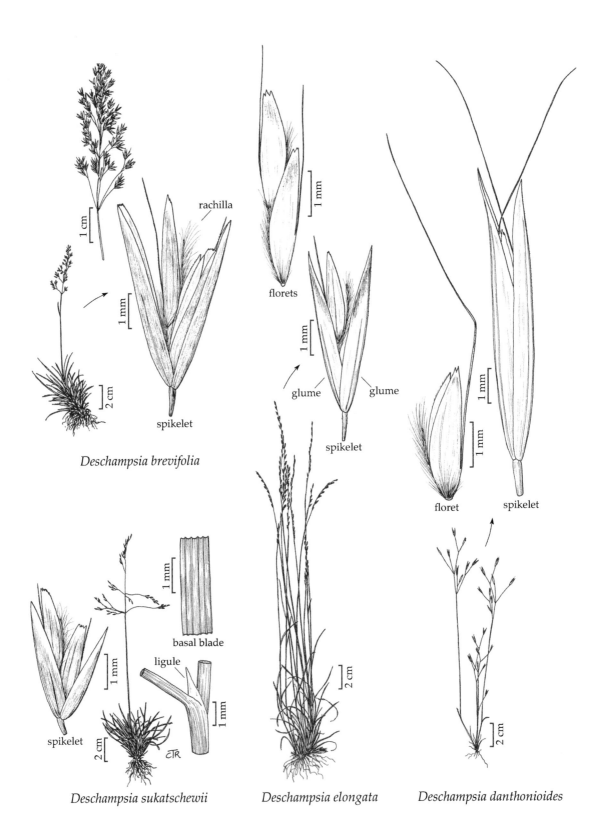

Deschampsia brevifolia

rachilla

spikelet

florets

glume   glume

spikelet

floret   spikelet

basal blade

ligule

spikelet

Deschampsia sukatschewii   *Deschampsia elongata*   *Deschampsia danthonioides*

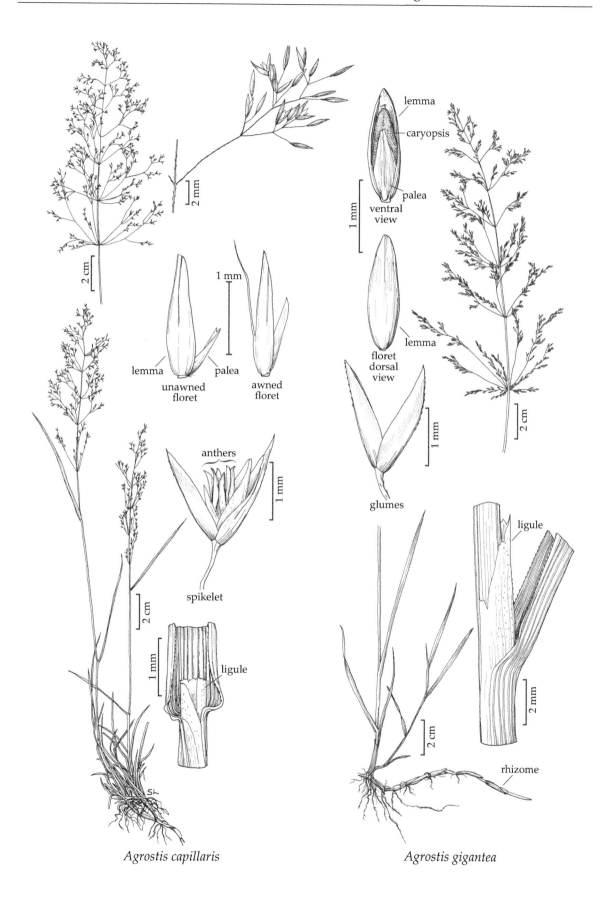

Agrostis capillaris

Agrostis gigantea

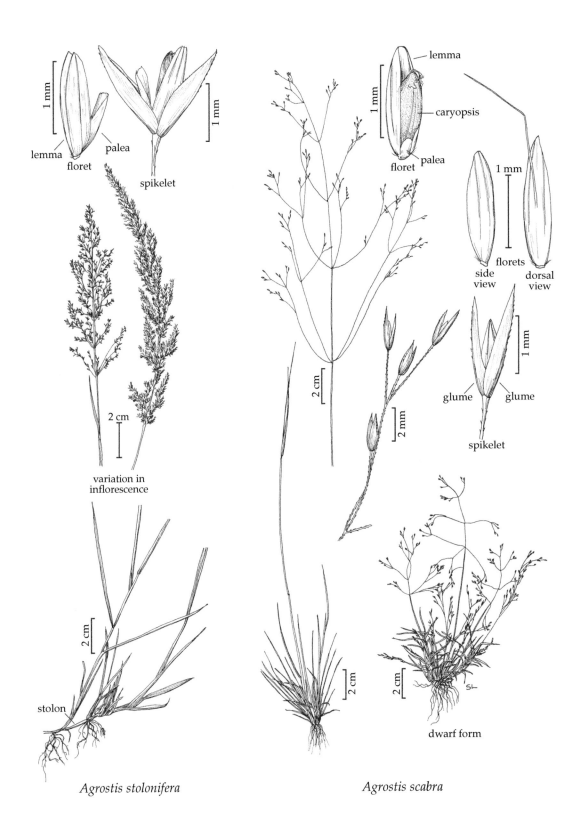

lemma

1 mm

lemma   palea
    floret

spikelet

variation in
inflorescence

2 cm

stolon

2 cm

Agrostis stolonifera

lemma

caryopsis

1 mm

floret   palea

1 mm

florets
side
view   dorsal
view

1 mm

glume   glume

spikelet

2 cm

2 mm

2 cm

2 cm

SL

dwarf form

Agrostis scabra

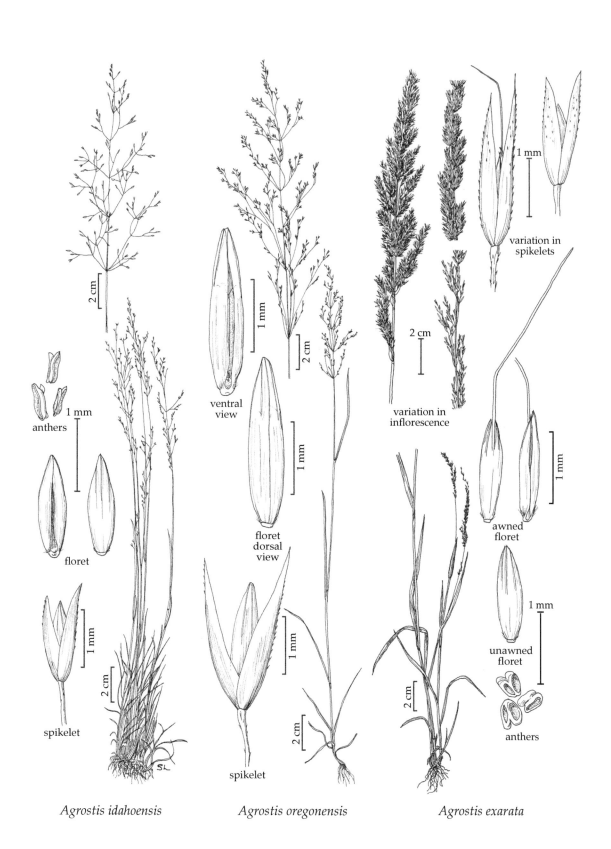

ventral
view

floret
dorsal
view

anthers

floret

spikelet

Agrostis idahoensis

spikelet

Agrostis oregonensis

variation in
spikelets

variation in
inflorescence

awned
floret

unawned
floret

anthers

Agrostis exarata

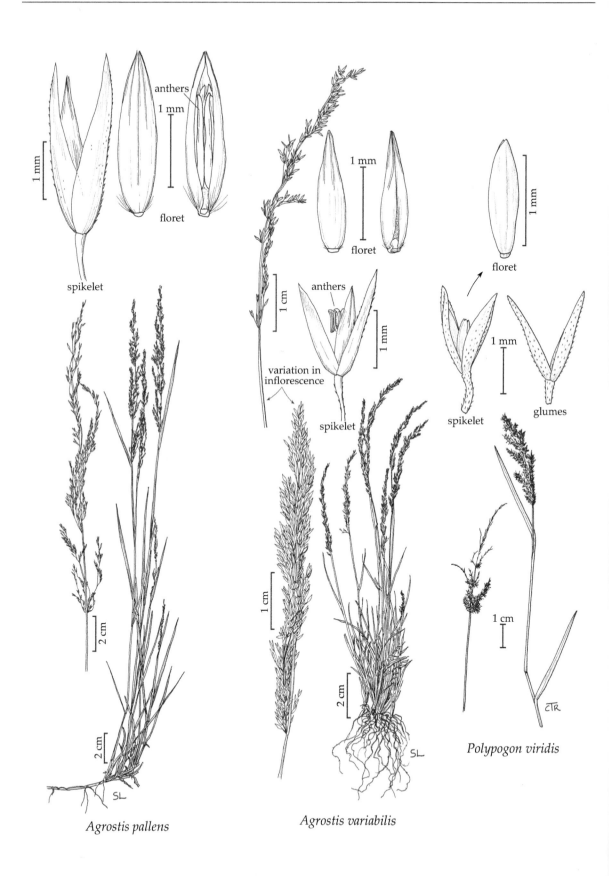

anthers

1 mm

1 mm

floret

spikelet

variation in inflorescence

anthers

1 mm

floret

1 mm

spikelet

1 mm

floret

1 mm

spikelet

1 mm

glumes

Polypogon viridis

1 cm

1 cm

2 cm

2 cm

2 cm

1 cm

SL

2 cm

SL

CTR

Agrostis pallens

Agrostis variabilis

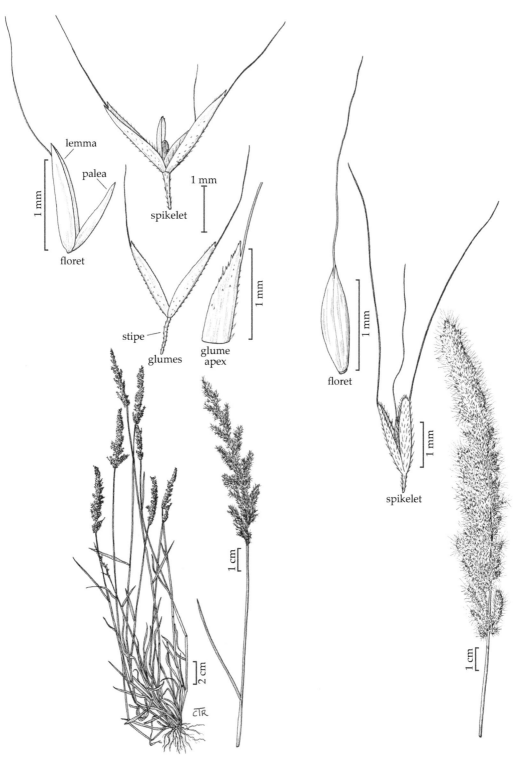

lemma

palea

1 mm

floret

spikelet

1 mm

stipe

glumes

glume apex

1 mm

1 mm

floret

1 mm

spikelet

1 mm

1 cm

2 cm

cTR

1 cm

Polypogon interruptus

Polypogon australis

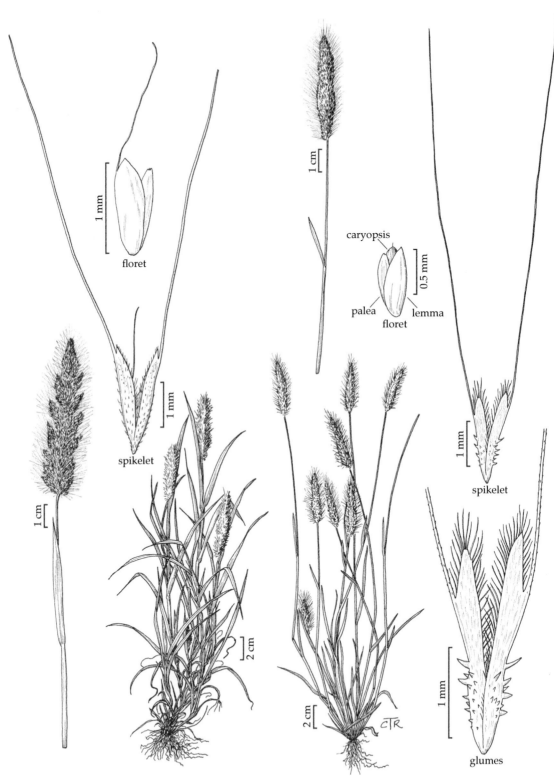

1 mm

floret

caryopsis

0.5 mm

palea   lemma

floret

1 cm

1 mm

spikelet

1 cm

spikelet

1 mm

1 mm

glumes

2 cm

2 cm

CTR

Polypogon monspeliensis   *Polypogon maritimus*

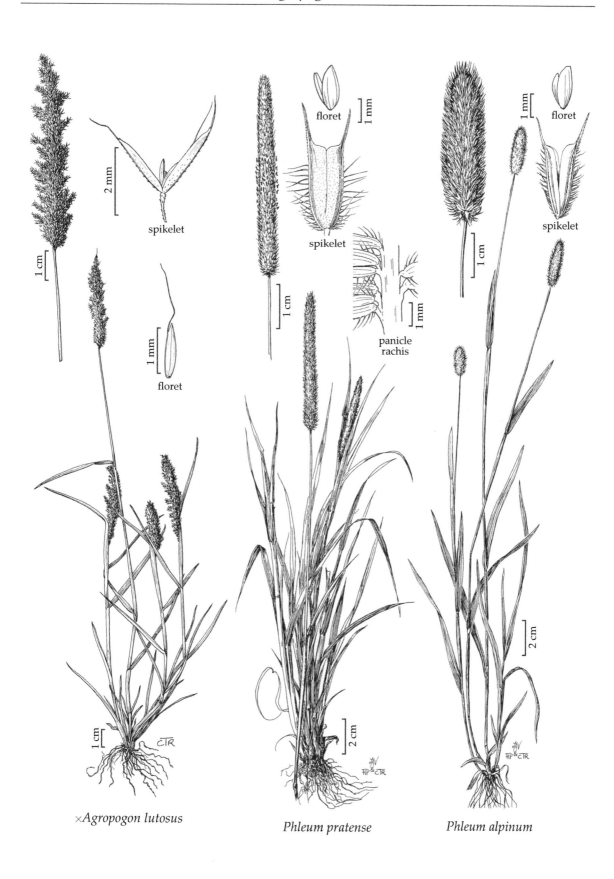

floret

spikelet

floret

×*Agropogon lutosus*

spikelet

floret

spikelet

panicle
rachis

Phleum pratense

floret

spikelet

Phleum alpinum

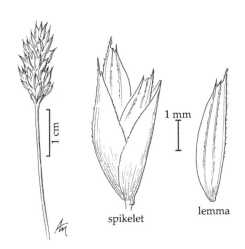

spikelet   lemma

Sesleria caerulea

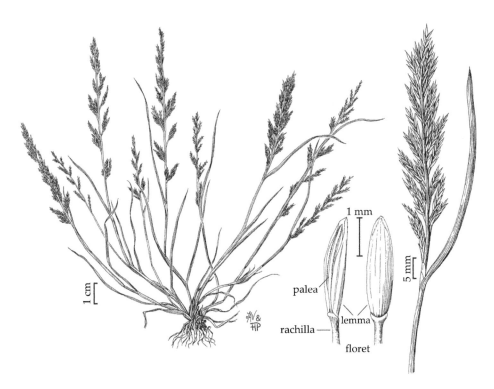

palea

lemma

rachilla

floret

Desmazeria rigida

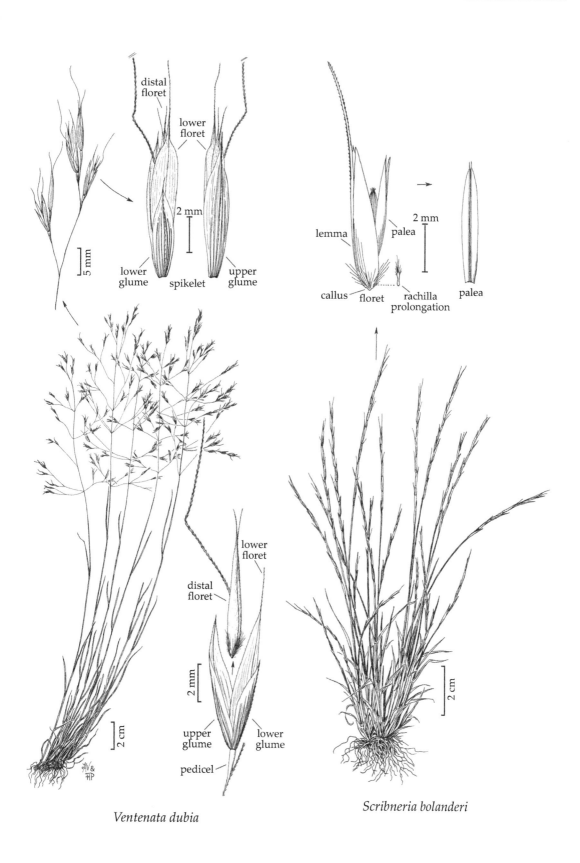

Ventenata dubia

Scribneria bolanderi

lemma

glumes

Vahlodea atropurpurea

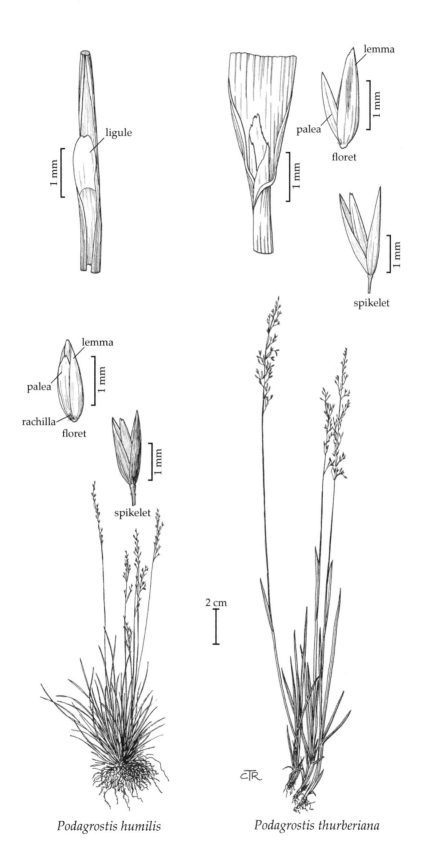

ligule

1 mm

lemma

palea

floret

1 mm

1 mm

spikelet

1 mm

lemma

palea

rachilla

floret

1 mm

spikelet

1 mm

2 cm

CTR

Podagrostis humilis   *Podagrostis thurberiana*

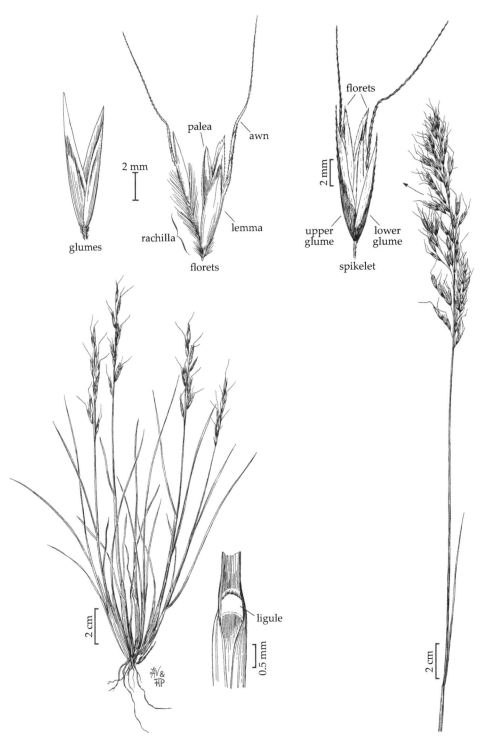

glumes

2 mm

palea

rachilla

florets

awn

lemma

florets

upper
glume

lower
glume

spikelet

2 mm

ligule

0.5 mm

2 cm

AV &
HP

2 cm

Helictotrichon mortonianum

Helictotrichon sempervirens

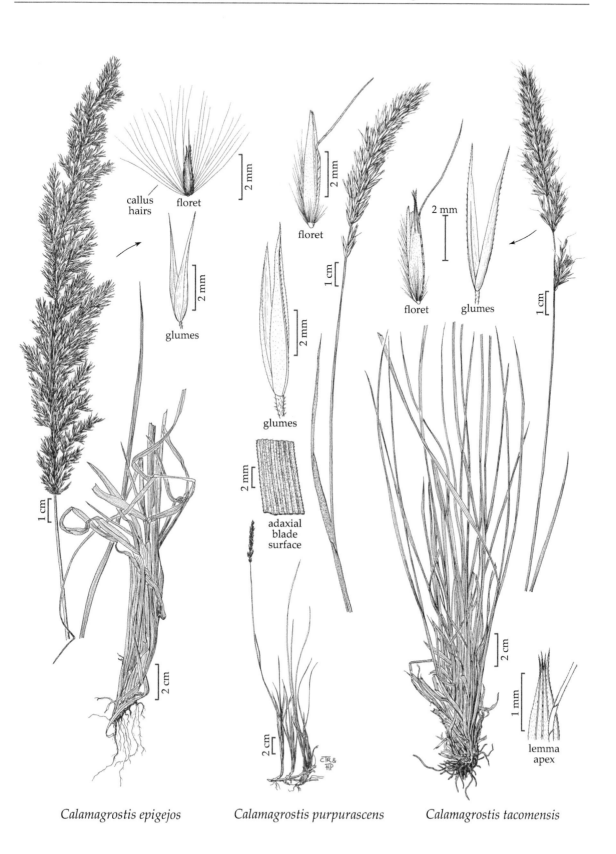

Calamagrostis epigejos   *Calamagrostis purpurascens*   *Calamagrostis tacomensis*

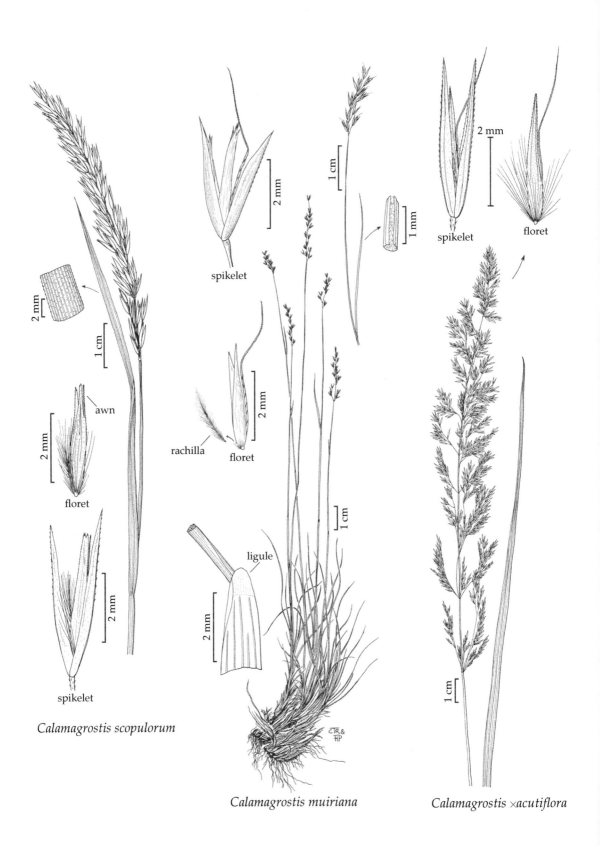

Calamagrostis scopulorum

Calamagrostis muiriana

Calamagrostis ×*acutiflora*

Calamagrostis koelerioides　　*Calamagrostis rubescens*　　*Calamagrostis montanensis*

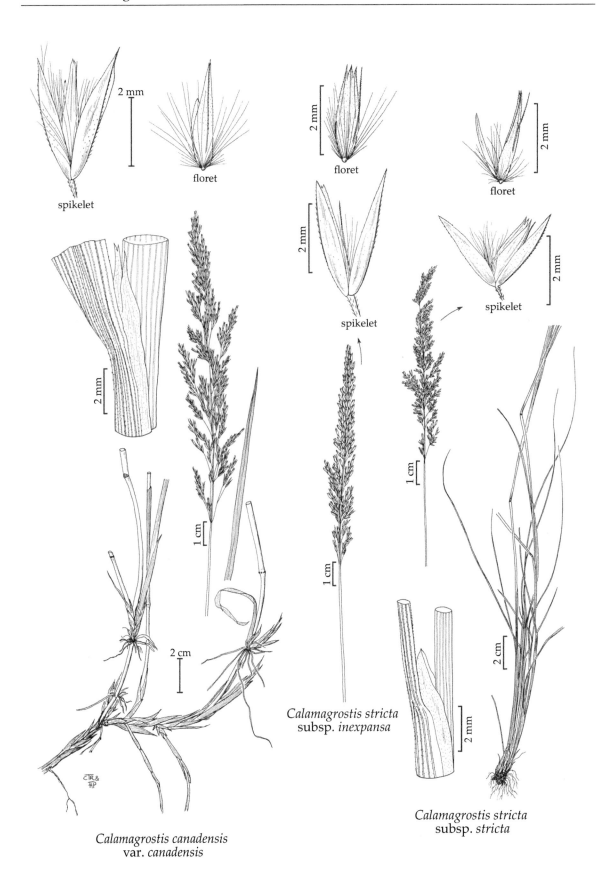

spikelet

2 mm

floret

floret

2 mm

floret

2 mm

2 mm

spikelet

2 mm

spikelet

2 mm

2 mm

1 cm

1 cm

1 cm

Calamagrostis stricta
subsp. *inexpansa*

2 mm

2 cm

2 cm

Calamagrostis canadensis
var. *canadensis*

Calamagrostis stricta
subsp. *stricta*

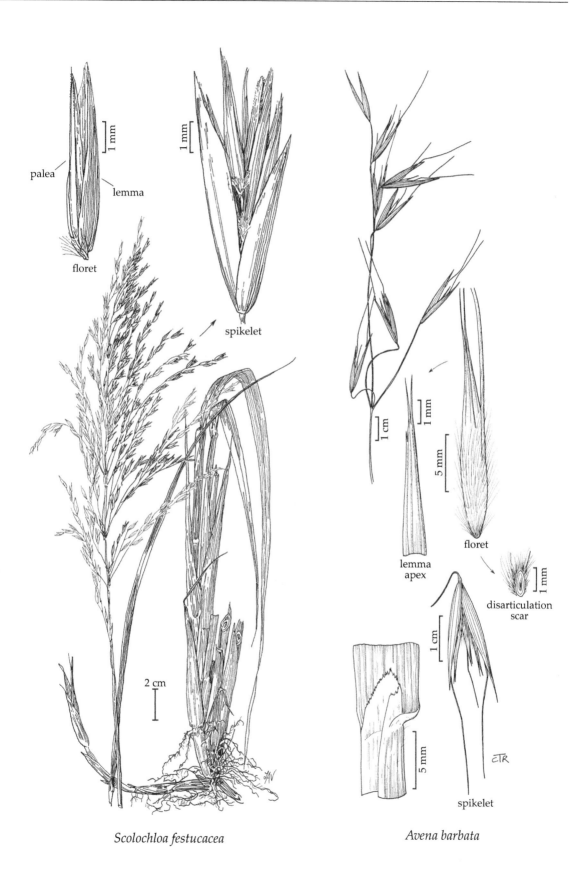

palea

lemma

floret

spikelet

Scolochloa festucacea

lemma apex

floret

disarticulation scar

spikelet

Avena barbata

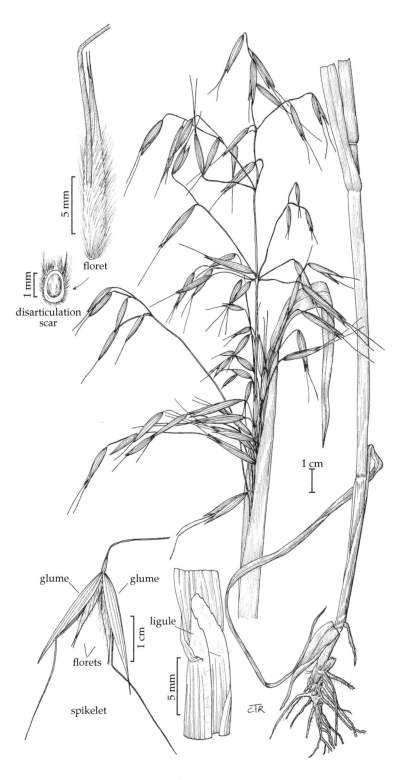

5 mm

floret

1 mm

disarticulation
scar

1 cm

glume   glume

ligule

1 cm

florets

5 mm

spikelet

CTR

Avena fatua

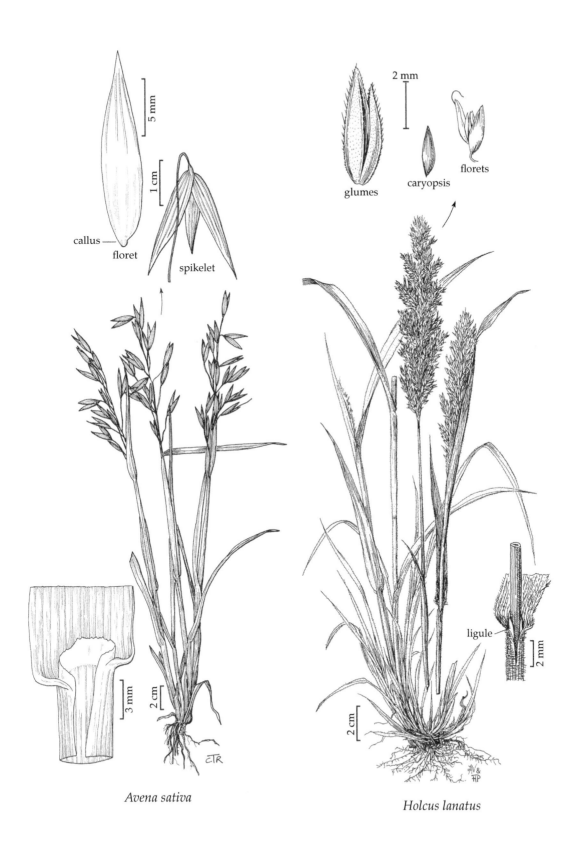

5 mm

1 cm

2 mm

glumes

caryopsis

florets

callus —

floret

spikelet

3 mm

2 cm

ligule

2 mm

2 cm

Avena sativa

Holcus lanatus

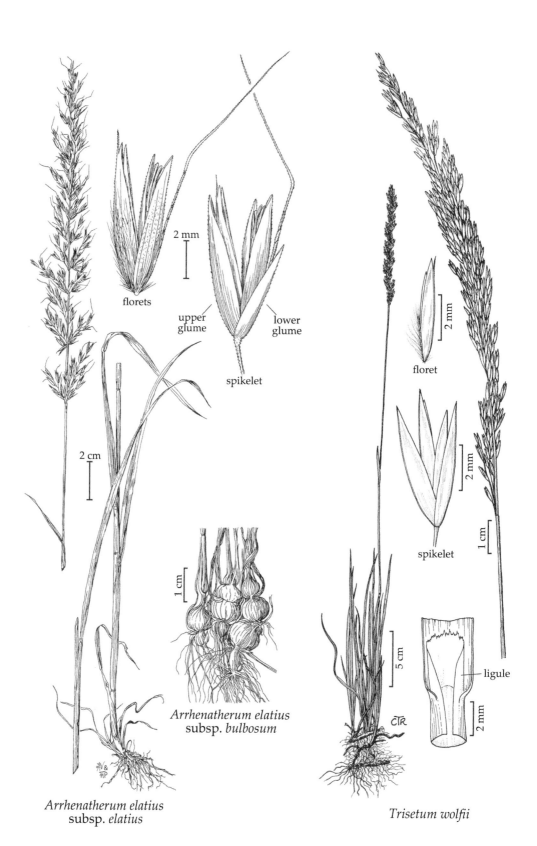

florets

2 mm

upper
glume

lower
glume

spikelet

2 cm

floret

2 mm

spikelet

2 mm

1 cm

Arrhenatherum elatius
subsp. *bulbosum*

1 cm

5 cm

ligule

2 mm

CTR

Arrhenatherum elatius
subsp. *elatius*

Trisetum wolfii

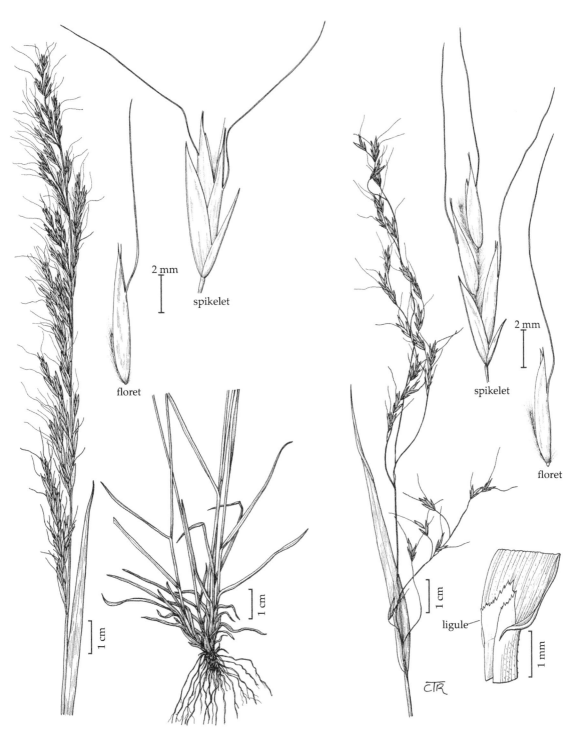

Trisetum canescens

Trisetum cernuum

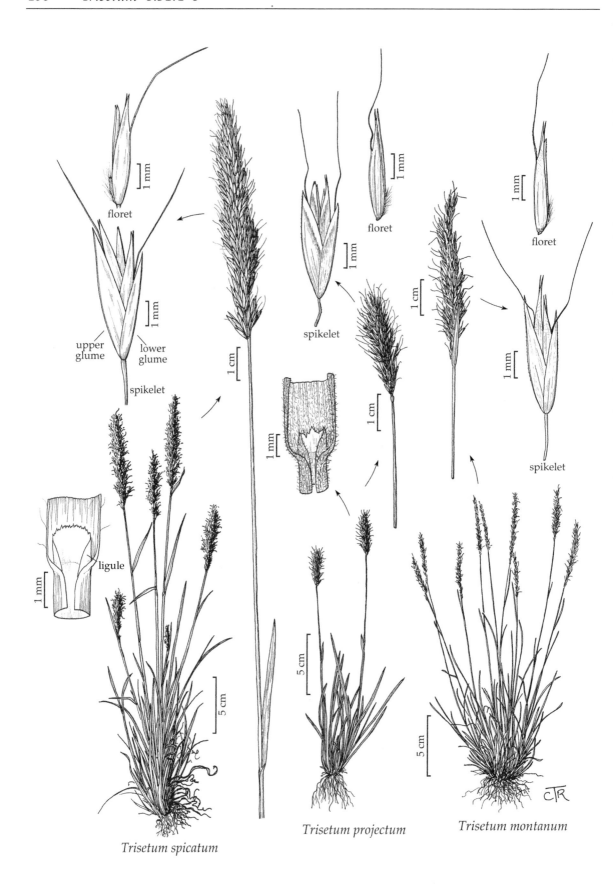

floret

upper
glume   lower
               glume

spikelet

ligule

Trisetum spicatum

floret

spikelet

Trisetum projectum

floret

spikelet

Trisetum montanum

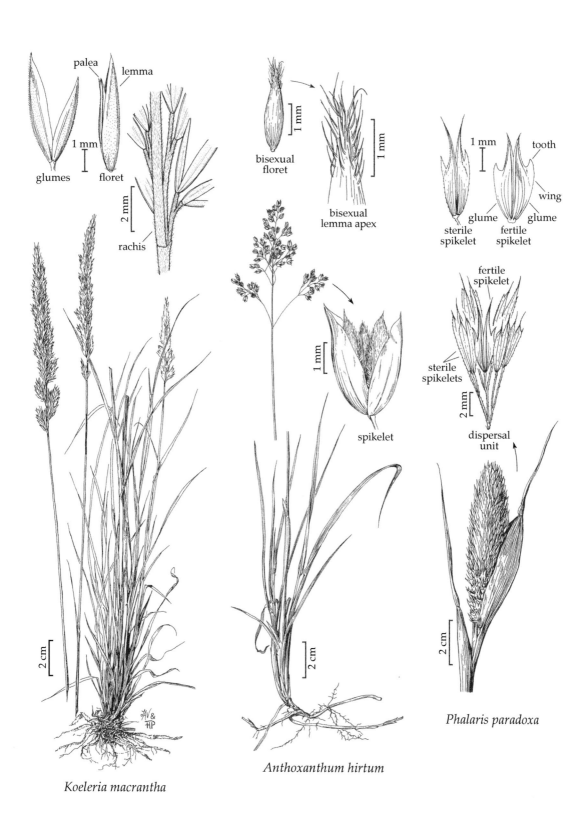

glumes

palea

lemma

floret

1 mm

2 mm

rachis

bisexual
floret

1 mm

1 mm

bisexual
lemma apex

1 mm

tooth

wing

glume

glume

sterile
spikelet

fertile
spikelet

fertile
spikelet

sterile
spikelets

2 mm

dispersal
unit

1 mm

spikelet

2 cm

2 cm

2 cm

Koeleria macrantha

Anthoxanthum hirtum

Phalaris paradoxa

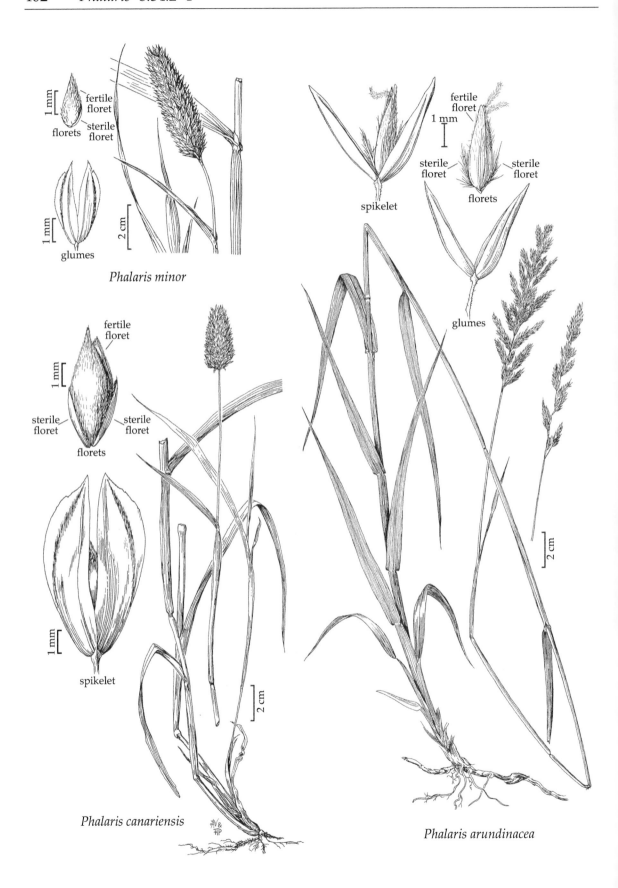

Phalaris minor

Phalaris canariensis

Phalaris arundinacea

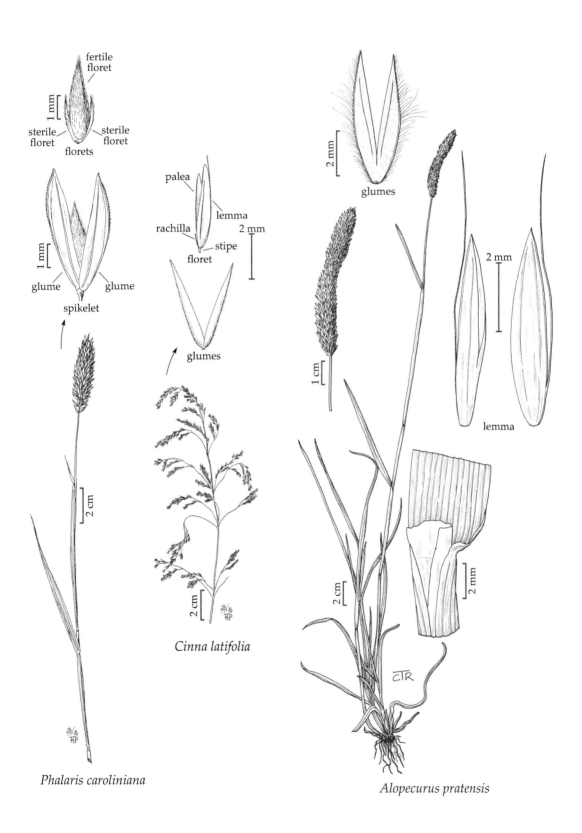

fertile
floret

1 mm

sterile
floret

sterile
floret

florets

palea

lemma

rachilla

2 mm

stipe

floret

1 mm

glume

glume

spikelet

glumes

2 mm

glumes

2 mm

glumes

2 mm

lemma

2 cm

1 cm

2 mm

2 cm

2 cm

2 mm

Cinna latifolia

Phalaris caroliniana

Alopecurus pratensis

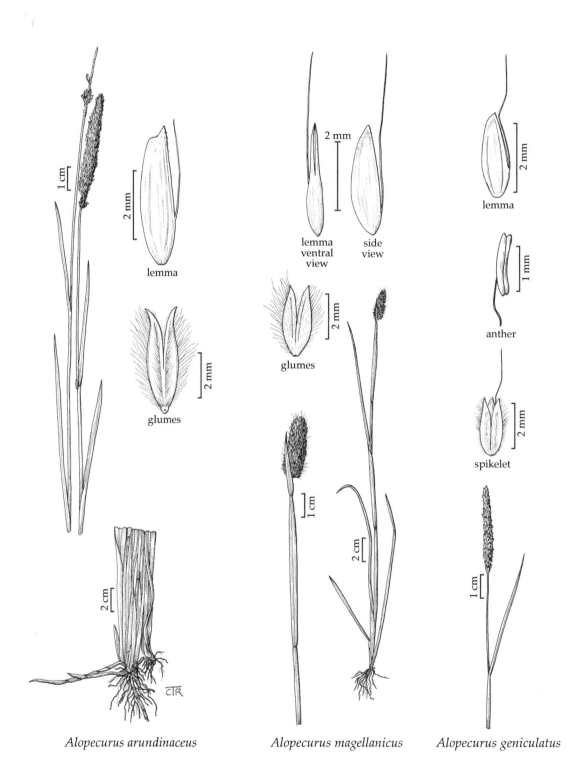

lemma

glumes

lemma
ventral
view

side
view

glumes

lemma

anther

spikelet

Alopecurus arundinaceus

Alopecurus magellanicus

Alopecurus geniculatus

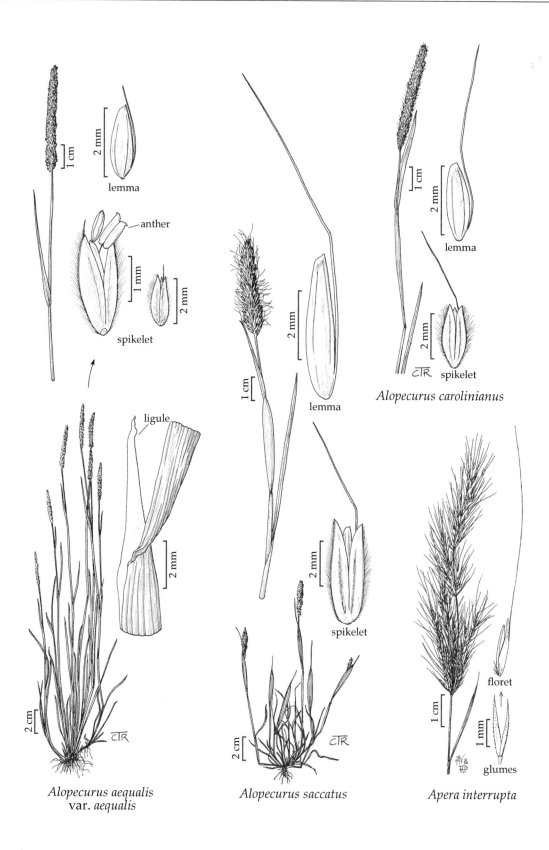

Alopecurus aequalis
var. *aequalis*

Alopecurus saccatus

Alopecurus carolinianus

Apera interrupta

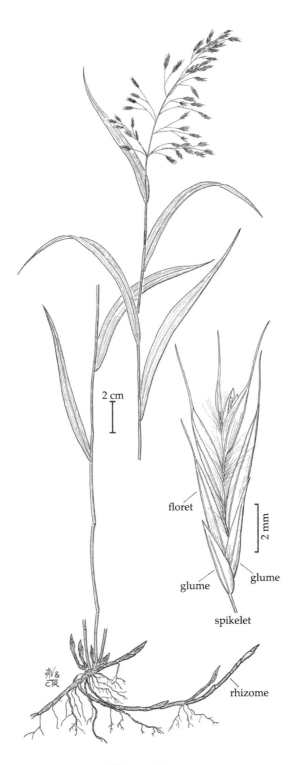

Hakonechloa macra

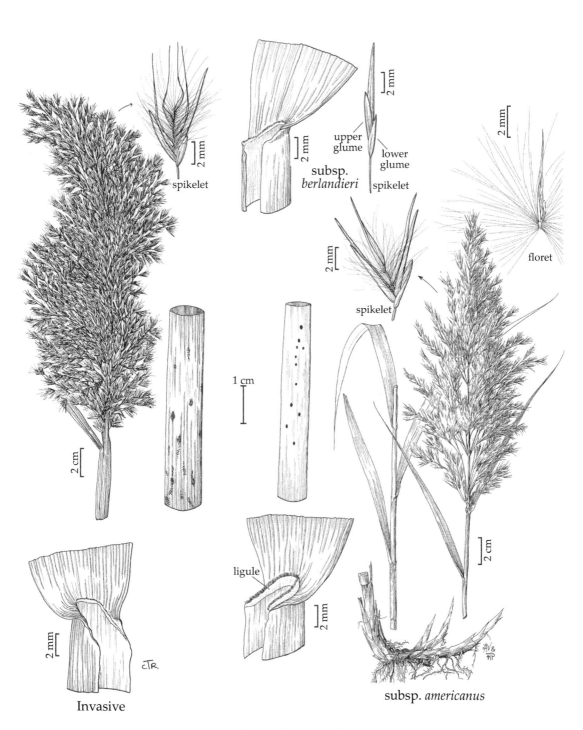

spikelet

subsp. *berlandieri*

upper glume

lower glume

spikelet

floret

spikelet

1 cm

Invasive

ligule

subsp. *americanus*

Phragmites australis

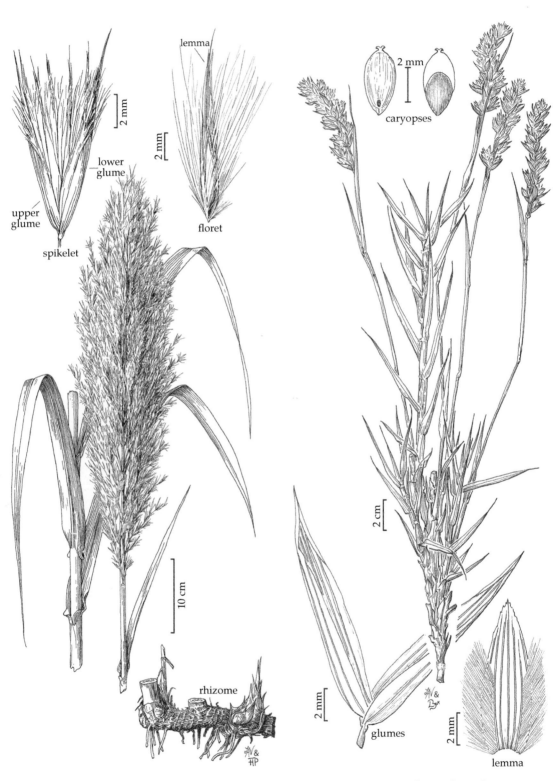

lemma

2 mm

lower
glume

upper
glume

spikelet

floret

2 mm

2 mm

caryopses

2 cm

10 cm

rhizome

2 mm

glumes

2 mm

lemma

Arundo donax

Swallenia alexandrae

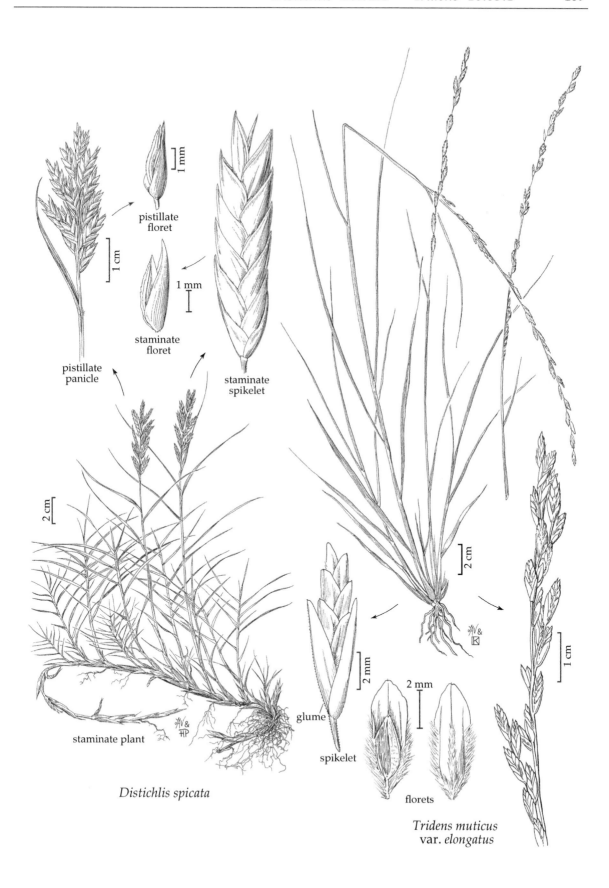

pistillate
floret

1 mm

staminate
floret

1 mm

pistillate
panicle

1 cm

staminate
spikelet

1 cm

2 cm

staminate plant

Distichlis spicata

glume

spikelet

2 mm

florets

2 mm

2 cm

1 cm

Tridens muticus
var. *elongatus*

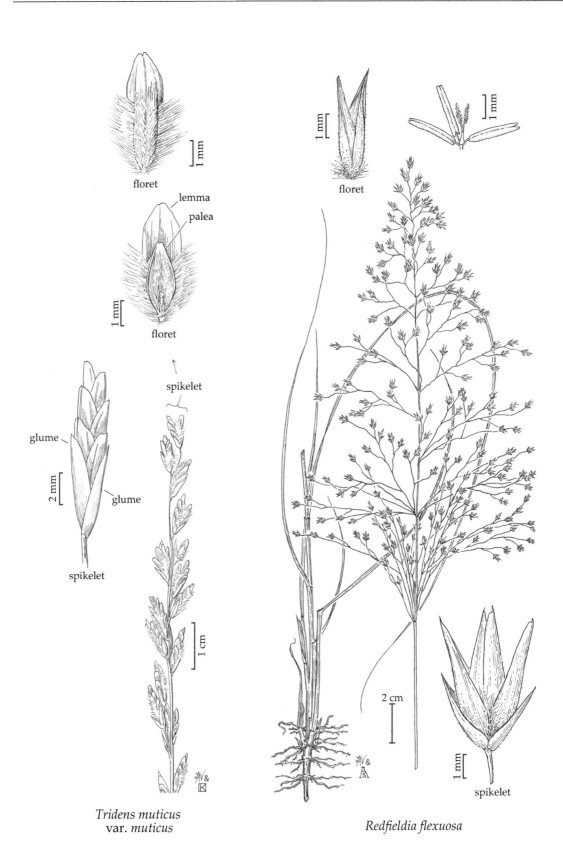

Tridens muticus
var. *muticus*

Redfieldia flexuosa

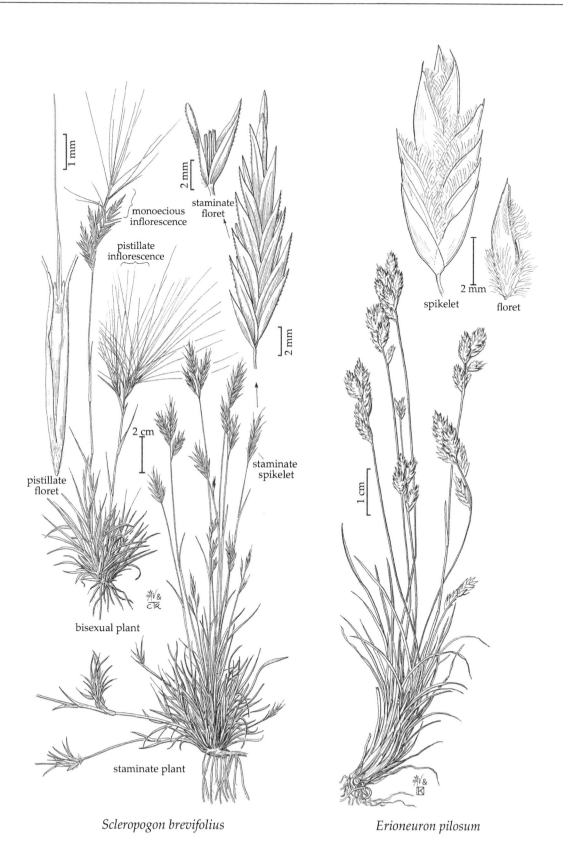

1 mm

2 mm

staminate
floret

monoecious
inflorescence

pistillate
inflorescence

2 mm

2 cm

pistillate
floret

staminate
spikelet

bisexual plant

staminate plant

spikelet

floret

2 mm

1 cm

Scleropogon brevifolius

Erioneuron pilosum

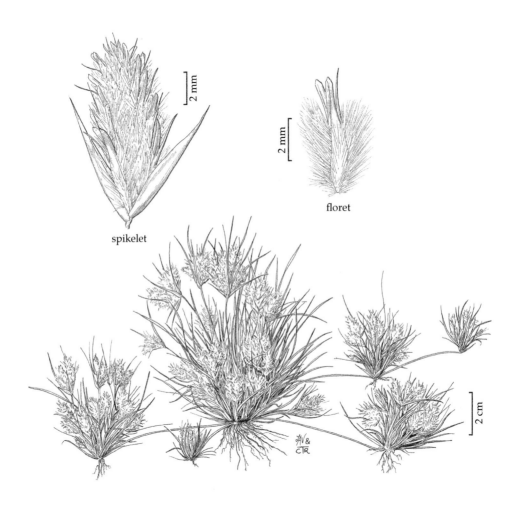

2 mm

spikelet

2 mm

floret

2 cm

Dasyochloa pulchella

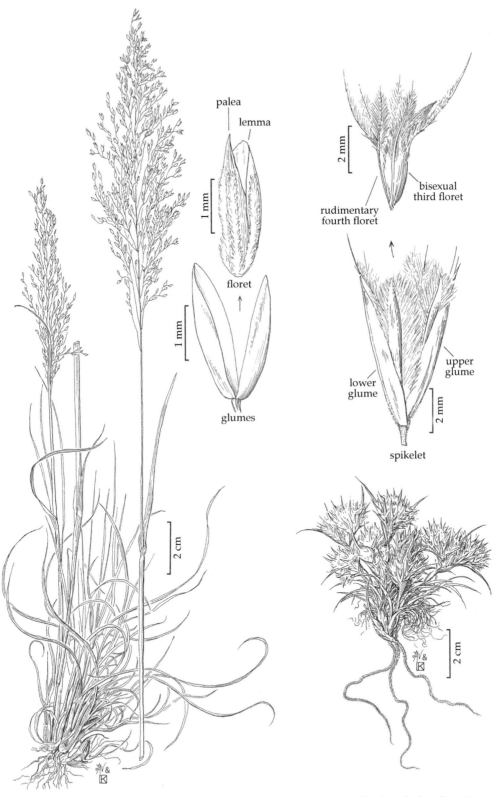

palea

lemma

1 mm

floret

1 mm

glumes

2 mm

bisexual
third floret

rudimentary
fourth floret

lower
glume

upper
glume

2 mm

spikelet

2 cm

2 cm

Blepharoneuron tricholepis

Blepharidachne kingii

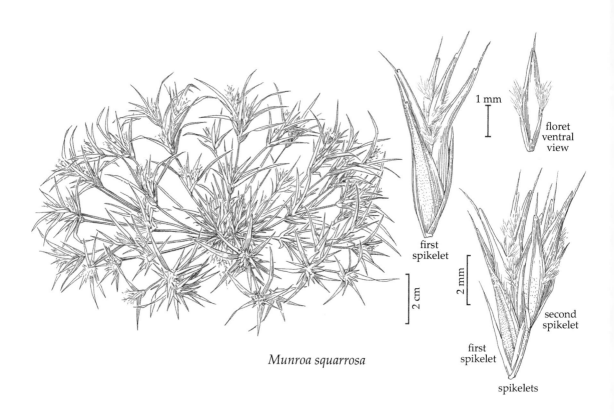

Munroa squarrosa

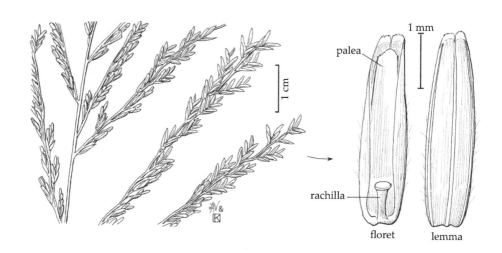

Leptochloa dubia

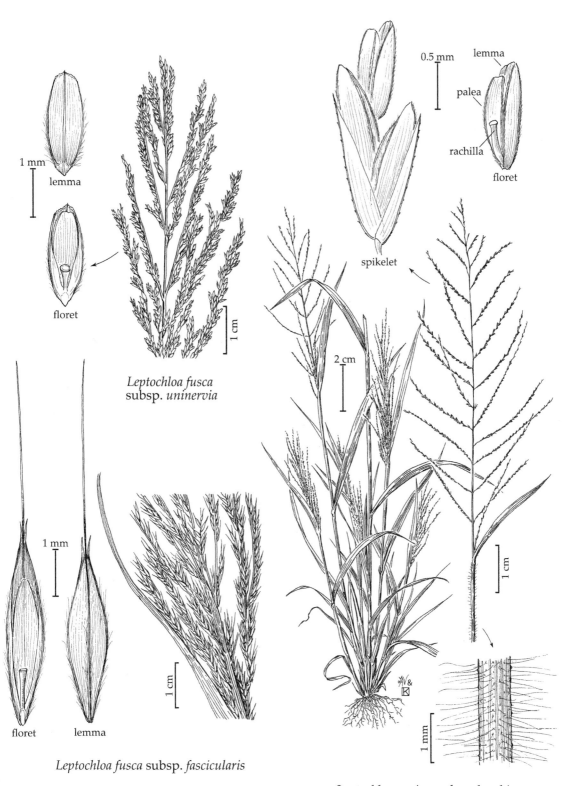

1 mm

lemma

floret

Leptochloa fusca
subsp. *uninervia*

1 cm

0.5 mm

lemma

palea

rachilla

floret

spikelet

2 cm

1 cm

1 mm

floret   lemma

Leptochloa fusca subsp. *fascicularis*

1 cm

1 mm

Leptochloa panicea subsp. *brachiata*

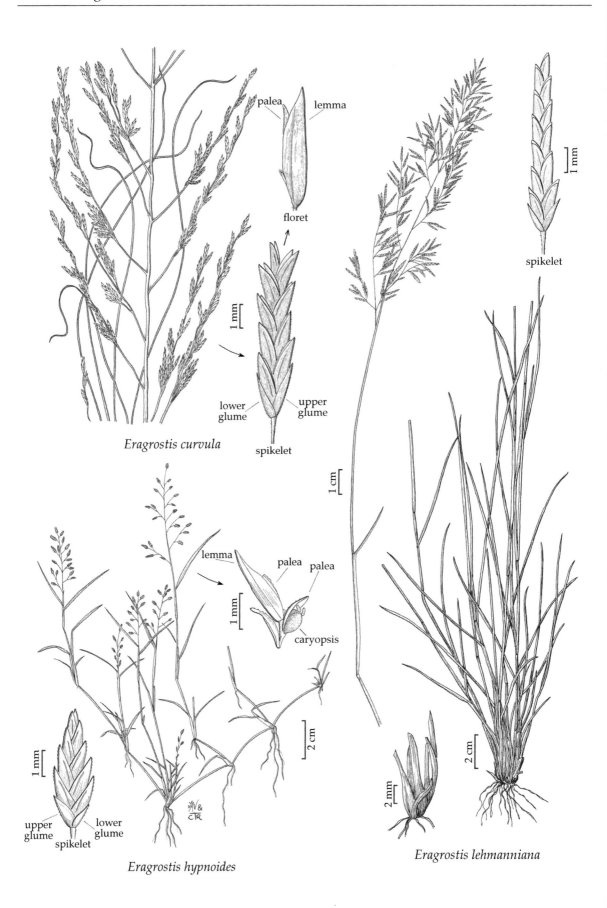

palea lemma

floret

1 mm

lower glume

upper glume

spikelet

Eragrostis curvula

spikelet

1 mm

1 cm

lemma palea palea

1 mm

caryopsis

2 cm

upper glume lower glume

spikelet

Eragrostis hypnoides

2 mm

2 cm

Eragrostis lehmanniana

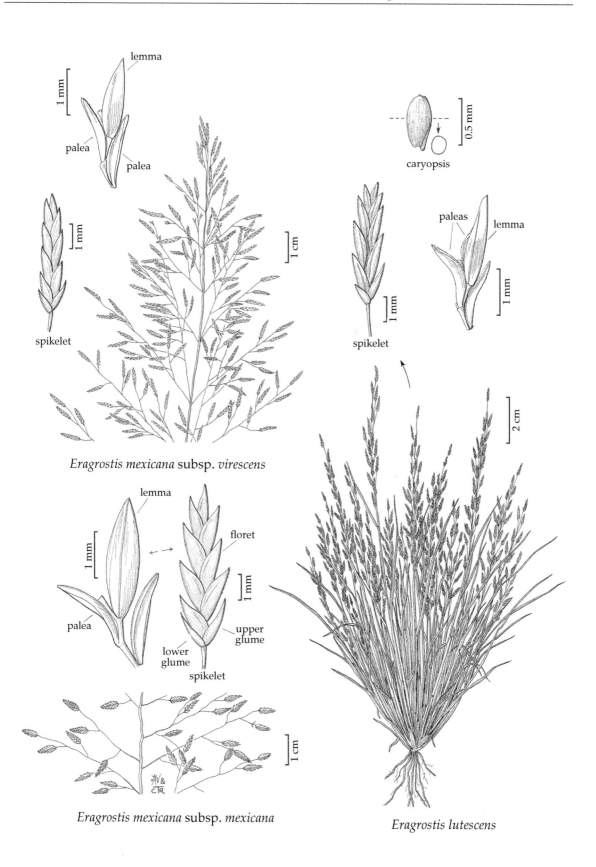

lemma

1 mm

palea

palea

1 mm

spikelet

caryopsis

0.5 mm

paleas

lemma

1 mm

1 mm

spikelet

1 cm

Eragrostis mexicana subsp. *virescens*

2 cm

lemma

floret

1 mm

palea

1 mm

lower
glume

upper
glume

spikelet

1 cm

ᴶᴬᴸ &
CᵀR

Eragrostis mexicana subsp. *mexicana*

Eragrostis lutescens

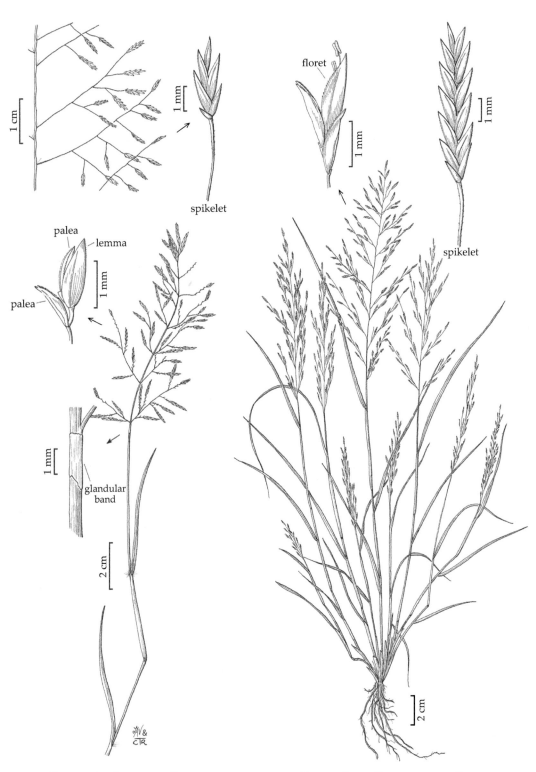

Eragrostis pilosa var. *pilosa*   *Eragrostis pectinacea* var. *pectinacea*

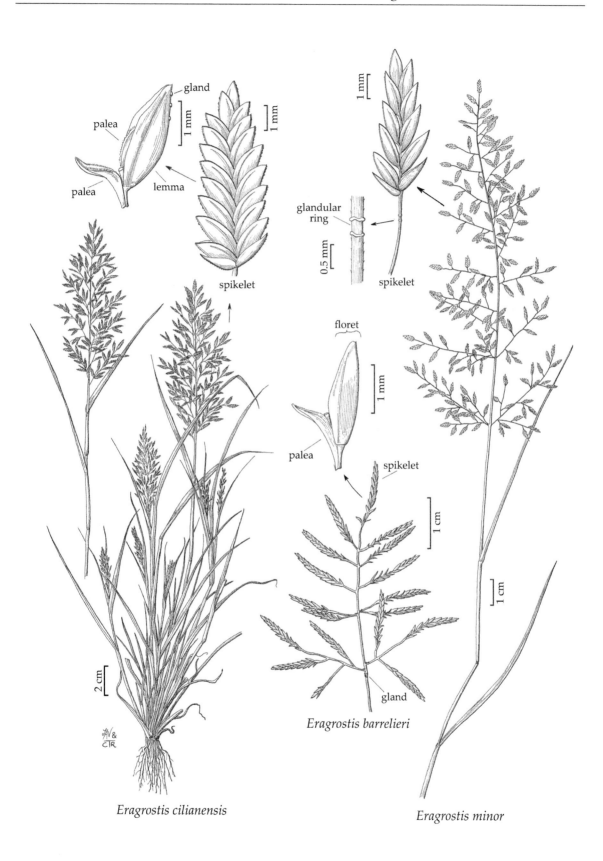

Eragrostis cilianensis

Eragrostis barrelieri

Eragrostis minor

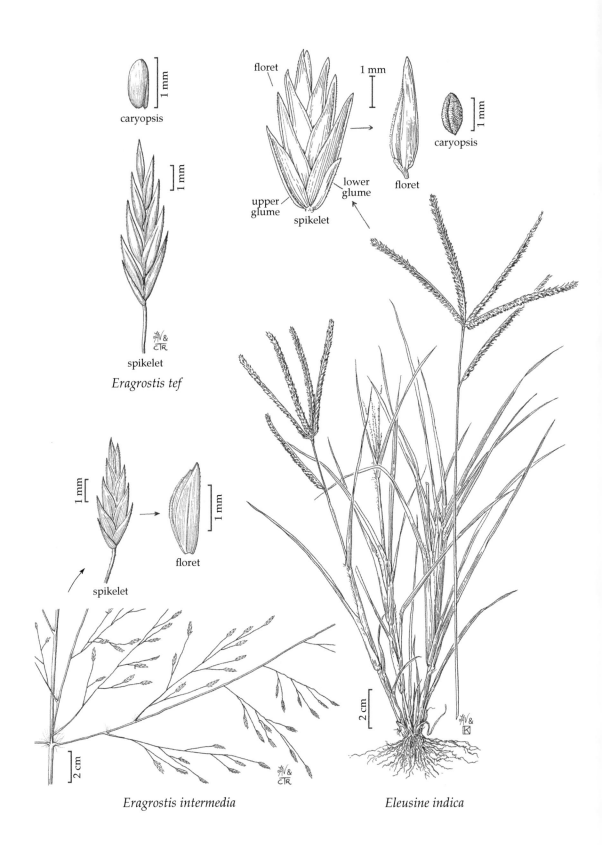

caryopsis

spikelet

Eragrostis tef

floret

upper glume

lower glume

spikelet

caryopsis

floret

spikelet

floret

Eragrostis intermedia

Eleusine indica

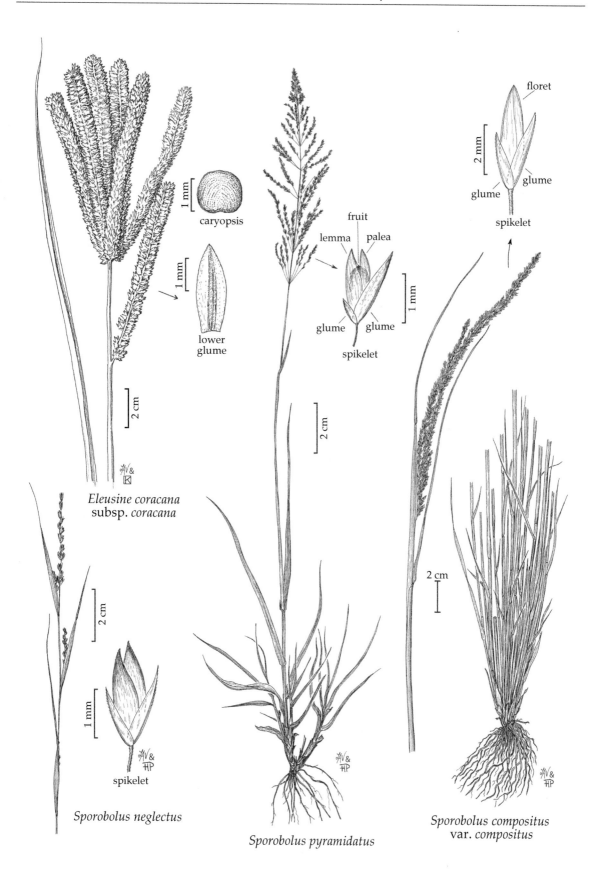

caryopsis

lower
glume

fruit

lemma palea

glume glume

spikelet

floret

glume glume

spikelet

Eleusine coracana
subsp. *coracana*

Sporobolus neglectus

spikelet

Sporobolus pyramidatus

Sporobolus compositus
var. *compositus*

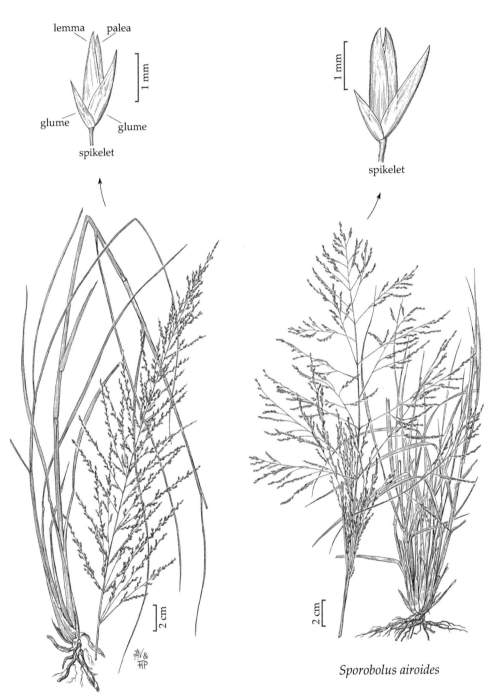

lemma   palea

1 mm

glume   glume

spikelet

1 mm

spikelet

2 cm

2 cm

Sporobolus airoides

Sporobolus wrightii

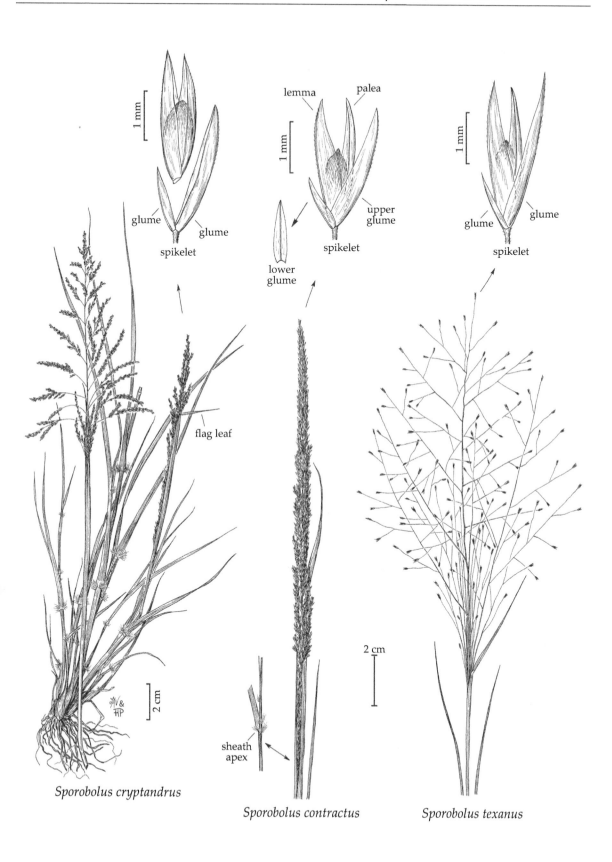

1 mm

lemma   palea

glume

glume

spikelet

1 mm

upper
glume

spikelet

lower
glume

1 mm

glume

glume

spikelet

flag leaf

2 cm

Sporobolus cryptandrus

sheath
apex

2 cm

Sporobolus contractus

Sporobolus texanus

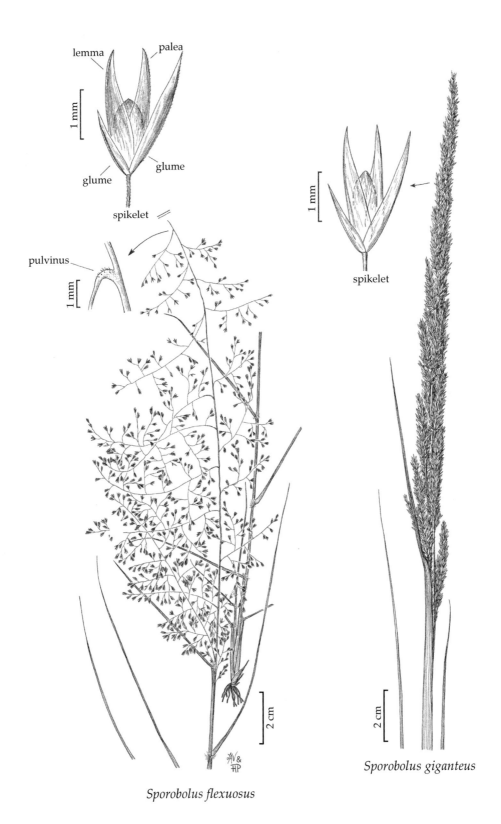

lemma   palea

1 mm

glume   glume

glume

spikelet

pulvinus

1 mm

1 mm

spikelet

2 cm

2 cm

Sporobolus flexuosus

Sporobolus giganteus

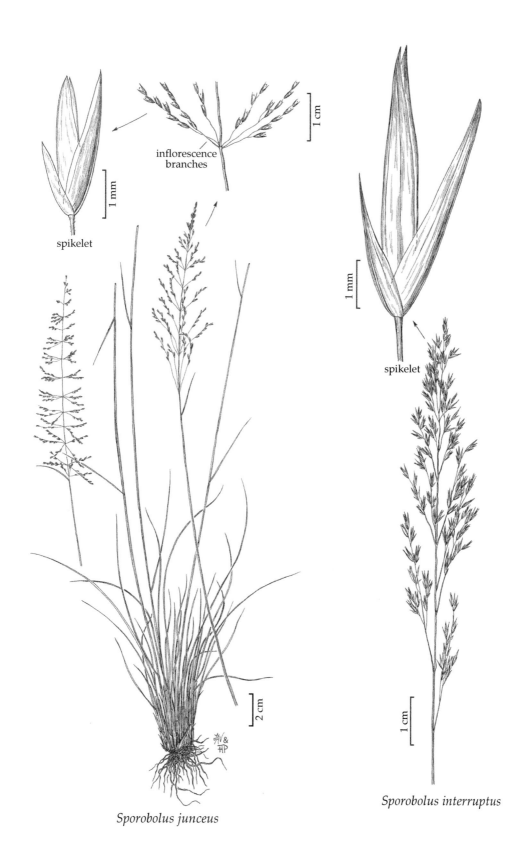

spikelet

inflorescence
branches

spikelet

Sporobolus junceus

Sporobolus interruptus

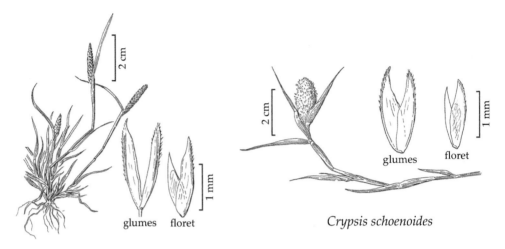

Crypsis alopecuroides

Crypsis schoenoides

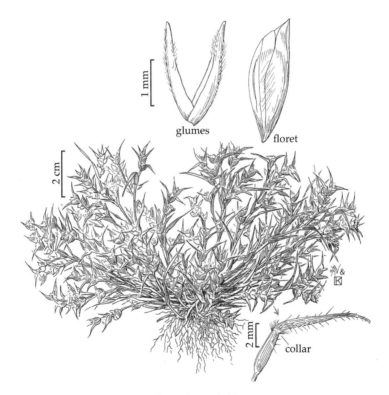

Crypsis vaginiflora

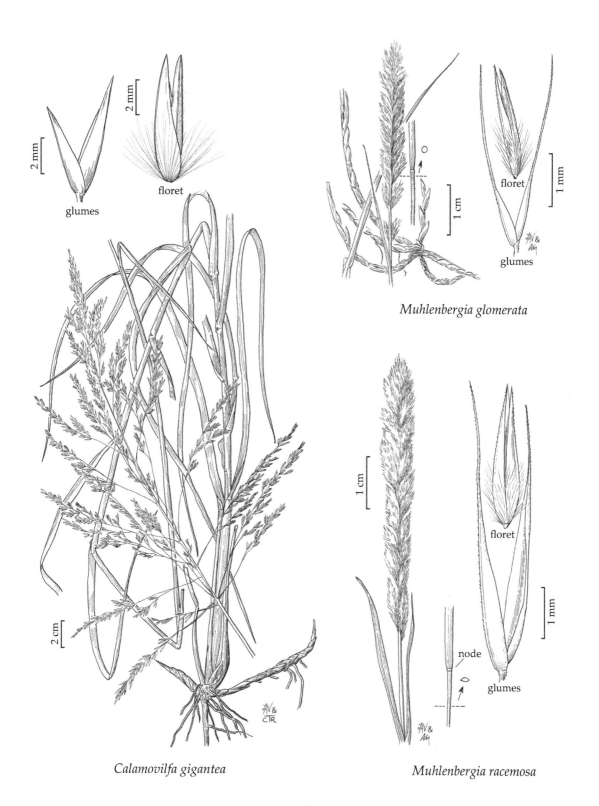

2 mm

glumes

2 mm

floret

Muhlenbergia glomerata

floret

1 mm

glumes

1 cm

2 cm

Calamovilfa gigantea

1 cm

floret

1 mm

node

glumes

Muhlenbergia racemosa

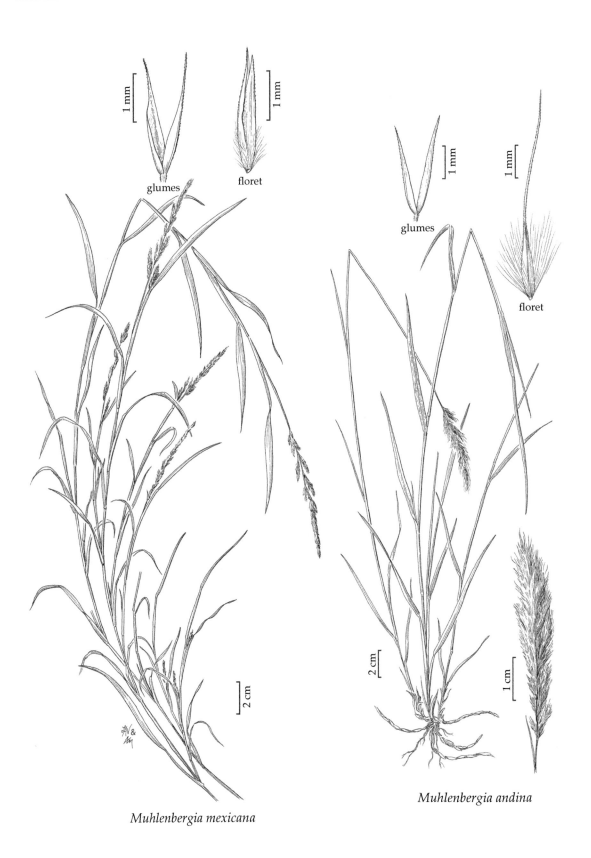

glumes

floret

glumes

floret

1 mm

1 mm

1 mm

1 mm

2 cm

2 cm

1 cm

Muhlenbergia mexicana

Muhlenbergia andina

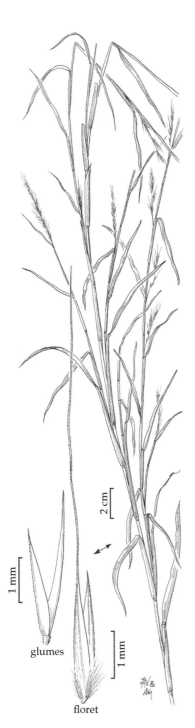

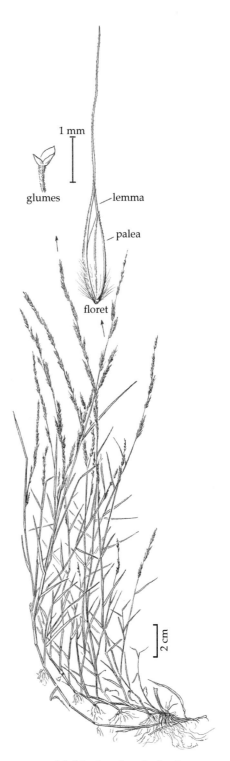

Muhlenbergia sylvatica   *Muhlenbergia schreberi*

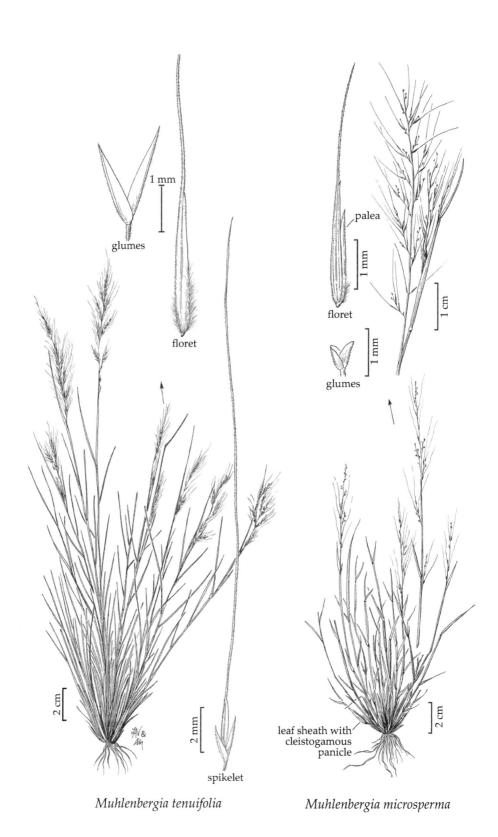

glumes

1 mm

floret

Muhlenbergia tenuifolia

2 cm

2 mm

spikelet

palea

1 mm

floret

1 mm

glumes

1 cm

leaf sheath with
cleistogamous
panicle

2 cm

Muhlenbergia microsperma

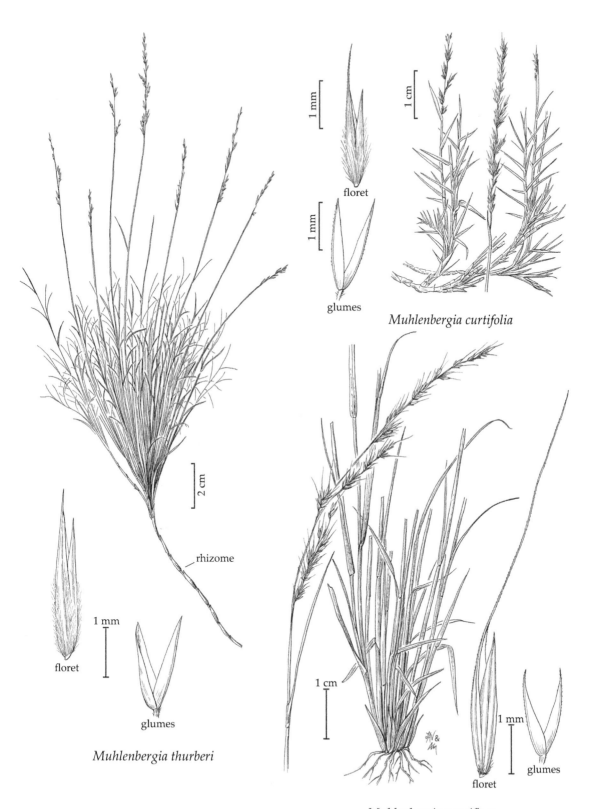

floret

1 mm

glumes

Muhlenbergia curtifolia

2 cm

rhizome

floret

1 mm

glumes

Muhlenbergia thurberi

1 cm

floret

1 mm

glumes

Muhlenbergia pauciflora

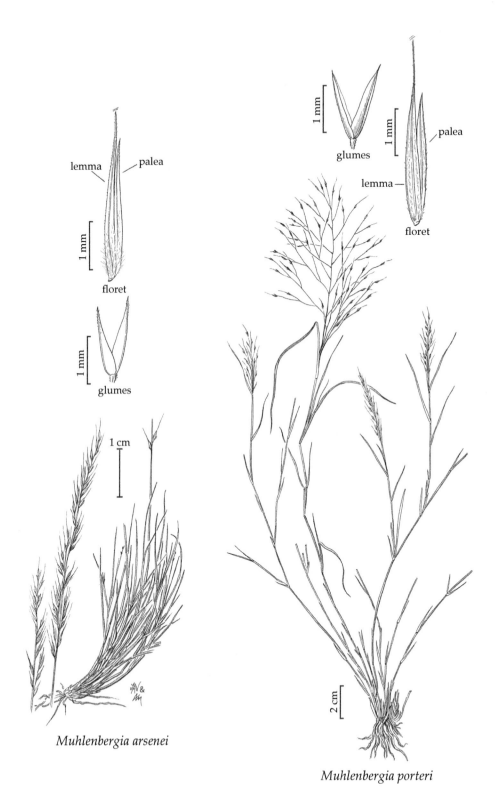

lemma   palea

1 mm

floret

1 mm

glumes

1 cm

Muhlenbergia arsenei

1 mm

glumes

palea

1 mm

lemma

floret

2 cm

Muhlenbergia porteri

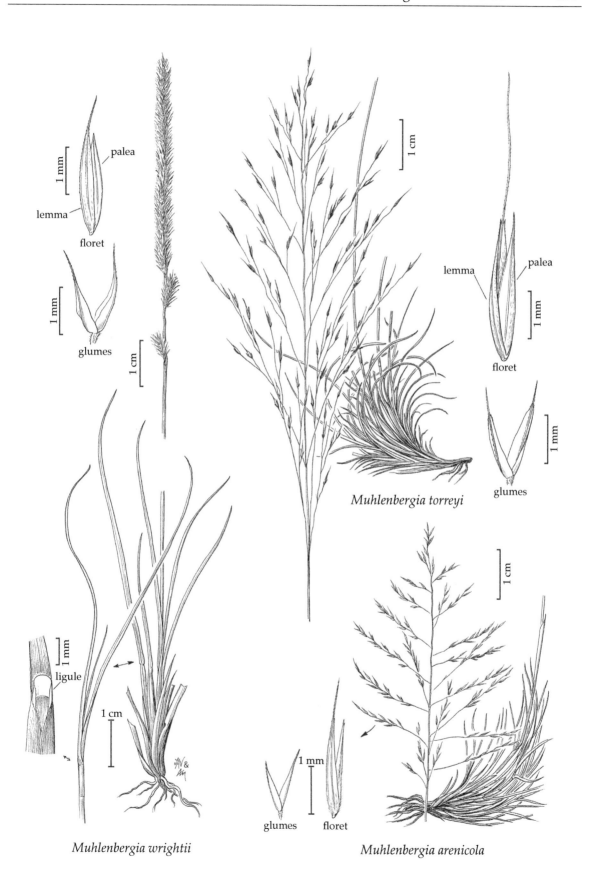

palea
lemma
floret

1 mm

glumes

1 mm

1 cm

Muhlenbergia torreyi

1 cm

lemma
palea

floret

1 mm

glumes

1 mm

ligule

1 mm

1 cm

Muhlenbergia wrightii

glumes   floret

1 mm

1 cm

Muhlenbergia arenicola

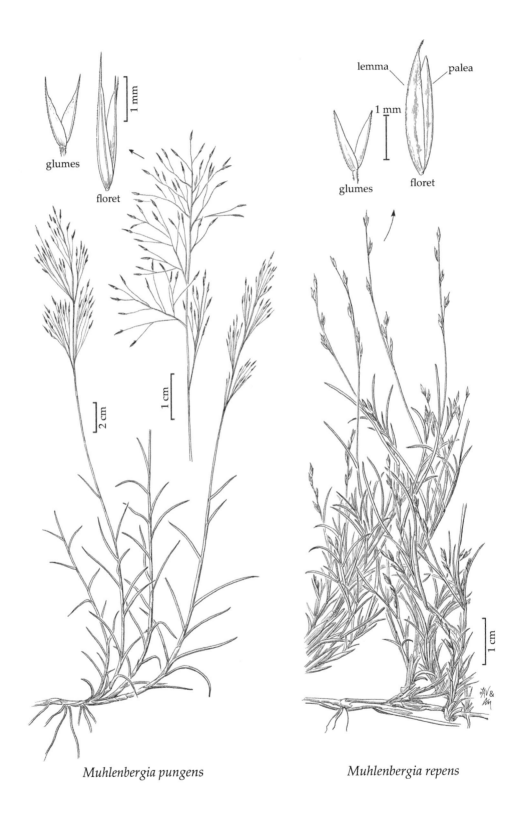

Muhlenbergia pungens   *Muhlenbergia repens*

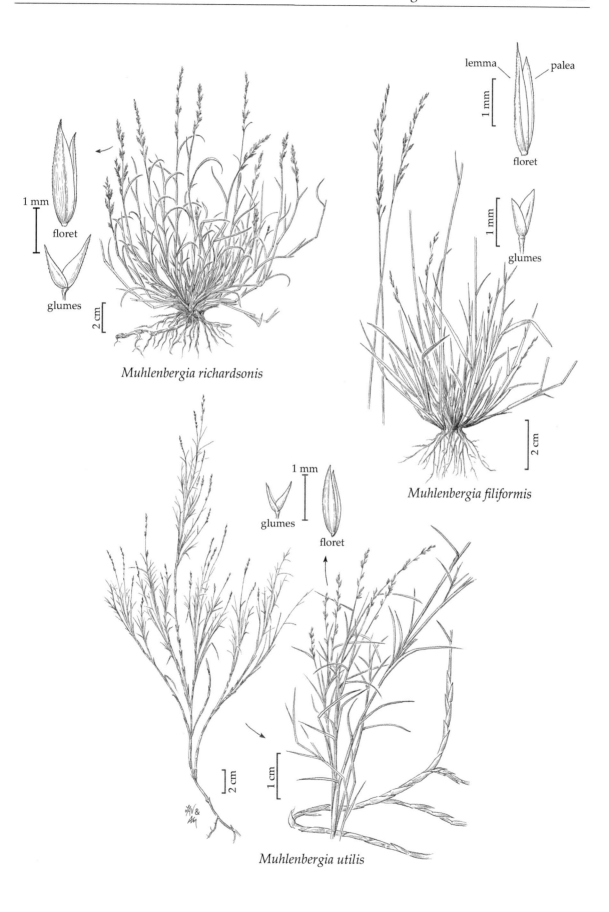

1 mm
floret

glumes

2 cm

Muhlenbergia richardsonis

lemma — palea

1 mm

floret

1 mm

glumes

2 cm

Muhlenbergia filiformis

1 mm

glumes

floret

2 cm

1 cm

Muhlenbergia utilis

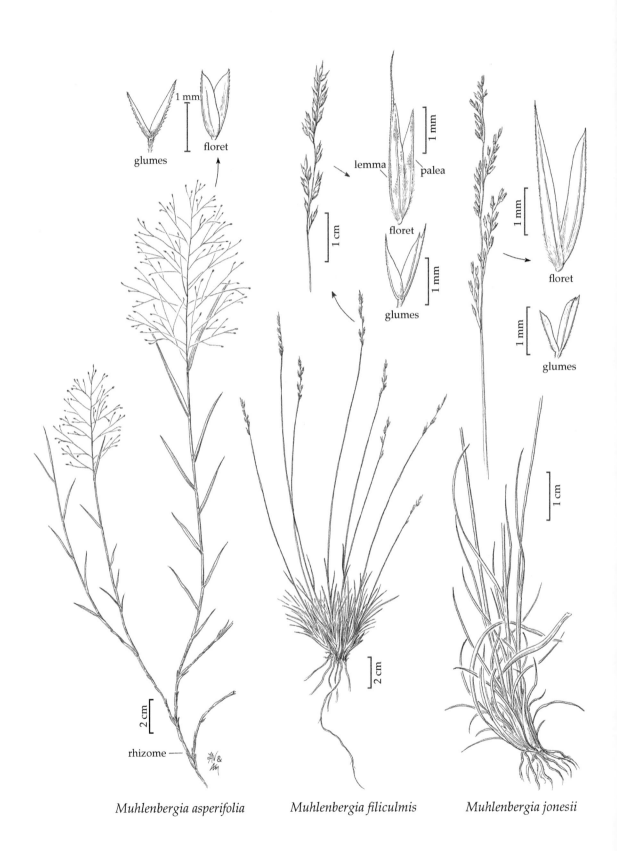

Muhlenbergia asperifolia   *Muhlenbergia filiculmis*   *Muhlenbergia jonesii*

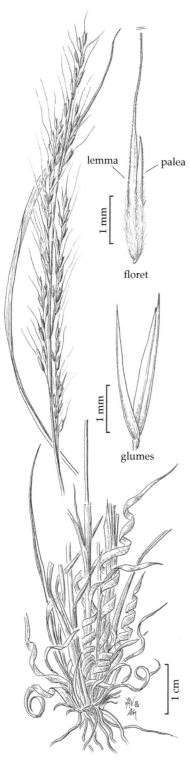

lemma   palea

1 mm

floret

1 mm

glumes

1 cm

Muhlenbergia straminea

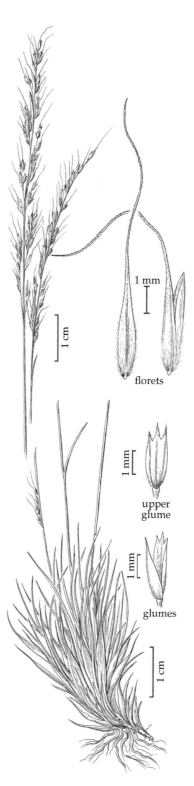

1 mm

florets

1 cm

1 mm

upper glume

1 mm

glumes

1 cm

Muhlenbergia montana

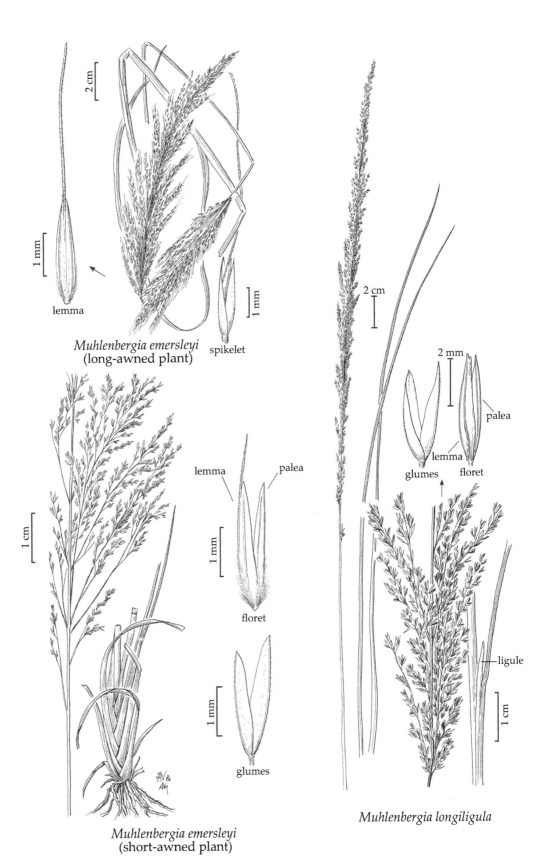

Muhlenbergia emersleyi
(long-awned plant)

lemma

spikelet

Muhlenbergia emersleyi
(short-awned plant)

lemma

palea

floret

glumes

Muhlenbergia longiligula

glumes

lemma

palea

floret

ligule

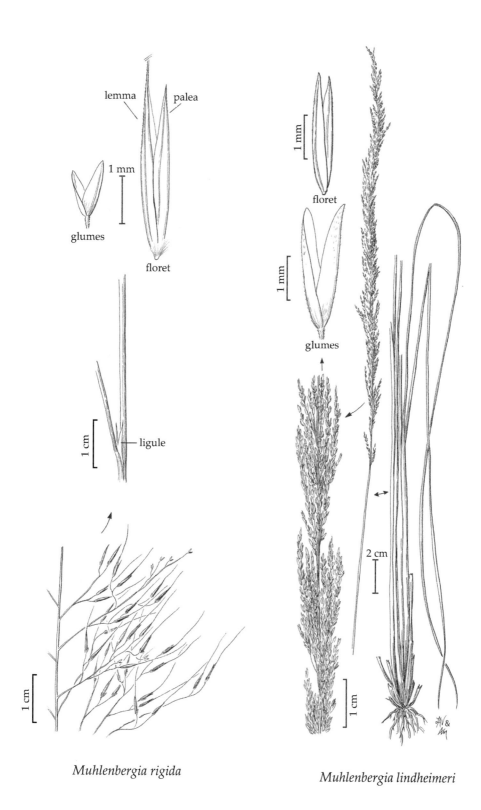

Muhlenbergia rigida

Muhlenbergia lindheimeri

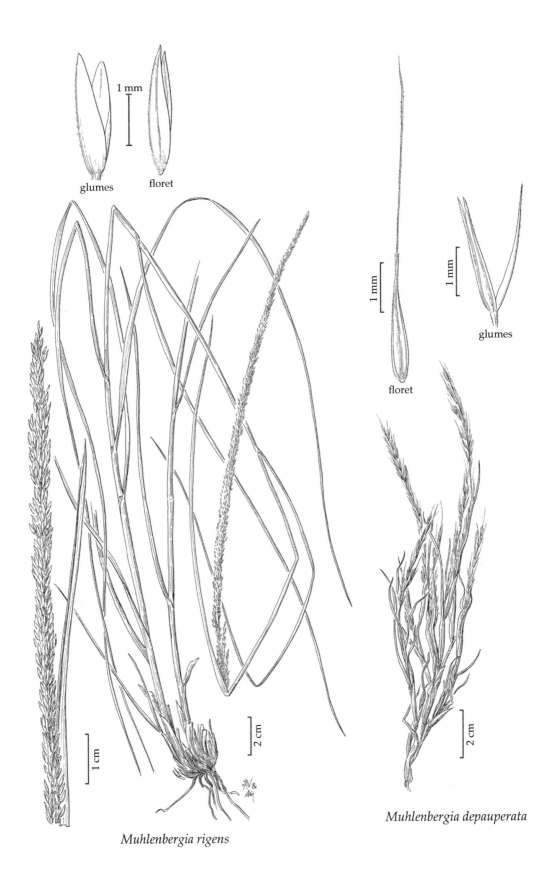

glumes

floret

1 mm

1 mm

floret

1 mm

glumes

2 cm

1 cm

2 cm

Muhlenbergia rigens

Muhlenbergia depauperata

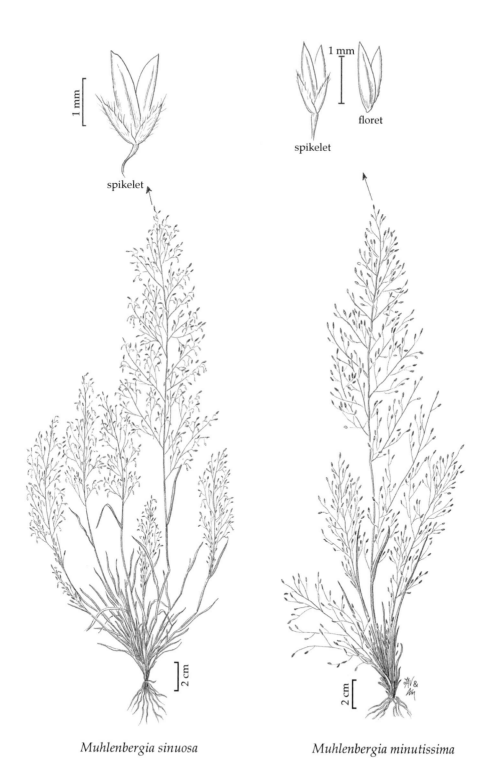

spikelet

1 mm

spikelet

floret

Muhlenbergia sinuosa

Muhlenbergia minutissima

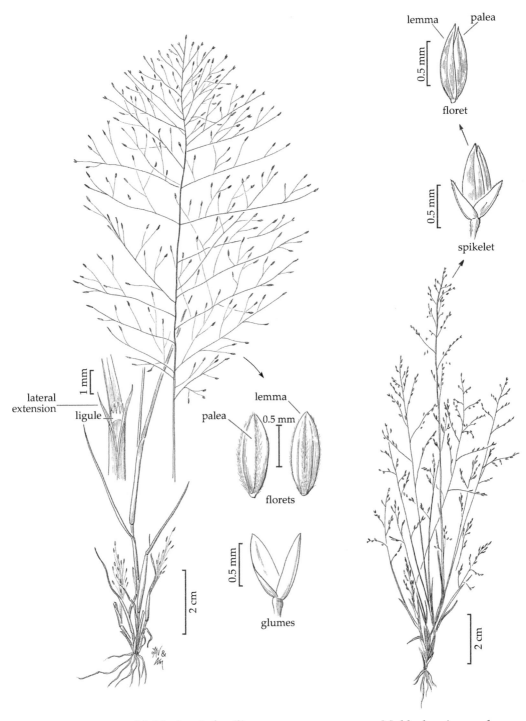

lemma   palea

0.5 mm

floret

0.5 mm

spikelet

lateral
extension

ligule

1 mm

lemma

palea

0.5 mm

florets

0.5 mm

glumes

2 cm

2 cm

Muhlenbergia fragilis   *Muhlenbergia ramulosa*

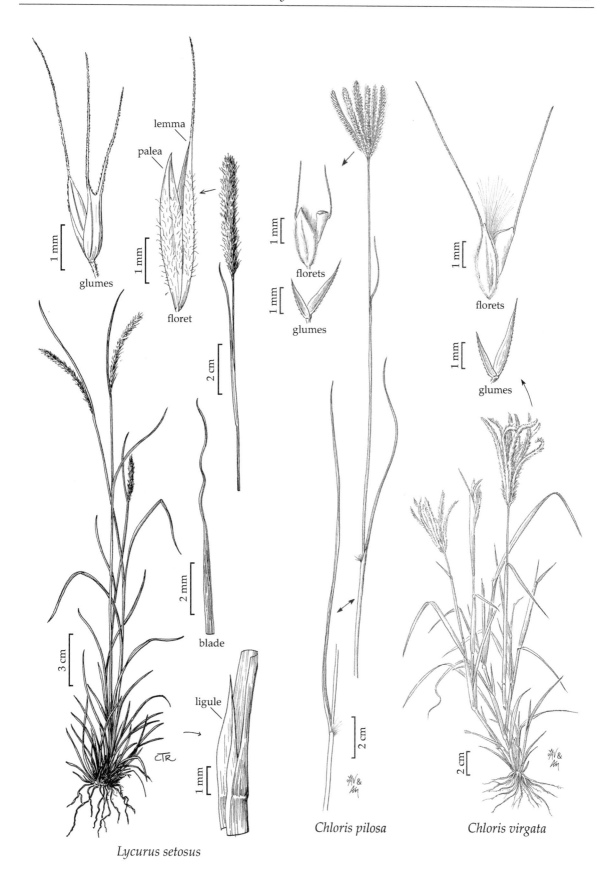

Lycurus setosus

Chloris pilosa

Chloris virgata

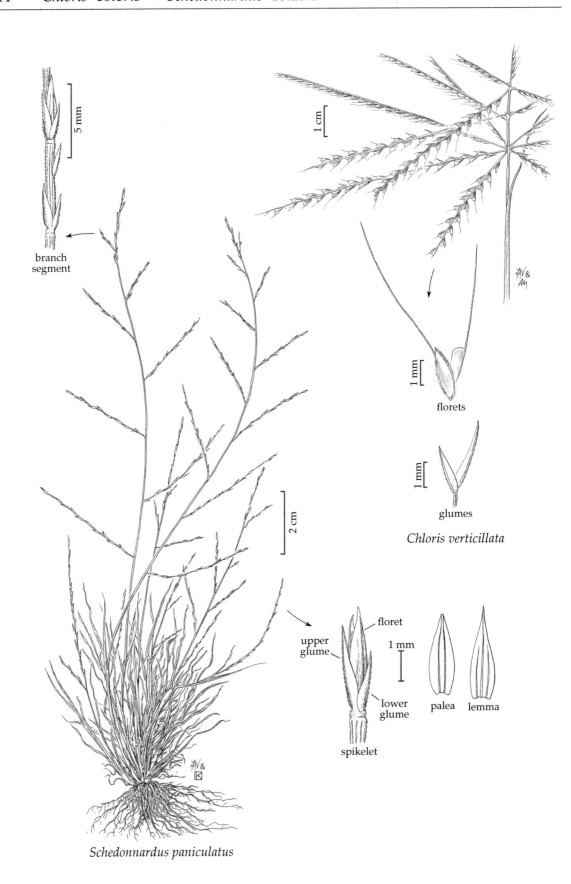

branch
segment

5 mm

1 cm

1 mm

florets

1 mm

glumes

Chloris verticillata

2 cm

floret

upper
glume

1 mm

lower
glume

palea   lemma

spikelet

Schedonnardus paniculatus

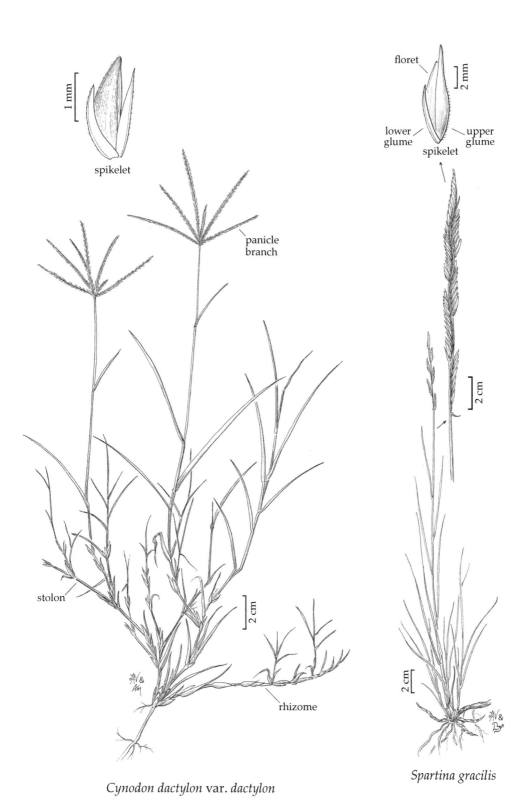

spikelet

floret

lower glume

upper glume

spikelet

panicle branch

stolon

rhizome

Cynodon dactylon var. *dactylon*

Spartina gracilis

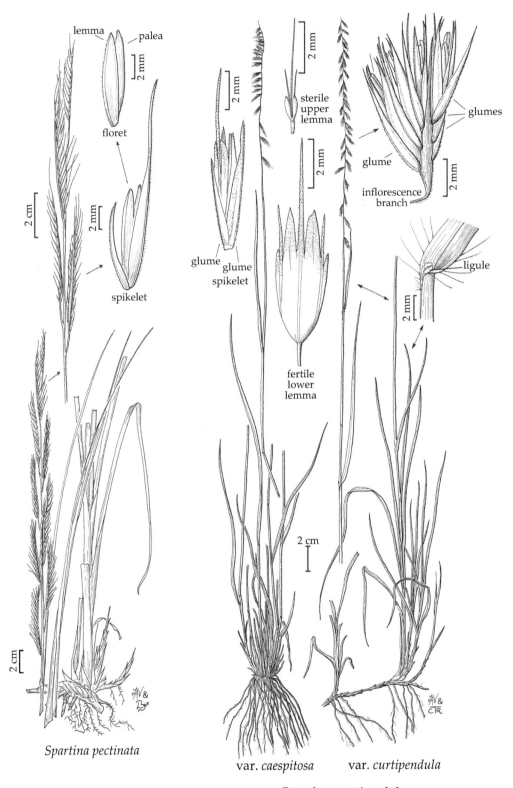

lemma   palea

floret

2 mm

2 cm   2 mm

spikelet

Spartina pectinata

2 mm

glume
glume
spikelet

2 mm

sterile
upper
lemma

2 mm

fertile
lower
lemma

glumes

glume

inflorescence
branch

2 mm

ligule

2 mm

2 cm

var. *caespitosa*   var. *curtipendula*

Bouteloua curtipendula

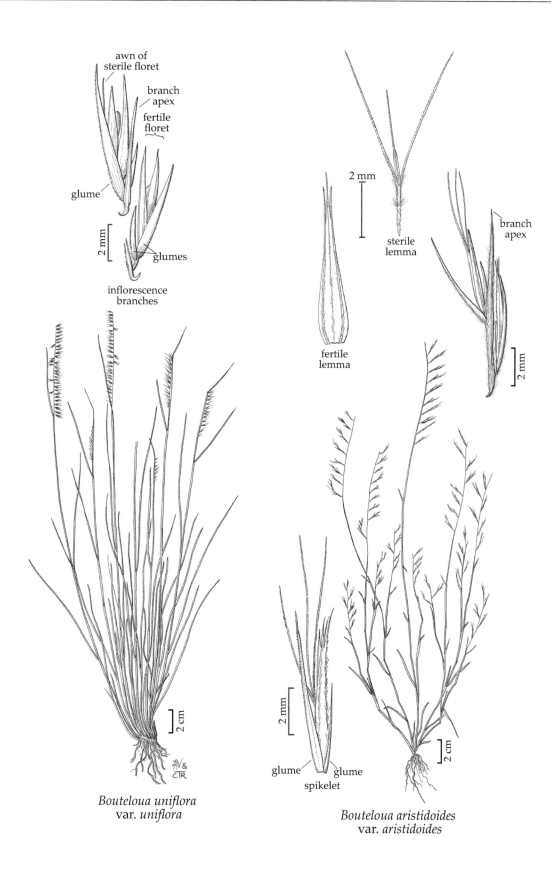

awn of
sterile floret

branch
apex

fertile
floret

glume

2 mm

glumes

inflorescence
branches

2 mm

sterile
lemma

branch
apex

fertile
lemma

2 mm

Bouteloua uniflora
var. *uniflora*

2 cm

2 mm

2 cm

glume   glume
spikelet

Bouteloua aristidoides
var. *aristidoides*

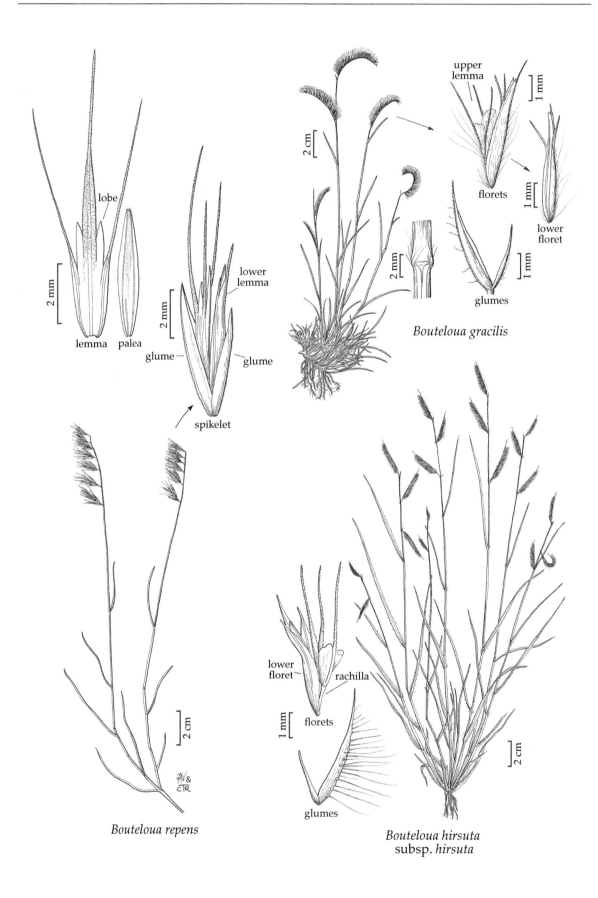

lobe

lemma   palea

2 mm

lower
lemma

glume   glume

spikelet

2 mm

Bouteloua repens

upper
lemma

florets

lower
floret

1 mm

1 mm

glumes

2 cm

2 mm

1 mm

Bouteloua gracilis

2 cm

lower
floret   rachilla

florets

1 mm

glumes

2 cm

Bouteloua hirsuta
subsp. *hirsuta*

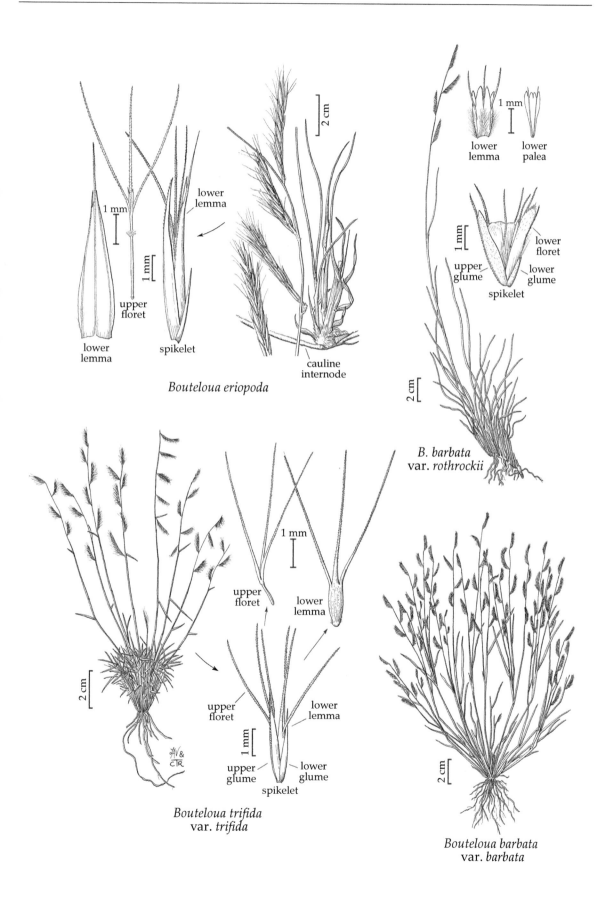

lower
lemma

upper
floret

lower
lemma

spikelet

cauline
internode

Bouteloua eriopoda

lower
lemma

lower
palea

upper
glume

lower
floret

lower
glume

spikelet

B. barbata
var. *rothrockii*

upper
floret

lower
lemma

upper
floret

upper
glume

lower
lemma

lower
glume

spikelet

Bouteloua trifida
var. *trifida*

Bouteloua barbata
var. *barbata*

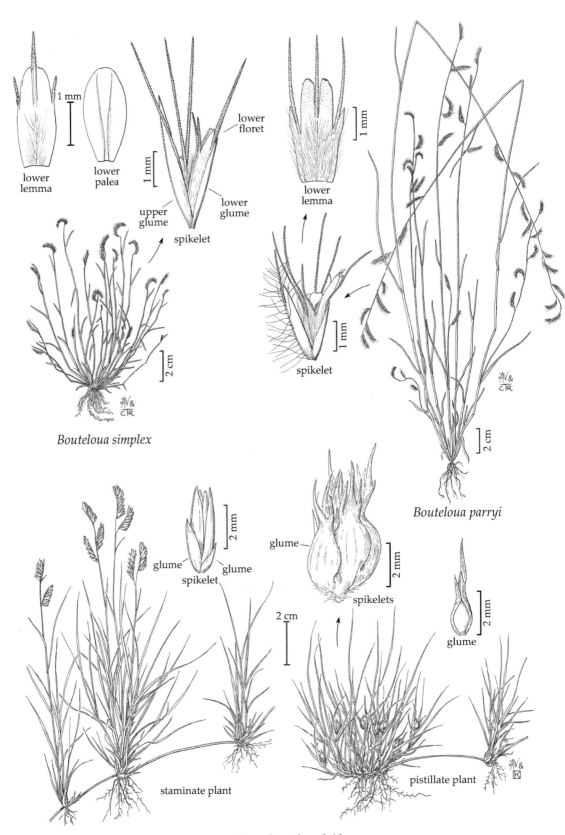

lower lemma

lower palea

lower floret

lower glume

upper glume

lower glume

spikelet

Bouteloua simplex

lower lemma

spikelet

Bouteloua parryi

glume

glume

spikelet

glume

spikelets

glume

staminate plant

pistillate plant

Bouteloua dactyloides

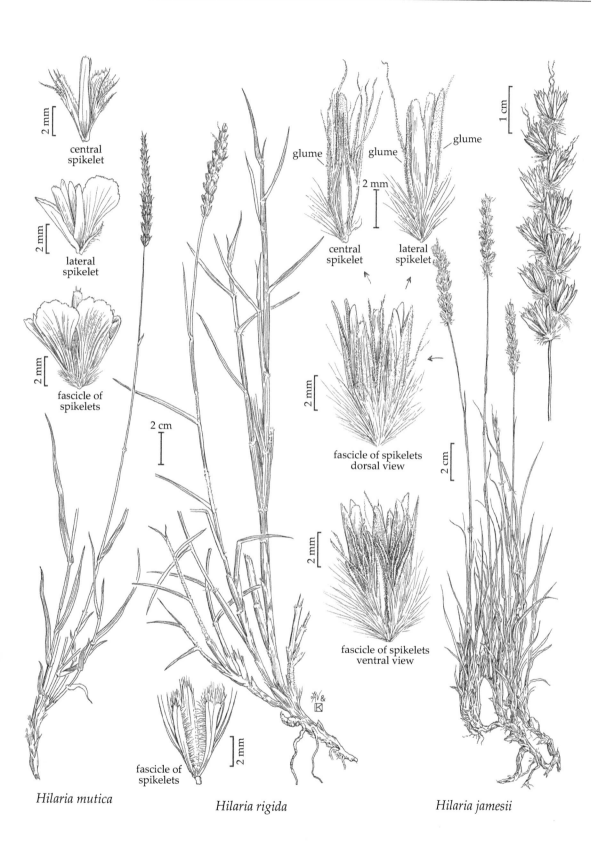

central
spikelet

2 mm

lateral
spikelet

fascicle of
spikelets

2 cm

glume

glume

glume

central
spikelet

lateral
spikelet

2 mm

fascicle of spikelets
dorsal view

2 mm

fascicle of spikelets
ventral view

2 mm

fascicle of
spikelets

2 mm

2 cm

1 cm

Hilaria mutica

Hilaria rigida

Hilaria jamesii

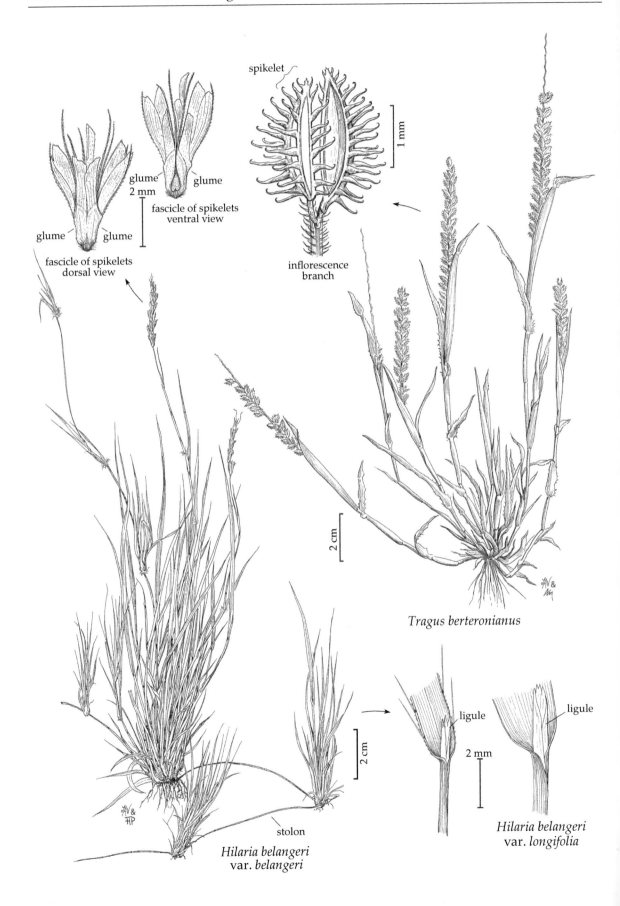

spikelet

glume

glume

2 mm

fascicle of spikelets
ventral view

glume

glume

fascicle of spikelets
dorsal view

1 mm

inflorescence
branch

Tragus berteronianus

2 cm

2 cm

ligule

ligule

2 mm

stolon

Hilaria belangeri
var. *belangeri*

Hilaria belangeri
var. *longifolia*

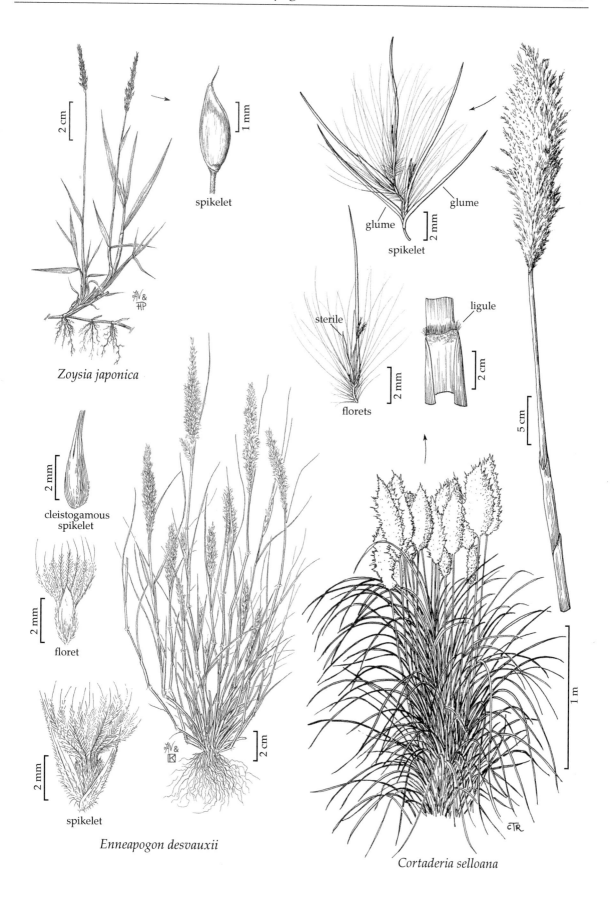

spikelet

Zoysia japonica

glume

glume

spikelet

sterile

florets

ligule

cleistogamous
spikelet

floret

spikelet

Enneapogon desvauxii

Cortaderia selloana

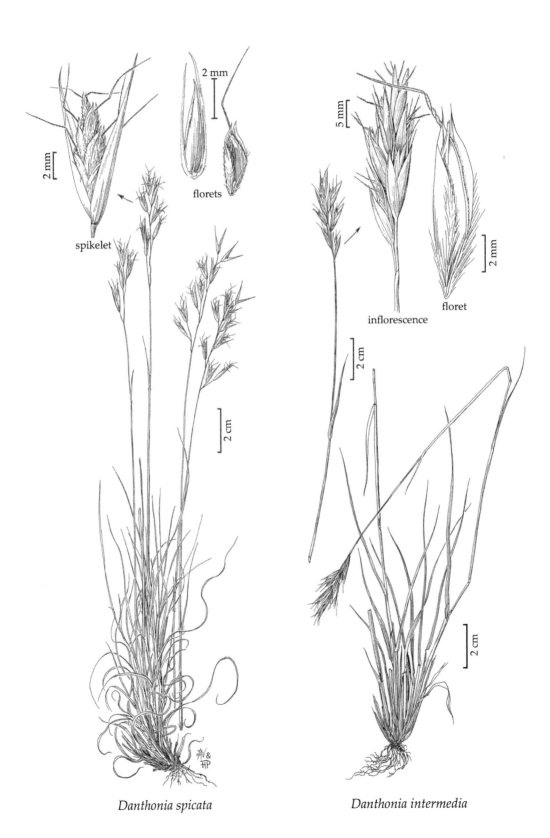

2 mm

florets

spikelet

2 mm

5 mm

floret

2 mm

inflorescence

2 cm

2 cm

2 cm

Danthonia spicata

Danthonia intermedia

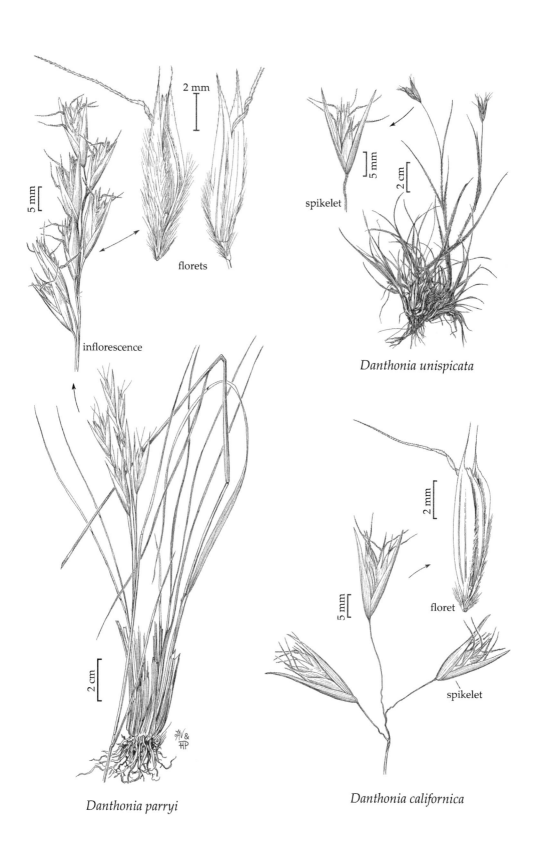

2 mm

florets

inflorescence

5 mm

spikelet

5 mm

2 cm

Danthonia unispicata

Danthonia parryi

2 cm

2 mm

floret

5 mm

spikelet

Danthonia californica

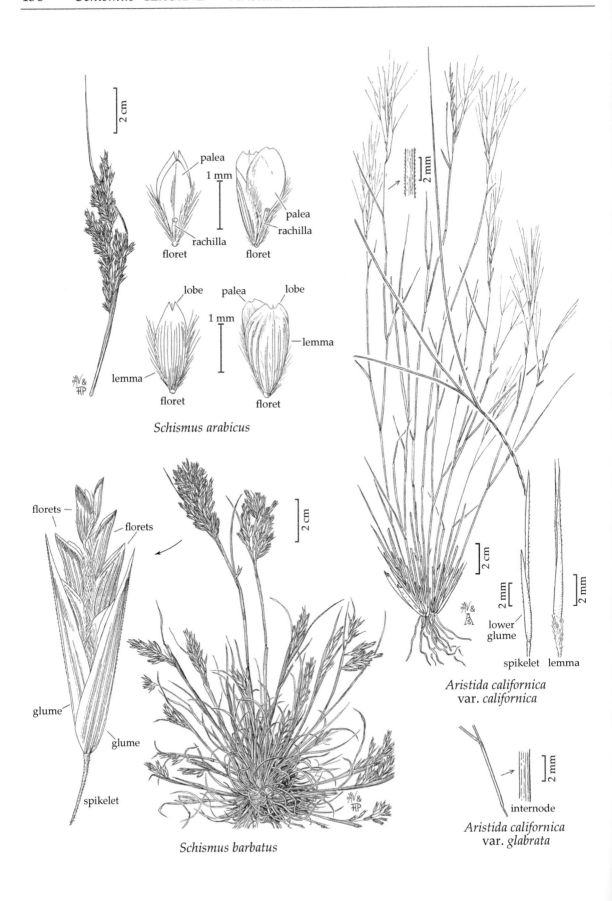

palea

1 mm

palea
rachilla

floret

rachilla

floret

lobe

palea   lobe

1 mm

lemma

lemma

floret

lemma

floret

Schismus arabicus

2 mm

2 cm

florets

florets

glume

glume

spikelet

2 cm

Schismus barbatus

2 mm

2 cm

2 mm

2 mm

lower
glume

spikelet   lemma

Aristida californica
var. *californica*

2 mm

internode

Aristida californica
var. *glabrata*

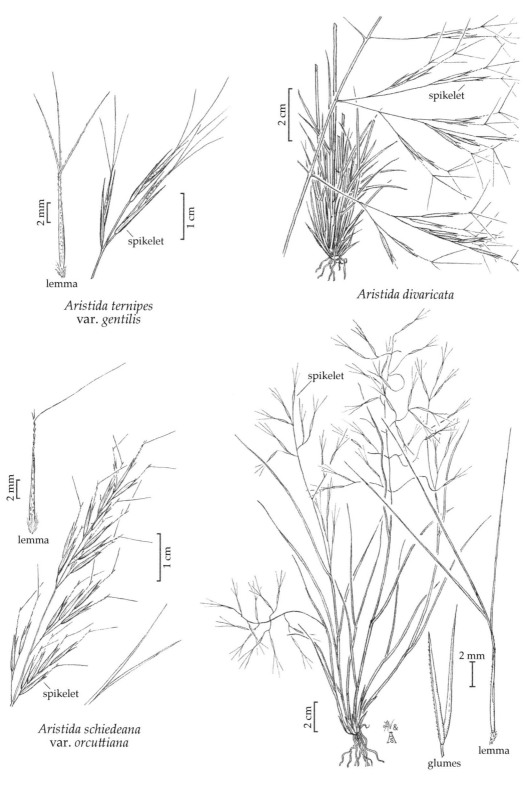

Aristida ternipes
var. gentilis

Aristida divaricata

Aristida schiedeana
var. orcuttiana

Aristida havardii

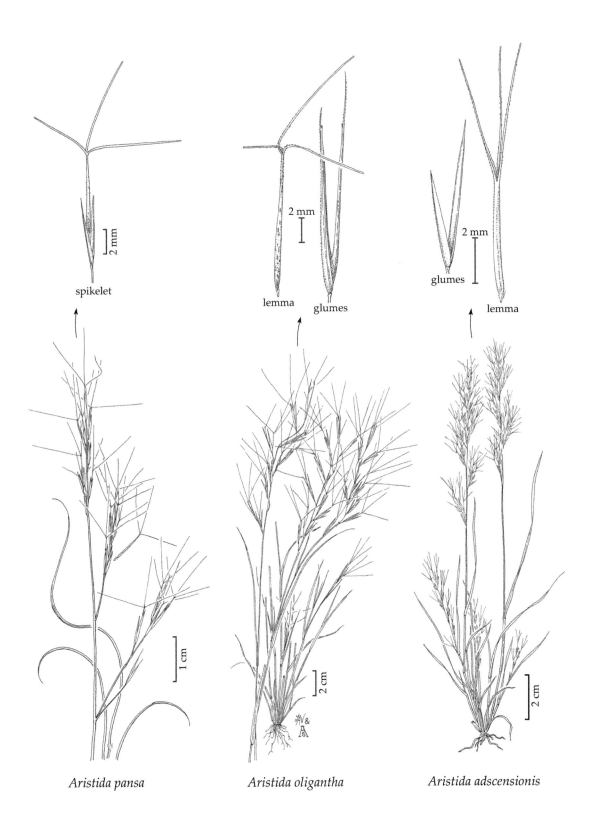

spikelet

2 mm

lemma   glumes

2 mm

glumes   lemma

2 mm

1 cm

2 cm

2 cm

Aristida pansa   *Aristida oligantha*   *Aristida adscensionis*

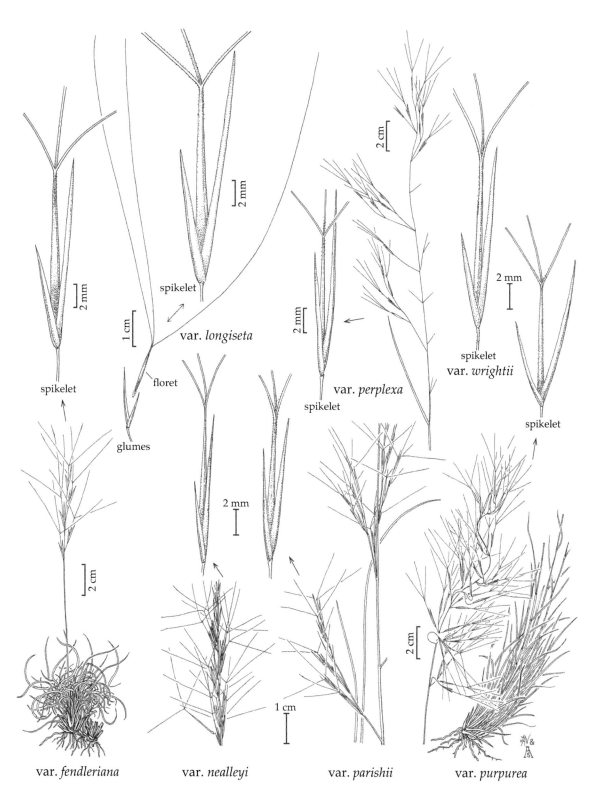

var. *fendleriana* var. *nealleyi* var. *parishii* var. *purpurea*

Aristida purpurea

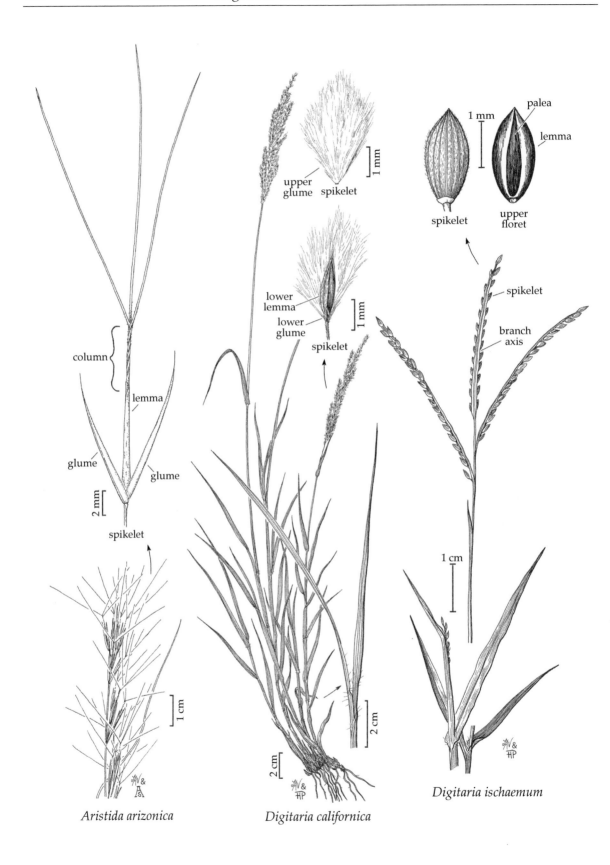

upper
glume   spikelet

1 mm

palea

1 mm   lemma

spikelet   upper
floret

column {

lemma

lower
lemma

lower
glume

spikelet

1 mm

spikelet

branch
axis

glume

glume

2 mm

spikelet

1 cm

2 cm

1 cm

2 cm

2 cm

Aristida arizonica

Digitaria californica

Digitaria ischaemum

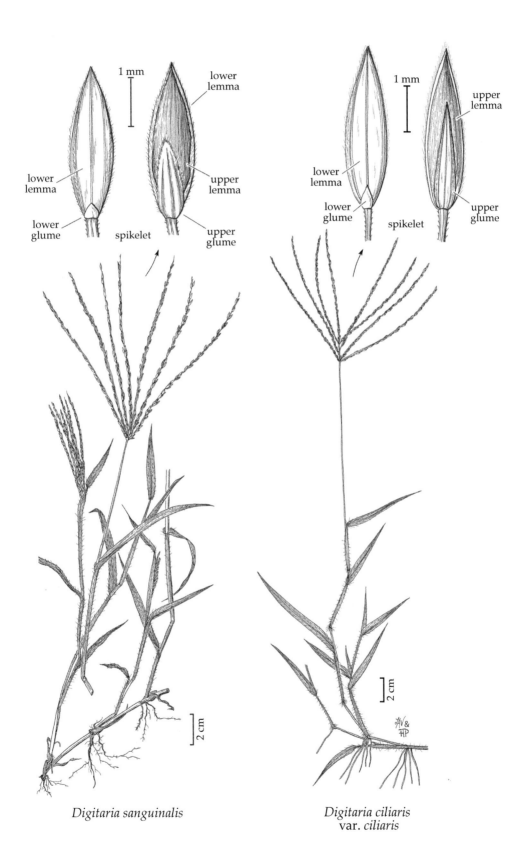

1 mm

lower
lemma

lower
lemma

lower
glume

spikelet

lower
lemma

upper
lemma

upper
glume

1 mm

lower
lemma

lower
glume

spikelet

upper
lemma

upper
glume

Digitaria sanguinalis

2 cm

Digitaria ciliaris
var. *ciliaris*

2 cm

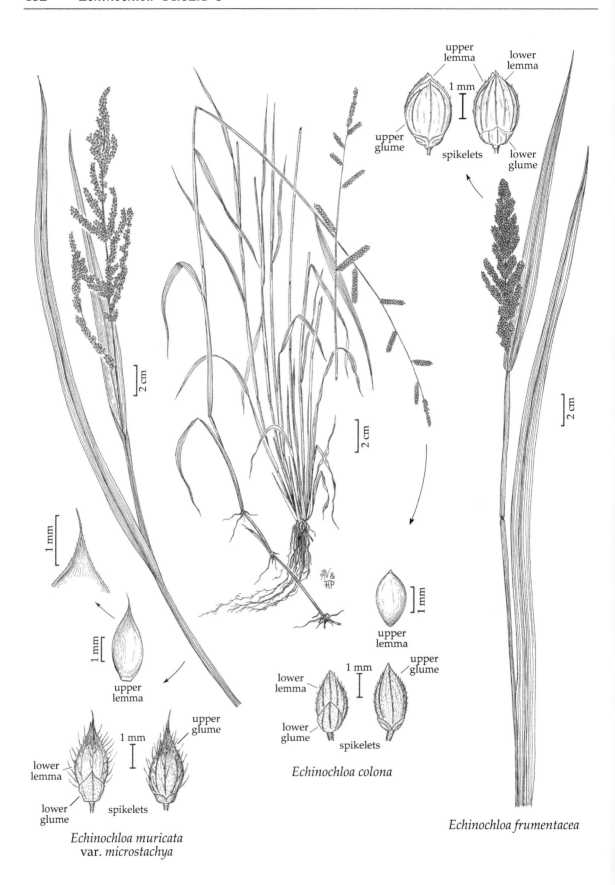

upper
lemma

lower
lemma

1 mm

upper
glume

spikelets

lower
glume

1 mm

2 cm

2 cm

1 mm

upper
lemma

1 mm

upper
lemma

1 mm

lower
lemma

upper
glume

lower
glume

spikelets

Echinochloa colona

1 mm

upper
lemma

1 mm

lower
lemma

upper
glume

lower
glume

spikelets

Echinochloa muricata
var. *microstachya*

Echinochloa frumentacea

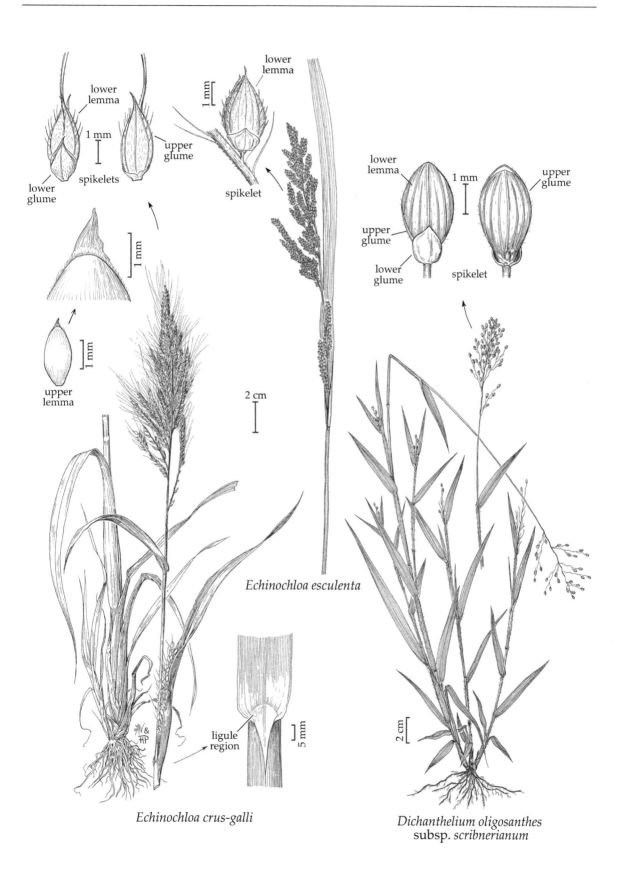

lower lemma

1 mm

lower glume

spikelets

upper glume

lower lemma

1 mm

spikelet

upper lemma

1 mm

1 mm

lower lemma

1 mm

upper glume

upper glume

lower glume

spikelet

2 cm

Echinochloa esculenta

ligule region

5 mm

2 cm

Echinochloa crus-galli

Dichanthelium oligosanthes subsp. *scribnerianum*

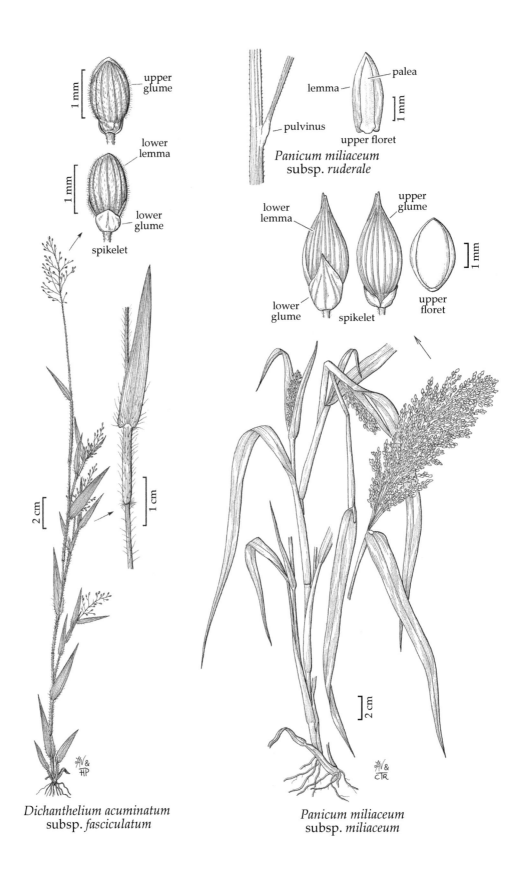

upper
glume

1 mm

lower
lemma

1 mm

lower
glume

spikelet

palea

lemma

pulvinus

upper floret

Panicum miliaceum
subsp. *ruderale*

lower
lemma

upper
glume

1 mm

lower
glume

spikelet

upper
floret

2 cm

1 cm

2 cm

Dichanthelium acuminatum
subsp. *fasciculatum*

Panicum miliaceum
subsp. *miliaceum*

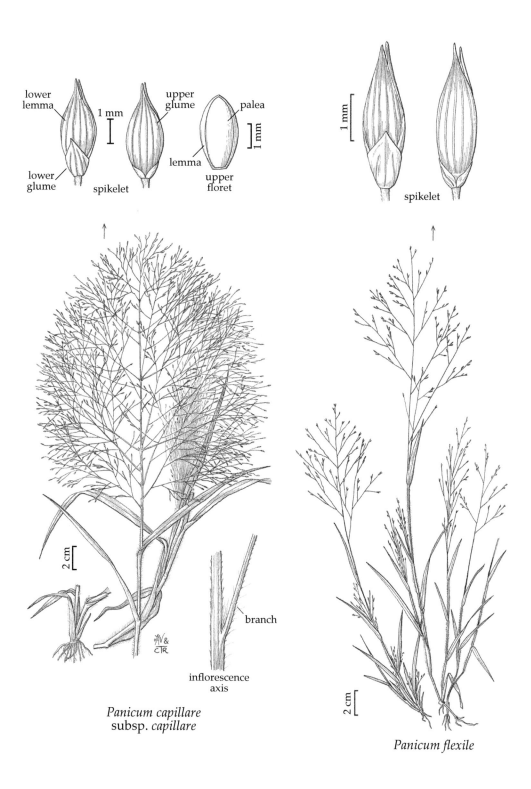

lower lemma

1 mm

lower glume

spikelet

upper glume

lemma

palea

1 mm

upper floret

1 mm

spikelet

branch

inflorescence axis

2 cm

Panicum capillare
subsp. *capillare*

2 cm

Panicum flexile

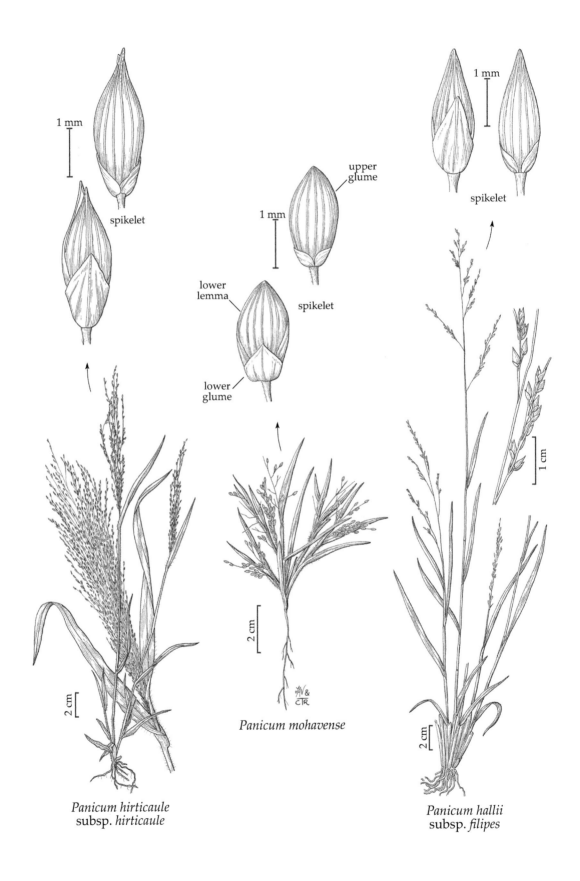

1 mm

spikelet

upper
glume

1 mm

spikelet

lower
lemma

lower
glume

1 mm

spikelet

Panicum mohavense

Panicum hirticaule
subsp. *hirticaule*

Panicum hallii
subsp. *filipes*

2 cm

2 cm

1 cm

2 cm

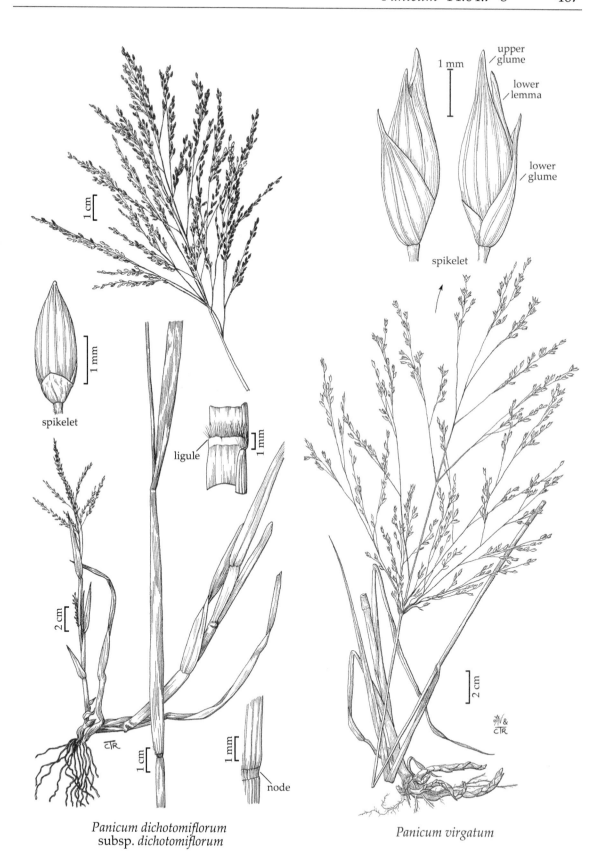

upper
/ glume

lower
/ lemma

lower
/ glume

1 mm

spikelet

spikelet

ligule

1 mm

1 cm

1 cm

1 mm

2 cm

1 mm

node

2 cm

CTR

Panicum dichotomiflorum
subsp. *dichotomiflorum*

Panicum virgatum

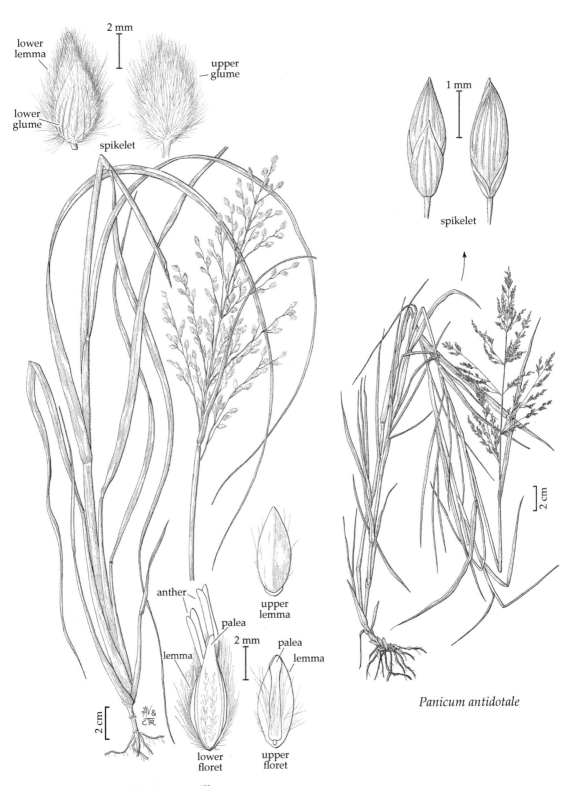

lower
lemma

2 mm

upper
glume

lower
glume

spikelet

1 mm

spikelet

anther

palea

upper
lemma

lemma

palea

2 mm

lemma

Panicum antidotale

2 cm

2 cm

lower
floret

upper
floret

Panicum urvilleanum

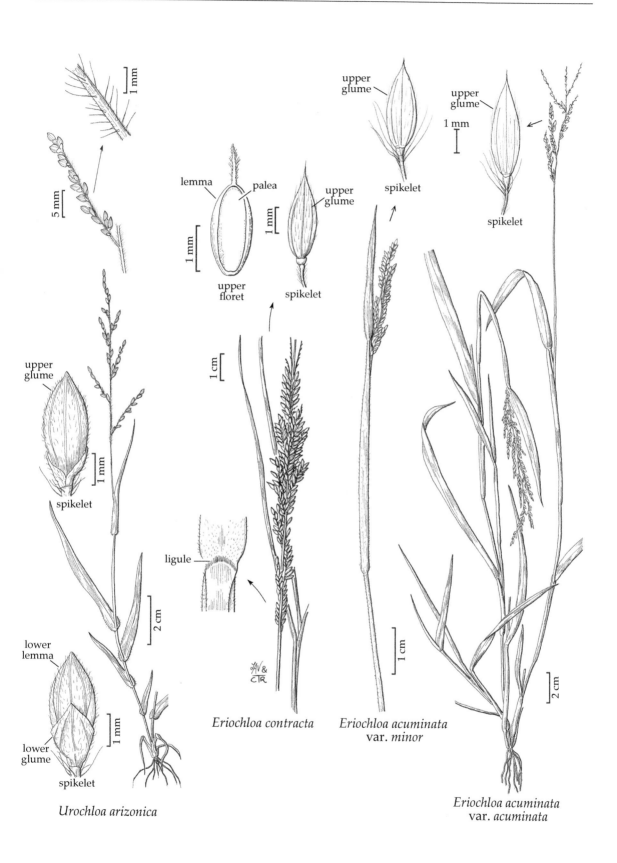

Urochloa arizonica

Eriochloa contracta

Eriochloa acuminata
var. *minor*

Eriochloa acuminata
var. *acuminata*

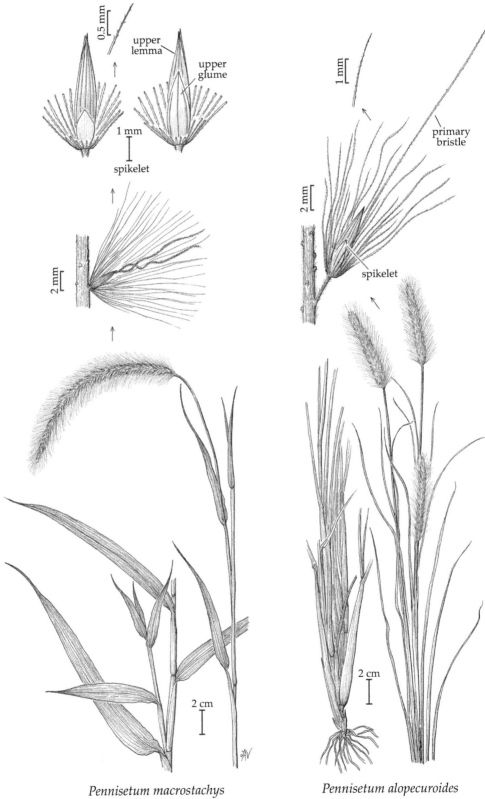

0.5 mm

upper
lemma

upper
glume

1 mm

spikelet

1 mm

primary
bristle

2 mm

2 mm

spikelet

2 mm

2 cm

2 cm

Pennisetum macrostachys   *Pennisetum alopecuroides*

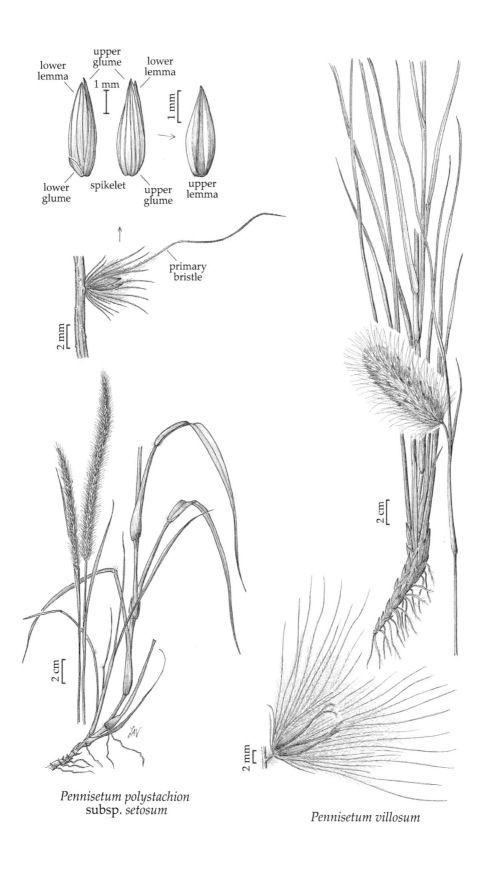

lower
lemma

upper
glume

lower
lemma

1 mm

1 mm

lower
glume

spikelet

upper
glume

upper
lemma

primary
bristle

2 mm

2 mm

Pennisetum polystachion
subsp. *setosum*

2 cm

Pennisetum villosum

2 cm

2 mm

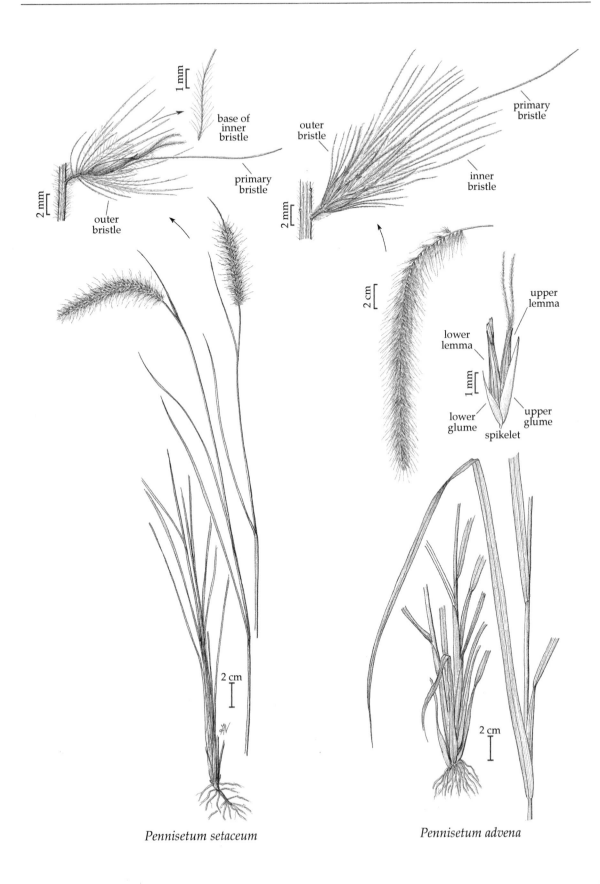

1 mm

base of
inner
bristle

primary
bristle

2 mm

outer
bristle

outer
bristle

inner
bristle

primary
bristle

2 mm

2 cm

upper
lemma

lower
lemma

1 mm

lower
glume

upper
glume

spikelet

2 cm

2 cm

Pennisetum setaceum

Pennisetum advena

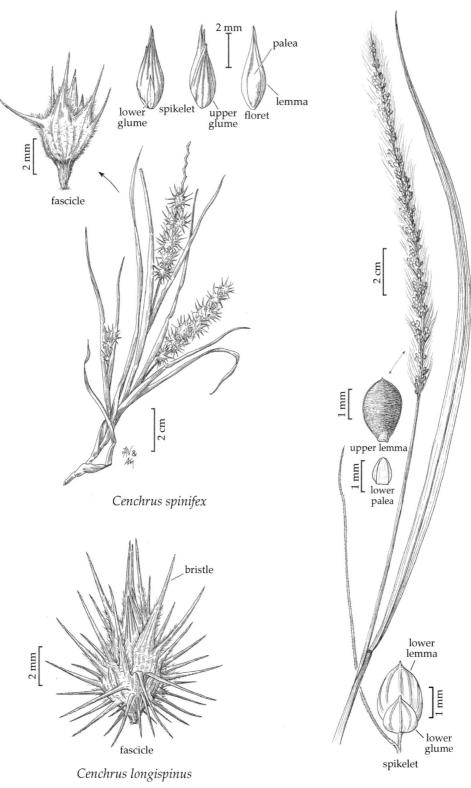

2 mm

palea

lower
glume

spikelet

upper
glume

lemma

floret

2 mm

fascicle

Cenchrus spinifex

2 cm

bristle

2 mm

fascicle

Cenchrus longispinus

2 cm

upper lemma

1 mm

lower
palea

1 mm

lower
lemma

lower
glume

spikelet

1 mm

Setaria macrostachya

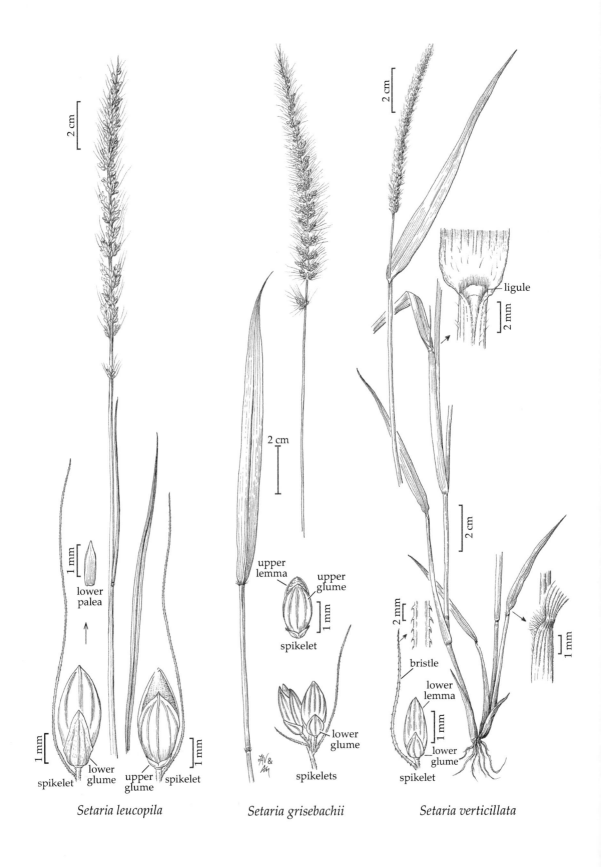

Setaria leucopila   *Setaria grisebachii*   *Setaria verticillata*

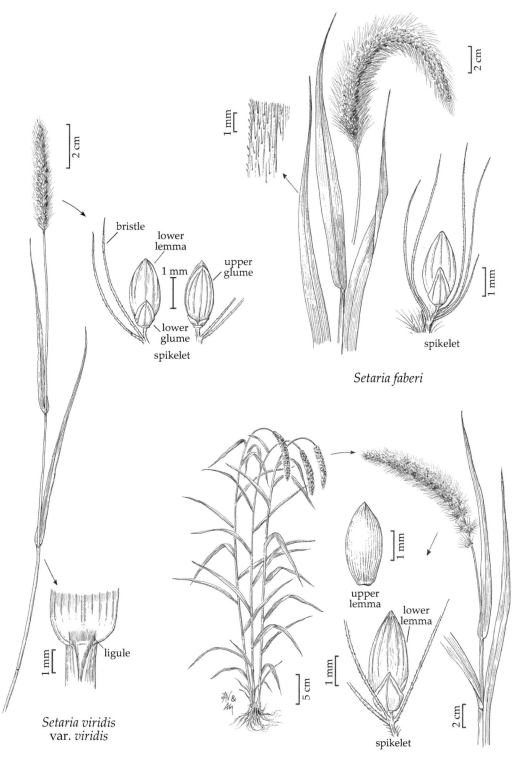

bristle

lower lemma

upper glume

1 mm

lower glume

spikelet

Setaria faberi

2 cm

1 mm

spikelet

ligule

1 mm

Setaria viridis
var. *viridis*

2 cm

upper lemma

1 mm

lower lemma

1 mm

spikelet

5 cm

2 cm

Setaria italica

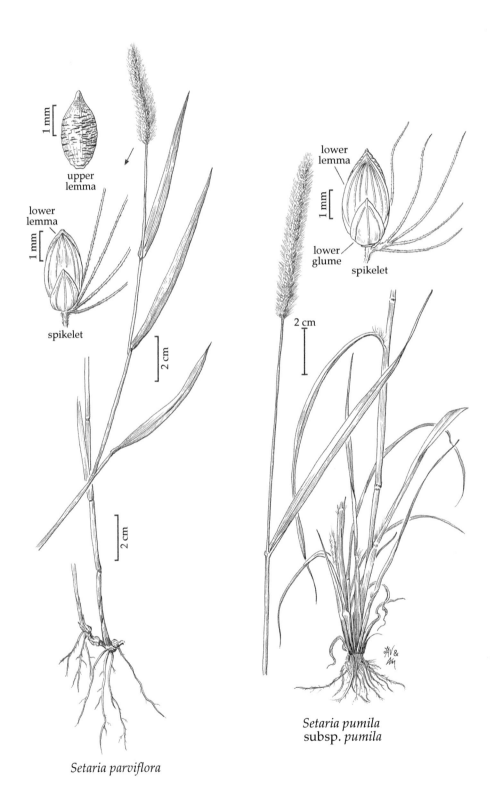

1 mm

upper
lemma

lower
lemma

1 mm

spikelet

2 cm

2 cm

Setaria parviflora

lower
lemma

1 mm

lower
glume

spikelet

2 cm

Setaria pumila
subsp. *pumila*

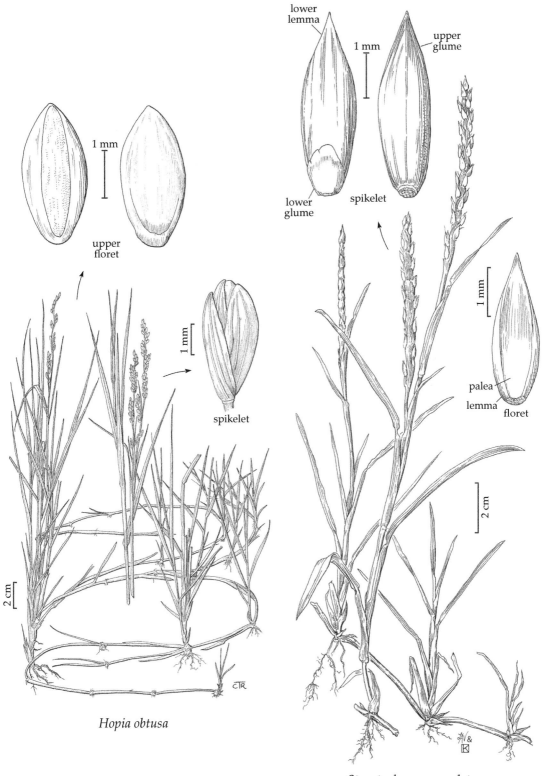

upper
floret

1 mm

lower
lemma

lower
glume

upper
glume

spikelet

1 mm

spikelet

1 mm

palea
lemma floret

1 mm

2 cm

2 cm

CTR

Hopia obtusa

Stenotaphrum secundatum

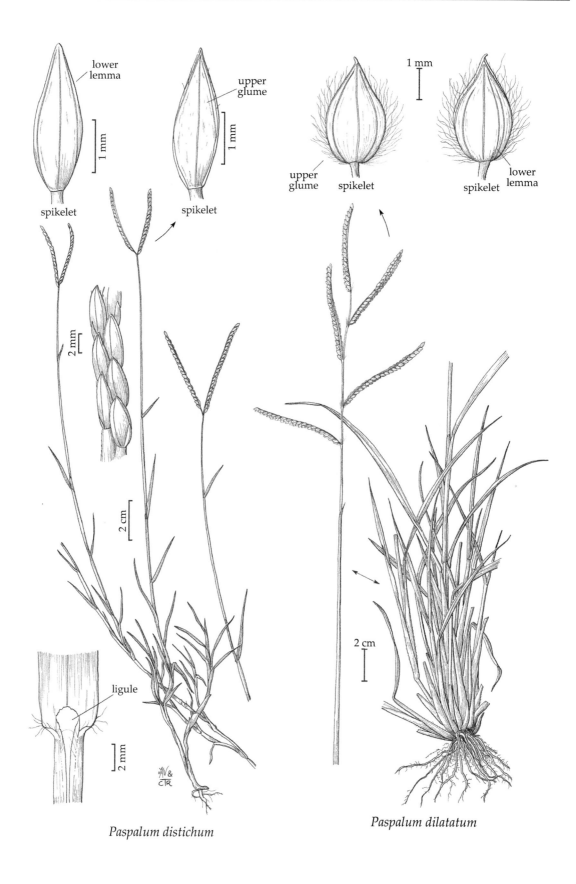

Paspalum distichum

Paspalum dilatatum

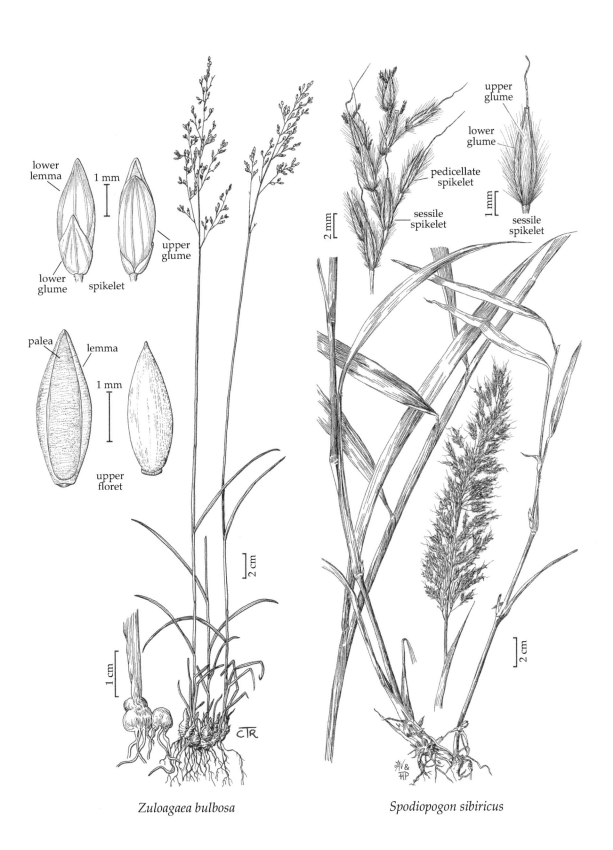

Zuloagaea bulbosa   *Spodiopogon sibiricus*

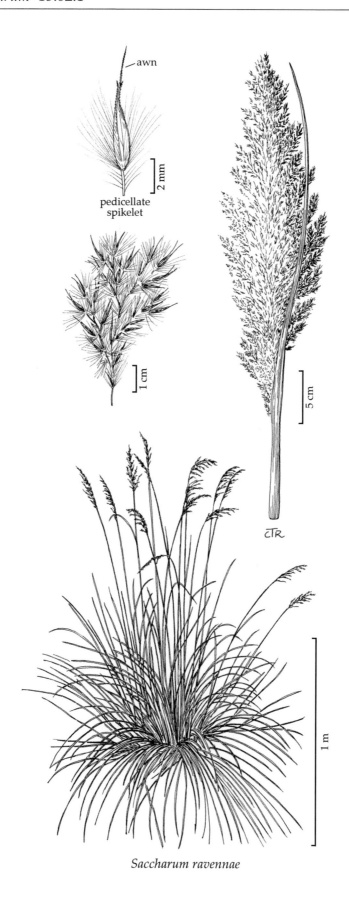

awn

pedicellate
spikelet

2 mm

1 cm

5 cm

CTR

1 m

Saccharum ravennae

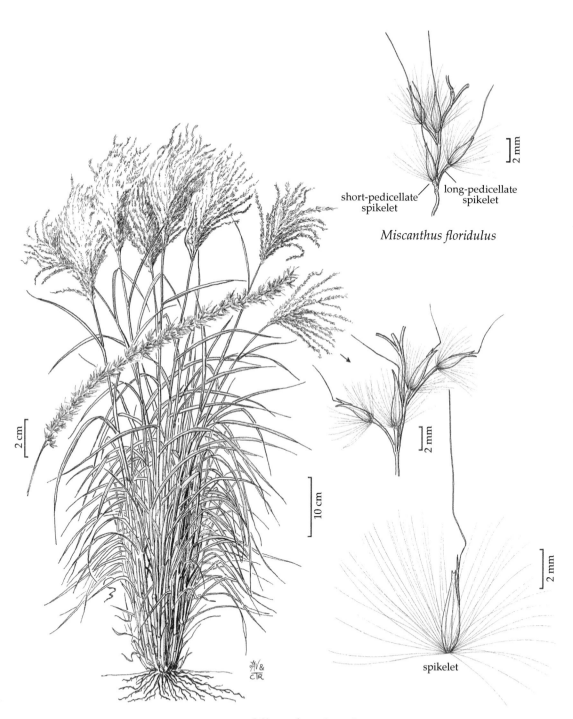

short-pedicellate spikelet

long-pedicellate spikelet

Miscanthus floridulus

2 mm

2 mm

2 mm

spikelet

2 cm

10 cm

Miscanthus sinensis

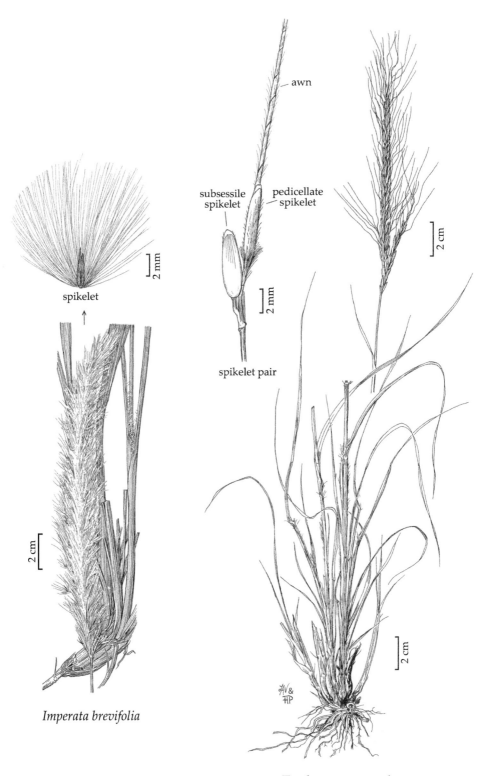

spikelet

2 mm

Imperata brevifolia

awn

subsessile
spikelet

pedicellate
spikelet

2 mm

spikelet pair

2 cm

2 cm

2 cm

Trachypogon secundus

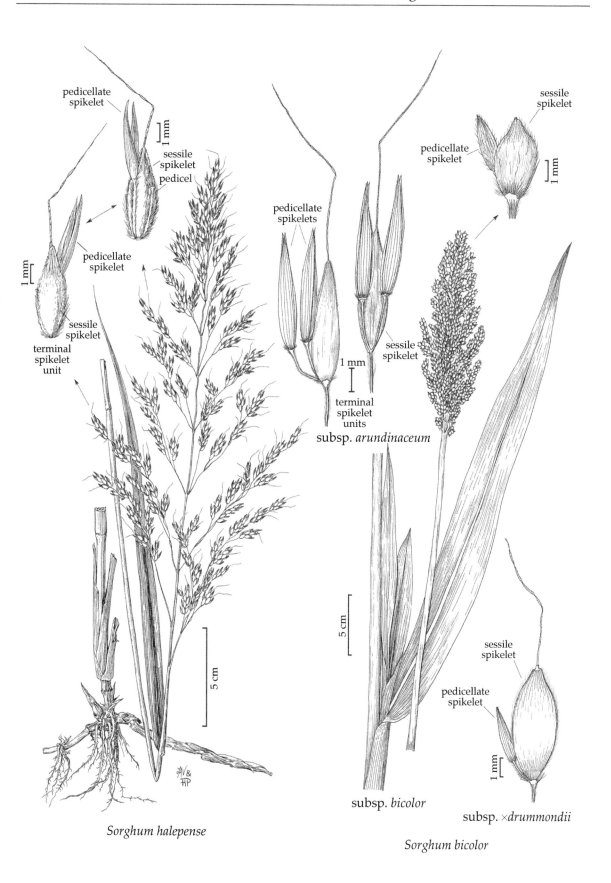

pedicellate
spikelet

1 mm

sessile
spikelet
pedicel

pedicellate
spikelet

1 mm

sessile
spikelet

terminal
spikelet
unit

pedicellate
spikelets

sessile
spikelet

1 mm

terminal
spikelet
units

subsp. *arundinaceum*

sessile
spikelet

pedicellate
spikelet

1 mm

5 cm

5 cm

sessile
spikelet

pedicellate
spikelet

1 mm

subsp. *bicolor*

subsp. ×*drummondii*

Sorghum halepense

Sorghum bicolor

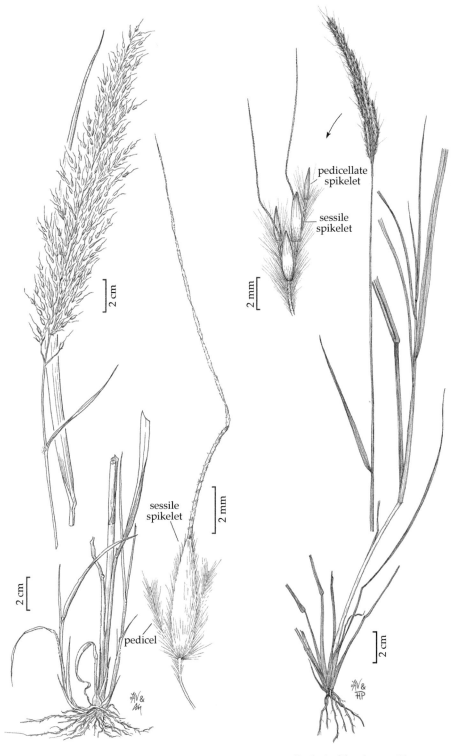

pedicellate
spikelet

sessile
spikelet

2 mm

2 cm

sessile
spikelet

2 mm

pedicel

2 cm

2 cm

Sorghastrum nutans

Bothriochloa laguroides
subsp. *torreyana*

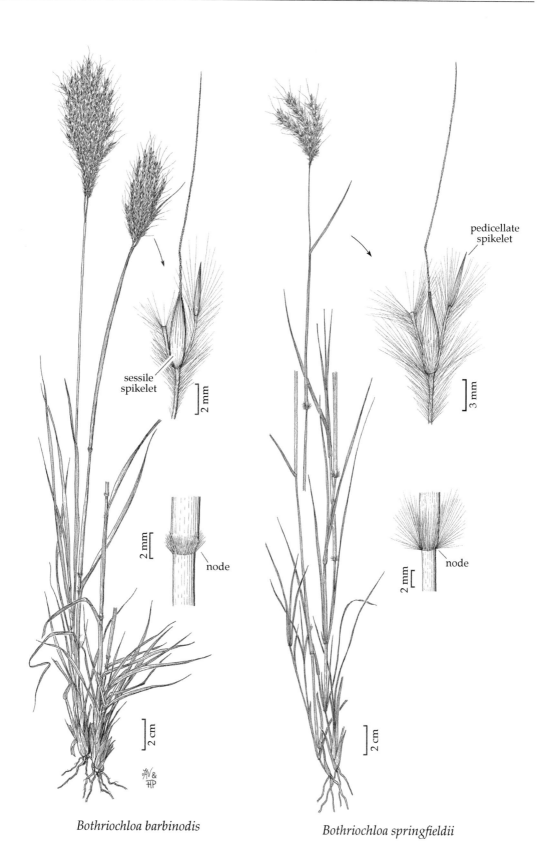

Bothriochloa barbinodis   *Bothriochloa springfieldii*

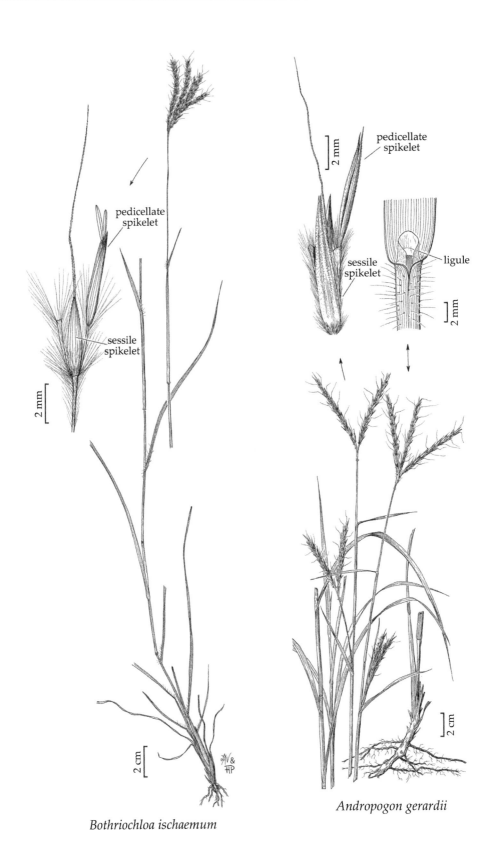

Bothriochloa ischaemum

Andropogon gerardii

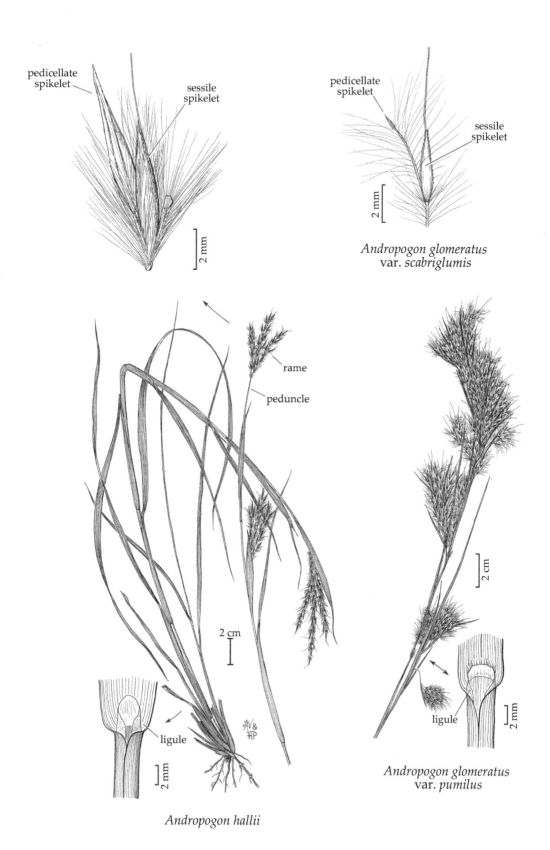

pedicellate spikelet

sessile spikelet

2 mm

pedicellate spikelet

sessile spikelet

2 mm

Andropogon glomeratus
var. *scabriglumis*

rame

peduncle

2 cm

ligule

2 mm

Andropogon hallii

2 cm

ligule

2 mm

Andropogon glomeratus
var. *pumilus*

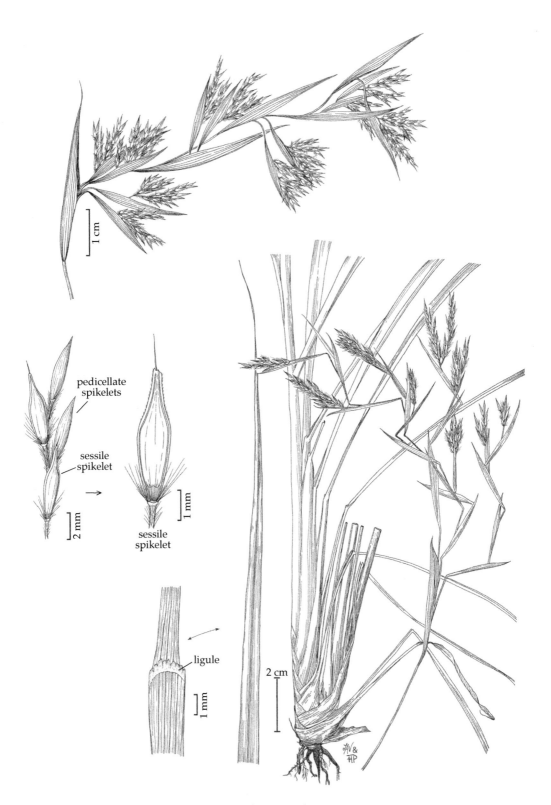

pedicellate
spikelets

sessile
spikelet

sessile
spikelet

ligule

Cymbopogon citratus

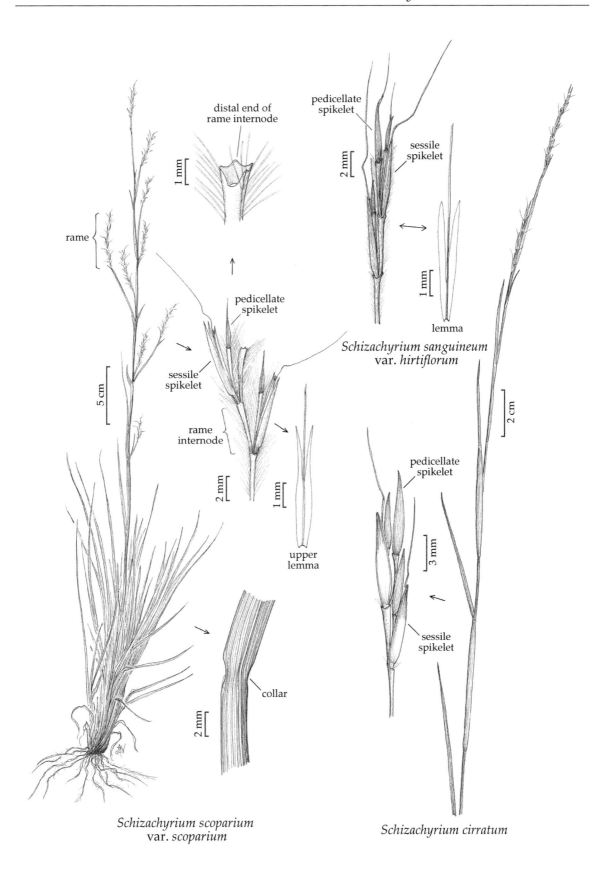

distal end of
rame internode

1 mm

rame

5 cm

pedicellate
spikelet

sessile
spikelet

rame
internode

2 mm

1 mm

upper
lemma

collar

2 mm

Schizachyrium scoparium
var. *scoparium*

pedicellate
spikelet

sessile
spikelet

2 mm

1 mm

lemma

Schizachyrium sanguineum
var. *hirtiflorum*

2 cm

pedicellate
spikelet

3 mm

sessile
spikelet

Schizachyrium cirratum

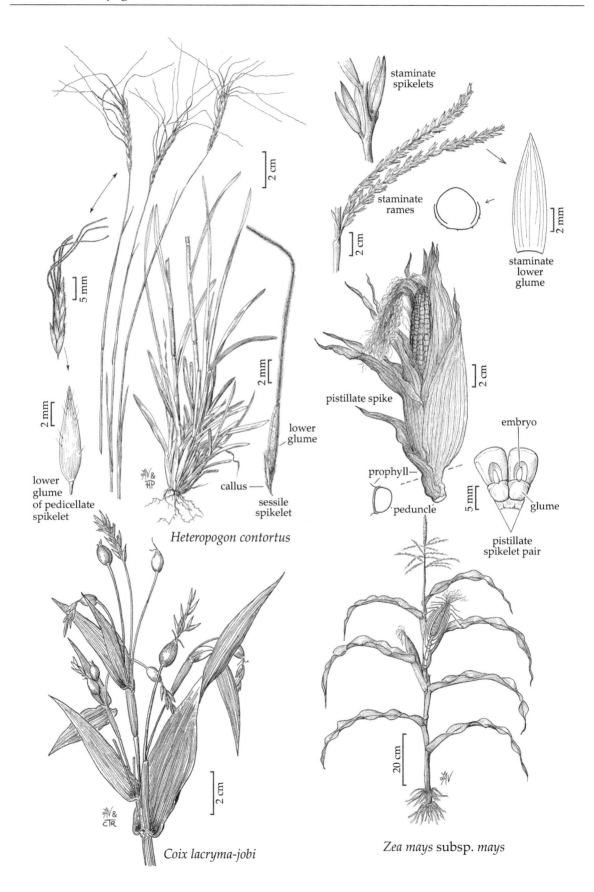

staminate
spikelets

staminate
rames

staminate
lower
glume

2 cm

5 mm

2 mm

lower
glume
of pedicellate
spikelet

2 mm

2 mm

lower
glume

callus

sessile
spikelet

Heteropogon contortus

pistillate spike

prophyll

peduncle

embryo

glume

5 mm

pistillate
spikelet pair

2 cm

20 cm

Coix lacryma-jobi

Zea mays subsp. *mays*

Distribution Maps

* Species with an asterisk are mapped within the Intermountain Region only in Mohave and/or Coconino counties, Arizona. If there is no dot on the map north of the Colorado River, the data were county-level only, and it was not possible to ascertain whether the species actually grows north of the Colorado River. These species were included in order to err on the side of caution.

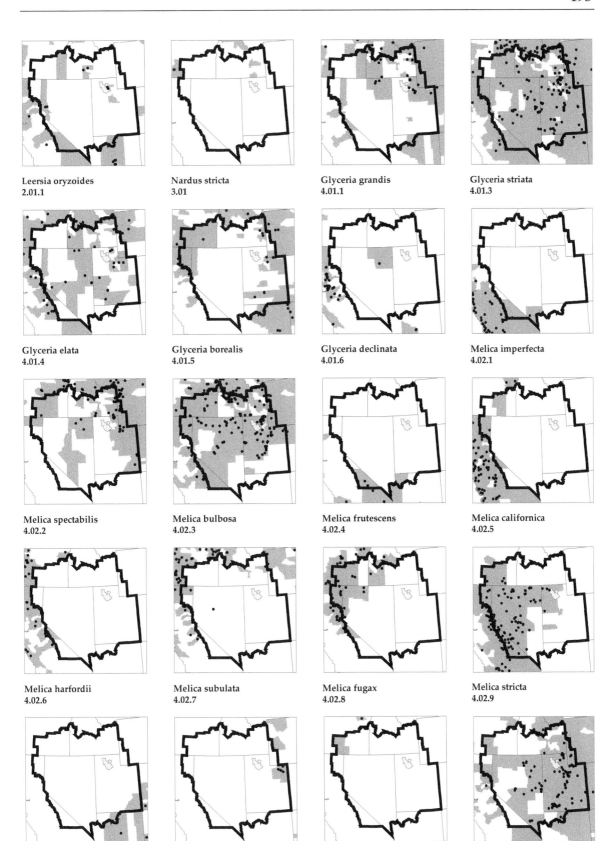

Leersia oryzoides
2.01.1

Nardus stricta
3.01

Glyceria grandis
4.01.1

Glyceria striata
4.01.3

Glyceria elata
4.01.4

Glyceria borealis
4.01.5

Glyceria declinata
4.01.6

Melica imperfecta
4.02.1

Melica spectabilis
4.02.2

Melica bulbosa
4.02.3

Melica frutescens
4.02.4

Melica californica
4.02.5

Melica harfordii
4.02.6

Melica subulata
4.02.7

Melica fugax
4.02.8

Melica stricta
4.02.9

Melica porteri
4.02.10

Schizachne purpurascens
4.03.1

Pleuropogon oregonus
4.04.1

Achnatherum lettermanii
5.01.1

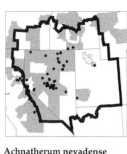

Achnatherum nevadense
5.01.2

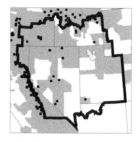

Achnatherum occidentale
5.01.3

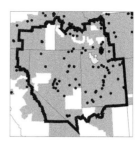

Achnatherum nelsonii
5.01.4

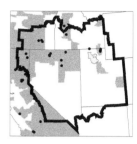

Achnatherum lemmonii
5.01.5

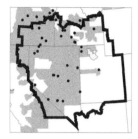

Achnatherum thurberianum
5.01.6

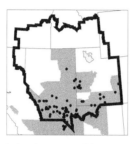

Achnatherum parishii
5.01.7

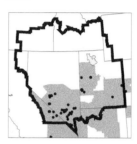

Achnatherum aridum
5.01.8

Achnatherum scribneri
5.01.9

Achnatherum perplexum
5.01.10

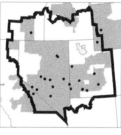

Achnatherum pinetorum
5.01.11

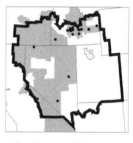

Achnatherum webberi
5.01.12

Achnatherum swallenii
5.01.13

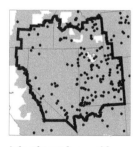

Achnatherum hymenoides
5.01.14

Achnatherum arnowiae
5.01.15

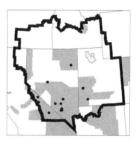

Achnatherum ×bloomeri
5.01.16

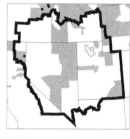

Piptatherum exiguum
5.02.1

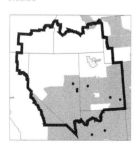

Piptatherum micranthum
5.02.2

Piptatherum shoshoneanum
5.02.3

Hesperostipa spartea
5.03.1

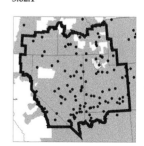

Hesperostipa comata
5.03.2

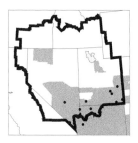

Hesperostipa neomexicana
5.03.3

Piptochaetium pringlei*
5.04.1

Oryzopsis asperifolia
5.05.1

Nassella pulchra
5.06.1

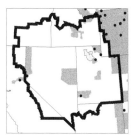

Nassella viridula
5.06.2

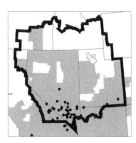

Pappostipa speciosa
5.07.1

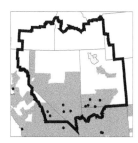

Bromus catharticus
6.01.1

Bromus sitchensis
6.01.2

Bromus arizonicus
6.01.3

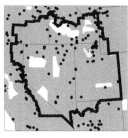

Bromus carinatus
6.01.4

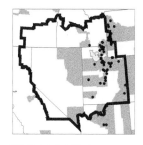

Bromus polyanthus
6.01.5

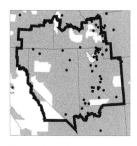

Bromus inermis
6.01.6

Bromus pumpellianus
6.01.7

Bromus laevipes
6.01.8

Bromus frondosus
6.01.9

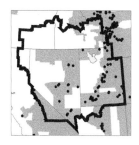

Bromus porteri
6.01.10

Bromus vulgaris
6.01.11

Bromus suksdorfii
6.01.12

Bromus mucroglumis
6.01.13

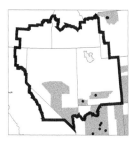

Bromus lanatipes
6.01.14

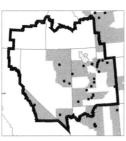

Bromus ciliatus
6.01.15

Bromus richardsonii
6.01.16

Bromus berteroanus
6.01.17

Bromus diandrus
6.01.18

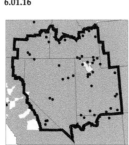

Bromus sterilis
6.01.19

Bromus tectorum
6.01.20

Bromus rubens
6.01.21

Bromus briziformis
6.01.22

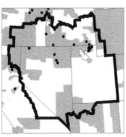

Bromus secalinus
6.01.23

Bromus commutatus
6.01.24

Bromus hordeaceus
6.01.25

Bromus racemosus
6.01.26

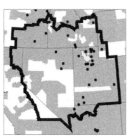

Bromus arenarius
6.01.27

Bromus japonicus
6.01.28

Bromus squarrosus
6.01.29

Hordeum pusillum
7.01.1

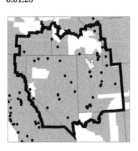

Hordeum depressum
7.01.2

Hordeum brachyantherum
7.01.3

Hordeum jubatum
7.01.4

Hordeum arizonicum
7.01.5

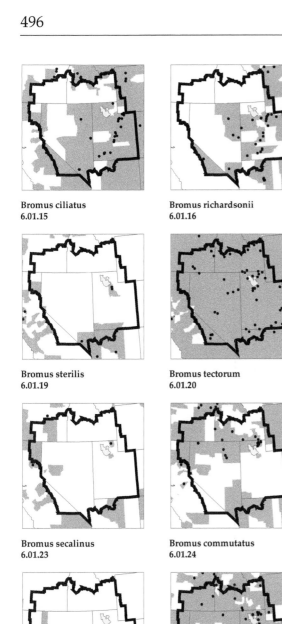

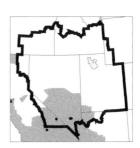
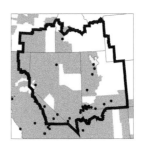
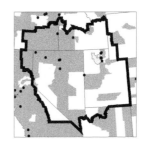

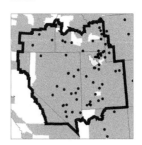
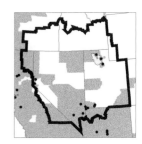
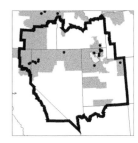
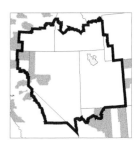
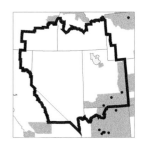

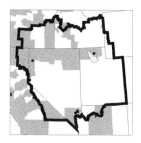

Hordeum geniculatum
7.01.6

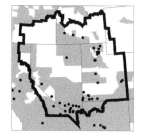

Hordeum murinum
7.01.7

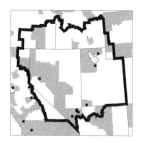

Hordeum vulgare
7.01.10

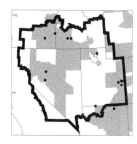

Eremopyrum triticeum
7.02.1

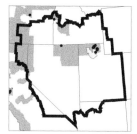

Taeniatherum caput-medusae
7.03.1

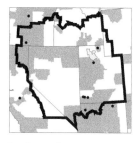

Secale cereale
7.04.1

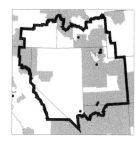

Aegilops cylindrica
7.06.1

Aegilops triuncialis
7.06.2

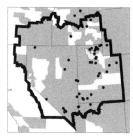

Agropyron cristatum
7.08.1

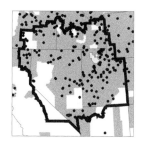

Pseudoroegneria spicata
7.09.1

Elymus curvatus
7.12.1

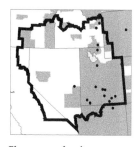

Elymus canadensis
7.12.2

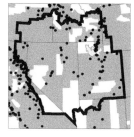

Elymus glaucus
7.12.3

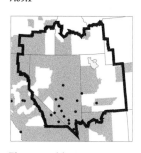

Elymus multisetus
7.12.5

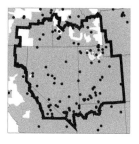

Elymus elymoides
7.12.6

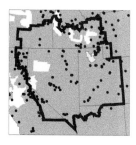

Elymus trachycaulus
7.12.7

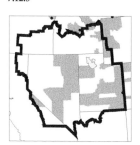

Elymus violaceus
7.12.8

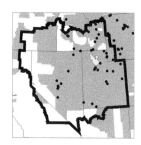

Elymus lanceolatus
7.12.9

Elymus stebbinsii
7.12.10

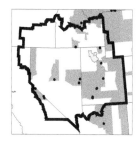

Elymus scribneri
7.12.11

Elymus sierrae
7.12.12

Elymus wawawaiensis
7.12.13

Elymus albicans
7.12.14

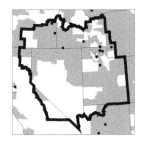

Elymus repens
7.12.15

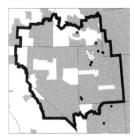

Pascopyrum smithii
7.14.1

Leymus racemosus
7.15.1

Leymus simplex
7.15.2

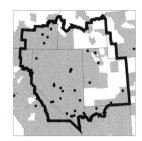

Leymus triticoides
7.15.3

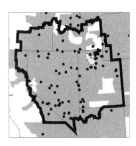

Leymus cinereus
7.15.5

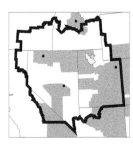

Leymus salina
7.15.6

Leymus flavescens
7.15.7

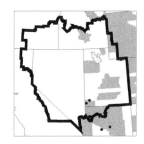

Psathyrostachys juncea
7.17.1

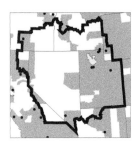

Thinopyrum intermedium
7.18.1

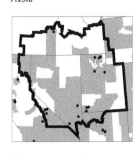

Thinopyrum ponticum
7.18.2

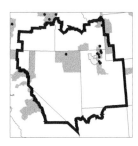

Festuca subulata
8.01.1

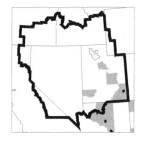

Festuca sororia
8.01.2

Festuca campestris
8.01.3

Festuca thurberi
8.01.4

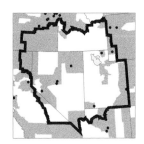

Festuca rubra
8.01.5

Festuca earlei
8.01.6

Festuca valesiaca
8.01.7

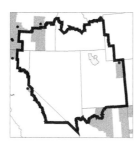

Festuca trachyphylla
8.01.11

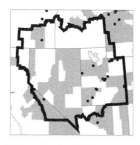

Festuca brachyphylla
8.01.12

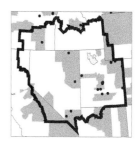

Festuca saximontana
8.01.13

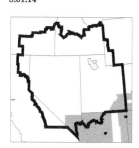

Festuca baffinensis
8.01.14

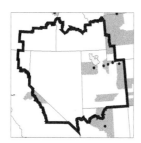

Festuca minutiflora
8.01.15

Festuca occidentalis
8.01.16

Festuca calligera
8.01.17

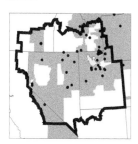

Festuca arizonica
8.01.18

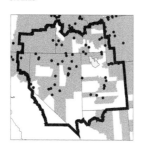

Festuca idahoensis
8.01.19

Festuca viridula
8.01.20

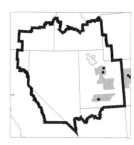

Festuca dasyclada
8.01.21

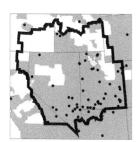

Leucopoa kingii
8.02.1

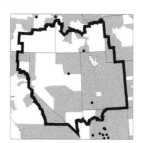

Schedonorus pratensis
8.03.1

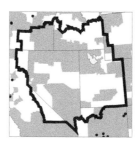

Schedonorus arundinaceus
8.03.2

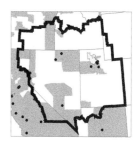

Vulpia myuros
8.04.1

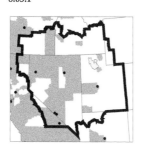

Vulpia octoflora
8.04.2

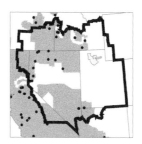

Vulpia bromoides
8.04.3

Vulpia microstachys
8.04.4

Vulpia ciliata
8.04.5

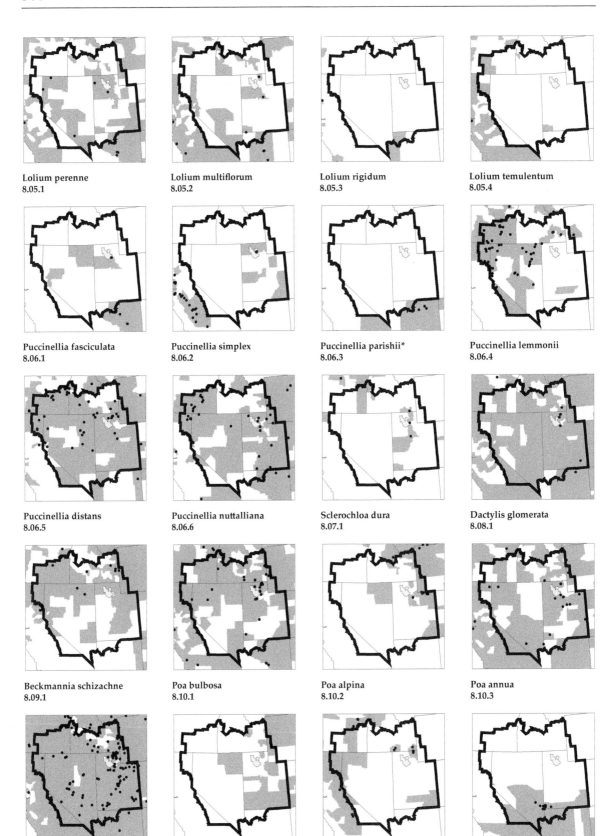

Lolium perenne
8.05.1

Lolium multiflorum
8.05.2

Lolium rigidum
8.05.3

Lolium temulentum
8.05.4

Puccinellia fasciculata
8.06.1

Puccinellia simplex
8.06.2

Puccinellia parishii*
8.06.3

Puccinellia lemmonii
8.06.4

Puccinellia distans
8.06.5

Puccinellia nuttalliana
8.06.6

Sclerochloa dura
8.07.1

Dactylis glomerata
8.08.1

Beckmannia schizachne
8.09.1

Poa bulbosa
8.10.1

Poa alpina
8.10.2

Poa annua
8.10.3

Poa pratensis
8.10.4

Poa arctica
8.10.5

Poa bolanderi
8.10.6

Poa bigelovii
8.10.7

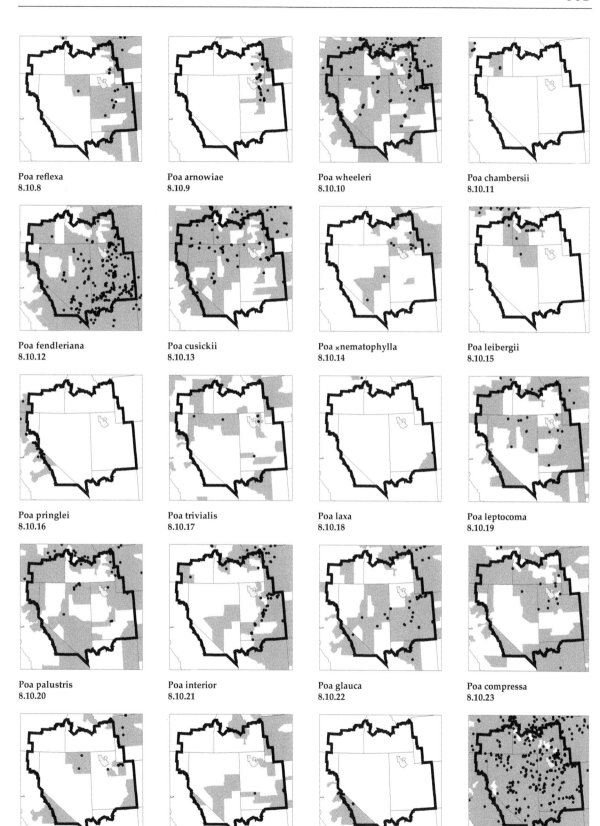

Poa reflexa
8.10.8

Poa arnowiae
8.10.9

Poa wheeleri
8.10.10

Poa chambersii
8.10.11

Poa fendleriana
8.10.12

Poa cusickii
8.10.13

Poa ×nematophylla
8.10.14

Poa leibergii
8.10.15

Poa pringlei
8.10.16

Poa trivialis
8.10.17

Poa laxa
8.10.18

Poa leptocoma
8.10.19

Poa palustris
8.10.20

Poa interior
8.10.21

Poa glauca
8.10.22

Poa compressa
8.10.23

Poa lettermanii
8.10.24

Poa abbreviata
8.10.25

Poa keckii
8.10.26

Poa secunda
8.10.27

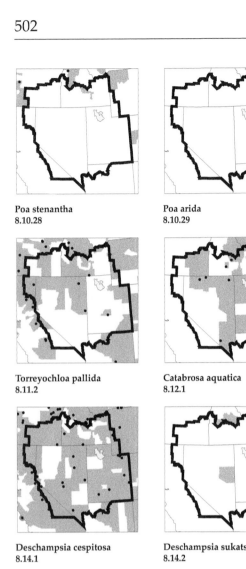

Poa stenantha
8.10.28

Poa arida
8.10.29

Poa ×limosa
8.10.30

Torreyochloa erecta
8.11.1

Torreyochloa pallida
8.11.2

Catabrosa aquatica
8.12.1

Sphenopholis obtusata
8.13.1

Sphenopholis intermedia
8.13.2

Deschampsia cespitosa
8.14.1

Deschampsia sukatschewii
8.14.2

Deschampsia brevifolia
8.14.3

Deschampsia elongata
8.14.4

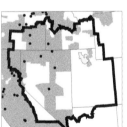

Deschampsia danthonioides
8.14.5

Agrostis capillaris
8.15.1

Agrostis gigantea
8.15.2

Agrostis stolonifera
8.15.3

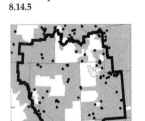

Agrostis scabra
8.15.4

Agrostis idahoensis
8.15.5

Agrostis oregonensis
8.15.6

Agrostis exarata
8.15.7

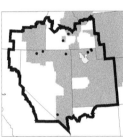

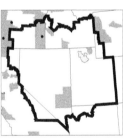
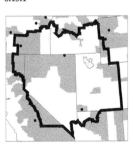
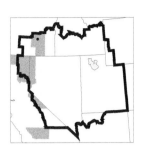
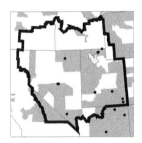
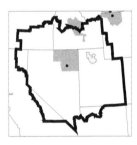
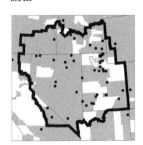
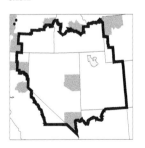

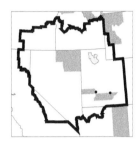
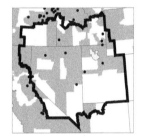
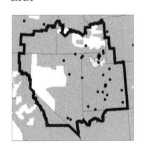
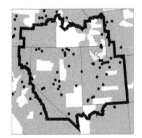

Agrostis pallens
8.15.8

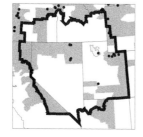

Agrostis variabilis
8.15.9

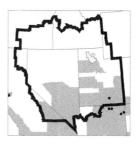

Polypogon viridis
8.16.1

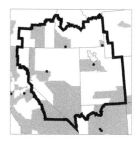

Polypogon interruptus
8.16.2

Polypogon australis
8.16.3

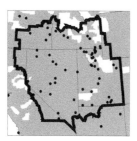

Polypogon monspeliensis
8.16.4

Polypogon maritimus
8.16.5

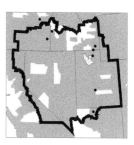

Phleum pratense
8.18.1

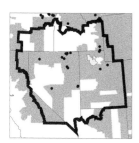

Phleum alpinum
8.18.2

Desmazeria rigida
8.20.1

Ventenata dubia
8.21.1

Scribneria bolanderi
8.22.1

Vahlodea atropurpurea
8.23.1

Podagrostis humilis
8.24.1

Podagrostis thurberiana
8.24.2

Helictotrichon mortonianum
8.25.1

Calamagrostis epigejos
8.26.1

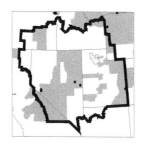

Calamagrostis purpurascens
8.26.2

Calamagrostis tacomensis
8.26.3

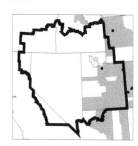

Calamagrostis scopulorum
8.26.4

Calamagrostis muiriana
8.26.5

Calamagrostis koelerioides
8.26.7

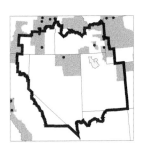

Calamagrostis rubescens
8.26.8

Calamagrostis montanensis
8.26.9

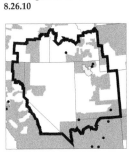

Calamagrostis canadensis
8.26.10

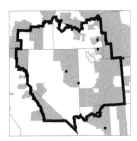

Calamagrostis stricta
8.26.11

Scolochloa festucacea
8.27.1

Avena barbata
8.28.1

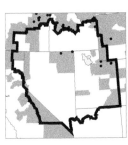

Avena fatua
8.28.2

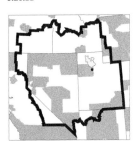

Avena sativa
8.28.3

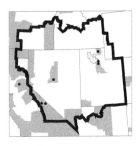

Holcus lanatus
8.29.1

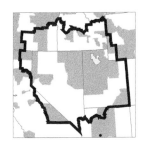

Arrhenatherum elatius
8.30.1

Trisetum wolfii
8.31.1

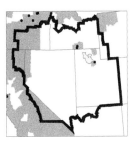

Trisetum canescens
8.31.2

Trisetum cernuum
8.31.3

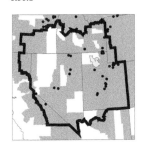

Trisetum spicatum
8.31.4

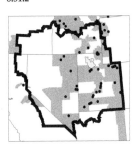

Trisetum projectum
8.31.5

Trisetum montanum
8.31.6

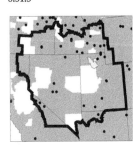

Koeleria macrantha
8.32.1

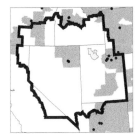

Anthoxanthum hirtum
8.33.1

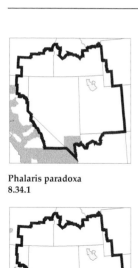

Phalaris paradoxa
8.34.1

Phalaris minor
8.34.2

Phalaris canariensis
8.34.3

Phalaris arundinacea
8.34.4

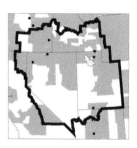

Phalaris caroliniana
8.34.5

Cinna latifolia
8.35.1

Alopecurus pratensis
8.36.1

Alopecurus arundinaceus
8.36.2

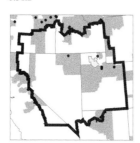

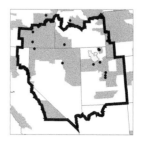

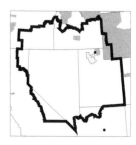

Alopecurus magellanicus
8.36.3

Alopecurus geniculatus
8.36.4

Alopecurus aequalis
8.36.5

Alopecurus saccatus
8.36.6

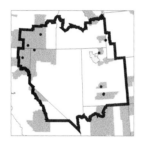

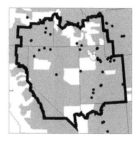

Alopecurus carolinianus
8.36.7

Apera interrupta
8.37.1

Phragmites australis
9.02.1

Arundo donax
9.03.1

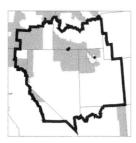

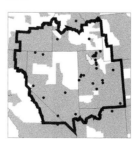

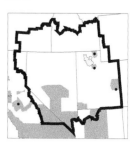

Swallenia alexandrae
10.01.1

Distichlis spicata
10.02.1

Tridens muticus
10.03.1

Redfieldia flexuosa
10.04.1

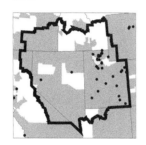

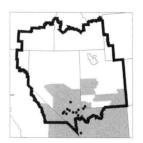

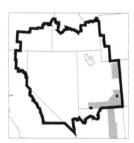

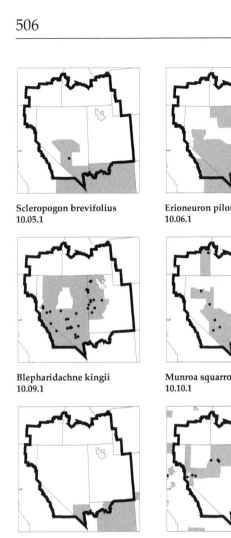

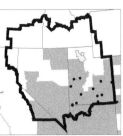

Scleropogon brevifolius
10.05.1

Erioneuron pilosum
10.06.1

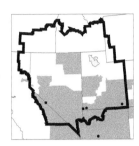

Dasyochloa pulchella
10.07.1

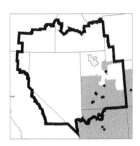

Blepharoneuron tricholepis
10.08.1

Blepharidachne kingii
10.09.1

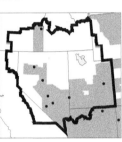

Munroa squarrosa
10.10.1

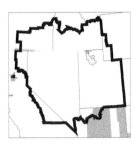

Leptochloa dubia*
10.11.1

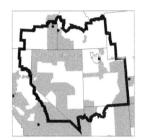

Leptochloa fusca
10.11.2

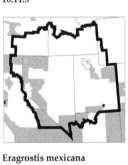

Leptochloa panicea
10.11.3

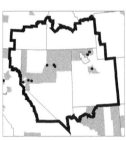

Eragrostis hypnoides
10.12.1

Eragrostis curvula
10.12.2

Eragrostis lehmanniana
10.12.3

Eragrostis mexicana
10.12.4

Eragrostis lutescens
10.12.5

Eragrostis pilosa
10.12.6

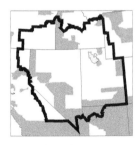

Eragrostis pectinacea
10.12.7

Eragrostis cilianensis
10.12.8

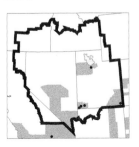

Eragrostis barrelieri
10.12.9

Eragrostis minor
10.12.10

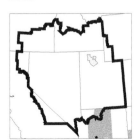

Eragrostis intermedia*
10.12.12

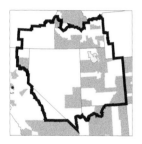

Eleusine indica
10.13.1

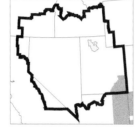

Sporobolus pyramidatus
10.14.1

Sporobolus neglectus
10.14.2

Sporobolus compositus
10.14.3

Sporobolus wrightii
10.14.4

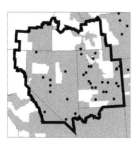

Sporobolus airoides
10.14.5

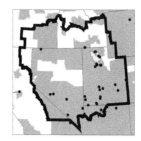

Sporobolus cryptandrus
10.14.6

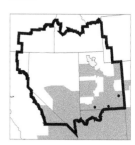

Sporobolus contractus
10.14.7

Sporobolus texanus
10.14.8

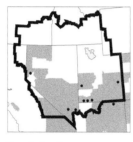

Sporobolus flexuosus
10.14.9

Sporobolus giganteus
10.14.10

Sporobolus junceus*
10.14.11

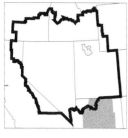

Sporobolus interruptus*
10.14.12

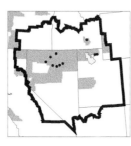

Crypsis alopecuroides
10.15.1

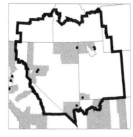

Crypsis schoenoides
10.15.2

Crypsis vaginiflora
10.15.3

Calamovilfa gigantea
10.16.1

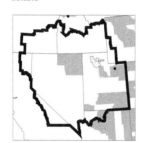

Muhlenbergia racemosa
10.17.1

Muhlenbergia glomerata
10.17.2

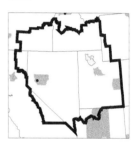

Muhlenbergia mexicana
10.17.3

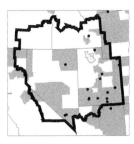

Muhlenbergia andina
10.17.4

Muhlenbergia sylvatica*
10.17.5

Muhlenbergia schreberi
10.17.6

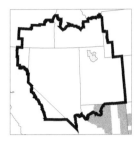

Muhlenbergia tenuifolia*
10.17.7

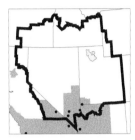

Muhlenbergia microsperma
10.17.8

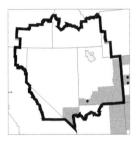

Muhlenbergia thurberi
10.17.9

Muhlenbergia curtifolia
10.17.10

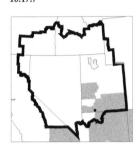

Muhlenbergia pauciflora
10.17.11

Muhlenbergia arsenei
10.17.12

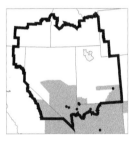

Muhlenbergia porteri
10.17.13

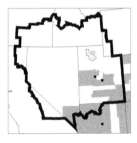

Muhlenbergia wrightii
10.17.14

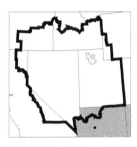

Muhlenbergia torreyi
10.17.15

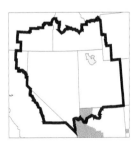

Muhlenbergia arenicola*
10.17.16

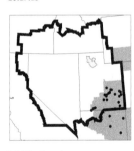

Muhlenbergia pungens
10.17.17

Muhlenbergia repens
10.17.18

Muhlenbergia utilis
10.17.19

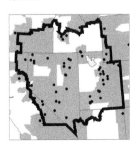

Muhlenbergia richardsonis
10.17.20

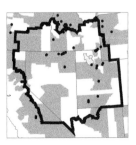

Muhlenbergia filiformis
10.17.21

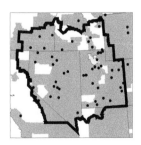

Muhlenbergia asperifolia
10.17.22

Muhlenbergia filiculmis
10.17.23

Muhlenbergia jonesii
10.17.24

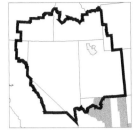

Muhlenbergia straminea*
10.17.25

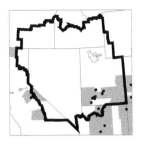

Muhlenbergia montana
10.17.26

Muhlenbergia emersleyi*
10.17.27

Muhlenbergia longiligula*
10.17.28

Muhlenbergia rigens
10.17.31

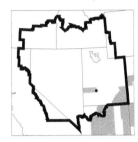

Muhlenbergia depauperata
10.17.32

Muhlenbergia sinuosa*
10.17.33

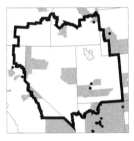

Muhlenbergia minutissima
10.17.34

Muhlenbergia fragilis
10.17.35

Muhlenbergia ramulosa*
10.17.36

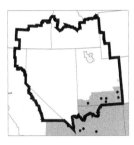

Lycurus setosus
10.18.1

Chloris pilosa
10.19.1

Chloris virgata
10.19.2

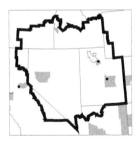

Chloris verticillata
10.19.3

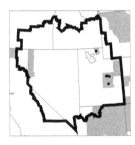

Schedonnardus paniculatus
10.20.1

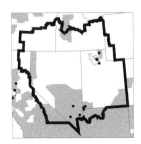

Cynodon dactylon
10.21.1

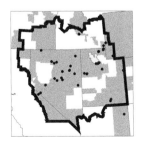

Spartina gracilis
10.22.1

Spartina pectinata
10.22.2

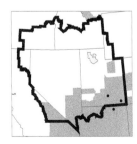

Bouteloua curtipendula
10.23.1

510

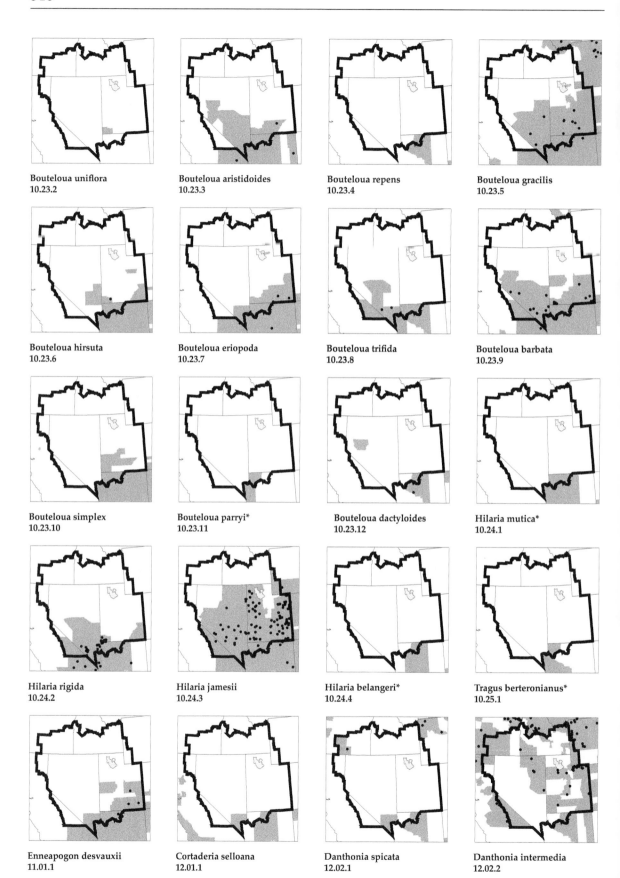

Bouteloua uniflora
10.23.2

Bouteloua aristidoides
10.23.3

Bouteloua repens
10.23.4

Bouteloua gracilis
10.23.5

Bouteloua hirsuta
10.23.6

Bouteloua eriopoda
10.23.7

Bouteloua trifida
10.23.8

Bouteloua barbata
10.23.9

Bouteloua simplex
10.23.10

Bouteloua parryi*
10.23.11

Bouteloua dactyloides
10.23.12

Hilaria mutica*
10.24.1

Hilaria rigida
10.24.2

Hilaria jamesii
10.24.3

Hilaria belangeri*
10.24.4

Tragus berteronianus*
10.25.1

Enneapogon desvauxii
11.01.1

Cortaderia selloana
12.01.1

Danthonia spicata
12.02.1

Danthonia intermedia
12.02.2

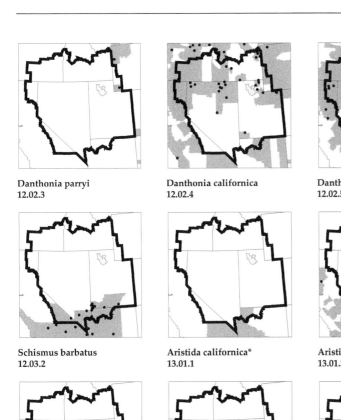

Danthonia parryi
12.02.3

Danthonia californica
12.02.4

Danthonia unispicata
12.02.5

Schismus arabicus
12.03.1

Schismus barbatus
12.03.2

Aristida californica*
13.01.1

Aristida ternipes*
13.01.2

Aristida schiedeana*
13.01.3

Aristida divaricata*
13.01.4

Aristida havardii
13.01.5

Aristida pansa*
13.01.6

Aristida oligantha
13.01.7

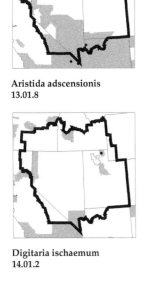

Aristida adscensionis
13.01.8

Aristida purpurea
13.01.9

Aristida arizonica
13.01.10

Digitaria californica*
14.01.1

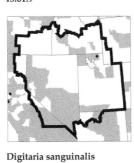

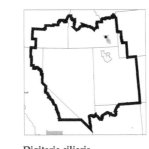

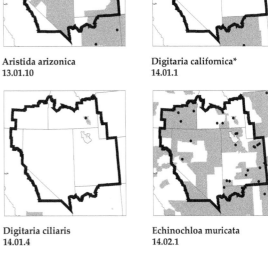

Digitaria ischaemum
14.01.2

Digitaria sanguinalis
14.01.3

Digitaria ciliaris
14.01.4

Echinochloa muricata
14.02.1

512

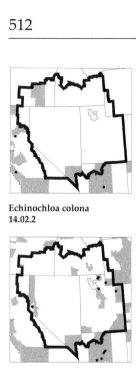

Echinochloa colona
14.02.2

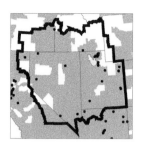

Echinochloa crus-galli
14.02.4

Dichanthelium oligosanthes
14.03.1

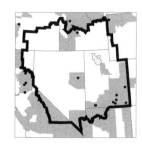

Dichanthelium acuminatum
14.03.2

Panicum miliaceum
14.04.1

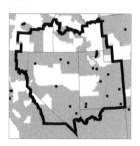

Panicum capillare
14.04.2

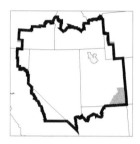

Panicum flexile
14.04.3

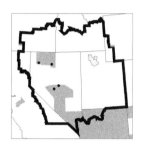

Panicum hirticaule
14.04.4

Panicum mohavense
14.04.5

Panicum hallii
14.04.6

Panicum dichotomiflorum
14.04.7

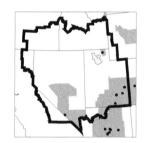

Panicum virgatum
14.04.8

Panicum urvilleanum
14.04.9

Panicum antidotale
14.04.10

Urochloa arizonica*
14.05.1

Eriochloa contracta
14.06.1

Eriochloa acuminata
14.06.2

Pennisetum polystachion
14.07.3

Pennisetum setaceum
14.07.5

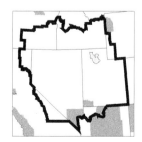

Cenchrus spinifex
14.08.1

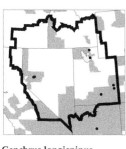

Cenchrus longispinus
14.08.2

Eriochloa acuminata
14.06.2

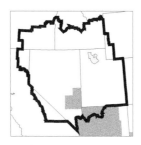

Setaria macrostachya
14.09.1

Setaria leucopila*
14.09.2

Setaria grisebachii*
14.09.3

Setaria verticillata
14.09.4

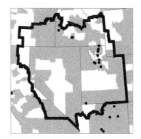

Setaria viridis
14.09.5

Setaria italica
14.09.6

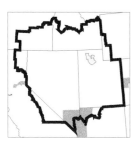

Setaria faberi
14.09.7

Setaria parviflora
14.09.8

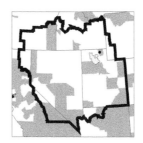

Setaria pumila
14.09.9

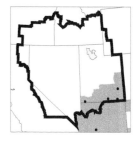

Hopia obtusa
14.10.1

Stenotaphrum secundatum
14.11.1

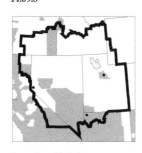

Paspalum distichum
14.12.1

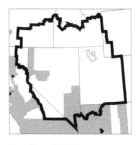

Paspalum dilatatum
14.12.2

Zuloagaea bulbosa
14.13.1

Imperata brevifolia
15.04.1

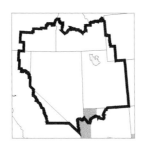

Trachypogon secundus*
15.05.1

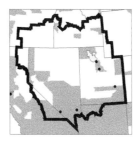

Sorghum halepense
15.06.1

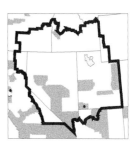

Sorghum bicolor
15.06.2

514

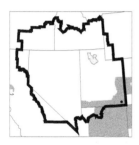

Sorghastrum nutans
15.07.1

Bothriochloa laguroides
15.08.1

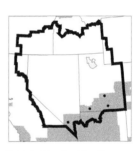

Bothriochloa barbinodis
15.08.2

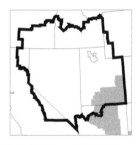

Bothriochloa springfieldii
15.08.3

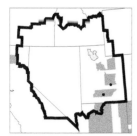

Bothriochloa ischaemum
15.08.4

Andropogon gerardii
15.09.1

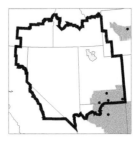

Andropogon hallii
15.09.2

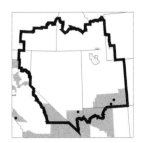

Andropogon glomeratus
15.09.3

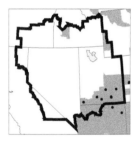

Schizachyrium scoparium
15.11.1

Schizachyrium sanguineum*
15.11.2

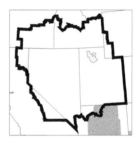

Schizachyrium cirratum*
15.11.3

Heteropogon contortus*
15.12.1

Literature Cited

Ainouche, M.L., R.J. Bayer, J.-P. Gourret, A. Defontaine, and M.-T. Misset. 1999. The allotetraploid invasive weed *Bromus hordeaceus* L. (Poaceae): Genetic diversity, origin and molecular evolution. Folia Geobot. 34:405–419. **[Bromus]**

Aliscioni, S.S., L.M. Giussani, F.O. Zuloaga, and E.A. Kellogg. 2003. A molecular phylogeny of *Panicum* (Poaceae: Paniceae): tests of monophyly and phylogenetic placement within the Panicoideae. Amer J. Bot. 90:796–821. **[Panicum]**

Anderson, D.E. 1974. Taxonomy of the genus *Chloris* (Gramineae). Brigham Young Univ. Sci. Bull., Biol. Ser. 19:1–133. **[Chloris]**

Arnow, L. A. 1987. Gramineae A. L. Jussieu, Grass Family. Pp. 684–788 *in* S. L. Welsh, N. D. Atwood, S. Goodrich, and L. C. Higgins (eds.). A Utah flora. Memoirs No. 9, Great Basin Naturalist.

Assadi, M. 1994. The genus *Elymus* L. (Poaceae) in Iran: Biosystematic studies and generic delimitation. Ph.D. dissertation, Lund University, Lund, Sweden. 104 pp. **[Thinopyrum]**

Baldini, R.M. 1995. Revision of the genus *Phalaris* L. (Gramineae). Webbia 49:265–329. **[Phalaris]**

Barber, J.C., S.S. Aliscioni, L.M. Giussani, J.D. Noll, M.R. Duvall, and E.A. Kellogg. 2002. Combined analyses of three independent datasets to investigate phylogeny of Poaceae subfamily Panicoideae. [Abstract.] http:// www.botany2002. org/. **[Paniceae, Panicoideae]**

Barker, N.P., H.P. Linder, and E.H. Harley. 1995. Polyphyly of Arundinoideae (Poaceae): Evidence from *rbcL* sequence data. Syst. Bot. 20:423–435. **[Arundinoideae, Chloridoideae]**

Barker, N.P., H.P. Linder, and E.H. Harley. 1998. Sequences of the grass-specific insert in the chloroplast *rpoC2* gene elucidate generic relationships of the Arundinoideae (Poaceae). Syst. Bot. 23:327–350. **[Arundinoideae]**

Barker, N.P., H.P. Linder, C.M. Morton, and M. Lyle. 2003. The paraphyly of *Cortaderia* (Danthonioideae: Poaceae): evidence from morphology and chloroplast and nuclear DNA sequence data. Ann. Missouri Bot. Gard. 90:1–24. **[Cortaderia]**

Barker, N.P., C.M. Morton, and H.P. Linder. 2000. The Danthonieae: Generic composition and relationships. Pp. 221–229 *in* S.W.L. Jacobs and J. Everett (eds.). Grasses: Systematics and Evolution. CSIRO Publishing, Collingwood, Victoria, Australia. 408 pp. **[Danthonioideae]**

Barkworth, M.E. 2000. Changing perceptions in the Triticeae. Pp. 110–120 *in* S.W.L. Jacobs and J. Everett (eds.). Grasses: Systematics and Evolution. CSIRO Publishing, Collingwood, Victoria, Australia. 408 pp. **[Elymus]**

Barkworth, M.E., M.O. Arriaga, J.F. Smith, S.W.L. Jacobs, J. Valdés-Reyna, and B.S. Bushman. 2008. Molecules and morphology in South American Stipeae (Poaceae). Syst. Bot. 33:719–731. **[Nassella]**

Barkworth, M.E. and R. von Bothmer. 2005. Twenty-one years later: Löve and Dewey's genomic classification proposal. Czech J. Genet. Pl. Breed. 41 (Special Issue):3–9. **[Elymus]**

Barkworth, M.E. and K.M. Capels. 2000. The Poaceae in North America: A geographic perspective. Pp. 327–346 *in* S.W.L. Jacobs and J. Everett (eds.). Grasses: Systematics and Evolution. CSIRO Publishing, Collingwood, Victoria, Australia. 408 pp. **[Chloridoideae, Panicoideae]**

Barkworth, M.E, K.M. Capels, S. Long, L.K. Anderton, and M.B. Piep (eds.). 2007. *Magnoliophyta: Commelinidae* (in part): *Poaceae*, part 1. Flora of North America North of Mexico, volume 24. Oxford University Press, New York and Oxford.

Barkworth, M.E., K.M. Capels, S. Long, and M.B. Piep (eds.). 2003. *Magnoliophyta: Commelinidae* (in part): *Poaceae*, part 2. *Flora of North America North of Mexico*, volume 25. Oxford University Press, New York. **[Cynodon, Zoysia]**

Barkworth, M.E. and M.A. Torres. 2001. Distribution and diagnostic characters of *Nassella* (Poaceae: Stipeae). Taxon 50:439–468. **[Nassella]**

Bess, E.C., A.N. Doust, G. Davidse, and E.A. Kellogg. 2006. *Zuloagaea*, a new genus of neotropical grass within the "Bristle Clade" (Poaceae: Paniceae). Syst. Bot. 31:656–670. **[Zuloagaea]**

Bödvarsdóttir, S.K. and K. Anamthawat-Jónsson. 2003. Isolation, characterization, and analysis of *Leymus*-specific DNA sequences. Genome 46:673–682. **[Leymus]**

Bothmer, R. von, C. Baden, and N.H. Jacobsen. 2007. *Hordeum*. Pp. 241–252 *in* Barkworth, M.E, K.M. Capels, S. Long, L.K. Anderton, and M.B. Piep (eds.). 2007. *Magnoliophyta: Commelinidae* (in part): *Poaceae*, part 1. Flora of North America North of Mexico, volume 24. Oxford University Press, New York and Oxford. **[Hordeum]**

Bothmer, R. von, N. Jacobsen, C. Baden, R.B. Jørgensen, and I. Linde-Laursen. 1995. An Ecogeographical Study of the Genus *Hordeum*, ed. 2. Systematic and Ecogeographic Studies on Crop Genepools No. 7. International Board of Plant Genetic Resources, Rome, Italy. 129 pp. **[Hordeum]**

Bouchenak-Khelladi, Y., N. Salamin, V. Savolainen, F. Forest, M. van der Bank, M.W. Chase, and T.R. Hodkinson. 2008. Large multi-gene phylogenetic trees of the grasses (Poaceae): Progress towards complete tribal and generic level sampling. Molec. Phylogenet. Evol. 47: 488–505. [Andropogoneae, Aristidoideae, Arundinoideae, Chloridoideae, Cynodonteae, Danthonioideae, Nardeae, Panicoideae, Schedonnardus]

Bowden, W.M. 1958. Natural and artificial ×Elymordeum hybrids. Canad. J. Bot. 36:101–123. [×Elyhordeum]

Bowden, W.M. 1967. Taxonomy of intergeneric hybrids of the tribe Triticeae from North America. Canad. J. Bot. 45:711–724. [×Elyleymus, ×Leydeum]

Boyle, W.S. 1945. A cytotaxonomic study of the North American species of Melica. Madroño 8:1–26. [Melica]

Brown, R. 1814. General remarks, geographical and systematical, on the botany of Terra Australis. Pp. 533–613 in M. Flinders (ed.). A Voyage to Terra Australis, vol. 2. G. and W. Nicol, London, England. 613 pp. [Paniceae, Panicoideae]

Campbell, C.S. 1985. The subfamilies and tribes of Gramineae (Poaceae) in the southeastern United States. J. Arnold Arbor. 66:123–199. [Chloridoideae, Cynodonteae]

Caro, J.A. 1981. Rehabilitación del género Dasyochloa (Gramineae). Dominguezia 2:1–17. [Dasyochloa]

Caro, J.A. and E.A. Sánchez. 1969. Las especies de Cynodon (Gramineae) de la República Argentina. Kurtziana 5:191–252. [Cynodon]

Catalán, P., P. Torrecilla, J.A.L. Rodriguez, and R.G. Olmstead, 2004. Phylogeny of the festucoid grasses of subtribe Loliinae and allies (Poeae, Pooideae) inferred from ITS and trnL–F sequences. Molec. Phylogenet. Evol. 31:517–541. [Leucopoa, Poeae, Pooideae]

Chambers, K.L. 1985. Pitfalls in identifying Ventenata dubia (Poaceae). Madroño 32:120–121. [Ventenata]

Church, G.L. 1949. Cytotaxonomic study of Glyceria and Puccinellia. Amer. J. Bot. 36:155–165. [Glyceria]

Church, G.L. 1952. The genus Torreyochloa. Rhodora 54:197–200. [Torreyochloa]

Cialdella, A.M. and M.O. Arriaga. 1998. Revisión de las especies sudamericanas del género Piptochaetium (Poaceae, Pooideae, Stipeae). Darwiniana 36:105–157. [Piptochaetium]

Clark, L.G., S. Dransfield, J. Triplett, and J. G. Sánchez-Ken. 2007. Phylogenetic relationships among the one-flowered, determinate genera of Bambuseae (Poaceae: Bambusoideae). Aliso 23: 315–332. [Bambusoideae]

Clayton, W.D. 1970. Flora of Tropical East Africa, Gramineae (Part 1). Crown Agents for Oversea Governments and Administrations, London, England. 176 pp. [Phragmites]

Clayton, W.D. and S.A. Renvoize. 1986. Genera Graminum: Grasses of the World. Kew Bull., Addit. Ser. 13. Her Majesty's Stationery Office, London, England. 389 pp. [Aristidoideae, Arundinoideae, Cynodonteae]

Clifford, H.T. 1996. Etymological Dictionary of Grasses, Version 1.0 (CD-ROM). Expert Center for Taxonomic Identification, Amsterdam, The Netherlands. [Cymbopogon, Eragrostis]

Columbus, J.T. 1999. An expanded circumscription of Bouteloua (Gramineae: Chlorideae): New combinations and names. Aliso 18:61–65. [Bouteloua]

Columbus, J.T., M.S. Kinney, R. Pant, and M.E. Siqueiros Delgado. 1998. Cladistic parsimony analysis of internal transcribed spacer region (nrDNA) sequences of Bouteloua and relatives (Gramineae: Chloridoideae). Aliso 17:99–130. [Hilaria]

Conert, H.J. 1987. Current concepts in the systematics of the Arundinoideae. Pp. 239–250 in T.R. Soderstrom, K.W. Hilu, C.S. Campbell, and M.E. Barkworth (eds.). Grass Systematics and Evolution. Smithsonian Institution Press, Washington, D.C., U.S.A. 473 pp. [Danthonioideae]

Conert, H.J. 1992b. Glyceria. Pp. 440–457 in G. Hegi. Illustrierte Flora von Mitteleuropa, ed. 3. Band I, Teil 3, Lieferung 6 (pp. 401–480). Verlag Paul Parey, Berlin and Hamburg, Germany. [Glyceria]

Cotton, R. and C.A. Stace. 1967. Taxonomy of the genus Vulpia (Gramineae): I. Chromosome numbers and geographical distribution of the Old World species. Genetica 46:235–255. [Vulpia]

Covas, G. 1949. Taxonomic observations on the North American species of Hordeum. Madroño 9:233–264. [×Leydeum]

Cronquist, A., A.H. Holmgren, N.H. Holmgren, and J.L. Reveal. 1972. Intermountain Flora. Vol. 1, Hafner Publ. Co. 270 pp. [Preface]

Dannhardt, G. and L. Steindl. 1985. Alkaloids of Lolium temulentum: Isolation, identification and pharmacological activity. Pl. Med. (Stuttgart) 1985:212–214. [Lolium]

Darbyshire, S.J. 1993. Realignment of Festuca subgenus Schedonorus with the genus Lolium. Novon 3:239–243. [Schedonorus]

Darbyshire, S.J. and J. Cayouette. 1989. The biology of Canadian weeds. 92. Danthonia spicata (L.) Beauv. in Roem. & Schult. Canad. J. Pl. Sci. 69:1217–1233. [Danthonia]

Darbyshire, S.J. and **L.E. Pavlick.** 2007. *Festuca.* Pp. 389–443 *in* Barkworth, M.E, K.M. Capels, S. Long, L.K. Anderton, and M.B. Piep (eds.). 2007. *Magnoliophyta: Commelinidae* (in part): Poaceae, part 1. Flora of North America North of Mexico, volume 24. Oxford University Press, New York and Oxford. **[Festuca]**

Darbyshire, S.J. and **S.I. Warwick.** 1992. Phylogeny of North American *Festuca* (Poaceae) and related genera using chloroplast DNA restriction site variation. Canad. J. Bot. 70:2415–2429. [1992 on title page; printed in 1993]. **[Leucopoa]**

Daubenmire, R.F. 1960. An experimental study of variation in the *Agropyron spicatum–A. inerme* complex. Bot. Gaz. 122:104–108. **[Pseudoroegneria]**

Dávila, P.D. 1994. Trachypogon Nees. Pp. 380–381 *in* G. Davidse, M. Sousa S., and A.O. Chater (eds.). Flora Mesoamericana, vol. 6: Alismataceae a Cyperaceae. Universidad Nacional Autónoma de México, Instituto de Biología, México, D.F., México. 543 pp. **[Trachypogon]**

Desvaux, E. 1854. Gramineas. Pp. 233–469 *in* C. Gay. Flora Chilena [Historia Fisica y Politica de Chile], vol. 6. Museo Historia Natural, Santiago, Chile. 551 pp. [1853 on title page; printed in March 1854]. **[Nassella]**

de Wet, J.M.J. 1978. Systematics and evolution of *Sorghum* sect. *Sorghum* (Gramineae). Amer. J. Bot. 65:477–484. **[Sorghum]**

de Wet, J.M.J. and **J.R. Harlan.** 1970. Biosystematics of *Cynodon* L.C. Rich. (Gramineae). Taxon 19:565–569. **[Cynodon]**

Dewey, D.R. 1965. Morphology, cytology, and fertility of synthetic hybrids of *Agropyron spicatum × Agropyron dasystachyum-riparium.* Bot. Gaz. 126:269–275. **[Elymus]**

Dewey, D.R. 1967a. Genome relations between *Agropyron scribneri* and *Sitanion hystrix.* Bull. Torrey Bot. Club 94:395–404. **[Elymus]**

Dewey, D.R. 1967b. Synthetic *Agropyron–Elymus* hybrids: II. *Elymus canadensis × Agropyron dasystachyum.* Amer. J. Bot. 54:1084–1089. **[Elymus]**

Dewey, D.R. 1968. Synthetic hybrids of *Agropyron dasystachyum × Elymus glaucus* and *Sitanion hystrix.* Bot. Gaz. 129:316–322. **[Elymus]**

Dewey, D.R. 1970. The origin of *Agropyron albicans.* Amer. J. Bot. 57:12–18. **[Elymus]**

Dewey, D.R. 1975a. Introgression between *Agropyron dasystachyum* and *A. trachycaulum.* Bot. Gaz. 136:122–128. **[Elymus]**

Dewey, D.R. 1975b. The origin of *Agropyron smithii.* Amer. J. Bot. 62:524–530. **[Pascopyrum]**

Dewey, D.R. 1976. Cytogenetics of *Agropyron pringlei* and its hybrids with *A. spicatum, A. scribneri, A. violaceum,* and *A. dasystachyum.* Bot. Gaz. 137:179–185. **[Elymus]**

Dewey, D.R. 1986. Taxonomy of the crested wheatgrasses (*Agropyron*). Pp. 31–42 *in* K.L. Johnson (ed.). Crested Wheatgrass: Its Values, Problems and Myths; Symposium Proceedings. Utah State University, Logan, Utah, U.S.A. 348 pp. **[Agropyron]**

Dewey, D.R. and **A.H. Holmgren.** 1962. Natural hybrids of *Elymus cinereus* and *Sitanion hystrix.* Bull. Torrey Bot. Club 89:217–228. **[×Elyleymus]**

Dillman, A.C. 1946. The beginnings of crested wheatgrass in North America. J. Amer. Soc. Agron. 38:237–250. **[Agropyron]**

Dore, W.G. and **J. McNeill.** 1980. Grasses of Ontario. Research Branch, Agriculture Canada Monograph No. 26. Canadian Government Publishing Centre, Hull, Québec, Canada. 568 pp. **[Danthonia, Oryzopsis]**

Döring, E., J.T. Albrecht, K.W. Hilu, and **M. Röser.** 2007. Phylogenetic relationships in the Aveneae/Poeae complex (Pooideae, Poaceae). Kew Bull. 62: 355–373. **[Pooideae]**

Edgar, E. 1995. New Zealand species of *Deyeuxia* P. Beauv. and *Lachnagrostis* Trin. (Gramineae:Aveneae). New Zealand J. Bot. 33:1–33. **[Calamagrostis]**

Edgar, E. and **H.E. Connor.** 2000. Flora of New Zealand, vol. 5. Manaaki Whenua Press, Lincoln, New Zealand. 650 pp. **[Ehrharteae, Schedonorus]**

Esen, A. and **K.W. Hilu.** 1991. Electrophoretic and immunological studies of prolamins in the Poaceae. II. Phylogenetic affinities of the Aristideae. Taxon 40:5–17. **[Aristidoideae]**

Finot, V.L., P.M. Peterson, R.J. Soreng, and **F.O. Zuloaga.** 2004. A revision of *Trisetum, Peyritschia,* and *Sphenopholis* (Poaceae: Pooideae: Aveninae) in Mexico and Central America. Ann. Missouri Bot. Gard. 91:1–30. **[Trisetum]**

Finot, V.L., P.M. Peterson, R.J. Soreng, and **F.O. Zuloaga.** 2005. A revision of *Trisetum* and *Graphephorum* (Poaceae: Pooideae: Aveninae) in North America north of Mexico. SIDA 21(3):1419–1453. **[Trisetum]**

Finot, V.L., P.M. Peterson, F.O. Zuloaga, R.J. Soreng, and **O. Matthei.** 2005. A revision of *Trisetum* (Poaceae: Pooideae: Aveninae) in South America. Ann. Missouri Bot. Gard. 92:533–568. **[Trisetum]**

Freckmann, R.W. and **M.G. Lelong.** 2002. Nomenclatural changes and innovations in *Panicum* and *Dichanthelium* (Poaceae: Paniceae). Sida 20:161–174. **[Panicum]**

Frederiksen, S. 1986. Revision of *Taeniatherum* (Poaceae). Nordic J. Bot. 6:389–397. **[Taeniatherum]**

Frederiksen, S. and G. Petersen. 1998. A taxonomic revision of *Secale* L. (Triticeae, Poaceae). Nordic J. Bot. 18:399–420. **[Secale]**

Gabel, M.L. 1989. Federal noxious weed identification bulletin. U.S. Dept. of Agriculture, Animal & Plant Health Inspection Service, Plant Protection & Quarantine Bull. 28:1–10. **[Imperata]**

Giussani, L.M., J.H. Cota-Sánchez, F.O. Zuloaga, and E.A. Kellogg. 2001. A molecular phylogeny of the grass subfamily Panicoideae (Poaceae) shows multiple origins of C_4 photosynthesis. Amer. J. Bot. 88:1993–2001. **[Andropogoneae, Panicoideae, Panicum]**

Goebel, K.I. 1882. Beitrage zur Entwickelungsgeschichte einiger Inflorescenzen. Jahrb. Wiss. Bot. 14:1–39. **[Pennisetum]**

Gómez-Martínez, R. and A. Culham. 2000. Phylogeny of the subfamily Panicoideae with emphasis on the tribe Paniceae: Evidence from the *trn*L-F cpDNA region. Pp. 136–140 *in* S.W.L. Jacobs and J. Everett (eds.). Grasses: Systematics and Evolution. CSIRO Publishing, Collingwood, Victoria, Australia. 408 pp. **[Panicoideae]**

Gould, F.W. 1979. The genus *Bouteloua* (Poaceae). Ann. Missouri Bot. Gard. 66:348–416. **[Bouteloua]**

Gould, F.W. and R.B. Shaw. 1983. Grass Systematics, ed. 2. Texas A&M University Press, College Station, Texas, U.S.A. 397 pp. **[Chloridoideae]**

Grass Phylogeny Working Group. 2001. Phylogeny and subfamilial classification of the grasses (Poaceae). Ann. Missouri Bot. Gard. 88:373–457. **[Arundinoideae, Chloridoideae, Cynodonteae, Danthonioideae, Meliceae, Nardeae]**

Greuter, W. 1968. Notulae nomenclaturales et bibliographicae, 1–4. Candollea 23:81–108. **[Koeleria]**

Hamilton, J.G. 1997. Changing perceptions of pre-European grasslands in California. Madroño 44:311–333. **[Nassella]**

Harlan, J.R. 1945a. Cleistogamy and chasmogamy in *Bromus carinatus* Hook. & Arn. Amer. J. Bot. 32:66–72. **[Bromus]**

Harlan, J.R. 1945b. Natural breeding structure in the *Bromus carinatus* complex as determined by population analyses. Amer. J. Bot. 32:142–147. **[Bromus]**

Harlan, J.R. and J.M.J. de Wet. 1972. A simplified classification of cultivated sorghum. Pers. Crop Sci. 12:172–176. **[Sorghum]**

Harrington, J., S. Reid, W. Black, and M. Brick. 2009. Was Rydberg right? Evidence for *Distichlis stricta* as a species distinct from *D. spicata*. Abstract for Botany and Mycology 2009. Accessible via http://2009.botanyconference.org/ **[Distichlis]**

Harvey, M.J. 2007a. *Agrostis*. Pp. 633–662 *in* Barkworth, M.E, K.M. Capels, S. Long, L.K. Anderton, and M.B. Piep (eds.). 2007. *Magnoliophyta: Commelinidae* (in part): *Poaceae*, part 1. Flora of North America North of Mexico, volume 24. Oxford University Press, New York and Oxford. **[Podgrostis]**

Harvey, M.J. 2007b. *Podagrostis*. Pp. 693–694 *in* Darkworth, M.E, K.M. Capels, S. Long, L.K. Anderton, and M.B. Piep (eds.). 2007. *Magnoliophyta: Commelinidae* (in part): *Poaceae*, part 1. Flora of North America North of Mexico, volume 24. Oxford University Press, New York and Oxford. **[Podagrostis]**

Hattersley, P.W. and L. Watson. 1976. C_4 grasses: an anotomical criterion for distinguishing between NADP-malic enzyme species and PCK or NAD-Malic enzyme species. Aust. J. Bot. 24:297–308. **[Poaceae]**

Hilu, K.W. and L.A. Alice. 2001. A phylogeny of the Chloridoideae (Poaceae) based on *matK* sequences. Syst. Bot. 26:386–405. **[Chloridoideae, Chloris, Cynodonteae, Eragrostis]**

Hilu, K.W. and A. Esen. 1990. Prolamin and immunological studies in the Poaceae. I. Subfamily Arundinoideae. Pl. Syst. Evol. 173:57–70. **[Arundinoideae]**

Hilu, K.W. and A. Esen. 1993. Prolamin and immunological studies in the Poacae: III. Subfamily Chloridoideae. Amer. J. Bot. 80:104–113. **[Chloridoideae]**

Hilu, K.W. and K. Wright. 1982. Systematics of Gramineae: A cluster analysis study. Taxon 31:9–36. **[Chloridoideae]**

Hitchcock, A.S. 1951. Manual of the Grasses of the United States, ed. 2, rev. A. Chase. U.S.D.A. Miscellaneous Publication No. 200. U.S. Government Printing Office, Washington, D.C., U.S.A. 1051 pp. [1950 on title page; printed in 1951] **[Bromus, Poa, Schizachyrium]**

Hitchcock, C.L. 1969. Gramineae. Pp. 384–725 *in* C.L. Hitchcock, A. Cronquist, and M. Ownbey. Vascular Plants of the Pacific Northwest, Part 1: Vascular Cryptogams, Gymnosperms, and Monocotyledons. University of Washington Press, Seattle, Washington, U.S.A. 914 pp. **[Calamagrostis]**

Holub, J. 1998. Reclassifications and new names in vascular plants 1. Preslia 70:97–122. **[Schedonorus]**

Hsiao, C., S.W.L. Jacobs, N.P. Barker, and N.J. Chatterton. 1998. A molecular phylogeny of the subfamily Arundinoideae (Poaceae) based on sequences of rDNA. Austral. Syst. Bot. 11:41–52. **[Arundinoideae]**

Hubbard, F.T. 1915. A taxonomic study of *Setaria italica* and its immediate allies. Amer. J. Bot. 2:169–198. **[Setaria]**

Hultén, E. 1942. Flora of Alaska and Yukon [in part]. Acta Univ. Lund, n.s., 38:1–281. **[Poa]**

Ingram, A.A. and J.J. Doyle. 2004. Is *Eragrostis* (Poaceae) monophyletic? Insights from nuclear and plastid sequence data. Syst. Bot. 29:545–552. **[Eragrostis]**

Jacobs, S.W.L. 1987. Systematics of the Chloridoid grasses. Pp. 871–903 *in* T.R. Soderstrom, K.W. Hilu, C.S. Campbell, and M.E. Barkworth (eds.) Grass Systematics and Evolution. Smithsonian Institution Press, Washington, D.C., U.S.A. 473 pp. **[Chloridoideae]**

Jacobs, S.W.L. 2001. The genus *Lachnagrostis* (Gramineae) in Australia. Telopea 9:439–448. **[Calamagrostis]**

Jacobs, S.W.L., R. Baker, J. Everett, M. Arriaga, M. Barkworth, A. Sabin-Badereau, A. Torres, F. Vázquez, and N. Bagnall. 2007. Systematics of the tribe Stipeae (Gramineae) using molecular data. Aliso 23:349–361. **[Nassella]**

Jakob, S.S., A. Ihlow, and F.R. Blattner. 2007. Combined ecological modelling and molecular phylogeography revealed the evolutionary history of *Hordeum marinum* (Poaceae) - niche differentiation, loss of genetic diversity, and speciation in Mediterranean Quaternary refugia. Molec. Ecol. 16:1713–1727. **[Hordeum]**

Jarvie, J.K. and M.E. Barkworth. 1992. Morphological variation and genome constitution in some perennial Triticeae. Bot. J. Linn. Soc. 108:167–189. **[Thinopyrum]**

Jensen, K.B. and K.H. Asay. 1996. Cytology and morphology of *Elymus hoffmannii* (Poaceae: Triticeae): A new species from the Erzurum Province of Turkey. Int. J. Pl. Sci. 157:750–758. **[Elymus]**

Jensen, K.B, S.L. Hatch, and J. Wipff. 1992. Cytogenetics and morphology of *Pseudoroegneria deweyi* (Poaceae: Triticeae), a new species from the Soviet Union. Canad. J. Bot. 70:900–909. **[Pseudoroegneria]**

Johnson, B.L. 1945. Natural hybrids between *Oryzopsis hymenoides* and several species of *Stipa*. Amer. J. Bot. 32:599–608. **[Achnatherum]**

Johnson, B.L. 1960. Natural hybrids between *Oryzopsis* and *Stipa*: I. *Oryzopsis hymenoides* × *Stipa speciosa*. Amer. J. Bot. 47:736–742. **[Achnatherum]**

Johnson, B.L. 1962a. Amphiploidy and introgression in *Stipa*. Amer. J. Bot. 49:253–262. **[Achnatherum]**

Johnson, B.L. 1962b. Natural hybrids between *Oryzopsis* and *Stipa*: II. *Oryzopsis hymenoides* × *Stipa nevadensis*. Amer. J. Bot. 49:540–546. **[Achnatherum]**

Johnson, B.L. 1963. Natural hybrids between *Oryzopsis* and *Stipa*: III. *Oryzopsis hymenoides* × *Stipa pinetorum*. Amer. J. Bot. 50:228–234. **[Achnatherum]**

Johnson, B.L. 1972. Polyploidy as a factor in the evolution and distribution of grasses. Pp. 18–35 *in* V.B. Youngner and C.M. McKell (eds.). The Biology and Utilization of Grasses. Academic Press, New York, New York, U.S.A. 426 pp. **[Achnatherum]**

Jozwik, F.X. 1966. A biosystematic analysis of the slender wheatgrass complex. Ph.D. dissertation, University of Wyoming, Laramie, Wyoming, U.S.A. 112 pp. **[Elymus]**

Judziewicz, E.J. 1990. Flora of the Guianas: 187. Poaceae (Gramineae). Series A: Phanerogams, Fascicle 8 (series ed. A.R.A. Görts-Van Rijn). Koeltz Scientific Books, Koenigstein, Germany. 727 pp. **[Trachypogon]**

Kearney, T.H. and R.H. Peebles. 1951. Arizona Flora. University of California Press, Berkeley and Los Angeles, California, U.S.A. 1032 pp. **[Muhlenbergia]**

Kellogg, E.A. 2000. Molecular and morphological evolution in the Andropogoneae. Pp. 149–158 *in* S.W.L. Jacobs and J. Everett (eds.). Grasses: Systematics and Evolution. CSIRO Publishing, Collingwood, Victoria, Australia. 408 pp. **[Andropogoneae]**

Kellogg, E.A. and C.S. Campbell. 1987. Phylogenetic analyses of the Gramineae. Pp. 310–322 *in* T.R. Soderstrom, K.W. Hilu, C.S. Campbell, and M.E. Barkworth (eds.) Grass Systematics and Evolution. Smithsonian Institution Press, Washington, D.C., U.S.A. 473 pp. **[Aristidoideae, Arundinoideae, Chloridoideae]**

Koyama, T. 1987. Grasses of Japan and Its Neighboring Regions: An Identification Manual. Kodansha, Ltd., Tokyo, Japan. 370 pp. **[Phragmites]**

Koyama, T. and S. Kawano. 1964. Critical taxa of grasses with North American and eastern Asiatic distribution. Canad. J. Bot. 42:859–864. **[Torreyochloa]**

Lawrence, W.E. 1945. Some ecotypic relations of *Deschampsia caespitosa*. Amer. J. Bot. 32:298–314. **[Deschampsia]**

Linder, H.P., G.A. Verboom, and N.P. Barker. 1997. Phylogeny and evolution in the *Crinipes* group of grasses (Arundinoideae: Poaceae). Kew Bull. 52:91–110. [Arundineae]

Mason-Gamer, R.J. 2001. Origin of North American *Elymus* (Poaceae:Triticeae) allotetraploids based on granule-bound starch synthase gene sequences. Syst. Bot. 26:757–768. [Elymus]

Mejia-Saulés, T. and F.A. Bisby. 2000. Preliminary views on the tribe Meliceae (Gramineae: Poöideae). Pp. 83–88 *in* S.W.L. Jacobs and J. Everett (eds.). Grasses: Systematics and Evolution. CSIRO Publishing, Collingwood, Victoria, Australia. 408 pp. [Meliceae]

Mejia-Saulés, T. and F.A. Bisby. 2003. Silica bodies and hooked papillae in lemmas of *Melica* species (Gramineae: Poöideae). Bot. J. Linn. Soc. 141:447–463. [Melica]

Merigliano, M.F. and P. Lesica. 1998. The native status of reed canarygrass (*Phalaris arundinacea* L.) in the inland northwest, USA. Nat. Areas J. 18:223–230. [Phalaris]

Moyer, J.R. and A.L. Boswall. 2002. Tall fescue or creeping foxtail suppresses foxtail barley. Canad. J. Pl. Sci. 82:89–92. [Alopecurus]

Nihsen, M.E., E.L. Piper, C.P. West, R.J. Crawford, Jr., T.M. Denard, Z.B. Johnson, C.A. Roberts, D.A. Spiers, and C.F. Rosenkrans, Jr. 2004. Growth rate and physiology of steers grazing tall fescue inoculated with novel endophytes. J. Animal Sci. 82:878–883. [Schedonorus]

Ogle, D. 2001. Intermediate wheatgrass, *Thinopyrum intermedium* (Host) Barkworth & D.R. Dewey. Plant Fact Sheet, U.S. Department of Agriculture, Natural Resource Conservation Service Plant Materials Program. http://plants.usda.gov/. [Thinopyrum]

Old, R.R. and R.H. Callihan. 1986. Distribution of *Ventenata dubia* in Idaho. Idaho Weed Control Rep. 1986:153. [Ventenata]

Peart, M.H. 1984. The effects of morphology, orientation, and position of grass diaspores on seedling survival. J. Ecol. 69:425–436. [Andropogoneae]

Peñailillo, P. 2002. El género *Jarava* Ruiz et Peterson, P.M. and J.T. Columbus. 2008. *Muhlenbergia alopecuroides* (Poaceae: Muhlenbergiinae), a new combination. Madroño 55:159–160. [Lycurus]

Peterson, P.M., R.J. Soreng, G. Davidse, T.S. Filgueras, F.O. Zuloaga, and E. Judziewicz. 2001. Catalogue of New World Grasses (Poaceae): II. Subfamily Chloridoideae. Contr. U.S. Nat. Herb. 41:1–255. [Cynodonteae]

Peterson, P.M., R.D. Webster and J. Valdés-Reyna. 1997. Genera of New World Eragrostideae (Poaceae: Chloridoideae). Smithsonian Contr. Bot. 87:1–50. [Sporobolus]

Phillips, S.M. and W.-L. Chen. 2003. Notes on grasses (Poaceae) for the Flora of China, I: *Deyeuxia*. Novon 13:318–321. [Calamagrostis]

Pizzolato, T.D. 1984. Vascular system of the fertile floret of *Anthoxanthum odoratum* L. Bot. Gaz. 145:358–371. [Anthoxanthum]

Prat, H. 1936. La systématique des Graminées. Ann. Sci. Nat., Bot., Ser. 10, 18:165–258. [Chloridoideae]

Quintanar, A., S. Castroviejo and P. Catalán. 2007. Phylogeny of the tribe Aveneae (Pooideae, Poaceae) inferred from plasted *trnT–F* and nuclear ITS sequences. Amer. J. Bot. 94:1554–1569. [Pooideae, Trisetum]

Riggins, R. 1977. A biosystematic study of the *Sporobolus asper* complex (Gramineae). Iowa State J. Res. 51:287–321. [Sporobolus]

Romaschenko, K., P.M. Peterson, R.J. Soreng, N. Garcia-Jacas, O. Futorna, and A. Susanna. 2008. Molecular phylogenetic analysis of the American Stipeae (Poaceae) resolves *Jarava sensu lato* polyphyletic: evidence for a new genus, *Pappostipa*. J. Bot. Res. Inst. Tex. 2:165–192. [Achnatherum, Nassella, Piptatherum, Stipeae]

Rumely, J.H. 2007. *Trisetum*. Pp. 744–753 *in* Barkworth, M.E, K.M. Capels, S. Long, L.K. Anderton, and M.B. Piep (eds.). 2007. *Magnoliophyta: Commelinidae* (in part): Poaceae, part 1. Flora of North America North of Mexico, volume 24. Oxford University Press, New York and Oxford. [Trisetum]

Sales, F. 1993. Taxonomy and nomenclature of *Bromus* sect. *Genea*. Edinburgh J. Bot. 50:1–31. [Bromus]

Saltonstall, K. 2002. Cryptic invasion by a non-native genotype of the common reed, *Phragmites australis*, into North America. Proc. Natl. Acad. Sci. U.S.A. 99:2445–2449. [Phragmites]

Saltonstall, K. 2003a. Genetic variation among North American Populations of *Phragmites australis*; implications for management. Estuaries 26:444–451. [Phragmites]

Saltonstall, K. 2003b. Microsatellite variation within and among North American lineages of *Phragmites australis*. Mol. Ecol. 12:1689–1702. [Phragmites]

Saltonstall, K., P.M. Peterson, and R. J. Soreng. 2004. Recognition of *Phragmites autralis* subsp. *americanus* (Poaceae: Arundinoideae) in North America: Evidence from morphological and genetic analyses. Sida 21:683–692. [Phragmites]

Savile, D.B.O. 1979. Fungi as aids in higher plant classification. Bot. Rev. 45:377–503. [Chloridoideae]

Schneider, J., E. Döring, K.W. Hilu, and Röser, M. [in press]. Phylogenetic structure of the grass subfamily Pooideae based on comparison of plastid *matK* gene-3'*trnK* exon and nuclear ITS sequences. [Pooideae]

Scholz, H. and N. Böhling. 2000. *Phragmites frutescens* (Gramineae) re-visited. The discovery of an overlooked, woody grass in Greece, especially Crete. Willdenowia 30:251–261. [Phragmites]

Schouten, Y. and J.F. Veldkamp. 1985. A revision of *Anthoxanthum* including *Hierochloë* (Gramineae) in Malesia and Thailand. Blumea 30:319–351. [Anthoxanthum]

Slageren, M.W. van. 1994. Wild Wheats: A Monograph of *Aegilops* L. and *Amblyopyrum* (Jaub. & Spach) Eig. Wageningen Agricultural University Papers 94–7. Wageningen Agricultural University and International Center for Agricultural Research in the Dry Areas (ICARDA), Wageningen, The Netherlands and Aleppo, Syria. 512 pp. [Aegilops]

Snow, N. 1997. Phylogeny and systematics of *Leptochloa* P. Beauv. *sensu lato* (Poaceae, Chloridoideae). Ph.D. dissertation, Washington University, St. Louis, Missouri, U.S.A. 506 pp. [Leptochloa]

Snow, N., P.M. Peterson, D. Gerald-Cañas. 2008. *Leptochloa* (Poaceae: Chloridoideae) in Colombia. J. Bot. Res. Inst. Texas 2:861–874. [Leptochloa]

Snyder, L.A. 1950. Morphological variability and hybrid development in *Elymus glaucus*. Amer. J. Bot. 37:628–636. [Elymus]

Snyder, L.A. 1951. Cytology of inter-strain hybrids and the probable origin of variability in *Elymus glaucus*. Amer. J. Bot. 38:195–202. [Elymus]

Sohns, E.R. 1955. *Cenchrus* and *Pennisetum*: Fascicle morphology. J. Wash. Acad. Sci. 45:135–143. [Pennisetum]

Soreng, R.J. 2003. *Podagrostis*. Contr. U.S. Natl. Herb. 48:581. [Agrostis]

Soreng, R.J. 2007. *Poa*. Pp. 486–600 *in* Barkworth, M.E, K.M. Capels, S. Long, L.K. Anderton, and M.B. Piep (eds.). *Magnoliophyta: Commelinidae* (in part): *Poaceae*, part 1. Flora of North America North of Mexico, volume 24. Oxford University Press, New York and Oxford. [Poa]

Soreng, R.J. and J.I. Davis. 2000. Phylogenetic structure in Poaceae subfamily Poöideae as inferred from molecular and morphological characters: Misclassification versus reticulation. Pp. 61–74 *in* S.W.L. Jacobs and J. Everett (eds.). Grasses: Systematics and Evolution. CSIRO Publishing, Collingwood, Victoria, Australia. 408 pp. [Leucopoa, Meliceae]

Soreng, R.J., J.I. Davis, and J.J. Doyle. 1990. A phylogenetic analysis of chloroplast DNA restriction site variation in Poaceae subfam. Poöideae. Pl. Syst. Evol. 172:83–97. [Catabrosa, Torreyochloa]

Soreng, R.J. and E.E. Terrell. 1997. Taxonomic notes on *Schedonorus*, a segregate genus from *Festuca* or *Lolium*, with a new nothogenus, ×*Schedololium*, and new combinations. Phytologia 83:85–88. [1997 on title page; printed in 1998]. [Schedonorus]

Spangler, R.E. 2000. Andropogoneae systematics and generic limits in *Sorghum*. Pp. 167–170 *in* S.W.L. Jacobs and J. Everett (eds.). Grasses: Systematics and Evolution. CSIRO Publishing, Collingwood, Victoria, Australia. 408 pp. [Andropogoneae, Sorghum]

Stebbins, G.L., Jr. 1947. The origin of the complex of *Bromus carinatus* and its phytogeographic implications. Contr. Gray Herb. 165:42–55. [Bromus]

Stebbins, G.L., Jr. 1957. The hybrid origin of microspecies in the *Elymus glaucus* complex. Pp. 336–340 *in* International Union of Biological Sciences (ed.) Proceedings of the International Genetics Symposia, 1956: Tokyo & Kyoto, September 1956. Organizing Committee, International Genetics Symposia, Science Council of Japan, Tokyo, Japan. 680 pp. [Elymus]

Stebbins, G.L. 1981. Chromosomes and evolution in the genus *Bromus* (Gramineae). Bot. Jahrb. 102:359–379. [Bromus]

Stebbins, G.L., Jr. and H.A. Tobgy. 1944. The cytogenetics of hybrids in *Bromus*: 1. Hybrids within the section *Ceratochloa*. Amer. J. Bot. 31:1–11. [Bromus]

Stebbins, G.L., H.A. Tobgy, and J.R. Harlan. 1944. The cytogenetics of hybrids in *Bromus*: II. *Bromus carinatus* and *Bromus arizonicus*. Proc. Cal. Acad. Sci. 25:307–321. [Bromus]

Stika, J. 2003. Sweetgrass ale. http://byo.com/feature/1067.html. [Anthoxanthum]

Sungkaew, S., C.M.A. Stapleton, N. Salamin, and T.R. Hodkinson. 2009. Non-monophyly of the woody bamboos (Bambuseae; Poaceae): a multi-gene region phylogenetic analysis of Bambusoideae s.s. J. Plant Res 122:95–108. [Bambusoideae]

Tercek, M.T., D.P. Hauber, and S.P. Darwin. 2003. Genetic and historical relationships among thermally adapted *Agrostis* (Bentgrass) of North America and Kamchatka: Evidence for a previously unrecognized, thermally adapted taxon. Amer. J. Bot. 90:1306–1312. [Agrostis]

Tsvelev, N.N. 1975. [On the possibility of despecialization by hybridogenesis for explaining the evolution of the Triticeae (Poaceae)]. Zhurn. Obshchei Biol. 36:90–99. [In Russian; translation of article by K. Gonzales, available at the Intermountain Herbarium, Utah State University, Logan, Utah 84322–5305, U.S.A.] **[Triticeae]**

Tsvelev, N.N. 1976. Zlaki SSSR. Nauka, Leningrad [St. Petersburg], Russia. 788 pp. **[Agropyron, Danthonia, Leymus]**

Tsvelev, N.N. 1993. Some notes on the grasses (Poaceae) of the Caucasus [in Russian]. Botaniceskij Zurnal (Moscow & Leningrad) 78(10): 83–95.

Walsh, N.G. 1994. *Cortaderia*. Pp. 546–548 *in* N.G. Walsh and T.J. Entwisle. Flora of Victoria, vol. 2: Ferns and Allied Plants, Conifers and Monocotyledons. Inkata Press, Melbourne, Australia. 946 pp. **[Cortaderia]**

Wang, R.R.-C. and K.B. Jensen. 1994. Absence of J genome in *Leymus* species (Poaceae: Triticeae): Evidence from DNA hybridization and meiotic pairing. Genome 37:231–235. **[Leymus]**

Watson, L., H.T. Clifford, and M.J. Dallwitz. 1985. The classification of the Poaceae: Subfamilies and supertribes. Austral. J. Bot. 33:433–484. **[Aristidoideae, Arundinoideae]**

Weimarck, G. 1971. Variation and taxonomy of *Hierochloë* (Gramineae) in the Northern Hemisphere. Bot. Not. 124:129–175. **[Anthoxanthum]**

Weimarck, G. 1987. *Hierochloë hirta* subsp. *praetermissa*, subsp. *nova* (Gramineae), an Asiatic–E. European taxon extending to N and C Europe in the Northern Hemisphere. Symb. Bot. Upal. 2:175–181. **[Anthoxanthum]**

Welsh, S.L., N.D. Atwood, S. Goodrich, and L.C. Higgins (eds.). 2003. A Utah Flora. 3rd ed. Provo (UT): Brigham Young University. 912 pp. **[Piptochaetium]**

Welsh, S.L., N.D. Atwood, S. Goodrich, and L.C. Higgins (eds.). 2008. 4th ed. A Utah Flora. Provo (UT): Brigham Young University. 1019 pp. **[Achnatherum, Echinochloa]**

Westerbergh, A. and J.F. Doebley. 2002. Morphological traits defining species differences in wild relatives of maize are controlled by multiple quantitative trait loci. Evolution 56:273–283. **[Zea]**

Willemse, L.P.M. 1982. A discussion of the Ehrharteae (Gramineae) with special reference to the Malesian taxa formerly included in *Microlaena*. Blumea 28:181–194. **[Ehrharteae]**

Wheeler, D.J.B., S.W.L. Jacobs, and R.D.B. Whalley. 2002. Grasses of New South Wales, ed. 3. University of New England, Armidale, New South Wales, Australia. 445 pp. **[Ehrharteae]**

Whipple, I.G., M.E. Barkworth, and B.S. Bushman. 2007. Molecular insights into the taxonomy of *Glyceria* (Poaceae: Meliceae) in North America. Amer. J. Bot. 94:551–557. **[Glyceria]**

Wilson, B.L., J. Kitzmiller, W. Rolle, and V.D. Hipkins. 2001. Isozyme variation and its environmental correlates in *Elymus glaucus* from the California floristic province. Canad. J. Bot. 79:139–153. **[Elymus]**

Wilson, F.D. 1963. Revision of *Sitanion* (Triticeae, Gramineae). Brittonia 15:303–323. **[Elymus]**

Wipff, J.K. 2001. Nomenclatural changes in *Pennisetum* (Poaceae: Paniceae). Sida:19:523–530. **[Pennisetum]**

Wolf, N.M. 1776. Genera Plantarum. [publisher unknown, Danzig, Germany]. 177 pp. **[Eragrostis]**

Van den Borre, A. 1994. A taxonomy of the Chloridoideae (Poaceae), with special reference to the genus *Eragrostis*. Ph.D. dissertation, Australian National University, Canberra, New South Wales, Australia. 313 pp. **[Chloris, Cynodonteae]**

Van den Borre, A. and L. Watson. 1997. On the classification of the Chloridoideae (Poaceae). Austral. Sys. Bot. 10:491–531. **[Chloridoideae, Cynodonteae]**

Van den Borre, A. and L. Watson. 2000. On the classification of the Chloridoideae: Results from morphological and leaf anatomical data analyses. Pp. 180–183 *in* S.W.L. Jacobs and J. Everett (eds.). Grasses: Systematics and Evolution. CSIRO Publishing, Collingwood, Victoria, Australia. 408 pp. **[Chloridoideae, Chloris, Eragrostis]**

Voss, E.G. 1972. Michigan Flora: A Guide to the Identification and Occurrence of the Native and Naturalized Seed-Plants of the State: Part I, Gymnosperms and Monocots. Cranbrook Institute of Science, Bloomfield Hills, Michigan, U.S.A. 488 pp. **[Glyceria]**

Yabuno, T. 1962. Cytotaxonomic studies on the two cultivated species and the wild relatives in the genus *Echinochloa*. Cytologia 27:296–305. **[Echinochloa]**

Yen, C., J.-L. Yang, and Y. Yen. 2005. Hitoshi Kihara, Áskell Löve and the modern genetic concept of the genera in the tribe Triticeae (Poaceae). Acta Phytotax. Sin. 43:82–93. **[Elymus]**

Zuloaga, F.O., L.M. Giussani, and O. Morrone. 2007. *Hopia*, a new monotypic genus segregated from *Panicum* (Poaceae). Taxon 56: 145-156. **[Hopia]**

Index

Names of accepted taxa are in **boldface**; names of other taxa are in *italics*; vernacular names are in SMALL CAPS. Page numbers in **boldface** refer to the primary treatment; page numbers in *italics* refer to the illustration; <u>underlined</u> page numbers refer to the distribution map; other page references are in regular type.

Infraspecific names are listed alphabetically, based on the *infraspecific epithet*; the rank (subspecies, variety, or form) is ignored. For those taxa in which the specific and infraspecific epithet are the same, authors are ignored as far as the alphabetic listing is concerned.

If the spellings are the same, vernacular names are listed after all the scientific names for that genus.

Abbreviations

abx . . . . abaxial
adx . . . . . adaxial, adaxially
ann . . . . annual
anth . . . . anthers
apc . . . . apices
asex . . . . asexual
aur . . . . auricles
bas . . . . basal, basally
bisex . . . . bisexual
bld . . . . blades
br . . . . branches
brchd . . . . branched
brchg . . . . branching
cal . . . . calluses
car . . . . caryopses
cent . . . . central
ces . . . . cespitose
clm . . . . culms
clstgn . . . . cleistogenes
col . . . . collars
dis . . . . disarticulation,
disarticulating
emb . . embryos, embryonic
emgt . . . . emarginate
epdm . . . . epidermes
exvag . . . . extravaginal
flt . . . . florets
fnctl . . . . functional
ftl . . . . fertile
glab . . . . glabrous
glm . . . . glumes
infl . . . . inflorescences

infvag . . . . infravaginal
intnd . . . . internodes
invag . . . . intravaginal
jnct . . . . junction
lat . . . . lateral, laterally
lf . . . . leaf
lfy . . . . leafy
lig . . . . ligules
lm . . . . lemmas
lo . . . . lower
lod . . . . lodicules
lvs . . . . leaves
memb . . . . . membranous
mid . . . . middle
mrg . . . . margins
mrgl . . . . marginal
nd . . . . nodes
occ . . . . occasionally
ov . . . . ovaries
pal . . . . paleas
pan . . . panicles, paniculate
ped . . . . pedicels
pedlt . . . . pedicellate
per . . . . perennial
pist . . . . pistillate
pl . . . . plants
pluricsp . . . . pluricespitose
pri . . . . primary
psdlig . . . . . . pseudoligules
psdpet . . . . pseudopetioles
pseudopetiolate
psdspklt . pseudospikelets

psdinvag . . . .
. . . . pseudointravaginal
rchl . . . . rachillas
rchs . . . . rachises
rcm . . . . racemes, racemose
rcmly . . . . racemosely
rdcd . . . . reduced
rdmt . . . . rudiments,
rudimentary
rebrchg . . . . . . rebranching
rhz . . . . rhizomes,
rhizomatous
sec . . . . secondary
shth . . . . sheaths
smt . . . . sometimes
spklt . . . . spikelets
sta . . . . stamens
stln . . stolons, stoloniferous
stmt . . . . staminate
strl . . . . sterile
sty . . . . styles
sex . . . . sexual, sexually
subglab . . . . . . subglabrous
subtm . . . . subterminal
tml . . . terminal, terminally
unisex . . . . unisexual
up . . . . upper
usu . . . . usually